Genomic and Epigenomic Biomarkers of Toxicology and Disease

Genomic and Epigenomic Biomarkers of Toxicology and Disease

Clinical and Therapeutic Actions

Edited by

Saura C. Sahu
Formerly of Center for Food Safety and Applied Nutrition,
US Food and Drug Administration,
MD, USA

WILEY

This edition first published 2022

Registered Office
John Wiley & Sons Ltd, The Atrium, Southern Gate, Chichester PO19 8SQ, UK

Editorial Office
The Atrium, Southern Gate, Chichester PO19 8SQ, UK

For details of our global editorial offices, customer services, and more information about Wiley products visit us at www.wiley.com.

Wiley also publishes its books in a variety of electronic formats and by print-on-demand. Some content that appears in standard print versions of this book may not be available in other formats.

A catalogue record for this book is available from the Library of Congress

Hardback ISBN: 9781119807674; ePub ISBN: 9781119807698; ePDF ISBN: 9781119807681;
Obook ISBN: 9781119807704

Cover image: © Ali Kahfi/Getty Images
Cover design by Wiley

Set in 9.5/12.5pt STIXTwoText by Integra Software Services Pvt. Ltd, Pondicherry, India
Printed and bound by CPI Group (UK) Ltd, Croydon, CR0 4YY

C9781119807674_120422

Contents

Dedication

This book is dedicated to my family:

My late parents, Gopinath and Ichhamoni, for their gift of life, love, and living examples

My wife, Jharana, for her life-long friendship, love, and support as well as for her patience and understanding of the long hours I spent at home on planning, writing, and editing this book

My children, Meghamala, Sudhir, Subir, and their spouses, for their love and care

My granddaughter, Naomi, and my grandson, Jonah, for their unconditional love, faith, and trust in me

Saura C. Sahu, PhD
Columbia, MD
August 2021

Preface

A biomarker is a measurable substance, found in body fluids or tissues as a result of exposure, infection, and/or disease, and it can be used for clinical practice. Genomics and epigenomics are modern scientific developments of the twenty-first century. They regulate gene expressions and play important roles in a wide range of human diseases, including cancer. The microRNAs are an important component of the molecular mechanisms of disease processes. Therefore they demonstrate potential for serving as biomarkers.

It is my sincere hope that the up-to-date information presented in this monograph will serve as a stimulus to investigators interested in genomic and epigenomic biomarkers research.

The importance of research on genomic and epigenomic biomarkers is evidenced by the growing number of contributions published each year. It becomes increasingly clear that developments in this field are moving so rapidly that new means are needed to report the status of ongoing research activities. The contributions presented in this monograph represent a collaborative effort by international experts in this emerging field of science.

The main purpose of this book is to assemble up-to-date, state-of-the-art information on genomic and epigenomic biomarkers presented by internationally recognized experts in a single edition. I sincerely hope that this book will provide an authoritative source of current information on biomarkers research and will prove useful to the scientists interested in this scientific discipline throughout the world. Moreover, it should be of interest to a variety of other scientific disciplines such as toxicology, medicine, and pharmacology, as well as to drug and food sciences.

Saura C. Sahu, PhD
Columbia, MD

Acknowledgements

I am indebted to the internationally recognized experts who shared my enthusiasm for this field of science and contributed generously to this book. They were selected from academia, industry, and government for their expertise in their own areas of research. Their work speaks for itself, and I am grateful to them for their strong commitment, cooperation, and excellent contributions in their areas of expertise.

I thank John Wiley & Sons Ltd, particularly Jenny Cossham and Katrina Maceda, for their excellent help, cooperation, support, and assistance with the timely publication of this book. Wiley published my first book, *Hepatotoxicity: From Genomics to* in vitro *and* in vivo *Models*, in 2007. Since then Wiley has published fifteen more of my books (including the current one), all in new, different, emerging, and developing areas of modern toxicology and medicine such as nanotoxicology, systems toxicology, toxicogenomics, epigenomics, microRNAs, stem cells—and I am still counting. I am amazed at Wiley's trust in me. I am grateful to John Wiley & Sons Ltd for the rest of my life!

Saura C. Sahu, PhD
Columbia, MD

List of Contributors

Eid Alshammari
Department of Biochemistry, Microbiology, and Immunology, Wayne State University School of Medicine, Detroit, MI, USA

Mayukh Banerjee
Department of Pharmacology and Toxicology, University of Louisville School of Medicine, Louisville, KY, USA

Robert Y.S. Cheng
Molecular Mechanisms Section, Cancer and Inflammation Program, Center for Cancer Research, National Cancer Institute, Frederick, MD, USA

Brian N. Chorley
Center for Computational Toxicology and Exposure, Office of Research and Development, U.S. Environmental Protection Agency, NC, USA

Carmen Elena Condrat
Carol Davila University of Medicine and Pharmacy, Bucharest, Romania;
Alessandrescu-Rusescu National Institute for Mother and Child Health, Fetal Medicine Excellence Research Center, Bucharest, Romania

Matei-Alexandru Cozma
Department of Gastroenterology, Colentina Clinical Hospital, 20125 Bucharest, Romania

Dragos Cretoiu
Alessandrescu-Rusescu National Institute for Mother and Child Health, Fetal Medicine Excellence Research Center, 020395 Bucharest, Romania;
Department of Cell and Molecular Biology and Histology, Carol Davila University of Medicine and Pharmacy, Bucharest, Romania

Sanda Maria Cretoiu
Department of Cell and Molecular Biology and Histology, Carol Davila University of Medicine and Pharmacy, Bucharest, Romania

Asim Bikas Das
Department of Biotechnology, National Institute of Technology Warangal, Warangal, Telangana, India

Elena-Codruța Dobrică
Department of Dermatology, "Elias" University Emergency Hospital, 011461 Bucharest, Romania

Ana P. Ferragut Cardoso
Department of Pharmacology and Toxicology, University of Louisville School of Medicine, Louisville, KY, USA

Jennifer L. Freeman
School of Health Sciences, Purdue University, West Lafayette, IN, USA

Yusuke Furukawa
Division of Cellular and Molecular Toxicology, Center for Biological Safety and Research, National Institute of Health Sciences (NIHS), Kawasaki, Japan

Mihnea-Alexandru Găman
Faculty of Medicine, "Carol Davila" University of Medicine and Pharmacy, Bucharest, Romania; Department of Hematology, Center of Hematology and Bone Marrow Transplantation, Fundeni Clinical Institute, Bucharest, Romania

David S. Hernández-Toledano
Department of Toxicology, Center for Research and Advanced Studies of the National Polytechnic Institute, Ciudad de México, México.

Yoko Hirabayashi
Center for Biological Safety and Research, National Institute of Health Sciences (NIHS), Kawasaki, Japan

Taisen Iguchi
Graduate School of Nanobioscience, Yokohama City University, Yokohama, Japan

Pius Joseph
Molecular Carcinogenesis Laboratory, Toxicology and Molecular Biology Branch, Health Effects Laboratory Division, National Institute for Occupational Safety and Health (NIOSH), Morgantown, WV, USA

Benjamin L. King
Department of Molecular and Biomedical Sciences, University of Maine, Orono, ME; Graduate School of Biomedical Science and Engineering, University of Maine, Orono, ME, USA

Satoshi Kitajima
Division of Cellular and Molecular Toxicology, Center for Biological Safety and Research, National Institute of Health Sciences (NIHS), Kawasaki, Japan

Makiko Kuwagata
Division of Cellular and Molecular Toxicology, Center for Biological Safety and Research, National Institute of Health Sciences (NIHS), Kawasaki, Japan

Deepak Mav
Sciome, Research Triangle Park, NC, USA

Moira Mccormick
Department of Environmental Health and Engineering, Johns Hopkins Bloomberg School of Public Health, Baltimore, MD, USA

B. Alex Merrick
Mechanistic Toxicology Branch, Division of the National Toxicology Program, National Institute of Environmental Health Sciences, NC, USA

Shinichi Miyagawa
Faculty of Advanced Engineering, Tokyo University of Science, Tokyo, Japan

Alexandra N. Nail
Department of Pharmacology and Toxicology, University of Louisville School of Medicine, Louisville, KY, USA

Tadaaki Nakajima
Department of Science, Yokohama City University, Yokohama, Japan

Mie Naruse
Division of Cellular and Molecular Toxicology, Center for Biological Safety and Research, National Institute of Health Sciences (NIHS), Kawasaki, Japan; Central Animal Division, National Cancer Center Research Institute, Tokyo, Japan

Gail M. Nelson
Center for Computational Toxicology and Exposure, Office of Research and

Development, U.S. Environmental Protection Agency, NC, USA

Takahiro Ochiya
Division of Cellular and Molecular Toxicology, Center for Biological Safety and Research, National Institute of Health Sciences (NIHS), Kawasaki, Japan; Division of Molecular and Cellular Medicine, Institute of Medical Science, Tokyo Medical University, Tokyo, Japan; Division of Molecular and Cellular Medicine, National Cancer Center Research Institute, Tokyo, Japan

Ryuichi Ono
Division of Cellular and Molecular Toxicology, Center for Biological Safety and Research, National Institute of Health Sciences (NIHS), Kawasaki, Japan

Dhiral P. Phadke
Sciome, Research Triangle Park, NC, USA

Jairus Pulczinski
Department of Environmental Health and Engineering, Johns Hopkins Bloomberg School of Public Health, Baltimore, MD, USA

Liz Saavedra Perez
Department of Molecular and Biomedical Sciences, University of Maine, Orono, ME, USA

Saura C. Sahu
Division of Toxicology, Center for Food Safety and Applied Nutrition, US Food and Drug Administration, Laurel, MD, USA

Athira Sairanthry Suku
Toxicology Division, Biomedical Technology Wing, Sree Chitra Tirunal Institute for Medical Sciences and Technology, Poojappura, Trivandrum, Kerala, India; School of Health Sciences, Purdue University, West Lafayette, IN, USA

Tomomi Sato
Graduate School of Nanobioscience, Yokohama City University, Yokohama, Japan

Ruchir R. Shah
Sciome, Research Triangle Park, NC

Jacob Sobota
Department of Biochemistry, Microbiology, and Immunology, Wayne State University School of Medicine, Detroit, MI, USA

J. Christopher States
Department of Pharmacology and Toxicology, University of Louisville School of Medicine, Louisville, KY, USA

Yuchen Sun
Department of Environmental and Occupational Health, University of Pittsburgh Graduate School of Public Health, Pittsburgh, PA, USA

Wan-yee Tang
Department of Environmental and Occupational Health, University of Pittsburgh Graduate School of Public Health, Pittsburgh, PA, USA

Erik J. Tokar
Mechanistic Toxicology Branch, Division of the National Toxicology Program, National Institute of Environmental Health Sciences, NC, USA

Parayanthala Valappil Mohanan
Toxicology Division, Biomedical Technology Wing, Sree Chitra Tirunal Institute for Medical Sciences and Technology, Poojappura, Trivandrum Kerala, India

Libia Vega
Center for Research and Advanced Studies of the National Polytechnic Institute, Department of Toxicology, Ciudad de México, México

Musa Watfa
Department of Environmental and Occupational Health, University of Pittsburgh Graduate School of Public Health, Pittsburgh, PA, USA

Genki Yamagishi
Faculty of Advanced Engineering, Tokyo University of Science, Tokyo, Japan

Zhe Yang
Department of Biochemistry, Microbiology, and Immunology, Wayne State University School of Medicine, Detroit, MI, USA

Yusuke Yoshioka
Division of Molecular and Cellular Medicine, Institute of Medical Science, Tokyo Medical University, Tokyo, Japan

Yingxue Zhang
Department of Biochemistry, Microbiology, and Immunology, Wayne State University School of Medicine, Detroit, MI, USA

1

Genomic and Epigenomic Biomarkers for Predictive Toxicity and Disease

Saura C. Sahu

Toxicity Testing in the Twenty-First Century

Consumer exposure to environmental chemicals is increasing. Therefore there is a need for developing short-term biomarkers for a rapid predictive toxicity screening of environmental chemicals. In 2007 the US National Academy of Sciences published its landmark report "Toxicity Testing in the 21st Century: A Vision and a Strategy" in order to reduce the costly and time-consuming classical animal studies required for safety risk analysis (Judson, Houck et al. 2016; Judson, Kavlock et al. 2013; Krewski et al. 2010; NAS 2007). In vitro cytotoxicity assays using human primary cells and cell lines in culture with focus on biological pathways is an important part of the twenty-first-century toxicity screening program. Alternative in vitro screening assays of this kind reduce, refine, and replace animal testing. One of the important goals of the twenty-first-century toxicity testing is the development of in vitro assays designed to predict the in vivo outcomes. In consequence, such asays can be used to predict starting doses for animal studies (Judson, Houck et al. 2016; Judson, Kavlock et al. 2013; Krewski et al. 2010; NAS 2007).

Biomarkers

The US National Institutes of Health (NIH) and Food and Drug Administration (FDA) define "biomarker" as "a defined characteristic that is measured as an indicator of normal biological processes, pathogenic processes or responses to an exposure or intervention" (Califf 2018). In practice, a biomarker is a measurable endpoint that can be found in body fluids or tissues as the indicator of exposure, infection, or disease. Biomarkers are used in clinical practice. They play important critical roles in the development of drugs and medical devices (Robb et al. 2016).

The new "omics" technology provides genomic and epigenomic tools for the development of biomarkers for predicting toxicity and disease. However, these biomarkers must be

Genomic and Epigenomic Biomarkers of Toxicology and Disease: Clinical and Therapeutic Actions,
First Edition. Edited by Saura C. Sahu.

validated, and that is not easy. Almost a decade ago, Janet Woodcock (2009) warned that "to deliver a new generation of biomarkers to clinical practice will be long and hard." In spite of her warning, a lot of progress has been made in the discovery of genomic and epigenomic biomarkers for toxicity and disease during the last decade. But unfortunately no single biomarker is expected to provide most of the information required for toxicity and disease. Therefore a battery of biomarkers is desirable for this purpose.

Genomic Biomarkers

A genome is an organism's complete set of DNA, including all of its genes. Each genome contains all of the information needed to build and maintain that organism.

(https://www.nih.gov)

The genome consists of a full set of chromosomes. The major component of a chromosome is the deoxyribonucleic acid (DNA), which carries the genetic information. The genetic information is transcribed into the messenger ribonucleic acid (RNA). The triplets of messenger RNA are translated into different amino acids, the building blocks of protein. The DNA, the RNA, and protein are the source of life. The genomic information is generated by profiling alterations in DNA, RNA and protein levels. Genetics is the totality of heritable changes in the gene expression profiles caused by the modification of the genomic DNA as a result of the changes in sequence of its bases. Various human diseases, including cancer, are caused by changes in the genomic DNA sequence.

The Human Genome Project, which started in 1990 and was completed in 2003, determined the whole genome sequences of all the estimated 20,000 to 25,000 human genes and made them accessible to the public. This genome sequence information led to analysis of the genome-wide expression of cellular genes through genomic technology. The simultaneous monitoring of thousands of nucleic acid sequences through microarray technology became possible by quantifying the amount of specific transcripts. The global analysis of gene expression profiles provides a comprehensive view of toxicity and disease. Besides genomics, other new technologies of this kind, for proteins (proteomics) and for metabolites (metabolomics), yield useful information on toxicity and disease.

One major application of genomics is to identifying the genes that predict the effects of exposure to environmental chemicals. Such genes are sensitive and informative indicators of time- and dose-dependent adverse effects of chemical exposure. Therefore they are predictive biomarkers of toxicity and disease. The microarrays and DNA sequencing technologies provide powerful tools; they can identify biomarkers from gene expression profiles that show altered expression levels in thousands of genes at a time. Genome-wide analysis through these technologies leads to an understanding of toxicological and pathological changes at the RNA or DNA level.

Epigenomic Biomarkers

Epigenomics is a new, emerging area of science. Humans are exposed daily to a mixture of chemicals present in their environment. Nearly every cell in the human body has the same set of genes, but different types of cells are affected differently by these chemicals. Gene expression in cells is controlled by several mechanisms. In genomic mechanisms of toxicity and disease, the genomic DNA sequence is altered through chemical exposure. Genomic

DNA sequences modified in this way are not cell- and tissue-specific. However, in some cases toxicity and diseases are caused by DNA modifications that are due to chemical exposure, but in the absence of any direct alteration in genomic DNA sequence. Such DNA modifications are known as epigenomics, where DNA methylation regulates gene expression without direct alterations in the genomic DNA sequence. In DNA methylation, gene expression occurs at the cytosine dinucleotide when a methyl group is added at position 5, producing methylcytosine (de Gannes et al. 2020; Jones 2012). Unlike genomic changes, epigenetic changes are cell- and tissue-specific. Epigenetic changes may be heritable and non-heritable. DNA methylation is associated with several human diseases, including cancer.

The epigenome is defined as heritable biological information contained outside the DNA sequence (Dolinoy and Jirtle 2008). It consists of DNA methylation, histone modifications, and microRNAs. Non-coding RNAs (ncRNAs) regulate gene expression at the transcriptional or post-translational levels without changing the genomic DNA sequence (Li and Cui 2018).

MicroRNAs

The microRNAs form a class of single-stranded, non-coding small RNAs with an average of 22 nucleotides. They are involved in the post-transcriptional regulation of gene expression. They were discovered in 1993 by the Victor Ambros research group (Lee et al. 1993), in a study on the development of *Caenorhabditis elegans*. These researchers discovered that the microRNA lin-4 regulated negatively the expression of another gene, lin-14. In subsequent studies, microRNAs were identified in multiple species, including plants, worms, insects, fish, rodents, and mammals. The sequences of microRNAs in mammalian species are highly conserved (Friedman et al. 2009). The microRNAs are found in body fluids such as saliva, urine, and plasma. Schueller et al. (2018) have demonstrated that the dysregulation of microRNAs is associated with toxicity and liver diseases. The microRNAs are considered new molecules related to disease classification and prediction (Otsuka et al. 2017) and can be used as a biomarker in the prediction of liver cancer (Livingstone et al. 2019). Changes in microRNA expression after chemical exposure can be determined by microarrays and next-generation sequencing technologies. It is believed that microRNA profiling is a valuable diagnostic tool. The microRNAs can serve as excellent biomarkers of toxicity and disease.

Differentially expressed microRNAs are the most common genomic biomarkers used for drug safety. Candidates for translational microRNA biomarkers for liver injury have been proposed (Li et al. 2021). Urinary microRNA biomarkers of drug-induced kidney injury have been suggested (Chorley et al. 2021). Recent studies have provided evidence that tumor-related genes regulated through DNA methylation and microRNA are involved in the occurrence of hepatocellular carcinoma (Lin et al. 2019).

Circular RNAs

Circular RNAs (circRNAs) are a class of closed, long, non-coding RNAs with stable expression in various tissues in different mammalian species. These most recently discovered ncRNAs are present abundantly in the transcriptome of eukaryotic cells. They play an important role in the regulation of gene expression (Shao and Jiang 2021) and in tumor development (Jeyaraman et al. 2019). Therefore they are considered potential biomarkers of toxicity and disease (Jeyaraman et al. 2019; Shao and Jiang 2021).

The Need for a Book on Genomic and Epigenomic Biomarkers for Predictive Toxicity and Disease

The rate of publications during the last few years has demonstrated that the developments of genomic and epigenomic biomarkers for predictive toxicity and disease are moving very rapidly, and therefore there is a need for new means to report the updated current status of this new area of research. As editor of *Genomic and Epigenomic Biomarkers for Toxicology and Disease: Clinical and Therapeutic Actions*, it gives me great pride, pleasure, and honor to introduce this unique book, which encompasses many aspects of genomic and epigenomic biomarker research never brought together in one publication before.

References

Califf, R.M. (2018). Biomarker definitions and their applications. *Exp. Biol. Med. (Maywood)* 243 (3): 213–221. Published online 2018 Feb 6. https://doi.org/10.1177/1535370217750088.

Chorley, B.N., Ellinger-Ziegelbauer, H., Tackett, M., Simutis, F.J., Harrill, A.H., McDuffie, J., Atabakhsh, E., Nassirpour, R., Whiteley, L.O., Léonard, J., Carswell, G.K., Harpur, E., Chen, C.L., and Gautier, J. (2021). Urinary miRNA biomarkers of drug-induced kidney injury and their site specificity within the nephron. *Toxicol. Sci.* 180 (1): 1–16. https://doi.org/10.1093/toxsci/kfaa181.

de Gannes, M., Ko, C., Zhang, X., Biesiada, J., Niu, L., Koch, S.E., Medvedovic, M., Rubinstein, J., and Puga, A. (2020). Dioxin disrupts dynamic DNA methylation patterns in genes that govern cardiomyocyte maturation. *Toxicol. Sci.* 178 (2): 325–337. https://doi.org/10.1093/toxsci/kfaa153.

Dolinoy, D.C. and Jirtle, R.L. (2008). Environmantal epigenomics in human health and disease. *Environ. Mol. Mutagen.* 49 (1): 4–8.

Friedman, R.C., Farh, K.K., Burge, C.B., and Bartel, D.P. (2009). Most mammalian mRNAs are conserved targets of microRNAs. *Genom. Res.* 19: 92–105.

Jeyaraman, S., Hanif, E.A.M., Ab Mutalib, N.S., Jamal, R., and Abu, N. (2019). Circular RNAs: Potential regulators of treatment resistance in human cancers. *Front. Genet.* 10: 1369.

Jones, P.A. (2012). Functions of DNA methylation: Islands, start sites, gene bodies and beyond. *Nat. Rev. Genet.* 13: 484–492.

Judson, R., Houck, K., Martin, M., Richard, A.M., Knudsen, T.B., Shah, I., Little, S., Wambaugh, J., Setzer, R.W., Kothya, P., Phuong, J., File, D., Smith, D., Reif, D., Rotroff, D., Kleinstreuer, N., Sipes, N., Xia, M., Huang|, R., Crofton, K., and Thomas, R.S. (2016). Analysis of the effects of cell stress and cytotoxicity on in vitro assay activity across a diverse chemical and assay space. *Toxicol. Sci.* 152 (2): 323–339.

Judson, R., Kavlock, R., Martin, M., Reif, D., Houck, K., Knudsen, T., Richard, A., Tice, R.R., Whelan, M., Xia, M., Huang, R., Austin, C., Daston, G., Hartung, T., Fowle, J.R., Wooge, W., Tong, W., and Dix, D. (2013). Perspectives on validation of high-throughput assays supporting 21st century toxicity testing. *ALTEX* 30 (1): 51–56.

Krewski, D., Acosta, D., Andersen, M., Anderson, H., Bailar, J.C., Boekelheide, K., Brent, R., Charnley, G., Cheung, V.G., Green, S., Kelsey, K.T., Kerkvliet, N.I., Li, A.A., McCray, L., Meyer, O., Patterson, R.D., Pennie, W., Scala, R.A., Solomon, G.M., Stephens, M., Yager, J., and Zeise, L. (2010). Toxicity testing in the 21st century: A vision and a strategy. *J. Toxicol. Environ. Health. B. Crit. Rev.* 2010 Feb 13 (0): 51–138. 10.1080/10937404.2010.483176.

Lee, R.C., Feinbaum, R.L., and Ambros, V. (1993). The *C. elegans* heterochronic gene lin-4 encodes small RNAs with antisense complementarity to lin-14. *Cell* 75: 843–854.

Li, C.Y. and Cui, J.Y. (2018). Regulation of protein-coding gene and long noncoding RNA pairs in liver of conventional and germ-free mice following oral PBDE exposure. *Plos One* 13: e0201387.

Li, D., Knox, B., Gong, B., Chen, S., Guo, L., Liu, Z., Tong, W., and Ning, B. (2021). Identification of translational microRNA biomarker candidates for ketoconazole-induced liver injury using next-generation sequencing. *Toxicol. Sci.* 179 (1): 31–43. https://doi.org/10.1093/toxsci/kfaa162.

Lin, C., Yuan, G., Hu, Z., Zeng, Y., Qiu, X., Yu, H., and He, S. (2019). Bioinformatics analysis of the interactions among lncRNA, miRNA and mRNA expression, genetic mutations and epigenetic modifications in hepatocellular carcinoma. *Mol. Med. Rep.* 19: 1356–1364. https://doi.org/10.3892/mmr.2018.9728.

Livingstone, M.C., Johnson, N.M., Roebuck, B.D., Kensler, T.W., and Groopman, J.D. (2019). Serum miR-182 is a predictive biomarker for dichotomization of risk of hepatocellular carcinoma in rats. *Mol. Carcinog.* 58: 2017–2025. https://doi.org/10.1002/mc.23093.

NAS (U.S. National Academy of Sciences). (2007). Toxicity Testing in the 21st Century: A Vision and A Strategy. Washington, DC: National Academy Press.

Otsuka, M., Kishikawa, T., Yoshikawa, T., Yamagami, M., Ohno, M., Takata, A., Shibata, C., Ishibashi, R., and Koike, K. (2017). MicroRNAs and liver disease. *J. Hum. Genet.* 62: 75–80. https://doi.org/10.1038/jhg.2016.53.

Robb, M.A., McInnes, P.M., and Califf, R.M. (2016). Biomarkers and surrogate endpoints: Developing common terminology and definitions. *JAMA* 2016 315: 1107–1108.

Schueller, F., Roy, S., Vucur, M., Trautwein, C., Lurdde, T., and Roderberg, C. (2018). The role of miRNA in the pathophyisiology of liver diseases and toxicity. *Int. J. Mol. Sci.* 19: 261–269.

Shao, Y. and Jiang, Y. (2021). Circular RNAs in Toxicology. *Toxicol. Sci.* 179 (2): 149–161. https://doi.org/10.1093/toxsci/kfaa173.

Woodcock, J. (2009). Chutes and ladders on the critical path: Comparative effectiveness, product value, and the use of biomarkers in drug development. *Clin. Pharmacol. Ther.* 86 (1): 12–14.

2

MicroRNAs as Non-invasive Biomarkers of Toxicity and Chemical Hazard

Genomic and Epigenomic Biomarkers of Toxicology and Disease

Gail M. Nelson and Brian N. Chorley

Center for Computational Toxicology and Exposure, Office of Research and Development, US Environmental Protection Agency, Research Triangle Park, NC 27709

Introduction

MicroRNAs (miRNAs) are short, non-coding RNAs whose primary known function is to regulate the transcription and translation of messenger RNA. Targeting occurs through sequence specificity; however, this interaction is not limited to a single gene. Rather, a particular miRNA can regulate hundreds of genes. In addition, a transcript may be targeted by multiple miRNAs. As a result, it is estimated that between 30% and 80% of all transcribed genes in mammals may be regulated by miRNAs, depending on the tissue or cell type (Clark et al. 2014; Gu and Kay 2010; Lu and Clark 2012). Thus, alterations in miRNA amounts, even if minute, could feasibly have a large impact on transcriptional profiles. The consequence is important in terms of biological effect, as well as for the identification of biomarkers of health impact.

The need for specific and sensitive biomarkers spans research and development, regulatory, safety, and clinical sectors. Biomarkers can indicate exposure, biological effect, and sensitivity. All three areas have important purposes and applications. Those that measure biological effect more directly link to biological function, and thus to the putative adverse health effects that a perturbation may mediate. A biomarker may be a measurable alteration—chemical, biochemical, physiological, behavioral, or of some other kind—within an organism (World Health Organization and International Programme on Chemical Safety 1993). miRNAs have the potential to be good biomarkers of biological effect because they are well defined, chemically uniform, restricted to a manageable number, and stable (not readily degraded). They also get released into extracellular matrices, where they are accessible and measurable. In these biofluids such as blood, urine, and sputum, miRNAs serve as unique biomarkers for a minimally invasive prediction of toxicant exposure. The altered biological pathways that is consequent upon (and due to) miRNA changes can therefore reflect the mechanisms of toxicant-related diseases. Measurements of these miRNAs in biofluids can therefore serve as biomarkers of effect.

Indeed, changes in intracellular miRNA expression due to toxicant exposure have been well documented in multiple tissues and model organisms (Yu and Cho 2015) and are regulated by both genetic and epigenetic mechanisms. DNA hypo- and hypermethylation and histone modifications are involved in the regulation of the expression of miRNA

Genomic and Epigenomic Biomarkers of Toxicology and Disease: Clinical and Therapeutic Actions,
First Edition. Edited by Saura C. Sahu.

promoters (Tomasetti et al. 2019). As a high proportion of miRNAs are embedded in CpG islands susceptible to methylation, miRNA genes are methylated more frequently than protein-coding genes (Kozomara et al. 2019; Morales et al. 2017). There is also evidence that some miRNAs can interact with other miRNAs, in a form of self-regulation (Hill and Tran 2021). In addition, mechanisms for the alteration of intracellular miRNA expression through exposure to environmental carcinogens have been proposed that involve the modulation or blocking of normal miRNA processing and maturation (Izzotti and Pulliero 2014). DNA damage can alter miRNA expression via p53-dependent mechanisms. p53 interaction with the Drosha/DGCR8 processing complex modulates the processing of pri-miRNAs to pre-miRNAs. In addition, Dicer, another processing unit that produces the final mature miRNA, is a direct transcriptional target for p53. Alternatively, the binding of electrophilic metabolites to nucleophilic sites of miRNA precursors can form miRNA adducts that are not able to access the catalytic pockets of Dicer, therefore arresting miRNA maturation. Further, toxicant metabolites binding to Dicer itself in the proximity of miRNA catalytic sites can block the maturation of miRNA precursors (Izzotti and Pulliero 2014).

The interest in miRNAs as biomarkers of exposure to environmental toxicants and resultant biological effects is bolstered by their stable detection in extracellular biofluids, where they can be non-invasively sampled. There are a number of mechanisms by which miRNAs are released from the cell into these matrices (Condrat et al. 2020) and, importantly, can be linked to tissue or cell type specificity (or both), as well as to mechanisms of biological perturbation that may relate to stress response, toxicity, and disease. In this chapter we review the promise of these putative biomarkers, together with the technical challenges that lie ahead if we want to establish them in the practice of toxicology and regulatory sciences.

Mechanisms that Contribute to Extracellular miRNA Release

When using the miRNAs present in biofluids as biomarkers of tissue perturbation, toxicity, or disease, one must understand the origin of these non-coding RNAs in the extracellular space. Most studies have focused on the miRNAs found in blood and urine, although extracellular miRNA has been noted in sputum, tears, amniotic fluid, cerebrospinal fluid, breast milk, and bronchial lavage fluid, among other sites (Arroyo et al. 2011; Chen et al. 2008; Turchinovich et al. 2011; Valadi et al. 2007; Vickers et al. 2011; Wang et al. 2010a; Weber et al. 2010). Although many miRNAs seem to be ubiquitously present in many characterized biofluids, blood plasma has the highest amount of uniquely present miRNAs (Weber et al. 2010). Blood is a complex liquid "tissue" that interacts with many cell types not accessible to other fluids, including those of hematopoietic residence—predominantly erythrocytes and reticulocytes. The contributions from these cell types are important to characterize, so that they can be separated, if needed, from the signals of other cell types or tissues. For example, in whole blood samples, miRs-486-5p and -451a are highly represented because they are derived from erythrocytes and can complicate or mask the evaluation of other putative biomarker miRNAs of lower abundance in the blood (Juzenas et al. 2020). Mechanistically, these potentially interfering factors can be removed by processing blood into serum or plasma, or, in cases where hemolysis may increase the proportion of erythrocyte-derived miRNAs in serum, blocking steps can reduce the measurement of these miRNAs (LaBelle et al. 2021).

Extracellular miRNAs are released by their cells of origin through a number of different mechanisms, which can simply be classified into "active" and "passive" release (Harrill et al. 2016). Those that are passively released are due to mechanisms of plasma membrane breakdown that lead to the spilling or "leaking" of intracellular contents into proximal biofluids. This can occur with cell death, necrosis, or toxicity. Importantly, some miRNAs have cell- or tissue-specific expression (Bailey and Glabb 2018). As such, these biomarkers may indicate specific tissue damage or disease. A number of studies have demonstrated the specificity of miRNAs and other non-coding RNAs for different tissues in clinical and non-clinical species such as human (miRNA TissueAtlas: Ludwig et al. 2016), rat (RATEmiRs database: Bushel et al. 2018; Smith et al. 2016), mouse (Isakova et al. 2020), and dog (Koenig et al. 2016). According to various analyses of these resources, those considered tissue-specific ranged from ~ 100 to ~ 400 miRNAs, depending on the species, and there was noted overlap among mammals that indicated putative biomarkers that could be used for cross-species profiling (Figure 2.1).

Packaging into Vesicles and Mechanisms of Release

Both prokaryotic and eukaryotic cells release membrane-enclosed microvesicles that package a variety of different cellular-derived components, including nucleic acids, proteins,

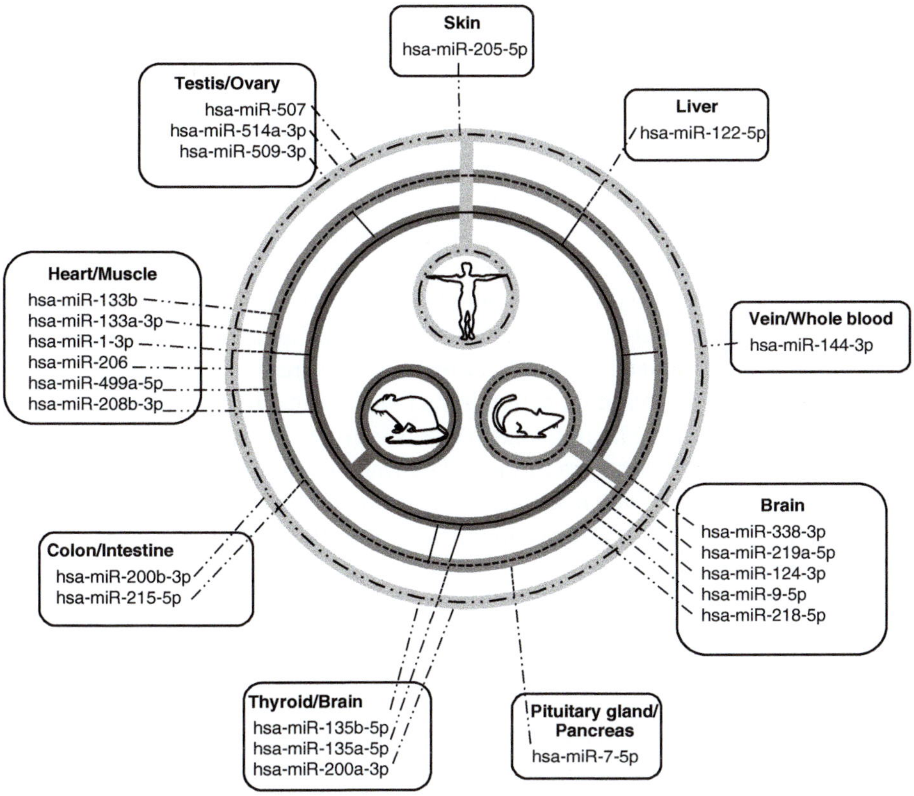

Figure 2.1 Tissue-specific miRNAs in mammals. Human tissue-specific miRNAs that were described as part of the human miRNA tissue atlas study (Ludwig et al. 2016) (light grey circle with dotted/dashed line) were cross-referenced with miRNA atlas studies that examine tissue-specific miRNAs in mouse (medium grey circle with dashed line) and rat (dark grey circle with solid line). Those miRNAs that transverse all three circles are ideal cross-species biomarkers for toxicological studies.

and lipids. The vesicles range in size (from small 50–150 nm exosomes to large 1,000–10,000 nm tumor cell-derived oncosomes), location of biogenesis, and content. These vesicles play important roles in cellular communications and their content is enriched in small non-coding RNAs (such as miRNAs, lncRNAs, snoRNAs, piRNA, snRNAs, etc.), but also in intact mRNAs and in fragmented tRNAs, mRNAs, lncRNAs, rRNAs, and other nucleic acids (Turchinovich et al. 2019). The roles of these packaged contents are not entirely clear for all RNA species; however, many studies have demonstrated that some of these RNAs can mediate differential responses in recipient cells. Many studies have focused on miRNAs in this paracrine role (reviewed in O'Brien et al. 2020) which therefore may make them serve as putative biomarkers of "active" release due to homeostatic, responsive, and perturbed cellular states. A number of characterization studies have been performed to determine the RNA contents of extracellular microvesicles isolated from human plasma, saliva, and urine, as well as cell lines of different lineages (see reviews in Amorim et al. 2017 and Turchinovich et al. 2019). In one of the first studies that use deep sequencing to profile extracellular vesicle (EV) RNAs, small RNAs were reported to dominate these "shuttle"-derived fractions; however, the distribution was different from that of the parent mouse dendritic cells co-cultured with cognate T cells (Nolte-'t Hoen et al. 2012). Importantly, whereas miRNAs dominated the intracellular small RNA content (~55% of the reads), miRNAs were only the fourth most populated of the EV-derived small RNAs, at around 10% of the total reads. Later studies have both supported and contradicted these findings. For example, Sork et al. (2018) also described different patterns of intracellular small, ribosomal, and other RNAs versus those found in EVs. Of the total reads, small RNAs were a minority; they were noted in the five cell cultures assessed, ranging from ~ 2% to ~ 40% of the reads. However, 58%–83% of the small RNA EV reads were attributed to miRNAs, many of these reflecting those found in the parent cell, with some notable exceptions such as miR-451a. The disagreements between different studies could be due to cell types, methods, and the analyses used. Overall, from a biomarker perspective, the content of these EVs reflects the cellular state and can be specific to a cell type by partially reflecting the transcriptome (Sork et al. 2018; Srinivasan et al. 2019); yet the EVs differ in intracellular RNA content from the source cell (Guduric-Fuchs et al. 2012; Zhang et al. 2010). Therefore packaging these EVs is not merely representative of the RNA cellular milieu; it is rather an active process that may be influenced by cellular responses to stressors.

Many recent studies have focused on determining mechanisms that are involved in EV formation and RNA packaging. The initial formation varies for the different types of EVs. For example, intraluminal vesicles form through the invagination of endosomes known as multivesicular bodies (MVBs), which are then guided by cytoskeletal components and degraded by giant intracellular lysosomes or released into the extracellular space after fusion with the plasma membrane (see reviews in Janas et al. 2015 and O'Brien et al. 2020). Other EVs, such as microvesicles, larger oncosomes, and apoptotic bodies, form directly from the plasma membrane itself. During these processes, RNAs are selectively packaged by the vesiculation machinery through different interactions. These vesicles are enriched with cholesterol, sphingomyelin, glycosphingolipids, and phosphatidylcholine comprised of saturated fatty acids when they are derived from raft-like regions of the MVBs and the plasma membrane. Ceramides, which preferentially locate on the outside rafts of MVBs, are thought to associate with ribonucleoprotein (RNP) complexes and may help selectively package RNA into EVs (Kosaka et al. 2013). Both RNA sequence and RNA secondary structure can mediate this selection (Janas et al. 2006). The sumoylated heterogenous nuclear RNP (specifically, hnRNPA2B1) can recognize the GGAG motif in miRNAs (e.g., miR-198 and -601) and subsequently transfer into exosomes (Villarroya-Beltri et al. 2013) without itself being packaged (Zhou et al. 2020). Sumoylated hnRNPA1 may work in a similar fashion (Li et al. 2004). Also, the synaptotagmin-binding cytoplasmic RNA-interacting protein (SYNCRIP) mediated the sorting of miRNAs bearing the GGCU motif in hepatocyte cell culture (Santangelo et al. 2016). These seed sequences, also termed "EXO-motifs," are located in the 3′ sites of miRNAs (Janas et al. 2006), whereas the sites that mediate mRNA targeting (and primary function) are located in the 5′ end. This suggests that the loading of miRNAs into vesicles could occur even if there were very different mRNA targeting functions. It has also been noted in B cells that the post-transcriptional addition of uracil at the 3′ end of the miRNA sequence preferentially sorts into exosomes, whereas adenylated miRNAs remain cell-bound (Koppers-Lalic et al. 2014). Other RNA-binding proteins implicated in preferential RNA loading include AGO2 (McKenzie et al. 2016), annexin A2 (Hagiwara et al. 2015), major vault protein (Statello et al. 2018), YBX1 (Kossinova et al. 2017; Shurtleff et al. 2016; Yanshina et al. 2018), lupus La protein (Temoche-Diaz et al. 2019), and Arc1 (Ashley et al. 2018) (see review in O'Brien et al. 2020).

Biological Interpretations Due to Mechanisms of Release

The difference between a "leakage" miRNA biomarker and one that is non-randomly packaged into an EV as a putative communication signal, is an important distinction for development of miRNA biomarkers that are indicative of health effects. For example, mammalian miR-122 has been heavily investigated as a leakage biomarker for liver toxicity, in the context of both non-clinical and clinical species. miR-122 accounts for approximately 72% of the total liver miRNAs in mice (Lagos-Quintana et al. 2002) and is highly specific for human, mouse, and rat livers (Landgraf et al. 2007; Smith et al. 2016). In a landmark study by Wang et al. (2009), mice were treated with hepatotoxic levels of acetominophen (APAP) and miRNA microarrays were used to measure difference in the plasma. Forty-four miRNAs were significantly altered, miR-122 showing a 470-fold increase twenty-four hours after APAP overdose, and these alterations could be observed as soon as one hour after exposure. This APAP overdose or acute liver injury-associated miR-122 release was later confirmed in human patients (Starkey Lewis et al. 2011). The observation that decreased liver-based miRNAs with APAP hepatotoxicity correlated with increases in serum plasma suggested that these miRNAs were released as packaged contents or leaked during hepatocyte necrosis; the leakage theory was supported by Arroyo et al. (2011), who demonstrated that miR-122 was primarily associated with a RNP complex (Ago2) and was not membrane-bound as in exosomes. However, later studies provided clues that the mechanism of miR-122 release from liver was context-dependent, that is, dependent on factors such as the condition of the liver, the timing of the injury or perturbation, the half-life of a miRNA, and baseline variability. Subtoxic exposures in vitro result primarily in hepatocyte-derived release of exosome-bound miR-122 (Holman et al. 2016; Mosedale et al. 2018), and alcoholic and inflammatory liver disease was associated with exosome-bound miR-122—as opposed to drug-induced injury, which was found in the protein-bound fraction of plasma (Bala et al. 2012). Further, the protection of miRNA from RNases present in biofluids may preferentially enrich exosome and protein-bound miRNAs (Arroyo et al. 2011; Koberle et al. 2013; Li et al. 2012; Turchinovich et al. 2011).

Therefore the decision whether to quantify miRNA biomarkers from unfractionated samples or from isolated EVs may depend on context and on intended use. Additionally, questions remain as to which one is more reliable in biomarker measurements; and some reports suggest that this may depend on the specific human disease. A disease-specific alteration in exosome release and clearance, or the administration of therapeutics that alter the volume of blood, can skew the yield (Buschmann et al. 2018; Hanna et al. 2019). The isolation of miRNAs from unfractionated biofluid samples should capture both miRNAs that originate from passive leakage into the sampled matrix and miRNAs that come from actively released membrane-bound vesicles lysed through the isolation procedure. This isolation avoids major drawbacks of vesicle purification, which lowers RNA yield and integrity and can result is greater experimental variability. Another challenge is that, depending on the method used, different populations of EVs are isolated; and this produces only partially overlapping RNA profiles (Buschmann et al. 2018). Recently, a comprehensive analysis of published results compared the utility of miRNAs isolated from EVs to the utitliy of those isolated from unfractionated whole serum or plasma for specific biomarker use (Nik Mohamed Kamal and Shahidan 2019). Twenty of the thirty-two studies cited suggested that EV-derived miRNAs may be preferable for the intended use examined. A commonly reported reason was the potential for increased stability that might be derived from being encapsulated in a vesicle. But it is also important to note that these intended uses varied, and included both in vitro and in vivo biomarker development for heart failure, chronic pain, septic shock, non-alcoholic fatty liver disease, immunization, and many types of cancer. So, while a slight majority of studies recommended vesicle fraction miRNA for biomarker measurement, specific research needs to be performed to provide the appropriate conditions and context of use.

MicroRNAs as Bioindicators of Toxicity and Disease

Much miRNA biomarker research has been focused on the early detection of drug-induced liver injury (DILI), a major roadblock in drug development. Alanine aminotransferase, aspartate aminotransferase, alkaline phosphatase, and total bilirubin are the currently approved DILI biomarkers in clinical practice (Robles-Diaz et al. 2016). These biomarkers are not specific to hepatotoxicity, as they increase in almost all abnormal liver conditions. Furthermore, they lack sensitivity, as they appear once liver damage has developed, and consequently have limited use for predicting potential liver injury at an early stage. For this reason, circulating miRNAs are being investigated for the monitoring of drug-induced tissue injury (tissue degeneration and necrosis) in cases where highly abundant tissue-specific miRNAs are passively released after cell injury. In incidences of DILI, studies have demonstrated that miR-122 presence in blood correlates tightly with pathological indications and clinical chemistries (i.e., ALT, AST) of liver degeneration (Krauskopf et al. 2015; Song et al. 2016; Wang et al. 2009). In combination with the liver injury biomarkers cytokeratin-18 and glutamate dehydrogenase, miR-122 outperformed individual biomarkers and had a higher than 0.92 correlation with alanine aminotransferase (ALT) in a validation cohort of individuals with acetominophen overdose (Llewellyn et al. 2021). In pre-clinical settings, miR-122 did mildly improve the diagnosis of DILI (4% increase in predictive accuracy using a multiparameter approach) and was highly correlated with necrosis, vacuolization, and hepatocellular hypertrophy (Sharapova et al. 2016), which indicated its usefulness as a leakage biomarker. However, the use of miR-122 has been hampered by a notably high baseline variability in human samples (> 100-fold interval; Vogt et al. 2019). This was likely due to differences in ethnic backgrounds (Church et al. 2019), undiagnosed states such as milder pathologies of non-alcoholic fatty liver disease (Cermelli et al. 2011), and short circulatory half-life (Thulin et al. 2017).

A review by Harrill et al. (2016) comparing biofluid-based miRNAs (e.g., blood, urine) to traditional biomarkers of tissue toxicity found that miRNA often outperformed conventional biomarkers in terms of a better correlation with tissue injury and earlier detection. Recent studies in rats have identified miRNA biomarkers specific to different stages (early, middle, late) and types (hepatocellular injury, cholestasis, steatosis) of DILI (Kagawa et al. 2018) and to different lesion sites in the nephron due to nephrotoxicants (Chorley et al. 2021b). Further study is needed to determine whether these biomarkers are translatable for the detection of specific organ injuries in humans. Potential miRNA biomarker candidates for the early detection of DILI were identified by examining hepatic miRNA expression in ketoconazole-treated rats with miRNA-sequencing (Li et al. 2021). The investigators then looked for the candidates evolutionarily conserved between rats and humans in the culture medium of ketoconazole-treated HepaRG cells with quantitative polymerase chain reaction (PCR) and identified miR-34a-5p, miR-331-3p, and miR-15b-3p as translational biomarker candidates for the early detection of ketoconazole-induced liver injury.

In addition to liver and kidney injury, circulating miRNAs are being investigated as mechanistic biomarkers of diseases such as cancer (Dutta et al. 2019; Jin et al. 2019; Lin et al. 2019; Pascut et al. 2019; Wang et al. 2018), cardiovascular disease (Zhou et al. 2018), metabolic syndrome (Huang et al. 2018), neurodegenerative disease (Juzwik et al. 2019; Sharma and Lu 2018), and chronic obstructive pulmonary disease (COPD) (Finicelli et al. 2020), all diseases whose etiology can include environmental toxicant exposure. Several recent reviews examine the dysregulation of miRNAs by environmental contaminants related to human health (Balasubramanian et al. 2020; Harrill et al. 2016; Qiao et al. 2019; Sollome et al. 2016; Tumolo et al. 2020; Wallace et al. 2020). Vrijens et al. (2015) summarized some of the miRNAs that respond to environmental exposure and their roles in human disease. The current knowledge linking cancer and neurodegenerative diseases to dysregulation of miRNAs after pesticide exposure was reviewed by Costa et al. (2020). A comprehensive literature review by Sima et al. 2011 (twenty-seven studies between 2012 and 2020) focused on miRNA expression in humans exposed to various air pollutants. Because air pollution is a factor in the development of lung cancer, they reported miRNAs commonly deregulated by both conditions, identifying twenty-five miRNAs that could serve as biomarkers of exposure to harmful pollutants that potentially contribute to lung cancer development. Several miRNAs were deregulated in multiple studies and may therefore be the most promising candidates: miRs-222, -21, -126-3p, -155 and -425.

Table 2.1 compiles the literature evidence from these and other studies in humans. The heterogeneity in sample types, methods of analysis, and study designs makes it difficult to compare the studies directly. Altered levels of some miRNAs are observed in several studies with the same pollutant category and sample type, but with different direction of expression (e.g., particulate matter (PM) exposure altered miRs-21-5p, -223-3p and -146; Bollati et al. 2010; Louwies et al. 2016; Motta et al. 2013; Rodosthenous et al. 2016). Additional miRNAs dysregulated by air pollution in more than one study included let-7g, miR-126 and miR-132. miR-146 was downregulated in both bronchial epithelial cells and induced sputum of smokers. miRs-21, -126 and -155 were dysregulated in both urine and blood–serum with arsenic exposure. Overall, miR-21, which is highly expressed and ubiquitous across most cell types, was most often reported as dysregulated across multiple categories of environmental contaminants. miR-21 plays a role in inflammation and is elevated in many different disease states, which suggests that it is commonly upregulated in a stress environment and is not specific to any one disease or exposure, as is probably the case for other ubiquitously expressed miRNA (Jenike and Halushka 2021).

Table 2.1 Human studies on environmental toxicant-induced changes in miRNA expression from accessible matrices.

Exposure	Sample type	miRNAs	Related Disease	References
Air Pollution	various biofluids	multiple, in particular miR-222, miR-21, miR-126-3p, miR-155 and miR-425	lung cancer	(Sima et al. 2021)
PM2.5, black carbon, organic carbon, sulfate	leukocytes	miR-126, miR-135a, miR-146a, miR-155, miR-9 ↓		(Fossati et al. 2014)
Coal fumes (miners)	blood lymphocytes	SNPs in pre-miRNA genes of miR-149	pneumoconiosus	(Wang et al. 2010b)
Black carbon and coal dust (urban traffic PM)	blood	let-7 g, miR-29, miR-146, miR-421 ↑		(Motta et al. 2013)
Air pollution PM10	plasma MV	miR-126 ↑		(Motta et al. 2013)
Metal rich fumes in steel industry (PM10)	blood leukocytes	miR-21 and miR-222 ↑		(Bollati et al. 2010)
Metal rich fumes in steel industry (PM10)	blood leukocytes	miR-146 ↓		(Bollati et al. 2010)
Metal rich PM	plasma	miR-128, miR-302c ↑		(Tumolo et al. 2020), (Bollati et al. 2015)
PM	blood	miR-21-5p, miR-223-3p ↓		(Tumolo et al. 2020), (Louwies et al. 2016)
PM	serum	miR-15a-5p, -19b-3p, -23a-3p, -93-5p, -126-3p, -130-3p, -142-3p, -146a-5p, -150-5p, -191-5p, -223-3p, let-7a-5p, let-7g-5p ↑		(Tumolo et al. 2020), (Rodosthenous et al. 2016)
PM	placenta	miR-20a-5p, miR-21-5p (1st trimester) ↑		(Tumolo et al. 2020), (Tsamou et al. 2018)
PM	placenta	miR-21-5p, miR-146a-5p, miR-222-3p (2nd trimester) ↓		(Tumolo et al. 2020), (Tsamou et al. 2018)
DEP	bronchial epithelial cells	miR-375 ↑		(Bleck et al. 2013)

(Continued)

Table 2.1 (Continued)

Exposure	Sample type	miRNAs	Related Disease	References
DEP	bronchial epithelial cells	miR-132-3p ↑		(Tumolo et al. 2020), (Rider et al. 2016)
DEP	bronchial epithelial cells	miR-183-5p ↓		
DEP	peripheral blood	miR-21-5p, miR-30e, miR-215, miR-144 ↑		(Tumolo et al. 2020), (Yamamoto et al. 2013)
TRAP	plasma	miR-1224-5p, miR-3127-5p ↑		(Tumolo et al. 2020), (Krauskopf et al. 2018)
TRAP	plasma	miR-27a-5p, miR-133a-3p, miR-145-5p, miR-193b-3p, miR-433-3p, miR-580-3p, miR-6716-3p ↓		(Tumolo et al. 2020), (Krauskopf et al. 2018)
O_3	induced sputum	miR-25-3p, miR-132-3p, miR-199a-3p, miR-222-3p, miR-434-5p, miR-582-5p ↑		(Tumolo et al. 2020), (Fry et al. 2014)
O_3	induced sputum	miR-143-3p, miR-223-3p, miR-145-5p, miR-199b-5p ↑		(Tumolo et al. 2020), (Fry et al. 2014)
Cigarette smoke				
cigarette smoke	bronchial epithelium	miR-181 ↑		(Izzotti and Pulliero 2014), (Schembri et al. 2009)
cigarette smoke	bronchial epithelium	miR-30a, miR-125b, miR-128, miR-146, miR-218, miR-223 and miR-500 ↓		(Izzotti and Pulliero 2014), (Schembri et al. 2009)
tobacco carcinogens	bronchial epithelial cells	miR-200b, miR-200c, miR-205↑		(Tellez et al. 2011)
smoking	induced sputum	let-7c, miR-146a, miR-150, miR-203, miR-340 ↓	COPD	(Van Pottelberge et al. 2011)
smoking	placenta	miR-16, miR-21, miR-146a ↓		(Maccani et al. 2010)
smoking	maternal and cord blood	miR-223 ↑		(Herberth et al. 2014)

(Continued)

Table 2.1 (Continued)

Exposure	Sample type	miRNAs	Related Disease	References
Organic chemicals				
PAHs	plasma	miR-142, miR-24, miR-27a, miR-28↓		(Deng et al. 2014)
PAHs	plasma	miR-150 ↑		
PCB-169	peripheral blood	miR-191 ↑		(Guida et al. 2013)
PCBs	serum	miR-122-5p, miR-192-5p, 99a-5p, 320a ↑	liver disease	*In journal review*
PCBs	serum	let-7d-5p, miR-17-5p, miR-24-3p, miR-197-3p, and miR-221-3p ↓	liver disease	*In journal review*
POPs (PCBs, HCB and DDT)	peripheral blood	miR-152, miR-31-5p, miR-34a-5p, miR-21-5p, miR-29a ↑		(Qiao et al. 2019), (Krauskopf et al. 2017)
POPs (PCBs, HCB and DDT)	peripheral blood	miR-320d, miR-320e, miR-486-5p, miR-324-3p, miR-331-3p, miR-501-3p, miR-532-3p ↓		(Qiao et al. 2019), (Krauskopf et al. 2017)
Benzene and toluene	maternal and cord blood	miR-155, miR-223 ↑		(Herberth et al. 2014)
Toluene, xylene, ethylbenzene	blood	microRNA profiles (>100)		(Song and Ryu 2015)
Toluene, xylene, ethylbenzene	blood	miR-6819-5p and miR-6778-5p ↑		(Sisto et al. 2019)
environmental pesticides	urine	miR-223, miR-518d-3p, miR-597, miR-133b, miR-517b, miR-28-5p ↑		(Qiao et al. 2019), (Weldon et al. 2016)
Methamidophos	peripheral blood	miR-133a, miR-214, miR-29b, miR-590 ↑		(Qiao et al. 2019), (Yuan et al. 2018)
Methamidophos	peripheral blood	miR-125b, miR-141-5p, miR-142a, miR-150 ↓		
Omethoate	peripheral blood lymphocytes	miR-145		(Qiao et al. 2019), (Wang et al. 2019)
PFOA	serum	miR-199a, miR-26b ↑		(Wang et al. 2012)
Organophosphates	serum	miR-155 ↓	Guillain-Barre Syndrome	(Costa et al. 2020), (Yuan et al. 2018)

(Continued)

Table 2.1 (Continued)

Exposure	Sample type	miRNAs	Related Disease	References
Heavy metals				
Arsenic	urine	miR-21, miR-126, miR-155, miR-221↑	albuminuria (miR-21)	(Kong et al. 2012)
Arsenic	urine	miR-205↓		(Gao et al. 2019)
Arsenic	peripheral blood mononuclear cells	miR-21↑	skin cancers	(Wallace et al. 2020), (Liu et al. 2018)
Arsenic	cord blood	has-let-7a, miR-107, miR-126, miR-16, miR-17, miR-195, miR-20a, miR-20b, miR-26b, miR-454, miR-96, miR-98 ↑		(Rager et al. 2014)
Arsenic	bronchial epithelial cells	miR-190 ↑		(Beezhold et al. 2011)
Arsenic	plasma	miR-423-5p, miR-423-5p, miR-142-5p -2, miR-454-5p	cardiometabolic disease	(Beck et al. 2018)
Arsenic	plasma	-320c-1, and -320c		
Arsenic	serum	miR-126, miR-155		(Wallace et al. 2020), (Ruiz-Vera et al. 2019)
Lead	placenta	miR-146a, miR-10a, miR-190b, miR-431 ↑		(Wallace et al. 2020), (Li et al. 2015)
Lead	placenta	miR-651 ↓		
Lead	blood	miR-21-5p, miR-122-5p ↑		(Wallace et al. 2020), (Guo et al. 2019)
Lead (battery factory workers)	blood	miR-520c-3p, miR-148a, miR-141,and miR-211↑.		(Wallace et al. 2020), (Xu et al. 2017)
Lead (battery factory workers)	blood	miR-572 and miR-130b↓		(Wallace et al. 2020), (Xu et al. 2017)

(Continued)

Table 2.1 (Continued)

Exposure	Sample type	miRNAs	Related Disease	References
Mercury (female factory workers)	blood	miR-92a-3p, miR-122-5p, miR-451a, and miR-486-5p↑		(Wallace et al. 2020), (Ding et al. 2016)
Mercury (female factory workers)	blood	miR-16-5p, miR-30c-3p,miR-181a-5p, and let-7e-5p↓		(Wallace et al. 2020), (Ding et al. 2016)
Other				
100 nm gold nanoparticles	cord blood	has-let-7a ↑		(Balansky et al. 2013)
Asbestos	peripheral blood	miR-103 ↑	mesothelioma	(Weber et al. 2012)
Occupational grain dust	serum	miR-18a-5p, miR-124-3p and miR-574-3p ↑	lung diseases	(Straumfors et al. 2020)
Occupational grain dust	serum	miR-19b-3p and miR-146a-5p ↓	lung diseases	(Straumfors et al. 2020)

Abbreviations: PM (particulate matter); SNPs (single nucleotide polymorphisms); MV (microvesicles); DEP (diethyl phthalate); TRAP (traffic-related air pollution); O3 (ozone); PAHs (polycyclic aromatic hydrocarbons); PCBs (polychlorinated biphenyls); POPs (persistent organic pollutants); HCB (hexachlorobenzene); DDT (dichloro-diphenyl-trichloroethane); PFOA (perfluorooctanoic acid)

The best miRNA biomarkers are likely to be those that indicate injury or perturbation to a specific cell type, such as the previously discussed hepatocyte-specific miR-122. Other cell-specific miRNAs that may be useful biomarkers of tissue injury include the myomiRs, miRs-133, -206, -208, and -499, which have shown promise through their ability to identify cardiac injury (Jenike and Halushka 2021). In addition, miRNA biomarkers that indicate the disruption of specific nuclear receptor or regulatory pathways as a result of chemical exposure would be desirable. Studies examining the changes in miRNA–mRNA interactions that are due to environmental exposures will be helpful in determining the mechanistic pathways involved. One such study applied a network approach, integrative Joint Random Forest, to investigate the effect of low-dose environmental chemical exposure on normal mammary gland development in rats (Petralia et al. 2017). The researchers detected miRNAs that regulated many mRNAs in the control group but not in the exposed groups, which suggested that the disruption of miRNA activity was due to chemical exposure. Messenger RNAs connecting with miR-200a and miR-375 in the control network only were enriched for the mammary gland development and gland morphogenesis pathways, and the effects of chemical exposure on these two miRNAs were confirmed in human breast cancer cell lines.

Song and Ryu (2015) identified characteristic miRNA profiles of human whole blood in workers exposed to volatile organic compounds (VOCs) and compared their effectiveness to the conventionally used metabolite markers of exposure. They were able to discern each VOC (toluene, xylene, and ethylbenzene) from the control group with higher accuracy, sensitivity, and specificity than the urinary biomarkers. In a recent study, serum miRNAs were associated with polychlorinated biphenyl (PCB) exposures and environmental liver disease in a residential cohort (Cave et al. 2022). This longitudinal cohort study enrolled residents of Anniston, Alabama that exhibited PCB levels twice and three times higher than found in the average population (Pavuk et al. 2014). A miRNA panel for liver toxicity and disease was assessed, and many of the miRNAs correlated with biomarkers of disease toxicity, metabolic changes, and inflammation. In addition, some of the miRNAs also correlated with levels of PCB measured in the serum. These miRNAs included miR-122-5p, -192-5p, and -99a-5p, which have been previously established to have important roles in liver disease and can serve as accessible biomarkers of disease progression (Brunetto et al. 2014; Gjorgjieva et al. 2019; Lopez-Sanchez et al. 2021). While more work needs to be done to establish the causative reasons why miRNAs in blood are correlating with both disease biomarkers and exposure, the indication is that they may serve as biomarkers of environmental pollutant health effects that include liver cell loss, change in function, and inflammatory response.

Application

The potential utility of miRNAs extends beyond their role as biomarkers of environmental toxicant exposure, encompassing applications in quantitative chemical risk assessment and in clinical diagnostics and therapeutics. Multiple studies have shown that the transcriptional benchmark dose modeling (BMD) of short-term chemical exposures can generate point-of-departure (POD) values or "breaking points" of chemical exposure that lead to disease or other adverse outcomes of regulatory concern. These next-generation toxicological assessments that use molecular measurements are consistent with traditional, longer-term apical endpoints and can inform the mode of action (MOA) via which the exposure imparts adverse outcomes (Lake et al. 2016; Thomas et al. 2007; Webster et al.

2015). The hepatic cell culture model, HepaRG, has been successfully leveraged for high-throughput transcriptomic studies that examine concentration–response relationships in order to predict liver injury (Ramaiahgari et al. 2019). Fewer studies have applied BMD assessments to miRNA expression levels but, because dysregulation of a small number of miRNAs can impact hundreds of gene targets, assessing miRNAs could reduce the complexity and variability associated with mRNA biomarkers. Rager et al. (2017) modeled multiple molecular endpoints from human cord blood after prenatal exposure to inorganic arsenic, including mRNA expression, protein expression, miRNA expression and DNA (CpG) methylation. They found twelve miRNAs significantly associated with exposure. The BMD values were 33 μg As/L urine (maternal exposure) for CpG methylation, 42 μg/L for miRNA, 45 μg/L for protein and 64 μg/L for mRNA. Interestingly, it was the epigenetic markers (including the miRNAs) that established the most conservative BMD values. Nevertheless, all endpoints reflected the epidemiological evidence of prenatal effects of inorganic arsenicals, all of which occurred below 100 μg/L urine. Stermer et al. (2019) used BMD-modeled miRNA sequencing reads and other small RNA molecular measurements of toxicity in rat sperm after exposure to the testicular toxicant ethylene glycol monomethyl ether (EMGE) and found them to be sensitive and predictive biomarkers of exposure. Calculations resulted in a BMD of 62 mg/kg using retained spermatid head (RSH), whereas percent miRNA resulted in a BMD of 59.2 mg/kg. A recent study in our laboratory (Chorley et al. 2020) indicated that miRNAs measured from mouse liver in response to DEHP exposure display early dose-responsive patterns linked to the predominant signaling pathway (PPARα); however, the BMD estimates based on these miRNAs were higher than for the target genes (163 vs 74 mg/kg-day). By contrast, another ongoing study sequenced miRNAs in liver tissue from B6C3F1 mice exposed to the rodent liver tumorigen furan and reported a mean miRNA BMD of 2.0 mg/kg/day, which was in close agreement with the 2-year determinations of hepatocyte cellular adenoma and human cellular carcinoma (HCA + HCC) BMD values of 2.3 mg/kg/day and more conservative than the most sensitive messenger RNA BMD of 2.8 mg/kg/day. miR-122 was also increased in the blood of treated mice (BMD 1.4 mg/kg/day). Overall, these studies indicate the potential usefulness of miRNA measurement in blood, tissue, and cell culture for quantitative risk assessment. Additional case studies are needed to examine a broad range of chemical MOA and whether the presence of these biomarkers in biofluids can consistently establish POD values reflective of cellular and tissue perturbation.

Clinical applications for miRNA biomarkers in biofluids are showing recent growth. In addition to their use in drug screening for DILI and drug-induced kidney injury (DIKI), they are finding utility as diagnostic and prognostic biomarkers for disease. A number of recent reviews have summarized the current miRNA diagnostic and therapeutic landscape (Bonneau et al. 2019; Chakraborty et al. 2021; Condrat et al. 2020; Hanna et al. 2019; Vandana Saini, 2021; Tribolet et al. 2020). These biomarkers can be used to indicate the presence of a pathology, or even the stage, progression, or genetic link of pathogenesis (Hanna et al. 2019). In certain situations, one miRNA biomarker may be sufficient to identify a health outcome; but more often a well-defined panel of miRNAs is necessary for increased diagnostic sensitivity or specificity. Despite the advantages of miRNA biomarkers for early detection, improved pathogen identification and personalized medicine, their translation from the research bench into the clinic has been slow. Many companies are successfully developing diagnostic miRNA panels and several products are already available to clinicians (Bonneau et al. 2019; Vandana Saini et al. 2021). In 2012, Rosetta Genomics in partnership with Precision Therapeutics launched miRview™ mets, a miRNA panel measuring the expression levels of sixty-four miRNAs that allow the identification of cancers of unknown or uncertain primary origin (CUPs), which account for up to 15% of newly diagnosed cancer cases in the United States every year. Interpace Diagnostics/Asuragen offers a product for the diagnosis of thyroid cancer that combines ThyraMIR®, a miRNA classifier including the quantification of 10 miRNAs, and ThyGeNEXT®, an oncogene panel for thyroid cancer stratification. This combination offers an interesting alternative to standard cytological evaluations, which fail to discriminate benign from malignant lesions in 15–30% of cases. In November 2018, the Blue Cross Blue Shield and the US Federal Employee Health Benefit Program agreed to include the ThyGeNEXT®/ThyraMIR® combined tests for their beneficiaries. TAmiRNA markets OsteomiR™, a product measuring the expression level of 19 blood-circulating miRNAs intended to calculate the risk of a first fracture in female patients with postmenopausal osteoporosis and type-2 diabetes. Similarly, TAmiRNA also provides the ThrombomiR™ panel (11 miRNAs), so as to assess platelet function in cardiovascular disease. In addition, miRNA-based therapeutics—including miRNA mimics or antagonists targeting miRNAs involved in hepatitis C, fibrosis, T-cell lymphoma, heart failure, mesothelioma, and ulcerative colitis—are in various phases of clinical trials (Vandana Saini et al. 2021).

Challenges and Future Focus

While there is considerable promise in the use of miRNA biomarkers in biofluids, the precise quantification of circulating miRNAs is challenging and the validation of introduced biomarkers is often unsuccessful. This is indicated by a decline in published biofluid-based miRNA biomarker research in the past couple of years, whereas between 2009 and 2015 yearly publications nearly doubled (Chorley et al. 2021a). This downward trajectory is partially attributed to technical obstacles for measuring miRNAs, some of which have been discussed in the preceding sections. Pre-analytical variables such as sample collection, processing, storage, and extraction are potential causes of inconsistency in miRNA measurements. Post-analytical challenges include the normalization of data and the interpretation of miRNA changes observed to be due to perturbation and release into biofluids. These uncertainties and challenges have inhibited full implementation in multiple sectors such as clinical, regulatory, and academic use.

For pre-analytical steps, the standardization of procedures and quality assurance will significantly reduce measurement variability. These considerations extend to consistency in sampling during normal cyclic patterns (circadian, menstrual, etc.), location and fraction of sampling (for example in blood: sublingual, aortic, jugular, etc. and whole blood, serum, plasma, or EV fraction), minimization of hemolysis of collected blood samples, and short-term and long-term storage considerations. Also, miRNAs in biofluids may vary in stability owing to interactions with RNA-binding proteins, EVs, and lipoproteins that may impart protection from high RNase activity in these matrices. In freshly isolated human serum, miRs-16, -21, and -142–3p demonstrated greater stability than miRs-122 and -1 after twenty-four hours at room temperature (Koberle et al. 2013). This difference was affected by RNase activity in the serum, whereas miRNAs in EVs were protected against this activity.

If isolating EVs, the techniques used can influence the content measured. First, there is confusion in the literature on how to distinguish different subclasses of EVs, and this subsequently complicates the identification of RNA cargo (Mathieu et al. 2019). Further, RNA can bind other types of carriers such as lipoprotein and ribonucleoproteins, which are difficult to separate from membrane-enclosed RNAs (O'Brien et al. 2020). Not surprisingly, vesicle purification can lower the RNA yield and integrity and, because of the lower RNA concentration, result in more interindividual variability and increased difficulty in measuring some miRNA biomarkers. To compound the issue, obtaining enough miRNA from vesicles is also a challenge, as studies have demonstrated an average of one or less than one miRNA per EV (Chevillet et al. 2014; Li et al. 2014). Therefore miRNA biomarkers released through active and passive mechanisms may have different stability levels in stored samples and recovery may be inherently low, requiring optimization and consistency in sample collection for specific miRNAs. There are a variety of methods that can be used to measure miRNAs in biofluids, and the selection depends on application, cost considerations, expertise, and available resources. These methods include lower-throughput techniques such as quantitative reverse transcriptase polymerase chain reaction (qRT-PCR) or digital drop PCR (ddPCR), medium-throughput techniques such as bead hybridization (e.g., NanoString™ or FirePlex™ assays), and higher-throughput techniques such as microarray and next generation sequencing. A comparative study that used qRT-PCR across eleven research institutions to examine miRNA biomarker alterations due to toxicant exposure found that, when outliers were removed, miRNA measurements were very consistent (only contributing 2% of the measured variance). This supports the idea that standardized protocols can lead to consistent measurements regardless of different site locations (Thompson et al. 2016). Comparison across methods shows consistent correlation when using fold change calculations: concordance ranges from 0.6 to 0.95 (Giraldez et al. 2018; Git et al. 2010; Mestdagh et al. 2014; Yauk et al. 2010).

While miRNAs can be consistently measured methodologically, a primary post-analytical challenge is data normalization. The normalization of measured miRNA data would theoretically correct for natural inter-individual fluctuations in fluid volume and rhythmic or cyclic release, as well as for technical variations attributed to miRNA extraction methods, assay inhibition due to biological factors, and other variables. This would allow precise site-to-site, temporal, and methodological comparisons, alongside increasing confidence in interindividual evaluations. Endogenous "housekeeping" calibrators measured in biofluids such as miRNAs or other nucleic acids would be ideal; however, no universally invariant calibrator has been found (Saliminejad et al. 2019). Where endogenous normalizers may not be available or optimized, synthetic exogenous normalizers or miRNAs "spiked-in" during or after small RNA extraction are commonly used to correct technical variations introduced during sample processing and measurement. A recent study analyzed the intra- and inter-individual variability of circulating miRNAs among healthy and cancer patients (Vigneron et al. 2016). Intra-individual variability significantly improved with a geometric mean of three exogenous measurements by comparison with endogenous normalization. These same normalizers—and, interestingly, the endogenous normalizer miR-16-5p—also improved cross-platform correlation of qRT-PCR and microarray measurements. Much like pre-analytical variables, normalization procedures must be optimized for each individual system and application to increase one's confidence in the measurements taken.

Recently the microRNA Biomarkers Working Group of the Health and Environmental Sciences Institute's (HESI) Committee on Emerging Systems Toxicology for the Assessment of Risk (eSTAR) examined the pre- and post-analytical variables that have impeded the development of miRNA biomarkers for toxicological and regulatory use (Chorley et al. 2021a). As a result of examining both the technical and the biological variables that influence the measurement of miRNA biomarkers in biofluids, the group made six recommendations for future research. These recommendations were: determining the pre-analytical stability of miRNAs; establishing standard operating procedures for collection and measurement; establishing a reporting framework; determining the mechanisms of action of extracellular release of miRNAs; determining the linkage of miRNA variants (also called isomeric miRNAs or "isomiRs") to adverse outcomes; and identifying the cellular or tissue source of extracellular miRNAs. Establishing these unknowns for a miRNA biomarker application strengthens one's confidence not only in the data produced but also in the mechanism by which the miRNA is released into the biofluid, thereby indicating clear linkages to the adverse outcome of interest and an increased specificity and sensitivity of the biomarker. Of the recommendations, the establishment of standard operating procedures was the highest priority. Procedures that follow standard guidelines for biofluid collection, storage, and handling (such as those outlined by the Early Detection Research Network (ERDN)), as well as the automation of procedures, where applicable, will ensure consistent, reproducible results (Farina et al. 2014).

Conclusions

The identification of miRNA biomarkers that can distinguish the biological effects of toxicant and other environmental chemical exposures is a much needed tool for toxicological and regulatory science. Studies have suggested that these biomarkers, which are stable and present in most biofluids, may indicate general toxicity in a tissue-specific manner. Further, the RNA content of EVs may reflect the type and the physiological or pathological state of the source cells. Extracellular vesicle-mediated functional transfer of miRNAs to recipient cells has been demonstrated both in vitro and in vivo for a variety of diseases and physiological states, but the mechanisms by which EVs are selectively packaged and "addressed" to different cell types are still largely unknown. In addition to specificity of tissue and MOA, in vitro and in vivo studies suggest early, dose-responsive alterations in expression that can indicate breaking points that may lead to adverse health outcomes. Recent development of miRNA biomarkers in biofluids for clinical indications provide promise that similar uses can be leveraged in toxicological sciences. Despite these qualities, there remain significant technical challenges to reducing variability in measurements and to increasing our confidence in the application of these biomarkers. A standardization of protocols and reporting efforts analogous to the Transcriptomics Reporting Framework (Gant et al. 2017), and large-scale database efforts to categorize EV contents relating to source cells will be important for biomarker identification and for linking biofluid measured miRNAs to specific tissues, cells, and biological effects (see exRNA Atlas: Murillo et al. 2019). These and other concerted efforts will ultimately lead to a consistent and widespread use of miRNAs in regulatory and toxicological sciences.

Acknowledgements

The authors wish to thank Drs. Sheau-Fung Thai, Chris Corton, Sid Hunter and Kimberly Slentz-Kesler for their critical reviews of this manuscript. This manuscript has been reviewed by the US Environmental Protection Agency and approved for publication. Approval does not signify that the contents reflect the views of the Agency, nor does the mentioning of trade names or commercial products constitute endorsement or recommendation of use.

References

Amorim, M.G., Valieris, R., Drummond, R.D., Pizzi, M.P., Freitas, V.M., Sinigaglia-Coimbra, R., Calin, G.A., Pasqualini, R., Arap, W., Silva, I.T., Dias-neto, E., and Nunes, D.N. (2017). A total transcriptome profiling method for plasma-derived extracellular vesicles: Applications for liquid biopsies. *Sci. Rep.* 7: 14395.

Arroyo, J.D., Chevillet, J.R., Kroh, E.M., Ruf, I.K., Pritchard, C.C., Gibson, D.F., Mitchell, P.S., Bennett, C.F., Pogosova-Agadjanyan, E.L., Stirewalt, D.L., Tait, J.F., and Tewari, M. (2011). Argonaute2 complexes carry a population of circulating microRNAs independent of vesicles in human plasma. *Proc. Natl. Acad. Sci. USA* 108: 5003–5008.

Ashley, J., Cordy, B., Lucia, D., Fradkin, L.G., Budnik, V., and Thomson, T. (2018). Retrovirus-like Gag protein Arc1 binds RNA and traffics across synaptic boutons. *Cell* 172 (262–274): e11.

Bailey, W.J. and Glaab, W.E. (2018). Accessible miRNAs as novel toxicity biomarkers. *Int. J. Toxicol.* 37: 116–120.

Bala, S., Petrasek, J., Mundkur, S., Catalano, D., Levin, I., Ward, J., Alao, H., Kodys, K., and Szabo, G. (2012). Circulating microRNAs in exosomes indicate hepatocyte injury and inflammation in alcoholic, drug-induced, and inflammatory liver diseases. *Hepatology* 56: 1946–1957.

Balansky, R., Longobardi, M., Ganchev, G., Iltcheva, M., Nedyalkov, N., Atanasov, P., Toshkova, R., De Flora, S., and Izzotti, A. (2013). Transplacental clastogenic and epigenetic effects of gold nanoparticles in mice. *Mutat. Res.* 751–752: 42–48.

Balasubramanian, S., Gunasekaran, K., Sasidharan, S., Jeyamanickavel Mathan, V., and Perumal, E. (2020). MicroRNAs and xenobiotic toxicity: An overview. *Toxicol. Rep.* 7: 583–595.

Beck, R., Bommarito, P., Douillet, C., Kanke, M., Del Razo, L.M., Garcia-Vargas, G., Fry, R.C., Sethupathy, P., and Styblo, M. (2018). Circulating miRNAs associated with arsenic exposure. *Environ. Sci. Technol.* 52: 14487–14495.

Beezhold, K., Liu, J., Kan, H., Meighan, T., Castranova, V., Shi, X., and Chen, F. (2011). miR-190-mediated downregulation of PHLPP contributes to arsenic-induced Akt activation and carcinogenesis. *Toxicol. Sci.* 123: 411–420.

Bleck, B., Grunig, G., Chiu, A., Liu, M., Gordon, T., Kazeros, A., and Reibman, J. (2013). MicroRNA-375 regulation of thymic stromal lymphopoietin by diesel exhaust particles and ambient particulate matter in human bronchial epithelial cells. *J. Immunol.* 190: 3757–3763.

Bollati, V., Angelici, L., Rizzo, G., Pergoli, L., Rota, F., Hoxha, M., Nordio, F., Bonzini, M., Tarantini, L., Cantone, L., Pesatori, A.C., Apostoli, P., Baccarelli, A.A., and Bertazzi, P.A. (2015). Microvesicle-associated microRNA expression is altered upon particulate matter exposure in healthy workers and in A549 cells. *J. Appl. Toxicol.* 35: 59–67.

Bollati, V., Marinelli, B., Apostoli, P., Bonzini, M., Nordio, F., Hoxha, M., Pegoraro, V., Motta, V., Tarantini, L., Cantone, L., Schwartz, J., Bertazzi, P.A., and Baccarelli, A. (2010). Exposure to metal-rich particulate matter modifies the expression of candidate microRNAs in peripheral blood leukocytes. *Environ. Health Perspect.* 118: 763–768.

Bonneau, E., Neveu, B., Kostantin, E., Tsongalis, G.J., and De Guire, V. (2019). How close are miRNAs from clinical practice? A perspective on the diagnostic and therapeutic market. *EJIFCC* 30: 114–127.

Brunetto, M.R., Cavallone, D., Oliveri, F., Moriconi, F., Colombatto, P., Coco, B., Ciccorossi, P., Rastelli, C., Romagnoli, V., Cherubini, B., Teilum, M.W., Blondal, T., and Bonino, F. (2014). A serum microRNA signature is associated with the immune control of chronic hepatitis B virus infection. *PLoS One* 9: e110782.

Buschmann, D., Kirchner, B., Hermann, S., Marte, M., Wurmser, C., Brandes, F., Kotschote, S., Bonin, M., Steinlein, O.K., Pfaffl, M.W., Schelling, G., and Reithmair, M. (2018). Evaluation of serum extracellular vesicle isolation methods for profiling miRNAs by next-generation sequencing. *J. Extracell Vesicles* 7: 1481321.

Bushel, P.R., Caiment, F., Wu, H., O'Lone, R., Day, F., Calley, J., Smith, A., and Li, J. (2018). RATEmiRs: The rat atlas of tissue-specific and enriched miRNAs database. *BMC Genom.* 19: 825.

Cave, M.C., Pinkston, C.M., Rai, S.N., Wahlang, B., Pavuk, M., Head, K.Z., Carswell, G.K., Nelson, G.M., Klinge, C.M., Bell, D.A., Birnbaum, L.S., and Chorley, B.N. (2022). Circulating microRNAs, polychlorinated biphenyls, and environmental liver disease in the Anniston Community Health Survey. *Environ. Health Perspect* 101: 17003.

Cermelli, S., Ruggieri, A., Marrero, J.A., Ioannou, G.N., and Beretta, L. (2011). Circulating microRNAs in patients with chronic hepatitis C and non-alcoholic fatty liver disease. *PLoS One* 6: e23937.

Chakraborty, C., Sharma, A.R., Sharma, G., and Lee, S.S. (2021). Therapeutic advances of miRNAs: A preclinical and clinical update. *J. Adv. Res.* 28: 127–138.

Chen, X., Ba, Y., Ma, L., Cai, X., Yin, Y., Wang, K., Guo, J., Zhang, Y., Chen, J., Guo, X., Li, Q., Li, X., Wang, W., Zhang, Y., Wang, J., Jiang, X., Xiang, Y., Xu, C., Zheng, P., Zhang, J., Li, R., Zhang, H., Shang, X., Gong, T., Ning, G., Wang, J., Zen, K., Zhang, J., and Zhang, C.Y. (2008). Characterization of microRNAs in serum: A novel class of biomarkers for diagnosis of cancer and other diseases. *Cell Res.* 18: 997–1006.

Chevillet, J.R., Kang, Q., Ruf, I.K., Briggs, H.A., Vojtech, L.N., Hughes, S.M., Cheng, H.H., Arroyo, J.D., Meredith, E.K., Gallichotte, E.N., Pogosova-Agadjanyan, E.L., Morrissey, C., Stirewalt, D.L., Hladik, F., Yu, E.Y., Higano, C.S., and Tewari, M. (2014). Quantitative and stoichiometric analysis of the microRNA content of exosomes. *Proc. Natl. Acad. Sci. USA* 111: 14888–14893.

Chorley, B.N., Atabakhsh, E., Doran, G., Gautier, J.C., Ellinger-Ziegelbauer, H., Jackson, D., Sharapova, T., Yuen, P.S.T., Church, R.J., Couttet, P., Froetschl, R., Mcduffie, J., Martinez, V., Pande, P., Peel, L., Rafferty, C., Simutis, F.J., and Harrill, A.H. (2021a). Methodological considerations for measuring biofluid-based microRNA biomarkers. *Crit. Rev. Toxicol.* 51 (3): 264–282264–282.

Chorley, B.N., Carswell, G.K., Nelson, G., Bhat, V.S., and Wood, C.E. (2020). Early microRNA indicators of PPARalpha pathway activation in the liver. *Toxicol. Rep.* 7: 805–815.

Chorley, B.N., Ellinger-Ziegelbauer, H., Tackett, M., Simutis, F.J., Harrill, A.H., Mcduffie, J., Atabakhsh, E., Nassirpour, R., Whiteley, L.O., Leonard, J.F., Carswell, G.K., Harpur, E., Chen, C.L., and Gautier, J.C. (2021b). Urinary miRNA biomarkers of drug-induced kidney injury and their site specificity within the nephron. *Toxicol. Sci.* 180: 1–16.

Church, R.J., Kullak-Ublick, G.A., Aubrecht, J., Bonkovsky, H.L., Chalasani, N., Fontana, R.J., Goepfert, J.C., Hackman, F., King, N.M.P., Kirby, S., Kirby, P., Marcinak, J., Ormarsdottir, S., Schomaker, S.J., Schuppe-koistinen, I., Wolenski, F., Arber, N., Merz, M., Sauer, J.M., Andrade, R.J., Van Bommel, F., Poynard, T., and Watkins, P.B. (2019). Candidate biomarkers for the diagnosis and prognosis of drug-induced liver injury: An international collaborative effort. *Hepatology* 69: 760–773.

Clark, P.M., Loher, P., Quann, K., Brody, J., Londin, E.R., and Rigoutsos, I. (2014). Argonaute CLIP-Seq reveals miRNA targetome diversity across tissue types. *Sci. Rep.* 4: 5947.

Condrat, C.E., Thompson, D.C., Barbu, M.G., Bugnar, O.L., Boboc, A., Cretoiu, D., Suciu, N., Cretoiu, S.M., and Voinea, S.C. (2020). miRNAs as biomarkers in disease: Latest findings regarding their role in diagnosis and prognosis. *Cells* 9 (2): 276.

Costa, C., Teodoro, M., Rugolo, C.A., Alibrando, C., Giambo, F., Briguglio, G., and Fenga, C. (2020). MicroRNAs alteration as early biomarkers for cancer and neurodegenerative diseases: New challenges in pesticides exposure. *Toxicol. Rep.* 7: 759–767.

Deng, Q., Huang, S., Zhang, X., Zhang, W., Feng, J., Wang, T., Hu, D., Guan, L., Li, J., Dai, X., Deng, H., Zhang, X., and Wu, T. (2014). Plasma microRNA expression and micronuclei frequency in workers exposed to polycyclic aromatic hydrocarbons. *Environ. Health Perspect.* 122: 719–725.

Ding, E., Zhao, Q., Bai, Y., Xu, M., Pan, L., Liu, Q., Wang, B., Song, X., Wang, J., Chen, L., and Zhu, B. (2016). Plasma microRNAs expression profile in female workers occupationally exposed to mercury. *J. Thorac. Dis.* 8: 833–841.

Dutta, R.K., Chinnapaiyan, S., and Unwalla, H. (2019). Aberrant microRNAomics in pulmonary complications: Implications in lung health and diseases. *Mol. Ther. Nucleic Acids* 18: 413–431.

Farina, N.H., Wood, M.E., Perrapato, S.D., Francklyn, C.S., Stein, G.S., Stein, J.L., and Lian, J.B. (2014). Standardizing analysis of circulating microRNA: Clinical and biological relevance. *J. Cell. Biochem.* 115: 805–811.

Finicelli, M., Squillaro, T., Galderisi, U., and Peluso, G. (2020). Micro-RNAs: Crossroads between the exposure to environmental particulate pollution and the obstructive pulmonary disease. *Int. J. Mol. Sci.* 21 (19): 7221.

Fossati, S., Baccarelli, A., Zanobetti, A., Hoxha, M., Vokonas, P.S., Wright, R.O., and Schwartz, J. (2014). Ambient particulate air pollution and microRNAs in elderly men. *Epidemiology* 25: 68–78.

Fry, R.C., Rager, J.E., Bauer, R., Sebastian, E., Peden, D.B., Jaspers, I., and Alexis, N.E. (2014). Air toxics and epigenetic effects: Ozone altered microRNAs in the sputum of human subjects. *Am. J. Physiol. Lung Cell Mol. Physiol.* 306: L1129–L1137.

Gant, T.W., Sauer, U.G., Zhang, S.D., Chorley, B.N., Hackermuller, J., Perdichizzi, S., Tollefsen, K.E., Van Ravenzwaay, B., Yauk, C., Tong, W., and Poole, A. (2017). A generic Transcriptomics Reporting Framework (TRF) foromics data processing and analysis. *Regul. Toxicol. Pharmacol.* 91 (Suppl 1): S36–S45.

Gao, S., Lin, P.I., Mostofa, G., Quamruzzaman, Q., Rahman, M., Rahman, M.L., Su, L., Hsueh, Y.M., Weisskopf, M., Coull, B., and Christiani, D.C. (2019). Determinants of arsenic methylation efficiency and urinary arsenic level in pregnant women in Bangladesh. *Environ Health* 18: 94.

Giraldez, M.D., Spengler, R.M., Etheridge, A., Godoy, P.M., Barczak, A.J., Srinivasan, S., De Hoff, P.L., Tanriverdi, K., Courtright, A., Lu, S., Khoory, J., Rubio, R., Baxter, D., Driedonks, T.A.P., Buermans, H.P.J., Nolte-'t Hoen, E.N.M., Jiang, H., Wang, K., Ghiran, I., Wang, Y.E., Van Keuren-Jensen, K., Freedman, J.E., Woodruff, P.G., Laurent, L.C., Erle, D.J., Galas, D.J., and Tewari, M. (2018). Comprehensive multi-center assessment of small RNA-seq methods for quantitative miRNA profiling. *Nat. Biotechnol.* 36: 746–757.

Git, A., Dvinge, H., Salmon-Divon, M., Osborne, M., Kutter, C., Hadfield, J., Bertone, P., and Caldas, C. (2010). Systematic comparison of microarray profiling, real-time PCR, and next-generation sequencing technologies for measuring differential microRNA expression. *RNA* 16: 991–1006.

Gjorgjieva, M., Sobolewski, C., Dolicka, D., Correia De Sousa, M., and Foti, M. (2019). miRNAs and NAFLD: From pathophysiology to therapy. *Gut* 68: 2065–2079.

Gu, S. and Kay, M.A. (2010). How do miRNAs mediate translational repression? *Silence* 1: 11.

Guduric-Fuchs, J., O'Connor, A., Camp, B., O'Neill, C.L., Medina, R.J., and Simpson, D.A. (2012). Selective extracellular vesicle-mediated export of an overlapping set of microRNAs from multiple cell types. *BMC Genom.* 13: 357.

Guida, M., Marra, M.L., Zullo, F., Guida, M., Trifuoggi, M., Biffali, E., Borra, M., De Mieri, G., D'Alessandro, R., and De Felice, B. (2013). Association between exposure to dioxin-like polychlorinated biphenyls and miR-191 expression in human peripheral blood mononuclear cells. *Mutat. Res.* 753: 36–41.

Guo, X., Yang, Q., Zhang, W., Chen, Y., Ren, J., and Gao, A. (2019). Associations of blood levels of trace elements and heavy metals with metabolic syndrome in Chinese male adults with microRNA as mediators involved. *Environ. Pollut.* 248: 66–73.

Hagiwara, K., Katsuda, T., Gailhouste, L., Kosaka, N., and Ochiya, T. (2015). Commitment of Annexin A2 in recruitment of microRNAs into extracellular vesicles. *FEBS Lett.* 589: 4071–4078.

Hanna, J., Hossain, G.S., and Kocerha, J. (2019). The potential for microRNA therapeutics and clinical research. *Front Genet.* 10: 478.

Harrill, A.H., Mccullough, S.D., Wood, C.E., Kahle, J.J., and Chorley, B.N. (2016). MicroRNA biomarkers of toxicity in biological matrices. *Toxicol. Sci.* 152: 264–272.

Herberth, G., Bauer, M., Gasch, M., Hinz, D., Roder, S., Olek, S., Kohajda, T., Rolle-Kampczyk, U., Von Bergen, M., Sack, U., Borte, M., Lehmann, I., and Lifestyle and Environmental Factors and Their Influence on Newborns Allergy Risk Study Group (2014). Maternal and cord blood miR-223 expression associates with prenatal tobacco smoke exposure and low regulatory T-cell numbers. *J. Allergy Clin. Immunol.* 133: 543–550.

Hill, M. and Tran, N. (2021). miRNA interplay: Mechanisms and consequences in cancer. *Dis. Model Mech.* 14 (4): dmm047662.

Holman, N.S., Mosedale, M., Wolf, K.K., Lecluyse, E.L., and Watkins, P.B. (2016). Subtoxic alterations in hepatocyte-derived exosomes: An early step in drug-induced liver injury? *Toxicol. Sci.* 151: 365–375.

Huang, Y., Yan, Y., Xv, W., Qian, G., Li, C., Zou, H., and Li, Y. (2018). A new insight into the roles of miRNAs in metabolic syndrome. *Biomed. Res. Int.* 2018: 7372636.

Isakova, A., Fehlmann, T., Keller, A., and Quake, S.R. (2020). A mouse tissue atlas of small noncoding RNA. *Proc. Natl. Acad. Sci. USA* 117: 25634–25645.

Izzotti, A. and Pulliero, A. (2014). The effects of environmental chemical carcinogens on the microRNA machinery. *Int. J. Hyg. Environ. Health* 217: 601–627.

Janas, T., Janas, M.M., Sapon, K., and Janas, T. (2015). Mechanisms of RNA loading into exosomes. *FEBS Lett.* 589: 1391–1398.

Janas, T., Janas, T., and Yarus, M. (2006). Specific RNA binding to ordered phospholipid bilayers. *Nucleic Acids Res.* 34: 2128–2136.

Jenike, A.E. and Halushka, M.K. (2021). miR-21: A non-specific biomarker of all maladies. *Biomark Res.* 9: 18.

Jin, Y., Wong, Y.S., Goh, B.K.P., Chan, C.Y., Cheow, P.C., Chow, P.K.H., Lim, T.K.H., Goh, G.B.B., Krishnamoorthy, T.L., Kumar, R., Ng, T.P., Chong, S.S., Tan, H.H., Chung, A.Y.F., Ooi, L., Chang, J.P.E., Tan, C.K., and Lee, C.G.L. (2019). Circulating microRNAs as potential diagnostic and prognostic biomarkers in hepatocellular carcinoma. *Sci. Rep.* 9: 10464.

Juzenas, S., Lindqvist, C.M., Ito, G., Dolshanskaya, Y., Halfvarson, J., Franke, A., and Hemmrich-Stanisak, G. (2020). Depletion of erythropoietic miR-486-5p and miR-451a improves detectability of rare microRNAs in peripheral blood-derived small RNA sequencing libraries. *NAR Genom. Bioinform.* 2: lqaa008.

Juzwik, C.A.S.S.D., Zhang, Y., Paradis-Isler, N., Sylvester, A., Amar-Zifkin, A., Douglas, C., Morquette, B., Moore, C.S., and Fournier, A.E. (2019). microRNA dysregulation in neurodegenerative diseases: A systematic review. *Prog. Neurobiol.* 182: 101664.

Kagawa, T., Shirai, Y., Oda, S., and Yokoi, T. (2018). Identification of specific microRNA biomarkers in early stages of hepatocellular injury, cholestasis, and steatosis in rats. *Toxicol. Sci.* 166: 228–239.

Koberle, V., Pleli, T., Schmithals, C., Augusto Alonso, E., Haupenthal, J., Bonig, H., Peveling-Oberhag, J., Biondi, R.M., Zeuzem, S., Kronenberger, B., Waidmann, O., and Piiper, A. (2013). Differential stability of cell-free circulating microRNAs: Implications for their utilization as biomarkers. *PLoS One* 8: e75184.

Koenig, E.M., Fisher, C., Bernard, H., Wolenski, F.S., Gerrein, J., Carsillo, M., Gallacher, M., Tse, A., Peters, R., Smith, A., Meehan, A., Tirrell, S., and Kirby, P. (2016). The beagle dog microRNA tissue atlas: Identifying translatable biomarkers of organ toxicity. *BMC Genom.* 17: 649.

Kong, A.P., Xiao, K., Choi, K.C., Wang, G., Chan, M.H., Ho, C.S., Chan, I., Wong, C.K., Chan, J.C., and Szeto, C.C. (2012). Associations between microRNA (miR-21, 126, 155 and 221), albuminuria and heavy metals in Hong Kong Chinese adolescents. *Clin. Chim. Acta.* 413: 1053–1057.

Koppers-Lalic, D., Hackenberg, M., Bijnsdorp, I.V., Van Eijndhoven, M.A.J., Sadek, P., Sie, D., Zini, N., Middeldorp, J.M., Ylstra, B., De Menezes, R.X., Wurdinger, T., Meijer, G.A., and Pegtel, D.M. (2014). Nontemplated nucleotide additions distinguish the small RNA composition in cells from exosomes. *Cell Rep.* 8: 1649–1658.

Kosaka, N., Iguchi, H., Hagiwara, K., Yoshioka, Y., Takeshita, F., and Ochiya, T. (2013). Neutral sphingomyelinase 2 (nSMase2)-dependent exosomal transfer of angiogenic microRNAs regulate cancer cell metastasis. *J. Biol. Chem.* 288: 10849–10859.

Kossinova, O.A., Gopanenko, A.V., Tamkovich, S.N., Krasheninina, O.A., Tupikin, A.E., Kiseleva, E., Yanshina, D.D., Malygin, A.A., Ven'Yaminova, A.G., Kabilov, M.R., and Karpova, G.G. (2017). Cytosolic YB-1 and NSUN2 are the only proteins recognizing specific motifs present in mRNAs enriched in exosomes. *Biochim. Biophys. Acta Proteins Proteom.* 1865: 664–673.

Kozomara, A., Birgaoanu, M., and Griffiths-Jones, S. (2019). miRBase: From microRNA sequences to function. *Nucleic Acids Res.* 47: D155–D162.

Krauskopf, J., Caiment, F., Claessen, S.M., Johnson, K.J., Warner, R.L., Schomaker, S.J., Burt, D.A., Aubrecht, J., and Kleinjans, J.C. (2015). Application of high-throughput sequencing to circulating microRNAs reveals novel biomarkers for drug-induced liver injury. *Toxicol. Sci.* 143: 268–276.

Krauskopf, J., Caiment, F., Van Veldhoven, K., Chadeau-Hyam, M., Sinharay, R., Chung, K.F., Cullinan, P., Collins, P., Barratt, B., Kelly, F.J., Vermeulen, R., Vineis, P., De Kok, T.M., and Kleinjans, J.C. (2018). The human circulating miRNome reflects multiple organ disease risks in association with short-term exposure to traffic-related air pollution. *Environ. Int.* 113: 26–34.

Krauskopf, J., De Kok, T.M., Hebels, D.G., Bergdahl, I.A., Johansson, A., Spaeth, F., Kiviranta, H., Rantakokko, P., Kyrtopoulos, S.A., and Kleinjans, J.C. (2017). MicroRNA profile for health risk assessment: Environmental exposure to persistent organic pollutants strongly affects the human blood microRNA machinery. *Sci. Rep.* 7: 9262.

LaBelle, J., Bowser, M., Brown, A., Farnam, L., Kho, A., Li, J., Mcgeachie, M., Chase, R., Piehl, S., Allen, K., Hobbs, B.D., Weiss, S.T., Hersh, C., Tantisira, K., and Amr, S.S. (2021). Commercially available blocking oligonucleotides effectively suppress unwanted hemolysis related miRNAs in a large whole blood RNA cohort. *J. Mol. Diagn. 23 (6): 671–682.*

Lagos-Quintana, M., Rauhut, R., Yalcin, A., Meyer, J., Lendeckel, W., and Tuschl, T. (2002). Identification of tissue-specific microRNAs from mouse. *Curr. Biol.* 12: 735–739.

Lake, A.D., Wood, C.E., Bhat, V.S., Chorley, B.N., Carswell, G.K., Sey, Y.M., Kenyon, E.M., Padnos, B., Moore, T.M., Tennant, A.H., Schmid, J.E., George, B.J., Ross, D.G., Hughes, M.F., Corton, J.C., Simmons, J.E., Mcqueen, C.A., and Hester, S.D. (2016). Dose and effect thresholds for early key events in a PPARalpha-mediated mode of action. *Toxicol. Sci.* 149: 312–325.

Landgraf, P., Rusu, M., Sheridan, R., Sewer, A., Iovino, N., Aravin, A., Pfeffer, S., Rice, A., Kamphorst, A.O., Landthaler, M., Lin, C., Socci, N.D., Hermida, L., Fulci, V., Chiaretti, S., Foa, R., Schliwka, J., Fuchs, U., Novosel, A., Muller, R.U., Schermer, B., Bissels, U., Inman, J., Phan, Q., Chien, M., Weir, D.B., Choksi, R., De Vita, G., Frezzetti, D., Trompeter, H.I., Hornung, V., Teng, G., Hartmann, G., Palkovits, M., Di Lauro, R., Wernet, P., Macino, G., Rogler, C.E., Nagle, J.W., Ju, J., Papavasiliou, F.N., Benzing, T., Lichter, P., Tam, W., Brownstein, M.J., Bosio, A., Borkhardt, A., Russo, J.J., Sander, C., Zavolan, M., and Tuschl, T. (2007). A mammalian microRNA expression atlas based on small RNA library sequencing. *Cell* 129: 1401–1414.

Li, D., Knox, B., Gong, B., Chen, S., Guo, L., Liu, Z., Tong, W., and Ning, B. (2021). Identification of translational microRNA biomarker candidates for ketoconazole-induced liver injury using next-generation sequencing. *Toxicol. Sci.* 179: 31–43.

Li, L., Zhu, D., Huang, L., Zhang, J., Bian, Z., Chen, X., Liu, Y., Zhang, C.Y., and Zen, K. (2012). Argonaute 2 complexes selectively protect the circulating microRNAs in cell-secreted microvesicles. *PLoS One* 7: e46957.

Li, M., Zeringer, E., Barta, T., Schageman, J., Cheng, A., and Vlassov, A.V. (2014). Analysis of the RNA content of the exosomes derived from blood serum and urine and its potential as biomarkers. *Philos. Trans. R. Soc. Lond. B. Biol. Sci.* 369 (1652): 20130502.

Li, Q., Kappil, M.A., Li, A., Dassanayake, P.S., Darrah, T.H., Friedman, A.E., Friedman, M., Lambertini, L., Landrigan, P., Stodgell, C.J., Xia, Y., Nanes, J.A., Aagaard, K.M., Schadt, E.E., Murray, J.C., Clark, E.B., Dole, N., Culhane, J., Swanson, J., Varner, M., Moye, J., Kasten, C., Miller, R.K., and Chen, J. (2015). Exploring the associations between microRNA expression profiles and environmental pollutants in human placenta from the National Children's Study (NCS). *Epigenetics* 10: 793–802.

Li, T., Evdokimov, E., Shen, R.F., Chao, C.C., Tekle, E., Wang, T., Stadtman, E.R., Yang, D.C., and Chock, P.B. (2004). Sumoylation of heterogeneous nuclear ribonucleoproteins, zinc finger proteins, and nuclear pore complex proteins: A proteomic analysis. *Proc. Natl. Acad. Sci. USA* 101: 8551–8556.

Lin, J., Ma, L., Zhang, D., Gao, J., Jin, Y., Han, Z., and Lin, D. (2019). Tumour biomarkers: Tracing the molecular function and clinical implication. *Cell Prolif.* 52: e12589.

Liu, J.Q., Niu, Q., Hu, Y.H., Li, Y., Wang, H.X., Xu, S.Z., Ding, Y.S., Li, S.G., and Ma, R.L. (2018). The bidirectional effects of arsenic on miRNA-21: A systematic review and meta-analysis. *Biomed. Environ. Sci.* 31: 654–666.

Llewellyn, H.P., Vaidya, V.S., Wang, Z., Peng, Q., Hyde, C., Potter, D., Wang, J., Zong, Q., Arat, S., Martin, M., Masek-Hammerman, K., Warner, R., Johnson, K., Kullak-Ublick, G.A., Aithal, G.P., Dear, J.W., and Ramaiah, S.K. (2021). Evaluating the sensitivity and Specificity of promising circulating biomarkers to diagnose liver injury in humans. *Toxicol. Sci.* 181: 23–34.

Lopez-Sanchez, G.N., Dominguez-Perez, M., Uribe, M., Chavez-Tapia, N.C., and Nuno-Lambarri, N. (2021). Non-alcoholic fatty liver disease and microRNAs expression, how it affects the development and progression of the disease. *Ann. Hepatol.* 21: 100212.

Louwies, T., Jaki Mekjavic, P., Cox, B., Eiken, O., Mekjavic, I.B., Kounalakis, S., and De Boever, P. (2016). Separate and combined effects of hypoxia and horizontal bed rest on retinal blood vessel diameters. *Invest Ophthalmol Vis Sci* 57: 4927–4932.

Lu, J. and Clark, A.G. (2012). Impact of microRNA regulation on variation in human gene expression. *Genome Res.* 22: 1243–1254.

Ludwig, N., Leidinger, P., Becker, K., Backes, C., Fehlmann, T., Pallasch, C., Rheinheimer, S., Meder, B., Stahler, C., Meese, E., and Keller, A. (2016). Distribution of miRNA expression across human tissues. *Nucleic Acids Res.* 44: 3865–3877.

Maccani, M.A., Avissar-whiting, M., Banister, C.E., Mcgonnigal, B., Padbury, J.F., and Marsit, C.J. (2010). Maternal cigarette smoking during pregnancy is associated with downregulation of miR-16, miR-21, and miR-146a in the placenta. *Epigenetics* 5: 583–589.

Mathieu, M., Martin-Jaular, L., Lavieu, G., and Thery, C. (2019). Specificities of secretion and uptake of exosomes and other extracellular vesicles for cell-to-cell communication. *Nat. Cell Biol.* 21: 9–17.

McKenzie, A.J., Hoshino, D., Hong, N.H., Cha, D.J., Franklin, J.L., Coffey, R.J., Patton, J.G., and Weaver, A.M. (2016). KRAS-MEK signaling controls ago2 sorting into exosomes. *Cell Rep.* 15: 978–987.

Mestdagh, P., Hartmann, N., Baeriswyl, L., Andreasen, D., Bernard, N., Chen, C., Cheo, D., D'Andrade, P., Demayo, M., Dennis, L., Derveaux, S., Feng, Y., Fulmer-smentek, S., Gerstmayer, B., Gouffon, J., Grimley, C., Lader, E., Lee, K.Y., Luo, S., Mouritzen, P., Narayanan, A., Patel, S., Peiffer, S., Ruberg, S., Schroth, G., Schuster, D., Shaffer, J.M., Shelton, E.J., Silveria, S., Ulmanella, U., Veeramachaneni, V., Staedtler, F., Peters, T., Guettouche, T., Wong, L., and Vandesompele, J. (2014). Evaluation of quantitative miRNA expression platforms in the microRNA quality control (miRQC) study. *Nat. Methods* 11: 809–815.

Morales, S., Monzo, M., and Navarro, A. (2017). Epigenetic regulation mechanisms of microRNA expression. *Biomol. Concepts* 8: 203–212.

Mosedale, M., Eaddy, J.S., Trask, O.J. JR, Holman, N.S., Wolf, K.K., Lecluyse, E., Ware, B.R., Khetani, S.R., Lu, J., Brock, W.J., Roth, S.E., and Watkins, P.B. (2018). miR-122 release in exosomes precedes overt tolvaptan-induced necrosis in a primary human hepatocyte micropatterned coculture model. *Toxicol. Sci.* 161: 149–158.

Motta, V., Angelici, L., Nordio, F., Bollati, V., Fossati, S., Frascati, F., Tinaglia, V., Bertazzi, P.A., Battaglia, C., and Baccarelli, A.A. (2013). Integrative analysis of miRNA and inflammatory gene expression after acute particulate matter exposure. *Toxicol. Sci.* 132: 307–316.

Murillo, O.D., Thistlethwaite, W., Rozowsky, J., Subramanian, S.L., Lucero, R., Shah, N., Jackson, A.R., Srinivasan, S., Chung, A., Laurent, C.D., Kitchen, R.R., Galeev, T., Warrell, J., Diao, J.A., Welsh, J.A., Hanspers, K., Riutta, A., Burgstaller-Muehlbacher, S., Shah, R.V., Yeri, A., Jenkins, L.M., Ahsen, M.E., Cordon-cardo, C., Dogra, N., Gifford, S.M., Smith, J.T., Stolovitzky, G., Tewari, A.K., Wunsch, B.H., Yadav, K.K., Danielson, K.M., Filant, J., Moeller, C., Nejad, P., Paul, A., Simonson, B., Wong, D.K., Zhang, X., Balaj, L., Gandhi, R., Sood, A.K., Alexander, R.P., Wang, L., Wu, C., Wong, D.T.W., Galas, D.J., Van Keuren-jensen, K., Patel, T., Jones, J.C., Das, S., Cheung, K.H., Pico, A.R., Su, A.I., Raffai, R.L., Laurent, L.C., Roth, M.E., Gerstein, M.B., and Milosavljevic, A. (2019). exRNA atlas analysis reveals distinct extracellular RNA cargo types and their carriers present across human biofluids. *Cell* 177: 463–477 e15.

Nik Mohamed Kamal, N. and Shahidan, W.N.S. (2019). Non-exosomal and exosomal circulatory microRNAs: Which are more valid as biomarkers? *Front Pharmacol.* 10: 1500.

Nolte-'t Hoen, E.N., Buermans, H.P., Waasdorp, M., Stoorvogel, W., Wauben, M.H., and T Hoen, P.A. (2012). Deep sequencing of RNA from immune cell-derived vesicles uncovers the selective incorporation of small non-coding RNA biotypes with potential regulatory functions. *Nucleic Acids Res.* 40: 9272–9285.

O'Brien, K., Breyne, K., Ughetto, S., Laurent, L.C., and Breakefield, X.O. (2020). RNA delivery by extracellular vesicles in mammalian cells and its applications. *Nat. Rev. Mol. Cell Biol.* 21: 585–606.

Pascut, D., Cavalletto, L., Pratama, M.Y., Bresolin, S., Trentin, L., Basso, G., Bedogni, G., Tiribelli, C., and Chemello, L. (2019). Serum miRNA are promising biomarkers for the detection of early hepatocellular carcinoma after treatment with direct-acting antivirals. *Cancers (Basel)* 11.

Pavuk, M., Olson, J.R., Sjodin, A., Wolff, P., Turner, W.E., Shelton, C., Dutton, N.D., Bartell, S., and Anniston Environmental Health Research Consortium (2014). Serum concentrations of polychlorinated biphenyls (PCBs) in participants of the Anniston Community Health Survey. *Sci. Total Environ.* 0: 473–474: 286–297.

Petralia, F., Aushev, V.N., Gopalakrishnan, K., Kappil, M.N.W.K., Chen, J., Teitelbaum, S.L., and Wang, P. (2017). A new method to study the change of miRNA-mRNA interactions due to environmental exposures. *Bioinformatics* 33: i199–i207.

Qiao, J., Du, Y., Yu, J., and Guo, J. (2019). MicroRNAs as potential biomarkers of insecticide exposure: A review. *Chem. Res. Toxicol.* 32: 2169–2181.

Rager, J.E., Auerbach, S.S., Chappell, G.A., Martin, E., Thompson, C.M., and Fry, R.C. (2017). Benchmark dose modeling estimates of the concentrations of inorganic arsenic that induce changes to the neonatal transcriptome, proteome, and epigenome in a pregnancy cohort. *Chem. Res. Toxicol.* 30: 1911–1920.

Rager, J.E., Bailey, K.A., Smeester, L., Miller, S.K., Parker, J.S., Laine, J.E., Drobna, Z., Currier, J., Douillet, C., Olshan, A.F., Rubio-Andrade, M., Styblo, M., Garcia-Vargas, G., and Fry, R.C. (2014). Prenatal arsenic exposure and the epigenome: Altered microRNAs associated with innate and adaptive immune signaling in newborn cord blood. *Environ. Mol. Mutagen* 55: 196–208.

Ramaiahgari, S.C., Auerbach, S.S., Saddler, T.O., Rice, J.R., Dunlap, P.E., Sipes, N.S., Devito, M.J., Shah, R.R., Bushel, P.R., Merrick, B.A., Paules, R.S., and Ferguson, S.S. (2019). The power of resolution: contextualized understanding of biological responses to liver injury chemicals using high-throughput transcriptomics and benchmark concentration modeling. *Toxicol. Sci.* 169: 553–566.

Rider, C.F., Yamamoto, M., Gunther, O.P., Hirota, J.A., Singh, A., Tebbutt, S.J., and Carlsten, C. (2016). Controlled diesel exhaust and allergen coexposure modulates microRNA and gene expression in humans: Effects on inflammatory lung markers. *J. Allergy Clin. Immunol.* 138: 1690–1700.

Robles-Diaz, M., Medina-Caliz, I., Stephens, C., Andrade, R.J., and Lucena, M.I. (2016). Biomarkers in DILI: One more step forward. *Front Pharmacol.* 7: 267.

Rodosthenous, R.S., Coull, B.A., Lu, Q., Vokonas, P.S., Schwartz, J.D., and Baccarelli, A.A. (2016). Ambient particulate matter and microRNAs in extracellular vesicles: A pilot study of older individuals. *Part Fibre Toxicol.* 13: 13.

Ruiz-Vera, T., Ochoa-Martinez, A.C., Zarazua, S., Carrizales-Yanez, L., and Perez-Maldonado, I.N. (2019). Circulating miRNA-126, -145 and -155 levels in Mexican women exposed to inorganic arsenic via drinking water. *Environ. Toxicol. Pharmacol.* 67: 79–86.

Saliminejad, K., Khorram Khorshid, H.R., and Ghaffari, S.H. (2019). Why have microRNA biomarkers not been translated from bench to clinic? *Future Oncol.* 15: 801–803.

Santangelo, L., Giurato, G., Cicchini, C., Montaldo, C., Mancone, C., Tarallo, R., Battistelli, C., Alonzi, T., Weisz, A., and Tripodi, M. (2016). The RNA-binding protein SYNCRIP is a component of the hepatocyte exosomal machinery controlling microRNA sorting. *Cell Rep.* 17: 799–808.

Schembri, F., Sridhar, S., Perdomo, C., Gustafson, A.M., Zhang, X., Ergun, A., Lu, J., Liu, G., Zhang, X., Bowers, J., Vaziri, C., Ott, K., Sensinger, K., Collins, J.J., Brody, J.S., Getts, R., Lenburg, M.E., and Spira, A. (2009). MicroRNAs as modulators of smoking-induced gene expression changes in human airway epithelium. *Proc. Natl. Acad. Sci. USA* 106: 2319–2324.

Sharapova, T., Devanarayan, V., Leroy, B., Liguori, M.J., Blomme, E., Buck, W., and Maher, J. (2016). Evaluation of miR-122 as a serum biomarker for hepatotoxicity in investigative rat toxicology studies. *Vet. Pathol.* 53: 211–221.

Sharma, S. and Lu, H.C. (2018). microRNAs in Neurodegeneration: Current findings and potential impacts. *J. Alzheimers Dis. Parkinsonism* 8.

Shurtleff, M.J., Temoche-diaz, M.M., Karfilis, K.V., Ri, S., and Schekman, R. (2016). Y-box protein 1 is required to sort microRNAs into exosomes in cells and in a cell-free reaction. *Elife* 5.

Sima, M., Rossnerova, A., Simova, Z., and Rossner, P., JR. (2021). The impact of air pollution exposure on the microRNA machinery and lung cancer development. *J. Pers. Med.* 11 (1): 60. doi: 10.3390/jpm11010060.

Sisto, R., Capone, P., Cerini, L., Sanjust, F., Paci, E., Pigini, D., Gatto, M.P., Gherardi, M., Gordiani, A., L'Episcopo, N., Tranfo, G., and Chiarella, P. (2019). Circulating microRNAs as potential biomarkers of occupational exposure to low dose organic solvents. *Toxicol. Rep.* 6: 126–135.

Smith, A., Calley, J., Mathur, S., Qian, H.R., Wu, H., Farmen, M., Caiment, F., Bushel, P.R., Li, J., Fisher, C., Kirby, P., Koenig, E., Hall, D.G., and Watson, D.E. (2016). The rat microRNA body atlas: Evaluation of the microRNA content of rat organs through deep sequencing and characterization of pancreas enriched miRNAs as biomarkers of pancreatic toxicity in the rat and dog. *BMC Genom.* 17: 694.

Sollome, J., Martin, E., Sethupathy, P., and Fry, R.C. (2016). Environmental contaminants and microRNA regulation: Transcription factors as regulators of toxicant-altered microRNA expression. *Toxicol. Appl. Pharmacol.* 312: 61–66.

Song, L., Zhang, Z., Zhang, J., Zhu, X., He, L., Shi, Z., Gao, L., and Feng, F. (2016). Ratio of microRNA-122/155 in isoniazid-induced acute liver injury in mice. *Exp. Ther. Med.* 12: 889–894.

Song, M.K. and Ryu, J.C. (2015). Blood miRNAs as sensitive and specific biological indicators of environmental and occupational exposure to volatile organic compound (VOC). *Int. J. Hyg. Environ. Health* 218: 590–602.

Sork, H., Corso, G., Krjutskov, K., Johansson, H.J., Nordin, J.Z., Wiklander, O.P.B., Lee, Y.X.F., Westholm, J.O., Lehtio, J., Wood, M.J.A., Mager, I., and El Andaloussi, S. (2018). Heterogeneity and interplay of the extracellular vesicle small RNA transcriptome and proteome. *Sci. Rep.* 8: 10813.

Srinivasan, S., Yeri, A., Cheah, P.S., Chung, A., Danielson, K., De Hoff, P., Filant, J., Laurent, C.D., Laurent, L.D., Magee, R., Moeller, C., Murthy, V.L., Nejad, P., Paul, A., Rigoutsos, I., Rodosthenous, R., Shah, R.V., Simonson, B., To, C., Wong, D., Yan, I.K., Zhang, X., Balaj, L., Breakefield, X.O., Daaboul, G., Gandhi, R., Lapidus, J., Londin, E., Patel, T., Raffai, R.L., Sood, A.K., Alexander, R.P., Das, S., and Laurent, L.C. (2019). Small RNA sequencing across diverse biofluids identifies optimal methods for exRNA isolation. *Cell* 177 (446–462): e16.

Starkey Lewis, P.J., Dear, J., Platt, V., Simpson, K.J., Craig, D.G., Antoine, D.J., French, N.S., Dhaun, N., Webb, D.J., Costello, E.M., Neoptolemos, J.P., Moggs, J., Goldring, C.E., and Park, B.K. (2011). Circulating microRNAs as potential markers of human drug-induced liver injury. *Hepatology* 54: 1767–1776.

Statello, L., Maugeri, M., Garre, E., Nawaz, M., Wahlgren, J., Papadimitriou, A., Lundqvist, C., Lindfors, L., Collen, A., Sunnerhagen, P., Ragusa, M., Purrello, M., Di Pietro, C., Tigue, N., and Valadi, H. (2018). Identification of RNA-binding proteins in exosomes capable of interacting with different types of RNA: RBP-facilitated transport of RNAs into exosomes. *PLoS One* 13: e0195969.

Stermer, A.R., Reyes, G., Hall, S.J., and Boekelheide, K. (2019). Small RNAs in rat sperm are a predictive and sensitive biomarker of exposure to the testicular toxicant ethylene glycol monomethyl ether. *Toxicol. Sci.* 169: 399–408.

Straumfors, A., Duale, N., Foss, O.A.H., and Mollerup, S. (2020). Circulating miRNAs as molecular markers of occupational grain dust exposure. *Sci. Rep.* 10: 11317.

Tellez, C.S., Juri, D.E., Do, K., Bernauer, A.M., Thomas, C.L., Damiani, L.A., Tessema, M., Leng, S., and Belinsky, S.A. (2011). EMT and stem cell-like properties associated with miR-205 and miR-200 epigenetic silencing are early manifestations during carcinogen-induced transformation of human lung epithelial cells. *Cancer Res.* 71: 3087–3097.

Temoche-Diaz, M.M., Shurtleff, M.J., Nottingham, R.M., Yao, J., Fadadu, R.P., Lambowitz, A.M., and Schekman, R. (2019). Distinct mechanisms of microRNA sorting into cancer cell-derived extracellular vesicle subtypes. *eLife* 8: e47544.

Thomas, R.S., Allen, B.C., Nong, A., Yang, L., Bermudez, E., and Clewell, H.J. 3RD, and Andersen, M.E. (2007). A method to integrate benchmark dose estimates with genomic data to assess the functional effects of chemical exposure. *Toxicol. Sci.* 98: 240–248.

Thompson, K.L., Boitier, E., Chen, T., Couttet, P., Ellinger-Ziegelbauer, H., Goetschy, M., Guillemain, G., Kanki, M., Kelsall, J., Mariet, C., De La Moureyre-Spire, C., Mouritzen, P., Nassirpour, R., O'Lone, R., Pine, P.S., Rosenzweig, B.A., Sharapova, T., Smith, A., Uchiyama, H., Yan, J., Yuen, P.S., and Wolfinger, R. (2016). Absolute measurement of cardiac injury-induced micrornas in biofluids across multiple test sites. *Toxicol. Sci.* 154: 115–125.

Thulin, P., Hornby, R.J., Auli, M., Nordahl, G., Antoine, D.J., Starkey Lewis, P., Goldring, C.E., Park, B.K., Prats, N., Glinghammar, B., and Schuppe-Koistinen, I. (2017). A longitudinal assessment of miR-122 and GLDH as biomarkers of drug-induced liver injury in the rat. *Biomarkers* 22: 461–469.

Tomasetti, M., Gaetani, S., Monaco, F., Neuzil, J., and Santarelli, L. (2019). Epigenetic regulation of mirna expression in malignant mesothelioma: MiRNAs as biomarkers of early diagnosis and therapy. *Front Oncol.* 9: 1293.

Tribolet, L., Kerr, E., Cowled, C., Bean, A.G.D., Stewart, C.R., Dearnley, M., and Farr, R.J. (2020). MicroRNA biomarkers for infectious diseases: From basic research to biosensing. *Front Microbiol.* 11: 1197.

Tsamou, M., Vrijens, K., Madhloum, N., Lefebvre, W., Vanpoucke, C., and Nawrot, T.S. (2018). Air pollution-induced placental epigenetic alterations in early life: A candidate miRNA approach. *Epigenetics* 13: 135–146.

Tumolo, M.R., Panico, A., De Donno, A., Mincarone, P., Leo, C.G., Guarino, R., Bagordo, F., Serio, F., Idolo, A., Grassi, T., and Sabina, S. (2020). The expression of microRNAs and exposure to environmental contaminants related to human health: A review. *Int. J. Environ. Health Res.* (online, May 12): 1–23. doi: 10.1016/j.hoc.2021.05.005.

Turchinovich, A., Drapkina, O., and Tonevitsky, A. (2019). Transcriptome of Extracellular Vesicles: State-of-the-Art. *Front Immunol.* 10: 202.

Turchinovich, A., Weiz, L., Langheinz, A., and Burwinkel, B. (2011). Characterization of extracellular circulating microRNA. *Nucleic Acids Res.* 39: 7223–7233.

Valadi, H., Ekstrom, K., Bossios, A., Sjostrand, M., Lee, J.J., and Lotvall, J.O. (2007). Exosome-mediated transfer of mRNAs and microRNAs is a novel mechanism of genetic exchange between cells. *Nat. Cell Biol.* 9: 654–659.

Van Pottelberge, G.R., Mestdagh, P., Bracke, K.R., Thas, O., Van Durme, Y.M., Joos, G.F., Vandesompele, J., and Brusselle, G.G. (2011). MicroRNA expression in induced sputum of smokers and patients with chronic obstructive pulmonary disease. *Am. J. Respir. Crit. Care Med.* 183: 898–906.

Vandana Saini, R.D, Suneja, S., Gangopadhyay, S., and Kaur, C. (2021). Can microRNA become next-generation tools in molecular diagnostics and therapeutics? A systematic review. *Egypt. J. Med. Hum. Genet.* 22: 1–9.

Vickers, K.C., Palmisano, B.T., Shoucri, B.M., Shamburek, R.D., and Remaley, A.T. (2011). MicroRNAs are transported in plasma and delivered to recipient cells by high-density lipoproteins. *Nat. Cell Biol.* 13: 423–433.

Vigneron, N., Meryet-Figuiere, M., Guttin, A., Issartel, J.P., Lambert, B., Briand, M., Louis, M.H., Vernon, M., Lebailly, P., Lecluse, Y., Joly, F., Krieger, S., Lheureux, S., Clarisse, B., Leconte, A., Gauduchon, P., Poulain, L., and Denoyelle, C. (2016). Towards a new standardized method for circulating miRNAs profiling in clinical studies: Interest of the exogenous normalization to improve miRNA signature accuracy. *Mol. Oncol.* 10: 981–992.

Villarroya-Beltri, C., Gutierrez-Vazquez, C., Sanchez-Cabo, F., Perez-Hernandez, D., Vazquez, J., Martin-Cofreces, N., Martinez-Herrera, D.J., Pascual-Montano, A., Mittelbrunn, M., and Sanchez-Madrid, F. (2013). Sumoylated hnRNPA2B1 controls the sorting of miRNAs into exosomes through binding to specific motifs. *Nat. Commun.* 4: 2980.

Vogt, J., Sheinson, D., Katavolos, P., Irimagawa, H., Tseng, M., Alatsis, K.R., and Proctor, W.R. (2019). Variance component analysis of circulating miR-122 in serum from healthy human volunteers. *PLoS One* 14: e0220406.

Vrijens, K., Bollati, V., and Nawrot, T.S. (2015). MicroRNAs as potential signatures of environmental exposure or effect: A systematic review. *Environ. Health Perspect.* 123: 399–411.

Wallace, D.R., Taalab, Y.M., Heinze, S., Tariba Lovakovic, B., Pizent, A., Renieri, E., Tsatsakis, A., Farooqi, A.A., Javorac, D., Andjelkovic, M., Bulat, Z., Antonijevic, B., and Buha Djordjevic, A. (2020). Toxic-metal-induced alteration in miRNA expression profile as a proposed mechanism for disease development. *cells* 9.

Wang, G.K., Zhu, J.Q., Zhang, J.T., Li, Q., Li, Y., He, J., Qin, Y.W., and Jing, Q. (2010a). Circulating microRNA: A novel potential biomarker for early diagnosis of acute myocardial infarction in humans. *Eur. Heart. J.* 31: 659–666.

Wang, J., Zhang, Y., Zhang, W., Jin, Y., and Dai, J. (2012). Association of perfluorooctanoic acid with HDL cholesterol and circulating miR-26b and miR-199-3p in workers of a fluorochemical plant and nearby residents. *Environ. Sci. Technol.* 46: 9274–9281.

Wang, K., Zhang, S., Marzolf, B., Troisch, P., Brightman, A., Hu, Z., Hood, L.E., and Galas, D.J. (2009). Circulating microRNAs, potential biomarkers for drug-induced liver injury. *Proc. Natl. Acad. Sci. USA* 106: 4402–4407.

Wang, M., Ye, Y., Qian, H., Song, Z., Jia, X., Zhang, Z., Zhou, J., and Ni, C. (2010b). Common genetic variants in pre-microRNAs are associated with risk of coal workers' pneumoconiosis. *J. Hum. Genet.* 55: 13–17.

Wang, T., Xu, H., Qi, M., Yan, S., and Tian, X. (2018). miRNA dysregulation and the risk of metastasis and invasion in papillary thyroid cancer: A systematic review and meta-analysis. *Oncotarget* 9: 5473–5479.

Wang, W., Zhang, H., Duan, X., Feng, X., Wang, T., Wang, P., Ding, M., Zhou, X., Liu, S., Li, L., Liu, J., Tang, L., Niu, X., Zhang, Y., Li, G., Yao, W., and Yang, Y. (2019). Association of genetic polymorphisms of miR-145 gene with telomere length in omethoate-exposed workers. *Ecotoxicol. Environ. Saf.* 172: 82–88.

Weber, D.G., Johnen, G., Bryk, O., Jockel, K.H., and Bruning, T. (2012). Identification of miRNA-103 in the cellular fraction of human peripheral blood as a potential biomarker for malignant mesothelioma–a pilot study. *PLoS One* 7: e30221.

Weber, J.A., Baxter, D.H., Zhang, S., Huang, D.Y., Huang, K.H., Lee, M.J., Galas, D.J., and Wang, K. (2010). The microRNA spectrum in 12 body fluids. *Clin. Chem.* 56: 1733–1741.

Webster, A.F., Chepelev, N., Gagne, R., Kuo, B., Recio, L., Williams, A., and Yauk, C.L. (2015). Impact of genomics platform and statistical filtering on transcriptional benchmark doses (BMD) and multiple approaches for selection of chemical point of departure (PoD). *PLoS One* 10: e0136764.

Weldon, B.A., Shubin, S.P., Smith, M.N., Workman, T., Artemenko, A., Griffith, W.C., Thompson, B., and Faustman, E.M. (2016). Urinary microRNAs as potential biomarkers of pesticide exposure. *Toxicol. Appl. Pharmacol.* 312: 19–25.

World Health Organization and International Programme on Chemical Safety (1993). *Biomarkers and Risk Assessment: Concepts and Principles, published under the joint sponsorship of the United Nations Environment Programme, the International Labour Organisation, and the World Health Organization*. World Health Organization. https://apps.who.int/iris/handle/10665/39037.

Xu, M., Yu, Z., Hu, F., Zhang, H., Zhong, L., Han, L., An, Y., Zhu, B., and Zhang, H. (2017). Identification of differential plasma miRNA profiles in Chinese workers with occupational lead exposure. *Biosci. Rep.* 37.

Yamamoto, M., Singh, A., Sava, F., Pui, M., Tebbutt, S.J., and Carlsten, C. (2013). MicroRNA expression in response to controlled exposure to diesel exhaust: Attenuation by the antioxidant N-acetylcysteine in a randomized crossover study. *Environ. Health Perspect.* 121: 670–675.

Yanshina, D.D., Kossinova, O.A., Gopanenko, A.V., Krasheninina, O.A., Malygin, A.A., Venyaminova, A.G., and Karpova, G.G. (2018). Structural features of the interaction of the 3′-untranslated region of mRNA containing exosomal RNA-specific motifs with YB-1, a potential mediator of mRNA sorting. *Biochimie.* 144: 134–143.

Yauk, C.L., Rowan-carroll, A., Stead, J.D., and Williams, A. (2010). Cross-platform analysis of global microRNA expression technologies. *BMC Genom.* 11: 330.

Yu, H.W. and Cho, W.C. (2015). The role of microRNAs in toxicology. *Arch. Toxicol.* 89: 319–325.

Yuan, H., Yuan, M., Tang, Y., Wang, B., and Zhan, X. (2018). MicroRNA expression profiling in human acute organophosphorus poisoning and functional analysis of dysregulated miRNAs. *Afr. Health Sci.* 18: 333–342.

Zhang, Y., Liu, D., Chen, X., Li, J., Li, L., Bian, Z., Sun, F., Lu, J., Yin, Y., Cai, X., Sun, Q., Wang, K., Ba, Y., Wang, Q., Wang, D., Yang, J., Liu, P., Xu, T., Yan, Q., Zhang, J., Zen, K., and Zhang, C.Y. (2010). Secreted monocytic miR-150 enhances targeted endothelial cell migration. *Mol. Cell* 39: 133–144.

Zhou, S.S., Jin, J.P., Wang, J.Q., Zhang, Z.G., Freedman, J.H., Zheng, Y., and Cai, L. (2018). miRNAS in cardiovascular diseases: Potential biomarkers, therapeutic targets and challenges. *Acta Pharmacol. Sin.* 39: 1073–1084.

Zhou, X., Wang, L., Zou, W., Chen, X., Roizman, B., and Zhou, G.G. (2020). hnRNPA2B1 Associated with recruitment of RNA into exosomes plays a key role in herpes simplex virus 1 release from infected cells. *J. Virol.* 94 (3): e00367-20.

3

EV (Extracellular Vesicle)-associated miRNAs as Biomarkers of Toxicity

Ryuichi Ono[1], Yusuke Yoshioka[2], Yusuke Furukawa[1], Mie Naruse[1,3], Makiko Kuwagata[1], Takahiro Ochiya[1,2,4], Satoshi Kitajima[1], and Yoko Hirabayashi[5]*

[1] *Division of Cellular and Molecular Toxicology, Center for Biological Safety and Research, National Institute of Health Sciences (NIHS)*
[2] *Division of Molecular and Cellular Medicine, Institute of Medical Science, Tokyo Medical University*
[3] *Central Animal Division, National Cancer Center Research Institute*
[4] *Division of Molecular and Cellular Medicine, National Cancer Center Research Institute*
[5] *Center for Biological Safety and Research, National Institute of Health Sciences (NIHS)*

Introduction

Exosomes were originally discovered in 1983, as membrane vesicles of approximately 100 nm in diameter with a lipid bilayer structure. These vesicles are secreted by reticulocytes (Harding and Stahl 1983; Pan and Johnstone 1983). They were initially thought to be a "trash bag" in which cells could eliminate excess proteins; however, from 2007 on great advances in exosome research were made, as it was shown that microRNAs (miRNAs) in exosomes could be transferred between cells and played important roles in many physiological and pathological processes (Valadi et al. 2007; Pegtel et al. 2010; Mittelbrunn et al. 2011; Hergenreider et al. 2012; Montecalvo et al. 2012).

Exosomes are secreted into body fluids from various types of cells and organs. Recent studies have demonstrated that exosomes secreted by tumor cells are transported to surrounding cells and participate in metastasis and infiltration. Since these tumor-derived exosomes containing cancer cell-specific miRNAs and mRNAs are secreted into body fluids such as blood, urine, ascites, amniotic fluid, bronchoalveolar lavage, and tears, the miRNAs and mRNAs in cancer cell-derived exosomes have come to be used as diagnostic biomarkers for early stage cancers (Thery et al. 2002; Logozzi et al. 2009; Lu et al. 2009; Rabinowits et al. 2009; Montecalvo et al. 2012; Peinado et al. 2012). In parallel, miRNAs contained in exosomes released into the blood from tissues and organs in response to adverse events, such as exposure to chemical substances and drugs, are expected to be useful as novel biomarkers for toxicity assessment.

In this chapter we will introduce the latest findings on exosomes, including the newly appreciated biological significance of exosomes, and will explain the potential use of exosomes as biomarkers of toxicity.

Genomic and Epigenomic Biomarkers of Toxicology and Disease: Clinical and Therapeutic Actions,
First Edition. Edited by Saura C. Sahu.

The Classification of Extracellular Vesicles (EVs)

Various particles are secreted by cells and released into body fluids such as blood, urine, ascites, amniotic fluid, bronchoalveolar lavage, and tears; the International Society for Extracellular Vesicles (ISEV) has named these particles "extracellular vesicles" (EVs) (Thery et al. 2002; Witwer et al. 2017). EVs are divided into three subtypes: exosomes, microvesicles (MVs), and apoptotic bodies. This classification is based on their size and on the biosynthetic pathways by which they are generated (see Figure 3.1 and Table 3.1; also Colombo et al. 2014).

Exosomes are thought to derive from intraluminal vesicles through the fusion of an intermediate endocytic compartment, the multivesicular body (MVB), with the plasma membrane (Witwer et al. 2013; Cufaro et al. 2019). The MVB contains vesicles that bind either to lysosomes, to degrade their contents, or to the plasma membrane. The latter produces the release of intraluminal vesicles defined as exosomes into extracellular space (Harding and Stahl 1983; Pan et al. 1985).

As already mentioned, exosomes are approximately 100 nm in diameter; hence they are smaller than MVs (Witwer et al. 2013). Exosome membrane proteins are enriched in heat shock proteins (HSP70, HSP90), integrins (LFA-1), proteins involved in MVB formation (ALIX, TSG101), tetraspanins (CD9, CD63, CD81, CD82, and CD151), immunostimulatory molecules (MHC class I/II proteins), lipid-related proteins, and phospholipases (Conde-vancells et al. 2008; Subra et al. 2010).

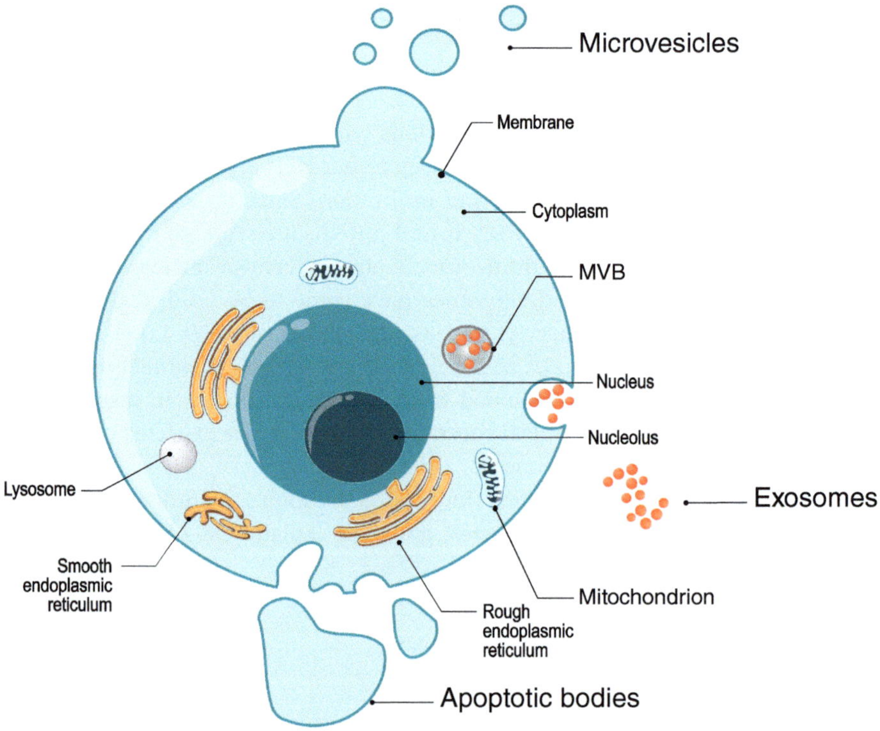

Figure 3.1 Schematic illustration of extracellular vesicles (EVs). EVs include exosomes, microvesicles and apoptotic bodies. MVB: multivesicular body.

Table 3.1 Main characteristics of exosomes, microvesicles (MV), and apoptotic bodies.

	Exosomes	Microvesicles	Apoptotic Bodies
Origin	Endocytic pathway	Plasma membrane	Plasma membrane
Size	40–120 nm	50–1000 nm	500–2000 nm
Function	Intercellular communication	Intercellular communication	Facilitate phagocytosis
Markers	Alix, Tsg101, tetraspanins	Selectins, integrins, CD proteins, DNA, RNA40 ligand	Histones, annexin V
Contents	Protein, DNA, RNA	Protein, DNA, RNA	Nuclear fractions, cell organelles

By contrast, MVs range from a few nanometers to a few microns in diameter and derive from outward budding of the plasma membrane (Witwer et al. 2013; Kalluri 2016). In some lipids and phosphatidylserine they are enriched.

Apoptotic bodies are released by apoptotic cells. They are the largest EVs, with a diameter of 1 to 5 μm, and contain several intracellular fragments, cellular organelles, membranes, and cytosolic contents (van der Pol et al. 2012; Cufaro et al. 2019). However, in practice it remains difficult to distinguish between the different subtypes of EVs. Therefore the ISEV recommends, by consensus, using the general term "EV" in the nomenclature (Gould and Raposo 2013; Thery et al. 2018). In this chapter we use this term to refer to all the subtypes of vesicles present in the extracellular space, alongside small EVs (sEVs) for exosomes.

Additional components are found in EVs, including a wide variety of genetic molecules such as DNA and coding or non-coding RNAs (Nawaz et al. 2018).

Components of Exosomes (EVs)

RNAs

Most of the RNA present in normal cells is ribosomal RNA, which is reported to account for 80–85% of total cellular RNA. Among its subunits, the 28S and 18S have the highest abundance; therefore, when electrophoresis of total RNA extracted from mammalian cells is performed, the two main bands represent the 28S and 18S subunits (Figure 3.2; see Imbeaud et al. 2005).

Both mRNAs and small RNAs in normal cells have low abundance by comparison with ribosomal RNAs. Small RNAs include small nuclear RNAs (snRNAs), miRNAs, small nucleolar RNAs (snoRNAs), and piwi-interacting RNAs (piRNAs) (Watson et al. 2019).

In contrast, small RNAs, especially miRNAs, are enriched in sEVs (Figure 3.2; see Valadi et al. 2007). MicroRNAs (miRNAs) were originally discovered in *Caenorhabditis elegans* and are found in most eukaryotes, including humans and mice. These molecules are 21–25 base (nt)-long single-stranded RNAs that are involved in the post-transcriptional regulation of gene expression in eukaryotes (Lee et al. 1993; Ambros 2004). To date, 1974 hairpin precursors of miRNAs and 2654 mature miRNA sequences have been annotated in the human genome (Kozomara et al. 2019). The latest information about miRNAs is available

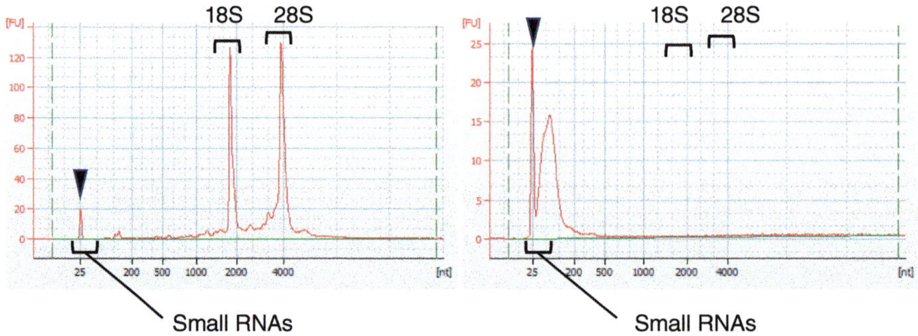

Figure 3.2 Gel electrophoresis of mouse liver total RNA (left panel) and mouse serum EV-associated RNAs (right panel). Black triangles represent size markers.

in the miRBase database (http://www.mirbase.org). Each miRNA binds to its target mRNA with incomplete homology, generally recognizes the 3′UTR of the target gene, and destabilizes the target mRNA or represses its translation by translation inhibition (Macfarlane and Murphy 2010).

Transcriptional repression mediated by miRNAs is known to play an important role in a wide variety of biological processes such as normal development, teratogenicity, cell proliferation and differentiation, apoptosis, and cancer formation (Alberti and Cochella 2017; Gueta et al. 2010; Mor et al. 2014; Macfarlane and Murphy 2010).

The standard miRNA production process consists of the following five steps: (1) transcriptional synthesis of primary miRNA (pri-miRNA); (2) cleavage of the pri-miRNA in the nucleus, to produce precursor miRNA (pre-miRNA); (3) nuclear export of the pre-miRNA; (4) cleavage of the pre-miRNA in the cytoplasm to produce miRNA duplexes; and (5) maturation of the miRNA.

These processes are collectively referred to as "miRNA processing" (see Figure 3.3).

1) *Transcriptional synthesis of primary miRNA (pri-miRNA).* The first miRNA precursor transcripts are produced in the nucleus through the transcription of the miRNA genes, primarily by RNA polymerase II and III, the same polymerases that transcribe those genes that are translated into proteins and polIII-driven repeat sequences (Lee et al. 2004; Borchert et al. 2006). Primary transcripts containing 60–70 nt RNA hairpin structures are called pri-miRNAs (Han et al. 2006). Most mammalian miRNAs are located within intergenic regions or within the introns of protein-encoding genes and non-coding RNAs. Less commonly, miRNA precursors may be located in exons of transcripts and in antisense transcripts.
2) *Cleavage of pri-miRNA in the nucleus to produce precursor miRNA (pre-miRNA).* The ribonuclease III endonuclease (RNase III), Drosha, along with DGCR8 and multiple cofactors, cleaves the hairpin base of the pri-miRNA in the nucleus to produce a 60–70-base pre-miRNA with a hairpin structure, which is an intermediate (Denli et al. 2004; Gregory et al. 2004; Han et al. 2004; Lee et al. 2003; Lee et al. 2002).
3) *Nuclear export of pre-miRNA.* Subsequent pre-miRNA processing occurs in the cytoplasm. The pre-miRNA is mainly transported from the nucleus to the cytoplasm by nucleocytoplasmic transporter containing exportin-5 (XPO5) and Ran-GTP, which prevents its degradation in the nucleus and facilitates its translocation into the cytoplasm (Bohnsack Czaplinski, Gorlich 2004; Okada et al. 2009; Yi et al. 2003; Zeng and Cullen 2004).

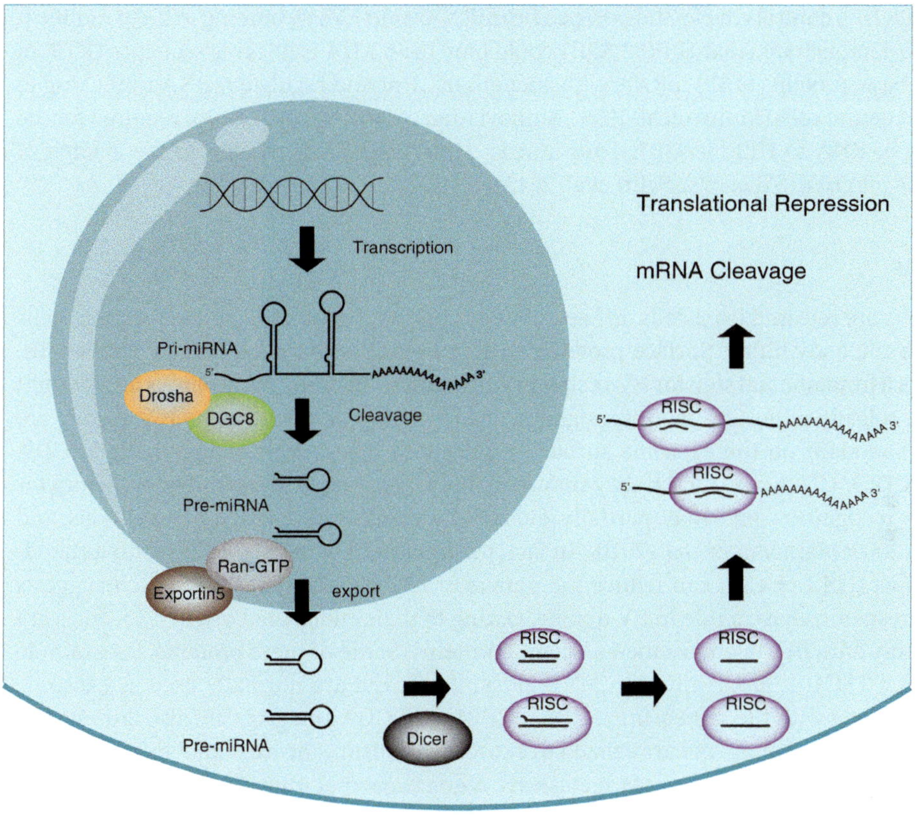

Figure 3.3 Canonical miRNA biogenesis pathway.

4) *Cleavage of pre-miRNA in the cytoplasm to produce miRNA duplexes.* The pre-miRNA undergoes a second cleavage event, whichis mediated by another RNase III endonuclease, Dicer, with its cofactor TRBP. The resulting small RNA is a 21-to-24-base miRNA duplex (miRNA-5p/miRNA-3p) (Hutvagner et al. 2001; Chendrimada et al. 2005).
5) *Maturation of miRNA.* A double-stranded miRNA duplex is incorporated into the Ago protein, and only one side of the RNA strand (the miRNA strand, guide strand, or mature miRNA) finally forms a stable complex with the Ago protein, forming an RNA-induced silencing complex (RISC) (Kwak and Tomari 2012; Iwasaki et al. 2010; Hammond et al. 2000). Finally, this single-stranded mature miRNA acts as a guide for gene expression regulation. RISC binds to its target mRNA, which has a sequence that is partially complementary to the miRNA incorporated into the RISC, and generally suppresses the output of the target mRNA to the protein (Jonas and Izaurralde 2015).

To date, there have been several reports on the mechanism of EV capture of miRNAs. Those miRNAs whose expression levels increase in T cells upon activation are not significantly increased in exosomes (Villarroya-Beltri et al. 2013). A similar trend was observed for mRNAs, which suggests that miRNA and mRNA loading into exosomes is not a passive process. Several miRNAs were more highly represented in EVs than in cells, for example miR-575, miR-451, miR-125a-3p, miR-198, miR-601 and miR-887. Analysis of these expression profiles revealed that miRNAs that are preferentially sorted to EVs contain specific short motifs, called "EXO motifs." The sumoylated heterogeneous nuclear ribonucleoprotein A2B1 (hnRNPA2B1) controls the loading of specific miRNAs into EVs by binding to these motifs. It was also demonstrated that HnRNPA2B1 could bind to an RNA trafficking sequence (RTS) of myelin basic protein (MBP) mRNA with a length of 21 nt and regulate mRNA trafficking to axons in neural cells (Munro et al. 1999). Sumoylation of HnRNPA2B1 is also required for the binding of RNA to HnRNPA2B1. Thus at least HnRNPA2B1 contributes to the loading of miRNAs into EVs (Villarroya-Beltri et al. 2013).

Proteins

After EVs are released from cells, EV-associated miRNAs will be delivered to recipient cells through the body fluids. Surface proteins on EVs help EVs to first adhere to target cells, which is a fundamental step for EV-recipient cell communication. Tetraspanins are thought to have roles in adhesion, motility, signal transduction, and cell activation; and they are highly abundant on the exosome surface (Figure 3.4). These tetraspanins include CD9, CD53, CD63, CD81, and CD82; they may contribute to the spatial assembly of factors for antigen recognition and may partially dictate the signal induced by EVs (Simons and Raposo 2009; Mathivanan et al. 2010). In fact, the treatment of recipient cells with antibodies against CD81 or CD9 can reduce the uptake of EVs by recipient cells, which suggests that tetraspanins have a role in EV uptake. Owing to their endosomal origin, exosomes are commonly enriched in endosome-associated proteins. Some of these proteins, for example ALIX and TSG101, are involved in MVB biosynthesis and are normally used as exosome markers along with the tetraspanins CD63, CD81, and CD9. Integrins are also enriched in the exosome membrane and are related to exosome tropism. The exosomal integrins $\alpha6\beta4$ and $\alpha6\beta1$ are associated with lung metastasis, while exosomal integrin $\alpha v\beta5$ was linked to liver metastasis. In fact, targeting the integrins $\alpha6\beta4$ and $\alpha v\beta5$ decreased respectively exosome uptake as well as lung and liver metastasis.

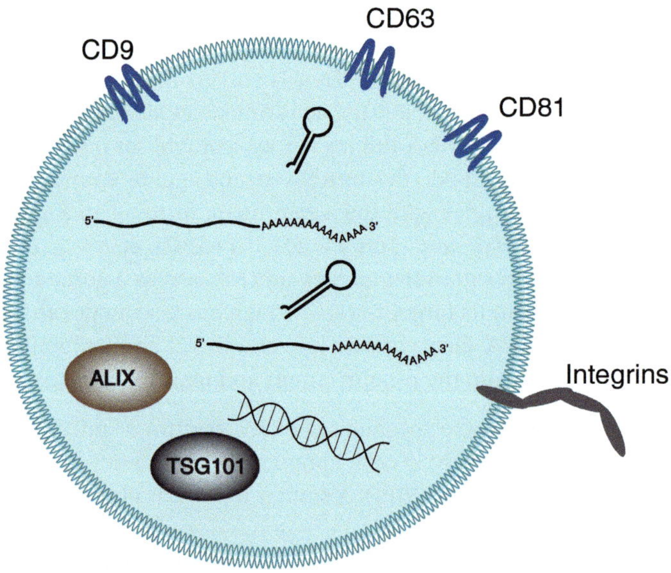

Figure 3.4 Schematic representation of the major components of exosomes. Common exosome markers include tetraspanins (CD9, CD63, and CD81), integrins, TSG101, and Alix. Exosomes also contain other proteins, different species of RNA, and DNA.

It is biologically important to determine the organ where metastasis might be facilitated by such a mechanism. Accumulating evidence suggests that exosome-mediated activities play an important role in various diseases, especially cancer.

Physiological and Cellular Functions of EVs

Angiogenesis is the physiological process through which new blood vessels are formed from pre-existing vessels (Birbrair et al. 2014; Birbrair et al. 2015). Angiogenesis is regulated by angiogenic factors, extracellular matrix components, and endothelial cells (ECs) (Carmeliet and Jain 2000).

Angiogenesis is also promoted by the uptake of cancer cell-derived EVs that are rich in matrix metalloproteinases, especially MMP-2, MMP-9 and MMP-13. These proteins have been found in glioblastoma-, melanoma-, myeloma- and nasopharyngeal carcinoma-derived exosomes (Skog et al. 2008; Ekstrom et al. 2014; You et al. 2015; Wang et al. 2016; Giusti et al. 2016; Chan et al. 2015). It has been recently reported that metastatic breast cancer-derived EVs, which express high levels of the proangiogenic protein annexin II (Anx II), promote angiogenesis and that head and neck squamous cell carcinoma-derived EVs regulate angiogenesis through ephrin-B reverse signaling (Sato et al. 2019).

EV-associated miRNAs are also known to promote angiogenesis. EV-associated miR-23a, derived from hypoxic lung cancer cells, promotes angiogenesis by targeting prolyl hydroxylase and the tight junction protein ZO-1 (Hsu et al. 2017). It also promotes angiogenesis in hypoxic HCC (Sruthi et al. 2018). Gastric cancer (GC) cell-derived EVs containing miR-155 promote angiogenesis by enhancing the expression of vascular endothelial growth factor (VEGF) (Deng et al. 2020). In this way EVs play an essential role in the promotion of angiogenesis, which suggests that they could promote carcinogenesis by promoting cancer metastasis. EVs are also important for the crosstalk between cancer cells and fibroblasts through EVs.

The interactions between cancer cells and cancer-associated fibroblasts (CAFs) vary and are considered to be important for carcinogenesis (Xouri and Christian 2010; Rasanen and Vaheri 2010; Kalluri 2016; Shiga et al. 2015; Kalluri and Zeisberg 2006; Tommelein et al. 2015). Moreover, the epithelial–mesenchymal transition (EMT), which enables tumor cells to acquire resistance to anticancer drugs, was reported to be facilitated by the presence of CAFs (Shan et al. 2017; Yu et al. 2014; Zhuang et al. 2015). There have been several reports that the interactions between cancer cells and CAFs are mediated by EVs. Cancer cells educate fibroblasts through EVs, which leads to the progression of metastasis (De Wever et al. 2014). Cancer-derived EVs are involved in the reprogramming of normal stromal fibroblasts to activated CAFs in chronic lymphocytic leukemia, hepatocellular carcinoma (HCC), and melanoma (Paggetti et al. 2015; Fang et al. 2018; Gener Lahav et al. 2019). These studies revealed that the CAF transition leads to metastatic niche formation and promotes cancer metastasis.

EVs facilitate not only cancer-related functions but also many other physiological activities. For example, it has been recently reported that EVs mediate horizontal gene transfer.

To date, there have been two reports of exosome-mediated trans-species horizontal gene transfer events in mammals: Ono et al. (2019) and Kawamura et al. (2019).

Approximately 40% of the mammalian genome is derived from retrotransposons, which have long been considered "junk." Recently accumulated evidence has demonstrated that mammalian ancestors acquire retrotransposons as endogenous genes. A chromodomain (chromatin organization modifier) is a protein structural domain. Chromodomains are highly conserved in chromoviruses, and SCAN domains might originate from GYPSYDR-1 retrotransposons. *Sirh*-family genes, which are conserved in mammals, contain a gag-like domain from the Ty3/Gypsy-type retrotransposon of fugu fish (Ono et al. 2001; Ono et al. 2003; Ono et al. 2006; Sekita et al. 2008; Naruse et al. 2014; Irie et al. 2015). In particular, in the case of *Sirh*-family genes, the mechanism of the horizontal gene transfer from fugu to the mammalian ancestor was a mystery (Ono et al. 2011). Recent *in silico* analyses demonstrated horizontal transfer of BovB (non-LTR retrotransposon from *Bos taurus*) and L1 retrotransposons (*Bos taurus*) in eukaryotes (Ivancevic et al. 2018).

In mouse NIH-3T3 cells, most DNA double-strand breaks (DSBs) introduced by CRISPR-Cas9 are repaired through non-homologous end joining (NHEJ) without homologous DNA sequences for homologous recombination (HR) (Wang et al. 2013). NHEJ-mediated repair of DSBs is prone to error, causing small indels (Wang et al. 2013). It was reported that DSBs introduced by CRISPR-Cas9 can be repaired through the capture of non-target (i.e., unintended) sequences (Ono et al. 2015). These non-target sequences may include retrotransposon sequences, reverse-transcribed spliced mRNA sequences, and CRISPR-Cas9 vector sequences (Figure 3.5a, b, c). Although most of the captured retrotransposons were derived from murine endogenous retrotransposons, horizontal gene transfer from bovines to mice was certainly confirmed (Figure 3.5d; see Ono et al. 2019).

As it was possible that these horizontal genes were transferred from the cell culture medium, we repeated these experiments using exosome-free 10% FBS (DMEM). Then most of the bovine DNA insertions were abolished by culture with exosome-free 10% FBS/DMEM (Figure 3.5e). Furthermore, it was reported that bovine retrotransposons were enriched in FBS-derived exosomes (Ono et al. 2019). These data demonstrate that trans-species gene transfer is mediated by exosomes (Ono et al. 2019).

It was also reported that LINE-1 retrotransposons underwent horizontal transfer mediated by exosomes by using the engineered L1-EGFP reporter assay (Figure 3.6a).

The human cancer cell line MDA-MB-231-D3H2LN (MM231) was transfected with an L1-EGFP reporter cassette as described previously (Farkash et al. 2006; Garcia-Perez et al. 2010; Ostertag et al. 2000; Coufal et al. 2009). In this construct, the EGFP reporter gene is interrupted by an intron and inserted in the opposite transcriptional orientation into the 3′UTR of a retrotransposition-competent human L1. Thus EGFP is expressed only when the L1 transcript is spliced, reverse-transcribed, and integrated into the chromosomal genome during a retrotransposition event (Ostertag et al. 2000). Successful L1 integration by retrotransposition can be detected by screening for EGFP expression or by performing a polymerase chain reaction (PCR) (Figure 3.6a).

To investigate the transfer of L1-EGFP RNA to recipient cells, MM231 L1-EGFP cells were co-cultured with MM231 wild-type (WT) cells as recipient cells (Figure 3.6b). Transwell inserts allow EVs, but not whole cells, to pass through.

After twenty-one days of co-culture, EGFP-expressing cells were detected in recipient cells that were exposed to L1-EGFP EVs (Figure 3.6), demonstrating that the L1-EGFP reporter was horizontally transferred from MM231 L1-EGFP cells to recipient cells and that retrotransposition events occurred in the recipient cells.

These reports support the occurrence of EV-mediated trans-species gene transfer. Because EVs are reportedly present in all fluids from living animals, the trans-species gene transfer events discovered in many species may be mediated by EVs.

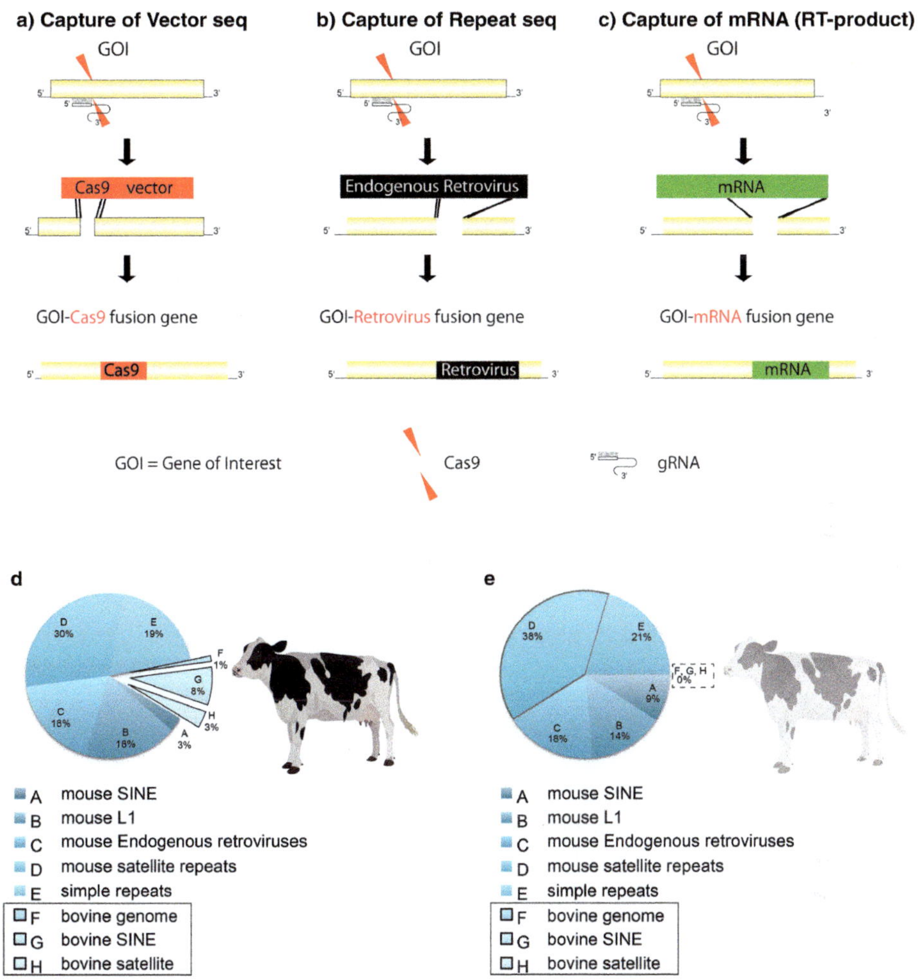

Figure 3.5 Horizontal gene transfer mediated by EVs. Three types of capture of non-target sequences associated with genome editing (a, b, c). Distribution of the captured retrotransposons at CRISPR-Cas9-induced DSB sites in NIH-3T3 cells cultured using DMEM containing 10% FBS; 12% of the reads that captured retrotransposon sequences, including genomic, Short interspersed nuclear element (SINE), and satellite DNA sequences (d), were from *Bos taurus* (bovine). Distribution of the captured retrotransposons at CRISPR-Cas9-induced DSB sites in NIH-3T3 cells cultured using DMEM containing 10% exosome-free FBS; none of the reads that captured retrotransposon sequences, including genomic, SINE, and satellite DNA sequences (e), were from *Bos taurus* (bovine).

Cancer Biomarkers

Cancer is one of the leading causes of death worldwide. Approximately 10 million people died prematurely as a result of cancer in 2018; that is, every sixth death in the world is due to cancer (Bray et al. 2018). To make progress in the prevention and treatment of cancer, many anticancer drugs are under active development worldwide. Because molecular targeting drugs and immune checkpoint inhibitors have been developed, improvements have been seen in cancer treatment outcomes (Welslau et al. 2014; Zhang 2016; Negrier et al.

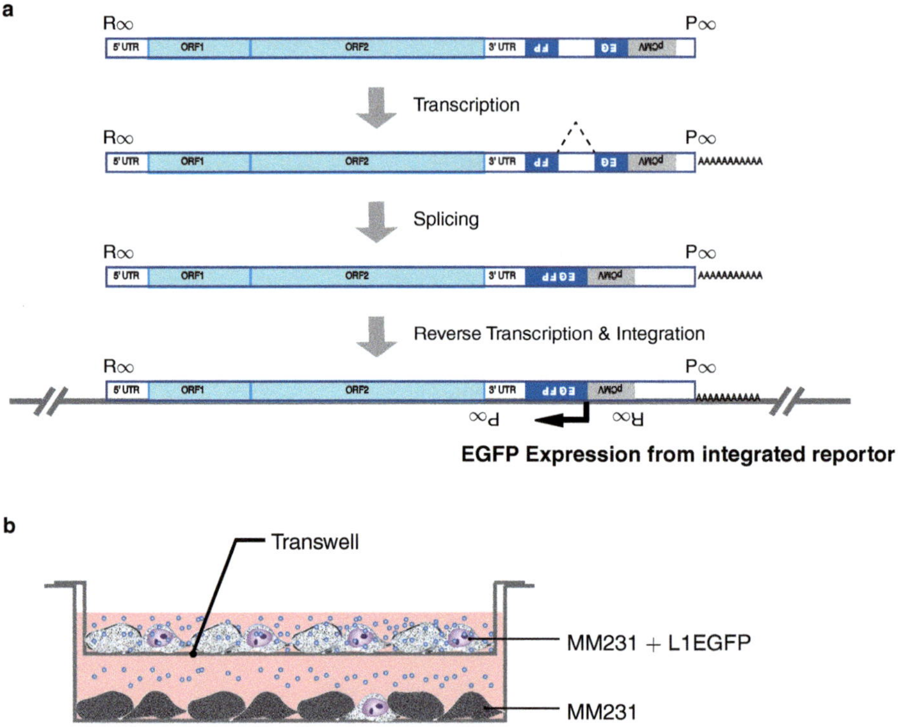

Figure 3.6 (a) Schematic representation of the engineered L1-EGFP reporter assay. (b) L1-EGFP RNA is transferred to recipient cells by using a co-culture system.

2011; Chen and Han 2015; Topalian et al. 2015; Topalian 2017; Sharma and Allison 2015). However, the prognosis of all patients with advanced cancer has not yet been significantly improved. It has been shown that early detection of cancers has a greater therapeutic effect than late detection. Therefore, to extend the healthy lifespans of cancer patients, it is important to detect cancers at early stages.

Liquid biopsies have been regarded as a powerful tool for screening and identifying tumors before symptoms appear. Historically, the detection of a single circulating protein (e.g., CA19–9, carcinoembryonic antigen (CEA), and prostate-specific antigen (PSA)) has achieved prominence due to its convenience and low cost. However, the limited specificity of such markers has been widely recognized. When a tumor marker is detected but the organ specificity of that marker is poor, it is not possible to determine which organ has cancer. For example, CEA is a tumor marker for breast cancer, but it is also expressed at high levels in colorectal cancer, stomach cancer, and pancreatic cancer (Vaidyanathan and Vasudevan 2012). Therefore further tests are required, which results in an increased burden on the patient. Because of these problems, screening for new molecular biomarkers to be used in liquid biopsy is underway: the aim is to develop more accurate methods for early cancer diagnosis.

The discovery of new molecular biomarkers overcomes the drawbacks, thanks to the intimate association of these se biomarkers with cancer-related genetic alterations. The proteins on the surfaces of exosomes are potential cancer biomarkers that can allow early tumor diagnosis (Table 3.2). Glypican-1 (GPC1) (+) exosomes could distinguish healthy subjects from patients with pancreatic cancer (Melo et al. 2015). The percentage of GPC1 (+) exosomes was markedly increased in plasma from patients with pancreatic cancer; however, they were decreased to normal levels after surgery. It was also reported that the exosome protein survivin-2B was a good biomarker for breast cancer diagnosis, and the level of TRIM3 protein in serum exosomes was decreased in gastric cancer patients (Khan et al. 2012; Fu et al. 2018).

Since exosomes are surrounded by a lipid bilayer membrane, they are not easily degraded even in those body fluids that contain many digestive enzymes. Therefore an EV-associated RNA is highly stable and makes an ideal noninvasive biomarker for cancer diagnosis. The most abundant type of RNA in exosomes is miRNA; consequently liquid biopsies using miRNAs in exosomes as cancer biomarkers have been developed in various cancer types (Table 3.3).

Increasing evidence has accumulated that EV-associated miRNAs may also prove clinically useful other than as cancer biomarkers—namely as biomarkers for non-cancer diseases. These diseases include pathogenic inflammation, infection, and chronic diseases. However, the disease-specific exosome cargo remains to be fully elucidated and validated.

Recently EV-associated miRNAs were identified as biomarkers for Alzheimer's disease and Parkinson's disease.

Table 3.2 Summary of exosomal proteins for early tumor diagnosis.

Type of cancer	Protein	Reference
Colorectal cancer	Copine III (CPNE3)	(Sun et al. 2019)
	CD147	(Tian et al. 2018)
	CD147	(Yoshioka et al. 2014)
Gastric cancer	HER-2/neu, EMMPRIN, MAGE-1, C-MET	(Baran et al. 2010)
	TRIM3	(Fu et al. 2018)
Lung adenocarcinoma	CD91, CD317, ITA2B	(Ueda et al. 2014)
	EGFR, KRAS, claudins and RAB-family proteins	(Clark et al. 2016)
	CD151, CD171 and tetraspanin 8	(Sandfeld-paulsen et al. 2016)
Melanoma	Caveolin	(Logozzi et al. 2009)
	Met	(Peinado et al. 2012)
Pancreatic ductal adenocarcinoma	GPC-1	(Melo et al. 2015)
Prostate cancer	PSA	(Mizutani et al. 2014)
	ephrinA2	(Li et al. 2018)
	Survivin	(Khan et al. 2012)
Renal cell carcinoma	MMP-9, DKP4, EMMPRIN, PODXL	(Raimondo et al. 2013)

Table 3.3 Summary of exosomal miRNAs for early tumor diagnosis.

Type of cancer	Protein	Reference
Breast cancer	miR-101, miR-372, miR-373	(Eichelser et al. 2014)
Cervical squamous cell carcinoma	miR-221-3p	(Zhou et al. 2019)
Colorectal carcinoma	miR-6803-5p	(Yan et al. 2018)
Endometrial cancer	miR-200c-3p	(Srivastava et al. 2018)
Glioblastoma multiforme	miR-320 and miR-574-3p	(Manterola et al. 2014)
Hepatocellular carcinoma	miR-21 and miR-144	(Pu et al. 2018)
Lung cancer	miR139-5p, miR-200b-5p, miR-378a, and miR-379	(Cazzoli et al. 2013)
	let-7g-5p, miR-24-3p, and miR-233-3p	(Rodriguez et al. 2014)
Ovarian cancer	miR-21, miR-141, miR-200a, miR-200c, miR203, miR-205, miR-214	(Taylor and Gercel-taylor 2008)
Pancreatic cancer	miR-1246, miR-3976, miR-4306, and miR-4644	(Madhavan et al. 2015)
	miR-17-5p and miR21	(Que et al. 2013)
Prostate cancer	miR-141	(Li et al. 2016)

Toxicology Biomarkers

Various chemical substances and drugs that are indispensable in daily life also carry a potential risk of harm to human health and the environment; therefore, in order to avoid health crises and maintain public safety, it is necessary to accurately estimate the risk carried by exposures to chemical substances and drugs.

Through the accumulation of toxicogenomics data collected during a single exposure to a chemical substance or drug, the highly accurate impact assessment of chemical substances and drugs on the basis of their molecular mechanisms is reaching the stage of practical use. However, these data are derived from specific organs (mainly the liver), and comprehensive toxicity evaluation at the individual level is both costly and labor-intensive.

Recent reports demonstrate that EVs circulating in various fluids could be used as diagnostic biomarkers for various cancers (Logozzi et al. 2009; Lu et al. 2009; Rabinowits et al. 2009; Choi et al. 2011). These EV-associated biomarkers are more sensitive and accurate than biomarkers that are currently widely used, such as CEA for adenocarcinoma and PSA for prostate cancer. Furthermore, microRNAs contained in EVs secreted from various cell types and human samples are being identified as specific biomarkers for chemically induced inflammation (Mobarrez et al. 2014; Li et al. 2010; Baek et al. 2016; Bala et al. 2012; Cho et al. 2017). In addition, EV-associated miRNAs are well protected owing to the lipid bilayer membrane of EVs, even in EVs that have been purified from the circulating bloodstream (Yanez-Mo et al. 2015).

Therefore "next-generation type" toxicity tests for chemical substances and drugs were developed by using EV-associated miRNAs in blood as biomarkers (Figure 3.7).

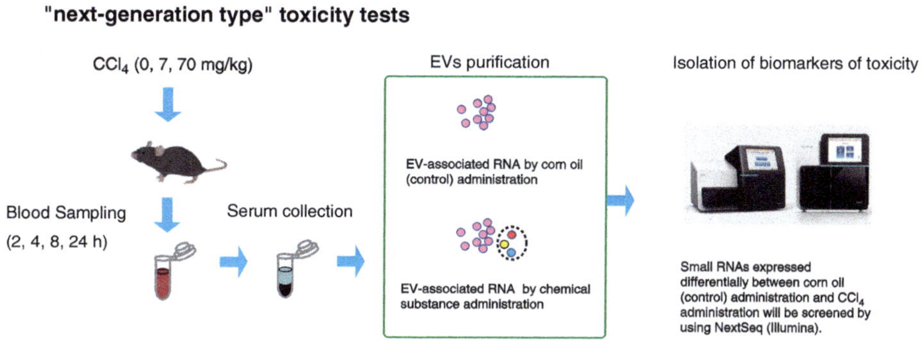

Figure 3.7 Schematic representation of toxicity testing using EVs as biomarkers.

Isolation and Characterization of EVs from Mouse Blood

Whole blood was collected from C57BL/6J male mice (twelve weeks), and the serum was carefully separated, as previously described. The serum was centrifuged at 10,000 × g for ten minutes to remove cellular debris and subsequently ultracentrifuged at 210,000 × g for thirty minutes. The pellet was washed and resuspended in PBS solution. To characterize EVs collected from whole blood, nanoparticle-tracking analysis (NTA) was performed by using NanoSight (Figure 3.8). The peak size of the EVs was 67 nm, which was comparable with that described in previous reports (Lobb et al. 2015; Urabe et al. 2017).

Carbon Tetrachloride (CCl_4) Administration and Histology

C57BL/6J male mice were orally dosed with carbon tetrachloride (CCl_4), because there are numerous reports that CCl_4 induces hepatotoxicity in many experimental animals (Chopra et al. 1972). Whole blood and liver samples were collected twenty-four hours after the administration of CCl_4 (0 mg/kg (vehicle control: corn oil), 7 mg/kg and 70 mg/kg) for the following experiments.

Hematoxylin and eosin (H&E) staining showed CCl_4 (70 mg/kg)-induced histopathological changes in the liver, with significant degeneration and necrosis of hepatocytes in the centrilobular region and perivenular inflammatory infiltrates twenty-four hours after administration, while histological tissue sections of mice in the vehicle control group and CCl_4 (7 mg/kg) showed a normal histological morphology of liver tissue samples (Figure 3.9).

Identification of Differentially Expressed miRNAs by RNA-Seq

To identify differentially expressed EV-associated miRNAs, we performed RNA-Seq on a size-selected EV-associated small RNA library for three doses of CCl_4, namely 0, 7, and 70 mg/kg. We obtained forty-five differentially expressed EV-associated miRNAs, including forty-five upregulated genes and no downregulated genes, between the corn oil and 70 mg/kg CCl_4 samples, while only one differentially expressed (upregulated) EV-associated was identified between the corn oil and 7 mg/kg CCl_4 samples (see Figure 3.10; also Ono et al. 2020).

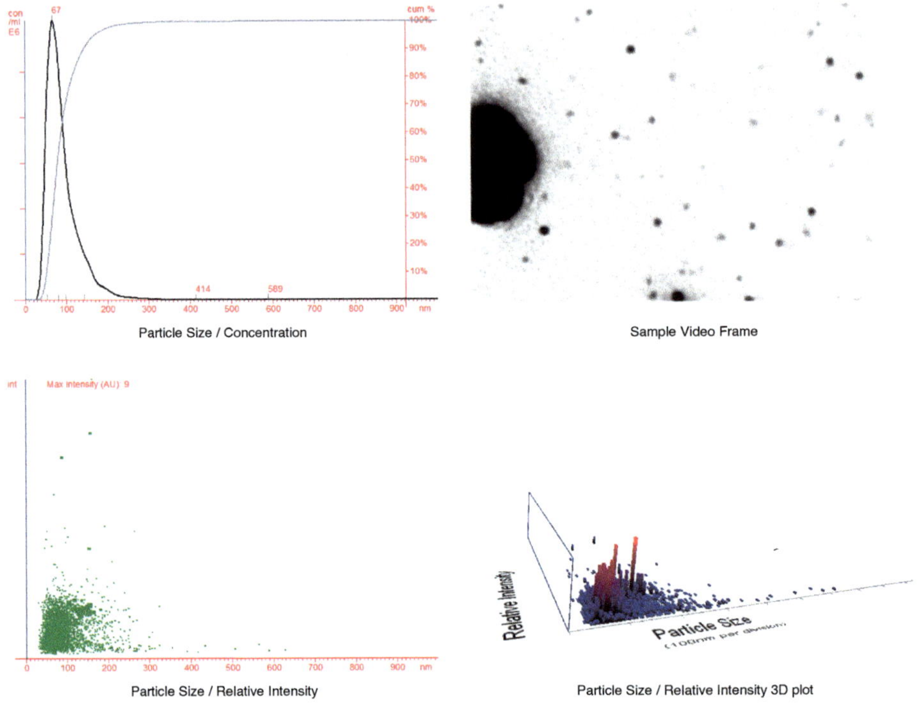

RESULTS:	
Size Distribution:	Mean: 93 nm, Mode: 67 nm, SD: 53 nm
Cumulative Data (nm):	D10: 53, D50: 80, D90: 144, D70: 99
User Lines:	0 nm, 0 nm
Total Concentration:	43.78 particles / frame, 3.87E8 particles / ml
Selected Concentration:	0.00 particles / frame, 0.00E8 particles / ml
Fitted Curve :	Mean: 0 nm, SD: 0
Completed Tracks:	3302
Drift Velocity:	370 nm/s

ANALYSIS SETTINGS:	
Frames Processed:	1499 of 1499
Frames per Second:	25.00
Calibration:	220 nm/pixel
Blur:	Auto
Detection Threshold:	7 Multi
Min Track Length:	Auto
Min Expected Size:	Auto - failed
Temperature:	24.70°C
Viscosity:	0.89 cP

Figure 3.8 Evaluation of EVs using NanoSight. Serum was separated after blood collection, and microparticle analysis by NanoSight was performed.

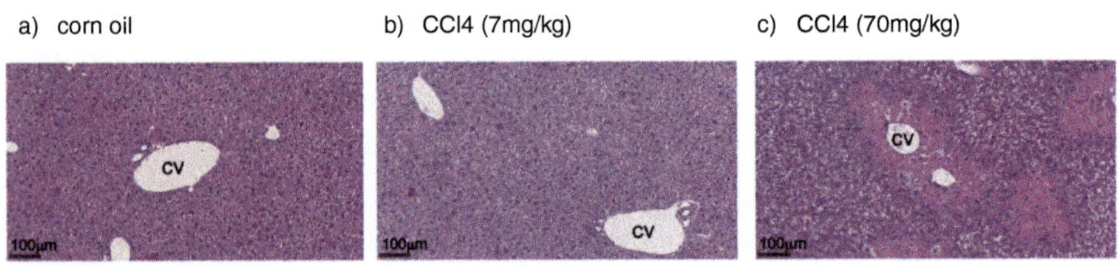

Figure 3.9 Representative H&E micrographs of liver tissues collected from mice treated with oral administration of corn oil (control) (a), 7 mg/kg CCl_4 (b), and 70 mg/kg CCl_4 (c). The control section shows the normal histological structure of the central vein (cv) and surrounding hepatocytes (a). Twenty-four hours after 7 mg/kg CCl_4 treatment, there were no histopathological changes by comparison with the control section (b). Twenty-four hours after 70 mg/kg CCl_4 treatment, the hepatocytes around the central veins (cv) were vacuolized and necrotic (c).

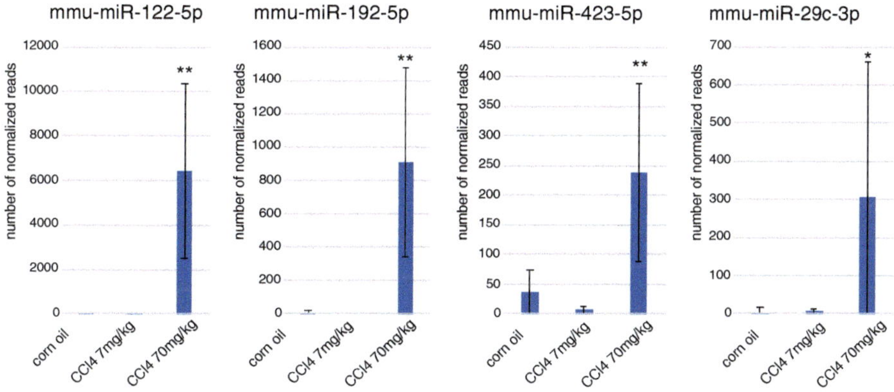

Figure 3.10 Differentially expressed EV-associated miRNAs. Individual normalized counts for four differentially expressed EV-associated miRNAs are shown. $^{**}P < 0.001$, $^{*}P < 0.01$ vs. control.

Conclusion

Liquid biopsy is a very powerful tool because it is rapid and non-invasive. In fact, AST and ALT are known to be very good biomarkers of liver damage; however, it is difficult to distinguish the cause of liver damage or the status of the liver only on the basis of elevated levels of AST and ALT. Thus, in some cases further examination is needed for detailed lesion assessment. Recently, using EV-associated miRNAs or circulating free miRNAs in serum as biomarkers has made it possible to diagnose thirteen types of cancer with an accuracy of 90% or more (Ogata-Kawata et al. 2014; Shimomura et al. 2016; Yokoi et al. 2018; Yao et al. 2019; Asano et al. 2019; Shiino et al. 2019; Usuba et al. 2019; Asakura et al. 2020). The development of next-generation toxicity tests using miRNA as a biomarker is therefore expected.

Our studies identified forty-two novel miRNAs—such as miR-122 and miR-192—as candidate liver damage biomarkers. Using these novel biomarkers, it may be possible to elucidate the mechanism of hepatotoxicity caused by the administration of drugs or chemical substances other than carbon tetrachloride. Since EVs in blood are secreted by a wide variety of cells, it is necessary to clarify the origin of the EV-associated miRNAs. In addition to their utility as markers of hepatotoxicity, EV-associated miRNAs are expected to be valuable as biomarkers of toxicity that targets other organs, such as the kidney, the lung, and the heart.

Although this study analyzed the effects of carbon tetrachloride administration at a single time point, twenty-four hours, evaluating the time courses of biomarkers in response to the repeated administration of carbon tetrachloride may be applicable if we want to shorten chronic toxicity tests and long-term carcinogenicity tests.

It may also be important to elucidate the function of the identified EV-associated miRNAs induced by hepatotoxicity in vivo.

Acknowledgments

The authors thank N. Moriyama, T. Momiyama, E. Tachihara, M. Uchiyama, and H. Aihara for excellent technical assistance. The images of mice, cows and cultured cells are from TogoTV (©2016 DBCLS TogoTV/CC-BY-4.0).

Competing Interests

The authors declare no competing interests.

Funding

This work was supported in part by Research on the Regulatory Science of Pharmaceuticals and Medical Devices (20mk0101163j0202), Research on the Development of New Drugs (20ak0101093j003) from the Japan Agency for Medical Research and Development, the Health Sciences Research Grants from the Ministry of Health, Labor, and Welfare, Japan, (H30-KAGAKU-IPPAN-002, 21KD1001), and JSPS KAKENHI (18K19315) to R.O.; Grant-in-Aid from the Research Program on Hepatitis from Japan Agency for Medical Research and Development (AMED: 16fk0310512h0005, 17fk0310101h0001 and 18fk0310101h0002) and Grants-in -Aid from a Project for Cancer research and Therapeutic Evolution (P-CREATE) to T. O.; JSPS KAKENHI (18K07053) to Y. H.

References

Alberti, C. and Cochella, L. (2017). A framework for understanding the roles of miRNAs in animal development. *Development* 144: 2548–2559.

Ambros, V. (2004). The functions of animal microRNAs. *Nature* 431: 350–355.

Asakura, K., Kadota, T., Matsuzaki, J., Yoshida, Y., Yamamoto, Y., Nakagawa, K., Takizawa, S., Aoki, Y., Nakamura, E., Miura, J., Sakamoto, H., Kato, K., Watanabe, S.I., and Ochiya, T. (2020). A miRNA-based diagnostic model predicts resectable lung cancer in humans with high accuracy. *Commun. Biol.* 3: 134.

Asano, N., Matsuzaki, J., Ichikawa, M., Kawauchi, J., Takizawa, S., Aoki, Y., Sakamoto, H., Yoshida, A., Kobayashi, E., Tanzawa, Y., Nakayama, R., Morioka, H., Matsumoto, M., Nakamura, M., Kondo, T., Kato, K., Tsuchiya, N., Kawai, A., and Ochiya, T. (2019). A serum microRNA classifier for the diagnosis of sarcomas of various histological subtypes. *Nat. Commun.* 10: 1299.

Baek, R., Varming, K., and Jorgensen, M.M. (2016). Does smoking, age or gender affect the protein phenotype of extracellular vesicles in plasma? *Transfus. Apher. Sci.* 55: 44–52.

Bala, S., Petrasek, J., Mundkur, S., Catalano, D., Levin, I., Ward, J., Alao, H., Kodys, K., and Szabo, G. (2012). Circulating microRNAs in exosomes indicate hepatocyte injury and inflammation in alcoholic, drug-induced, and inflammatory liver diseases. *Hepatology* 56: 1946–1957.

Baran, J., Baj-Krzyworzeka, M., Weglarczyk, K., Szatanek, R., Zembala, M., Barbasz, J., Czupryna, A., Szczepanik, A., and Zembala, M. (2010). Circulating tumour-derived microvesicles in plasma of gastric cancer patients. *Cancer Immunol. Immunother.* 59: 841–850.

Birbrair, A., Zhang, T., Wang, Z.M., Messi, M.L., Mintz, A., and Delbono, O. (2015). Pericytes at the intersection between tissue regeneration and pathology. *Clin. Sci. (Lond)* 128: 81–93.

Birbrair, A., Zhang, T., Wang, Z.M., Messi, M.L., Olson, J.D., Mintz, A., and Delbono, O. (2014). Type-2 pericytes participate in normal and tumoral angiogenesis. *Am. J. Physiol. Cell. Physiol.* 307: C25–38.

Bohnsack, M.T., Czaplinski, K., and Gorlich, D. (2004). Exportin 5 is a RanGTP-dependent dsRNA-binding protein that mediates nuclear export of pre-miRNAs. *RNA* 10: 185–191.

Borchert, G.M., Lanier, W., and Davidson, B.L. (2006). RNA polymerase III transcribes human microRNAs. *Nat. Struct. Mol. Biol.* 13: 1097–1101.

Bray, F., Ferlay, J., Soerjomataram, I., Siegel, R.L., Torre, L.A., and Jemal, A. (2018). Global cancer statistics 2018: GLOBOCAN estimates of incidence and mortality worldwide for 36 cancers in 185 countries. *CA Cancer J. Clin.* 68: 394–424.

Carmeliet, P. and Jain, R.K. (2000). Angiogenesis in cancer and other diseases. *Nature* 407: 249–257.

Cazzoli, R., Buttitta, F., Di Nicola, M., Malatesta, S., Marchetti, A., Rom, W.N., and Pass, H.I. (2013). microRNAs derived from circulating exosomes as noninvasive biomarkers for screening and diagnosing lung cancer. *J. Thorac. Oncol.* 8: 1156–1162.

Chan, Y.K., Zhang, H., Liu, P., Tsao, S.W., Lung, M.L., Mak, N.K., Ngok-Shun Wong, R., and Ying-Kit Yue, P. (2015). Proteomic analysis of exosomes from nasopharyngeal carcinoma cell identifies intercellular transfer of angiogenic proteins. *Int. J. Cancer* 137: 1830–1841.

Chen, L. and Han, X. (2015). Anti-PD-1/PD-L1 therapy of human cancer: Past, present, and future. *J. Clin. Invest.* 125: 3384–3391.

Chendrimada, T.P., Gregory, R.I., Kumaraswamy, E., Norman, J., Cooch, N., Nishikura, K., and Shiekhattar, R. (2005). TRBP recruits the Dicer complex to Ago2 for microRNA processing and gene silencing. *Nature* 436: 740–744.

Cho, Y.E., Kim, S.H., Lee, B.H., and Baek, M.C. (2017). Circulating plasma and exosomal microRNAs as indicators of drug-induced organ injury in rodent models. *Biomol. Ther. (Seoul)* 25: 367–373.

Choi, D.S., Park, J.O., Jang, S.C., Yoon, Y.J., Jung, J.W., Choi, D.Y., Kim, J.W., Kang, J.S., Park, J., Hwang, D., Lee, K.H., Park, S.H., Kim, Y.K., Desiderio, D.M., Kim, K.P., and Gho, Y.S. (2011). Proteomic analysis of microvesicles derived from human colorectal cancer ascites. *Proteomics* 11: 2745–2751.

Chopra, P., Roy, S., Ramalingaswami, V., and Nayak, N.C. (1972). Mechanism of carbon tetrachloride hepatotoxicity. An in vivo study of its molecular basis in rats and monkeys. *Lab. Invest.* 26: 716–727.

Clark, D.J., Fondrie, W.E., Yang, A., and Mao, L. (2016). Triple SILAC quantitative proteomic analysis reveals differential abundance of cell signaling proteins between normal and lung cancer-derived exosomes. *J. Proteomics* 133: 161–169.

Colombo, M., Raposo, G., and Thery, C. (2014). Biogenesis, secretion, and intercellular interactions of exosomes and other extracellular vesicles. *Annu. Rev. Cell. Dev. Biol.* 30: 255–289.

Conde-Vancells, J., Rodriguez-Suarez, E., Embade, N., Gil, D., Matthiesen, R., Valle, M., Elortza, F., Lu, S.C., Mato, J.M., and Falcon-Perez, J.M. (2008). Characterization and comprehensive proteome profiling of exosomes secreted by hepatocytes. *J. Proteome. Res.* 7: 5157–5166.

Coufal, N.G., Garcia-Perez, J.L., Peng, G.E., Yeo, G.W., Mu, Y., Lovci, M.T., Morell, M., O'Shea, K.S., Moran, J.V., and Gage, F.H. (2009). L1 retrotransposition in human neural progenitor cells. *Nature* 460: 1127–1131.

Cufaro, M.C., Pieragostino, D., Lanuti, P., Rossi, C., Cicalini, I., Federici, L., De Laurenzi, V., and Del Boccio, P. (2019). Extracellular vesicles and their potential use in monitoring cancer progression and therapy: The contribution of proteomics. *J. Oncol.* 2019: 1639854.

De Wever, O., Van Bockstal, M., Mareel, M., Hendrix, A., and Bracke, M. (2014). Carcinoma-associated fibroblasts provide operational flexibility in metastasis. *Semin. Cancer Biol.* 25: 33–46.

Deng, T., Zhang, H., Yang, H., Wang, H., Bai, M., Sun, W., Wang, X., Si, Y., Ning, T., Zhang, L., Li, H., Ge, S., Liu, R., Lin, D., Li, S., Ying, G., and Ba, Y. (2020). Exosome miR-155 derived from gastric carcinoma promotes angiogenesis by targeting the c-MYB/VEGF axis of endothelial cells. *Mol. Ther. Nucleic Acids* 19: 1449–1459.

Denli, A.M., Tops, B.B., Plasterk, R.H., Ketting, R.F., and Hannon, G.J. (2004). Processing of primary microRNAs by the microprocessor complex. *Nature* 432: 231–235.

Eichelser, C., Stuckrath, I., Muller, V., Milde-Langosch, K., Wikman, H., Pantel, K., and Schwarzenbach, H. (2014). Increased serum levels of circulating exosomal microRNA-373 in receptor-negative breast cancer patients. *Oncotarget* 5: 9650–9663.

Ekstrom, E.J., Bergenfelz, C., Von Bulow, V., Serifler, F., Carlemalm, E., Jonsson, G., Andersson, T., and Leandersson, K. (2014). WNT5A induces release of exosomes containing pro-angiogenic and immunosuppressive factors from malignant melanoma cells. *Mol. Cancer* 13: 88.

Fang, T., Lv, H., Lv, G., Li, T., Wang, C., Han, Q., Yu, L., Su, B., Guo, L., Huang, S., Cao, D., Tang, L., Tang, S., Wu, M., Yang, W., and Wang, H. (2018). Tumor-derived exosomal miR-1247-3p induces cancer-associated fibroblast activation to foster lung metastasis of liver cancer. *Nat. Commun.* 9: 191.

Farkash, E.A., Kao, G.D., Horman, S.R., and Prak, E.T. (2006). Gamma radiation increases endonuclease-dependent L1 retrotransposition in a cultured cell assay. *Nucleic Acids Res.* 34: 1196–1204.

Fu, H., Yang, H., Zhang, X., Wang, B., Mao, J., Li, X., Wang, M., Zhang, B., Sun, Z., Qian, H., and Xu, W. (2018). Exosomal TRIM3 is a novel marker and therapy target for gastric cancer. *J. Exp. Clin. Cancer Res.* 37: 162.

Garcia-Perez, J.L., Morell, M., Scheys, J.O., Kulpa, D.A., Morell, S., Carter, C.C., Hammer, G.D., Collins, K.L., O'Shea, K.S., Menendez, P., and Moran, J.V. (2010). Epigenetic silencing of engineered L1 retrotransposition events in human embryonic carcinoma cells. *Nature* 466: 769–773.

Gener Lahav, T., Adler, O., Zait, Y., Shani, O., Amer, M., Doron, H., Abramovitz, L., Yofe, I., Cohen, N., and Erez, N. (2019). Melanoma-derived extracellular vesicles instigate proinflammatory signaling in the metastatic microenvironment. *Int. J. Cancer* 145: 2521–2534.

Giusti, I., Delle Monache, S., Di Francesco, M., Sanita, P., D'Ascenzo, S., Gravina, G.L., Festuccia, C., and Dolo, V. (2016). From glioblastoma to endothelial cells through extracellular vesicles: Messages for angiogenesis. *Tumour Biol.* 37: 12743–12753.

Gould, S.J. and Raposo, G. (2013). As we wait: Coping with an imperfect nomenclature for extracellular vesicles. *J. Extracell Vesicles* 2. doi: 10.3402/jev.v2i0.20389.

Gregory, R.I., Yan, K.P., Amuthan, G., Chendrimada, T., Doratotaj, B., Cooch, N., and Shiekhattar, R. (2004). The microprocessor complex mediates the genesis of microRNAs. *Nature* 432: 235–240.

Gueta, K., Molotski, N., Gerchikov, N., Mor, E., Savion, S., Fein, A., Toder, V., Shomron, N., and Torchinsky, A. (2010). Teratogen-induced alterations in microRNA-34, microRNA-125b and microRNA-155 expression: Correlation with embryonic p53 genotype and limb phenotype. *BMC Dev. Biol.* 10: 20.

Hammond, S.M., Bernstein, E., Beach, D., and Hannon, G.J. (2000). An RNA-directed nuclease mediates post-transcriptional gene silencing in drosophila cells. *Nature* 404: 293–296.

Han, J., Lee, Y., Yeom, K.H., Kim, Y.K., Jin, H., and Kim, V.N. (2004). The Drosha-DGCR8 complex in primary microRNA processing. *Genes Dev.* 18: 3016–3027.

Han, J., Lee, Y., Yeom, K.H., Nam, J.W., Heo, I., Rhee, J.K., Sohn, S.Y., Cho, Y., Zhang, B.T., and Kim, V.N. (2006). Molecular basis for the recognition of primary microRNAs by the Drosha-DGCR8 complex. *Cell* 125: 887–901.

Harding, C. and Stahl, P. (1983). Transferrin recycling in reticulocytes: PH and iron are important determinants of ligand binding and processing. *Biochem. Biophys. Res. Commun.* 113: 650–658.

Hergenreider, E., Heydt, S., Treguer, K., Boettger, T., Horrevoets, A.J., Zeiher, A.M., Scheffer, M.P., Frangakis, A.S., Yin, X., Mayr, M., Braun, T., Urbich, C., Boon, R.A., and Dimmeler, S. (2012). Atheroprotective communication between endothelial cells and smooth muscle cells through miRNAs. *Nat. Cell Biol.* 14: 249–256.

Hsu, Y.L., Hung, J.Y., Chang, W.A., Lin, Y.S., Pan, Y.C., Tsai, P.H., Wu, C.Y., and Kuo, P.L. (2017). Hypoxic lung cancer-secreted exosomal miR-23a increased angiogenesis and vascular permeability by targeting prolyl hydroxylase and tight junction protein ZO-1. *Oncogene* 36: 4929–4942.

Hutvagner, G., Mclachlan, J., Pasquinelli, A.E., Balint, E., Tuschl, T., and Zamore, P.D. (2001). A cellular function for the RNA-interference enzyme Dicer in the maturation of the let-7 small temporal RNA. *Science* 293: 834–838.

Imbeaud, S., Graudens, E., Boulanger, V., Barlet, X., Zaborski, P., Eveno, E., Mueller, O., Schroeder, A., and Auffray, C. (2005). Towards standardization of RNA quality assessment using user-independent classifiers of microcapillary electrophoresis traces. *Nucleic Acids Res.* 33: e56.

Irie, M., Yoshikawa, M., Ono, R., Iwafune, H., Furuse, T., Yamada, I., Wakana, S., Yamashita, Y., Abe, T., Ishino, F., and Kaneko-Ishino, T. (2015). Cognitive function related to the Sirh11/Zcchc16 gene acquired from an ltr retrotransposon in eutherians. *PLoS Genet.* 11: e1005521.

Ivancevic, A.M., Kortschak, R.D., Bertozzi, T., and Adelson, D.L. (2018). Horizontal transfer of BovB and L1 retrotransposons in eukaryotes. *Genome Biol.* 19: 85.

Iwasaki, S., Kobayashi, M., Yoda, M., Sakaguchi, Y., Katsuma, S., Suzuki, T., and Tomari, Y. (2010). Hsc70/Hsp90 chaperone machinery mediates ATP-dependent RISC loading of small RNA duplexes. *Mol. Cell* 39: 292–299.

Jonas, S. and Izaurralde, E. (2015). Towards a molecular understanding of microRNA-mediated gene silencing. *Nat. Rev. Genet.* 16: 421–433.

Kalluri, R. (2016). The biology and function of exosomes in cancer. *J. Clin. Invest.* 126: 1208–1215.

Kalluri, R. and Zeisberg, M. (2006). Fibroblasts in cancer. *Nat. Rev. Cancer* 6: 392–401.

Kawamura, Y., Sanchez Calle, A., Yamamoto, Y., Sato, T.A., and Ochiya, T. (2019). Extracellular vesicles mediate the horizontal transfer of an active LINE-1 retrotransposon. *J. Extracell Vesicles* 8: 1643214.

Khan, S., Jutzy, J.M., Valenzuela, M.M., Turay, D., Aspe, J.R., Ashok, A., Mirshahidi, S., Mercola, D., Lilly, M.B., and Wall, N.R. (2012). Plasma-derived exosomal survivin, a plausible biomarker for early detection of prostate cancer. *PLoS One* 7: e46737.

Kozomara, A., Birgaoanu, M., and Griffiths-Jones, S. (2019). miRBase: From microRNA sequences to function. *Nucleic Acids Res.* 47: D155–D162.

Kwak, P.B. and Tomari, Y. (2012). The N domain of Argonaute drives duplex unwinding during RISC assembly. *Nat. Struct. Mol. Biol.* 19: 145–151.

Lee, R.C., Feinbaum, R.L., and Ambros, V. (1993). The C. elegans heterochronic gene lin-4 encodes small RNAs with antisense complementarity to lin-14. *Cell* 75: 843–854.

Lee, Y., Ahn, C., Han, J., Choi, H., Kim, J., Yim, J., Lee, J., Provost, P., Radmark, O., Kim, S., and Kim, V.N. (2003). The nuclear RNase III Drosha initiates microRNA processing. *Nature* 425: 415–419.

Lee, Y., Jeon, K., Lee, J.T., Kim, S., and Kim, V.N. (2002). MicroRNA maturation: Stepwise processing and subcellular localization. *EMBO J.* 21: 4663–4670.

Lee, Y., Kim, M., Han, J., Yeom, K.H., Lee, S., Baek, S.H., and Kim, V.N. (2004). MicroRNA genes are transcribed by RNA polymerase II. *EMBO J.* 23: 4051–4060.

Li, M., Yu, D., Williams, K.J., and Liu, M.L. (2010). Tobacco smoke induces the generation of procoagulant microvesicles from human monocytes/macrophages. *Arterioscler Thromb Vasc. Biol.* 30: 1818–1824.

Li, S., Zhao, Y., Chen, W., Yin, L., Zhu, J., Zhang, H., Cai, C., Li, P., Huang, L., and Ma, P. (2018). Exosomal ephrinA2 derived from serum as a potential biomarker for prostate cancer. *J. Cancer* 9: 2659–2665.

Li, Z., Ma, Y.Y., Wang, J., Zeng, X.F., Li, R., Kang, W., and Hao, X.K. (2016). Exosomal microRNA-141 is upregulated in the serum of prostate cancer patients. *Onco. Targets Ther.* 9: 139–148.

Lobb, R.J., Becker, M., Wen, S.W., Wong, C.S., Wiegmans, A.P., Leimgruber, A., and Moller, A. (2015). Optimized exosome isolation protocol for cell culture supernatant and human plasma. *J. Extracell Vesicles* 4: 27031.

Logozzi, M., De Milito, A., Lugini, L., Borghi, M., Calabro, L., Spada, M., Perdicchio, M., Marino, M.L., Federici, C., Iessi, E., Brambilla, D., Venturi, G., Lozupone, F., Santinami, M., Huber, V., Maio, M., Rivoltini, L., and Fais, S. (2009). High levels of exosomes expressing CD63 and caveolin-1 in plasma of melanoma patients. *PLoS One* 4: e5219.

Lu, Q., Zhang, J., Allison, R., Gay, H., Yang, W.X., Bhowmick, N.A., Frelix, G., Shappell, S., and Chen, Y.H. (2009). Identification of extracellular delta-catenin accumulation for prostate cancer detection. *Prostate* 69: 411–418.

Macfarlane, L.A. and Murphy, P.R. (2010). MicroRNA: Biogenesis, function and role in cancer. *Curr. Genom.* 11: 537–561.

Madhavan, B., Yue, S., Galli, U., Rana, S., Gross, W., Muller, M., Giese, N.A., Kalthoff, H., Becker, T., Buchler, M.W., and Zoller, M. (2015). Combined evaluation of a panel of protein and miRNA serum-exosome biomarkers for pancreatic cancer diagnosis increases sensitivity and specificity. *Int. J. Cancer* 136: 2616–2627.

Manterola, L., Guruceaga, E., Gallego Perez-Larraya, J., Gonzalez-Huarriz, M., Jauregui, P., Tejada, S., Diez-Valle, R., Segura, V., Sampron, N., Barrena, C., Ruiz, I., Agirre, A., Ayuso, A., Rodriguez, J., Gonzalez, A., Xipell, E., Matheu, A., Lopez De Munain, A., Tunon, T., Zazpe, I., Garcia-Foncillas, J., Paris, S., Delattre, J.Y., and Alonso, M.M. (2014). A small noncoding RNA signature found in exosomes of GBM patient serum as a diagnostic tool. *Neuro Oncol* 16: 520–527.

Mathivanan, S., Ji, H., and Simpson, R.J. (2010). Exosomes: Extracellular organelles important in intercellular communication. *J. Proteomics* 73: 1907–1920.

Melo, S.A., Luecke, L.B., Kahlert, C., Fernandez, A.F., Gammon, S.T., Kaye, J., Lebleu, V.S., Mittendorf, E.A., Weitz, J., Rahbari, N., Reissfelder, C., Pilarsky, C., Fraga, M.F., Piwnica-Worms, D., and Kalluri, R. (2015). Glypican-1 identifies cancer exosomes and detects early pancreatic cancer. *Nature* 523: 177–182.

Mittelbrunn, M., Gutierrez-Vazquez, C., Villarroya-Beltri, C., Gonzalez, S., Sanchez-Cabo, F., Gonzalez, M.A., Bernad, A., and Sanchez-Madrid, F. (2011). Unidirectional transfer of microRNA-loaded exosomes from T cells to antigen-presenting cells. *Nat. Commun.* 2: 282.

Mizutani, K., Terazawa, R., Kameyama, K., Kato, T., Horie, K., Tsuchiya, T., Seike, K., Ehara, H., Fujita, Y., Kawakami, K., Ito, M., and Deguchi, T. (2014). Isolation of prostate cancer-related exosomes. *Anticancer Res.* 34: 3419–3423.

Mobarrez, F., Antoniewicz, L., Bosson, J.A., Kuhl, J., Pisetsky, D.S., and Lundback, M. (2014). The effects of smoking on levels of endothelial progenitor cells and microparticles in the blood of healthy volunteers. *PLoS One* 9: e90314.

Montecalvo, A., Larregina, A.T., Shufesky, W.J., Stolz, D.B., Sullivan, M.L., Karlsson, J.M., Baty, C.J., Gibson, G.A., Erdos, G., Wang, Z., Milosevic, J., Tkacheva, O.A., Divito, S.J., Jordan, R., Lyons-Weiler, J., Watkins, S.C., and Morelli, A.E. (2012). Mechanism of transfer of functional microRNAs between mouse dendritic cells via exosomes. *Blood* 119: 756–766.

Mor, E., He, L., Torchinsky, A., and Shomron, N. (2014). MicroRNA-34a is dispensable for p53 function as teratogenesis inducer. *Arch. Toxicol.* 88: 1749–1763.

Munro, T.P., Magee, R.J., Kidd, G.J., Carson, J.H., Barbarese, E., Smith, L.M., and Smith, R. (1999). Mutational analysis of a heterogeneous nuclear ribonucleoprotein A2 response element for RNA trafficking. *J. Biol. Chem.* 274: 34389–34395.

Naruse, M., Ono, R., Irie, M., Nakamura, K., Furuse, T., Hino, T., Oda, K., Kashimura, M., Yamada, I., Wakana, S., Yokoyama, M., Ishino, F., and Kaneko-Ishino, T. (2014). Sirh7/Ldoc1 knockout mice exhibit placental P4 overproduction and delayed parturition. *Development* 141: 4763–4771.

Nawaz, M., Shah, N., Zanetti, B.R., Maugeri, M., Silvestre, R.N., Fatima, F., Neder, L., and Valadi, H. (2018). Extracellular vesicles and matrix remodeling enzymes: the emerging roles in extracellular matrix remodeling, progression of diseases and tissue repair. *Cells* 7 (10): 167.

Negrier, S., Gravis, G., Perol, D., Chevreau, C., Delva, R., Bay, J.O., Blanc, E., Ferlay, C., Geoffrois, L., Rolland, F., Legouffe, E., Sevin, E., Laguerre, B., and Escudier, B. (2011). Temsirolimus and bevacizumab, or sunitinib, or interferon alfa and bevacizumab for patients with advanced renal cell carcinoma (TORAVA): A randomised phase 2 trial. *Lancet. Oncol.* 12: 673–680.

Ogata-Kawata, H., Izumiya, M., Kurioka, D., Honma, Y., Yamada, Y., Furuta, K., Gunji, T., Ohta, H., Okamoto, H., Sonoda, H., Watanabe, M., Nakagama, H., Yokota, J., Kohno, T., and Tsuchiya, N. (2014). Circulating exosomal microRNAs as biomarkers of colon cancer. *PLoS One* 9: e92921.

Okada, C., Yamashita, E., Lee, S.J., Shibata, S., Katahira, J., Nakagawa, A., Yoneda, Y., and Tsukihara, T. (2009). A high-resolution structure of the pre-microRNA nuclear export machinery. *Science* 326: 1275–1279.

Ono, R., Ishii, M., Fujihara, Y., Kitazawa, M., Usami, T., Kaneko-Ishino, T., Kanno, J., Ikawa, M., and Ishino, F. (2015). Double strand break repair by capture of retrotransposon sequences and reverse-transcribed spliced mRNA sequences in mouse zygotes. *Sci. Rep.* 5: 12281.

Ono, R., Kobayashi, S., Wagatsuma, H., Aisaka, K., Kohda, T., Kaneko-Ishino, T., and Ishino, F. (2001). A retrotransposon-derived gene, PEG10, is a novel imprinted gene located on human chromosome 7q21. *Genomics* 73: 232–237.

Ono, R., Kuroki, Y., Naruse, M., Ishii, M., Iwasaki, S., Toyoda, A., Fujiyama, A., Shaw, G., Renfree, M.B., Kaneko-Ishino, T., and Ishino, F. (2011). Identification of tammar wallaby SIRH12, derived from a marsupial-specific retrotransposition event. *DNA Res.* 18: 211–219.

Ono, R., Nakamura, K., Inoue, K., Naruse, M., Usami, T., Wakisaka-Saito, N., Hino, T., Suzuki-Migishima, R., Ogonuki, N., Miki, H., Kohda, T., Ogura, A., Yokoyama, M., Kaneko-Ishino, T., and Ishino, F. (2006). Deletion of Peg10, an imprinted gene acquired from a retrotransposon, causes early embryonic lethality. *Nat. Genet.* 38: 101–106.

Ono, R., Shiura, H., Aburatani, H., Kohda, T., Kaneko-Ishino, T., and Ishino, F. (2003). Identification of a large novel imprinted gene cluster on mouse proximal chromosome 6. *Genome Res.* 13: 1696–1705.

Ono, R., Yasuhiko, Y., Aisaki, K.I., Kitajima, S., Kanno, J., and Hirabayashi, Y. (2019). Exosome-mediated horizontal gene transfer occurs in double-strand break repair during genome editing. *Commun. Biol.* 2: 57.

Ono, R., Yoshioka, Y., Furukawa, Y., Naruse, M., Kuwagata, M., Ochiya, T., Kitajima, S., and Hirabayashi, Y. (2020). Novel hepatotoxicity biomarkers of extracellular vesicle (EV)-associated miRNAs induced by CCl4. *Toxicol. Rep.* 7: 685–692.

Ostertag, E.M., Prak, E.T., Deberardinis, R.J., Moran, J.V., and Kazazian, H.H.J.R. (2000). Determination of L1 retrotransposition kinetics in cultured cells. *Nucleic Acids Res.* 28: 1418–1423.

Paggetti, J., Haderk, F., Seiffert, M., Janji, B., Distler, U., Ammerlaan, W., Kim, Y.J., Adam, J., Lichter, P., Solary, E., Berchem, G., and Moussay, E. (2015). Exosomes released by chronic lymphocytic leukemia cells induce the transition of stromal cells into cancer-associated fibroblasts. *Blood* 126: 1106–1117.

Pan, B.T. and Johnstone, R.M. (1983). Fate of the transferrin receptor during maturation of sheep reticulocytes in vitro: Selective externalization of the receptor. *Cell* 33: 967–978.

Pan, B.T., Teng, K., Wu, C., Adam, M., and Johnstone, R.M. (1985). Electron microscopic evidence for externalization of the transferrin receptor in vesicular form in sheep reticulocytes. *J. Cell Biol.* 101: 942–948.

Pegtel, D.M., Cosmopoulos, K., Thorley-Lawson, D.A., Van Eijndhoven, M.A., Hopmans, E.S., Lindenberg, J.L., De Gruijl, T.D., Wurdinger, T., and Middeldorp, J.M. (2010). Functional delivery of viral miRNAs via exosomes. *Proc. Natl. Acad. Sci. USA* 107: 6328–6333.

Peinado, H., Aleckovic, M., Lavotshkin, S., Matei, I., Costa-Silva, B., Moreno-Bueno, G., Hergueta-Redondo, M., Williams, C., Garcia-Santos, G., Ghajar, C., Nitadori-Hoshino, A., Hoffman, C., Badal, K., Garcia, B.A., Callahan, M.K., Yuan, J., Martins, V.R., Skog, J., Kaplan, R.N., Brady, M.S., Wolchok, J.D., Chapman, P.B., Kang, Y., Bromberg, J., and Lyden, D. (2012). Melanoma exosomes educate bone marrow progenitor cells toward a pro-metastatic phenotype through MET. *Nat. Med.* 18: 883–891.

Pu, C., Huang, H., Wang, Z., Zou, W., Lv, Y., Zhou, Z., Zhang, Q., Qiao, L., Wu, F., and Shao, S. (2018). Extracellular vesicle-associated miR-21 and miR-144 are markedly elevated in serum of patients with hepatocellular carcinoma. *Front. Physiol.* 9: 930.

Que, R., Ding, G., Chen, J., and Cao, L. (2013). Analysis of serum exosomal microRNAs and clinicopathologic features of patients with pancreatic adenocarcinoma. *World J. Surg. Oncol.* 11: 219.

Rabinowits, G., Gercel-Taylor, C., Day, J.M., Taylor, D.D., and Kloecker, G.H. (2009). Exosomal microRNA: A diagnostic marker for lung cancer. *Clin. Lung. Cancer* 10: 42–46.

Raimondo, F., Morosi, L., Corbetta, S., Chinello, C., Brambilla, P., Della Mina, P., Villa, A., Albo, G., Battaglia, C., Bosari, S., Magni, F., and Pitto, M. (2013). Differential protein profiling of renal cell carcinoma urinary exosomes. *Mol. Biosyst.* 9: 1220–1233.

Rasanen, K. and Vaheri, A. (2010). Activation of fibroblasts in cancer stroma. *Exp. Cell Res.* 316: 2713–2722.

Rodriguez, M., Silva, J., Lopez-Alfonso, A., Lopez-Muniz, M.B., Pena, C., Dominguez, G., Garcia, J.M., Lopez-Gonzalez, A., Mendez, M., Provencio, M., Garcia, V., and Bonilla, F. (2014). Different exosome cargo from plasma/bronchoalveolar lavage in non-small-cell lung cancer. *Genes. Chromosomes Cancer* 53: 713–724.

Sandfeld-Paulsen, B., Jakobsen, K.R., Baek, R., Folkersen, B.H., Rasmussen, T.R., Meldgaard, P., Varming, K., Jorgensen, M.M., and Sorensen, B.S. (2016). Exosomal proteins as diagnostic biomarkers in lung cancer. *J. Thorac. Oncol.* 11: 1701–1710.

Sato, S., Vasaikar, S., Eskaros, A., Kim, Y., Lewis, J.S., Zhang, B., Zijlstra, A., and Weaver, A.M. (2019). EPHB2 carried on small extracellular vesicles induces tumor angiogenesis via activation of ephrin reverse signaling. *JCI Insight* 4.

Sekita, Y., Wagatsuma, H., Nakamura, K., Ono, R., Kagami, M., Wakisaka, N., Hino, T., Suzuki-Migishima, R., Kohda, T., Ogura, A., Ogata, T., Yokoyama, M., Kaneko-Ishino, T., and Ishino, F. (2008). Role of retrotransposon-derived imprinted gene, Rtl1, in the feto-maternal interface of mouse placenta. *Nat. Genet.* 40: 243–248.

Shan, T., Chen, S., Chen, X., Lin, W.R., Li, W., Ma, J., Wu, T., Ji, H., Li, Y., Cui, X., and Kang, Y. (2017). Prometastatic mechanisms of CAF-mediated EMT regulation in pancreatic cancer cells. *Int. J. Oncol.* 50: 121–128.

Sharma, P. and Allison, J.P. (2015). The future of immune checkpoint therapy. *Science* 348: 56–61.

Shiga, K., Hara, M., Nagasaki, T., Sato, T., Takahashi, H., and Takeyama, H. (2015). Cancer-associated fibroblasts: Their characteristics and their roles in tumor growth. *Cancers (Basel)* 7: 2443–2458.

Shiino, S., Matsuzaki, J., Shimomura, A., Kawauchi, J., Takizawa, S., Sakamoto, H., Aoki, Y., Yoshida, M., Tamura, K., Kato, K., Kinoshita, T., Kitagawa, Y., and Ochiya, T. (2019). Serum miRNA-based prediction of axillary lymph node metastasis in breast cancer. *Clin. Cancer Res.* 25: 1817–1827.

Shimomura, A., Shiino, S., Kawauchi, J., Takizawa, S., Sakamoto, H., Matsuzaki, J., Ono, M., Takeshita, F., Niida, S., Shimizu, C., Fujiwara, Y., Kinoshita, T., Tamura, K., and Ochiya, T. (2016). Novel combination of serum microRNA for detecting breast cancer in the early stage. *Cancer Sci.* 107: 326–334.

Simons, M. and Raposo, G. (2009). Exosomes–vesicular carriers for intercellular communication. *Curr. Opin. Cell Biol.* 21: 575–581.

Skog, J., Wurdinger, T., Rijn, V.A.N., Meijer, S., Gainche, D.H., Sena-Esteves, L., Curry, W.T., Jr., Carter, B.S., Krichevsky, A.M., and Breakefield, X.O. (2008). Glioblastoma microvesicles transport RNA and proteins that promote tumour growth and provide diagnostic biomarkers. *Nat. Cell Biol.* 10: 1470–1476.

Srivastava, A., Moxley, K., Ruskin, R., Dhanasekaran, D.N., Zhao, Y.D., and Ramesh, R. (2018). A non-invasive liquid biopsy screening of urine-derived exosomes for miRNAs as biomarkers in endometrial cancer patients. *AAPS J.* 20: 82.

Sruthi, T.V., Edatt, L., Raji, G.R., Kunhiraman, H., Shankar, S.S., Shankar, V., Ramachandran, V., Poyyakkara, A., and Kumar, S.V.B. (2018). Horizontal transfer of miR-23a from hypoxic tumor cell colonies can induce angiogenesis. *J. Cell Physiol.* 233: 3498–3514.

Subra, C., Grand, D., Laulagnier, K., Stella, A., Lambeau, G., Paillasse, M., De Medina, P., Monsarrat, B., Perret, B., Silvente-Poirot, S., Poirot, M., and Record, M. (2010). Exosomes account for vesicle-mediated transcellular transport of activatable phospholipases and prostaglandins. *J. Lipid. Res.* 51: 2105–2120.

Sun, B., Li, Y., Zhou, Y., Ng, T.K., Zhao, C., Gan, Q., Gu, X., and Xiang, J. (2019). Circulating exosomal CPNE3 as a diagnostic and prognostic biomarker for colorectal cancer. *J. Cell. Physiol.* 234: 1416–1425.

Taylor, D.D. and Gercel-Taylor, C. (2008). MicroRNA signatures of tumor-derived exosomes as diagnostic biomarkers of ovarian cancer. *Gynecol. Oncol.* 110: 13–21.

Thery, C., Witwer, K.W., Aikawa, E., Alcaraz, M.J., Anderson, J.D., Andriantsitohaina, R., Antoniou, A., Arab, T., Archer, F., Atkin-Smith, G.K., Ayre, D.C., Bach, J.M., Bachurski, D., Baharvand, H., Balaj, L., Baldacchino, S., Bauer, N.N., Baxter, A.A., Bebawy, M., Beckham, C., Bedina Zavec, A., Benmoussa, A., Berardi, A.C., Bergese, P., Bielska, E., Blenkiron, C., Bobis-Wozowicz, S., Boilard, E., Boireau, W., Bongiovanni, A., Borras, F.E., Bosch, S., Boulanger, C.M., Breakefield, X., Breglio, A.M., Brennan, M.A., Brigstock, D.R., Brisson, A., Broekman, M.L., Bromberg, J.F., Bryl-Gorecka, P., Buch, S., Buck, A.H., Burger, D., Busatto, S., Buschmann, D., Bussolati, B., Buzas, E.I., Byrd, J.B., Camussi, G., Carter, D.R., Caruso, S., Chamley, L.W., Chang, Y.T., Chen, C., Chen, S., Cheng, L., Chin, A.R., Clayton, A., Clerici, S.P., Cocks, A., Cocucci, E., Coffey, R.J., Cordeiro da Silva, A., Couch, Y., Coumans, F.A., Coyle, B., Crescitelli, R., Criado, M.F., D'Souza-Schorey, C., Das, S., Datta Chaudhuri, A., De Candia, P., De Santana, E.F., De Wever, O., Del Portillo, H.A., Demaret, T., Deville, S., Devitt, A., Dhondt, B., Di Vizio, D., Dieterich, L.C., Dolo, V., Dominguez Rubio, A.P., Dominici, M., Dourado, M.R., Driedonks, T.A., Duarte, F.V., Duncan, H.M., Eichenberger, R.M., Ekstrom, K., El Andaloussi, S., Elie-Caille, C., Erdbrugger, U., Falcon-Perez, J.M., Fatima, F., Fish, J.E., Flores-Bellver, M., Forsonits, A., Frelet-Barrand, A. et al. (2018). Minimal information for studies of extracellular vesicles 2018 (MISEV2018): A position statement of the International Society for Extracellular Vesicles and update of the MISEV2014 guidelines. *J. Extracell Vesicles* 7: 1535750.

Thery, C., Zitvogel, L., and Amigorena, S. (2002). Exosomes: Composition, biogenesis and function. *Nat. Rev. Immunol.* 2: 569–579.

Tian, Y., Ma, L., Gong, M., Su, G., Zhu, S., Zhang, W., Wang, S., Li, Z., Chen, C., Li, L., Wu, L., and Yan, X. (2018). Protein profiling and sizing of extracellular vesicles from colorectal cancer patients via flow cytometry. *ACS Nano* 12: 671–680.

Tommelein, J., Verset, L., Boterberg, T., Demetter, P., Bracke, M., and De Wever, O. (2015). Cancer-associated fibroblasts connect metastasis-promoting communication in colorectal cancer. *Front. Oncol.* 5: 63.

Topalian, S.L. (2017). Targeting immune checkpoints in cancer therapy. *JAMA* 318: 1647–1648.

Topalian, S.L., Drake, C.G., and Pardoll, D.M. (2015). Immune checkpoint blockade: A common denominator approach to cancer therapy. *Cancer Cell* 27: 450–461.

Ueda, K., Ishikawa, N., Tatsuguchi, A., Saichi, N., Fujii, R., and Nakagawa, H. (2014). Antibody-coupled monolithic silica microtips for highthroughput molecular profiling of circulating exosomes. *Sci. Rep.* 4: 6232.

Urabe, F., Kosaka, N., Yoshioka, Y., Egawa, S., and Ochiya, T. (2017). The small vesicular culprits: The investigation of extracellular vesicles as new targets for cancer treatment. *Clin. Transl. Med.* 6: 45.

Usuba, W., Urabe, F., Yamamoto, Y., Matsuzaki, J., Sasaki, H., Ichikawa, M., Takizawa, S., Aoki, Y., Niida, S., Kato, K., Egawa, S., Chikaraishi, T., Fujimoto, H., and Ochiya, T. (2019). Circulating miRNA panels for specific and early detection in bladder cancer. *Cancer Sci.* 110: 408–419.

Vaidyanathan, K. and Vasudevan, D.M. (2012). Organ specific tumor markers: What's new? *Indian J. Clin. Biochem.* 27: 110–120.

Valadi, H., Ekstrom, K., Bossios, A., Sjostrand, M., Lee, J.J., and Lotvall, J.O. (2007). Exosome-mediated transfer of mRNAs and microRNAs is a novel mechanism of genetic exchange between cells. *Nat. Cell Biol.* 9: 654–659.

van der Pol, E., Boing, A.N., Harrison, P., Sturk, A., and Nieuwland, R. (2012). Classification, functions, and clinical relevance of extracellular vesicles. *Pharmacol. Rev.* 64: 676–705.

Villarroya-Beltri, C., Gutierrez-Vazquez, C., Sanchez-Cabo, F., Perez-Hernandez, D., Vazquez, J., Martin-Cofreces, N., Martinez-Herrera, D.J., Pascual-Montano, A., Mittelbrunn, M., and

Sanchez-Madrid, F. (2013). Sumoylated hnRNPA2B1 controls the sorting of miRNAs into exosomes through binding to specific motifs. *Nat. Commun.* 4: 2980.

Wang, H., Yang, H., Shivalila, C.S., Dawlaty, M.M., Cheng, A.W., Zhang, F., and Jaenisch, R. (2013). One-step generation of mice carrying mutations in multiple genes by CRISPR/Cas-mediated genome engineering. *Cell* 153: 910–918.

Wang, J., De Veirman, K., Faict, S., Frassanito, M.A., Ribatti, D., Vacca, A., and Menu, E. (2016). Multiple myeloma exosomes establish a favourable bone marrow microenvironment with enhanced angiogenesis and immunosuppression. *J. Pathol.* 239: 162–173.

Watson, C.N., Belli, A., and Pietro, D.I. (2019). Small non-coding RNAs: New class of biomarkers and potential therapeutic targets in neurodegenerative disease. *Front. Genet.* 10: 364.

Welslau, M., Dieras, V., Sohn, J.H., Hurvitz, S.A., Lalla, D., Fang, L., Althaus, B., Guardino, E., and Miles, D. (2014). Patient-reported outcomes from EMILIA, a randomized phase 3 study of trastuzumab emtansine (T-DM1) versus capecitabine and lapatinib in human epidermal growth factor receptor 2-positive locally advanced or metastatic breast cancer. *Cancer* 120: 642–651.

Witwer, K.W., Buzas, E.I., Bemis, L.T., Bora, A., Lasser, C., Lotvall, J., Nolte-'t Hoen, E.N., Piper, M.G., Sivaraman, S., Skog, J., Thery, C., Wauben, M.H., and Hochberg, F. (2013). Standardization of sample collection, isolation and analysis methods in extracellular vesicle research. *J. Extracell. Vesicles* 2.

Witwer, K.W., Soekmadji, C., Hill, A.F., Wauben, M.H., Buzas, E.I., Di Vizio, D., Falcon-Perez, J.M., Gardiner, C., Hochberg, F., Kurochkin, I.V., Lotvall, J., Mathivanan, S., Nieuwland, R., Sahoo, S., Tahara, H., Torrecilhas, A.C., Weaver, A.M., Yin, H., Zheng, L., Gho, Y.S., Quesenberry, P., and Thery, C. (2017). Updating the MISEV minimal requirements for extracellular vesicle studies: Building bridges to reproducibility. *J Extracell Vesicles* 6: 1396823.

Xouri, G. and Christian, S. (2010). Origin and function of tumor stroma fibroblasts. *Semin. Cell Dev. Biol.* 21: 40–46.

Yan, S., Jiang, Y., Liang, C., Cheng, M., Jin, C., Duan, Q., Xu, D., Yang, L., Zhang, X., Ren, B., and Jin, P. (2018). Exosomal miR-6803-5p as potential diagnostic and prognostic marker in colorectal cancer. *J. Cell Biochem.* 119: 4113–4119.

Yanez-Mo, M., Siljander, P.R., Andreu, Z., Zavec, A.B., Borras, F.E., Buzas, E.I., Buzas, K., Casal, E., Cappello, F., Carvalho, J., Colas, E., Cordeiro da Silva, A., Fais, S., Falcon-Perez, J.M., Ghobrial, I.M., Giebel, B., Gimona, M., Graner, M., Gursel, I., Gursel, M., Heegaard, N.H., Hendrix, A., Kierulf, P., Kokubun, K., Kosanovic, M., Kralj-Iglic, V., Kramer-Albers, E.M., Laitinen, S., Lasser, C., Lener, T., Ligeti, E., Line, A., Lipps, G., Llorente, A., Lotvall, J., Mancek-Keber, M., Marcilla, A., Mittelbrunn, M., Nazarenko, I., Nolte-'t Hoen, E.N., Nyman, T.A., O'Driscoll, L., Olivan, M., Oliveira, C., Pallinger, E., Del Portillo, H.A., Reventos, J., Rigau, M., Rohde, E., Sammar, M., Sanchez-Madrid, F., Santarem, N., Schallmoser, K., Ostenfeld, M.S., Stoorvogel, W., Stukelj, R., Van Der Grein, S.G., Vasconcelos, M.H., Wauben, M.H., and De Wever, O. (2015). Biological properties of extracellular vesicles and their physiological functions. *J. Extracell. Vesicles.* 4: 27066.

Yao, B., Qu, S., Hu, R., Gao, W., Jin, S., Liu, M., and Zhao, Q. (2019). A panel of miRNAs derived from plasma extracellular vesicles as novel diagnostic biomarkers of lung adenocarcinoma. *FEBS Open. Bio.* 9: 2149–2158.

Yi, R., Qin, Y., Macara, I.G., and Cullen, B.R. (2003). Exportin-5 mediates the nuclear export of pre-microRNAs and short hairpin RNAs. *Genes. Dev.* 17: 3011–3016.

Yokoi, A., Matsuzaki, J., Yamamoto, Y., Yoneoka, Y., Takahashi, K., Shimizu, H., Uehara, T., Ishikawa, M., Ikeda, S.I., Sonoda, T., Kawauchi, J., Takizawa, S., Aoki, Y., Niida, S.,

Sakamoto, H., Kato, K., Kato, T., and Ochiya, T. (2018). Integrated extracellular microRNA profiling for ovarian cancer screening. *Nat. Commun.* 9: 4319.

Yoshioka, Y., Kosaka, N., Konishi, Y., Ohta, H., Okamoto, H., Sonoda, H., Nonaka, R., Yamamoto, H., Ishii, H., Mori, M., Furuta, K., Nakajima, T., Hayashi, H., Sugisaki, H., Higashimoto, H., Kato, T., Takeshita, F., and Ochiya, T. (2014). Ultra-sensitive liquid biopsy of circulating extracellular vesicles using ExoScreen. *Nat. Commun.* 5: 3591.

You, Y., Shan, Y., Chen, J., Yue, H., You, B., Shi, S., Li, X., and Cao, X. (2015). Matrix metalloproteinase 13-containing exosomes promote nasopharyngeal carcinoma metastasis. *Cancer Sci.* 106: 1669–1677.

Yu, Y., Xiao, C.H., Tan, L.D., Wang, Q.S., Li, X.Q., and Feng, Y.M. (2014). Cancer-associated fibroblasts induce epithelial-mesenchymal transition of breast cancer cells through paracrine TGF-beta signalling. *Br. J. Cancer* 110: 724–732.

Zeng, Y. and Cullen, B.R. (2004). Structural requirements for pre-microRNA binding and nuclear export by Exportin 5. *Nucleic Acids Res.* 32: 4776–4785.

Zhang, H. (2016). Three generations of epidermal growth factor receptor tyrosine kinase inhibitors developed to revolutionize the therapy of lung cancer. *Drug. Des. Devel. Ther.* 10: 3867–3872.

Zhou, C.F., Ma, J., Huang, L., Yi, H.Y., Zhang, Y.M., Wu, X.G., Yan, R.M., Liang, L., Zhong, M., Yu, Y.H., Wu, S., and Wang, W. (2019). Cervical squamous cell carcinoma-secreted exosomal miR-221-3p promotes lymphangiogenesis and lymphatic metastasis by targeting VASH1. *Oncogene* 38: 1256–1268.

Zhuang, J., Lu, Q., Shen, B., Huang, X., Shen, L., Zheng, X., Huang, R., Yan, J., and Guo, H. (2015). TGFbeta1 secreted by cancer-associated fibroblasts induces epithelial-mesenchymal transition of bladder cancer cells through lncRNA-ZEB2NAT. *Sci. Rep.* 5: 11924.

4

Circulating miRNAs as Biomarkers of Toxic Heavy Metal Exposure

Alexandra N. Nail, Ana P. Ferragut Cardoso, Mayukh Banerjee, and J. Christopher States

Department of Pharmacology and Toxicology, University of Louisville, Louisville, KY

Metals and the Focus of This Chapter

Metals are defined as elements that readily form cations and are composed of a lattice of positive ions surrounded by a cloud of delocalized electrons. Heavy metals have played an important role in human civilization for thousands of years. Commonly defined by their specific density (greater than 5.0 g/cm^3), heavy metals occur naturally in the environment. Five heavy metals, namely arsenic (which is classified as a metalloid but is often grouped within heavy metals), lead, mercury, cadmium, and chromium, are of considerable public health importance because of their high degree of toxicity (Jarup 2003). In fact arsenic, lead, mercury, and cadmium are ranked within the top seven substances in the Agency for Toxic Substances and Disease Registry's *Substance Priority List*, which considers a combination of frequency, toxicity, and potential for human exposure when classifying substances of concern (ATSDR 2019). While metals often occur naturally in the environment, anthropogenic activities such as mining, smelting, industrial production, and agriculture can lead to their increased mobilization and can alter their chemical form, thereby also altering their toxicity. Human activities thus contribute to environmental and occupational exposure to arsenic, lead, mercury, cadmium, and chromium. Exposure to these metals can cause a myriad of health problems such as multiple organ damage, cellular damage, cardiovascular and metabolic diseases, and cancer (Rehman et al. 2018).

In general, concentrations of metals in biological samples are used as biomarkers for assessing the level of concern with metal intoxication. As technologies continue to improve in molecular toxicology, biomarker possibilities have expanded. This chapter discusses circulating miRNAs as potential biomarkers for heavy metal(s) exposure and adverse health outcomes associated with types of heavy metal exposure. The discussion highlights current biofluids used for determining individual heavy metals (arsenic, lead, mercury, cadmium, chromium) and mixed metals exposure, miRNA biogenesis, and studies that have investigated circulating miRNAs associated with heavy metal exposure.

Genomic and Epigenomic Biomarkers of Toxicology and Disease: Clinical and Therapeutic Actions, First Edition. Edited by Saura C. Sahu.

Exposure to Metals

Heavy metal contamination of water, soil, and air can occur as a result of natural phenomena such as soil erosion, atmospheric deposition, and volcanic eruptions (Tchounwou et al. 2012). Furthermore, modern industrial practices have resulted in increased human exposure to heavy metal through mining, smelting, agriculture, industrial processing, by-products, and effluents, as well as in domestic use of metals and metal-containing compounds (Tchounwou et al. 2012). Heavy metal exposure can occur through inhalation, ingestion, or contact with the skin (See Figure 4.1). The effects and the severity of metal toxicity can vary considerably according to the specific metal, its chemical form, its physio-chemical properties, the duration of exposure (acute or chronic), the route of exposure, and individual susceptibility as shaped by genetics, lifestyle factors, and nutritional status (Rajkumar and Gupta 2021). Depending on the dosage and duration of exposure, heavy metals can exert both acute and chronic effects, causing multi-organ maladies such as gastrointestinal and kidney dysfunction, nervous system disorders, skin lesions, vascular damage, immune system dysfunction, birth defects, and cancer (Balali-Mood et al. 2021).

Biomarkers

The term "biomarker" refers to a characteristic that can be measured objectively, accurately, and reproducibly as an indicator of normal biological processes, pathogenic processes, or pharmacologic response to a therapeutic intervention (Califf 2018; Strimbu and Tavel 2010). Biomarkers should be separated from clinical outcome assessment (COA), which encompasses direct measures of how a patient feels, functions, or survives (Califf 2018). This distinction is important, because therapeutics must meet regulatory standards that rely on COAs for measuring the outcomes. Biomarkers, on the other hand, can serve various purposes, such as predicting a COA on the basis of a measurement. To be used in a clinical setting, biomarkers must undergo rigorous scrutiny and be validated for yielding

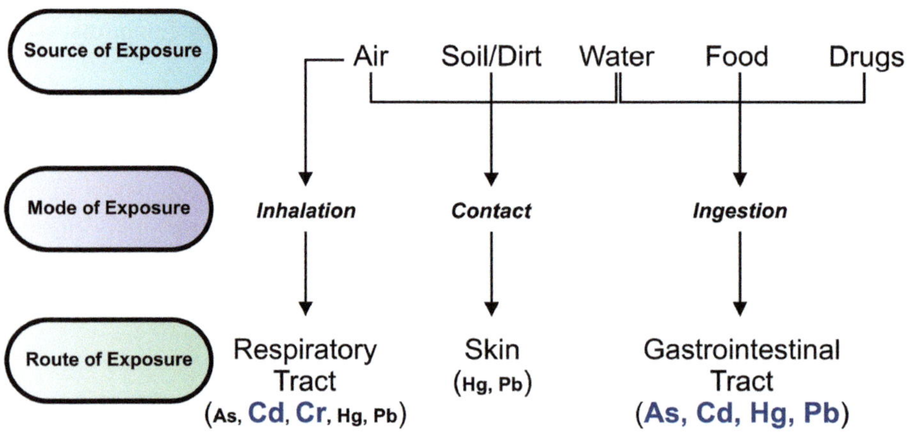

Figure 4.1 Major routes of exposure for heavy metals. The most common sources, modes, and routes of exposure for arsenic (As), cadmium (Cd), chromium (Cr), mercury (Hg), and lead (Pb) are depicted. Large bold blue text indicates the predominant route of exposure for individual metals, while black small text indicates additional exposure routes. For Cd, ingestion is the predominant exposure route in the non-smoking population, while inhalation of cigarette smoke is the predominant exposure source in smokers.

reliable and reproducible results across multiple populations. Subtypes of biomarkers have been defined according to their applications and classified as diagnostic biomarkers, monitoring biomarkers, pharmacodynamic or response biomarkers, predictive biomarkers, prognostic biomarkers, and safety biomarkers (Califf 2018; see also Table 4.1). Importantly, a biomarker may fall into one or more biomarker subtypes.

Biomarkers of Heavy Metal Exposure

Biomarkers that are typically used to assess heavy metal exposure are concentrations of heavy metals in blood, urine, bone, nail, hair, and other tissues and fall into both monitoring and safety biomarker subtypes. These traditional biomarkers are practical and reliable measures of exposure, can be vital when signs and symptoms of metal exposure are lacking, and can be useful for assessing the risk for disease development (Gil and Pla 2001). Typically, blood and urine are used to assess recent exposure, whereas bone, nail, and hair can be used to indicate chronic or past exposure. However, the type of biological tissue examined is highly dependent on the biological half-life of the metal, or the time it takes to excrete the metal from the body. For instance, lead can be retained in bone decades after exposure, whereas the half-life of lead in blood is only a few weeks. However, these traditional biomarkers of exposure are useful only in defining the severity and the duration of exposure (which can be either acute or chronic, depending on the biological sample and metal tested for) and do not have the capacity to determine disease susceptibility or whether adverse health effects can be mitigated. Additionally, metal toxicity can present non-descript and non-specific symptoms, which do not pinpoint a specific metal exposure if the type of exposure is unknown (Hackenmueller et al. 2019). Therefore an immediate need exists to find additional, specific, and reliable biomarkers that play dual roles in identifying exposures to types of heavy metals (i.e., in monitoring and safety) and may be used as early indicators of disease as a result of heavy metal exposure (i.e., for diagnostic purposes). In the last decade or so, ribonucleic acid (RNA) has emerged as a major target for diagnostics and therapeutics

Table 4.1 Description of biomarker subtypes.

Biomarker Subtypes	Description
Diagnostic	Confirms the presence of a disease or condition with precision and reliability
Monitoring	Measured over a span of time and monitors the status of a disease or medical condition due to exposure of an individual to medicinal or environmental agents
Pharmacodynamic/Response	Changes in response to exposure to a medicinal or environmental agent
Predictive	Predicts whether favourable or unfavourable outcomes are more likely to be experienced by an individual or group of individuals
Prognostic	Identifies the likelihood of a clinical event, disease recurrence, or disease progression
Safety	Measures before or after exposure to a medicinal or environmental agent to indicate likelihood, presence or extent of toxicity as an adverse event

This table has been created on the basis of information from Califf (2018).

(Zampetaki et al. 2018), including in heavy metal exposure. This chapter focuses on microRNAs (miRNAs) as potential biomarkers in heavy metal toxicity and disease.

miRNA Biogenesis

RNAs can be broadly classified into two groups: coding RNAs, which code for proteins and are known as messenger RNAs (mRNAs); and non-coding RNAs (ncRNAs), which comprise the largest class of RNAs and include ribosomal RNAs (rRNAs), transfer RNAs (tRNAs), long non-coding RNAs (lncRNAs, > 200 nucleotides) and small non-coding RNAs (< 200 nucleotides). miRNAs constitute the best studied class of small non-coding RNAs and act post-transcriptionally to regulate gene expression (Ebert and Sharp 2012). Mature miRNAs occur as single-stranded RNAs, are typically between 21 and 24 nucleotides in length, and are estimated to regulate about 50% of all mammalian protein-coding genes (Krol et al. 2010).

The biogenesis of miRNA has been comprehensively reviewed (Fabian and Sonenberg 2012; Ha and Kim 2014; Hammond 2015; Libri et al. 2013) and is shown diagrammatically in Figure. 4.2. The majority of miRNAs are encoded within introns or exons of non-coding RNAs or introns of pre-mRNA (Saliminejad et al. 2019). In the canonical pathway, miRNAs are transcribed by RNA polymerase II (pol II) into large primary miRNAs (pri-miRNAs); these contain embedded stem loop structures that are roughly 70 nucleotides long. Pri-miRNAs are cleaved within the nucleus by the microprocessor complex to form stem-loop

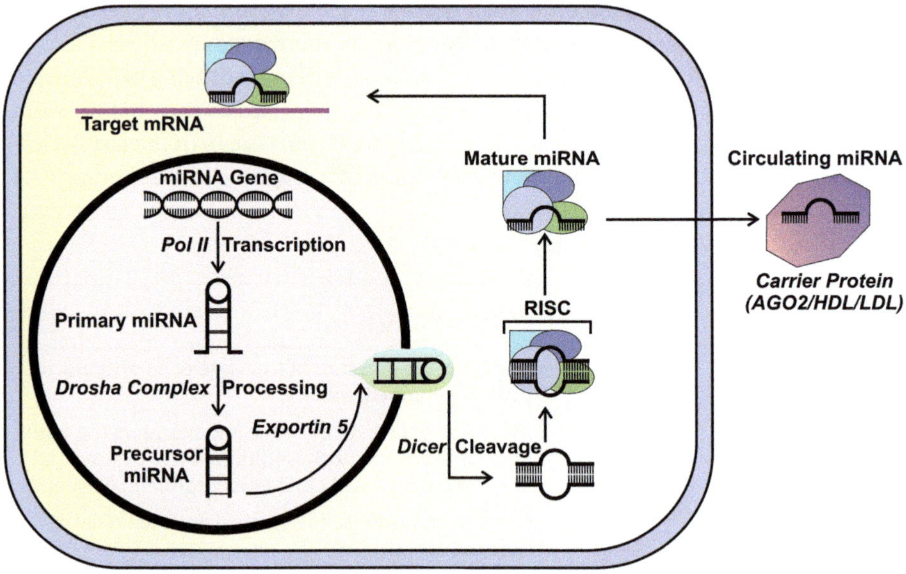

Figure 4.2 Simplified overview of miRNA biogenesis and release of miRNA into the circulation. In the canonical pathway, miRNAs encoded within genes are transcribed by RNA Polymerase II (Pol II Transcription) into large primary miRNAs. Primary miRNAs are cleaved within the nucleus by the Drosha Microprocessing Complex (Drosha Complex Processing), which gives rise to precursor miRNAs. Precursor miRNAs are exported to the cytoplasm by Exportin 5. They are next cleaved by Dicer, followed by loading into RISC. The RISC facilitates removal of the passenger strand RNA to generate mature miRNA. Mature miRNAs can be actively secreted into circulation in association with RNA-binding carrier proteins (AGO2/HDL/LDL), or by packaging into microvesicles/exosomes (not shown). Alternatively, mature miRNA targets cognate mRNA in the cytoplasm through complimentary base pairing, which leads to regulation of gene expression.

structures called precursor miRNAs (pre-miRNAs). Pre-miRNAs are exported into the cytoplasm by Exportin 5 (XPO5) and are cleaved by Dicer to form a miRNA duplex (see Figure 4.2). miRNA duplexes are then loaded onto Argonaute protein (Ago), which stimulates the assembly of the RNA-induced silencing complex (RISC) (Ha and Kim 2014). The RISC facilitates the removal of the passenger strand RNA in order to generate mature miRNA (Michlewski and Caceres 2019).

Mature miRNAs regulate their target mRNAs by binding to 3′ untranslated regions (3′UTR) of mRNAs; however, miRNA binding sequences have also been identified within 5′ UTRs (Friedman et al. 2009). Mature miRNAs are guided to their cytosolic target mRNAs by complementary base pairing, and miRNA targeting is reliant on base pairing of the miRNA seed region, nucleotides 2-7, to mRNA targets. Perfect miRNA-mRNA base pairing leads to the degradation of the target mRNA, whereas imperfect base pairing leads to decreased mRNA translation, which can be restored once the repressing miRNA is degraded. Intriguingly, it has also been shown that miRNA-mRNA interaction with the 5′ UTR can increase the translation of the mRNA, and that some miRNAs can target promoters or enhancers (Fabbri 2018; Odame et al. 2021; Vasudevan et al. 2007). Biological functions of separate miRNAs have been extensively studied using a variety of miRNA-knockout/knockdown models and transgenic overexpression experiments (Hammond 2015; Saliminejad et al. 2019).

Circulating miRNAs as Biomarkers for Disease

miRNA expression profiling has demonstrated dysregulation of miRNA expression in a wide range of human diseases (Bonneau et al. 2019). Dysregulated miRNA expression has been shown to affect a large variety of cellular processes, for example transcription, signal transduction, cellular proliferation, differentiation, apoptosis, and epithelial-mesenchymal transition (Aleckovic and Kang 2015). Alterations in miRNA expression signatures offer great potential for use as clinical biomarkers, because miRNAs are stable in biological samples, including formalin fixed paraffin sections (FFPE) and plasma (Amini et al. 2017; Aryani and Denecke 2015; Heneghan et al. 2010; Kotorashvili et al. 2012; Wimmer et al. 2018). Additionally, miRNAs can be reliably detected in multiple bodily fluids such as blood, serum, amniotic fluid, breast milk, bronchial lavage, cerebrospinal fluid, colostrum, peritoneal fluid, plasma, pleural fluid, saliva, seminal fluid, tears, and urine (De Guire et al. 2013; Hammond 2015; Lawrie et al. 2008; Weber et al. 2010). But the total RNA concentration and the total number of distinct miRNAs can differ vastly between various body fluids (Weber et al. 2010). Likewise, specific circulating miRNAs can be enriched within specific fluids.

Circulating miRNAs are released into circulation in one of two forms: they are either bound to specific RNA-binding proteins (Ago; high density lipoprotein, HDL; or low-density lipoprotein, LDL) or encapsulated in microvesicles along with proteins, lipids, and other nucleic acids (Aleckovic and Kang 2015; Cui et al. 2019; Mori et al. 2019; see Figure 4.2). The release mechanisms of protein- or lipid-bound miRNAs is largely unknown, whereas exosomal miRNAs are known to be selectively recruited and actively secreted in a regulated manner (Aleckovic and Kang 2015). When injected intravenously, the exosomes can remain in the circulation for up to two hours, which implies a large range in the bioavailability of circulating miRNAs (Mori et al. 2019).

The potential use of circulating miRNAs as non-invasive diagnostic and prognostic biomarkers for disease status in biological fluids was first realized in the field of cancer biology because cancer diagnosis and prognosis rely on invasive tissue biopsies (Correia et al. 2017; Iorio and Croce 2012). More recently, changes in circulating miRNA profiles upon heavy metal exposure have been described. In consequence, this chapter strives to provide a comprehensive resource, which will highlight studies that have identified circulating miRNAs as a result of heavy metal exposure and assist with rigor and reproducibility in future studies. We have chosen to limit our focus to the five metals (arsenic, lead, mercury, cadmium, chromium) ranked currently at the top of the ASTDR substance priority list on the basis of their significant toxicity and of the high potential for human exposure to them.

Circulating miRNAs Associated with Arsenic Exposure

Arsenic (As) is a naturally occurring metalloid ubiquitously distributed throughout the earth's crust and groundwater. It is also found in the atmosphere and in food, particularly in cereal and cereal products (Cubadda et al. 2017; Zhang et al. 2020). Arsenic is classified as a Group I human carcinogen by the International Agency for Research on Cancer, because chronic exposure to arsenic is strongly associated with an increased risk for cancer development in skin, urinary bladder, and lung in humans (IARC 1980, 2012; Straif et al. 2009). Therefore arsenic contamination is considered a serious global public health issue.

Arsenic contamination affects several countries, for example Bangladesh, India, Taiwan, China, Ghana, United States, Argentina, Vietnam, and Chile (Hunt et al. 2014). An estimated 225 million people are chronically exposed to arsenic from drinking water, at concentrations that exceed the maximum contaminant level (MCL) of 10 μg/L recommended by the World Health Organization (WHO) (Podgorski and Berg 2020). In Bangladesh alone, arsenic levels exceeding 50 μg/L have been reported, affecting between 21% and 48% of the total population (Rahman et al. 2001). In the United States, approximately 3 million people relying on private domestic wells for water usage are exposed to high concentrations of arsenic (Ayotte et al. 2017). In addition to naturally occurring arsenic, human-made arsenic-based compounds constitute other sources of arsenic exposure. Arsenic-based pesticides are still widely used in agriculture (Li et al. 2016); in this way they produce significant arsenic contamination in the environment and contribute to a number of acute and subacute arsenic intoxication cases (Armstrong et al. 1984; Li et al. 2016).

The main forms of arsenic present in the environment are arsenate (iAs^{V}), arsenite (iAs^{III}), and other organic arsenicals such as methylated arsenicals—monomethyl arsenic acid (MMA^{V}) and dimethyl arsenic acid (DMA^{V}) (Thomas 2015; Tsuji et al. 2019). After ingestion, arsenate or arsenite are readily absorbed in the gastrointestinal tract and transported through the blood to other parts of the body. In the liver, around 90% of arsenic undergoes a series of sequential reduction and oxidative-methylation reactions. iAs^{V} is reduced to iAs^{III}, with subsequent methylation to MMA^{V}, which is then reduced to monomethylarsonous acid (MMA^{III}). MMA^{III} is oxidatively methylated to DMA^{V}, which can be reduced to dimethylarsinic acid (DMA^{III}) (Tsuji et al. 2019; Zhou and Xi 2018). In rodents and humans exposed to extremely high concentrations of arsenic, DMA^{III} can be further methylated to trimethyl arsenic oxide ($TMA^{V}O$) and, in rodents, it can be subsequently reduced to volatile trimethylarsine (TMA^{III}) (Cohen et al. 2013). The reduction reactions are catalyzed by glutathione-S-transferase omega (GSTO); and oxidative methylation is mainly performed by the arsenic-3-methyl transferase enzyme (AS3MT) (Naranmandura et al. 2006; Tsuji et al. 2019; Zhou and Xi 2018).

The clinical features of acute arsenic poisoning include, but are not limited to, nausea, vomiting, diarrhea, severe abdomen pain, skin rash, and seizures (Ratnaike 2003). Depending on the amount ingested, arsenic can induce severe systemic toxicity and death. Although acute exposures to high levels of arsenic are occasionally reported, chronic exposure to low levels over a long period of time is more common and currently of great concern. The long-term effects of arsenic exposure differ between individuals, population groups, and geographical areas and clinical outcomes include diabetes, pulmonary and cardiovascular disease, skin lesions, hyperkeratosis, and, as discussed earlier, skin, urinary bladder, and lung cancers (IARC 2012; States 2015; WHO 2018). The mechanisms by which arsenic causes cancer remains elusive, but some of the proposed mechanisms of action are inhibition of DNA repair, oxidative stress, aneuploidy, aberrant DNA methylation, and miRNA dysregulation (Hughes 2002; Tam et al. 2020).

Because of arsenic's multisystem toxicity and carcinogenic potential, it is important to prevent further exposure to this substance by providing safe water supplies for drinking, cooking, and the irrigation of food crops. It is also important to monitor high-risk populations for early signs of arsenic toxicity, which is mainly characterized by skin lesions (WHO 2018). Recently, studies have identified circulating miRNAs associated with arsenic exposure and arsenic-induced disease development. In a study conducted in Mexico, serum levels of miR-155 and miR-126 were found, up- and downregulated respectively, in women with high arsenic levels in urine (Ochoa-Martinez et al. 2021; Ruiz-Vera et al. 2019). Downregulation of miR-126 was also observed in the plasma of children exposed to arsenic from the same area, but the levels of miR-155 were unchanged (Perez-Vazquez et al. 2017). In the plasma of a Chinese population exposed to arsenic, 56 upregulated miRNAs (miR-21, -141, -148a, -145, -155, -191, -218 and -491 presenting the largest fold change increase), and 18 downregulated miRNAs (miR-200b, -200 c, -26, and -34 c levels with the most decreased expression) were observed using miRNA microarray (Sun et al. 2017). Further RT-qPCR validation in larger population-based studies in the same area validated the expression levels of miR-21, miR-145, miR-155 and miR-191 (Sun et al. 2017; Xu et al. 2020; Zeng et al. 2019). Increased levels of miR-21 and miR-145 were seen in patients who presented skin alterations and liver damage (Zeng et al. 2019). Additionally, miR-191 was upregulated in patients with skin (Zeng et al. 2019) and renal alterations (Xu et al. 2020; Zeng et al. 2019) . Finally, miR-155 was found to be increased in patients with skin manifestations of arsenic toxicity (Zeng et al. 2019). Therefore circulating levels of these miRNAs could be important for monitoring arsenic-induced skin, liver, and kidney damage. Arsenic-related skin lesions (precancerous and cancerous) were also associated with 202 dysregulated circulating miRNAs in the plasma (199 miRNAs up- and 3 downregulated) of an exposed population in West Bengal, India (Banerjee et al. 2019). RT-qPCR validated microarray data for miR-21, miR-23a, miR-619, miR-126, miR-3613 (upregulated miRNAs) and miR-1282 and miR-4530 (downregulated miRNAs) (Banerjee et al. 2019). Circulating levels of miR-21 in blood were, again, found to be increased in arsenic-exposed individuals in West Bengal (Banerjee et al. 2017). Within the arsenic-exposed population, a much higher miR-21 expression was found in individuals who presented with arsenic-induced skin lesions (malignant squamous cell carcinoma and basal cell carcinoma) than in individuals without skin lesions (Banerjee et al. 2017). In another study, the levels of miR-200c and miR-205 in urine were inversely associated with arsenic exposure (Michailidi et al. 2015). However, the arsenic-exposed and non-exposed cohorts selected for this study were from different populations (Bangladesh versus Baltimore), which could add confounding factors and bias to the analysis. miR-205 is expressed in epithelial tissues and is up- or downregulated in different epithelial cancers (Ferrari and Gandellini 2020). Rager et al. (2014) identified miRNAs in newborn cord blood samples by miRNA array in order to assess prenatal arsenic exposure in Mexico. Increased expression of miRNAs known to have roles in cancer and inflammatory response—for example let-7a, miR-107, miR-126, miR-16, miR-17, miR-195, miR-20a, miR-20b, miR-26b, miR-454, miR-96, and miR-98—was associated with maternal urinary arsenic (Rager et al. 2014). Two studies evaluated the association between circulating miRNAs and arsenic and its metabolites. In a population-based study conducted in Mexico, plasma MMA^{III} significantly correlated with plasma levels of miR-423-5p, miR-142-5p-2, miR-423-5p+1, miR-320c-1, miR-320c-2, and miR-454-5p, while no associations were found for plasma inorganic arsenic or DMA^{V} (Beck et al. 2018). miR-142-5p is a common circulating miRNA in type 1, type 2, and gestational diabetes (Collares et al. 2013). Blood levels of miR-548c-3p were also negatively correlated with iAs, MMA, and DMA concentrations in the urine of a Chinese cohort (Cheng et al. 2018). Although published research highlights several circulating miRNAs as potential responders to arsenic exposure and disease development, few miRNAs were consistently dysregulated between studies. miR-126 was downregulated in two studies performed in Mexico and miR-21 was constantly found upregulated in arsenic-exposed individuals from China and India. These might represent good candidates for further validation.

Circulating miRNAs Associated with Lead Exposure

Elemental lead (Pb) is found in lead ore deposits that are distributed throughout the world (Abadin et al. 2007). Lead does not degrade in the environment and is dispersed worldwide as a result of anthropogenic activities such as mining and smelting ore, the manufacture of lead-containing products (lead-containing gasoline, paints, batteries, radiation shields, water pipes, ammunition, ceramics and metal containers), the combustion of coal and oil, and waste incineration (Tokar et al. 2013). Lead is listed as one of the topmost toxic substances, and its toxic properties have been recognized for over 2,000 years (Abadin et al. 2007). However, in the last few decades, there has been increased awareness about the detrimental health effects of low-level lead exposure, particularly in children. This awareness has prompted changes in public health policies, and the result the is phasing out of lead from several sources, including gasoline and paints. Nonetheless, exposure to lead still occurs, primarily from lead-containing paint and water pipes present in older housing, from residues along roadways, or in the form of occupational exposure.

For the general population, lead exposure occurs orally for the most part; however, inhalation and dermal contact can also occur, albeit rarely, in occupational settings. The health effects caused by lead exposure are diverse (neurological, cardiovascular, renal, hematological, reproductive, developmental, respiratory, hepatic, endocrine, gastrointestinal, musculoskeletal, ocular, and cancer) and depend on numerous factors including age, nutritional status, and life stage (e.g., *in utero*).

The blood lead reference value was set at 5 μg/dL in 2012; elevated lead in blood is an indication of excessive exposure (CDC 2012). The most common metric for lead exposure is the concentration of lead in blood, although other measurements of lead in urine, bone, and hair can be used to quantitate exposure (Abadin et al. 2007). Levels of lead in plasma and semen are difficult to measure because lead concentrations in these fluids are often near the lower limits of detection, and lead measurements in saliva and sweat show inconsistent results when compared to blood lead.

Circulating miRNAs associated with lead exposure have been identified in humans (Mitra et al. 2021; Xu et al. 2017, de Araujo et al. 2021). In a large study that analyzed 1,130 Chinese workers with occupational lead exposure, miRNA microarray analyses from plasma identified three miRNAs that were significantly downregulated (miR-520c-3p, miR-148a, and miR-211) and one miRNA that was significantly upregulated (miR-572) in workers chronically exposed to lead who also had high blood lead levels (BLLs) (51.35 ± 6.86 μg/dL; see Xu et al. 2017). However, workers with high BLLs were compared to workers with low BLLs (8.93 ± 1.53 μg/dL), which still contain levels above the blood lead reference values set by the Centers for Disease Control (CDC). Thus it is unclear whether the changes in circulating miRNA expression are a function of lead exposure or happen only when a specific exposure threshold is breached. In another study, which analyzed workers who suffered from occupational lead exposure in Northern India, expression levels of circulating miR-20b-5, miR-221-3p (miR-221), miR-155-5p (miR-155), miR-20b from serum were quantified by RT-qPCR and compared to expression levels in individuals without any history of occupational lead exposure (Mitra et al. 2021). Circulating miR-221 and miR-155 were significantly higher in lead-exposed workers than in non-exposed individuals. In a third study, circulating miRNA expression profiles were measured in blood using RT-qPCR from battery plant workers from Paraná State, Brazil and revealed an association of DNA methylation status and miR-148a expression in lead-exposed workers (de Araujo et al. 2021). However, no unexposed control population was included in this study. Although the de Araujo study correlated circulating miR-148a to DNA methylation status in lead-exposed workers only, DNA methylation increased with the lead dose and miR-148a expression was inversely correlated to DNA methylation. Taken together, circulating miRNA analyses from two small populations of lead-exposed workers revealed that several circulating miRNAs are differentially expressed as a result of lead exposure; thus, pending further validation, these miRNAs hold promise for being used as biomarkers.

Circulating miRNAs Associated with Mercury Exposure

Mercury is present in the environment in several forms, for example as inorganic mercury—elemental mercury (Hg^0), mercurous mercury (Hg^{1+}), mercuric salts (Hg^{2+})—and in organic compounds (methylated mercury). In its elemental form, metallic mercury is the only metal that exists in a liquid state at ambient temperatures. It is also known as quicksilver (Clarkson and Magos 2006). Mercury vapors released from elemental mercury occur naturally in the environment as a consequence of the removal of gases from the earth's crust, volcanic eruptions, and evaporation from oceans and soils. Anthropogenic sources such as metal mining, smelting (with mercury, gold, copper, and zinc), coal combustion, municipal incinerators, and the chloralkaline industry contribute significantly to atmospheric mercury, and mercury vapors are stable within the atmosphere for about one year (Tokar et al. 2013). Once released into the environment, the various forms of mercury undergo complex oxidation-reduction and methylation-demethylation reactions known as the mercury cycle, which further gives rise to inorganic or organic species that become globally distributed.

Exposure to inorganic mercury can occur through inhalation and through oral or dermal routes, but inhalation is the most biologically relevant owing to its high absorbance potential in the lungs (up to 80%; see Risher et al. 2003). Inorganic mercury poisoning often occurs as a result of the inhalation of mercury vapor from industrial sources, dental amalgams, or the continued use of mercury compounds in consumer products (Clarkson and Magos 2006). The inhalation of mercury vapor from the use of liquid mercury can follow from occupational exposure in the chloroalkali industry, in the manufacture of scientific instruments, in fluorescent lightbulb manufacturing, in small-scale gold mining, and in dentistry where mercury amalgams are used as fillings for tooth decay (Clarkson and

Magos 2006; Risher et al. 2003). Mercury poisoning can also result from inorganic mercury, which is most commonly found as calomel or mercurous chloride (Hg_2Cl_2) and was used widely in diuretics, antiseptics, skin products, laxatives, and teething powder in the twentieth century (Clarkson and Magos 2006; Klaassen 2013). Additionally, the use of phenylmercuric compounds as antifungal agents in paints, although banned in the United States, still continues in other countries (Risher et al. 2003).

Ding et al. (2016) investigated the plasma miRNA expression profile for female workers in eastern China who were occupationally exposed to inorganic mercury and found that four miRNAs (miR-16-5p, miR-30c-3p, miR-181a-5p, and let-7e-5p) were downregulated and four miRNAs (miR-92a-3p, miR-122-5p, miR-451a, and miR-486-5p) were upregulated, as measured by microarray. Validation of these miRNAs by TaqMan-based RT-qPCR revealed that two miRNAs, namely miR-92a-3p and miR-486-5p, were consistently upregulated when measured by both methods.

Both metallic and organic mercury compounds are oxidized to inorganic mercury in the blood and in the liver, which plays a key role in animal toxicity (Tchounwou et al. 2003). However, organomercurial compounds are generally more toxic to animals than inorganic mercury on account of their high bioaccumulation potential, and methyl mercury (MeHg) is toxicologically the most important organic form. In humans, consumption of methyl mercury-contaminated fish is the predominant route of exposure to organomercury, but other consumables such as drinking water, cereals, vegetables, and meat can also be sources of exposure (Holmes et al. 2009; Karagas et al. 2012; Yang et al. 2020). Early stages of life are generally more sensitive to mercury, such that exposure to high-level elemental, inorganic, and organic mercury results in severe developmental and neurological defects, depending on the length and dose of this exposure (Yang et al. 2020). However, to date, no literature exists that has examined circulating miRNAs that are due to the consumption of methylmercury-containing food sources.

Circulating miRNAs Associated with Cadmium Exposure

Cadmium (Cd) is a naturally occurring toxic metal of environmental and occupational concern. The main source of cadmium contamination is related to its use in industries, for example in the production of nickel–cadmium batteries, pigments, chemical stabilizers, and metal coatings (Genchi et al. 2020). Fossil fuel combustion and the use of phosphate fertilizers also contribute to cadmium accumulation in the environment. Human exposure to cadmium occurs mainly through the ingestion of contaminated food and water and through inhalation; cigarette smoking is the most significant source of human exposure to cadmium (Bernhoft 2013; Friberg 1983). Once absorbed, cadmium is bound by metallothionein (MT) within liver cells and is released into circulation in the form of MT-cadmium complexes. These complexes are filtered through the renal glomeruli and reabsorbed by the proximal tubular cells, processes that make the kidney a primary target organ of toxicity (Yuan et al. 2020). Since cadmium remains tightly bound to MT and is almost completely reabsorbed in the renal tubules, excretion in urine is very low, resulting in a twenty-five–thirty-year half-life for cadmium in the body (Genchi et al. 2020). Additionally, women with lower iron status are at risk for greater absorption of cadmium after oral exposure (Olsson et al. 2002). Cadmium can be measured in blood, urine, feces, hair, nails, saliva, and milk. Cadmium in blood indicates recent exposure and its acceptable level ranges between 0.03 and 0.12 μg/dl in blood (Alli 2015). In smokers, cadmium concentrations are typically above 3 μg/L, whereas in the general population the concentration varies between 0.1 and 1.0 μg/L (Fowler et al. 2015). Acute exposure to high levels of cadmium can result from inhalation or ingestion and initial symptoms of inhaled cadmium exposure (8.63 mg/m^3 for five hours) include chills, fever, and myalgias. Later, individuals can develop chest pain, cough, and dyspnea (Newman-Taylor 1998). Gastrointestinal symptoms such as nausea, vomiting, diarrhea, abdominal cramps and pain have also been described after the ingestion of cadmium-contaminated food or drinks (Fowler et al. 2015). Long-term or chronic exposure to low levels of cadmium is also of concern. The most common health effects of low-dose chronic exposure include cancers, cardiovascular disease, diabetes, osteoporosis-related fractures, and kidney dysfunction, the kidney being the critical target for cadmium toxicity (Jarup and Akesson 2009). Recent studies have shown adverse kidney effects even at doses lower than the provisional tolerable weekly intake of 1 μg/kg body weight/day (Kobayashi et al. 2006; Noonan et al. 2002; WHO 1993). Hence it is crucial to identify potential biomarkers for the early detection of cadmium exposure.

Few studies have addressed cadmium-induced alterations in miRNA expression in humans. miR-122-5p and miR-326-3p were found to be upregulated in the serum of a cadmium-exposed population from China (Yuan et al. 2020). The commonly used biomarkers for cadmium exposure, urinary β2-microglobulin and retinol-binding protein, remained in the normal range, which suggests that these two miRNAs exhibited greater sensitivity to cadmium exposure than the urinary markers; and this makes them suitable candidates for the early detection of exposure (Yuan et al. 2020). In another study, levels of miRNAs were measured from the serum and urine of patients in China diagnosed with occupational chronic cadmium poisoning (Chen et al. 2021). In the urine, 16 miRNAs were found to be upregulated and 36 downregulated, while in the serum 46 miRNAs were upregulated and 131 were downregulated. An overlap of 59 abnormally expressed miRNAs was found between serum and urine samples of occupational chronic cadmium poisoning patients; these included miR-122-5p, miR-363-3p, miR-129-5p, miR-204-3p and miR-361-3p (Chen et al. 2021). In individuals exposed to low-levels of cadmium in China, the levels of miR-21 and miR-29b in plasma were upregulated. Serum levels of miR-30 were decreased in Chinese chronic obstructive pulmonary disease (COPD) patients with elevated levels of cadmium in serum by comparison to the levels in healthy control subjects. Additionally, an epithelial-mesenchymal transition was observed in the lung tissue of these patients (Zheng et al. 2021). In another study, individuals in India who were occupationally exposed to cadmium presented elevated levels of serum miR-221 and IL-17, a pro-inflammatory cytokine (Goyal et al. 2021; Zenobia and Hajishengallis 2015). Overall, several studies have identified differentially expressed miRNAs in the urine, serum, and plasma of individuals exposed to cadmium only in the Chinese population.

Circulating miRNAs Associated with Chromium Exposure

Chromium (Cr) is the seventh most abundant element on the earth. As a transition metal, chromium can occur in any of the oxidation states from -II to +VI, but it is mostly found in oxidation states 0, +III, and +VI (Sperling 2005). In the environment, chromium can occur in hexavalent (Cr(VI)) or trivalent (Cr(III)) species (Nath et al. 2009). Cr(VI) has been classified as Group I human carcinogen (Proctor et al. 2014; Straif et al. 2009). Hexavalent chromium is used extensively in industry, for instance in leather tanning, in textile and pigment manufacturing, in stainless steel production, and in welding, polishing, and grinding (Wang et al. 2017). Occupational exposure is thus a major concern, primarily because of inhalation. Once absorbed, chromium circulates in the blood, getting bound to the β-globulin plasma fraction, and is transported to tissues after being complexed with transferrin or other proteins (Pechova and Pavlata 2007). Within cells, hexavalent chromium compounds undergo reduction by ascorbate and non-protein thiols, cysteine, and glutathione, and this leads to the formation of transient and highly reactive Cr(V) and Cr(IV) species, organic radicals, and, finally, Cr(III) products (Manning et al. 1994; Salnikow and Zhitkovich 2008; Wang et al. 2017). Intermediate chromium forms are mostly responsible for oxidative stress and DNA damage, which they do by inducing chromium–DNA adducts, DNA strand breaks, and DNA–protein cross-links, all known to contribute to mutagenesis and carcinogenesis (O'Brien et al. 2003; Wang et al. 2017; Wise et al. 2002).

The acute effects of inhaled hexavalent chromium include shortness of breath, coughing, and gastrointestinal and neurological effects (IARC 1990; USEPA 1998). The chronic inhalation of hexavalent chromium causes ulcerations of the septum, bronchitis, pneumonia, decreased pulmonary function, and increased risk of developing respiratory system cancers, primarily bronchogenic and nasal (IARC 1990; USEPA 1998). The chronic inhalation of high levels of hexavalent chromium induces liver, kidney, and gastrointestinal damage (IARC 1990; USEPA 1998).

Several studies have investigated chromium-induced circulating miRNA dysregulation. miRNA profiles from Chinese workers involved in the production of chromate revealed lower levels of miR-3940-5p associated with high levels of hexavalent chromium in blood (Li et al. 2014). In addition, plasma levels of miR-3940-5p were positively correlated with micronuclei frequency in the hexavalent chromium-exposed group (Li et al. 2014). The micronuclei test has been recognized as a reliable biomarker of genotoxicity in occupationally chromium-exposed individuals (IARC 1990; Sudha et al. 2011). Therefore miR-3940-5p could in principle be a biomarker of chromium exposure linked to genetic damage. A very recent study observed forty-five significant differentially expressed miRNAs in the plasma of hexavalent chromium-exposed workers from China (Jia et al. 2020). RT-qPCR validated both the decreased levels of miR-19a-3p, miR-19b-3p, and miR-142-3p and the increased levels of miR-590-3p and miR-941 in the hexavalent chromium-exposed group (Jia et al. 2020). Expression levels of miR-143 were decreased, whereas Interleukin 6 (IL-6) was increased in the blood samples of hexavalent chromium-exposed workers by comparison with the samples of unexposed workers (Wang et al. 2019). Only one study attempted to better understand the linkage between chromium, miRNAs, and metabolic and cardiovascular diseases. Blood levels of forty-three miRNAs were negatively associated with chromium in the urine of obese subjects in Italy (Dioni et al. 2017). By RT-qPCR, nine miRNAs (miR-451, miR-301, miR-15b, miR-21, miR-26a, miR-362-3p, miR-182, miR-183 and miR-486-3p) were downregulated in association with chromium in urine. miR-451 was further associated with glycated hemoglobin, whereas miR-486-3p was associated with both diastolic and systolic blood pressure in obese subjects who presented urinary chromium (Dioni et al. 2017). These two circulating miRNAs can be considered putative biomarkers of chromium exposure, and their relation with glycated hemoglobin and blood pressure may give insights into chromium-induced metabolic and cardiovascular diseases (Dioni et al. 2017). In summary, studies so far demonstrate a differential expression of several miRNAs upon chromium exposure, albeit with no overlap of a single miRNA across all these studies. This inconsistency is a major caveat, given that all the studies have been conducted on the Chinese population and further substantiation of these findings across other exposed populations is required.

Circulating miRNAs Associated with Mixed Metals Exposure

Often populations are exposed to a mixture of metals released into the environment either through natural processes or as an outcome of anthropogenic activities (Cory-Slechta 2005). While it is important to understand the effect of each metal on human health individually, this does not adequately elucidate the molecular events operative within the environmental reality of mixed metal exposures (Cory-Slechta 2005). This is particularly important, given that exposure to one metal can affect the uptake, metabolism, disposition, and elimination of other metals and their metabolites (Young et al. 2019).

Recently studies have started analyzing how expression of miRNA(s) are altered by exposure to multimetal mixtures. In a study of welders exposed to mixed metal fumes, plasma miR-21 and miR-155 levels were lower than in matched unexposed control populations (Amrani et al. 2020). Furthermore, miR-21 and miR-155 expression levels were demonstrated to be positively associated with urinary chromium levels, while miR-146a was positively associated with urinary nickel levels (Amrani et al. 2020). In another study on sixty-three healthy male steel plant workers occupationally exposed to mixed metal particulate matter, blood leukocyte expression of miR-222 was positively correlated with lead exposure, while that of miR-146a was negatively correlated with lead and cadmium exposures (Bollati et al. 2010). Importantly, only miR-222 was found to be significantly upregulated in workers at the end of the workweek by comparison with the situation at the beginning of the workweek, after two days' break from exposure to particulate matter. In a subsequent study from the same group, the leukocyte expression of four upregulated miRNAs (miR-421, miR-146a, miR-29a, and let-7g) was found to regulate the downstream expression of eleven target candidate inflammatory genes in a population of ten steel plant workers exposed to mixed metal particulate matter (Motta et al. 2013). In another study, urinary expression of miR-200c and miR-423 was positively associated with chromium concentration in individuals in the upper tertile of urinary chromium levels, but not with arsenic levels in a cohort of Mexican children (Cardenas-Gonzalez et al. 2016). In a study of healthy male coke oven workers in southern China, decreased plasma expression levels of miR-27a-3p were associated with lead exposure (Deng et al. 2019). In another study, on Hong Kong Chinese adolescents, expressions of urinary miR-21 and miR-221 were found to be negatively correlated both with urinary arsenic and lead (Kong et al. 2012).

Multiple heavy metal exposures often interact in a myriad of ways, modulating toxicity and health outcomes (Wah Chu and Chow 2002; Young et al. 2019). No studies thus far have examined whether there is any additive, synergistic, or antagonistic effect between the different metals in terms of miRNA expression modulation. This lack of investigation represents a considerable knowledge gap in the field—and one that needs to be addressed.

Summary of Potential Circulating miRNA Biomarkers for Heavy Metals

Figure 4.3 depicts the candidate circulating miRNA biomarker expression status across different metals (or metal mixtures), stratified by body fluids (plasma, blood, serum and urine). Plasma and blood are by far the most heavily investigated biofluids; and they are followed by urine and serum. However, regardless of the body fluids assessed, very few miRNAs are represented as putative biomarkers across multiple metal exposures. Presently miR-21 seems to be the most promising biomarker candidate and the one with the

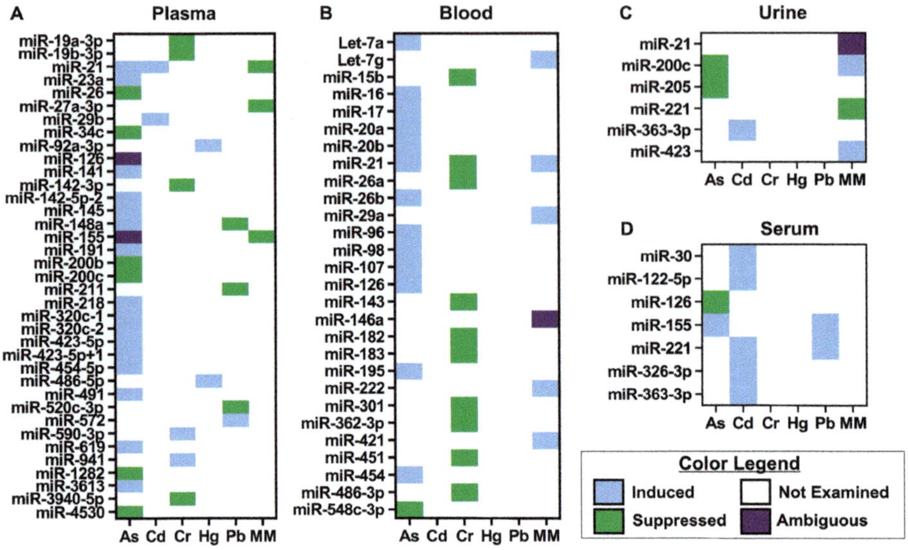

Figure 4.3 Differential expression map of circulating miRNAs dysregulated after heavy metal or mixed metal exposure, stratified by body fluid. Circulating miRNAs dysregulated as a result of heavy metal exposure to arsenic (As), cadmium (Cd), chromium (Cr), mercury (Hg), lead (Pb), and mixed metals (MM) across all studies examined in this chapter are presented within specific biofluids analysed: (A) plasma, (B) whole blood, (C) urine, and (D) serum. Induced circulating miRNAs are represented in blue, suppressed or lower circulating miRNAs are represented in green, miRNAs that were not examined in a particular biofluid for specific metals are represented in white, and miRNAs that showed conflicting results in multiple studies within a biofluid are shown in purple.

strongest experimental evidence. Circulating miR-21 is higher in both the plasma and the blood of arsenic-exposed individuals from India and China. Additionally, plasma levels of miR-21 are suppressed in chromium-exposed individuals but induced in cadmium-exposed individuals. Circulating miR-21 has also been detected in the urine of individuals exposed to metal mixtures, although the nature of its expression varied depending on the population (Cardenas-Gonzalez et al. 2016; Kong et al. 2012).

Figure 4.4 represents the circulating miRNA biomarker expression status across different body fluids for each metal (or metal mixture) exposure, individually. Even for the same metal exposure, the nature of dysregulated miRNA expression varies considerably within the different body fluids tested so far, with little or no overlap. Furthermore, even if the same miRNA is differentially expressed in multiple body fluids for a particular metal exposure, often the direction of expression change is not consistent across the different fluids. For example, miR-126 is dysregulated upon arsenic exposure in blood, plasma, and serum. However, while it is induced in blood, it is found to be suppressed in both plasma and serum (Figure 9.4), which suggests that the induction is in circulating cells. In summary, despite all the studies, there is still no circulating miRNA that is unequivocally representative of exposure to a specific heavy metal, either across all biofluids or across all heavy metals for any specific biofluid. This inconsistency is possibly due to the fact that the number of studies that are not on arsenic exposure is so limited. Such inconsistencies are further complicated by differences between study designs, some of which employ end points and statistical measures (e.g., point estimates versus interval estimates) that make it difficult to unify all the results and gain a holistic understanding.

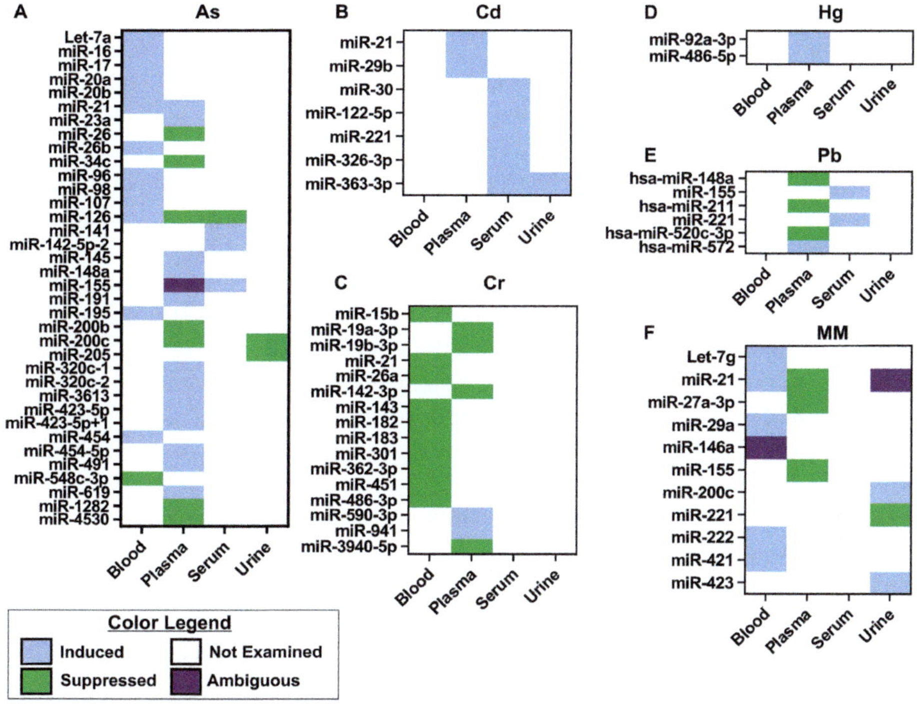

Figure 4.4 Differential expression map of circulating miRNAs dysregulated after a specific heavy metal or mixed metal exposure across different body fluids. Circulating miRNAs dysregulated in different biofluids (blood, plasma, serum and urine) upon exposure to (A) arsenic (As), (B) cadmium (Cd), (C) chromium (Cr), (D) mercury (Hg), (E) lead (Pb), and (F) mixed metals (MM). Induced circulating miRNAs are represented in blue, suppressed or lower circulating miRNAs are represented in green, miRNAs that were not examined in a particular biofluid for specific metals are represented in white, and miRNAs that showed conflicting results in multiple studies within a biofluid are shown in purple.

Future Avenues of Research

Importantly, all circulating miRNAs associated with metal exposure are currently in stages of discovery and development and have not yet been validated; nor are they considered biomarkers. Biomarkers must be validated in accordance with well-established protocols that involve multiple independent qualitative and quantitative steps, before they can be employed for diagnosis or monitoring (Califf 2018). Consequently, there is a significant knowledge gap when it comes to identifying a circulating miRNA that can be used consistently or can be used as a unique biomarker for heavy metal exposure or for disease outcome(s) of such exposure. Therefore future studies should first find out whether there are differences between heavy metal exposures for the circulating miRNAs summarized in Figure 4.3. Furthermore, the assessment of these candidate biomarkers and of other circulating miRNAs should be validated in other fluids that do not rely on blood draw and contain high levels of miRNAs, for instance sweat, tears, saliva, semen, or breast milk (Barcelo et al. 2019; Karvinen et al. 2020; Rubio et al. 2018; Setti et al. 2020; Weber et al. 2010).

Other areas that need to be addressed are things such as limited sample sizes and the inclusion of geographically restricted populations. Both factors need to be taken into account when suggesting the use of a circulating miRNA as a putative biomarker. Ideally, future studies should be performed with these candidate circulating miRNAs across larger populations from numerous geographic regions. This approach will ultimately enable multiple stratifications across biological and socioeconomic parameters in order to investigate whether initial patterns discovered for circulating miRNAs are both consistent and associated with these additional factors.

Although causality is not a required criterion for a biomarker, it would be important and interesting to examine whether any of these dysregulated miRNAs plays a causal role in the etiology of any metal exposure-induced diseases. A few studies have been conducted in this direction. As consistent with human studies, miR-21 was increased in immortalized human keratinocytes (HaCaT) exposed to 500 nM arsenite for four weeks (Gonzalez et al. 2015) and in human umbilical vein endothelial cells (HUVEC) exposed to 20 μM arsenite for twenty-four hours (Li et al. 2012). A recent systematic review and meta-analysis suggests that arsenic-induced miR-21 expression suppresses phosphatase and tensin homolog (PTEN) and protein sprouty homolog 1 (Spry1) levels, leading to epithelial–mesenchymal transition (EMT) and malignant transformation (Liu et al. 2018). miR-21 is a well-conserved miRNA, frequently found upregulated in numerous types of cancer (Feng and Tsao 2016). Furthermore, in cadmium-exposed individuals, miR-21 was associated with renal dysfunction, characterized by increased excretion of the low molecular weight protein N-Acetyl-beta-(D)-glucosaminidase in urine (Lei et al. 2019). Thus it is important to consider circulating miR-21 as a potential biomarker for heavy metal exposure and to investigate its potential mechanistic role in heavy metal exposure-induced disease outcomes. Similarly, upregulation of miR-92a-3p and miR-486-5p after mercury exposure has been recapitulated in vitro by exposing HEK-293 and HUVEC human cell lines to mercuric chloride (Ding et al. 2017).

Techniques and Challenges for Identifying Circulating miRNAs

Although using circulating miRNA as biomarkers has its advantages, since this is a relatively non-invasive technique and also one that can in principle be used in the early diagnosis of diseases, miRNA profiling in biofluids is still in its infancy. Therefore it is necessary to highlight limitations that may lead to inconsistent findings. The identification of potential confounding factors will also help optimize the reproducibility of miRNA future biomarker studies used in metal toxicology.

Perhaps the biggest challenge to the successful identification of circulating biomarkers lies in the technical constraints on isolating and purifying the samples for analysis. Methods used by different groups to isolate miRNAs from blood vary greatly, which may contribute to reproducibility issues (Mariner et al. 2018). For instance, previous studies have shown that the differential expression analysis of circulating miRNA from plasma can be confounded by the presence of platelets, which require additional steps of centrifugation in order to accurately quantify differential circulating miRNA expression profiles (Cheng et al. 2013). However, one protocol to address this issue has already been developed (Mariner et al. 2018). Additionally, the type of RNA extraction protocol can vary from lab to lab. Traditional approaches to extracting RNA use phenol:chloroform steps to separate RNAs from proteins, then precipitation of RNA with alcohol and pelleting by centrifugation (Mariner et al. 2018). But the size and quantity of circulating miRNAs may make them difficult to pellet, and this leads to greater sample variability. Furthermore, commercialized kits that use column-based purification techniques such as RT-qPCR, next-generation sequencing, or microarray approaches may inhibit the most common downstream miRNA quantification as a result of the incomplete removal of denaturants.

Other challenges—for instance the normalization of data to a housekeeping circulating miRNA or other genes, low concentration of circulating miRNAs, and identification of acceptable ranges for certain circulating miRNAs in profiles of normal individuals—can contribute to data variability or impede meaningful interpretations (Cui et al. 2019). One study has addressed the effect of different normalization strategies to quantify the circulating miRNAs in human plasma and has found that miR-320d may be the most reliable endogenous circulating miRNA to use for normalization (Faraldi et al. 2019). Alternatively, others have used exogenous, synthetic miRNA mimic from *C. elegans* as normalization controls (Farina et al. 2014). Although according to some studies storage conditions do not seem to play a role in affecting the stability of circulating miRNAs, multiple freeze thaw cycles should be avoided in order to reduce the degradation of limited circulating miRNA species, and samples should kept at -80 C for long-term storage (Farina et al. 2014; Tiberio et al. 2015). Lastly, no studies to date have addressed the matter of a normal physiological range for circulating miRNAs. However, on the basis of individual variability in circulating miRNAs, these studies are needed before the use of circulating miRNAs as biomarkers in a clinical setting.

Acknowledgement

The authors were supported by the National Institute Of Environmental Health Sciences of the National Institutes of Health under Award Numbers R01ES027778, R21ES030334, P30ES030283, and T32ES011564. The content is solely the responsibility of the authors and does not necessarily represent the official views of the National Institutes of Health.

References

Abadin, H., Ashizawa, A., Stevens, Y.W., Llados, F., Diamond, G., Sage, G., Citra, M., Quinones, A., Bosch, S.J., and Swarts, S.G. (2007). *Toxicological Profile for Lead*. Atlanta, GA: Agency for Toxic Substances and Disease Registry.

Aleckovic, M. and Kang, Y. (2015). Regulation of cancer metastasis by cell-free miRNAs. *Biochim. Biophys. Acta* 1855: 24–42.

Alli, L.A. (2015). Blood level of cadmium and lead in occupationally exposed persons in Gwagwalada, Abuja, Nigeria. *Interdiscip. Toxicol.* 8: 146–150.

Amini, P., Ettlin, J., Opitz, L., Clementi, E., Malbon, A., and Markkanen, E. (2017). An optimised protocol for isolation of RNA from small sections of laser-capture microdissected FFPE tissue amenable for next-generation sequencing. *BMC Mol. Biol.* 18: 22.

Amrani, I., Haddam, N., Garat, A., Allorge, D., Zerimech, F., Schraen, S., Taleb, A., Merzouk, H., Edme, J.L., and Lo-Guidice, J.M. (2020). Exposure to metal fumes and circulating miRNAs in Algerian welders. *Int. Arch. Occup. Environ. Health* 93: 553–561.

Armstrong, C.W., Stroube, R.B., Rubio, T., Siudyla, E.A., and Miller, G.B. Jr. (1984). Outbreak of fatal arsenic poisoning caused by contaminated drinking water. *Arch. Environ. Health* 39: 276–279.

Aryani, A. and Denecke, B. (2015). In vitro application of ribonucleases: Comparison of the effects on mRNA and miRNA stability. *BMC Res. Notes* 8: 164.

ATSDR. 2019. *ATSDR's Substance Priority List* [Online]. ATSDR. https://www.atsdr.cdc.gov/spl/index.html (accessed June 21, 2021).

Ayotte, J.D., Medalie, L., Qi, S.L., Backer, L.C., and Nolan, B.T. (2017). Estimating the high-arsenic domestic-well population in the conterminous United States. *Environ. Sci. Technol.* 51: 12443–12454.

Balali-Mood, M., Naseri, K., Tahergorabi, Z., Khazdair, M.R., and Sadeghi, M. (2021). Toxic mechanisms of five heavy metals: Mercury, lead, chromium, cadmium, and arsenic. *Front Pharmacol.* 12: 643972.

Banerjee, N., Bandyopadhyay, A.K., Dutta, S., Das, J.K., Roy Chowdhury, T., Bandyopadhyay, A., and Giri, A.K. (2017). Increased microRNA 21 expression contributes to arsenic induced skin lesions, skin cancers and respiratory distress in chronically exposed individuals. *Toxicology* 378: 10–16.

Banerjee, N., Das, S., Tripathy, S., Bandyopadhyay, A.K., Sarma, N., Bandyopadhyay, A., and Giri, A.K. (2019). MicroRNAs play an important role in contributing to arsenic susceptibility in the chronically exposed individuals of West Bengal, India. *Environ. Sci. Pollut. Res. Int.* 26: 28052–28061.

Barcelo, M., Castells, M., Bassas, L., Vigues, F., and Larriba, S. (2019). Semen miRNAs contained in exosomes as non-invasive biomarkers for prostate cancer diagnosis. *Sci. Rep.* 9: 13772.

Beck, R., Bommarito, P., Douillet, C., Kanke, M., Del Razo, L.M., Garcia-Vargas, G., Fry, R.C., Sethupathy, P., and Styblo, M. (2018). Circulating miRNAs associated with arsenic exposure. *Environ. Sci. Technol.* 52: 14487–14495.

Bernhoft, R.A. (2013). Cadmium toxicity and treatment. *Sci. World J.* 7: 394652.

Bollati, V., Marinelli, B., Apostoli, P., Bonzini, M., Nordio, F., Hoxha, M., Pegoraro, V., Motta, V., Tarantini, L., Cantone, L., Schwartz, J., Bertazzi, P.A., and Baccarelli, A. (2010). Exposure to metal-rich particulate matter modifies the expression of candidate microRNAs in peripheral blood leukocytes. *Environ. Health Perspect.* 118: 763–768.

Bonneau, E., Neveu, B., Kostantin, E., Tsongalis, G.J., and Guire, D.E. (2019). How close are miRNAs from clinical practice? A perspective on the diagnostic and therapeutic market. *EJIFCC* 30: 114–127.

Califf, R.M. (2018). Biomarker definitions and their applications. *Exp. Biol. Med. (Maywood)* 243: 213–221.

Cardenas-Gonzalez, M., Osorio-Yanez, C., Gaspar-Ramirez, O., Pavkovic, M., Ochoa-Martinez, A., Lopez-Ventura, D., Medeiros, M., Barbier, O.C., Perez-Maldonado, I.N., Sabbisetti, V.S., Bonventre, J.V., and Vaidya, V.S. (2016). Environmental exposure to arsenic and chromium in children is associated with kidney injury molecule-1. *Environ. Res.* 150: 653–662.

CDC. (2012). Response to Advisory Commitee on Childhood Lead Poisoning Prevention recommendations in Low Level Lead Exposure Harms Children: A Renewed Call for Primary Prevention. https://www.cdc.gov/nceh/lead/acclpp/cdc_response_lead_exposure_recs.pdf.

Chen, J., Lai, W., Deng, Y., Liu, M., Dong, M., Liu, Z., Wang, T., Li, X., Zhao, Z., Yin, X., Yang, J., Yu, R., and Liu, L. (2021). MicroRNA-363-3p promotes apoptosis in response to cadmium-induced renal injury by down-regulating phosphoinositide 3-kinase expression. *Toxicol. Lett.* 345: 12–23.

Cheng, H., Hu, P., Wen, W., and Liu, L. (2018). Relative miRNA and mRNA expression involved in arsenic methylation. *PLoS One* 13: e0209014.

Cheng, H.H., Yi, H.S., Kim, Y., Kroh, E.M., Chien, J.W., Eaton, K.D., Goodman, M.T., Tait, J.F., Tewari, M., and Pritchard, C.C. (2013). Plasma processing conditions substantially influence circulating microRNA biomarker levels. *PLoS One* 8: e64795.

Clarkson, T.W. and Magos, L. (2006). The toxicology of mercury and its chemical compounds. *Crit. Rev. Toxicol.* 36: 609–662.

Cohen, S.M., Arnold, L.L., Beck, B.D., Lewis, A.S., and Eldan, M. (2013). Evaluation of the carcinogenicity of inorganic arsenic. *Crit. Rev. Toxicol.* 43: 711–752.

Collares, C.V., Evangelista, A.F., Xavier, D.J., Rassi, D.M., Arns, T., Foss-Freitas, M.C., Foss, M.C., Puthier, D., Sakamoto-Hojo, E.T., Passos, G.A., and Donadi, E.A. (2013). Identifying common and specific microRNAs expressed in peripheral blood mononuclear cell of type 1, type 2, and gestational diabetes mellitus patients. *BMC Res. Notes* 6: 491.

Correia, C.N., Nalpas, N.C., Mcloughlin, K.E., Browne, J.A., Gordon, S.V., Machugh, D.E., and Shaughnessy, R.G. (2017). Circulating microRNAs as potential biomarkers of infectious disease. *Front Immunol.* 8: 118.

Cory-Slechta, D.A. (2005). Studying toxicants as single chemicals: Does this strategy adequately identify neurotoxic risk? *Neurotoxicology* 26: 491–510.

Cubadda, F., Jackson, B.P., Cottingham, K.L., Van Horne, Y.O., and Kurzius-Spencer, M. (2017). Human exposure to dietary inorganic arsenic and other arsenic species: State of knowledge, gaps and uncertainties. *Sci. Total Environ.* 579: 1228–1239.

Cui, M., Wang, H., Yao, X., Zhang, D., Xie, Y., Cui, R., and Zhang, X. (2019). Circulating microRNAs in cancer: Potential and challenge. *Front Genet.* 10: 626.

de Araujo, M.L., Gomes, B.C., Devoz, P.P., Duarte, N.A.A., Ribeiro, D.L., De Araujo, A.L., Batista, B.L., Antunes, L.M.G., Barbosa, F., JR., Rodrigues, A.S., Rueff, J., and Barcelos, G.R.M. (2021). Association between miR-148a and DNA methylation profile in individuals exposed to lead (Pb). *Front Genet.* 12: 620744.

De Guire, V., Robitaille, R., Tetreault, N., Guerin, R., Menard, C., Bambace, N., and Sapieha, P. (2013). Circulating miRNAs as sensitive and specific biomarkers for the diagnosis and monitoring of human diseases: Promises and challenges. *Clin. Biochem.* 46: 846–860.

Deng, Q., Dai, X., Feng, W., Huang, S., Yuan, Y., Xiao, Y., Zhang, Z., Deng, N., Deng, H., Zhang, X., Kuang, D., Li, X., Zhang, W., Zhang, X., Guo, H., and Wu, T. (2019). Co-exposure to metals and polycyclic aromatic hydrocarbons, microRNA expression, and early health damage in coke oven workers. *Environ. Int.* 122: 369–380.

Ding, E., Guo, J., Bai, Y., Zhang, H., Liu, X., Cai, W., Zhong, L., and Zhu, B. (2017). MiR-92a and miR-486 are potential diagnostic biomarkers for mercury poisoning and jointly sustain NF-kappaB activity in mercury toxicity. *Sci. Rep.* 7: 15980.

Ding, E., Zhao, Q., Bai, Y., Xu, M., Pan, L., Liu, Q., Wang, B., Song, X., Wang, J., Chen, L., and Zhu, B. (2016). Plasma microRNAs expression profile in female workers occupationally exposed to mercury. *J. Thorac. Dis.* 8: 833–841.

Dioni, L., Sucato, S., Motta, V., Iodice, S., Angelici, L., Favero, C., Cavalleri, T., Vigna, L., Albetti, B., Fustinoni, S., Bertazzi, P., Pesatori, A., and Bollati, V. (2017). Urinary chromium is associated with changes in leukocyte miRNA expression in obese subjects. *Eur. J. Clin. Nutr.* 71: 142–148.

Ebert, M.S. and Sharp, P.A. (2012). Roles for microRNAs in conferring robustness to biological processes. *Cell* 149: 515–524.

Fabbri, M. (2018). MicroRNAs and miRceptors: A new mechanism of action for intercellular communication. *Philos. Trans. R. Soc. Lond., B, Biol. Sci.* 373 (1737): 20160486.

Fabian, M.R. and Sonenberg, N. (2012). The mechanics of miRNA-mediated gene silencing: A look under the hood of miRISC. *Nat. Struct. Mol. Biol.* 19: 586–593.

Faraldi, M., Gomarasca, M., Sansoni, V., Perego, S., Banfi, G., and Lombardi, G. (2019). Normalization strategies differently affect circulating miRNA profile associated with the training status. *Sci. Rep.* 9: 1584.

Farina, N.H., Wood, M.E., Perrapato, S.D., Francklyn, C.S., Stein, G.S., Stein, J.L., and Lian, J.B. (2014). Standardizing analysis of circulating microRNA: Clinical and biological relevance. *J. Cell. Biochem.* 115: 805–811.

Feng, Y.H. and Tsao, C.J. (2016). Emerging role of microRNA-21 in cancer. *Biomed. Rep.* 5: 395–402.

Ferrari, E. and Gandellini, P. (2020). Unveiling the ups and downs of miR-205 in physiology and cancer: Transcriptional and post-transcriptional mechanisms. *Cell Death Dis.* 11: 980.

Fowler, B.A., Alexander, J., and Oskarsson, A. (2015). Toxic metals in food. Ch. 6 in *Handbook on the Toxicology of Metals*, 4th edn., G.F. Nordberg, B.A. Fowler, and M. Nordberg eds., San Diego: Academic Press.

Friberg, L. (1983). Cadmium. *Annu. Rev. Public Health* 4: 367–373.

Friedman, R.C., Farh, K.K., Burge, C.B., and Bartel, D.P. (2009). Most mammalian mRNAs are conserved targets of microRNAs. *Genome Res.* 19: 92–105.

Genchi, G., Sinicropi, M.S., Lauria, G., Carocci, A., and Catalano, A. (2020). The effects of cadmium toxicity. *Int. J. Environ. Res. Public Health* 17 (11): 3782.

Gil, F. and Pla, A. (2001). Biomarkers as biological indicators of xenobiotic exposure. *J. Appl. Toxicol.* 21: 245–255.

Gonzalez, H., Lema, C., Kirken, R.A., Maldonado, R.A., Varela-Ramirez, A., and Aguilera, R.J. (2015). Arsenic-exposed keratinocytes exhibit differential microRNAs expression profile: Potential implication of miR-21, miR-200a and miR-141 in melanoma pathway. *Clin. Cancer Drugs* 2: 138–147.

Goyal, T., Mitra, P., Singh, P., Ghosh, R., Sharma, S., and Sharma, P. (2021). Association of microRNA expression with changes in immune markers in workers with cadmium exposure. *Chemosphere* 274: 129615.

Ha, M. and Kim, V.N. (2014). Regulation of microRNA biogenesis. *Nat. Rev. Mol. Cell Biol.* 15: 509–524.

Hackenmueller, S.A., Gherasim, C., Walden, J.Q., Law, C.L., and Strathmann, F.G. (2019). Unrecognized elevations of toxic elements in urine and blood highlight the potential need for a broader approach to exposure assessment. *J. Anal Toxicol.* 43: 284–290.

Hammond, S.M. (2015). An overview of microRNAs. *Adv. Drug Deliv. Rev.* 87: 3–14.

Heneghan, H.M., Miller, N., and Kerin, M.J. (2010). MiRNAs as biomarkers and therapeutic targets in cancer. *Curr. Opin. Pharmacol.* 10: 543–550.

Holmes, P., James, K.A., and Levy, L.S. (2009). Is low-level environmental mercury exposure of concern to human health? *Sci. Total Environ.* 408: 171–182.

Hughes, M.F. (2002). Arsenic toxicity and potential mechanisms of action. *Toxicol. Lett.* 133: 1–16.

Hunt, K.M., Srivastava, R.K., Elmets, C.A., and Athar, M. (2014). The mechanistic basis of arsenicosis: Pathogenesis of skin cancer. *Cancer Lett.* 354: 211–219.

IARC. (1980). Some metals and metallic compounds. *IARC Monogr. Eval. Carcinog Risk Chem Hum* 23: 1–415.

IARC. (1990). *Chromium, Nickel and Welding*. Lyon: WHO.

IARC. (2012). Arsenic, metals, fibres, and dusts. *IARC Monogr Eval Carcinog Risks Hum* 100: 11–465.

Iorio, M.V. and Croce, C.M. (2012). MicroRNA dysregulation in cancer: Diagnostics, monitoring and therapeutics. A comprehensive review. *EMBO. Mol. Med.* 4: 143–159.

Jarup, L. (2003). Hazards of heavy metal contamination. *Br. Med. Bull.* 68: 167–182.

Jarup, L. and Akesson, A. (2009). Current status of cadmium as an environmental health problem. *Toxicol. Appl. Pharmacol.* 238: 201–208.

Jia, J., Li, T., Yao, C., Chen, J., Feng, L., Jiang, Z., Shi, L., Liu, J., Chen, J., and Lou, J. (2020). Circulating differential miRNAs profiling and expression in hexavalent chromium exposed electroplating workers. *Chemosphere* 260: 127546.

Karagas, M.R., Choi, A.L., Oken, E., Horvat, M., Schoeny, R., Kamai, E., Cowell, W., Grandjean, P., and Korrick, S. (2012). Evidence on the human health effects of low-level methylmercury exposure. *Environ. Health Perspect.* 120: 799–806.

Karvinen, S., Sievanen, T., Karppinen, J.E., Hautasaari, P., Bart, G., Samoylenko, A., Vainio, S.J., Ahtiainen, J.P., Laakkonen, E.K., and Kujala, U.M. (2020). MicroRNAs in extracellular vesicles in sweat change in response to endurance exercise. *Front Physiol.* 11: 676.

Klaassen, C.D. (2013). *Toxicology the Basic Science of Poisons.* New York: Mc Graw Hill.

Kobayashi, E., Suwazono, Y., Uetani, M., Kido, T., Nishijo, M., Nakagawa, H., and Nogawa, K. (2006). Tolerable level of lifetime cadmium intake estimated as a benchmark dose low, based on excretion of beta2-microglobulin in the cadmium-polluted regions of the Kakehashi River Basin, Japan. *Bull Environ. Contam. Toxicol.* 76: 8–15.

Kong, A.P., Xiao, K., Choi, K.C., Wang, G., Chan, M.H., Ho, C.S., Chan, I., Wong, C.K., Chan, J.C., and Szeto, C.C. (2012). Associations between microRNA (miR-21, 126, 155 and 221), albuminuria and heavy metals in Hong Kong Chinese adolescents. *Clin. Chim. Acta* 413: 1053–1057.

Kotorashvili, A., Ramnauth, A., Liu, C., Lin, J., Ye, K., Kim, R., Hazan, R., Rohan, T., Fineberg, S., and Loudig, O. (2012). Effective DNA/RNA co-extraction for analysis of microRNAs, mRNAs, and genomic DNA from formalin-fixed paraffin-embedded specimens. *PLoS One* 7: e34683.

Krol, J., Loedige, I., and Filipowicz, W. (2010). The widespread regulation of microRNA biogenesis, function and decay. *Nat. Rev. Genet.* 11: 597–610.

Lawrie, C.H., Gal, S., Dunlop, H.M., Pushkaran, B., Liggins, A.P., Pulford, K., Banham, A.H., Pezzella, F., Boultwood, J., Wainscoat, J.S., Hatton, C.S., and Harris, A.L. (2008). Detection of elevated levels of tumour-associated microRNAs in serum of patients with diffuse large B-cell lymphoma. *Br. J. Haematol.* 141: 672–675.

Lei, L.J., Zhang, Z., Guo, J.Y., Shi, X.J., Zhang, G.Y., Kang, H., Gao, Y.Y., Hu, X.Q., Wang, T., and Mu, L.N. (2019). MiR-21 as a potential biomarker for renal dysfunction induced by cadmium exposure. *Int. J. Clin. Exp. Med.* 12: 1631–1639.

Li, X., Shi, Y., Wei, Y., Ma, X., Li, Y., and Li, R. (2012). Altered expression profiles of microRNAs upon arsenic exposure of human umbilical vein endothelial cells. *Environ. Toxicol. Pharmacol.* 34: 381–387.

Li, Y., Li, P., Yu, S., Zhang, J., Wang, T., and Jia, G. (2014). miR-3940-5p associated with genetic damage in workers exposed to hexavalent chromium. *Toxicol. Lett.* 229: 319–326.

Li, Y., Ye, F., Wang, A., Wang, D., Yang, B., Zheng, Q., Sun, G., and Gao, X. (2016). Chronic arsenic poisoning probably caused by arsenic-based pesticides: Findings from an investigation study of a household. *Int. J. Environ. Res. Public Health* 13 (1): 133.

Libri, V., Miesen, P., Rij, V.A.N.R.P, and Buck, A.H. (2013). Regulation of microRNA biogenesis and turnover by animals and their viruses. *Cell. Mol. Life Sci.* 70: 3525–3544.

Liu, J.Q., Niu, Q., Hu, Y.H., Li, Y., Wang, H.X., Xu, S.Z., Ding, Y.S., Li, S.G., and Ma, R.L. (2018). The bidirectional effects of arsenic on miRNA-21: A systematic review and meta-analysis. *Biomed. Environ. Sci.* 31: 654–666.

Manning, F.C., Blankenship, L.J., Wise, J.P., Xu, J., Bridgewater, L.C., and Patierno, S.R. (1994). Induction of internucleosomal DNA fragmentation by carcinogenic chromate: Relationship to DNA damage, genotoxicity, and inhibition of macromolecular synthesis. *Environ. Health Perspect.* 102 (Suppl 3): 159–167.

Mariner, P.D., Korst, A., Karimpour-Fard, A., Stauffer, B.L., Miyamoto, S.D., and Sucharov, C.C. (2018). Improved detection of circulating miRNAs in serum and plasma following rapid heat/freeze cycling. *Microrna* 7: 138–147.

Michailidi, C., Hayashi, M., Datta, S., Sen, T., Zenner, K., Oladeru, O., Brait, M., Izumchenko, E., Baras, A., Vandenbussche, C., Argos, M., Bivalacqua, T.J., Ahsan, H., Hahn, N.M., Netto, G.J., Sidransky, D., and Hoque, M.O. (2015). Involvement of epigenetics and EMT-related miRNA in arsenic-induced neoplastic transformation and their potential clinical use. *Cancer Prev. Res. (Phila)* 8: 208–221.

Michlewski, G. and Caceres, J.F. (2019). Post-transcriptional control of miRNA biogenesis. *RNA* 25: 1–16.

Mitra, P., Goyal, T., Singh, P., Sharma, S., and Sharma, P. (2021). Assessment of circulating miR-20b, miR-221, and miR-155 in occupationally lead-exposed workers of North-Western India. *Environ. Sci. Pollut. Res. Int.* 28: 3172–3181.

Mori, M.A., Ludwig, R.G., Garcia-Martin, R., Brandao, B.B., and Kahn, C.R. (2019). Extracellular miRNAs: From biomarkers to mediators of physiology and disease. *Cell Metab.* 30: 656–673.

Motta, V., Angelici, L., Nordio, F., Bollati, V., Fossati, S., Frascati, F., Tinaglia, V., Bertazzi, P.A., Battaglia, C., and Baccarelli, A.A. (2013). Integrative analysis of miRNA and inflammatory gene expression after acute particulate matter exposure. *Toxicol. Sci.* 132: 307–316.

Naranmandura, H., Suzuki, N., and Suzuki, K.T. (2006). Trivalent arsenicals are bound to proteins during reductive methylation. *Chem. Res. Toxicol.* 19: 1010–1018.

Nath, K., Singh, D., Shyam, S., and Sharma, Y.K. (2009). Phytotoxic effects of chromium and tannery effluent on growth and metabolism of *Phaseolus mungo* Roxb. *J. Environ. Biol.* 30: 227–234.

Newman-Taylor, A. (1998). Cadmium. In *Environmental and Occupational Medicine*, W.N. Rom ed., Philadelphia, PA: Lippincott-Raven.

Noonan, C.W., Sarasua, S.M., Campagna, D., Kathman, S.J., Lybarger, J.A., and Mueller, P.W. (2002). Effects of exposure to low levels of environmental cadmium on renal biomarkers. *Environ. Health Perspect.* 110: 151–155.

O'Brien, T.J., Ceryak, S., and Patierno, S.R. (2003). Complexities of chromium carcinogenesis: Role of cellular response, repair and recovery mechanisms. *Mutat. Res.* 533: 3–36.

Ochoa-Martinez, A.C., Araiza-Gamboa, Y., Varela-Silva, J.A., Orta-Garcia, S.T., Carrizales-Yanez, L., and Perez-Maldonado, I.N. (2021). Effect of gene-environment interaction (arsenic exposure: PON1 Q192R polymorphism) on cardiovascular disease biomarkers in Mexican population. *Environ. Toxicol. Pharmacol.* 81: 103519.

Odame, E., Chen, Y., Zheng, S., Dai, D., Kyei, B., Zhan, S., Cao, J., Guo, J., Zhong, T., Wang, L., Li, L., and Zhang, H. (2021). Enhancer RNAs: Transcriptional regulators and workmates of NamiRNAs in myogenesis. *Cell. Mol. Biol. Lett.* 26: 4.

Olsson, I.M., Bensryd, I., Lundh, T., Ottosson, H., Skerfving, S., and Oskarsson, A. (2002). Cadmium in blood and urine: Impact of sex, age, dietary intake, iron status, and former smoking: Association of renal effects. *Environ. Health Perspect.* 110: 1185–1190.

Pechova, A. and Pavlata, L. (2007). Chromium as an essential nutrient: A review. *Vet Med (Praha)* 52: 1–18.

Perez-Vazquez, M.S., Ochoa-Martinez, A.C., Ruiz-Vera, T., Araiza-Gamboa, Y., and Perez-Maldonado, I.N. (2017). Evaluation of epigenetic alterations (miR-126 and miR-155 expression levels) in Mexican children exposed to inorganic arsenic via drinking water. *Environ. Sci. Pollut. Res. Int.* 24: 28036–28045.

Podgorski, J. and Berg, M. (2020). Global threat of arsenic in groundwater. *Science* 368: 845–850.

Proctor, D.M., Suh, M., Campleman, S.L., and Thompson, C.M. (2014). Assessment of the mode of action for hexavalent chromium-induced lung cancer following inhalation exposures. *Toxicology* 325: 160–179.

Rager, J.E., Bailey, K.A., Smeester, L., Miller, S.K., Parker, J.S., Laine, J.E., Drobna, Z., Currier, J., Douillet, C., Olshan, A.F., Rubio-Andrade, M., Styblo, M., Garcia-Vargas, G., and Fry, R.C. (2014). Prenatal arsenic exposure and the epigenome: Altered microRNAs associated with innate and adaptive immune signaling in newborn cord blood. *Environ. Mol. Mutagen.* 55: 196–208.

Rahman, M.M., Chowdhury, U.K., Mukherjee, S.C., Mondal, B.K., Paul, K., Lodh, D., Biswas, B.K., Chanda, C.R., Basu, G.K., Saha, K.C., Roy, S., Das, R., Palit, S.K., Quamruzzaman, Q., and Chakraborti, D. (2001). Chronic arsenic toxicity in Bangladesh and West Bengal, India: A review and commentary. *J. Toxicol. Clin. Toxicol.* 39: 683–700.

Rajkumar, V. and Gupta, V. (2021). *Heavy Metal Toxicity*. Treasure Island, FL: StatPearls Publishing LLP. https://www.ncbi.nlm.nih.gov/books/NBK560920

Ratnaike, R.N. (2003). Acute and chronic arsenic toxicity. *Postgrad. Med. J.* 79: 391–396.

Rehman, K., Fatima, F., Waheed, I., and Akash, M.S.H. (2018). Prevalence of exposure of heavy metals and their impact on health consequences. *J. Cell. Biochem.* 119: 157–184.

Risher, J.F., Nickle, R.A., and Amler, S.N. (2003). Elemental mercury poisoning in occupational and residential settings. *Int. J. Hyg. Environ. Health* 206: 371–379.

Rubio, M., Bustamante, M., Hernandez-Ferrer, C., Fernandez-Orth, D., Pantano, L., Sarria, Y., Pique-Borras, M., Vellve, K., Agramunt, S., Carreras, R., Estivill, X., Gonzalez, J.R., and Mayor, A. (2018). Circulating miRNAs, isomiRs and small RNA clusters in human plasma and breast milk. *PLoS One* 13: e0193527.

Ruiz-Vera, T., Ochoa-Martinez, A.C., Zarazua, S., Carrizales-Yanez, L., and Perez-Maldonado, I.N. (2019). Circulating miRNA-126, -145 and -155 levels in Mexican women exposed to inorganic arsenic via drinking water. *Environ. Toxicol. Pharmacol.* 67: 79–86.

Saliminejad, K., Khorshid, K.H.O.R.R.A.M., Soleymani Fard, H.R., and Ghaffari, S.H. (2019). An overview of microRNAs: Biology, functions, therapeutics, and analysis methods. *J. Cell. Physiol.* 234: 5451–5465.

Salnikow, K. and Zhitkovich, A. (2008). Genetic and epigenetic mechanisms in metal carcinogenesis and cocarcinogenesis: Nickel, arsenic, and chromium. *Chem. Res. Toxicol.* 21: 28–44.

Setti, G., Pezzi, M.E., Viani, M.V., Pertinhez, T.A., Cassi, D., Magnoni, C., Bellini, P., Musolino, A., Vescovi, P., and Meleti, M. (2020). Salivary microRNA for diagnosis of cancer and systemic diseases: A systematic review. *Int. J. Mol. Sci.* 21 (3): 907.

Sperling, M. (2005). Chromium. In *Encyclopedia of Analytical Science*, 2nd edn., P. Worsfold, A. Townshend, and C. Poole, eds., Oxford: Elsevier.

States, J.C. (2015). *Arsenic: Exposure Sources, Health Risks, and Mechanisms of Toxicity*. Hoboken, NJ: John Wiley & Sons.

Straif, K., Benbrahim-Tallaa, L., Baan, R., Grosse, Y., Secretan, B., El Ghissassi, F., Bouvard, V., Guha, N., Freeman, C., Galichet, L., Cogliano, V., and the WHO International Agency for Research on Cancer Monograph Working Group. (2009). A review of human carcinogens. Part C: Metals, arsenic, dusts, and fibres. *Lancet Oncol.* 10: 453–454.

Strimbu, K. and Tavel, J.A. (2010). What are biomarkers? *Curr. Opin. HIV AIDS* 5: 463–466.

Sudha, S., Kripa, S.K., Shibily, P., and Shyn, J. (2011). Elevated frequencies of micronuclei and other nuclear abnormalities of chrome plating workers occupationally exposed to hexavalent chromium. *Iran J. Cancer Prev.* 4: 119–124.

Sun, B., Xue, J., Li, J., Luo, F., Chen, X., Liu, Y., Wang, Q., Qi, C., Zou, Z., Zhang, A., and Liu, Q. (2017). Circulating miRNAs and their target genes associated with arsenism caused by coal-burning. *Toxicol. Res. (Camb)* 6: 162–172.

Tam, L.M., Price, N.E., and Wang, Y. (2020). Molecular mechanisms of arsenic-induced disruption of DNA repair. *Chem. Res. Toxicol.* 33: 709–726.

Tchounwou, P.B., Ayensu, W.K., Ninashvili, N., and Sutton, D. (2003). Environmental exposure to mercury and its toxicopathologic implications for public health. *Environ. Toxicol.* 18: 149–175.

Tchounwou, P.B., Yedjou, C.G., Patlolla, A.K., and Sutton, D.J. (2012). Heavy metal toxicity and the environment. *Exp. Suppl.* 101: 133–164.

Thomas, D.J. (2015). The chemistry and metabolism of arsenic. In *Arsenic: Exposure Sources, Health Risks and Mechanisms of Toxicity*, J.C. States, ed., New York: Wiley.

Tiberio, P., Callari, M., Angeloni, V., Daidone, M.G., and Appierto, V. (2015). Challenges in using circulating miRNAs as cancer biomarkers. *Biomed. Res. Int.* 2015: 731479.

Tokar, E.J., Boyd, W.A., Freedman, J.H., and Waalkes, M.P. (2013). Toxic effects of metals. In *Casarett and Doull's Toxicology: The Basic Science of Poisons*, (8th edn.), edited by C. D. Klaassen. New York: McGraw Hill.

Tsuji, J.S., Chang, E.T., Gentry, P.R., Clewell, H.J., Boffetta, P., and Cohen, S.M. (2019). Dose-response for assessing the cancer risk of inorganic arsenic in drinking water: The scientific basis for use of a threshold approach. *Crit. Rev. Toxicol.* 49: 36–84.

USEPA. (1998). *Toxicological Review of Hexavalent Chromium: Cas. No. 18540-29-9.* Washington, DC: US Environmental Protection Agency.

Vasudevan, S., Tong, Y., and Steitz, J.A. (2007). Switching from repression to activation: MicroRNAs can up-regulate translation. *Science* 318: 1931–1934.

Wah Chu, K. and Chow, K.L. (2002). Synergistic toxicity of multiple heavy metals is revealed by a biological assay using a nematode and its transgenic derivative. *Aquat. Toxicol.* 61: 53–64.

Wang, L., Qiu, J.G., He, J., Liu, W.J., Ge, X., Zhou, F.M., Huang, Y.X., Jiang, B.H., and Liu, L.Z. (2019). Suppression of miR-143 contributes to overexpression of IL-6, HIF-1alpha and NF-kappaB p65 in Cr(VI)-induced human exposure and tumor growth. *Toxicol. Appl. Pharmacol.* 378: 114603.

Wang, Y., Su, H., Gu, Y., Song, X., and Zhao, J. (2017). Carcinogenicity of chromium and chemoprevention: A brief update. *Onco. Targets Ther.* 10: 4065–4079.

Weber, J.A., Baxter, D.H., Zhang, S., Huang, D.Y., Huang, K.H., Lee, M.J., Galas, D.J., and Wang, K. (2010). The microRNA spectrum in 12 body fluids. *Clin. Chem.* 56: 1733–1741.

WHO. (1993). Evaluation of certain food additives and contaminants: Forty-first report of the Joint FAO/WHO Expert Committee on Food Additives. *World Health Organ. Tech. Rep. Ser.* 837: 1–53.

WHO. (2018). *Arsenic* [Online]. https://www.who.int/news-room/fact-sheets/detail/arsenic (accessed May 30 2021).

Wimmer, I., Troscher, A.R., Brunner, F., Rubino, S.J., Bien, C.G., Weiner, H.L., Lassmann, H., and Bauer, J. (2018). Systematic evaluation of RNA quality, microarray data reliability and pathway analysis in fresh, fresh frozen and formalin-fixed paraffin-embedded tissue samples. *Sci. Rep.* 8: 6351.

Wise, J.P., Sr., Wise, S.S., and Little, J.E. (2002). The cytotoxicity and genotoxicity of particulate and soluble hexavalent chromium in human lung cells. *Mutat. Res.* 517: 221–229.

Xu, M., Yu, Z., Hu, F., Zhang, H., Zhong, L., Han, L., An, Y., Zhu, B., and Zhang, H. (2017). Identification of differential plasma miRNA profiles in Chinese workers with occupational lead exposure. *Biosci. Rep.* 37.

Xu, Y., Zou, Z., Liu, Y., Wang, Q., Sun, B., Zeng, Q., Liu, Q., and Zhang, A. (2020). miR-191 is involved in renal dysfunction in arsenic-exposed populations by regulating inflammatory response caused by arsenic from burning arsenic-contaminated coal. *Hum Exp Toxicol* 39: 37–46.

Yang, L., Zhang, Y., Wang, F., Luo, Z., Guo, S., and Strahle, U. (2020). Toxicity of mercury: Molecular evidence. *Chemosphere* 245: 125586.

Young, J.L., Yan, X., Xu, J., Yin, X., Zhang, X., Arteel, G.E., Barnes, G.N., States, J.C., Watson, W.H., Kong, M., Cai, L., and Freedman, J.H. (2019). Cadmium and high-fat diet disrupt renal, cardiac and hepatic essential metals. *Sci. Rep.* 9: 14675.

Yuan, W., Liu, L., Liang, L., Huang, K., Deng, Y., Dong, M., Chen, J., Wang, G., and Zou, F. (2020). MiR-122-5p and miR-326-3p: Potential novel biomarkers for early detection of cadmium exposure. *Gene* 724: 144156.

Zampetaki, A., Albrecht, A., and Steinhofel, K. (2018). Long non-coding RNA structure and function: Is there a link? *Front Physiol.* 9: 1201.

Zeng, Q., Zou, Z., Wang, Q., Sun, B., Liu, Y., Liang, B., Liu, Q., and Zhang, A. (2019). Association and risk of five miRNAs with arsenic-induced multiorgan damage. *Sci. Total Environ.* 680: 1–9.

Zenobia, C. and Hajishengallis, G. (2015). Basic biology and role of interleukin-17 in immunity and inflammation. *Periodontol 2000*, 69: 142–159.

Zhang, L., Gao, Y., Wu, S., Zhang, S., Smith, K.R., Yao, X., and Gao, H. (2020). Global impact of atmospheric arsenic on health risk: 2005 to 2015. *Proc. Natl. Acad. Sci. USA* 117: 13975–13982.

Zheng, L., Jiang, Y.L., Fei, J., Cao, P., Zhang, C., Xie, G.F., Wang, L.X., Cao, W., Fu, L., and Zhao, H. (2021). Circulatory cadmium positively correlates with epithelial-mesenchymal transition in patients with chronic obstructive pulmonary disease. *Ecotoxicol. Environ. Saf.* 215: 112164.

Zhou, Q. and Xi, S. (2018). A review on arsenic carcinogenesis: Epidemiology, metabolism, genotoxicity and epigenetic changes. *Regul. Toxicol. Pharmacol.* 99: 78–88.

5

MicroRNA Biomarkers of Malignant Mesothelioma

Lijin Zhu, Fangfang Zhang, Min Zhang, Hailing Xia, Xiuyuan Yuan, and Yanan Gao

Hangzhou Medical College

Pleural malignant mesothelioma (MM), which arises from the cells that line the lung and the chest cavity (pleura), is a highly aggressive tumor with a high recurrence rate after surgical resection, and is insensitive to chemotherapy and radiotherapy. The median survival time commonly does not exceed 12–18 months after diagnosis (Wright et al. 2013). Therefore biomarkers for early detection are imperative even for experienced pathologists (Wu et al. 2013; Zhang et al. 2014). Approximately 80% of the cases of pleural MM are attributed to asbestos exposure, and the latency after exposure could be 20–60 years (Rascoe et al. 2012). Asbestos exposure, a genetic basis, and other factors are likely to contribute to the etiology of pleural MM (Robinson and Lake 2005).

MicroRNA (miRNA) is a kind of highly conserved, non-coding, single-stranded small RNA with a length of 18–25 nucleotides (Kirschner et al. 2011). About 35,828 mature miRNAs have been found in human genome, which can be divided into 223 species; these miRNAs regulate one third of human genes. The same miRNA can regulate single or multiple target genes and, in turn, the same gene can be regulated by multiple miRNAs (Luo et al. 2010) . Here mature miRNAs bind to untranslated sequences at the 3′-end of the target mRNA through base complementary pairing, to inhibit mRNA translation or degradation, thereby regulating gene expression and playing a role in promoting or inhibiting cancer.

During the two decades since Calin et al. (2002) first discovered two miRNAs (miR-15 and miR-16) closely related to the occurrence of chronic lymphoblastic leukemia, miRNAs have been proved to play a key role in cancer progression, treatment response, and diagnosis (Berindan-Neagoe et al. 2014; Hata and Lieberman 2015). The tumor-promoting or tumor-suppressing effects of miRNAs in various cancers depend on their expression levels. More and more researchers have applied miRNAs to the diagnosis and treatment of malignant tumors—for example the miR-200 family, miR-9, miR-34, miR-21, and miR-340 in the prognosis of pancreatic cancer (Zöller 2013) and miR-140 and miR-145 in the diagnosis and treatment of ovarian cancer (Banno et al., 2014).

The earliest study on miRNA in relation to MM began with Guled's research in 2009 (Guled et al. 2009). In this study, the miRNA expression profiles of seventeen freshly frozen MM tissue samples and normal pericardium were analyzed using miRNA microarray, and multiple differentially expressed miRNAs between MM tissues and adjacent tissues were found. Moreover, different tissue subtypes of MM expressed specific miRNAs: for example, epithelial MM expressed miR-135b, miR-181a-2*, miR-499-5p, miR-517b, miR-519d, miR-615-5p,

Genomic and Epigenomic Biomarkers of Toxicology and Disease: Clinical and Therapeutic Actions,
First Edition. Edited by Saura C. Sahu.

and miR-624, biphasic MM expressed miR-218-2*, miR-346, miR-377*, miR-485-5p, and miR-525-3p, and sarcomatous MM expressed miR-301b, miR-433, and miR-543. In this study, patients' exposure to smoking and asbestos were also considered, and some miRNAs were specifically expressed in smoking patients (miR-379, miR-301a, miR-299-3p, miR-455-3p, and miR-127-3p); No miRNAs specifically expressed in asbestos exposure samples were found, but this absence may be related to different methods of asbestos exposure assessment. Other study found that most of these miRNAs were located in abnormal chromosomal positions of MM patients and that their target genes, such as CDKN2A, NF2, JUN, HGF, and PDGFA, were most likely to affect the occurrence and development of MM. This pioneering research has opened up a new kind of approach in the field. Since its publication in 2009, more studies have discussed the role of miRNA in MM in detail (Truini et al. 2014).

Methods for Detecting the Expression of miRNAs

As a biomarker of human diseases, ideally the detection method of miRNA should not need expensive reagents and instruments and be easy to operate. Such a method should have good specificity for distinguishing miRNAs with similar sequences. At the same time, it should have sufficient sensitivity for quantitative analysis, even for micro clinical samples, and it should be able to detect multiple samples in parallel (de Planell-Saguer and Rodicio 2011; Van Roosbroeck, Pollet, and Calin 2013).

The Gold Standard Conference believes that the method for meeting these requirements and for detecting miRNA in clinical laboratories is quantitative reverse transcription polymerase chain reaction (PCR) (qRT-PCR). miRNA microarray is more expensive than qRT-PCR and is normally used in the discovery stage of biomarkers.

However, these methods all need to extract miRNAs from tumor samples of patients, which include non-tumor matrix and inflammatory cells in addition to malignant tumor cells. If the goal is to detect the specific expression of miRNA in malignant tumor cells, it is recommended to use flow cytometry to sort liquid tumors or laser capture microscopy to cut solid tumors. In situ hybridization technology (Sempere and Korc 2013) can also be used; it can detect target miRNA in different types of cells (malignant tumor cells or microenvironment cells) that constitute tumors. This method provides additional miRNA subcellular localization information. At present, the latest method for detecting miRNA expression is next-generation sequencing. This technique is highly sensitive and specific, can be used for high-throughput analysis, and can discover new miRNAs. Next-generation sequencing produces a large amount of complex data, which need to be analyzed by a trained bioinformatician; besides, the cost of single RNA sequencing is too high, so this method is not suitable for diagnosis but can still be considered as a screening method for miRNAs of interest (Berindan-Neagoe et al. 2014).

Early Screening for Malignant Mesothelioma

In view of the high degree of malignancy and difficult treatment of MM, early screening is particularly important in the prevention and treatment of the disease. At present, there have been studies designed to improve the detection rate of early MM patients by identifying the expression level of specific miRNAs in serum or plasma (Mujoomdar et al. 2010).

Santarelli et al. (2011) found that miR-126 is downregulated in the tissues of MM patients. They noticed that one of the target genes of miR-126 is vascular endothelial growth factor (VEGF), which encodes a protein that stimulates angiogenesis and increases the VEGF content in the serum of MM patients. In addition, miR-126 expression is downregulated and VEGF expression is upregulated in lung cancer patients, which suggests that miR-126 plays a role in tumor inhibition (Liu et al. 2009). Moreover, there is a correlation between miR-126 and the serum MM-specific marker soluble mesothelin-related protein (SMRP); and the level of SMRP is related to the high risk of MM development. These kinds of evidence indicate that the correlation between miR-126 and SMRP can be used as an early detection marker of MM (Santarelli et al. 2015). Further research on serum samples from forty-five untreated MM and twenty non-small cell lung cancer (NSCLC) patients showed that the circulating miR-126 level in MM patients decreased and could be significantly different from the level in non-small cell lung cancer and in healthy control group samples. Although miR-126 can be considered as a marker of MM, it lacks tumor specificity owing to its low expression in various tumors, which indicates that miR-126 will be used only to bind to other MM-specific biomarkers such as mesothelin (Tomasetti et al. 2012).

Weber et al. (2014) verified the expression level of miR-103 in the plasma of MM patients and of an asbestos-exposed control population, with sensitivity and specificity of 85% and 63% respectively. However, when combined with mesothelitin, sensitivity reached 95% and specificity reached 81%.

Therefore miRNA can be used as a biomarker in the early detection of MM, especially in combination with other specific protein markers. It can also be used as a serum-monitoring indicator for asbestos-exposed population.

Diagnosis of Malignant Mesothelioma

Studies have shown that the expression of some miRNAs in cancer tissues or body fluids of patients with MM is dysregulated and affects the expression of target genes, which may play an important role in the occurrence and development of malignant mesothelioma. In addition, some differentially expressed miRNAs have a certain specificity, which can distinguish MM from lung adenocarcinoma and other tumors. In consequence, these miRNAs can be used as biomarkers of MM for clinical diagnosis and differential diagnosis (Kirschner et al. 2012).

Differential Expression of miRNAs in Cancer Tissues

At present, the diagnosis of MM is mainly based on histopathology, but it is difficult to distinguish some cases from metastatic carcinoma only by pathology. Some scholars have found that the expression of miR-145 (Casarsa et al. 2011; Cioce et al. 2014; Schramm et al. 2010) and miR-16 (Reid et al. 2013) in MM tissues is significantly downregulated; and in vitro experiments have also found that the expression of these two miRNAs in MM cell lines is significantly downregulated. Further study found that miR-16 can, specifically, bind to the 3 UTR of BCL-2 and CCND-1, induce apoptosis of MM cells, and inhibit cell proliferation and clone formation by inhibiting the expression of its target genes BCL-2 and CCND-1. This indicates that miR-16 may have the effect of inhibiting the growth of MM.

Significant downregulation of miR-126, miR-143, miR-145, and miR-652 can be used to distinguish malignant pleural mesothelioma (MPM) from reactive mesothelial hyperplasia. Logistic regression analysis has shown that it had high sensitivity and specificity (0.96 area under the curve) to distinguish MPM from non-tumor mesothelium, and the overall accuracy reached 94% (Andersen et al. 2014).

This being the case, the expression of miRNAs may become a new clinical diagnostic method, to be used as a supplement to the current immunohistochemical methods, and may improve the diagnosis rate.

Differential Expression of MicroRNAs in Serum and Plasma

A large number of studies have shown that free miRNAs in serum may be an important marker for cancer diagnosis (Santarelli et al. 2011). Tomasetti et al. (2012) found that miR-126 in the serum of patients with MM was significantly downregulated, and patients with MM could be distinguished from the control population according to the expression level of miR-126. Studies by Kirschner et al. (2012) show that serum miR-625-3p in patients with MM is significantly increased. Receiver operating curve (ROC) analysis shows that the accuracy rate of distinguishing patients with MM from controls by serum miR-625-3p expression level is 82.4%; sensitivity and specificity are 73.33% and 78.58% respectively. When the digital methylation-specific PCR (MSP) method was used to detect quantitatively the methylation level of miR-34b/c in circulating serum from MPM patients, from benign asbestos pleurisy patients, and from healthy people, it was found that the methylation level of miR-34b/c in the serum of MPM patients was significantly increased. Through ROC analysis it was found that the accuracy of detecting the methylation of miR-34b/c in the serum to diagnose MPM was 77%, and the sensitivity and specificity were 67% and 77% (Muraoka et al. 2013) .

The expression level of circulating miR-132-3p in the plasma of MM patients and asbestos-exposed people is different. Its sensitivity and specificity for MM diagnosis are 86% and 61%, and it is not affected by age and smoking. The sensitivity and specificity of miR-126 combined with miR-132-3p in the diagnosis of malignant mesothelioma were 86% and 77%, respectively (Weber et al. 2017). Plasmatic extracellular vesicles-associated miR-103a-3p and miR-30e-3p are able to discriminate MPM subjects from past asbestos exposure (PAE) subjects (Cavalleri et al. 2017).

Serum or plasma miRNA is a rapid, convenient, and relatively non-invasive marker. If multiple miRNAs can be used in combination for diagnosis, the sensitivity and specificity may be further improved, which will be helpful for early diagnosis of MM.

Differential Expression of miRNAs in Pleural Effusion

Cytological miRNA examination of pleural effusion is a very challenging task: to distinguish MM from reactive mesothelial cells (RMCS), even if combined with in situ hybridization staining or immunofluorescence, is quite difficult. Cappellesso et al. (2016) screened and verified the differential expression of miRNA-19a (miR-19a, miR-19b), miR-21, and miR-126 in MM cell lines and mesothelioma patient tissue samples. The diagnostic sensitivity and specificity are both over 80%. The combined use of miR-21 and miR-126 can make the diagnostic sensitivity and specificity reach 86% and 87% respectively. In pleural effusion cells, combining miR-143, miR-210, and miR-200c could differentiate MPM with an area under the curve (AUC) of 0.92 (Birnie et al. 2019).

MicroRNAs Associated with Malignant Mesothelioma Tissue Subtypes

Different histopathological subtypes have great influence on the curative effect and prognosis of MM, so correctly distinguishing histopathological subtypes is also an important part of the diagnosis of MM. Researchers have studied the difference of miRNAs expression in different tissue subtypes of MM, hoping to apply it to the tissue classification of MM.

Busacca et al. (2010) analyzed miRNAs expression in two MM cell lines (MPP-89 and REN) and formalin-immobilized tumor samples using microarray and RT-qPCR methods. They found that there were seven kinds of miRNAs (miR-17-5p, miR-21, miR-29a, miR-30c, miR-30e-5p, miR-106a, and miR-143) with significant differences in expression, and it can distinguish the three tissue subtypes of MM tumors. The specific expression levels are high expression in epithelial MM, biphasic MM, and low expression in sarcomatous MM. Guled et al. (2009) also compare and analyze that expression of miRNAs in different subtypes of MM tissue. Seven miRNAs were specifically expressed in epithelial MM (miR-135b, miR-181a-2*, miR-499-5p, miR-517b, miR-519d, miR-615-5p, and miR-624), five miRNAs were specifically expressed in biphasic MM (miR-218-2*, miR-346, miR-377*, miR-485-5p, and miR-525-3p), and three miRNAs were specifically expressed in sarcomatous MM (miR-301b, miR-433, and miR-543). Tese miRNAs were specifically expressed in tissue subtypes.

Differential Diagnosis

The diagnostic performances of miR-130a expression analysis and immunohistochemistry appear to be similar. miR-130a quantification could be used reliably as a second-level diagnostic tool to differentiate MM from lung adenocarcinoma in pleural effusion cytology, mainly in those cases with ambiguous or negative immunohistochemistry (Cappellesso et al. 2017).

Multiple miRNAs were combined to diagnose a MM and can better distinguish it from other tumors or from pleural metastatic carcinoma.

Benjamin et al. (2010) were the first to find that eleven miRNAs were differentially expressed by gene chip in MPM patients and in twelve tumor patients—namely patients with bladder cancer, breast cancer, intestinal cancer, endometrial cancer, esophageal cancer, renal cancer, liver cancer, lung cancer, ovarian cancer, pancreatic cancer, prostate cancer, and gastric cancer. Among these miRNAs, four (in the miR-193 family, which includes miR-193a-3p/5p and -193b and miR-152) are highly expressed in MPM tissues, whereas seven (in the miR-200 family, which includes miR-141, miR-200a/b/c and miR-429, miR-192, and miR-194) are highly expressed in other tumor tissues. Further validation results showed that the combination of miR-200c/193a-3p was sufficient to distinguish MPM from most tumors except renal cell carcinoma, while miR-192 could be used to distinguish MPM from renal cell carcinoma, liver cancer, and gastrointestinal cancer. Subsequent double-blind experiments showed that the combination of miR-200c/193a-3p could make the accuracy of distinguishing MPM from breast cancer, intestinal cancer, endometrial cancer, renal cancer, lung cancer, ovarian cancer, and other tumors reach 95%. In pleural effusion cells, combining miR-143, miR-210, and miR-200c could discriminate MPM from a further adenocarcinoma with an AUC of 0.9887 (Birnie et al. 2019).

Gee et al. (2010) also found that miR-141, miR-200a/b/c, and miR-429 were downregulated in MPM samples and that miR-203 and miR-205 were downregulated as well. All seven miRNA markers can effectively distinguish MPM from lung adenocarcinoma, and their sensitivity or specificity exceeds the 80% recommended by the International Mesothelioma

Interest Group. All members of the miR-200 family had AUCs above 0.9, while the remaining markers had AUCs between 0.83 and 0.89. In addition, these seven miRNA markers are only suitable for differential diagnosis, but cannot classify histological subtypes.

MicroRNAs Related to the Occurrence and Development of Mesothelioma

Because different miRNAs may have either carcinogenic or antitumor effects, researchers tried to clarify their role in the occurrence and development of MM. In vitro experiments and animal experiments have found that some miRNAs can inhibit tumor growth by regulating cell cycle, proliferation, clone formation, migration, invasion, apoptosis, and other biological processes. Some studies have also shown that miRNAs can directly inhibit the growth of tumors implanted in experimental mice.

MicroRNAs Related to Cell Cycle Regulation

Uncontrolled cell cycle is one of the key steps in tumorigenesis. Almost all tumor cells have cell cycle regulation defects, and MM is no exception. MicroRNAs can regulate the process of cell cycle by targeting key regulators that promote or inhibit cell cycle, such as cyclin-cyclin-dependent protein kinase (CDK) complex or other cell growth regulatory genes. Kubo et al. (2011) and Maki et al. (2012) found that the expression level of miR-34b/c in MM cell lines (H2052 and H28) decreased. After the level of miR-34 in cells was increased by transfection, the expression of the cell cycle-related proteins CDK4, CDK6, and CCND1 was inhibited and the proportion of G1 phase cells was significantly increased. When normal pleural cells (LP-9) were transfected with miR-34 inhibitor, the number of G1 phase cells decreased significantly and the expression levels of miR-34 target genes c-MET and Bcl-2 were significantly upregulated. This indicates that miR-34b/c plays an important role in regulating G1 cell cycle arrest (Tanaka et al. 2013). Some studies have also found that tumor cells are inhibited and in S phase by transfecting miR-31 into MM cell HP-1 or inhibiting the expression of endogenous miR-31 in MM cell H2461 (Ivanov et al. 2010).

MicroRNAs Related to Proliferation and Clone Formation

Abnormal cell proliferation is an important feature of tumors, and miRNA can regulate genes related to cell proliferation. Therefore the targeted regulation of miRNAs involved in cell proliferation may be helpful to the treatment of MM, and related studies have been reported. Some studies have transfected miR-31 analogues into MM cells and induced miR-31 precursor expression, and found that cell proliferation and the clone formation ability were significantly weakened. Similarly, there have been studies on the decrease of intracellular target protein expression after the transfection of miR-16 analogue and miR-145 activator into tumor cell lines, and the decrease of target protein expression is positively correlated with miRNA transfection concentration, while cell proliferation and clone formation are inhibited. However, the proliferation of normal mesothelial cells transfected with related miRNAs was not affected (Cioce et al. 2014; Reid et al. 2013). In addition, it has been reported that butterfly element (EphA2) inhibits the growth of MM cells by promoting the expression of let-7a1 and by inhibiting the proto-oncogene RAS. When MM cells are transfected with the precursor of miRNAlet-7, the expression of proto-oncogene RAS is inhibited, and cell proliferation is also inhibited (Khodayari et al. 2011).

MicroRNAs Associated with Migration and Invasion

Malignant tumors have the characteristics of metastasis and invasiveness. They can metastasize from the primary site of the tumor to distant ones and invade adjacent tissues or organs. At the cellular level, they show the ability of cell migration and invasion. Identifying metastasis-related factors and understanding their role in the metastasis mechanism are helpful processes for the treatment of MM. Studies have found that the upregulation or deletion of some specific miRNAs makes cancer cells have potential metastasis ability. Fassina et al. (2012) used biphasic MM cell line MSTO-211H for migration and invasion experiments, and found that overexpression of miR-205 can inhibit the migration and invasion of tumor cells. Similarly, an in vitro chemotaxis test—a Boyden chamber assay—has been used to detect that the migration and invasion ability of LP-9 cells transfected with miR-34 inhibitor is significantly enhanced (Tanaka et al. 2013); but some tumor cell lines (H28, H290 and H2052) have increased their cell migration and invasion ability after miR-34 methylation (Muraoka et al. 2013). When miR-34b/c is transfected into tumor cell lines, the ability of tumor cells to form clones is obviously weakened, and the migration and invasion ability of cells is also obviously inhibited (Santarelli et al. 2011). However, through the transwell cell migration test and the cell scratch test, the migration ability of MSTO-211H and NCI-H2052 cell lines transfected with miR-145 analogue was compared with that of non-transfected miR-145 analogue cells, and it was found that migration was significantly weakened in the transfected group and was not found at all in the NCI-H28 cell line (Cioce et al. 2014).

Apoptosis-related miRNAs

In the process of tumor formation and malignant transformation, tumor cells escape the monitoring system by avoiding apoptosis and survive in the microenvironment of tumor growth. Studies have found that miRNA may play a role in inducing apoptosis. The expression levels of miR-1(Kirschner et al. 2012) and miR-34b/c(Kubo et al. 2011) decreased in MM cell lines. After transfection of miR-1 and adenovirus carrying miR-34b/c into tumor cell lines, the number of early and late apoptosis of tumor cells increased significantly. It indicated that miR-1 and miR-34b/c could promote the apoptosis of MM cell.

Inhibition of miRNAs Associated with Explant Tumor

Apart from the experimental studies in vitro already mentioned, there are also many in vivo experimental studies that demonstrate the role of miRNA in the treatment of MM. Ueno et al. (2014) injected adenovirus-carried miR-34b/c into explants of MM tissues from mice and found that the expression of miR-34b/c in tumor cells increased. When compared with the lentivirus-carried luciferase treatment, the volume of tumor did not change for six to ten days after injection, which indicated that miR-34b/c has a certain inhibitory effect on tumor growth. Another study transplanted tumors in nude mice, and then injected into these mice, intravenously, micro cells containing miR-16 analogues. It was found that the degree of inhibition of tumor growth became more obvious as the dose increased; and DNA fragment analysis found that miR-16 analogues also damaged intracellular DNA. When MSTO-211H or MM 05 cell line is transfected with miR-16 analogue, the sensitivity of tumor cells to chemotherapy drugs increases and the efficacy of these drugs is enhanced (Reid et al. 2013). Cioce et al. (2014) inoculated the experimental mice with MSTO-211H. The results showed that eight out of eight mice in the positive control group that were not transfected with miR-145 analogues had solid tumors, while only two out of eight mice in the experimental group that were transfected with miR-145 analogues had tumors; and the tumor volume in the experimental group was significantly smaller than in the positive control group. These experimental results on animals provide more powerful evidence for the usefulness of miRNA in the treatment of MM.

miRNA is able not only to regulate cell cycle, proliferation, clone formation, migration, invasion ability, and apoptosis or inhibit the growth of explant tumors, but also to improve the sensitivity of tumor cells to traditional radiotherapy by regulating the expression level of some miRNAs in cells. Increased expression of miR-34b/c in tumor cells can destroy DNA double-strand damage repair and inhibit the expression of tumor cell growth-related proteins. It can also enhance the sensitivity of MM cells to radiation, which indicates that the combination of increased expression level of miR-34b/c in cells and radiotherapy may be used for the treatment of MM (Maki et al. 2012). At present, on the basis of this research, the application of miRNA to the treatment of MM produces the following three ideas. First, miRNA antagonist is used to inhibit the effect of carcinogenic miRNA, block endogenous miRNA from being processed by RISC, or lead to the degradation of endogenous miRNA. The second idea is to enhance the expression level of the endogenous tumor suppressor miRNA and inhibit the expression of its target proto-oncogene (Benjamin et al. 2010). The third idea is to use the artificial miRNA (amiRNA) expression vector that targets genes related to the malignant tumor phenotype. These methods could become a new way of treating MM, namely by interfering with the changes in miRNA expression level related to the occurrence and development of the tumor.

Exosomal miRNA as a Target for Malignant Mesothelioma

Exosomes provide the opportunity to deliver therapeutic cargo to cancer stroma. The cells were treated with exosome-enriched miR-126. The reduced miR-126 content in fibroblasts in favor of endothelial cells reduced angiogenesis and suppressed cell growth in an miR-126-sensitive environment. Conversely, the accumulation of miR-126 in fibroblasts and the reduced level of miR-126 in endothelial cells induced tube formation in a miR-126-resistant environment via VEGF/EGFL7 upregulation and IRS1-mediated cell proliferation. These findings suggest that the transfer of miR-126 via exosomes represents a novel strategy to inhibit angiogenesis and cell growth in MM (Monaco et al. 2019).

Munson et al. (2019) employed small molecule inhibitors to block exosome secretion, thereby reducing miR-16-5p exosome loss and replenishing cellular miR-16-5p. These processes led to reduced tumorigenic capacity and miR-16-5p target oncoproteins CCND1 and BCL2. Additionally, the researchers force-fed MM tumor exosomes back to MM tumor cells, which caused cell death and a reduction in the same oncoproteins.

MicroRNAs Related to Prognosis of Malignant Mesothelioma

Patients with MM have poor prognosis, and their average survival time is generally seven to twenty-four months. There are many factors affecting the survival time or prognosis of patients. . For example, Fassina et al. (2012) analyzed the average survival time of an epithelial MM: the survival curve is thirteen months, the average survival time of a biphasic MM is fourteen months, and the survival time of a sarcomatous MM is only six months. The COX model also shows that a sarcomatous MM has the worst prognosis. It is not only tissue subtypes are related to prognosis; studies have shown that the expression level of some specific miRNAs is related to prognosis too. Pass et al. (2010) found that miR-29c* in tissue samples of MM patients with good prognosis was significantly downregulated. After comparing it with tissue subtypes, they found that miR-29c* expression in epithelial MM tissues was significantly upregulated and that the prognosis of patients was better (Jean et al. 2012). Some studies have also found that, when the expression of miR-17 and miR-30c in sarcomatous MM is down-regulated (Cappellesso et al. 2016; Jean et al. 2012; Pass et al. 2010), the prognosis of patients is also better. Another study found, through Kaplan-Meier analysis, that the low expression of miR-126 in the serum of patients with MM has a strong correlation with poor prognosis (Tomasetti et al. 2012). Kaplan-Meier analysis showed that hsa-miR-2053 is an independent prognostic factor in MPM (Matboli et al. 2019). Also, a 2-miRNA prognostic signature (Let-7c-5p and miR-151a-5p) related to hypoxia and energy metabolism was identified (De Santi et al. 2017). The expression of miRNA-16 in plasma and tissue was positively related with cumulative survival in MPM patients (Fennell 2017; Mozzoni et al. 2017). miR-215-5p is a poor prognosis miRNA, downregulated in MPM tissues (Singh et al. 2019). Another study found that miR-299-3p, -301,-379 and-455-3p were differentially expressed in smokers and non-smokers, but related miRNAs not significantly different were found in asbestos-exposed and -non-exposed patients (Price 2011). It can be seen that specific miRNA can be used not only to distinguish tumor types but also to predict prognosis for patients (Jean et al. 2012). These results can help clinicians to use tissue or serum samples better, so as to predict the progression and outcome of MMs.

To sum up, although some progress has been made in the research on miRNA in branches of oncology such as the study of MM, there are still many problems. The sensitivity and specificity of miRNA as a biomarker in the diagnosis and prognosis of MM need to be further improved. At present, no miRNA marker has been found that can be used to distinguish between the various pathological tissue subtypes and clinical stages of MMs. Research on miRNA in the treatment of MM is limited to the observation of short-term curative effects displayed by the cell model and the rat tumor model, while research on long-term treatments and their side effects is relatively scarce. However, with the deepening of research, the application of miRNA to the diagnosis and treatment of MM may make substantial progress.

References

Andersen, M., Grauslund, M., Ravn, J., Sørensen, J.B., Andersen, C.B., and Santoni-Rugiu, E. (2014). Diagnostic potential of miR-126, miR-143, miR-145, and miR-652 in malignant pleural mesothelioma. *J Mol Diagn* 16 (4): 418–430.

Banno, K., Yanokura, M., Iida, M. et al. (2014). Application of microRNA in diagnosis and treatment of ovarian cancer. *Biomed. Res. Int.* 2014: 232817.

Benjamin, H., Lebanony, D., Rosenwald, S. et al. (2010). A diagnostic assay based on microRNA expression accurately identifies malignant pleural mesothelioma. *J Mol Diagn* 12 (6): 771–779.

Berindan-Neagoe, I., Monroig Pdel, C., Pasculli, B., and Calin, G.A. (2014). MicroRNAome genome: A treasure for cancer diagnosis and therapy. *CA Cancer J. Clin.* 64 (5): 311–336.

Birnie, K.A., Prêle, C.M., Musk, A. et al. (2019). MicroRNA signatures in malignant pleural mesothelioma effusions. *Dis. Markers* 2019: 8628612.

Busacca, S., Germano, S., De Cecco, L. et al. (2010). MicroRNA signature of malignant mesothelioma with potential diagnostic and prognostic implications. *Am. J. Respir. Cell Mol. Biol.* 42 (3): 312–319.

Calin, G.A., Dumitru, C.D., Shimizu, M. et al. (2002). Frequent deletions and down-regulation of microRNA genes miR15 and miR16 at 13q14 in chronic lymphocytic leukemia. *Proc. Natl. Acad. Sci. USA* 99 (24): 15524–15529.

Cappellesso, R., Galasso, M., Nicolè, L., Dabrilli, P., Volinia, S., and Fassina, A. (2017). miR-130A as a diagnostic marker to differentiate malignant mesothelioma from lung adenocarcinoma in pleural effusion cytology. *Cancer Cytopathol.* 125 (8): 635–643.

Cappellesso, R., Nicolè, L., Caroccia, B. et al. (2016). Young investigator challenge: microRNA-21/microRNA-126 profiling as a novel tool for the diagnosis of malignant mesothelioma in pleural effusion cytology. *Cancer Cytopathol.* 124 (1): 28–37.

Casarsa, C., Bassani, N., Ambrogi, F. et al. (2011). Epithelial-to-mesenchymal transition, cell polarity and stemness-associated features in malignant pleural mesothelioma. *Cancer Lett.* 302 (2): 136–143.

Cavalleri, T., Angelici, L., Favero, C. et al. (2017). Plasmatic extracellular vesicle microRNAs in malignant pleural mesothelioma and asbestos-exposed subjects suggest a 2-miRNA signature as potential biomarker of disease. *PloS One* 12 (5): e0176680.

Cioce, M., Ganci, F., Canu, V. et al. (2014). Protumorigenic effects of miR-145 loss in malignant pleural mesothelioma. *Oncogene* 33 (46): 5319–5331.

de Planell-Saguer, M. and Rodicio, M.C. (2011). Analytical aspects of microRNA in diagnostics: A review. *Anal. Chim. Acta* 699 (2): 134–152.

De Santi, C., Melaiu, O., Bonotti, A. et al. (2017). Deregulation of miRNAs in malignant pleural mesothelioma is associated with prognosis and suggests an alteration of cell metabolism. *Sci. Rep.* 7 (1): 3140.

Fassina, A., Cappellesso, R., Guzzardo, V. et al. (2012). Epithelial-mesenchymal transition in malignant mesothelioma. *Mod. Pathol. Off. J. US Can. Acad. Pathol. Inc* 25 (1): 86–99.

Fennell, D. (2017). miR-16: Expanding the range of molecular targets in mesothelioma. *Lancet Oncol.* 18 (10): 1296–1297.

Gee, G.V., Koestler, D.C., Christensen, B.C. et al. (2010). Downregulated microRNAs in the differential diagnosis of malignant pleural mesothelioma. *Int. J. Cancer* 127 (12): 2859–2869.

Guled, M., Lahti, L., Lindholm, P.M. et al. (2009). CDKN2A, NF2, and JUN are dysregulated among other genes by miRNAs in malignant mesothelioma -A miRNA microarray analysis. *Genes Chromosom. Cancer* 48 (7): 615–623.

Hata, A. and Lieberman, J. (2015). Dysregulation of microRNA biogenesis and gene silencing in cancer. *Sci. Signal* 8 (368): re3.

Ivanov, S.V., Goparaju, C.M., Lopez, P. et al. (2010). Pro-tumorigenic effects of miR-31 loss in mesothelioma. *J. Biol. Chem.* 285 (30): 22809–22817.

Jean, D., Daubriac, J., Le Pimpec-Barthes, F., Galateau-Salle, F., and Jaurand, M.C. (2012). Molecular changes in mesothelioma with an impact on prognosis and treatment. *Arch. Pathol. Lab. Med.* 136 (3): 277–293.

Khodayari, N., Mohammed, K.A., Goldberg, E.P., and Nasreen, N. (2011). EphrinA1 inhibits malignant mesothelioma tumor growth via let-7 microRNA-mediated repression of the RAS oncogene. *Cancer Gene. Ther.* 18 (11): 806–816.

Kirschner, M.B., Cheng, Y.Y., Badrian, B. et al. (2012). Increased circulating miR-625-3p: A potential biomarker for patients with malignant pleural mesothelioma. *J. Thoracic Oncol. Off. Publ. Int. Assoc. Study Lung Cancer* 7 (7): 1184–1191.

Kirschner, M.B., Kao, S.C., Edelman, J.J. et al. (2011). Haemolysis during sample preparation alters microRNA content of plasma. *PloS One* 6 (9): e24145.

Kubo, T., Toyooka, S., Tsukuda, K. et al. (2011). Epigenetic silencing of microRNA-34b/c plays an important role in the pathogenesis of malignant pleural mesothelioma. *Clin. Cancer Res. Off. J. Am. Assoc. Cancer Res.* 17 (15): 4965–4974.

Liu, B., Peng, X.C., Zheng, X.L., Wang, J., and Qin, Y.W. (2009). MiR-126 restoration down-regulate VEGF and inhibit the growth of lung cancer cell lines in vitro and in vivo. *Lung Cancer J. Int. Assoc. Study Lung Cancer* 66 (2): 169–175.

Luo, L., Shi, H.Z., Liang, Q.L., Jiang, J., Qin, S.M., and Deng, J.M. (2010). Diagnostic value of soluble mesothelin-related peptides for malignant mesothelioma: A meta-analysis. *Respir. Med.* 104 (1): 149–156.

Maki, Y., Asano, H., Toyooka, S. et al. (2012). MicroRNA miR-34b/c enhances cellular radiosensitivity of malignant pleural mesothelioma cells. *Anticancer Res.* 32 (11): 4871–4875.

Matboli, M., Shafei, A.E., Ali, M.A. et al. (2019). Clinical significance of serum DRAM1 mRNA, ARSA mRNA, hsa-miR-2053 and lncRNA-RP1-86D1.3 axis expression in malignant pleural mesothelioma. *J. Cell. Biochem.* 120 (3): 3203–3211.

Monaco, F., Gaetani, S., Alessandrini, F. et al. (2019). Exosomal transfer of miR-126 promotes the anti-tumour response in malignant mesothelioma: Role of miR-126 in cancer-stroma communication. *Cancer Lett.* 463: 27–36.

Mozzoni, P., Ampollini, L., Goldoni, M. et al. (2017). MicroRNA expression in malignant pleural mesothelioma and asbestosis: A pilot study. *Dis. Markers* 2017: 9645940.

Mujoomdar, A.A., Tilleman, T.R., Richards, W.G., Bueno, R., and Sugarbaker, D.J. (2010). Prevalence of in vitro chemotherapeutic drug resistance in primary malignant pleural mesothelioma: Result in a cohort of 203 resection specimens. *J. Thorac. Cardiovasc. Surg.* 140 (2): 352–355.

Munson, P.B., Hall, E.M., Farina, N.H., Pass, H.I., and Shukla, A. (2019). Exosomal miR-16-5p as a target for malignant mesothelioma. *Sci. Rep.* 9 (1): 11688.

Muraoka, T., Soh, J., Toyooka, S. et al. (2013). The degree of microRNA-34b/c methylation in serum-circulating DNA is associated with malignant pleural mesothelioma. *Lung Cancer J. Int. Assoc. Study Lung Cancer* 82 (3): 485–490.

Pass, H.I., Goparaju, C., Ivanov, S. et al. (2010). hsa-miR-29c* is linked to the prognosis of malignant pleural mesothelioma. *Cancer Res.* 70 (5): 1916–1924.

Price, A. (2011). What is the role of radiotherapy in malignant pleural mesothelioma. *The Oncologist* 16 (3): 359–365.

Rascoe, P.A., Jupiter, D., Cao, X., Littlejohn, J.E., and Smythe, W.R. (2012). Molecular pathogenesis of malignant mesothelioma. *Expert Rev. Mol. Med.* 14: e12.

Reid, G., Pel, M.E., Kirschner, M.B. et al. (2013). Restoring expression of miR-16: A novel approach to therapy for malignant pleural mesothelioma. *Ann. Oncol. Off. J. Eur. Soc. Med. Oncol.* 24 (12): 3128–3135.

Robinson, B.W. and Lake, R.A. (2005). Advances in malignant mesothelioma. *N. Engl. J. Med.* 353 (15): 1591–1603.

Santarelli, L., Staffolani, S., Strafella, E. et al. (2015). Combined circulating epigenetic markers to improve mesothelin performance in the diagnosis of malignant mesothelioma. *Lung Cancer J. Int. Assoc. Study Lung Cancer* 90 (3): 457–464.

Santarelli, L., Strafella, E., Staffolani, S. et al. (2011). Association of miR-126 with soluble mesothelin-related peptides, a marker for malignant mesothelioma. *PloS One* 6 (4): e18232.

Schramm, A., Opitz, I., Thies, S. et al. (2010). Prognostic significance of epithelial-mesenchymal transition in malignant pleural mesothelioma. *Eur. J. Cardiothorac. Surg: Off. J. Eur. Assoc. Cardiothorac. Surg.* 37 (3): 566–572.

Sempere, L.F. and Korc, M. (2013). A method for conducting highly sensitive microRNA in situ hybridization and immunohistochemical analysis in pancreatic cancer. *Methods Mol. Biol.* 980: 43–59.

Singh, A., Bhattacharyya, N., Srivastava, A. et al. (2019). MicroRNA-215-5p treatment suppresses mesothelioma progression via the MDM2-p53-signaling axis. *Mol. Ther. J. Am. Soc. Gene Ther.* 27 (9): 1665–1680.

Tanaka, N., Toyooka, S., Soh, J. et al. (2013). Downregulation of microRNA-34 induces cell proliferation and invasion of human mesothelial cells. *Oncol. Rep.* 29 (6): 2169–2174.

Tomasetti, M., Staffolani, S., Nocchi, L. et al. (2012). Clinical significance of circulating miR-126 quantification in malignant mesothelioma patients. *Clin. Biochem.* 45 (7-8): 575–581.

Truini, A., Coco, S., Alama, A. et al. (2014). Role of microRNAs in malignant mesothelioma. *CMLS* 71 (15): 2865–2878.

Ueno, T., Toyooka, S., Fukazawa, T., et al. (2014). Preclinical evaluation of microRNA-34b/c delivery for malignant pleural mesothelioma. *Acta Med Okayama.* 68 (1): 23–26.

Van Roosbroeck, K., Pollet, J., and Calin, G.A. (2013). miRNAs and long noncoding RNAs as biomarkers in human diseases. *Expert Rev. Mol. Diagn.* 13 (2): 183–204.

Weber, D.G., Casjens, S., Johnen, G. et al. (2014). Combination of miR-103a-3p and mesothelin improves the biomarker performance of malignant mesothelioma diagnosis. *PloS One* 9 (12): e114483.

Weber, D.G., Gawrych, K., Casjens, S. et al. (2017). Circulating miR-132-3p as a candidate diagnostic biomarker for malignant mesothelioma. *Dis. Markers* 2017: 9280170.

Wright, C.M., Kirschner, M.B., Cheng, Y.Y. et al. (2013). Long non coding RNAs (lncRNAs) are dysregulated in malignant pleural mesothelioma (MPM). *PloS One* 8 (8): e70940.

Wu, C., Jin, B., Chen, L. et al. (2013). MiR-30d induces apoptosis and is regulated by the Akt/FOXO pathway in renal cell carcinoma. *Cell. Signal.* 25 (5): 1212–1221.

Zhang, P., Garnett, J., Creighton, C.J. et al. (2014). EZH2-miR-30d-KPNB1 pathway regulates malignant peripheral nerve sheath tumour cell survival and tumourigenesis. *J. Pathol.* 232 (3): 308–318.

Zöller, M. (2013). Pancreatic cancer diagnosis by free and exosomal miRNA. *World J. Gastrointest. Pathophysiol.* 4 (4): 74–90.

6

Role of Non-coding RNAs in Innate Immune Responses Perturbed by Environmental Arsenic Exposure

Liz Saavedra Perez[1] *and Benjamin L. King*[1,2]

[1] *Department of Molecular and Biomedical Sciences, University of Maine, Orono, Maine*
[2] *Graduate School of Biomedical Science and Engineering, University of Maine, Orono, Maine*

Introduction

The immune system is essential for human health, and dysregulation of the immune function increases one's susceptibility to several human diseases. Diseases linked to the immune function include autoimmune diseases (e.g., multiple sclerosis), single-gene disorders (e.g., cystic fibrosis), pro-inflammatory disorders (e.g., asthma), and cardiovascular disease. The immune system is complex and has both innate and adaptive components that involve many effector cell types. Many genes function as part of effector cell development and the signaling pathways used to respond to infection and injury appropriately. Among these genes are non-coding ones, which are transcribed as RNA. These lack long reading frames and function to regulate other genes, including protein-coding ones. Non-coding RNAs include microRNAs (miRNAs) (Bartel 2018), which negatively regulate other genes, and long non-coding RNAs (lncRNAs), which regulate genes in *cis* or *trans* (Yao et al. 2019). Both protein-coding genes and non-coding RNAs function as part of complex gene regulatory networks that control the immune system's function. Beyond these genetic factors, environmental factors also influence immune function. For example, environmental toxicants such as arsenic negatively impact this function (see, e.g., Goodale et al. 2017). Immunotoxicological studies provide valuable insight into the fields of immunology and toxicology, as they are complementary to genetic studies (Vos 1977). For example, a toxicant can be used to modify the function of genes or cells in a temporal and temporary fashion, as needed. Furthermore, there may be interactions between genetic and environmental factors whereby the impact on the immune function is observed with a specific combination of genetic and environmental perturbations. Some of these combinations of factors may not be intuitive; for example, a response could be greater at a moderate level of exposure than at a low or high level (or dose).

In this chapter, we review studies of gene function in the context of environmental exposure to arsenic, focusing on immune function and non-coding RNAs. These gene-by-environment interactions are inexplicitly linked to the immune function. To limit the scope of our review, we focus on the innate immune system, since it mounts the initial response to infection and injury and is fundamental to understanding host–pathogen interactions. Our review seeks to call attention to the importance of studying these complex interactions and thereby to advance the fields of immunotoxicology and environmental functional genomics (see Feder and Mitchell-Olds 2003).

Genomic and Epigenomic Biomarkers of Toxicology and Disease: Clinical and Therapeutic Actions,
First Edition. Edited by Saura C. Sahu.

The Innate Immune System

The innate immune system functions to defend the organism against infection and injury, but environmental and genetic factors can alter its response (Goering et al. 2018). The innate immune system has several defense mechanisms, including lysozyme, complement, and phagocytic cells. Phagocytic cells consist of monocytes, macrophages, neutrophils, dendritic cells, and mast cells, all of which secrete antimicrobial factors and phagocytize pathogens and cellular debris. Macrophages and neutrophils are the two main cell types responsible for phagocytosis. Macrophages reside in tissues and originate in monocytes within the vascular system and within bone marrow promonocytes (Simon et al. 2015). In humans, neutrophils originate in the bone marrow and reside within it or circulate in the vascular system. In addition to performing phagocytosis, macrophages and neutrophils function by migrating to an injury or infection and then recruiting additional cells through cytokines, chemokines, and reactive oxygen species (ROS) (Arango Duque and Descoteaux 2014; Butterfield et al. 2006; Tecchio et al. 2014). Neutrophils secrete pro-inflammatory and anti-inflammatory cytokines such as interferon gamma (IFNG), tumor necrosis factor (TNF), interleukin 4 (IL4), and interleukin 10 (IL10). The cytokines secreted by macrophages include TNF, interleukin 1 (IL1), interleukin 6 (IL6), interleukin 8 (IL8), and interleukin 12 (IL12). These cytokines activate signaling networks required to initiate innate and adaptive immune processes by regulating inflammation, immunity, tissue repair, hematopoiesis, and angiogenesis (Feldmann 2008). Neutrophils use cytokines to initiate phagocytosis, degranulation, formation of neutrophil extracellular traps (NETs), and ROS. Macrophages use cytokines to activate phagocytosis, autophagy, vesicle trafficking, antigen presentation for T-cell recognition, and subsequent activation of adaptive immune responses.

The innate immune system initiates the response to infection and injury through pathogen-associated molecular pattern receptors (PAMPs) and damage-associated molecular patterns (DAMPs) that use phagocytic cell pattern recognition receptors (PRRs). Toll-like receptors (TLRs) are a prominent family of PRRs expressed on phagocytic cells, and they can activate an array of signaling pathways that control essential innate immune system processes. The activation of TLR pathways can initiate inflammation, immunity, repair, hematopoiesis, and angiogenesis by expressing cytokines and chemokines that regulate the immune response. TLR pathways also induce an acute-phase response by activating the complement system, which includes inflammasomes. Cross talk between TLR and other pathways helps activate the adaptive immune system.

TLR pathways are activated after the ligand-induced dimerization of the TLR ectodomains, which bring the TLRs cytoplasmic Toll/IL1 Receptor (TIR) domains closer together and allow for their interaction with cytoplasmic adaptor molecules (Xu et al. 2000). Different adaptors are used for specific innate immune responses. Four adaptors for mammalian TLRs are myeloid differentiation factor 88 (MYD88), MYD88 adaptor-like (MAL), TLR adaptor molecule 1 (TICAM1), and TLR adaptor molecule 2 (TICAM2) (Takeda and Akira 2004; Yamamoto and Akira 2005). MYD88 is a signal transduction adaptor that can bind to the TLR TIR domain with the amino terminus death domain and the carboxyl terminus TIR domain of MYD88. Binding of the TLR TIR domain and the MYD88 death domain recruits and activates two serine-threonine protein kinases, interleukin 1 receptor-associated kinase 4 (IRAK4) and interleukin 1 receptor-associated kinase 1 (IRAK1) (Murphy and Weaver 2016). The activated complex recruits TNF receptor-associated factor 6 (TRAF6) to interact with mitogen-activated protein kinase kinase kinase 7 (MAP3K7) and to create a scaffold of polyubiquitin chains on TRAF6 and the inhibitor of NF-kappa-B kinase regulator subunit gamma (IKBKG) (Sun et al. 2004). IKBKG recruits and activates MAP3K7 either to activate mitogen-activated protein kinases (MAPKs)—such as c-Jun terminal kinase (JNK) and mitogen-activated protein kinase 14 (MAPK14), which further activates the AP-1 family transcription factors, Jun proto-oncogene (JUN), and Fos proto-oncogene (FOS)—or to phosphorylate and activate the IκB kinase (IKK) complex. IKK is composed of IKKα, IKKβ, and IKKγ. IKK activation phosphorylates IκB, releasing NFκB, which then translocates to the nucleus to transcribe several pro-inflammatory cytokines, including TNF, interleukin 1 beta (IL1B), and IL6. Activation of NFκB by the TLR pathway is a crucial part of the innate immune response to mount a defense against invading pathogens.

The complement system functions in extracellular spaces to coat microbes with proteins that facilitate their recognition and removal by phagocytes. Once neutrophils are recruited and activated, they can initiate robust inflammatory responses. Neutrophils can generate superoxides by membrane-associated NADPH oxidases upon activation. The process starts when phagocyte activation occurs owing to the actions of phagocytic receptors and to signaling by *N*-formylmethionyl-leucyl-phenylalanine (fMLP) or complement component 5, factor a (C5a) receptors that activate Rac2, a gene involved in the generation of ROS (Roberts et al. 1999; Yang et al. 2001). These interactions cause the movement of cytosolic subunits—p40 (encoded by *NCF4*), p47 (encoded by *NCF1*), and p67 (encoded by *NCF2*)—closer to the membrane-associated p91 (encoded by *CYBB*) and p22 (encoded by *CYBA*) complex to create a complete and functional NADPH oxidase in the phagolysosome membrane that produces the respiratory burst response. ROS, hydrolytic enzymes, and membranes disrupt peptides that are released during the respiratory burst response, causing extensive tissue damage.

The interferon (IFN) signaling pathway can further modulate neutrophils and macrophages in the innate immune response. IFNs are extracellular glycoproteins that can activate a large set of IFN-regulated genes (IRGs). IRGs have many functions, including the prevention of microbial replication via inhibition of translation initiation and assistance in RNA degradation (Frucht et al. 2001; Kak et al. 2018). The three classes of IFNs are IFNα, IFNβ, and IFNγ. Macrophages and neutrophils express all three, but IFNγ is more strongly expressed by macrophages and recognized as a link between the innate and the adaptive immune system (Ethuin et al. 2004; Rodrigues et al. 2014). The mechanism employed by the IFN activating pathway upon encountering microbes depends on the induction of IFNγ to activate macrophages into M1 (pro-inflammatory) or M2 (anti-inflammatory) classes. The receptor used is composed of a ligand-binding alpha subunit (IFNγR1) and a signal transduction beta subunit (IFNγR2), which initiates a signaling cascade that produces an array of responses. One response is the activation of the Janus kinases, JAK1 and JAK2, via phosphorylation. These activated kinases phosphorylate the inactive cytosolic transcription factor signal transducer and activator of transcription 1 (STAT1) that translocates to the nucleus to bind IFNγ-activated sequence elements within the promoter region. STAT1 binding mediates the transcription of genes needed in innate and adaptive immune responses. Additionally, STAT1 phosphorylation also causes the activation of TNFβ that is responsible for T cell polarization and commitment to the Type 1 T helper (TH1) cell lineage. The IFNγ-stimulated macrophages have enhanced activity. They are capable of enhanced antigen processing and antigen presentation, increased ROS and nitric oxide synthase (NOS) production, autophagy, increased secretion of pro-inflammatory cytokines, activation of natural killer cells, regulation of antibody production to modulate B cell responses, and facilitation of leukocyte migration to sites of infection.

Negative Impacts of Arsenic on Human Health

Arsenic is a naturally occurring metalloid that can be found in soil, air, and water. It occurs in two states: a non-toxic organic state and a toxic inorganic state, which exists in trivalent (iAs^{III}) and pentavalent (iAs^{V}) forms. The trivalent form, also called arsenite, is the most toxic one (Cervantes et al. 1994). Ingestion of arsenic-contaminated drinking water has a negative impact on the health of an estimated ~ 200 million people worldwide (Podgorski and Berg 2020; World Health Organization 2011).

Arsenic exposure has been linked to lung, bladder, and skin cancers, diabetes, liver and birth disorders, cardiovascular disease, and neurogenesis (Carlin et al. 2016). The incidence, prevalence, and severity of these adverse effects vary depending on the concentration of arsenic, but even low concentrations have been shown to impact human health. To limit exposure, the World Health Organization established a maximum arsenic level of ten parts per billion (ppb), that is, ten micrograms per liter, for drinking water (USEPA 2018; World Health Organization., 2011). The US Environmental Protection Agency also recommends a maximum containment level of 10 ppb for drinking water. Additional research on the health effects of low-level (< 10 ppb) arsenic exposure produced even lower limits for drinking water. The New Hampshire Birth Cohort study has linked arsenic exposure greater than 5 ppb to increased risk of gestational diabetes, preeclampsia, and adverse birth outcomes, including decreased fetal growth (Farzan et al. 2013, 2016). Furthermore, gene expression studies of lung cells showed that arsenic concentrations as low as 10 ppb impact the immune function and increase the risk of infection and chronic obstructive pulmonary disease (Goodale et al. 2017). On the basis of these and other studies on the health effects of low-level arsenic exposure, the states of New Hampshire and New Jersey have lowered the limit of arsenic in drinking water to 5 ppb.

Negative Impacts of Arsenic on the Immune Function

Arsenic exposure can negatively impact the function of the immune system. The innate immune system is of particular importance, because it represents the first line of defense against infection and injury and any modifications can have a multitude of adverse effects. Early-life exposure to arsenic, especially during the development of the innate immune system in humans and mice, can be a concern, as it increases the risk of cancer and non-cancerous genetic dysregulation; but the mechanisms remain largely unknown (Abernathy et al. 2003).

The Comparative Toxicogenomics Database (CTD; http://ctdbase.org) is a curated knowledge base that uses peer-reviewed literature and transitive inferences to show how chemicals such as arsenic impact health (Davis et al. 2015; King et al. 2012). A total of 4,839 genes had chemical-gene interactions with arsenic in the February 2022 release of CTD. Using this database, we exported the list of the top ten most enriched pathways represented among the 4,903 genes (Table 6.1). The most enriched pathway was "Immune System" (R-HSA-168256) from Reactome (Jassal et al. 2020): it included 766 of the 4,903 genes. Other enriched Reactome pathways associated with immune function among the top ten were "Innate Immune System" (R-HSA-168249) with 450 genes, "Cytokine Signaling in Immune System" (R-HSA-1280215) with 341 genes, "Signaling by Interleukins" (R-HSA-449147) with 251 genes, and "Adaptive Immune System" (R-HSA-1280218) with 290 genes. Analysis of the 450 genes from the Reactome "Innate Immune System" pathway showed that 32 genes were part of the KEGG "microRNAs in cancer" (hsa:05206) pathway (Figure 6.1). This suggests that the genes with arsenic interactions from CTD may also be regulated

Table 6.1 Comparative Toxicogenomics Database's top ten Reactome pathways significantly enriched among genes that interact with arsenic or its descendants. Summary of pathways and the current number of annotated genes with an adjusted p-value < 0.01.

Top Ten Pathways Affected by Exposure to Arsenic	Number of Genes (adj. p-value)
Immune System (R-HSA-168256)	766 (0)
Signal Transduction (R-HSA-162582)	725 (2.70 E-255)
Metabolism (R-HSA-1430728)	600 (9.81 E-204)
Innate Immune System (R-HSA-168249)	450 (4.59 E-193)
Cytokine Signaling in Immune System (R-HSA-1280215)	341 (2.86 E-186)
Developmental Biology (R-HSA-1266738)	390 (8.27 E-174)
Metabolism of Proteins (R-HSA-392499)	445 (6.79 E-146)
Signaling by Interleukins (R-HSA-449147)	251 (5.77 E-143)
Gene Expression (R-HSA-74160)	469 (8.64 E-142)
Adaptive Immune System (R-HSA-1280218)	290 (2.14 E-124)

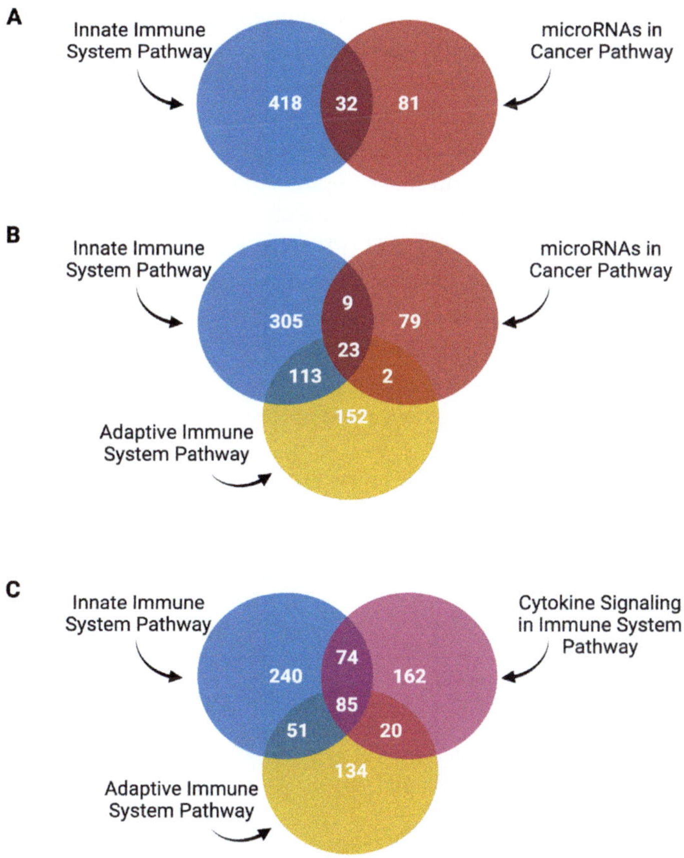

Figure 6.1 Comparison of innate immune system genes and other genes with chemical-gene interactions with arsenic from the Comparative Toxicogenomics Database. (A) Comparison of 450 Innate Immune System pathway (Reactome: R-HSA-168249) genes and 113 microRNAs in Cancer pathway (KEGG:hsa05206) genes that all have chemical-gene interactions with arsenic. Thirty-two genes overlap between the two gene sets, indicating that these innate immune system genes also regulated by miRNAs in cancer. (B) Comparison of 450 innate immune system genes, 113 genes in the microRNAs in Cancer pathway and 290 Adaptive Immune System pathway (Reactome: R-HSA-1280218) genes that all have chemical-gene interactions with arsenic. Twenty-three genes overlap between the three gene sets, indicating that these genes are regulated by miRNAs; 113 genes have both innate and adaptive immune system function. (C) Comparison of 450 innate immune system genes, 341 genes in the Cytokine Signaling in Immune System (Reactome: R-HSA-1280215) pathway and 290 Adaptive Immune System pathway genes that all have chemical-gene interactions with arsenic. Eighty-five genes overlap between the three subsets, indicating that arsenic alters cytokine signaling in the context of the innate and adaptive immune systems.

by miRNAs in the context of innate immunity. Of these 32 genes, 23 are also part of the Reactome pathway "Adaptive Immune Response" (R-HSA-1280218). Similarly, 159 genes of the same 450 "Innate Immune System" pathway genes were part of the "Cytokine Signaling in Immune System" (R-HSA-1280215) pathway, and 85 of these 159 were also part of the "Adaptive Immune Response" pathway. Taken together, the results of the pathway enrichment analysis provided by CTD constitute strong evidence that the immune function is impacted by arsenic.

Several studies have detailed how arsenic exposure alters genes that function in specific signaling pathways involved in immune system function (Table 6.2). Arsenic exposure can perturb the function of the innate immune system by generating oxidative stress through the production of ROS, by altering DNA methylation, and by causing genotoxic damage and even chromosomal abnormalities (Ren et al. 2011). The impact of arsenic exposure on gene expression during early development was extensive (Fry et al. 2007; Rager et al. 2014). Prenatal arsenic exposure was associated with altered gene expression and dampened cytokine expression during innate immune responses. Furthermore, newborns exposed to arsenic in utero were shown to dysregulate the expression of DUSP1 (Fry et al. 2007), a regulator of broad and local inflammatory responses that influences the MAPK and JNK function (Lang et al. 2006; Liu et al. 2007). Depending on the level and duration of exposure, genes associated with the immune function, for instance with cytokine activity, inflammatory response, and the activation of NFκB, were differentially expressed (Ahmed et al. 2012, 2011; Farzan et al. 2016; Rahman et al. 2011; Raqib et al. 2009). Genes that participate in the innate and adaptive immune response, for instance in inflammation, were downregulated. Some of the genes affected were integrin subunit alpha 5 (*ITGA5*), which functions as part of triggering receptor expressed on myeloid cells-1 (TREM1) signaling; dynamin 2 (*DNM2*), from endocytosis signaling; toll-like receptor 5 (*TLR5*) from TLR signaling; *MAP3K3* and NFκB inhibitor alpha (*NFKBIA*) from NFκB signaling and protein kinase C theta (PRKCQ) signaling in T lymphocytes; and IFN-induced protein 35 (*IFI35*) from IFN signaling (Rager et al. 2014). Combined, these adverse immune-related outcomes were shown to reduce thymic size and function in newborns, which resulted in decreased levels of the trophic factors lactotransferrin (LTF) and interleukin 7 (IL7), reduced placental T-cell counts, induced placental inflammatory cytokines, and increased susceptibility to infection during infancy, especially infection in the lower respiratory tract.

Table 6.2 Immune system pathways perturbed by exposure to low levels of inorganic arsenic. 1 ppb = 1 mg/L; ppb; parts per billion.

	Type I IFN Signaling	**Arsenic Concentration (ppb)**	**Reference**
Mx, Il1b, Tnfa	Exposure to low-levels of arsenic inhibited clearance of viral and bacterial infection in *Danio rerio*. It dampened the respiratory burst response because of a delayed induction of antiviral and antibacterial cytokines. Furthermore, it affected the typical bimodal expression pattern for *Il1b* and *Tnfa*. This caused an ineffective antibacterial immune response to bacterial infection that in turn determined an ineffective recruitment of phagocytes at the site of infection.	2, 10	Nayak et al. 2007

(Continued)

Table 6.2 (Continued)

Ifn1	Low levels of arsenic dysregulated the establishment of microbial communities in *Danio rerio*. An increase in the abundance of genera such as *Acinetobacter*, *Sediminibacterium*, *Methylotenera,Legionellales*, and *Janthinobacterium* was observed along with a decrease in genera such as *Bdellovibrio* and *Pseudomonas*. Furthermore, arsenic influenced the spread of antibiotic resistance by affecting the abundance of microbial variants relevant to antimicrobial resistance. Increasing concentrations of arsenic contributed to the abundance of type I interferon in the microbiome, but the increase was not associated with any specific taxa.	10,50,100	Dahan et al. 2018
	JAK-STAT Signaling		
Il2, Il3, Il4, Il6, Il11, Socs3b, Mhc1uja	Exposure to high concentrations of arsenic played a critical role in the function of the JAK-STAT pathway in *Danio rerio* through the downregulation of *Socs3b* and the suppression of cytokine signaling (*Il2, Il3, Il4, Il6, Il11, erythroprotein, Ifng*, and *Ifna*). *Ifng* expression was upregulated and correlated to an increase in the expression of *Mhc1uja* upon exposure to arsenic.	20, 40, 80	Ray et al., 2017
Complement Signaling, Immune Cell Migration, and the Respiratory Burst Response			
GPx, Cat, Nrf2, Keap1, and *MnSod* (mRNA antioxidant genes), *Cox1, Ucp2*, and *Bcl2* (inner mitochondrial membrane genes)	Exposure to low levels of arsenic for 60–90 days induced histopathological alterations and oxidative stress in the liver and kidney of *Danio rerio*. Low levels of arsenic altered the expression of gluthathione (GSH), glutathione (GST), glutathione peroxidase (GPx), and glucocorticoid reductase (GR) antioxidant genes, and the expression of inner mitochondrial membrane genes.	0.05	Sarkar et al., 2017
C3, Fn1, Foxo5, Notch1a, Notch1b, Akt2, Nfkb2, Plg	Low levels of arsenic downregulated the expression of genes critical for initiating the immune response during early stages of embryonic development in *Danio rerio* at 48 hours post fertilization.	10, 100	Mattingly et al. 2009
	ATP Hydrolysis		
Entpd2_mg, Entpd_mq, Nt5e (Ecto-5′-nucleotidase), *As3mt*	Exposure to low levels of arsenic throughout development led to the development of fatty liver disease (FLD) in over 50% of *Danio rerio* larvae by 5 days post fertilization. The oxidative stress response and the unfolded protein response (UPR) caused by endoplasmic reticulum (ER) stress were found in models of FLD. The combination of low levels of arsenic and ethanol potentiated the expression of UPR-associated genes, inducing FLD (steatosis) while reducing the expression of *As3mt*, and increasing the concentrations of inorganic arsenic in the liver.	0.05, 5, 15	Baldissarelli et al., 2012

Arsenic exposure has also been shown to impact innate immune system function at later life stages by dampening gene expression and the overall response to infection. Seven-week-old C57BL/6 J mice exposed to 100 ppb arsenic and then infected with Influenza A virus (IAV) (Puerto Rico/8/34; H1N1) showed decreased survival by comparison with controls (Kozul et al. 2009). Arsenic-exposed mice had fewer dendritic cells in the mediastinal lymph nodes during early stages of infection, and this compromised their immune response to infection. Studies have shown that low-level arsenic exposure can alter the expression of genes with immune function (Goodale et al. 2017; Hsu et al. 2020). Cultured human bronchial epithelial cells exposed to arsenic presented altered gene expression responses to the opportunistic pathogen *Pseudomonas aeruginosa* when compared to the responses of controls (Goodale et al. 2017). Analysis of gene expression data from 0, 5, 10, and 50 ppb resulted in the identification of 102 differentially expressed genes. Two nuclear factors, erythroid 2 like 2 (NFE2L2) oxidative stress response pathway genes, heme oxygenase 1 (*HMOX1*), and ferritin heavy chain 1 (*FTH1*), were among the forty-four genes downregulated with arsenic. Other downregulated genes were lysozyme (*LYZ*), cysteine rich secretory protein 3 (*CRISP3*), nitric oxide synthase 2 (*NOS2*), and interleukin 19 (*IL19*). Gene expression patterns during viral infection were found to be altered by arsenic exposure. In a follow-up study to that of Kozul et al. (2009), single-cell RNA sequencing was used to identify possible mechanisms through which arsenic exposure increased the lung inflammation and the tissue damage they observed after IAV infection (Hsu et al. 2020). Significantly, the authors demonstrated that the functions of monocyte-derived macrophages, neutrophils, alveolar macrophages, and natural killer cells were altered by 100 ppb arsenic exposure. Both studies provide evidence that arsenic alters the function of several innate immune system components.

The Zebrafish Model for Toxicological Responses to Arsenic

The zebrafish (*Danio rerio*) offers a vertebrate model that is important for understanding toxicological responses to arsenic. The fully sequenced genome showed that 70% of the protein-coding genes are conserved between humans and zebrafish (Howe et al. 2013). Zebrafish develop from fertilized embryos into free-swimming larvae quite rapidly, in five days. Several genetic tools have been developed and used to generate mutant lines, including transgenic florescent reporter lines. Since development is external and embryos are optically transparent during the first few days of their development, florescent confocal imaging can be used to study specifically tagged cells in vivo. They also have a high fecundity and are relatively inexpensive to maintain. Embryos can be easily exposed to toxins such as arsenic by adding the toxins to the embryo water. Developmental phenotypes can be observed effortlessly, through various imaging modalities.

Using the zebrafish model, Mattingly et al. (2009) studied how gene expression was altered through low-dose arsenic exposure (0, 10, 100 ppb) and found that several immune system pathways were dysregulated. This study of gene expression was conducted on embryos at twenty-four and forty-eight hours post fertilization (hpf), using microarrays and qRT-PCR analysis. The researchers found that the expression of immune system genes was altered by arsenic depending on the stage of development (24 or 48 hpf). qRT-PCR analysis showed a significant downregulation of the genes that function in immune system development and immunity at 24 hpf. This pattern showed a reversed profile in microarray analysis data at 48 hpf. Two genes previously reported to be predictive of transplacental arsenate exposure in infants were significantly differentially expressed: forkhead box 3b (*foxo3b*, the zebrafish ortholog of human *FOXO3*) and phosphoinositide-3-kinase regulatory subunit 1 (*pik3r1*). Genes that were differentially downregulated at both 10 and 100 ppb exposure were fibronectin 1 (*fn1*), notch receptors 1a and 1b (*notch1a*, *notch1b*), and *pik3r1*. Genes downregulation at 10 ppb, but not at 100 ppb, extended to AKT serine/threonine kinase 2 (*akt2*), *foxo5*, and nuclear factor kappa B subunit 2 (*nfkb2*). Gene Ontology and pathway enrichment analyses demonstrated that arsenic perturbed genes and networks involved in immune response during these stages of zebrafish development. These pathways included acute-phase response signaling, phosphatase and tensin homolog (PTEN) signaling, phosphatidylinositol-4,5-bisphosphate 3-kinase (PI3K)/AKT signaling, and interleukin 15 (IL15) signaling. Two subsequent studies focused on the specific mechanisms involved in the dysregulation of the genes involved in the innate immune response to arsenic exposure. These studies showed that multiple genes were dysregulated in the livers of adult female and male zebrafish at 10, 50, and 500 ppb arsenic, by comparison with 0 ppb dysregulation in controls (Carlson and Van Beneden 2014; Xu et al. 2013). These genes function in cell cycle regulation and in the DNA damage response through AKT serine/threonine kinase 1 (Akt1) signaling, as well as in lipid transport and metabolism. Breast cancer DNA repair-associated 2 (*brca2*) were downregulated in both female and male zebrafish at 500 ppb, at seven and twenty-one days post exposure. WEE1 kinase (*wee1*) was downregulated in females after seven days of 500 ppb exposure. Carnitine O-octanoyltransferase (*crot*), the lipid metabolism gene, was downregulated in female livers with 10 and 500 ppb exposure for seven days. Another lipid metabolism gene, 3-hydroxy-3-methylglutaryl-CoA synthase 1 (*hmgcs1*), was downregulated in males at 10 ppb and in females at 500 ppb, after seven days of exposure. Genome-wide analyses of gene expression in the liver of adult zebrafish exposed to 15 parts per million (ppm) arsenic showed that oxidative stress, carcinogenesis, and metabolism genes were dysregulated (Xu et al. 2013). Twenty-seven genes were validated through qRT-PCR analysis as potential biomarkers of arsenic exposure. Four genes were further validated as potential biomarkers by assaying the expression of homologs in Japanese medaka (*Oryzias latipes*) exposed to arsenic. These potential biomarkers were ferrintin H3 (*frh3*), microsomal glutathione S-transferase-like (*mgst1*), carboxymethylenebutenolidase homolog (*cmbl*), and solute carrier family 40, member 1 (*slc40a1*). Taken together, these cases indicate that arsenic exposure affects the expression of immune response and of metabolism genes that may have several negative consequences on health.

Additional studies have further examined the effects on gene expression and zebrafish development. Survival and gene expression studies have shown that exposure to low levels of arsenic lowers the survival of zebrafish larvae while causing oxidative stress through the direct modulation of expression levels of genes and proteins required for the detoxification of free radical superoxides and protection against oxidative tissue injury, and in two biomarkers for responses to environmental stress (Sun et al. 2019). The expression of superoxide dismutase 1 (*sod1*), superoxide dismutase 2 (*sod2*), methallothionein 1 (*mt1*), and heat shock protein 70 (*hsp70*) was studied in larvae exposed to 0, 25, 50, 75, and 150 ppb arsenic. Decreased survival was observed in zebrafish larvae exposed to 50, 75, and 150 ppb arsenic. A separate study on the embryonic development of zebrafish showed that arsenic exposure caused a significant dysregulation of the genes involved in development and retinogenesis (Babich and Van Beneden 2019). Dysregulation of gene expression had a negative effect on the retinal pigmented epithelium (RPE), which protects visual integrity and facilitates retina–sclera communication. Exposure to arsenic caused the dysregulation of eye development genes, including paired box 6a (*pax6a*), paired box 2a (*pax2a*), neurogenin 1 (*neurog1*), SRY-box transcription factor 2 (*sox2*), and sonic hedgehog signaling molecule a (*shha*). Arsenic exposure also altered eye development. Morphological defects included increased eye diameter and a thinning of the retinal RPE. The dysregulation of gene expression during early developmental phases could in principle explain the improper formation of the RPE, a process that leads to inability of the sclera to maintain the shape of the eye and improper retina–sclera communication.

The Zebrafish Model of Innate Immune Responses

The zebrafish is a valuable vertebrate model in the study of the immune system function as well. Adult zebrafish have both innate and adaptive immune systems with the same immune cell types shared by other vertebrates, including humans. Their innate immune system is functional two days post fertilization (dpf), but their adaptive immune system is not functional until six to seven weeks post fertilization. Additionally, several transgenic fluorescent reporter lines are available that allow for in vivo imaging of immune cells, since embryos, as we mentioned, are optically transparent during the first several days of development (Table 6.3). Zebrafish have been used to study several human bacterial (Phennicie et al. 2010), fungal (Brothers et al. 2011) and viral (Gabor et al. 2014; Sullivan et al. 2021) pathogens. Genetic tools available for the zebrafish allow for functional studies of host–pathogen interactions.

Using the zebrafish to model how the response to infection or injury can be altered through arsenic has yielded valuable information regarding the function of the innate immune system. Studies have shown that low-level arsenic exposure dampens the immune response to viral and bacterial infection by dysregulating the induction of essential cytokines and modifying microbiome composition. One of the first studies addressing arsenic's effect on the response to infection in zebrafish examined how it decreased the induction of type I IFN and the IFN-response gene, myxovirus resistance A (*mxa*) (Nayak et al. 2007). Dysregulation of IFN correlated directly with the inability to clear viral infection after arsenic exposure. In contrast, an increased induction of interleukin 1 beta (*il1b*) and tumor necrosis factor alpha (*tnfa*) correlated with the inability to clear bacterial infections. Exposure to low levels of arsenic during the initial establishment of the zebrafish microbiome correlates with an abundance of genera such as *Acinetobacter*, *Sediminibacterium*, *Methylotenera*, *Legionellales*, and *Janthinobacterium* and a decrease in *Bdellovibrio* and *Pseudomonas* (Dahan et al. 2018). These microbiome modifications influence the spread of antibiotic resistance by affecting the abundance of microbial variants relevant to antimicrobial resistance. Specific arsenic concentrations like 10 ppb and 50 ppb lower microbial diversity, while a concentration of 100 ppb increases the abundance of type I IFN in the zebrafish microbiome.

Table 6.3 *Danio rerio* macrophage and neutrophil specific reporter lines. *Tg(coro1a:eGFP;lyz:dsRed)* is a double transgenic line that marks both macrophages and neutrophils.

Macrophage		Neutrophil	
Tg(mpeg1:eGFP)^{gl22}	*GFP Reporter*	*Tg(mpx:GFP)*^{i114}	GFP Reporter
Tg(mpeg1:YFP)^{w200}	*YFP Reporter*	*Tg(mpx:GFP)*^{uwm1}	
Tg(mpeg1:mCherry)^{gl23}	*RFP Reporter*	*Tg(lyz:eGFP)*^{nz117}	
Tg(mpeg1:mCherry-F;tnfa:eGFP-F)		*Tg(mpx:mCherry)*^{uwm7}	RFP Reporter
		Tg(lyz:dsRed2)^{nz50}	
Tg(coro1a:eGF;lyz:dsRed)		GFP Reporter	RFP Reporter

Additionally, dysregulation in the induction of essential cytokines during early development can cause acute responses detrimental to survival (Xu et al. 2018). An increased induction of neutrophils, associated with an acute inflammatory response, was observed in zebrafish larvae exposed to arsenic. To further understand this potential mechanism of immunotoxicity, Islam and Takeyama (2021) showed that exposure to low concentrations of arsenic (20 ppb) significantly affects the neutrophils' formation of NETs, phagocytosis, and the generation of ROS. The release of NET components in polymorphonuclear neutrophils (PMN) was inhibited in zebrafish through exposure to 20 ppb arsenic for twelve hours, and lower concentrations had a similar effect. Superoxide, the phorbol myristate acetate (PMA)-induced NET formation often used to induce ROS, was also inhibited.

Role of Non-coding RNAS in Innate Immunity

Overview of Non-coding RNAs

The complex defense mechanisms of the innate immune system require genetic regulation of biological processes to prevent infection or to respond to injury while adapting to multiple environments. Both protein-coding and -non-coding RNA genes regulate the immune function. The function of non-coding RNAs has not yet been well characterized by comparison to that of protein-coding genes. Non-coding RNAs transcripts lack long open reading frames and have been shown to be capable of regulating genes in *cis* or *trans* (Statello et al. 2021). Some non-coding RNAs function in the nucleus to regulate gene expression by interacting with chromatin, altering transcription, splicing, transcript-processing, or exporting to the cytoplasm. Non-coding RNAs can also function in the cytoplasm to alter translation and transcriptional degradation and have been associated with organelles, including mitochondria and exosomes, depending on the non-coding RNA.

Two prominent classes of non-coding RNAs with demonstrated regulatory roles in immune function are lncRNAs and miRNAs. lncRNAs are transcripts greater than 200 nucleotides in length that lack long open reading frames. Ensembl (version 105) annotation of the current human genome assembly (GRCh38) includes 16,896 lncRNAs. The regulatory function of the few lncRNAs characterized encompasses epigenetic and post-transcriptional regulation of gene expression in immune responses. For example, the mouse's lncRNA, *Ptgs2os2*, was shown to regulate the expression of several proinflammatory genes, chemokines, chemokine receptors, and IFN-stimulated genes in both *cis* and *trans* specimens (Carpenter et al. 2013). The expression of *Ptgs2os2* was induced through TLR activation, in a MYD88- and NFκB-dependent manner, after stimulation by lipopolysaccharide (LPS) in mouse bone marrow-derived dendritic cells and in Pam_3CSK_4 bone marrow-derived macrophages. Short-hairpin RNA (shRNA) knockdown of *Ptgs2os2* in a mouse macrophage cell line stimulated with Pam_3CSK_4 resulted in downregulation of proinflammatory genes (*Tlr1*, *Il6*, and *Il23a*) and in upregulation of the chemokines *Ccl5* and *Cx3cl1*, the chemokine receptor *Ccr1*, and IFN-stimulated genes *Irf7*, *Oas1a*, *Oas1l*, *Oas2*, *Ifi204* and *Isg15*. Additional studies on the roles that lncRNAs have in immune function are needed.

The second class, miRNAs, is made up of highly conserved small RNAs that regulate gene expression and play critical roles in stress responses in various organisms. Functional mature miRNAs are single-stranded RNAs, 20–26 nucleotides in length (Fromm et al. 2015). Two mature products are cleaved from imperfect loop structures in RNA precursors. One of the products is functional, the other is degraded. The functional product is incorporated into the RNA-induced silencing complex and negatively regulates the expression of target genes on the basis of sequence complementarity of the seed sequence (positions 2–8) between the miRNA and the target transcript (Bartel and Bartel 2003; Ebert and Sharp 2012; Lai 2003). Multiple transcripts may be targeted by a single miRNA, depending on the complementarity of the seed sequence. Conversely, multiple miRNAs may regulate a single transcript. In vertebrates, binding of the RNA-induced silencing complex (RISC) to target transcripts results in degradation of the transcript (Bartel 2018). The diverse functions of miRNAs include the regulation of development, regeneration, cancer, and immune function (Carrington and Ambros 2003; Small and Olson 2011).

miRNAs function as part of complex gene regulatory networks where one miRNA or more can regulate a biological process (King et al. 2018; King and Yin 2016). Multiple modes of such regulation have been described (Small and Olson 2011). miRNAs may target multiple genes in a biological process in order to modulate the overall function of that process. A single miRNA may target genes that function as part of a single biological process, or a set of miRNAs may target those genes cooperatively or redundantly. miRNAs can also function as physiological buffers by negatively regulating both the activators and the inhibitors of a biological process.

Studies of miRNA families across animal taxa have shown that they are highly conserved; there is gradual accrual across evolutionary time and rare secondary loss (Heimberg et al. 2008; Sempere et al. 2006; Tarver et al. 2013). miRNA families consist of genes that encode miRNAs that share similar seed sequences. Increasing organismal and immune system complexity correlates with a higher number of cell types and a higher number of miRNA families (Tarver et al. 2013). Humans (*Homo sapiens*) have 556 miRNA genes, and there are 447 in mice (*Mus musculus*), 390 in the zebrafish (*Danio rerio*), and 8 in the sponge (Fromm et al. 2015; Tarver et al. 2013). A total of 79 families are conserved between the most often used model systems in immunology: human cell lines and primary cells, mice, and zebrafish (Sullivan et al. 2021). It should be noted that each family contains different numbers of miRNA genes respectively in humans, mice, and zebrafish.

MiRNAs Regulate Response to Injury and Infection Involving Multiple Cell Types

Knowledge of the conservation of miRNAs across taxa and of how they negatively regulate gene function has guided research toward investigating the role of miRNAs in development and in the innate immune function (Table 6.4)—or, more specifically, the effect of miRNAs on the development, recruitment, and function of macrophages and neutrophils, two main effector cells of the innate immune system.

One of the best known studies on the role of miRNAs in development was carried out by Kenneth Poss' laboratory (Yin et al. 2008). Their work is relevant to our understanding of the immune function, since complex tissue regeneration requires a functional immune system. Expression of miR-133 in the zebrafish was shown to be lower during caudal fin regeneration than in uninjured controls. Using a dominant negative fibroblast growth factor receptor 1 (*fgfr1*) in a zebrafish line where caudal fin regeneration is blocked, the authors showed that miR-133 expression was not decreased, as observed in wild-type controls. Blastema formation was rescued and even accelerated when expression of the miR-133 target gene, ttk protein kinase (*ttk*), was inhibited (Yin et al. 2008). The researchers showed how miR-133 depletion was required for complex tissue regeneration. This and other studies of miRNA function led to more questions about the potential roles of miRNAs in regulatory mechanisms for other biological processes, such as the innate immune system and its effector cells.

Table 6.4 miRNAs associated with the innate immune function and their predicted target genes.

miRNA	Immune System Target Genes	Immune System Function	Reference
miR-223	*stat1/3/5a, nlrp3, traf6, cul1 scf, ikkα*	Neutrophilic inflammation was augmented in miR-223-deficient *Danio rerio* because of an elevated activation of the NFκB pathway in epithelial cells. miR-223-deficient zebrafish were capable of regulating tumorigenesis at multiple levels by suppressing the inflammatory tumor microenvironment and therefore modulating cancer cells development and function.	Jeffries et al. 2019
	Il1β, Nlrp3 Inflammasomes	MiR-223 deficiency led to an increase in Ly6G + cells in the peripheral blood of mice. Deficiency inhibited the differentiation of monocyte precursors into Ly6G + cells, but an opposing effect was seen with increased miR-223 expression. Furthermore, miR-223 inhibited the translation of the NLRP3 inflammasome in bone marrow-derived neutrophils and increased the production of IL1β.	Feng et al., 2017
	Il6, CEBPA	MiR-223 deficiency in mice was associated with enhanced neutrophil activation and interlukin-6 (Il6) expression. Wounds infected with *Staphylococcus aureus* (*S. aureus*) in miR-223-deficient mice showed enhanced wound repair. The expression of miR-223 in human neutrophils was regulated by CCAAT/enhancer binding protein alpha (CEBPA) after exposure to *S. aureus* peptides during infection.	De Kerckhove et al. 2018
let-7, miR-223, miR-16, miR-142-5p, miR-15a, miR-16	*IKKα, NFκB target genes, P52*	The miRNAs present in human monocytes with a predicted target sequence in the 3′UTR of the gene *IKKα* are let-7, miR-223, miR-16, miR-142-5p, miR-15a, miR-16. The target site for miR-223, miR-15a, and miR-16 are ~ 35 nucleotides apart, potentially allowing cooperation between target sites, especially by the same miRNA. miR-223, miR-15a, and miR-16 are downregulated during granulocyte-macrophage colony-stimulating factor (GM-CSF) differentiation. Their downregulation corresponded to an increase in *IKKα* protein expression, an increase in *P52* downstream, and P52 binding to nuclear DNA in macrophages, allowing repression of transcription. Therefore it is estimated that miR-223, miR-15a, and miR-16 downregulation increased *IKKα* protein expression and led to the repression of basal gene expression of *NFκB* target genes.	Li et al., 2010
miR-722	*rac2*	Overexpression of miR-722 downregulated *rac2* transcript levels, leading to a reduction in neutrophil motility and chemotaxis at the site of injury and infection in *Danio rerio*.	Hsu et al. 2017
miR-142-3p	*ifnγ, irf1b, irf9, crfb17*	miR-142-3p deficiency resulted in abnormal reduction and hypermutation of neutrophils and in the impairment of the inflammatory migration of neutrophils during early stages of development. An abnormal activation of the *ifnγ* signaling pathway was observed along with hypermutation of neutrophils in adult *Danio rerio*.	Fan et al. 2014

(Continued)

Table 6.4 (Continued)

miRNA	Immune System Target Genes	Immune System Function	Reference
miR-199	*Pdcd4*	In mice, miR-199a-3p targeted hepatic programmed cell death 4 (*Pdcd4*), a promoter for apoptosis. miR-199a-3p suppressed hepatocyte apoptosis and hepatic injury by targeting and suppressing *Pdcd4*. A deficiency in miR-199a-3p in hepatocytes could aggravate hepatic injury and HCC development.	Li et al., 2020
	cdk2	miR-199 reduced neutrophil migration and reduced systemic inflammation, which prevented a severe immunocompromised state in *Danio rerio*. It directly suppresses cyclin-dependent kinase 2 (*cdk2*), reducing neutrophil migration.	Hsu et al. 2019
	Il1β, *Il6*, *Tnfα*, *Sirt1*	Upon LPS stimulation, miR-199a was downregulated in mouse alveolar macrophages that protected against sepsis-induced acute respiratory distress syndrome (ARDS). Reduction of miR-199a derepressed its target gene, sirtuin 1 (*Sirt1*). The increased expression of *Sirt1* reduced *Il1β*, *Il6* and *Tnfα* expression, which reduced inflammation and inhibited apoptosis in lung tissue.	Liu et al. 2018
miR-205-3p, miR-141-5p, miR-200a-5p, miR-92a-2-5p, miR-192, miR-1788	Many target genes that participate in several biological processes and pathways	A group of 37 *Danio rerio* miRNAs were differentially expressed during infection; 26 of these miRNA have 107 predicted target genes. These targets were associated with specific Gene Ontology (GO) terms such as peripheral nervous system neuron axonogenesis, organophosphate metabolic process, heme binding, protein binding, tetrapyrrole binding, protein dimerization activity, and aromatase activity, which regulate nerve conduction, energy metabolism, hematopoiesis, and protein synthesis. They were also associated with Kyoto Encyclopedia of Genes and Genomes (KEGG) pathways such as phototransduction, tryptophan metabolism, notch signaling, and purine metabolism.	Ji et al., 2019
miR-146	*myd88*, *traf6*, *tlr4*	miR-146 regulated lipid metabolism during the inflammatory response to infection by *Salmonella typhimurium* in *Danio rerio*.	Ordas et al. 2013

(Continued)

Table 6.4 (Continued)

miRNA	Immune System Target Genes	Immune System Function	Reference
miR-205-5p	*ifn1*, *trim25*, *rig-I*, *mavs*, *traf3*, *irf3*, *ifn1*	*Danio rerio* infected with red spotted grouper nervous necrosis virus (RGNNV) were found to overexpress miR-205-5p. Increased miR-205-5p expression led to reduced expression of *ifn1* and its downstream antiviral genes which promoted RGNNV replication. Overexpression of miR-205-5p downregulated its target gene, *trim25*, and interferon signaling through the trim25-mediated retinoic acid inducible gene-I (RIG-I)-like receptors (RLRs) pathway.	Liu et al., 2020
miR-124, miR–144, miR–148, miR–155, miR–19a, miR–223	Many target genes that participate in several biological processes and pathways	Prediction-based work identified miRNAs with potential roles in the regulation of processes triggered by nanoparticle induction. Identified miRNAs can be used as starting points and further analyzed via validation experiments.	Hu et al., 2019
miR-133	*ttk*	miR-133 acted as a regulator of regeneration in zebrafish. In a normal regenerative response, miR-133 expression is reduced via fibroblast growth factor (fgf) signaling. miR-133 targets ttk protein kinase (*ttk*), which directs blastemal proliferation and tissue renewal.	Yin et al. 2008

As discussed earlier, neutrophils and macrophages serve as essential effector cells whose migration and subsequent recognition of pathogens or injured tissues lead to a wide range of innate immune system responses. These responses include phagocytosis, degranulation, formation of NETs, release of ROS, secretion of inflammatory cytokines and chemokines, and recruitment of other effector cells. The role of miRNAs in neutrophil and macrophage function remains understudied.

MiRNAs Regulate Macrophage Function

MiRNAs have essential roles in regulating the recruitment and function of macrophages after infection and injury. Studies have focused on miRNAs that are part of macrophage PRR involved in pathway signaling responsible for recruitment and activation of additional cytokines. These studies provide insight into how specific miRNAs function as part of the antibacterial and antiviral roles of macrophages.

O'Connell et al. (2007) employed microarray analyses to investigate the relationship between inflammation, innate immunity, and miRNA expression during viral infection (O'Connell et al. 2007). Their study characterized the response of mature murine bone marrow macrophages, stimulated either by a synthetic viral intermediate, polyriboinosinic :polyribocytidylic acid (poly(I:C)) double-stranded RNA, or by the antiviral response cytokine, interferon beta 1 (Ifnb1). miR-155 was the most highly upregulated miRNA after stimulation by both methods. Monitoring of the kinetics of miR-155 induction for forty-eight hours after stimulation further revealed that the regulation of miR-155 by poly (I:C) and Ifnb1 requires an upregulation of the Mir155 host gene (*Mir155hg*). Several TLR ligands also stimulated miR-155 expression through MYD88 and TRIF pathways. Lastly, the authors demonstrated that the JNK pathway was involved in miR-155 activation, as the JNK inhibitor, sp600125, blocked poly(I:C) and Ifnb1 activation.

Studies have also investigated how miRNAs regulate macrophage function during the pathogenesis of specific diseases. In cardiovascular disease, inflammatory macrophages are recognized as essential effector cells during the progress of atherosclerotic plaque formation and rupture, two central disease stages. A study of potential miRNA targets in atherosclerosis-prone apolipoprotein E-deficient (*ApoE*$^{-/-}$) mice has been used to further understand the regulators of inflammatory macrophages that underlie cardiovascular disease (Wang et al. 2015). The expression of miR-223 was increased within atherosclerotic aortic lesions. After exposure to *Escherichia coli* (serotype 055:B5), LPS caused the activation of macrophages and a subsequent decrease in miR-223 expression, in a time- and LPS dose-dependent manner. The overexpression of miR-223 was responsible for the attenuation of foam cell formation after the recruitment of macrophages to atherosclerotic lesion sites, lipid accumulation, and pro-inflammatory cytokine production. The opposite effect was observed with the downregulation of miR-223. This attenuating effect was hypothesized to be due to the negative regulatory roles of miR-223 on the activation of the toll-like receptor 4-nuclear factor kappa b (TLR4-NFkb) pathway. Furthermore, miR-223 upregulation was shown to lead to the activation of PI3K/AKT. Inhibition of miR-223 leads to an increase in lipid deposition and in inflammatory cytokine production. Together, these results suggest that miR-223 may be a potential target for cardiovascular disease by inhibiting atherosclerosis through the activation of the PI3K/AKT pathway and subsequent inhibition of the TLR4 signaling pathway.

Tao et al. (2021) studied the effect of bacterial infections on the activation and expression of miRNAs with immune roles in antibacterial defenses. Using cytokine-induced killer (CIK) cells infected by *Aeromonas hydrophila* (*A. hydrophila*), the authors showed that infection caused an inhibition of miR-142a-3p expression, leading to an upregulation in the expression of the miRNA's target genes, *tnfaip2* and *glut3*. Additionally, it was shown that CIK viability or apoptosis rates were reduced over time in cells that overexpressed miR-142a-3p, while the expression of inflammatory factors *tnfa*, *il1b*, and *cxcl8a* was downregulated. An opposite effect occurred in cells where the expression of miR-142a-3p was downregulated. The expression of anti-inflammatory factors *il10* and *tgfb1a* by miR-142a-3p was enhanced in Kupffer cells (macrophage; *Ctenopharyngodon idella* liver cell line) after an *A. hydrophila* infection. These results demonstrated that miR-142a-3p has multiple antibacterial immune response roles in the host.

MiRNAs Regulate Neutrophil Function

Ongoing studies have found significant connections between various miRNAs and the recruitment of neutrophils to sites of infection and injury (Hsu et al. 2017; Zhou et al. 2018). Some of these miRNAs and their target genes have been demonstrated to function in the innate immune system. For example, Hsu et al. (2017) provided evidence that miR-223, miR-199, and miR-722 function as negative regulators of neutrophil migration to sites of infection and injury in zebrafish. They showed that neutrophil migration required the downregulation of miR-223, miR-199, and miR-722. Functional studies of these miRNAs were used to understand their roles. Overexpression in neutrophils was used to inhibit their function and to identify target genes.

The overexpression of miR-722 by neutrophils attempting to regulate, via downregulation of Rac2 transcripts, the inflammatory response to infection given by opportunistic pathogens like *Pseudomonas aeruginosa* was found to increase zebrafish survival (Hsu et al. 2017). The recruitment of neutrophils was also negatively regulated by a decrease in the expression of miR-223 within neutrophils and basal and epithelium surfaces (Zhou et al. 2018). Furthermore, neutrophilic inflammation was augmented in miR-223 deficient zebrafish, in part because of the elevated activation of the NF-κB pathway in epithelial cells (Jeffries et al. 2019). In mice, miR-223 deficiency was associated with enhanced neutrophilic activation, interleukin-6 (*Il6*) expression, and enhanced repair in wounds infected by *Staphylococcus aureus* (de Kerckhove et al. 2018). These studies corroborated the effect of reducing or enhancing neutrophil motility and chemotaxis to the site of injury or infection and the upregulation of anti-inflammatory cytokines in the defense against detrimental systemic inflammation based on the pattern of miR-223 expression. Furthermore, Hsu et al. (2019) discovered that the migration of neutrophils to sites of infection required the expression of miR-199 by targeting cyclin-dependent kinase 2 (*cdk2*) in zebrafish. Overexpression of miR-199 reduced neutrophil migration and controlled systemic inflammation, which prevented a severely immunocompromised state through the direct suppression of *cdk2*—a process that reduced the quantity of migrating neutrophils. The downregulation of miR-199a in mice protects lung tissue against sepsis-induced acute respiratory distress syndrome (ARDS) after LPS stimulation (Liu et al. 2018). The downregulation of miR-199a in alveolar macrophages reduces inflammation and apoptosis by derepressing its target gene, sirtuin 1 (*Sirt1*), which results in reduced *Il1b*, *Il6*, and *Tnfa* expression.

With the help of the zebrafish model, both miR-142 and miR-146 have been demonstrated to have roles in neutrophil recruitment and function. Zebrafish miR-142a/b null double mutants that were shown not to express miR-142-3p had abnormal neutrophil phenotypes (Fan et al. 2014). During definitive myelopoiesis, the number of neutrophils was reduced, and the neutrophils that developed had an increased cell size and a decreased nucleocytoplasmic ratio. In mutant larvae, neutrophils had impaired inflammatory migration and abnormal activation of the interferon γ (IFN-γ) signaling pathway. Both miR-146a and miR-146b were upregulated after infection by *Salmonella typhimurium* in zebrafish, owing to the disruption of the Myd88-Traf6 pathway. This finding was significant, because the Myd88-Traf6 pathway is responsible for the transduction of TLR signals and cytokine responses, essential components of the immune system. Interestingly, infection in zebrafish deficient in miR-146 did not significantly affect the Myd88-Traf6 pathway. Instead, the main effect is a regulation of the lipid metabolism, as a result of an increased induction of members of the apolipoprotein gene family (Ordas et al. 2013). Although pro-inflammatory genes and transcription factors are not affected in the absence or decreased expressions of miR-146a and miR-146b, miRNAs of this family are significant because of their ability to induce inflammatory pathways. The insight gained by studying these miRNAs in zebrafish should inform the design of experiments in mammalian models, given that both the miR-142 and the miR-146 families are conserved among mammals.

These studies have provided insight into the regulation of neutrophils and macrophages by miRNAs, but more studies are needed if we want to understand the regulation of the innate immune function even further. Understanding the relationship between genetic and environmental factors and miRNA expression is also needed in order to assess interactions that dysregulate the mechanisms of the immune system.

The Expression of Non-coding RNAs Altered by Arsenic

Several studies in vertebrates have analyzed the relationship between arsenic exposure and the dysregulation of miRNA expression using whole-genome miRNA expression or their predicted target genes (Table 6.5). These studies used human cells, mice, zebrafish, or Atlantic killifish (*Fundulus heteroclitus*) exposed to varying concentrations of arsenic. Studies in which exposure to low concentrations of arsenic are used often allow for the investigation of short- and long-term effects to the immune system, especially on innate immune defense mechanisms. Exposure to high arsenic concentrations can lead to immediate and drastic adverse effects. For example, if arsenic concentrations are high enough to elicit an oxidative stress response that exceeds the immune system's antioxidant removal capabilities, an increase in lipid hydroperoxides can cause cells damage and, potentially, death (Fukai and Ushio-Fukai 2011). Low concentrations can also have adverse effects. If accumulated over time, immune system suppression can lead to an inability to clear infections through the dysregulation of natural oxidation, a reduction in protein homeostasis, and a reduced production of pro-inflammatory factors (Jomova et al. 2011).

Several in vitro studies have described how arsenic exposure results in changes in miRNA and how those changes are dependent on arsenic concentration. For example, studies on immortalized human keratinocytes (HaCAT cells) exposed to 25 nM sodium arsenite (~13 ppb) showed dysregulation of expressed miRNAs after a twenty-four-week treatment (Zhou et al. 2017). Significantly upregulated miRNAs extended to miR-181b and miR-100. The targets of these miRNAs include genes from epigenetic and metabolic pathways, which suggests that arsenic exposure produces epigenetic and metabolic alterations in these cells. Furthermore, the study of low arsenic exposure by Zhou et al. showed that four essential proteins, typically associated with stress responses, were differentially expressed. These proteins (and their coding genes) were heat shock protein-beta-1 (*HSPB1*), peroxiredoxin-2 (*PRDX2*), adenpsylhomocysteinase (*AHCY*), and leukocyte elastase inhibitor (*SERPINB1*). Some of these genes were directly or indirectly affected by the expression of one or more of the differentially expressed miRNAs.

Table 6.5 Arsenic perturbation of miRNAs associated with innate immune function and their predicted target genes.

miRNA	Immune System Gene Tragets	Arsenic Concentration (ppb)	Immune System Function	Reference
let-7a, miR-107, miR-126, miR-16, miR-17*, miR-195, miR-20a, miR-20b, miR-26b, miR-454, miR-96, miR-98	*DUSP1*, *DNM2*, *IFI35*, *ITGA5*, *NFKBIA*, *TLR5*, *TLR9*	6.2 -320	Prenatal arsenic exposure in humans was associated with altered mRNA expression and dampened cytokine expression during innate immune responses. It is also associated with prenatal arsenic exposure and with altered expression levels of miRNAs. Microarray analysis revealed that prenatal exposure to arsenic was associated with the increased expression of twelve miRNAs in newborn cord blood. These are let-7a, miR-107, miR-126, miR-16, miR-17*, miR-195, miR-20a, miR-20b, miR-26b, miR-454, miR-96, and miR-98. The expression of dual specificity phosphatase 1 (*DUSP1*), a potential gene biomarker of prenatal arsenic exposure, was found to be associated with prenatal arsenic exposure. Furthermore, predictive analyses showed that decreased expression of dynamin-2 (*DNM2*), interferon-induced protein 35 (*IFI35*), integrin alpha 5 (*ITGA5*), NFκBIA, *TLR5*, and *TLR9* was associated with significant over-representation of innate immune response signaling pathways such as TREM1, clathrin-mediated endocytosis, toll-like receptor, NFκB, and interferon.	Rager et al. 2014
miR-6739-5p, miR-4521, miR-181b-5p, miR-100-5p, miR-3919, miR-513a-5p, miR-1273g-3p	*HSPB1*, *PRDX2*, *AHCY*, *ILEU*	52	Low levels of arsenic induced HaCaT cell malignant transformation and differential expression of miRNAs. Network analysis showed an interaction between differentially expressed miRNAs and three differentially expressed proteins: heat HSPB1, PRDX2, and AHCY. These interactions suggest an association with oxidative stress, glutathione metabolism, metabolic cycle, and DNA methylation.	Zhou et al. 2017

(Continued)

Table 6.5 (Continued)

miRNA	Immune System Gene Tragets	Arsenic Concentration (ppb)	Immune System Function	Reference
miR-663, miR-222, miR-638, miR-30d, miR-221, miR-181a, miR-150	*SP1*, *CYP1A1*, *RNF4*, *TGFB1*	~ 2	The expression of miRNAs was modulated in Jurkat cells after exposure to low levels of arsenic. Arsenic exposure caused the induction of cell cycle progression and the failure of apoptosis. These findings support the idea of inorganic arsenic's carcinogenic activity. Network analyses showed that upregulation of miR-221, miR-222, and miR-638 corresponded to a decrease in the expression of their putative target gene, the Ring Finger Protein 4 (*RNF4*). MiR-663 expression was negatively correlated with its putative target gene, *TGFB1*. Furthermore, the expression of miR-181a and miR-150 corresponded to the opposite expression of their respective putative target genes, the transcription factor Sp1 (*SP1*) and cytochrome P450, family 1, subfamily A, polypeptide 1 (*CYP1A1*).	Sturchio et al. 2014
miR-135b	*kcnb2*, *kcnj14*, *slc9a1*, *slc16a4*, *tor1a*, *mrpl14*, *fkbp7*, *rangap1*, *and tmem68 homologs*	100	miRNA expression profiling in the Atlantic killifish (*Fundulus heteroclitus*) gills identified a small group of highly expressed, well-conserved miRNAs as well as sixteen novel miRNAs not yet identified in other organisms. In response to changes in salinity, gills undergo reversible remodeling to acclimate. Gill remodeling was inhibited by exposure to arsenic. After analysis of the individual and combined effects of salinity and arsenic on miRNA expression in killifish gill, miR-135b was found to be differentially expressed in response to arsenic and twenty-four hours after transfer to salt water. miR-135b predicted targets included genes involved in ion transport, cell motility and migration, GTPase mediated signal transduction and organelle assembly.	Goodale et al. 2019

Sturchio et al. (2014) characterized the expression of miRNAs after arsenic exposure in Jurkat immortalized human T lymphocyte cells. When Jurkat cells were exposed to 260 ppb arsenic for 24 hours, they remained at the G2/M phase thereby preventing the segregation of damaged chromosomes. In this study, miR-30d, miR-221, miR-222, miR-638 and miR-663 were upregulated after arsenic treatment for 24 and 144 hours. MiR-150, miR-181a, and miR-142-5p were also significantly upregulated at 24 hours post exposure, but were downregulated at 144 hours. The let-7 family members, let-7a, let-7 c and let-7e, were downregulated at 24 hours. Sturchio et al. then assayed the expression of several miRNA target genes by quantitative real time PCR (qRT-PCR) at 24 and 144 hours post exposure. Ring finger protein 4 (*RNF4*), a predicted target gene for the upregulated miRNAs miR-221, miR-222, and miR-638, had decreased expression. Transforming growth factor beta 1 (*TGFB1*), a target gene for the upregulated miR-663, was downregulated. One target gene of the upregulated miRNAs miR-150 and miR-181a was the Sp1 transcription factor (*SP1*), and it was downregulated at 24 hours. BTB domain and CNC homolog 1 (*BACH1*), a target gene of the let-7 family, was upregulated. A major function of *BACH1* is in the oxidative stress response, as it regulates Heme oxygenase 1 (*HMOX1*), an important enzyme in the antioxidant response. *HMOX1* and the oxidative stress response have been shown to be induced by sodium arsenite (Hashimoto et al. 2017).

Studies of arsenic toxicity in animal models complement in vitro models, as physiological phenotypes can be correlated with exposure. The mouse is a well-established mammalian animal model with high genetic conservation with humans. States et al. (2012) combined predicted target data with in vivo miRNA profile analyses in order to study the effect of arsenic. Profiles of miRNA expression were assayed in the liver of mice pups that were exposed to 49 parts per million (ppm) arsenic in utero. Mice were exposed to arsenic from gestational day 8 to term. miRNA expression was assayed at birth and at ten weeks. The expression of miR-361, miR-148a and miR-467a was induced, and let-7d, miR-719, miR-679, miR-592, miR-222, miR-497, miR-211, and miR-188 were suppressed in the livers of arsenic-exposed mice at the day of birth. At ten weeks, the expression of miR-211, miR-291a-5p, miR-291b-5p, miR-294, miR-302, and miR-464 were induced, while miR-1, miR-10a, miR-15b, miR-214a, miR-130a, miR-149, miR-186, miR-193, miR-218, miR-337, miR-376a, miR-412, miR-476b, miR-681, and miR-715 were suppressed. Furthermore, analysis of Gene Ontology annotations of the target genes suggested that glycolysis/gluconeogenesis, protein metabolism processes such as protein export and ribosome, and inflammatory processes such as complement, antigen processing, and coagulation would be regulated by these differentially expressed miRNAs. Collectively, these findings show that arsenic exposure during early development leads to pro-inflammatory responses postnatally.

Goodale et al. (2019) studied the effect of low-level arsenic exposure on miRNA expression profiles in the Atlantic killifish (*Fundulus heteroclitus*) gill. The Atlantic killifish, native to estuarine areas of the Atlantic coast in the United States, has become a valuable ecotoxicological model because of its ability to acclimate to rapid environmental changes and adapt to polluted habitats (Reid et al. 2017, 2016). Killifish respond to large fluctuations in salinity with rapid changes in gene expression and protein trafficking in order to maintain osmotic balance; this response is followed by a secondary phase of gene and protein expression changes, which enable remodeling of the gills. Arsenic exposure (100 ppb) was previously shown to inhibit gene expression responses in the killifish gill and the ability of killifish to acclimate to a rapid increase in salinity (Shaw et al. 2007, 2014). Goodale et al.'s (2019) was one of the first studies of the potential roles miRNA have in environmental acclimation. The study profiled miRNA expression in killifish gill for the first time and identified a small group of highly expressed, well-conserved miRNAs, as well as sixteen novel miRNAs not yet identified in other organisms. After analyzing the individual and combined effects of salinity and arsenic on miRNA expression in killifish gill, the researchers found miR-135b to be differentially expressed in response to arsenic and at twenty-four hours after transfer to saltwater. Predicted targets of miR-135b were involved in ion transport, cell motility and migration, GTPase mediated signal transduction, and organelle assembly. In agreement with previous studies of these two environmental stressors, there was a significant interaction (i.e., arsenic-dependent salinity effect), whereby killifish exposed to arsenic exhibited an opposite response in miR-135b expression at twenty-four hours after the hyperosmotic challenge by comparison with controls. When examining the mRNA expression of predicted miRNA targets during salinity acclimation and arsenic exposure, miR-135b targets were found to be significantly more likely to decrease during salinity acclimation than non-targets. The identification of a significant interaction effect of arsenic and salinity on miR-135b expression supports the hypothesis that arsenic alters upstream regulators of stress response networks, which may adversely affect the response of killifish to osmotic stress. The characterization of miRNAs in this ecotoxicological model will be a valuable resource for future studies that investigate the role of miRNAs in response to environmental stress.

Discussion

Innate immunity uses various induced effector cells to combat infection, repair injured tissue, and maintain homeostasis until the more robust adaptive immune responses are activated (Goering et al. 2018). Both genetic and environmental factors impact the function of these effector cells (Figure 6.2). The mechanisms by which genetic and environmental factors influence innate immune function remain understudied. For example, low-level arsenic exposure has been shown to perturb the innate immune system. This has been done by using human cell lines, mouse models, and zebrafish models. The studies demonstrated an increase in the cells' susceptibility to infection and disease. Arsenic impacts the function of

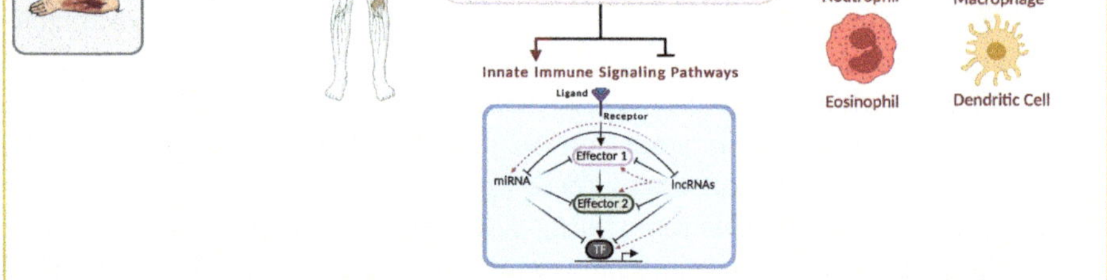

Figure 6.2 Regulation of innate immune system responses and signaling pathways by non-coding RNAs. Genetic and environmental factors can influence the function of the innate immune response to infection and injury. Immune cell recruitment and activation rely upon several signaling pathways that are regulated in part by non-coding RNAs. Created with BioRender.com.

phagocytic cells, which were characterized by decreased recruitment and a dysregulated ROS production. However, the mechanisms behind these changes remain primarily uncharacterized (Islam and Takeyama 2021; Mattingly et al. 2009). Further studies on low-dose arsenic exposure are needed to identify what developmental defects are induced and by what mechanisms the innate immune function and its regulation are altered. Ongoing work on non-coding RNAs, especially miRNAs, has shown strong promise to address some of these challenges.

As discussed earlier, miRNAs regulate biological processes by negatively regulating innate immune genes and pathways and thereby modulating the innate immune system's function. miRNAs are estimated to modulate the expression of about 60% of all protein-coding genes by directly regulating multiple genes that can participate in many biological functions (Small and Olson 2011). Numerous studies have shown the potential of miRNAs as innovative tools to investigate innate immune responses and to develop gene-targeted therapies. But we must understand how genetic and environmental factors contribute to the complex mechanisms of innate immunity if we want to advance our understanding of the immune function and develop new therapies.

Gene-by-Environment Interactions

Many studies have focused on examining how one independent gene or environmental factor affects innate immune responses. Such studies can be used to evaluate whether genetic factors are more important than environmental factors. It is now recognized that both kinds play significant roles, and studies are needed to investigate how the immune system's responses are affected by multiple factors. Gene–environment interactions have become a focal point in the field (Dick et al. 2015; Ottman 1996). The concept of interaction focuses on characterizing how interactions between genetic and environmental factors work together to impact phenotypes, for instance in the form of an increased susceptibility to disease. Is the innate immune response to infection altered by genetic or environmental factors, as seen in patients with autosomal recessive or dominant disorders who are exposed to low-dose arsenic? Is the immune function more severely affected in these patients when they are exposed to arsenic in an additive fashion, or are the interactions more complex? These questions cannot be investigated with single-factor studies. Furthermore, do these interactions affect the expression of miRNAs and, if so, are there any specific miRNA target genes that correlate with or influence the phenotypes? These studies could provide evidence about how miRNAs may alter specific innate immune pathways.

Functional Studies of Non-coding RNAs Using the Zebrafish Model

The zebrafish is a powerful vertebrate model in the study of the function of miRNAs. It allows us to address directly host–pathogen interactions that occur during infection and can be translated into observations of human disease. The model offers a variety of genetic tools and strategies for studying the function of miRNAs. These strategies include targeted knockouts (e.g., CRISPR), miRNA anti-sense RNAs (e.g., locked nucleic acid (LNA) anti-sense RNAs) knockdown, and transgenics where miRNA precursors (pre-miRNAs) or their target genes can be overexpressed in specific cell types. Engineered endonucleases (EENs) such as the clustered, regularly interspaced, short palindromic repeats (CRISPR) are used to target genes by cleaving the zebrafish genome (Huang et al. 2016; Xiao et al. 2013b; Zu et al. 2013). Their products offer disruption of a specific target gene's function and modification of the target locus (e.g., specific donor templates) (Cade et al. 2012). Newer approaches have achieved the induction of insertion–deletion mutations, the target of interest for non-coding RNA genes, using the Cas9/gRNA system (Huang et al. 2013; Xiao et al. 2013a). The system has a simple design, which is easier to engineer because the synthesis of a specific guide RNA (gRNA) oligo of fewer than 100 nucleotides is required for each target sequence, and the Cas9 protein for different target sites is universal. The Cas9/gRNA system produces DNA double-strand breaks (DSBs) in the target sequence, which induce insertion–deletion mutations and could be further used to characterize the function of miRNAs and other non-coding RNAs.

miRNA function can be altered, apart from EENs, by overexpressing miRNA precursors (pre-miRNAs) using transgenic lines or by injecting miRNA anti-sense RNAs (anti-miRs) to knock down one or more mature miRNA. For example, transgenic lines were developed to overexpress miR-722 (Hsu et al. 2017) and miR-199 (Hsu et al. 2019) pre-miRNAs in neutrophils and were used to demonstrate how these miRNAs regulate neutrophil migration. Overexpression of the pre-miRNA was driven by a transgene that contained the lysozyme C (*lyz*) promoter; Tol2 transgenesis was used to specifically express the pre-miRNA in neutrophils. Increased expression of the pre-miRNA led to an increased expression of the mature miRNA. Conversely, the expression of a mature miRNA may be inhibited with anti-sense RNA. For example, zebrafish embryos at one-cell or two-cell stages can be injected with an anti-sense RNA morpholino (Kimmel et al. 1995; Lehmann et al. 2008; Leucht et al. 2008) that targets a pre-miRNA in embryos. LNA anti-miRs was shown to knock down miR-122 in non-human primates (Elmen et al. 2008). When combined with gene expression studies, overexpression or knockdown studies can provide information about the effect of either the upregulation or the down-regulation of specific miRNAs after infection or chemical exposure (e.g., to low-dose arsenic). Subsequently, gene expression networks containing both miRNAs and their target genes can be constructed (King et al. 2018; King and Yin 2016). Using EENs, pre-miRNA overexpression lines, and anti-miRs in combination could be advantageous, because each can be tested independently and all can be used in experiments of gene–environment interaction that investigate the innate immune system's pathways. Such studies may help identify the regulatory mechanisms for these pathways, including the roles of non-coding RNAs.

Future Studies on Gene-by-Environment Interactions

The discovery that non-coding RNAs such as miRNAs are essential regulators of biological processes has opened a new perspective on our understanding of the innate immune system's pathways and their outcomes. Moreover, the identification of target genes and the characterization of the multiple mechanisms by which miRNAs regulate them may continue to disclose parts of the innate immune system, with potential for the future development of traditional and novel therapeutics. Although studies that characterize and validate how non-coding RNAs and their target genes function in the context of gene–environment interactions are still in progress and have not reached final conclusions, the growing interest in the subject and the increasing number of such studies will significantly shape future research. Hopefully future discoveries will throw light on some of the most complex areas—molecular biology, genetics, and immunology.

Acknowledgments

We acknowledge the University of Maine System Research Reinvestment Fund for supporting this work.

References

Abernathy, C.O., Thomas, D.J., and Calderon, R.L. (2003). Health effects and risk assessment of arsenic. *J. Nutr.* 133 (5 Suppl 1): 1536S–1538S.

Ahmed, S., Ahsan, K.B., Kippler, M., Mily, A., Wagatsuma, Y., Hoque, A.M., Ngom, P.T., El Arifeen, S., Raqib, R., and Vahter, M. (2012). In utero arsenic exposure is associated with impaired thymic function in newborns possibly via oxidative stress and apoptosis. *Toxicol. Sci.* 129 (2): 305–314.

Ahmed, S., Mahabbat-e Khoda, S., Rekha, R.S., Gardner, R.M., Ameer, S.S., Moore, S., Ekström, E.C., Vahter, M., and Raqib, R. (2011). Arsenic-associated oxidative stress, inflammation, and immune disruption in human placenta and cord blood. *Environ. Health Perspect.* 119 (2): 258–264.

Arango Duque, G. and Descoteaux, A. (2014). Macrophage cytokines: Involvement in immunity and infectious diseases. *Front Immunol.* 5: 491.

Babich, R. and Van Beneden, R.J. (2019). Effect of arsenic exposure on early eye development in zebrafish (*Danio rerio*). *J. Appl. Toxicol.* 39 (6): 824–831.

Baldissarelli, L.A., Capiotti, K.M., Bogo, M.R., Ghisleni, G., and Bonan, C.D. (2012). Arsenic alters behavioral parameters and brain ectonucleotidases activities in zebrafish (*Danio rerio*). *Comp Biochem Physiol C Toxicol Pharmacol.* 155 (4): 566–572.

Bartel, B. and Bartel, D.P. (2003). MicroRNAs: At the root of plant development? *Plant Physiol.* 132 (2): 709–717.

Bartel, D.P. (2018). Metazoan MicroRNAs. *Cell* 173 (1): 20–51.

Brothers, K.M., Newman, Z.R., and Wheeler, R.T. (2011). Live imaging of disseminated candidiasis in zebrafish reveals role of phagocyte oxidase in limiting filamentous growth. *Eukaryot. Cell* 10 (7): 932–944.

Butterfield, T.A., Best, T.M., and Merrick, M.A. (2006). The dual roles of neutrophils and macrophages in inflammation: A critical balance between tissue damage and repair. *J. Athl. Train* 41 (4): 457–465.

Cade, L., Reyon, D., Hwang, W.Y., Tsai, S.Q., Patel, S., Khayter, C., Joung, J.K., Sander, J.D., Peterson, R.T., and Yeh, J.R. (2012). Highly efficient generation of heritable zebrafish gene mutations using homo- and heterodimeric TALENs. *Nucleic Acids Res.* 40 (16): 8001–8010.

Carlin, D.J., Naujokas, M.F., Bradham, K.D., Cowden, J., Heacock, M., Henry, H.F., Lee, J.S., Thomas, D.J., Thompson, C., Tokar, E.J., Waalkes, M.P., Birnbaum, L.S., and Suk, W.A. (2016). Arsenic and environmental health: State of the science and future research opportunities. *Environ. Health Perspect.* 124 (7): 890–899.

Carlson, P. and Van Beneden, R.J. (2014). Arsenic exposure alters expression of cell cycle and lipid metabolism genes in the liver of adult zebrafish (*Danio rerio*). *Aquat. Toxicol.* 153: 66–72.

Carpenter, S., Aiello, D., Atianand, M.K., Ricci, E.P., Gandhi, P., Hall, L.L., Byron, M., Monks, B., Henry-Bezy, M., Lawrence, J.B., O'Neill, L.A., Moore, M.J., Caffrey, D.R., and Fitzgerald, K.A. (2013). A long noncoding RNA mediates both activation and repression of immune response genes. *Science* 341 (6147): 789–792.

Carrington, J.C. and Ambros, V. (2003). Role of microRNAs in plant and animal development. *Science* 301 (5631): 336–338.

Cervantes, C., Ji, G., Ramirez, J.L., and Silver, S. (1994). Resistance to arsenic compounds in microorganisms. *FEMS Microbiol. Rev.* 15 (4): 355–367.

Dahan, D., Jude, B.A., Lamendella, R., Keesing, F., and Perron, G.G. (2018). Exposure to arsenic alters the microbiome of larval zebrafish. *Front Microbiol.* 9: 1323.

Davis, A.P., Grondin, C.J., Lennon-Hopkins, K., Saraceni-Richards, C., Sciaky, D., King, B.L., Wiegers, T.C., and Mattingly, C.J. (2015). The Comparative Toxicogenomics Database's 10th year anniversary: Update 2015. *Nucleic Acids Res.* 43 (Database issue): D914–920.

de Kerckhove, M., Tanaka, K., Umehara, T., Okamoto, M., Kanematsu, S., Hayashi, H., Yano, H., Nishiura, S., Tooyama, S., Matsubayashi, Y., Komatsu, T., Park, S., Okada, Y., Takahashi, R., Kawano, Y., Hanawa, T., Iwasaki, K., Nozaki, T., Torigoe, H., Ikematsu, K., Suzuki, Y., Tanaka, K., Martin, P., Shimokawa, I., and Mori, R. (2018). Targeting miR-223 in neutrophils enhances the clearance of *Staphylococcus aureus* in infected wounds. *EMBO Mol. Med.* 10 (10).

Dick, D.M., Agrawal, A., Keller, M.C., Adkins, A., Aliev, F., Monroe, S., Hewitt, J.K., Kendler, K.S., and Sher, K.J. (2015). Candidate gene-environment interaction research: Reflections and recommendations. *Perspect Psychol. Sci.* 10 (1): 37–59.

Ebert, M.S. and Sharp, P.A. (2012). Roles for microRNAs in conferring robustness to biological processes. *Cell* 149 (3): 515–524.

Elmen, J., Lindow, M., Schutz, S., Lawrence, M., Petri, A., Obad, S., Lindholm, M., Hedtjarn, M., Hansen, H.F., Berger, U., Gullans, S., Kearney, P., Sarnow, P., Straarup, E.M., and Kauppinen, S. (2008). LNA-mediated microRNA silencing in non-human primates. *Nature* 452 (7189): 896–899.

Ethuin, F., Gérard, B., Benna, J.E., Boutten, A., Gougereot-Pocidalo, M.A., Jacob, L., and Chollet-Martin, S. (2004). Human neutrophils produce interferon gamma upon stimulation by interleukin-12'. *Lab. Invest.* 84 (10): 1363–1371.

Fan, H.B., Liu, Y.J., Wang, L., Du, T.T., Dong, M., Gao, L., Meng, Z.Z., Jin, Y., Chen, Y., Deng, M., Yang, H.T., Jing, Q., Gu, A.H., Liu, T.X., and Zhou, Y. (2014). miR-142-3p acts as an essential modulator of neutrophil development in zebrafish. *Blood* 124 (8): 1320–1330.

Farzan, S.F., Korrick, S., Li, Z., Enelow, R., Gandolfi, A.J., Madan, J., Nadeau, K., and Karagas, M.R. (2013). In utero arsenic exposure and infant infection in a United States cohort: A prospective study. *Environ. Res.* 126: 24–30.

Farzan, S.F., Li, Z., Korrick, S.A., Spiegelman, D., Enelow, R., Nadeau, K., Baker, E., and Karagas, M.R. (2016). Infant infections and respiratory symptoms in relation to in utero arsenic exposure in a US cohort. *Environ. Health Perspect.* 124 (6): 840–847.

Feder, M.E. and Mitchell-Olds, T. (2003). Evolutionary and ecological functional genomics. *Nat. Rev. Genet.* 4 (8): 651–657.

Feldmann, M. (2008). Many cytokines are very useful therapeutic targets in disease. *J. Clin. Invest.* 118 (11): 3533–3536.

Feng, Z., Qi, S., Zhang, Y., Qi, Z., Yan, L., Zhou, J., He, F., Li, Q., Yang, Y., Chen, Q., Xiao, S. and Chen, Y. (2017). Ly6G+ neutrophil-derived miR-223 inhibits the NLRP3 inflammasome in mitochondrial DAMP-induced acute lung injury. *Cell Death Dis.* 8 (11): e3170.

Fromm, B., Billipp, T., Peck, L.E., Johansen, M., Tarver, J.E., King, B.L., Newcomb, J.M., Sempere, L.F., Flatmark, K., Hovig, E., and Peterson, K.J. (2015). A uniform system for the annotation of vertebrate microRNA genes and the evolution of the human microRNAome. *Annu. Rev. Genet.* 49: 213–242.

Frucht, D.M., Fukao, T., Bogdan, C., Schindler, H., O'Shea, J.J., and Koyasu, S. (2001). IFN-gamma production by antigen-presenting cells: Mechanisms emerge. *Trends Immunol.* 22 (10): 556–560.

Fry, R.C., Navasumrit, P., Valiathan, C., Svensson, J.P., Hogan, B.J., Luo, M., Bhattacharya, S., Kandjanapa, K., Soontararuks, S., Nookabkaew, S., Mahidol, C., Ruchirawat, M., and Samson, L.D. (2007). Activation of inflammation/NF-kappaB signaling in infants born to arsenic-exposed mothers. *PLoS Genet.* 3 (11): e207.

Fukai, T. and Ushio-Fukai, M. (2011). Superoxide dismutases: Role in redox signaling, vascular function, and diseases. *Antioxid. Redox Signal.* 15 (6): 1583–1606.

Gabor, K.A., Goody, M.F., Mowel, W.K., Breitbach, M.E., Gratacap, R.L., Witten, P.E., and Kim, C.H. (2014). Influenza A virus infection in zebrafish recapitulates mammalian infection and sensitivity to anti-influenza drug treatment. *Dis. Model Mech.* 7 (11): 1227–1237.

Goering, R., Dockrell, H., Zuckerman, M., and Chiodini, P.L. (2018). *Mims' Medical Microbiology E-Book*. Edinburgh: Elsevier.

Goodale, B.C., Hampton, T.H., Ford, E.N., Jackson, C.E., Shaw, J.R., Stanton, B.A., and King, B.L. (2019). Profiling microRNA expression in Atlantic killifish (*Fundulus heteroclitus*) gill and responses to arsenic and hyperosmotic stress. *Aquat. Toxicol.* 206: 142–153.

Goodale, B.C., Rayack, E.J., and Stanton, B.A. (2017). Arsenic alters transcriptional responses to *Pseudomonas aeruginosa* infection and decreases antimicrobial defense of human airway epithelial cells. *Toxicol. Appl. Pharmacol.* 331: 154–163.

Hashimoto, K., Majumdar, R., and Tsuji, Y. (2017). Nuclear lamins and progerin are dispensable for antioxidant Nrf2 response to arsenic and cadmium. *Cell. Signal.* 33: 69–78.

Heimberg, A.M., Sempere, L.F., Moy, V.N., Donoghue, P.C., and Peterson, K.J. (2008). MicroRNAs and the advent of vertebrate morphological complexity. *Proc. Natl. Acad. Sci. USA* 105 (8): 2946–2950.

Howe, K., Clark, M.D., Torroja, C.F., Torrance, J., Berthelot, C., Muffato, M., Collins, J.E., Humphray, S., McLaren, K., Matthews, L., McLaren, S., Sealy, I., Caccamo, M., Churcher, C., Scott, C., Barrett, J.C., Koch, R., Rauch, G.J., White, S., Chow, W., Kilian, B., Quintais, L.T., Guerra-Assuncao, J.A., Zhou, Y., Gu, Y., Yen, J., Vogel, J.H., Eyre, T., Redmond, S., Banerjee, R., Chi, J., Fu, B., Langley, E., Maguire, S.F., Laird, G.K., Lloyd, D., Kenyon, E., Donaldson, S., Sehra, H., Almeida-King, J., Loveland, J., Trevanion, S., Jones, M., Quail, M., Willey, D., Hunt, A., Burton, J., Sims, S., McLay, K., Plumb, B., Davis, J., Clee, C., Oliver, K., Clark, R., Riddle, C., Elliot, D., Threadgold, G., Harden, G., Ware, D., Begum, S., Mortimore, B., Kerry, G., Heath, P., Phillimore, B., Tracey, A., Corby, N., Dunn, M., Johnson, C., Wood, J., Clark, S., Pelan, S., Griffiths, G., Smith, M., Glithero, R., Howden, P., Barker, N., Lloyd, C., Stevens, C., Harley, J., Holt, K., Panagiotidis, G., Lovell, J., Beasley, H., Henderson, C., Gordon, D., Auger, K., Wright, D., Collins, J., Raisen, C., Dyer, L., Leung, K., Robertson, L., Ambridge, K., Leongamornlert, D., McGuire, S., Gilderthorp, R., Griffiths, C., Manthravadi, D., Nichol, S., Barker, G., Whitehead, S., Kay, M., Brown, J., Murnane, C., Gray, E., Humphries, M., Sycamore, N., Barker, D., Saunders, D., Wallis, J., Babbage, A., Hammond, S., Mashreghi-Mohammadi, M., Barr, L., Martin, S., Wray, P., Ellington, A., Matthews, N., Ellwood, M., Woodmansey, R., Clark, G., Cooper, J., Tromans, A., Grafham, D., Skuce, C., Pandian, R., Andrews, R., Harrison, E., Kimberley, A., Garnett, J., Fosker, N., Hall, R., Garner, P., Kelly, D., Bird, C., Palmer, S., Gehring, I., Berger, A., Dooley, C.M., Ersan-Urun, Z., Eser, C., Geiger, H., Geisler, M., Karotki, L., Kirn, A., Konantz, J., Konantz, M., Oberlander, M., Rudolph-Geiger, S., Teucke, M., Lanz, C., Raddatz, G., Osoegawa, K., Zhu, B., Rapp, A., Widaa, S., Langford, C., Yang, F., Schuster, S.C., Carter, N.P., Harrow, J., Ning, Z., Herrero, J., Searle, S.M., Enright, A., Geisler, R., Plasterk, R.H., Lee, C., Westerfield, M., De Jong, P.J., Zon, L.I., Postlethwait, J.H., Nusslein-Volhard, C., Hubbard, T.J., Roest Crollius, H., Rogers, J., and Stemple, D.L. (2013). The zebrafish reference genome sequence and its relationship to the human genome. *Nature* 496 (7446): 498–503.

Hsu, A.Y., Wang, D., Gurol, T., Zhou, W., Zhu, X., Lu, H.Y., and Deng, Q. (2017). Overexpression of microRNA-722 fine-tunes neutrophilic inflammation by inhibiting Rac2 in zebrafish. *Dis. Model Mech.* 10 (11): 1323–1332.

Hsu, A.Y., Wang, D., Liu, S., Lu, J., Syahirah, R., Bennin, D.A., Huttenlocher, A., Umulis, D.M., Wan, J., and Deng, Q. (2019). Phenotypical microRNA screen reveals a noncanonical role of CDK2 in regulating neutrophil migration. *Proc. Natl. Acad. Sci. USA* 116 (37): 18561–18570.

Hsu, K.S., Goodale, B.C., Ely, K.H., Hampton, T.H., Stanton, B.A., and Enelow, R.I. (2020). Single-cell RNA-seq analysis reveals that prenatal arsenic exposure results in long-term, adverse effects on immune gene expression in response to influenza A infection. *Toxicol. Sci.* 176 (2): 312–328.

Hu, M., Jovanović, B., and Palić, D. (2019). In silico prediction of microRNA role in regulation of zebrafish (*Danio rerio*) responses to nanoparticle exposure. *Toxicol in Vitro* 60: 187–202.

Huang, P., Xiao, A., Tong, X., Lin, S., and Zhang, B. (2016). Targeted mutagenesis in zebrafish by TALENs. *Methods Mol. Biol.* 1338: 191–206.

Huang, P., Xu, L., Liang, W., Tam, C.I., Zhang, Y., Qi, F., Zhu, Z., Lin, S., and Zhang, B. (2013). Genomic deletion induced by Tol2 transposon excision in zebrafish. *Nucleic Acids Res.* 41 (2): e36.

Islam, M.M. and Takeyama, N. (2021). Inorganic arsenic administration suppresses human neutrophil function. *Hum. Exp. Toxicol.* 40 (5): 725–734.

Jassal, B., Matthews, L., Viteri, G., Gong, C., Lorente, P., Fabregat, A., Sidiropoulos, K., Cook, J., Gillespie, M., Haw, R., Loney, F., May, B., Milacic, M., Rothfels, K., Sevilla, C., Shamovsky, V., Shorser, S., Varusai, T., Weiser, J., Wu, G., Stein, L., Hermjakob, H., and D'Eustachio, P. (2020). The reactome pathway knowledgebase. *Nucleic Acids Res.* 48 (D1): D498–D503.

Jeffries, J., Zhou, W., Hsu, A.Y., and Deng, Q. (2019). miRNA-223 at the crossroads of inflammation and cancer. *Cancer Lett.* 451: 136–141.

Ji, C., Guo, X., Ren, J., Zu, Y., Li, W., and Zhang, Q. (2019). Transcriptomic analysis of microRNAs-mRNAs regulating innate immune response of zebrafish larvae against *Vibrio parahaemolyticus* infection. *Fish Shellfish Immunol.* 91: 333–342.

Jomova, K., Jenisova, Z., Feszterova, M., Baros, S., Liska, J., Hudecova, D., Rhodes, C.J., and Valko, M. (2011). Arsenic: Toxicity, oxidative stress and human disease. *J. Appl. Toxicol.* 31 (2): 95–107.

Kak, G., Raza, M., and Tiwari, B.K. (2018). Interferon-gamma (IFN-γ): Exploring its implications in infectious diseases. *Biomol Concepts* 9 (1): 64–79.

Kimmel, C.B., Ballard, W.W., Kimmel, S.R., Ullmann, B., and Schilling, T.F. (1995). Stages of embryonic development of the zebrafish. *Dev. Dyn.* 203 (3): 253–310.

King, B.L., Davis, A.P., Rosenstein, M.C., Wiegers, T.C., and Mattingly, C.J. (2012). Ranking transitive chemical-disease inferences using local network topology in the comparative toxicogenomics database. *PLoS One* 7 (11): e46524.

King, B.L., Rosenstein, M.C., Smith, A.M., Dykeman, C.A., Smith, G.A., and Yin, V.P. (2018). RegenDbase: A comparative database of noncoding RNA regulation of tissue regeneration circuits across multiple taxa. *NPJ Regen Med.* 3: 10.

King, B.L. and Yin, V.P. (2016). A conserved microRNA regulatory circuit is differentially controlled during limb/appendage regeneration. *PLoS One* 11 (6): e0157106.

Kozul, C.D., Ely, K.H., Enelow, R.I., and Hamilton, J.W. (2009). Low-dose arsenic compromises the immune response to influenza A infection in vivo. *Environ. Health Perspect.* 117 (9): 1441–1447.

Lai, E.C. (2003). microRNAs: Runts of the genome assert themselves. *Curr. Biol.* 13 (23): R925–36.

Lang, R., Hammer, M., and Mages, J. (2006). DUSP meet immunology: Dual specificity MAPK phosphatases in control of the inflammatory response. *J. Immunol.* 177 (11): 7497–7504.

Lehmann, U., Hasemeier, B., Christgen, M., Müller, M., Römermann, D., Länger, F., and Kreipe, H. (2008). Epigenetic inactivation of microRNA gene hsa-miR-9-1 in human breast cancer. *J. Pathol.* 214 (1): 17–24.

Leucht, C., Stigloher, C., Wizenmann, A., Klafke, R., Folchert, A., and Bally-Cuif, L. (2008). MicroRNA-9 directs late organizer activity of the midbrain-hindbrain boundary. *Nat. Neurosci.* 11 (6): 641–648.

Li, T., Morgan, M.J., Choksi, S., Zhang, Y., Kim, Y.S., and Liu, Z.G. (2010). MicroRNAs modulate the noncanonical transcription factor NF-kappaB pathway by regulating expression of the kinase IKKalpha during macrophage differentiation. *Nat Immunol.* 11 (9): 799–805.

Li, Z., Zhou, Y., Zhang, L., Jia, K., Wang, S., Wang, M., Li, N., Yu, Y., Cao, X., and Hou, J. (2020). microRNA-199a-3p inhibits hepatic apoptosis and hepatocarcinogenesis by targeting PDCD4. *Oncogenesis* 9 (10): 95.

Liu, Y., Guan, H., Zhang, J.L., Zheng, Z., Wang, H.T., Tao, K., Han, S.C., Su, L.L., and Hu, D. (2018). Acute downregulation of miR-199a attenuates sepsis-induced acute lung injury by targeting SIRT1. *Am. J. Physiol., Cell Physiol.* 314 (4): C449–C455.

Liu, Y., Shepherd, E.G., and Nelin, L.D. (2007). MAPK phosphatases–regulating the immune response. *Nat. Rev. Immunol.* 7 (3): 202–212.

Liu, W., Jin, Y., Zhang, W., Xiang, Y., Jia, P., Yi, M., and Jia, K. (2020). MiR-202-5p inhibits RIG-I-dependent innate immune responses to RGNNV infection by targeting TRIM25 to mediate RIG-I ubiquitination. *Viruses* 12 (3): 261.

Mattingly, C.J., Hampton, T.H., Brothers, K.M., Griffin, N.E., and Planchart, A. (2009). Perturbation of defense pathways by low-dose arsenic exposure in zebrafish embryos. *Environ. Health Perspect.* 117 (6): 981–987.

Murphy, K. and Weaver, C. (2016). *Janeway's Immunobiology*. New York: W. W. Norton.

Nayak, A.S., Lage, C.R., and Kim, C.H. (2007). Effects of low concentrations of arsenic on the innate immune system of the zebrafish (*Danio rerio*). *Toxicol. Sci.* 98 (1): 118–124.

O'Connell, R.M., Taganov, K.D., Boldin, M.P., Cheng, G., and Baltimore, D. (2007). MicroRNA-155 is induced during the macrophage inflammatory response. *Proc. Natl. Acad. Sci. USA* 104 (5): 1604–1609.

Ordas, A., Kanwal, Z., Lindenberg, V., Rougeot, J., Mink, M., Spaink, H.P., and Meijer, A.H. (2013). MicroRNA-146 function in the innate immune transcriptome response of zebrafish embryos to Salmonella typhimurium infection. *BMC Genom.* 14: 696.

Ottman, R. (1996). Gene-environment interaction: Definitions and study designs. *Prev. Med.* 25 (6): 764–770.

Phennicie, R.T., Sullivan, M.J., Singer, J.T., Yoder, J.A., and Kim, C.H. (2010). Specific resistance to *Pseudomonas aeruginosa* infection in zebrafish is mediated by the cystic fibrosis transmembrane conductance regulator. *Infect. Immun.* 78 (11): 4542–4550.

Podgorski, J. and Berg, M. (2020). Global threat of arsenic in groundwater. *Science* 368 (6493): 845–850.

Rager, J.E., Bailey, K.A., Smeester, L., Miller, S.K., Parker, J.S., Laine, J.E., Drobná, Z., Currier, J., Douillet, C., Olshan, A.F., Rubio-Andrade, M., Stýblo, M., García-Vargas, G., and Fry, R.C. (2014). Prenatal arsenic exposure and the epigenome: Altered microRNAs associated with innate and adaptive immune signaling in newborn cord blood. *Environ. Mol. Mutagen.* 55 (3): 196–208.

Rahman, A., Vahter, M., Ekström, E.C., and Persson, L. (2011). Arsenic exposure in pregnancy increases the risk of lower respiratory tract infection and diarrhea during infancy in Bangladesh. *Environ. Health Perspect.* 119 (5): 719–724.

Raqib, R., Ahmed, S., Sultana, R., Wagatsuma, Y., Mondal, D., Hoque, A.M., Nermell, B., Yunus, M., Roy, S., Persson, L.A., Arifeen, S.E., Moore, S., and Vahter, M. (2009). Effects of in utero arsenic exposure on child immunity and morbidity in rural Bangladesh. *Toxicol. Lett.* 185 (3): 197–202.

Ray, A., Bhaduri, A., Srivastava, N., and Mazumder, S. (2017). Identification of novel signature genes attesting arsenic-induced immune alterations in adult zebrafish (*Danio rerio*). *J Hazard Mater.* 321: 121–131.

Reid, N.M., Jackson, C.E., Gilbert, D., Minx, P., Montague, M.J., Hampton, T.H., Helfrich, L.W., King, B.L., Nacci, D.E., Aluru, N., Karchner, S.I., Colbourne, J.K., Hahn, M.E., Shaw, J.R., Oleksiak, M.F., Crawford, D.L., Warren, W.C., and Whitehead, A. (2017). The landscape of extreme genomic variation in the highly adaptable Atlantic killifish. *Genome Biol Evol* 9 (3): 659–676.

Reid, N.M., Proestou, D.A., Clark, B.W., Warren, W.C., Colbourne, J.K., Shaw, J.R., Karchner, S.I., Hahn, M.E., Nacci, D., Oleksiak, M.F., Crawford, D.L., and Whitehead, A. (2016). The genomic landscape of rapid repeated evolutionary adaptation to toxic pollution in wild fish. *Science* 354 (6317): 1305–1308.

Ren, X., McHale, C.M., Skibola, C.F., Smith, A.H., Smith, M.T., and Zhang, L. (2011). An emerging role for epigenetic dysregulation in arsenic toxicity and carcinogenesis. *Environ. Health Perspect.* 119 (1): 11–19.

Roberts, A.W., Kim, C., Zhen, L., Lowe, J.B., Kapur, R., Petryniak, B., Spaetti, A., Pollock, J.D., Borneo, J.B., Bradford, G.B., Atkinson, S.J., Dinauer, M.C., and Williams, D.A. (1999). Deficiency of the hematopoietic cell-specific Rho family GTPase Rac2 is characterized by abnormalities in neutrophil function and host defense. *Immunity* 10 (2): 183–196.

Rodrigues, D.R., Fernandes, R.K., Balderramas, H.E.A., Penitenti, M., Bachiega, T.F., Calvi, S.A., Dias-Melicio, L.A., Ikoma, M.R., and Soares, Â. (2014). Interferon-gamma production by human neutrophils upon stimulation by IL-12, IL-15 and IL-18 and challenge with Paracoccidioides brasiliensis. *Cytokine* 69 (1): 102–109.

Sarkar, S., Mukherjee, S., Chattopadhyay, A., and Bhattacharya, S. (2017). Differential modulation of cellular antioxidant status in zebrafish liver and kidney exposed to low dose arsenic trioxide. *Ecotoxicol Environ Saf.* 135: 173–182.

Sempere, L.F., Cole, C.N., McPeek, M.A., and Peterson, K.J. (2006). The phylogenetic distribution of metazoan microRNAs: Insights into evolutionary complexity and constraint. *J. Exp. Zool. B Mol. Dev. Evol.* 306 (6): 575–588.

Shaw, J.R., Gabor, K., Hand, E., Lankowski, A., Durant, L., Thibodeau, R., Stanton, C.R., Barnaby, R., Coutermarsh, B., Karlson, K.H., Sato, J.D., Hamilton, J.W., and Stanton, B.A. (2007). Role of glucocorticoid receptor in acclimation of killifish (Fundulus *heteroclitus*) to seawater and effects of arsenic. *Am. J. Physiol. Regul. Integr. Comp. Physiol.* 292 (2): R1052–1060.

Shaw, J.R., Hampton, T.H., King, B.L., Whitehead, A., Galvez, F., Gross, R.H., Keith, N., Notch, E., Jung, D., Glaholt, S.P., Chen, C.Y., Colbourne, J.K., and Stanton, B.A. (2014). Natural selection canalizes expression variation of environmentally induced plasticity-enabling genes. *Mol. Biol. Evol.* 31 (11): 3002–3015.

Simon, A.K., Hollander, G.A., and McMichael, A. (2015). Evolution of the immune system in humans from infancy to old age. *Proc. Biol. Sci.* 282 (1821): 20143085.

Small, E.M. and Olson, E.N. (2011). Pervasive roles of microRNAs in cardiovascular biology. *Nature* 469 (7330): 336–342.

Statello, L., Guo, C.J., Chen, L.L., and Huarte, M. (2021). Gene regulation by long non-coding RNAs and its biological functions. *Nat. Rev. Mol. Cell Biol.* 22 (2): 96–118.

States, J.C., Singh, A.V., Knudsen, T.B., Rouchka, E.C., Ngalame, N.O., Arteel, G.E., Piao, Y., and Ko, M.S. (2012). Prenatal arsenic exposure alters gene expression in the adult liver to a proinflammatory state contributing to accelerated atherosclerosis. *PLoS One* 7 (6): e38713.

Sturchio, E., Colombo, T., Boccia, P., Carucci, N., Meconi, C., Minoia, C., and Macino, G. (2014). Arsenic exposure triggers a shift in microRNA expression. *Sci. Total Environ.* 472: 672–680.

Sullivan, C., Soos, B.L., Millard, P.J., Kim, C.H., and King, B.L. (2021). Modeling virus-induced inflammation in zebrafish: A balance between infection control and excessive inflammation. *Front Immunol* 12: 636623.

Sun, H.J., Zhang, J.Y., Wang, Q., Zhu, E., Chen, W., Lin, H., Chen, J., and Hong, H. (2019). Environmentally relevant concentrations of arsenite induces developmental toxicity and oxidative responses in the early life stage of zebrafish. *Environ. Pollut.* 254 (Pt A): 113022.

Sun, L., Deng, L., Ea, C.K., Xia, Z.P., and Chen, Z.J. (2004). The TRAF6 ubiquitin ligase and TAK1 kinase mediate IKK activation by BCL10 and MALT1 in T lymphocytes. *Mol. Cell* 14 (3): 289–301.

Takeda, K. and Akira, S. (2004). TLR signaling pathways. *Semin. Immunol.* 16 (1): 3–9.

Tao, L., Pang, Y., Wang, A., Li, L., Shen, Y., Xu, X., and Li, J. (2021). Functional miR-142a-3p induces apoptosis and macrophage polarization by targeting tnfaip2 and glut3 in grass carp (*Ctenopharyngodon idella*). *Front Immunol.* 12 (2481).

Tarver, J.E., Sperling, E.A., Nailor, A., Heimberg, A.M., Robinson, J.M., King, B.L., Pisani, D., Donoghue, P.C., and Peterson, K.J. (2013). miRNAs: Small genes with big potential in metazoan phylogenetics. *Mol. Biol. Evol.* 30 (11): 2369–2382.

Tecchio, C., Micheletti, A., and Cassatella, M.A. (2014). Neutrophil-derived cytokines: Facts beyond expression. *Front Immunol* 5: 508.

USEPA [Environment Protection Agency]. UUÜ (2018). *Drinking water requirements for states and public water systems.* https://www.epa.gov/dwreginfo/chemical-contaminant-rules (accessed July 15 2021).

Vos, J.G. (1977). Immune suppression as related to toxicology. *CRC Crit. Rev. Toxicol.* 5 (1): 67–101.

Wang, J., Bai, X., Song, Q., Fan, F., Hu, Z., Cheng, G., and Zhang, Y. (2015). miR-223 inhibits lipid deposition and inflammation by suppressing toll-like receptor 4 signaling in macrophages. *Int. J. Mol. Sci.* 16 (10): 24965–24982.

World Health Organization. (2011). *Guidelines for Drinking-water Quality*, 4th edn. Geneva: World Health Organization.

Xiao, A., Wang, Z., Hu, Y., Wu, Y., Luo, Z., Yang, Z., Zu, Y., Li, W., Huang, P., Tong, X., Zhu, Z., Lin, S., and Zhang, B. (2013a). Chromosomal deletions and inversions mediated by TALENs and CRISPR/Cas in zebrafish. *Nucleic Acids Res.* 41 (14): e141.

Xiao, A., Wu, Y., Yang, Z., Hu, Y., Wang, W., Zhang, Y., Kong, L., Gao, G., Zhu, Z., Lin, S., and Zhang, B. (2013b). EENdb: A database and knowledge base of ZFNs and TALENs for endonuclease engineering. *Nucleic Acids Res.* 41 (Database issue): D415–422.

Xu, H., Lam, S.H., Shen, Y., and Gong, Z. (2013). Genome-wide identification of molecular pathways and biomarkers in response to arsenic exposure in zebrafish liver. *PLoS One* 8 (7): e68737.

Xu, H., Zhang, X., Li, H., Li, C., Huo, X.J., Hou, L.P., and Gong, Z. (2018). Immune response induced by major environmental pollutants through altering neutrophils in zebrafish larvae. *Aquat. Toxicol.* 201: 99–108.

Xu, Y., Tao, X., Shen, B., Horng, T., Medzhitov, R., Manley, J.L., and Tong, L. (2000). Structural basis for signal transduction by the Toll/interleukin-1 receptor domains. *Nature* 408 (6808): 111–115.

Yamamoto, M. and Akira, S. (2005). TIR domain-containing adaptors regulate TLR signaling pathways. *Adv. Exp. Med. Biol.* 560: 1–9.

Yang, F.C., Atkinson, S.J., Gu, Y., Borneo, J.B., Roberts, A.W., Zheng, Y., Pennington, J., and Williams, D.A. (2001). Rac and Cdc42 GTPases control hematopoietic stem cell shape, adhesion, migration, and mobilization. *Proc. Natl. Acad. Sci. USA* 98 (10): 5614–5618.

Yao, R.W., Wang, Y., and Chen, L.L. (2019). Cellular functions of long noncoding RNAs. *Nat. Cell Biol.* 21 (5): 542–551.

Yin, V.P., Thomson, J.M., Thummel, R., Hyde, D.R., Hammond, S.M., and Poss, K.D. (2008). Fgf-dependent depletion of microRNA-133 promotes appendage regeneration in zebrafish. *Genes Dev.* 22 (6): 728–733.

Zhou, W., Pal, A.S., Hsu, A.Y., Gurol, T., Zhu, X., Wirbisky-Hershberger, S.E., Freeman, J.L., Kasinski, A.L., and Deng, Q. (2018). MicroRNA-223 suppresses the canonical NF-kappaB pathway in basal keratinocytes to dampen neutrophilic inflammation. *Cell Rep.* 22 (7): 1810–1823.

Zhou, Y., Wang, Y., Su, J., Wu, Z., Wang, C., Zhong, W., Liu, X., Cui, L., Zhou, X., Ma, Y., Xin, Y., Zhang, J., Wu, L., Hu, X., Chen, X., Peng, C., and Gao, M. (2017). Integration of microRNAome, proteomics and metabolomics to analyze arsenic-induced malignant cell transformation. *Oncotarget* 8 (53): 90879–90896.

Zu, Y., Tong, X., Wang, Z., Liu, D., Pan, R., Li, Z., Hu, Y., Luo, Z., Huang, P., Wu, Q., Zhu, Z., Zhang, B., and Lin, S. (2013). TALEN-mediated precise genome modification by homologous recombination in zebrafish. *Nat. Methods* 10 (4): 329–331.

7

Transcriptomics

Applications in Toxicology and Medicine

Pius Joseph

Molecular Carcinogenesis Laboratory, Toxicology and Molecular Biology Branch, Health Effects Laboratory Division, National Institute for Occupational Safety and Health (NIOSH)

Introduction

Materials that are foreign to a biological system are commonly referred as xenobiotics. Chemicals that are synthesized for specific applications constitute a major class among xenobiotics. Nanomaterials with several desirable features such as high aspect ratio, light weight, and high tensile strength constitute yet another class of xenobiotics, though a relatively new one among them. Drugs, developed and marketed by the pharmaceutical industry to treat various diseases, are also considered to be xenobiotics. The number of new xenobiotics introduced for various applications has steadily increased over the past several decades. For example, since the establishment of the Toxic Substances Control Act (TSCA) Inventory in 1976, nearly 42,000 new chemical submissions or Premanufacture Notices (PMNs) have been reviewed by the United States Environmental Protection Agency (EPA 2021). Similarly, the global market size of nanomaterials and products enabled by them, valued at 8.5 billion US dollars in 2019, is expected to grow at a compound rate of 13.1% from 2020 to 2027, primarily owing to demands from sectors such as healthcare, cosmetics, electronics, paint and coating, energy and power, automotive, and aerospace (Market Analysis Report 2020).

Xenobiotics, whether naturally occurring or artificial, have numerous applications that create an opportunity for human exposure to these substances and for their potential entry into the body. Occupational exposure to xenobiotics may occur during the manufacture and use of various products. Similarly, activities such as mining and drilling may result in significant occupational exposure to xenobiotics such as crystalline silica (Prajapati et al. 2021), asbestos (Silvestri et al. 2020), and various metals (Moon et al. 2021) that are present in significant quantities in the earth's crust. Significant amounts of xenobiotics are also present as contaminants in the air and in water, which potentially causes human exposure to them. Next, xenobiotics may enter the food chain and be widely distributed, a process whose outcome is, again, the entry of these substances into the human body. Both prescription and over-the-counter medicines are employed to treat various diseases, as well as to improve the well-being of humans. Therefore human exposure to xenobiotics is almost unavoidable, primarily on account of their omnipresence in the food, in the environment, and at the workplace, especially given their intentional use as medicines.

Genomic and Epigenomic Biomarkers of Toxicology and Disease: Clinical and Therapeutic Actions,
First Edition. Edited by Saura C. Sahu.

Exposure to xenobiotics, be it intentional or non-intentional, may result in their entry into the human body: both artificial and naturally occurring xenobiotics may enter the human body after exposure to them. The potential for many xenobiotics to introduce or raise toxicity or to have adverse effects on health is well established. The human body is capable of detoxifying and thus eliminating the toxic materials in order to defend itself against the toxicity. However, too much exposure to xenobiotics may exceed the detoxification and defense capacity of the organism. Virtually every organ and tissue in the human body has been identified as a target for toxicity and adverse health effects that may result from exposure to various xenobiotics. For example, such exposure has been identified as a major etiological factor in several adverse cardiovascular (Lippman 2014), renal (Mascarenhas et al. 2017), pulmonary (Peretz et al. 2006), neurological (Racette et al. 2012), reproductive (Cavallini et al. 2016), and dermal effects (Kim et al. 2017) and in diseases that affect the immune system (Pollard 2016) and the endocrine system (den Braver-Sewradj et al. 2020). Many xenobiotics are also classified as human carcinogens (Lamm et al. 2015). So, to summarize, an association is known to exist between exposure to many xenobiotics and the incidence of adverse health effects in humans, including potentially life-threatening diseases (Lamm et al. 2015; Peretz et al. 2006).

At the same time, as already mentioned, human exposure to xenobiotics is almost unavoidable, given their presence everywhere. Therefore it is very important to be able to prevent the onset and progression of adverse health effects associated with human exposure to xenobiotics, especially those effects that can have serious consequences. The prevention of xenobiotic-induced toxicity and its resulting adverse effects on health, a priority for several public health agencies, may be achieved by detecting the xenobiotic-induced toxicity and adverse health effects at an early stage, when these effects are reversible or even preventable. In addition, a proper understanding of the mechanisms that underlie xenobiotic-induced toxicity has merits on its own, for the future prevention of the potentially life-threatening effects of human exposure to toxic xenobiotics. Holistic approaches are valuable precisely because they are capable of both detecting toxicity at an early (and maybe reversible) stage and providing key information on the mechanisms that underlie toxicity.

Significant progress has been made in the past several decades in determining the potential for xenobiotics to cause toxicity after exposure to them, as well as in elucidating the possible mechanisms behind their toxicity. Whether a xenobiotic is toxic or not can be determined by employing a variety of approaches or techniques. It is also possible to determine the potential mechanisms underlying the toxicity of xenobiotics quite efficiently. The main goal of an efficient and successful toxicity test is to detect or predict with accuracy the potential of a xenobiotic to have an adverse health effect on human beings exposed to it. The most desired features of an ideal toxicity test are sensitivity, specificity, and reproducibility. The test should be highly sensitive, so that subtle or preclinical toxicity may be detected long before the onset of any clinical symptom of target organ toxicity. Specificity minimizes the detection of false positives and false negatives. In addition, the test should be non-invasive or minimally invasive in nature, so that it may be easily applied in situations of monitoring for ill-health effects. Being able to accurately detect and predict the potential of xenobiotics to cause health problems immediately after exposure to them, in combination with an understanding of the mechanisms involved in that causal process, will help us develop strategies designed to prevent the onset of toxicity and to protect the exposed individuals from developing conditions that otherwise would result in significant morbidity and mortality.

In the past, various in vitro (cell/tissue culture) and in vivo (experimental animals) toxicity models have been developed and employed to determine the toxicity that could result from exposure to xenobiotic(s). These tests rely heavily on various endpoints pertaining to structural and functional damage to cells and organs associated with exposure to xenobiotics. While the traditional toxicity tests employed can detect xenobiotic-induced toxicity, many of them lack one or more of the desirable features that would make them ideal for predicting or determining toxicity. For example, live animals can be exposed to xenobiotics, and the histological–biochemical changes that take place in organs or in biological fluids can be determined as indicators of target organ toxicity. However, it needs to be emphasized that the biochemical and histological changes detectable in the body fluids and target organs of the exposed animals may not be early predictors of subtle toxicity, but rather indicators of an injury or damage that has already taken place in response to exposure. Such changes are noticeable only when significant damage to the target organs or tissues, including structural and functional damage, has already taken place. Therefore many of the traditional toxicity tests lack the high sensitivity required to detect early, subtle toxicity (toxicity not yet associated with clinical symptoms) after exposure to very low concentrations of a xenobiotic (see Figure 7.1). This deficit makes many of the currently available toxicity tests less desirable in situations such as monitoring for possible health effects of chronic exposure to low concentrations of xenobiotics—such as those that take place routinely in the environment or in the workplace.

The start of this century witnessed a major paradigm shift in toxicology that was mainly due to the realization that manifestations of xenobiotic-induced toxicity are the result of changes in molecules such as genes, proteins, lipids, and metabolites at the cellular level. A new branch in toxicology, referred as toxicogenomics, soon emerged, showing great promise. The availability of powerful revolutionary techniques capable of capturing the

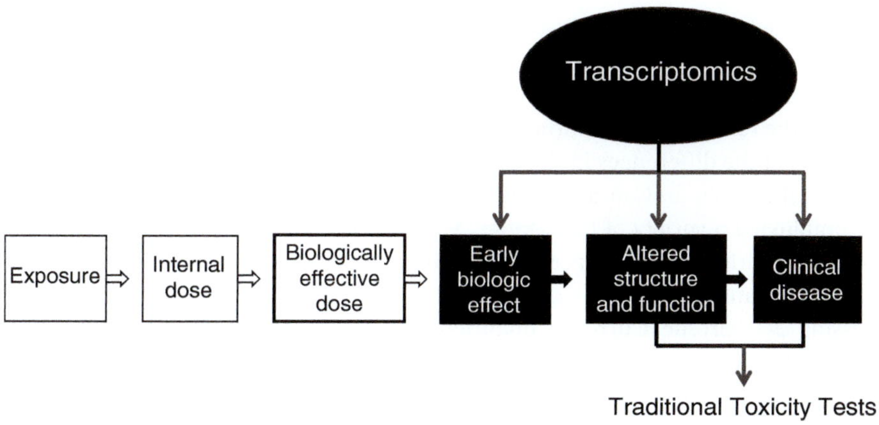

Figure 7.1 Xenobiotic exposure to the disease development model. Xenobiotics may enter the body after their exposure and may produce a biologic effect when the internal or target dose reaches a certain threshold. In the majority of cases, the early, subtle effects of exposure are reversible and may not result in any significant toxicity. Failure to detect and prevent toxicity at the early stage(s) may result later on in significant and advanced toxicity, characterized by structural and functional changes in the target organ(s). Transcriptome analysis, given its superior sensitivity, may detect the early and reversible stage(s) in toxicity development, which potentially leads to the prevention of toxicity and of the adverse health effect(s) associated with it. Traditional toxicity tests, due to their poor sensitivity, may not detect toxicity in its early and reversible stages, and this potentially results in the development of serious adverse health effect(s).

"global" changes collectively referred to as "-ome" that take place in a biological system in various molecules, in response to the xenobiotic exposure, played a significant role in toxicogenomics. The US National Research Council defines toxicogenomics as "combin(ing) toxicology with information-dense genomic technologies to integrate toxicant-specific alterations in gene, protein, and metabolite expression patterns with phenotypic response of cells, tissues, and organisms" (NRC 2007). Toxicogenomics facilitates the interrogation of cellular response to a xenobiotic exposure at the global, molecular level (Alexander-Dann et al. 2018). Recent developments in biology, especially molecular biology, in computer science, in mathematics, and in bioinformatics contributed enormously to the development of various branches of toxicogenomics such as genomics (the study of the genome or DNA), transcriptomics (the study of the transcriptome or transcripts), proteomics (the study of the proteome or proteins), lipidomics (the study of the lipidome or lipids), and metabolomics or metabonomics (the study of the metabolome or metabolites, that is, small molecules). The "-omics" technologies enable the simultaneous measurement of all definable entities of an "-ome." Proteomics, for example, enables the simultaneous detection and analysis of all the proteins expressed in a biological system at a given time.

Whereas the various -omics techniques are helpful and important if we want to understand the toxicity potential of a xenobiotic, the scope of this chapter is restricted to transcriptomics. An overview of techniques currently employed to profile the transcriptome will be presented. The potential applications of transcriptomics in toxicology will be presented, with special emphasis on the prevention of the adverse health effects of human exposure to a specific xenobiotic: crystalline silica. Furthermore, attempts will be made to demonstrate the applicability of transcriptomics in medicine, especially to the detection of diseases, using Alzeimer's disease as an example.

Transcriptomics: Technical Aspects

As already mentioned, the objective of transcriptomics or transcriptome analysis in toxicology is to determine the differential gene expression (DGE)—the changes in the expression profile of the transcripts present in a biological system in response to its exposure to a xenobiotic—and the relevance or importance of DGE in the ensuing toxicity. This involves the quantification of the differences in the abundance of transcripts between a xenobiotic-exposed biological sample and an appropriate control (Figure 7.2). This process can be achieved by employing a variety of techniques. Many of the techniques employed earlier, such as northern hybridization (Bergeron et al. 2001), serial analysis of gene expression (Taulan et al. 2004), and differential display (Tomita et al. 2002), are no longer routinely used in assessing DGE profile. Currently quantitative real time polymerase chain reaction (QRT-PCR), microarray, and next-generation sequence (NGS) analysis or, more specifically, RNA sequencing (RNA-seq) are most frequently employed in profiling a transcriptome. Even though all three techniques currently employed start with the isolation of a high-quality RNA with significant integrity, the subsequent steps taken to determine the DGE profile differ significantly. While QRT-PCR is relatively simple, RNA-seq is the most complex and technically challenging of the three techniques.

QRT-PCR consists of a reverse transcription of mRNAs, undertaken in order to synthesize cDNAs and DNA polymerase-catalyzed amplification, detection, and quantification of the cDNAs that are annealed to specific oligonucleotide primers. While QRT-PCR is quantitatively more accurate than microarray and RNA-seq techniques in determining DGE,

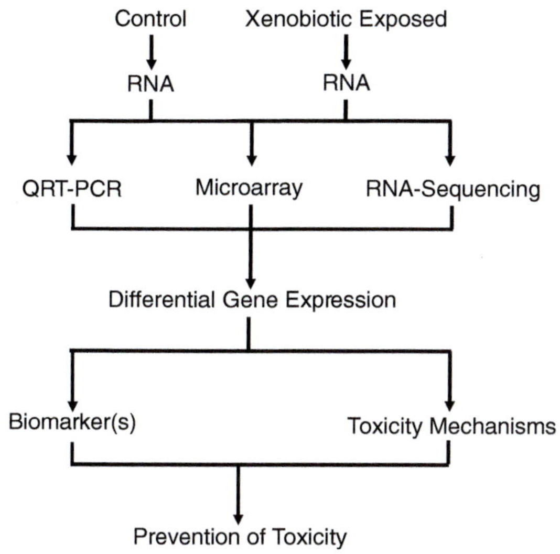

Figure 7.2 Steps involved in the transcriptome analysis potentially leading to the prevention of toxicity. DGE profile in a biological system, in response to exposure to a xenobiotic, is determined by techniques such as quantitative real-time polymerase chain reaction (QRT-PCR), microarray analysis, and RNA-sequencing. Bioinformatic analysis of the DGE data may facilitate the prevention of toxicity through the identification of biomarker(s) for the prediction and detection of toxicity, as well as of the molecular mechanism(s) underlying the toxicity.

the major drawback of this technique is its low throughput. It is not practical to determine the abundance of all, or even most, transcripts present in the transcriptome by QRT-PCR. In addition, on a global basis, QRT-PCR is prohibitively costly, considering that only a limited number of transcripts can be detected simultaneously by this technique. Nevertheless, QRT-PCR is frequently employed, especially in the validation of the DGE profile detected by both microarray and RNA-seq analysis, as well as in situations where the objective of profiling the transcriptome is to determine the expression profile of a limited number of genes that are involved in a specific mechanism, pathway, or network.

The major advantage of employing microarray and RNA-seq, as opposed to QRT-PCR, in transcriptome analysis is their high throughput. Microarray technique is capable of simultaneously detecting tens of thousands of the transcripts present in the transcriptome—or most of them. This technique consists in a sequence-specific hybridization of the transcripts (targets) prepared from a biological sample with the oligonucleotides or cDNAs (probes) arrayed on the microarray. The various steps involved in a microarray analysis of the DGE profile are as follows: reverse transcription of the mRNAs in order to synthesize cDNAs; their labeling with fluorescent dye(s) in order to prepare the targets; sequence-specific hybridization of the labeled targets with oligonucleotides or cDNAs arrayed on the microarray; washing designed to remove non-specific binding of the target, while probes and the free labels potentially contribute to background signal; scanning of the arrays in order to detect the fluorescence intensity associated with the hybridized target–probe complex; and comparison of the signal intensity with appropriate control and xenobiotic-exposed samples, in order to detect and quantify the DGE profile (Joseph 2017). Since its inception in the early 1990s (Bumgarner 2013), the microarray technology has been employed in studying the toxicity of xenobiotics, which affects virtually every organ or system in the body in a variety of species.

RNA-sequence (RNA-seq) analysis is the latest advancement in profiling the transcriptome for DGE. Since RNA-seq is based on *de novo* sequence detection, the technique can detect virtually all transcripts that are present in the transcriptome. RNA-seq consists of multiple steps such as preparation of sequencing libraries from control and xenobiotic-exposed biological samples, sequencing the libraries, and quantification of the transcripts to determine the DGE between the control and the xenobiotic-exposed samples. Since RNA molecules such as rRNA and globin mRNA, which are present in very high abundance in biological samples, are known to result in their overrepresentation during RNA-seq, complicating the detection of less abundant transcripts and their differential expression, it is important to selectively remove those abundant RNA species before or during library preparation. The next step in RNA-seq, as in QRT-PCR and microarray analysis, is the reverse transcription of RNA in order to synthesize cDNAs that are ligated to adaptors that serve as barcodes to identify the cDNAs prepared from a sample. The adaptor-ligated cDNA library samples are subsequently subjected to high-throughput sequencing, typically to a read depth of 10–30 million reads per sample, and the data are further analyzed to determine the differences in the gene expression profile between the xenobiotic-exposed and the appropriately matched control samples.

Analysis of the raw data generated through the techniques employed in DGE profiling is a critical step in the identification and quantification of differentially expressed genes. Data analysis can be simple or very complex, depending on the technique employed to generate the gene expression data. Among the three currently employed techniques, data analysis is least complicated one in the case of QRT-PCR, and consists in normalizing the expression of the genes of interest (target genes) to the expression of one or more house-keeping genes and calculating the fold changes in gene expression between the xenobiotic-exposed and the matched control samples. Data analysis in the case of microarray is not as simple as in the case of QRT-PCR and typically consists in normalizing the hybridization signal intensities and calculating the fold changes in the expression of all the detected genes (tens of thousands in the majority of the cases) represented on the microarray between the sample sets. Data analysis is much more complex in RNA-seq than in QRT-PCR and microarray, and requires significant expertise and computational resources (Stark et al. 2019). The first step in RNA-seq data analysis is alignment or mapping the sequence reads generated to a known transcriptome. Various alignment tools, such as TopHat (Kim et al. 2013), STAR (Dobin et al. 2013), or HISAT (Kim et al. 2015), are available and routinely employed. These tools facilitate the spliced alignment of the cDNA sequences generated by RNA-seq with the exon boundaries of the genes. The next step in data analysis is assignment of the sequences to genes or transcripts in order to determine their counts or abundance. Some of the common quantification tools employed in the RNA-seq analysis are RSEM (Li and Dewey 2011), CuffLinks (Trapnell et al. 2012), MMSeq (Turro et al. 2011), and HTSeq (Anders et al. 2015). Various factors such as read depth, expression patterns of genes, GC content, and technical biases are known to contribute to differences in the quantification of transcripts (Risso et al. 2011; Wagner et al. 2012), and therefore the data need to be normalized and filtered to improve quality. Quantile or median normalization (Bourgon et al. 2010) is commonly employed for this purpose. The final step in RNA-seq data analysis is differential expression modeling, which determines the transcripts that may have changed in expression and the quantification of changes between the sample sets. Some of the tools commonly employed in this step of the RNA-seq analysis are edgeR (Robinson et al. 2010) and limma + voom (Law et al. 2014). Each of the steps involved in RNA-seq data analysis may significantly influence the DGE data output; in consequence, great care should be exercised while analyzing the RNA-seq data to obtain biologically or toxicologically meaningful gene expression data.

Despite the challenges, especially those associated with data analysis, RNA-seq has several advantages by comparison with the microarray analysis in high-throughput DGE studies. It is more sensitive, and therefore capable of detecting transcripts that are present in very low abundance (Wang et al. 2014). Since microarray is based on hybridization between the labeled transcripts (probes) expressed in the biological sample and the targets that are arrayed on the microarray platform, unknown or novel transcripts are not detected by microarray technique. RNA-seq, on the other hand, involves *de novo* sequencing of all the transcripts present in the biological specimen employed and for this reason can detect novel transcripts and RNA species other than mRNA such as miRNA, long non-coding RNA (lncRNA), and small nuclear RNA (snRNA) (Stark et al. 2019). RNA-seq, unlike microarray, can also detect splice variants of the transcripts. Despite being the newcomer in the field, RNA-seq is being increasingly used in DGE analysis in toxicogenomics (Sager et al. 2020; Joseph et al. 2021).

The next step in toxicogenomics, after the determination of the gene expression profile in the xenobiotic-exposed and in the appropriately matched control samples, is to determine the relationship, if any, that may exist between the differentially expressed genes and the toxicity that could result from exposure to the xenobiotic. Transcriptome is a very dynamic system and, as described so far, the various steps involved in determining the DGE profile are complex and may introduce errors, including false positives and false negatives. Similarly, often the gene expression data are very noisy. Therefore it is important to establish that the DGE determined in response to exposure to a xenobiotic is distinct from the background noise and truly related to that xenobiotic's toxicity. Statistical analysis of the gene expression data (Anjum et al. 2016) is helpful in removing the background noise and thus facilitating the identification of genes whose expressions are significantly different across the xenobiotic-exposed and control groups of samples. The most frequently employed approach is to apply a cutoff in fold change in expression, along with a correction for false discovery rate (Ritchie et al. 2015; Sellamuthu et al. 2012; Sager et al. 2020; Joseph et al. 2021). It is also very important to establish the relationship between gene expression changes and toxicity phenotypes (i.e., the phenotypic anchoring of DGE) such as histological or biochemical indicators of toxicity, as demonstrated in the case of crystalline silica-induced pulmonary toxicity and changes in the lung gene expression profile (Sellamuthu et al. 2011b, 2012), or acetaminophen-induced hepatotoxicity and DGE in the liver (Hamadeh et al. 2002; Heinloth et al. 2004). The most meaningful, if cumbersome, approach is the use of transgenic models to validate the relationship between toxicity and DGE. Cell lines overexpressing transgenes can be developed and tested in order to confirm the involvement of specific genes in xenobiotic-induced toxicity (Valentine et al. 1996). Transgenic approaches facilitating the suppression of specific genes' expression can also be employed to demonstrate the involvement of genes in toxicity (Salganik et al. 2015). Transgenic animal models have also been successfully employed to demonstrate the functional significance of xenobiotic-induced DGE in the toxicity (Radjendirane et al. 1998).

The downstream applications of DGE data in toxicogenomics, from the perspective of the prevention of xenobiotic-induced adverse health effects, include prediction and detection of the toxicity potential of a xenobiotic and determination of the molecular mechanisms that underlie xenobiotic-induced toxicity. Both these applications have significant implications for the prevention of adverse health effects potentially associated with xenobiotic exposure. This is further discussed in the following sections, where silica-induced lung toxicity will be developed as a specific example.

Superior Sensitivity of Gene Expression Changes to Detect Target Organ Toxicity

The early detection of target organ toxicity is important for the prevention of serious and potentially life-threatening conditions associated with exposure to xenobiotics. This operation will require techniques with superior sensitivity, capable of catching the phenomenon of target organ toxicity in its earliest stages, when development might still be prevented (Figure 7.1). Whether transcriptome profiling is superior in this respect to traditional toxicity endpoints such as histology and clinical chemistry was investigated by employing animal models of hepatotoxicity in the early 00s. In a classic study, Heinloth et al. (2004), employed acetaminophen, a prototype hepatotoxic xenobiotic, to investigate the potential application of transcriptome profiling as a highly sensitive technique of detecting target organ toxicity. Rats were administered acetaminophen at either subtoxic (50 and 150 mg/kg body weight (bw)) or toxic (1,500 mg/kg bw) doses. After acetaminophen administration, at time intervals of six, twenty-four, and forty-eight hours, groups of rats were euthanized and hepatotoxicity was assessed on the basis of liver histopathology and blood chemistry in the control and in the acetaminophen-administered rats. The total RNA isolated from the liver was used to determine global gene expression profiles by microarray analysis at each of the post-exposure time intervals selected in the study. On the basis of the results of liver histology and marker enzyme activities in the blood, no hepatotoxicity was detected at any of the post-exposure time intervals analyzed in either group of the rats administered doses of acetaminophen. With respect to liver histology and blood chemistry, the normal results observed in the rats administered the subtoxic doses may indicate an inability of these traditional hepatotoxicity endpoints to detect the subtle toxicity induced by acetaminophen because of their poor sensitivity. Normal liver histology and blood chemistry were also found in the rats administered the overtly toxic dose of acetaminophen (1,500 mg/kg bw) at the earliest post-exposure time interval of six hours. By contrast, significant alterations in liver histology and blood clinical chemistry were noticed at the later intervals of twenty-four and forty-eight hours in the rats administered the same amount (1,500 mg acetaminophen/kg bw); and these alteration suggested the induction of hepatotoxicity. At these later post-exposure intervals, microarray analysis of global gene expression profiling identified in the liver of the rats administered the toxic dose many genes whose expressions were significantly different from those of the vehicle-administered controls. Bioinformatic analysis of the differentially expressed genes suggested the potential involvement of energy metabolism and stress response as key molecular mechanisms underlying the development of acetaminophen-induced hepatotoxicity in the rats. Interestingly, many of the genes whose expressions were significantly affected by the administration of the toxic dose of acetaminophen at the post-exposure intervals of twenty-four and forty-eight hours and are therefore considered to be indicators of acetaminophen-induced hepatotoxicity were also found, differentially expressed, in the rats that had been administered the same toxic dose, when analyzed at the six-hour post-exposure interval. Similarly, significant differential expressions of many of the genes involved in energy metabolism and stress response were detected in the liver of the rats administered the subtoxic doses of acetaminophen, in spite of the absence of histological or biochemical changes indicative of hepatotoxicity. Taken together, these results demonstrated the superior sensitivity of gene expression changes to detect hepatotoxicity by comparison with the sensitivity of traditional histological and clinical markers of hepatotoxicity. A transcriptome profiling sensitivity superior to that of the traditional hepatotoxicity end points has also been reported by Hamadeh et al. (2002) in connection with detecting hepatotoxicity in rats.

In another early toxicogenomic study, Waring et al. (2001) administered multiple well-characterized hepatotoxic chemicals to rats. The hepatotoxicity effects of the chemicals selected for administration included necrosis, DNA damage, cirrhosis, hypertrophy, and hepatic carcinoma. The toxicity parameters employed by investigators in this study to determine hepatotoxicity induction, just like the parameters used by Heinloth et al. (2004), included liver histopathology and the clinical chemistry evaluation of the blood. Global gene expression profiles in the liver of the rats were also determined by microarray analysis. The authors reported a good correlation between histopathology, clinical chemistry, and the gene expression profiles induced in rats through the administration of the hepatotoxic agents of their choice. The authors also attempted to cluster the various hepatotoxic compounds employed in the study according to their hepatotoxicity (histopathology and clinical chemistry results) and liver gene expression profiles. The results of the cluster analysis demonstrated that the various hepatotoxic compounds employed in the study clustered together, generally in good agreement, on the basis of the changes in clinical chemistry, histopathology, and gene expression profile detected in the rats. In some cases, liver gene expression profiles outperformed the traditional hepatotoxicity endpoints (histopathology of liver and blood clinical chemistry) in clustering the hepatotoxic chemicals. On the basis of the results of their study, the authors concluded that global gene expression profiling could be employed as a highly sensitive and reliable technique in identifying hepatotoxic chemicals. Gene expression profiling has been recently reported in the early identification of target organ toxicity induced in a variety of organs and tissues such as liver, kidney, heart, and skeletal muscle (Glaab et al. 2021).

Silica Exposure and Toxicity

Silica is present in abundance in the earth's crust and, additionally, a wide range of materials and products contain substantial amount of silica. Silica exists primarily in amorphous or crystalline forms. Between the amorphous and the crystalline forms of the silica present, the crystalline form is more important with respect to the potential to result in adverse health effects, including cancer and silicosis. The generation of dust that contains crystalline silica particles of respirable size takes place routinely in various occupational settings. Approximately 2.3 million workers in the United States, mainly those employed in the construction industry, are estimated to be exposed to crystalline silica (Doney et al. 2020). Mining, drilling, and hydraulic fracturing are some of the other occupations where significant exposure to crystalline silica is known to take place among workers.

Because of the potential of crystalline silica to induce toxicity, excessive exposure to it is often associated with adverse health outcomes among workers. Lungs, in addition to being the principal route of human exposure to dust that contains crystalline silica of respirable size, is the main target organ for the health effects of silica. The International Agency for Research on Cancer has classified crystalline silica (quartz and cristobalite forms) as a Group 1 human lung carcinogen (IARC 1997). Crystalline silica exposure is also associated with cardiopulmonary failure, mycobacterial infection, renal diseases and renal failures, and autoimmune diseases (Cooper et al. 2002; IARC 1997; Vupputuri et al. 2012). Another serious adverse health outcome associated with crystalline silica exposure is silicosis, an incurable and irreversible but preventable progressive lung fibrotic disease (Greenberg et al. 2007). Silicosis is an interstitial lung disease characterized by alveolar proteinosis and diffuse fibrosis, which result in progressively restrictive lung function and death (Castranova and Vallyathan 2000). Currently there is no effective treatment for silicosis.

Protecting workers through the prevention of the adverse health effects potentially associated with crystalline silica exposure is a priority for the US National Institute for Occupational Safety and Health (NIOSH) and the US Occupational Safety and Health Administration (OSHA). NIOSH (1996) recommends limiting workers' exposure to crystalline silica at 50 $\mu g/m^3$, for up to a ten-hour workday and forty-hour workweek, for their protection. NIOSH also recommends medical monitoring through chest X-rays and pulmonary function testing, respiratory protection, employee training, and work practices and dust control procedures as additional measures designed to protect workers from developing silicosis. OSHA enforces the presence of airborne crystalline silica in workplaces at a permissible exposure limit (PEL), which is currently set at 50 $\mu g/m^3$ (OSHA 2013). The past few decades have registered significant progress in protecting workers from developing silicosis in the United States, mainly through regulation and control of the amount of crystalline silica to which one is exposed in the workplace; this amount was kept below the PEL. Thus, whereas in 1970 more than 1,000 deaths due to silicosis were reported (Bell and Mazurek 2020), in 2005 that figure fell to approximately 200 (Bell and Mazurek 2020). Currently less than 100 deaths per year are attributed to silicosis in the United States (Bell and Mazurek 2020).

Despite significant progress in the prevention of silicosis made in the United States over the past decades recent reports indicate that workers in specific occupations are exposed to crystalline silica at levels much higher than the PEL. For example, it has been reported that, among workers employed in hydraulic fracturing, the level of exposure to crystalline silica is 100 times higher than the PEL (Esswein et al. 2013). Exposure to crystalline silica at levels significantly above the PEL has also been reported among construction workers (Doney et al. 2019, 2020). Occupational exposure to crystalline silica, especially at levels higher than the PEL, is alarming, considering the potential effects of silica on health, and mainly silicosis.

A major limitation in the prevention of silicosis, besides the capacity to limit workers' exposure to dust that contains crystalline silica below the PEL, is the inability to detect the disease at the early or preclinical stage. Currently silicosis is diagnosed by detecting lung abnormalities via chest X-ray. Determination of lung function by spirometry is often employed, too, as a supplemental approach. Both chest X-ray and spirometry are based on the detection of structural and functional abnormalities of the lungs that represent relatively advanced stages in silicosis. Since silicosis is a progressive and irreversible disease, diagnosis at an advanced stage is not significantly helpful for its prevention. On the other hand, the preventable nature of silicosis strongly supports the need to develop tests or techniques sensitive enough to predict the disease or to detect it at a very early, pre-clinical, and potentially preventable stage, in agreement with the NIOSH (2002) recommendation to develop highly sensitive and non-invasive or minimally invasive tests by which silicosis can be caught before it produces any structural or functional impairment of the lungs. Similarly, an understanding of the molecular mechanisms involved in the initiation and progression of silicosis may be helpful for managing the disease. Several studies have been conducted investigating the application of transcriptome profiling to the lung's response to crystalline silica exposure and the ensuing toxicity.

In Vitro Cell Culture Studies

It has been well recognized that the crystalline form of silica is biologically more active and therefore more pathogenic than the amorphous form (Fubini and Hubbard 2003; Johnston et al. 2000). In vitro cell culture models have been employed to determine whether a relationship exists between cytotoxicity and changes in the transcriptome in response to exposure to silica (Perkins et al. 2012). The cytotoxicity that resulted in the human bronchial epithelial cell line (BEAS2B) and primary human bronchial epithelial cells (NHBE) from exposure to crystalline silica was higher than the toxicity induced in the same cells through exposure to the amorphous form. In both the BEAS2B and the NHBE cells, transcriptome profiling revealed more gene expression changes in response to exposure to crystalline silica than in response to the amorphous form. Both the number of the significantly differentially expressed genes (SDEGs) and the fold changes in their differential expression were higher in the cells exposed to the crystalline form of silica than in the cells exposed to the amorphous form. In conclusion, these findings suggested that changes in the transcriptome in the cell culture models employed in the study were relevant and corresponded to the biological activity and pathogenicity of the silica particles employed.

In another in vitro study, Sellamuthu et al. (2011a) employed the human lung epithelial cell line A549 to investigate the relationship between cytotoxicity and alterations in the transcriptome in response to exposure to crystalline silica. The A549 cells in culture were exposed to Min-U-Sil® 5 crystalline silica at concentrations of 15, 30, 60, 120, and 240 μg/cm^2 cell culture area for six hours. The A549 cells in culture were also exposed to Min-U-Sil® 5 crystalline silica at a concentration of 60 μg/m^3 culture area for time intervals of two, six, and twenty-four hours. At the end of the Min-U-Sil® 5 crystalline silica exposure, both in the dose-response studies and in the time-course studies, cytotoxicity and global gene expression profile were determined in the cells respectively by lactate dehydrogenase (LDH) assay and microarray analysis. The number of the SDEGs detected in the cells correlated well with the crystalline silica-induced cytotoxicity; the correlation coefficients (r^2 values) were 0.89 in the dose-response and 0.98 in the time-course study. This further supported the hypothesis of a strong relationship between toxicity and changes in the transcriptome in response to crystalline silica exposure (Figure 7.3). In the same study, the authors also conducted bioinformatic analysis of the genes whose expressions were significantly affected by exposure of the cells to Min-U-Sil® 5 crystalline silica. The enrichment in several biological functions that are relevant to silica-induced cytotoxicity also exhibited the same relationship that was noticed between cytotoxicity and DGE profile in the cells (Figure 7.4). Again, these results suggested that the changes in the transcriptome detected in the cell culture model were relevant to the mechanisms that underlie the crystalline silica-induced toxicity.

In Vivo Animal Studies

Several investigators have employed animal models to study the lung's response to crystalline silica exposure. Silicosis is characterized by fibrotic changes that culminate in a significant reduction in lung function, potentially causing death. Inflammation plays a critical role in this disease (Castranova 2004). Rats exhibit a significant similarity to humans when it comes to the initiation and progression of silicosis and for this reason have been used in our laboratory as the preferred animal model for studying human silicosis. In a series of studies conducted in our laboratory, rats were exposed, through inhalation, to air or an aerosol that contained crystalline silica and induced lung toxicity. Lungs and blood samples collected from both controls and silica-exposed rats were analyzed in order to determine the application of transcriptomics in the lung's response to crystalline silica exposure. The major findings of these studies are described in the next sections.

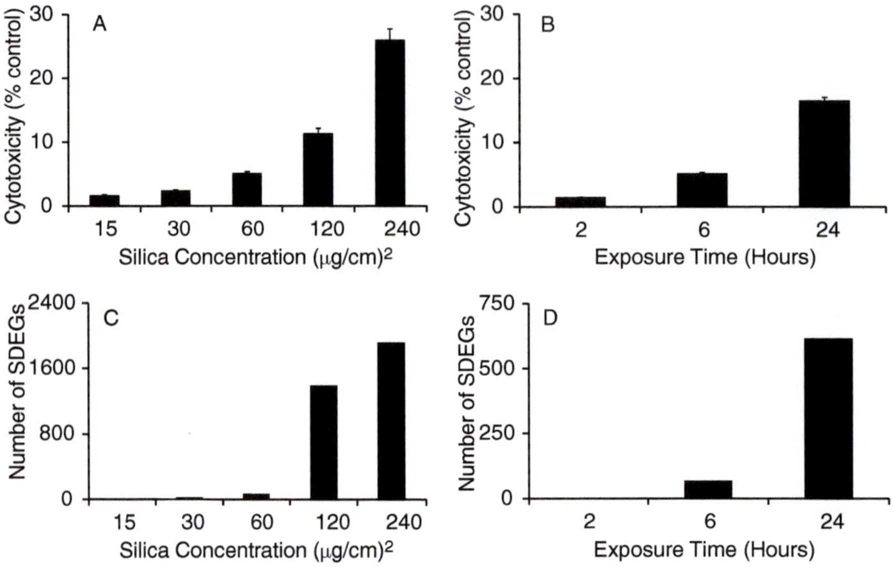

Figure 7.3 Cytotoxicity and DGE profile in an in vitro silica toxicity model. Exponentially growing human lung epithelial cells (A549) were exposed to Min-U-Sil® 5 crystalline silica particles at concentrations of 15, 30, 60, 120, and 240 μg/cm^2 for 6-hours (A and C) or 60 μg/cm^2 for time intervals of 2, 6, or 24 hours (B and D). Cytotoxicity (A and B) and global gene expression profile (C and D) were determined by LDH assay and microarray analysis respectively, as described in our original publication (Sellamuthu et al. 2011a). Cytotoxicity (percentage of dead cells in the silica-exposed, by comparison with the control) and the number of SDEGs were calculated with respect to the corresponding control and presented. Reproduced with permission from *Inhalation Toxicology* 2011 (23): 927–937.

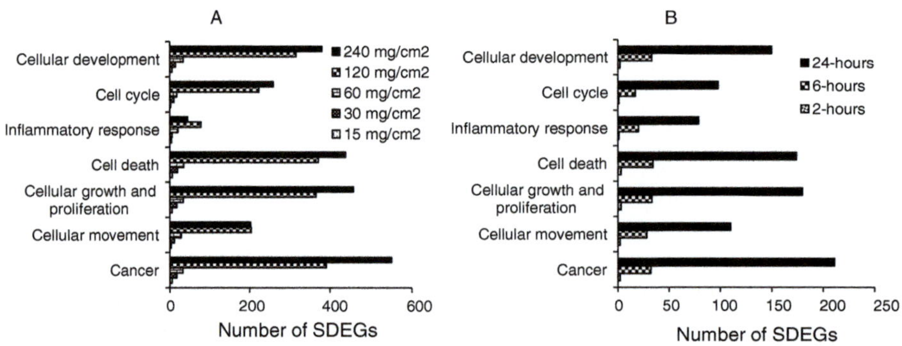

Figure 7.4 Biological functions enriched in response to crystalline silica exposure in an in vitro toxicity model. Exponentially growing human lung epithelial cells (A549) were exposed to Min-U-Sil® 5 crystalline silica particles at concentrations of 15, 30, 60, 120, and 240 μg/cm^2 for 6 hours (A) or 60 μg/cm^2 for time intervals of 2, 6, or 24 hours (B). DGE profiles in the silica-exposed cells, compared with profiles in the corresponding controls, were determined by microarray analysis, and the gene expression data were further analyzed to identify the biological functions significantly enriched in response to silica exposure, as described in the original publication (Sellamuthu et al. 2011a). Reproduced with permission from *Inhalation Toxicology* 2011 (23): 927–937.

Lung Response to Crystalline Silica Exposure in Rats

Healthy male Fisher 344 rats were exposed, though whole-body inhalation, to air or an aerosol containing Min-U-Sil® 5 crystalline silica (US Silica, Berkley Springs, WV, US). The crystalline silica exposure was performed at a concentration of 15 mg/m^3, six hours a day for five consecutive days. At post-exposure time intervals of 1, 2, 4, 8, 16, and 32 weeks, both the controls and the silica-exposed rats were euthanized and their pulmonary responses to the exposures were assessed on the basis of the biochemical and histological changes found in the lungs. Biochemical parameters of toxicity in the bronchoalveolar lavage fluid (BALF), such as LDH activity and levels of proinflammatory cytokines, were significantly affected in the silica-exposed rat lungs by comparison with the time-matched controls (Figure 7.5). The fold changes in the selected BALF parameters of toxicity corresponded to the post-exposure time intervals at which the rats were euthanized in order for the analyses to be performed. The changes were lowest during the initial and highest during the last post-exposure interval analyzed, intermediate changes being noticed during the intervening intervals. The lung histological changes detected in the rats exhibited a trend, just like the changes detected in the biochemical parameters determined in the BALF samples. No significant lung histological changes were detected during the early post-exposure intervals up to the fourth week. During the eighth week, the major histological change detected in the lungs of silica-exposed rats was the infiltration of macrophages and neutrophils. During the sixteenth, in addition to a further increase in the infiltration of the phagocytes, type II pneumocyte hyperplasia was detected in the lungs of the same rats (Sellamuthu et al. 2011b). Finally, during the thirty-second week, a further progression was detected in all the histological changes noticed in the rats' lungs in the previous weeks. In the same week fibrosis, as evidenced by the presence of collagen fibers, was also found in the lungs of the silica-exposed rats (Sellamuthu et al. 2012). In summary, the results demonstrated a steady progression in the lung's response to crystalline silica exposure in the model employed in the study.

Lung Gene Expression Changes in Response to Crystalline Silica Exposure

The total RNA was isolated from the unlavaged apical lung lobe obtained from the air- and silica-exposed rats and employed to determine the global gene expression profile through

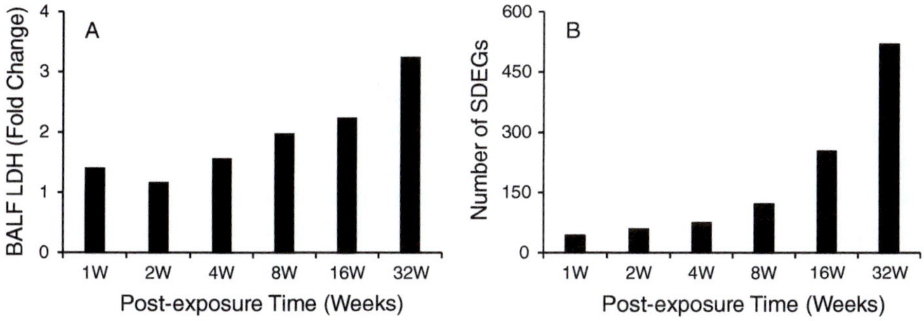

Figure 7.5 Crystalline silica-induced lung toxicity and DGE profile in an in vivo rat pulmonary toxicity model. Rats were exposed to crystalline silica (15 mg/m^3, six hours a day for five days) and the BALF LDH activity and the number of SDEGs in the lungs were determined, as described in detail in the original publications (Sellamuthu et al. 2011b, 2012, 2013). The fold changes in the BALF LDH activity (A) and the number of genes significantly differentially expressed (B) in the silica exposed rat lungs, compared with the corresponding time-matched controls at the post-exposure time intervals analyzed [(W = Week(s)], are presented. Reproduced with permission from *Toxicological Sciences* 2011 (122): 253–264; *Inhalation Toxicology* 2012 (24): 570–579; and *Journal of Applied Toxicology* 2013 (33): 301–312.

microarray analysis (Sellamuthu et al. 2012, 2013). SDEGs were identified in the lungs of the silica-exposed rats by comparison with the time-matched controls. Many genes were found to be significantly differentially expressed in the lungs of the rats in response to their inhalation exposure to crystalline silica. The number of SDEGs identified in the rat lungs increased steadily during the post-exposure intervals and correlated well with the lung toxicity parameters determined; the correlation coefficients (r^2 values) ranged between 0.776 and 0.927 (Figure 7.5). These results suggested that the lung's transcriptomic changes were related to the lung's response to crystalline silica exposure in the rat model employed.

Blood Gene Expression Changes in Response to Crystalline Silica Exposure

The investigators (Sellamuthu et al. 2011b) also isolated the total RNA from the blood samples of the silica-exposed rats and the corresponding time-matched controls and performed a microarray analysis to determine changes in the blood transcriptome. Significant changes in the blood gene expression profiles were detected in the rats in response to their inhalation exposure to the aerosol that contained crystalline silica. The number of SDEGs detected in the blood samples analyzed correlated well with the results of the lung toxicity parameters determined in the rats; the correlation coefficients (r^2 values) ranged between 0.831 and 0.958. In fact the number of SDEGs detected in the blood samples, when compared with the number of SDEGs detected in the lungs, correlated slightly better with the changes in the lung toxicity parameters. Thus the findings of the blood transcriptome analysis suggested that the blood gene expression profiling is a highly sensitive and toxicologically relevant end point in determining the lung response to crystalline silica exposure in the rat model employed.

Blood Gene Expression Signature to Predict/Detect Silica Exposure/Lung Toxicity

As described here, silicosis, a serious adverse health outcome associated with crystalline silica exposure, is an irreversible, incurable, and potentially fatal disease. Silicosis results mostly from low-level chronic exposures to silica such as those that typically take place in most occupations and is characterized by a long latency period, during which the disease progresses asymptomatically, and therefore without being diagnosed. Chest radiography, the most popular among the techniques currently employed in its diagnosis, shows the structural changes that occur in the lungs in response to crystalline silica exposure. But subtle changes, which occur during the initial phase of silicosis development, cannot be detected through chest X-ray because of its poor sensitivity. Therefore, in the majority of the cases, silicosis progresses undetected until the manifestation of clinical symptoms, which are typically associated with advanced and incurable stages in the development of this disease.

As already pointed out, a characteristic feature of the transcriptome is its superior sensitivity and capacity to detect lower levels of the toxicity that follows exposure to xenobiotics. It is entirely possible that gene expression changes that are relevant to xenobiotic-induced toxicity precede the alterations perceived in other parameters, which are commonly employed in the detection of toxicity—for instance biochemical changes in tissues or biofluids, and histological changes in target organs. Furthermore, it has been recognized that changes in the transcriptome that take place in surrogate tissues are relevant to target organ toxicity (Sellamuthu et al. 2012). Since the blood gene expression changes that occurred in rats in response to inhalation exposure to crystalline silica corresponded to the progression of their lungs' response (Sellamuthu et al. 2011b), a rat model was employed to develop and test a blood gene expression signature predictive of crystalline silica-induced lung toxicity. Rats were exposed to air or an aerosol that contained crystalline silica at concentrations of 1, 2, or 15 mg/m^3, six hours a day, for five consecutive days. Approximately sixteen hours after the last exposure, the rats were euthanized and their lung response to crystalline silica exposure was determined through lung histology and BALF parameters of pulmonary toxicity. The total RNA isolated from the unclotted blood samples obtained from the rats was used to determine global gene expression changes through microarray analysis. The rats' exposure, through inhalation, to the highest concentration of crystalline silica employed in the study (15 mg/m^3) resulted in significant pulmonary toxicity, as evidenced by changes in the BALF parameters. The rats exposed to crystalline silica at the concentration of 2 mg/m^3 exhibited mild lung toxicity, as indicated by low but statistically significant changes in the BALF parameters. On the other hand, the rats exposed to crystalline silica at the concentration of 1 mg/m^3 did not exhibit any pulmonary toxicity detectable by the BALF parameters analyzed.

Global gene expression profiles in the RNA isolated from the blood samples of the rats were analyzed using microarray. By employing the blood gene expression data obtained from the rats in the highest dose group (15 mg/m^3) as the training set, the researchers developed a blood gene expression signature designed to predict crystalline silica-induced lung toxicity (Sellamuthu et al. 2011b). The gene expression signature thus developed was tested in the other two groups of rats, exposed to lower concentrations of crystalline silica (the 2 and 1 mg/m^3 test sets). The predictive gene expression signature that was employed identified correctly seven out of the eight rats (87.5% accuracy) that exhibited mild lung toxicity in response to their exposure to 2 mg/m^3 crystalline silica. In the remaining test set, that of the rats exposed to 1 mg/m^3 crystalline silica with no detectable lung toxicity, six out of the eight silica-exposed rats (75% accuracy) were correctly identified through the blood gene expression signature. These results therefore demonstrated the applicability of blood transcriptome profiling as a highly sensitive technique to the detection of exposure to crystalline silica and of the resulting lung toxicity in the rat model. Importantly, this includes subtle toxicity, undetectable through the application of histology or biochemical parameters. It is worth mentioning that the gene expression signature developed and tested for crystalline silica-induced lung toxicity in the rats was quite distinct from the one that was employed to predict hepatotoxicity with a very high degree of accuracy (Bushel et al. 2007), despite the predominance of genes involved in an inflammatory response in both signatures. Similarly, the blood gene expression signature identified in the rat model for silica-induced lung toxicity was distinctly different from those successfully employed in the detection of pulmonary response to cigarette smoke (Charlesworth et al. 2010) and diesel exhaust (Peretz et al. 2007) exposures, despite the fact that inflammation was the prominent mechanism involved in the lung toxicity induced by all three agents investigated. These results may therefore indicate a specificity of the blood gene expression signatures developed and employed to detect or predict lung toxicity in response to exposure to distinct pulmonary toxic agents.

Molecular Mechanisms of Lung Response to Silica Exposure

The gene expression data obtained by the microarray analysis of the lung samples, isolated from the Min-U-Sil® 5 crystalline silica-exposed rats (15 mg/m^3, six hours a day for five days) at post-exposure time intervals of one, two, four, eight, and sixteen weeks, were subjected to further bioinformatic analysis, in order for the molecular mechanisms potentially involved in the production of lung toxicity in response to the exposure to be understood. The initiation and progression of silicosis is primarily driven by inflammation (Castranova 2004). Furthermore, silicosis is characterized by fibrosis (Rimal et al. 2005), which compromises the lung function. Therefore the discussion of the molecular mechanisms that underlie the lung's response to silica exposure in the rats is focused on the induction of inflammation and fibrosis.

Crystalline Silica and Lung Inflammation

After entering the lungs, crystalline silica particles are engulfed by phagocytes, primarily alveolar macrophages, which attempt to eliminate them. The interaction of inhaled silica particles with the resident alveolar macrophages may result in damage to the macrophages, and this triggers their activation. The transcript for the gene *RETNLA*, a marker for macrophage activation (Jablonski et al. 2015), was significantly overexpressed in the lungs of the silica exposed rats, and the fold changes in its overexpression corresponded to the progression in lung toxicity during the post-exposure time intervals (Table 7.1). The damaged or activated macrophages release various molecules that act as signaling mediators, playing key roles in the lung's response to silica exposure. Molecules that are involved in inflammation represent a major class among those released by the activated or damaged macrophages in response to crystalline silica exposure.

Table 7.1 Fold changes in expressions of genes related to inflammation in the lungs of silica exposed rats.

Gene	Fold change in expression				
	1 week	2 weeks	4 weeks	8 weeks	16 weeks
Chemokine (C-C motif) ligand 2 (CCl2)	2.36	1.99	2.30	4.28	6.22
Chemokine (C-C motif) ligand 3 (CCl3)	1.84	1.53	1.58	2.24	2.25
Chemokine (C-C motif) ligand 4 (CCl4)	1.21	1.12	1.23	1.50	1.50
Chemokine (C-C motif) ligand 7 (CCl7)	1.39	1.29	1.35	2.37	4.32
Chemokine (C-X-C motif) ligand 11 (CXCl1)	3.05	2.26	2.76	3.20	2.71
Chemokine (C-X-C motif) ligand 12 CXCl2	1.46	1.12	1.22	1.57	1.42
Chemokine (C-X-C motif) ligand 15 (CXCl5)	2.52	2.06	3.39	4.32	3.96
Chemokine (C-X-C motif) ligand 19 (CXCl9)	1.09	1.29	1.31	2.08	2.91
Chemokine (C-X-C motif) ligand 10 (CXCl10)	1.00	1.05	1.04	1.04	1.52
Chemokine (C-X-C motif) ligand 11 (CXCl11)	1.15	1.04	1.05	1.40	2.63
Interleukin 1 beta (IL1β)	1.21	1.16	1.22	1.48	1.73
Interleukin 1 receptor antagonist, transcript variant 2 (ILR2)	1.45	1.21	1.21	2.27	2.80
Resistin-like alpha (RETNLA)	2.46	2.52	2.42	3.57	8.46

(Continued)

Table 7.1 (Continued)

S100 calcium binding protein A8 (S100A8)	1.26	1.16	1.42	2.50	3.59
Triggering receptor expressed on myeloid cells 1 (TREM1)	1.30	1.08	1.10	1.46	1.72
Triggering receptor expressed on myeloid cells 2 (TREM2)	1.50	1.28	1.27	1.63	1.64
Lipocalin 2 (LCN2)	3.24	3.34	3.58	5.53	5.96
Chitinase 3-like 1 (CHI3L1)	1.89	2.15	2.17	2.75	3.34
Secreted phosphoprotein1 (SPP1)	1.34	1.01	1.12	1.85	6.27

Rats were exposed to Min-U-Sil® 5 crystalline silica (15 mg/m^3, six hours a day for five days) and the lung gene expression profile was determined by microarray analysis at the post-exposure time intervals indicated in the table. The gene expression data were further analyzed, and genes belonging to the IPA category inflammation were identified. Fold changes in expressions of the genes, compared with corresponding changes in the time-matched controls, are presented.
Modified with permission from Sellamuthu et al., *J. Appl. Toxicol.* 2013 (33): 301–312.

Numerous studies have demonstrated a prominent role for inflammation in the lung's response to crystalline silica exposure, and especially in silicosis (Castranova 2004). Further evidence for the involvement of inflammation in this disease comes from the observation that the treatment of patients with the anti-inflammatory agent prednisolone produced a significant improvement in lung function (Sharma et al. 1991). A significant induction of inflammation was detected in the lungs of the rats used in the investigation summarized here of the pulmonary response to exposure to crystalline silica (Sellamuthu et al. 2011b). The pulmonary inflammation induced in the rats involved an increase in the number of PMNs that infiltrated the lungs, as well as an increase in the BALF levels of inflammatory cytokines. Like the lung toxicity parameters, namely LDH activity and lung histology, the infiltration of PMNs into the lungs and the increases in the BALF levels of the cytokines steadily progressed in the rats during post-exposure intervals, and this suggested the onset and progression of pulmonary inflammation in correlation with the progression of toxicity from silica exposure.

As we will see later, microarray analysis of the global gene expression profile in the lungs of the silica-exposed rats identified significant differential expressions of many genes involved in inflammation. Bioinformatic analysis of the SDEGs detected in the lungs of these rats found a significant enrichment in many biological functions that are involved in an inflammatory response. Some of the top-ranking biological functions significantly enriched in the rats' lungs were related to respiratory diseases, inflammatory diseases, cellular movement, cell-to-cell signaling and interaction, and inflammatory response (Figure 7.6). The enrichment in these biological functions, all clustered around inflammation, also exhibited a steady progression during the post-exposure intervals analyzed and correlated with the progression of lung injury. All this suggested the involvement of inflammation in this injury, which constituted the lungs' response to the presence of crystalline silica in them. A similar progression, this time in the enrichment of canonical pathways and networks related to inflammation, was detected in the lungs of the rats in response to their exposure to and inhalation of crystalline silica. This further supports the involvement of inflammation in the induction and progression of lung toxicity in the rats.

A list of the genes that were found to be significantly differentially expressed in the lungs of these rats by comparison with the time-matched controls, along with their fold changes in

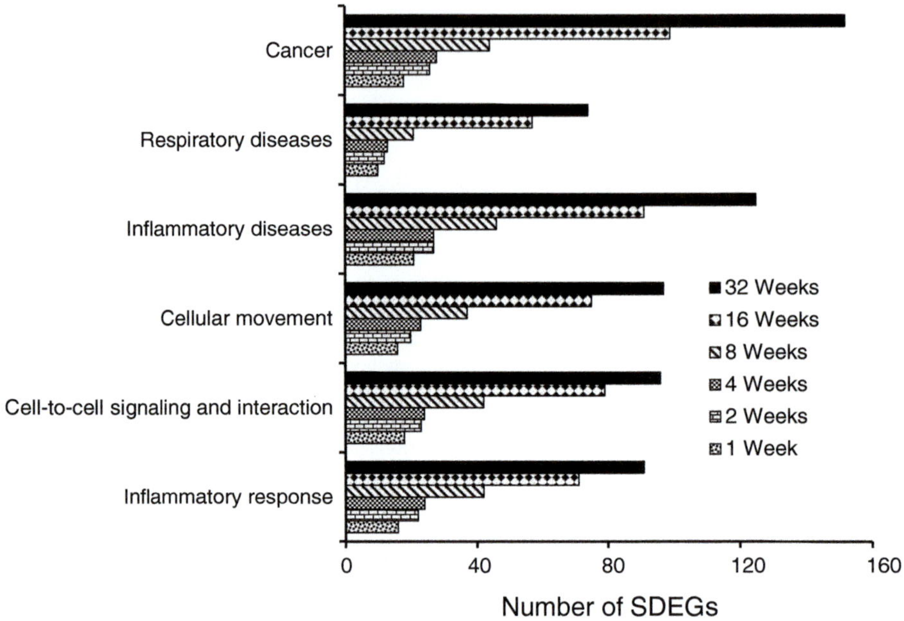

Figure 7.6 Biological functions enriched in response to crystalline silica exposure in an in vivo rat lung toxicity model. Rats were exposed to Min-U-Sil® 5 crystalline silica (15 mg/m^3, six hours a day for five days) and the lung gene expression profile was determined by microarray analysis, as described in detail in the original publications (Sellamuthu et al. 2012, 2013). The microarray data were subjected to bioinformatics analysis, and the number of SDEGs in the crystalline silica-exposed rat lungs belonging to the six top ranking IPA biological functions at each of the post-exposure time interval analyzed is presented. Modified with permission from *Inhalation Toxicology* 2012 (24): 570–579 and *Journal of Applied Toxicology* 2013 (33): 301–312.

expressions, is presented in Table 7.1. The induction of pulmonary inflammation in response to exposure to toxic particles of substances such as crystalline silica is characterized by the migration of neutrophils from blood to the site of lung injury. The migration of neutrophils and the ensuing inflammatory response are complex processes regulated by the expressions of many genes. Myelin-related proteins (MRPs) play an active role in the migration of neutrophils from the blood to organs; and it is this migration that results in an inflammatory response (Ryckman et al. 2003). The transcript for the *S100A8*, a calcium-binding MRP involved in neutrophil migration (Wang et al. 2018), was significantly overexpressed in the lungs of the silica-exposed rats. The correlation between the number of infiltrating neutrophils and the fold changes in the expression of *S100A8* transcript detected in the rats' lungs during the post-exposure intervals analyzed supported the hypothesis of the potential involvement of MRPs in the silica-induced neutrophil infiltration and in the induction of pulmonary inflammation in the rats. The transcripts for several inflammatory response cytokines and chemokines, namely *CCl2*, *CCl3*, *CCl4*, *CCl7*, *CCl11*, *CXCl1*, *CXCl2*, *CXCl5*, *CXCl9*, *CXCl10*, *CXCl11*, *IL1β*, and *ILR2*, were significantly overexpressed in the lungs of the silica-exposed rats (Table 7.1). The protein products of the cytokine and chemokine genes often lead to the recruitment and activation of immune cells as well as to the release of additional cytokines and chemokines, which results in the induction of inflammation (Kany et al. 2019). The overexpressions of the various cytokine mRNAs detected in the rats' lungs progressed along with the pulmonary inflammation and the toxicity found during the post-exposure intervals. These observations may therefore suggest the potential contribution of these overexpressions to the pulmonary inflammation and injury found in the rats in response to silica exposure. *Osteopontin* or *secreted phosphoprotein 1* (*SPP1*) is a multifunctional gene whose involvement in the lung's inflammatory response to toxic particles has been well characterized (Dong and Ma 2017). The significant overexpression of the *SPP1* mRNA (Table 7.1) in the lungs of rats exposed to silica, along with the progression of the injury, may suggest that the *SPP1* gene contributes to lung injury by inducing inflammation. The other genes that exhibited significant overexpression in the rats' lungs in correlation with the progression of lung injury and have established roles in inflammation were *TREM1* and *TREM2* (Ford and McVicar 2009), *LCN2* (Zhang et al. 2008), and *CHI3L1* (Eurich et al. 2009). Significant overexpression of genes and enrichment in the pathways related to inflammation, in agreement with the report from our laboratory (Sellamuthu et al. 2013), has been reported in the lungs of mice (Chen et al. 2018), rats (Zhao et al. 2020), and humans (Pang et al. 2021) in response to silica exposure. The gene expression data therefore suggested the potential involvement of inflammation through the transcriptional regulation of expressions of multiple genes, in lung injuries detected in response to exposure to crystalline silica.

Crystalline Silica and Lung Fibrosis

Silicosis, perhaps the most serious health outcome of inhalation exposure to crystalline silica, is characterized by pulmonary fibrosis accompanied by a reduction in lung function. Crystalline silica-induced lung fibrosis involves the release of fibrogenic factors by alveolar macrophages, the proliferation of the fibroblasts, and an increased production of collagen by pulmonary fibroblasts—a process that culminates in fibrosis. In the rat model employed by Sellamuthu et al. (2012), which investigates the relationship between alterations in the transcriptome and the mechanisms of lung toxicity induced through crystalline silica exposure, an increase in Mason's trichrome staining indicative of lung fibrosis was detected when the lungs were analyzed thirty-two weeks after exposure. The absence of trichrome staining during the earlier intervals analyzed and up to the sixteenth week may suggest that silica exposure, as employed in the studies (Sellamuthu et al. 2011b), did not cause lung fibrosis during those intervals. Alternatively, the results may indicate that trichrome staining was not highly sensitive and therefore failed to pick up the subtle lung fibrosis that could have occurred during the earlier intervals. It is generally accepted that the unresolved chronic inflammation and oxidative stress induced by crystalline silica may result in lung injury, including scar formation and fibrosis (Kawasaki 2015). The degradation and rebuilding of the extracellular matrix (ECM), the key events involved in lung fibrosis, are regulated by the expressions of several genes. Several lines of evidence suggest a prominent role for the *CCL2* gene, which encodes the MCP-1 protein, a multifunctional pro-inflammatory chemokine, in fibrosis. This evidence includes the up-regulation of the *MCP-1* gene, which is detected in the lungs of pulmonary fibrotic patients (Distler et al. 2001; Yamamoto and Nishioka 2003) and implicates the involvement of chemokine in lung fibrosis. More direct evidence for the involvement of *MCP-1* gene in fibrosis comes from the observation that collagen fibrils exhibiting decreased diameter and irregular arrangement, characteristic features of fibrosis, were more abundant in bleomycin-induced dermal lesions in the $MCP1^{+/+}$ mice than in $MCP1^{-/-}$ mice (Ferreira et al. 2006). Additional mechanisms of *MCP1*-induced fibrosis included the increased differentiation of fibroblasts to myofibroblasts, a rich source for the synthesis of collagen and profibrogenic mediators, and an increase in the synthesis of collagen—a prominent component of the ECM (Gabbiani 2003). The findings that the expression of the *CCL2* transcript (Table 7.2) was significantly higher in the lungs of the crystalline silica-exposed rats than in the lungs of controls suggested the involvement of the *CCL2* gene, through enhanced ECM formation, in

Table 7.2 Fold changes in expressions of genes related to fibrosis and tissue remodeling in the lungs of rats exposed to silica.

Gene	Fold change in expression				
	1 week	2 weeks	4 weeks	8 weeks	16 weeks
Matrix metallopeptidase 8 (MMP8)	1.09	1.13	1.22	1.33	1.50
Matrix metallopeptidase 12 (MMP12)	3.78	3.49	3.87	4.34	4.96
Secreted phosphoprotein 1 (SPP1)	1.34	1.01	1.11	1.85	6.27
Haptoglobin (HP)	1.49	1.56	1.78	2.11	2.42
Hemopexin (HPX)	1.24	1.05	1.46	1.51	1.56
Platelet factor 4 (PF4)	1.34	1.45	1.52	1.99	1.55
Arginase 1 (ARG1)	1.59	1.29	1.38	1.93	2.28
Chemokine C-X-C motif 9 (CXCL9)	1.09	1.29	1.31	2.08	2.91
Chemokine (C-C motif) ligand 2 (CCl2)	2.36	1.99	2.30	4.28	6.22
Chemokine (C-C motif) ligand 7 (CCl7)	1.39	1.29	1.35	2.37	4.32
Complement component 3 (C3)	2.50	2.14	2.26	2.88	3.02
Complement component 4 binding protein alpha (C4BPA)	2.14	2.19	2.23	2.90	3.55
Complement component 5 (C5)	1.19	1.24	1.31	1.533	1.62
Complement factor B (CFB)	1.32	1.18	1.22	1.37	1.5
Complement factor 1(CFI)	1.28	1.47	1.65	2.09	4.18

Rats were exposed to Min-U-Sil® 5 crystalline silica (15 mg/m^3, six hours a day for five days) and the lung gene expression profile was determined by microarray analysis at the post-exposure time intervals indicated in the table. The gene expression data were further analyzed and a list of the genes that belonged to the fibrosis and tissue remodeling category was drawn. Fold changes in expressions of the genes, compared with corresponding changes in the time-matched controls, are presented.
Modified with permission from Sellamuthu et al., *J. Appl. Toxicol.* 2013 (33): 301–312.

silica-induced lung fibrosis in the rats. It is noteworthy that a further overexpression of the *CCl2* transcript was detected during the thirty-second week after exposure, when a positive result to trichrome staining (which is indicative of fibrosis) was detected in the lungs of rats (Sellamuthu et al. 2012). However, there was a steady increase in the overexpression of *CCl2* transcript during the intervals after exposure—including the earlier ones, when fibrosis by trichrome staining was undetected (Table 7.2).

Lung remodeling, which involves the degradation and synthesis of the ECM, takes place under normal conditions as well as in response to the injury that results from exposure to toxic agents such as crystalline silica (Scabilloni et al. 2005). By cleaving ECM and non-ECM molecules (Page-McCaw et al. 2007), the matrix metalloproteinases (MMPs) play a central role in tissue remodeling. The transcripts for two of the MMPs, namely *MMP8* and *MMP12*, were overexpressed in the lungs of the silica-exposed rats (Table 7.2). The fold changes in the overexpression of *MMP8* and *MMP12* corresponded to the severity of lung toxicity detected in the silica-exposed rats. *MMP12* codes for a protein that exhibits elastolytic activity and is secreted by activated macrophages (Wagner et al. 2016). The activity of *MMP12* is increased in children with cystic fibrosis (Trojanek et al. 2014), and the sputum levels of *MMP12* correlate with emphysema severity in chronic obstructive pulmonary disease (COPD) patients (Chaudhuri et al. 2012). These facts signal its role in lung diseases. Furthermore, genetic deletion of the *MMP12* gene and the pharmacological inhibition of *MMP12* activity significantly blocked emphysema (Trojanek et al. 2014). The underlying mechanism responsible for the role of *MMP12* in lung damage is thought to relate to its proteolytic activity toward ECM, antiproteases, and pro-inflammatory cytokines (Churg and Wright 2005; Hautamaki et al. 1997; Zheng et al. 2000).

Differential expressions of several other genes known to be involved in fibrosis were also detected in the lungs of the crystalline silica-exposed rats. The *SPP1* gene, in addition to its established role in the induction of inflammation (Dong and Ma 2017), is also involved in cellular functions that underlie fibrosis. In a mouse pulmonary toxicity model, *SPP1* promoted the differentiation of fibroblasts into myofibroblasts (Dong and Ma 2017), a key event involved in lung fibrosis. The *ARG1* gene transcript, which was significantly and progressively overexpressed in the silica-exposed lung samples (Table 7.2), is associated with bleomycin-induced pulmonary fibrosis in mice (Endo et al. 2003). The significant overexpressions of the profibrotic chemokines *CCl2* and *CCl7* observed in the lungs of crystalline silica-exposed rats (Table 7.2) are also known for their involvement in pulmonary fibrosis (Mercer et al. 2009; Moore and Hogaboam 2008). This view is further supported by the significant overexpression of these chemokines in tuberculosis, a human infectious fibrotic lung disease (Nau et al. 1997). The involvement of the *RETNLA* gene in the induction of pulmonary fibrosis through the promotion of the differentiation of myoblasts that mediate collagen deposition has been suggested (Liu et al. 2004). The *RETNLA* gene transcript was highly overexpressed in the silica-exposed rats' lungs (Table 7.1). Since a definite relationship is known to exist between unresolved pulmonary inflammation and fibrosis (Reynolds 2005), it is reasonable to assume that the significant overexpression of the multiple pro-inflammatory genes detected in the rats' lungs might have contributed to the crystalline silica-induced pulmonary fibrosis by facilitating an unresolved pulmonary inflammation. The overexpression of the genes that are known to be involved in tissue remodeling and in fibrosis steadily increased in magnitude, in parallel with the progression of crystalline silica-induced pulmonary toxicity in the rats; this suggests the role of these genes in crystalline silica-induced pulmonary fibrosis and possibly in causing silicosis (Sellamuthu et al. 2012, 2013, and 2017). Significant overexpression of genes and enrichment in pathways related to fibrosis have been reported in the lungs of mice (Chen et al. 2018), rats (Zhao et al. 2020), and humans (Pang et al. 2021) in response to silica exposure. These reports, in agreement with ours (Sellamuthu et al. 2013), further support the application of transcriptome analysis as a relevant approach to investigating the mechanisms that underlie silicosis.

Transcriptomics and Alzeimer's Disease

Alzeimer's disease (AD), the most common among age-related and progressive neurodegenerative diseases, is characterized by dementia (Jellinger et al. 2008). As the world's population ages, the number of people with AD is expected to increase dramatically. It is estimated that the number of global cases of AD will increase from 24 million in 2001 to 81 million in 2040 (Ferri et al. 2005). It is also estimated that, by 2050, AD will affect 1 out of

85 people globally (Brookmeyer et al. 2007). Histologically, AD is characterized by extracellular amyloid plaques and by tangles of intracellular neurofibrils. According to the "amyloid cascade hypothesis," an imbalance between the production and the clearance of amyloid beta (Aβ) in the brain is the initiating event that ultimately results in neuronal or synaptic degeneration and dementia (Hardy and Selkoe 2002). The deposition of aggregated Aβ fibrils in the brain triggers an inflammatory response that plays an important role in the pathogenesis of AD (Pellicano et al. 2012). Currently the disease is diagnosed clinically through a combination of physical, mental, and neuropsychological examinations that are time-consuming and costly. Positron emission tomography (PET) scan of brain and analysis of cerebrospinal fluid have also been used to identify cases of AD. Decreased Aβ and increased tau levels in the cerebrospinal fluid have been employed to distinguish AD patients from people not affected by the disease, as well as from those with mild cognitive impairment. While these tests are noted for their relatively high accuracy, they are costly, and therefore not an attractive option for screening large numbers of individuals for AD. Primarily because of the rapid increase in cases of AD globally and because of the associated burden on society, there is a strong and urgent need to develop more cost- and time-effective tools for the reliable detection of AD.

Blood, a rich source of various molecules, has been employed to develop biomarkers for early detection of AD. In a study conducted by Nakamura et al. (2018), the plasma Aβ level was tested as a biomarker for AD. While the authors were successful in their effort to detect AD on the basis of plasma Aβ levels in patients, the value of these findings was questioned, mainly on account of to the non-specificity of Aβ as a biomarker for AD. Another study, by Kiddle et al. (2014), reviewed the findings of twenty-one previous studies, which had investigated 163 proteins in the blood as possible biomarkers for the detection of AD. Four proteins, namely α-1-antitrypsin, α-2-microglobulin, apolipoprotein E, and complement C3, were identified as blood biomarkers for AD. However, their utility for this function was questionable, since the blood level of the same proteins was also found to be affected by other diseases, including Parkinson's disease and schizophrenia (Chan et al. 2014).

Blood transcriptome was investigated in several studies, with a view to developing gene expression changes as minimally invasive biomarkers for AD. The initial studies focused on the identification of individual genes whose expressions were different between AD patients and normal individuals. While individual genes exhibiting differential expressions between AD patients and normal individuals were identified, their potential use as biomarkers for the disease was limited, mainly owing to the lack of specificity and reproducibility of the findings (Han et al. 2013; Rye et al. 2011). Subsequently, global gene expression profiling was applied in combination with machine learning in order to develop gene expression signatures as blood biomarkers for AD. In a study conducted by Voyle et al. (2016), a training set consisting of 110 AD patients and 118 controls was used to obtain blood gene expression-based AD biomarkers. The biomarkers developed were validated in an independent test study cohort consisting of 118 AD patients and 107 controls. The authors were able to identify the AD patients and the controls successfully, on the basis of the blood transcriptome biomarker. The blood biomarkers exhibited a sensitivity of 56%, a specificity of 74.6%, an accuracy of 66%, a positive predictive power of 69.1%, and a negative predictive power of 63%. Overall, the predictive power of the blood biomarkers (gene expression profile) was lower than the 90% predictive power desired for a clinical diagnostic test (Huynh and Mohan 2017).

A significant improvement in the detection of AD by employing biomarkers developed on the basis of blood gene expression profile was made recently by Patel et al. (2020). The authors employed publicly available blood gene expression data to develop AD biomarkers.

They used two different classification models for their purpose. The first one consisted of 160 AD and 127 healthy controls. In the second model, in addition to the microarray data obtained from the same 160 AD patients, they put together data from a mixed population of individuals with Parkinson's disease, multiple sclerosis, amyotrophic lateral sclerosis, bipolar disorder, schizophrenia, coronary artery disease, rheumatoid arthritis, and chronic obstructive pulmonary disease alongside cognitively healthy subjects. The two classification models developed were tested in an independent cohort, which consisted of 127 AD patients and 687 mixed controls. The authors found that the use of a mixed population instead of a healthy population as the control group had a profound effect on the AD detection. The first model (AD versus healthy controls) exhibited a sensitivity of 48.7% and a specificity of 41.9% and had a positive predictive value of 13.6% and a negative predictive value of 81.1%. On the other hand, the second model (AD versus mixed control) exhibited a sensitivity of 40.8% and a specificity of 95.3% and had a positive predictive value of 61.4% and a negative predictive value of 89.7%. Despite the need for further improvement to sensitivity, the study demonstrated the benefits of incorporating a mixed population with related disorders as the control group for improving the ability of blood transcriptome-based biomarkers designed to distinguish AD patients from non-AD patients.

References

Alexander-Dann, B., Pruteanu, L.L., Oerton, E. et al. (2018). Developments in toxicogenomics: Understanding and predicting compound-induced toxicity from gene expression data. *Mol. Omics* 14 (4): 218–236.

Anders, S., Pyl, P.T., and Huber, W. (2015). HTSeq–a Python framework to work with high-throughput sequencing data. *Bioinformatics* 31 (2): 166–169.

Anjum, A., Jaggi, S., Varghese, E., Lall, S., Bhowmik, A., and Rai, A. (2016). Identification of differentially expressed genes in RNA-seq data of *Arabidopsis thaliana*: A compound distribution approach. *J. Comput. Biol.* 23 (4): 239–247.

Bell, J.L. and Mazurek, J.M. (2020). Trends in pneumoconiosis deaths: United States, 1999–2018. *MMWR Morb. Mortal. Wkly. Rep.* 69 (23): 693–698.

Bergeron, M., Honkaniemi, J., and Sharp, F.R. (2001). Northern blot analysis of RNA. *Curr. Protoc. Toxicol.* 7, Appendix 3: A3E 1–13.

Bourgon, R., Gentleman, R., and Huber, W. (2010). Independent filtering increases detection power for high-throughput experiments. *Proc. Natl. Acad. Sci. USA* 107 (21): 9546–9551.

Brookmeyer, R., Johnson, E., Ziegler-Graham, K., and Arrighi, H.M. (2007). Forecasting the global burden of Alzheimer's disease. *Alzheimers Dement* 3 (3): 186–191.

Bumgarner, R. (2013). Overview of DNA microarrays: Types, applications, and their future. *Curr. Protoc. Mol. Biol.* 101 (1). doi: 10.1002/0471142727.mb2201s101.

Bushel, P.R., Heinloth, A.N., Li, J. et al. (2007). Blood gene expression signatures predict exposure levels. *Proc. Natl. Acad. Sci. USA* 104 (46): 18211–18216.

Castranova, V. (2004). Signaling pathways controlling the production of inflammatory mediators in response to crystalline silica exposure: Role of reactive oxygen/nitrogen species. *Free Radic. Biol. Med.* 37 (7): 916–925.

Castranova, V. and Vallyathan, V. (2000). Silicosis and coal workers' pneumoconiosis. *Environ. Health Perspect.* 108 (Suppl 4): 675–684.

Cavallini, A., Lippolis, C., Vacca, M. et al. (2016). The effects of chronic lifelong activation of the AHR pathway by industrial chemical pollutants on female human reproduction. *PLoS One* 11 (3): e0152181.

Chan, M.K., Gottschalk, M.G., Haenisch, F. et al. (2014). Applications of blood-based protein biomarker strategies in the study of psychiatric disorders. *Prog. Neurobiol.* 122: 45–72.

Charlesworth, J.C., Curran, J.E., Johnson, M.P. et al. (2010). Transcriptomic epidemiology of smoking: The effect of smoking on gene expression in lymphocytes. *BMC Med. Genom* 3: 29.

Chaudhuri, R., McSharry, C., Brady, J. et al. (2012). Sputum matrix metalloproteinase-12 in patients with chronic obstructive pulmonary disease and asthma: Relationship to disease severity. *J. Allergy Clin. Immunol.* 129 (3): 655–663. e658.

Chen, J., Yao, Y., Su, X. et al. (2018). Comparative RNA-Seq transcriptome analysis on silica induced pulmonary inflammation and fibrosis in mice silicosis model. *J. Appl. Toxicol.* 38 (5): 773–782.

Churg, A. and Wright, J.L. (2005). Proteases and emphysema. *Curr. Opin. Pulm. Med.* 11 (2): 153–159.

Cooper, G.S., Miller, F.W., and Germolec, D.R. (2002). Occupational exposures and autoimmune diseases. *Int. Immunopharmacol.* 2 (2-3): 303–313.

den Braver-Sewradj, S.P., Van Spronsen, R., and Hessel, E.V.S. (2020). Substitution of bisphenol A: A review of the carcinogenicity, reproductive toxicity, and endocrine disruption potential of alternative substances. *Crit. Rev. Toxicol.* 50 (2): 128–147.

Distler, O., Pap, T., Kowal-Bielecka, O. et al. (2001). Overexpression of monocyte chemoattractant protein 1 in systemic sclerosis: Role of platelet-derived growth factor and effects on monocyte chemotaxis and collagen synthesis. *Arthritis Rheum.* 44 (11): 2665–2678.

Dobin, A., Davis, C.A., Schlesinger, F. et al. (2013). STAR: Ultrafast universal RNA-seq aligner. *Bioinformatics* 29 (1): 15–21.

Doney, B.C., Blackley, D., Hale, J.M. et al. (2019). Respirable coal mine dust in underground mines, United States, 1982–2017. *Am. J. Ind. Med.* 62 (6): 478–485.

Doney, B.C., Miller, W.E., Hale, J.M., and Syamlal, G. (2020). Estimation of the number of workers exposed to respirable crystalline silica by industry: Analysis of OSHA compliance data (1979–2015). *Am. J. Ind. Med.* 63 (6): 465–477.

Dong, J. and Ma, Q. (2017). Osteopontin enhances multi-walled carbon nanotube-triggered lung fibrosis by promoting TGF-beta1 activation and myofibroblast differentiation. *Part Fibre Toxicol* 14 (1): 18.

Endo, M., Oyadomari, S., Terasaki, Y. et al. (2003). Induction of arginase I and II in bleomycin-induced fibrosis of mouse lung. *Am. J. Physiol. Lung Cell Mol. Physiol.* 285 (2): L313–321.

EPA [Environmental Protection Agency]. 2021. Reviewing new chemicals under the Toxic Substances Control Act (TSCA). Available at https://www.epa.gov/reviewing-new-chemicals-under-toxic-substances-control-act-tsca). Accessed on May 25, 2021.

Esswein, E.J., Breitenstein, M., Snawder, J., Kiefer, M., and Sieber, W.K. (2013). Occupational exposures to respirable crystalline silica during hydraulic fracturing. *J. Occup. Environ. Hyg.* 10 (7): 347–356.

Eurich, K., Segawa, M., Toei-Shimizu, S., and Mizoguchi, E. (2009). Potential role of chitinase 3-like-1 in inflammation-associated carcinogenic changes of epithelial cells. *World J. Gastroenterol.* 15 (42): 5249–5259.

Ferreira, A.M., Takagawa, S., Fresco, R., Zhu, X., Varga, J., and DiPietro, L.A. (2006). Diminished induction of skin fibrosis in mice with MCP-1 deficiency. *J. Invest. Dermatol.* 126 (8): 1900–1908.

Ferri, C.P., Prince, M., Brayne, C. et al. (2005). Global prevalence of dementia: A Delphi consensus study. *Lancet* 366 (9503): 2112–2117.

Ford, J.W. and McVicar, D.W. (2009). TREM and TREM-like receptors in inflammation and disease. *Curr. Opin. Immunol.* 21 (1): 38–46.

Fubini, B. and Hubbard, A. (2003). Reactive oxygen species (ROS) and reactive nitrogen species (RNS) generation by silica in inflammation and fibrosis. *Free Radic. Biol. Med.* 34 (12): 1507–1516.

Gabbiani, G. (2003). The myofibroblast in wound healing and fibrocontractive diseases. *J. Pathol.* 200 (4): 500–503.

Glaab, W.E., Holder, D., He, Y.D. et al. (2021). Universal toxicity gene signatures for early identification of drug-induced tissue injuries in rats. *Toxicol. Sci.* 181 (2): 148–159.

Greenberg, M.I., Waksman, J., and Curtis, J. (2007). Silicosis: A review. *Dis. Mon.* 53 (8): 394–416.

Hamadeh, H.K., Bushel, P.R., Jayadev, S. et al. (2002). Prediction of compound signature using high density gene expression profiling. *Toxicol. Sci.* 67 (2): 232–240.

Han, G., Wang, J., Zeng, F. et al. (2013). Characteristic transformation of blood transcriptome in Alzheimer's disease. *J. Alzheimers Dis.* 35 (2): 373–386.

Hardy, J. and Selkoe, D.J. (2002). The amyloid hypothesis of Alzheimer's disease: Progress and problems on the road to therapeutics. *Science* 297 (5580): 353–356.

Hautamaki, R.D., Kobayashi, D.K., Senior, R.M., and Shapiro, S.D. (1997). Requirement for macrophage elastase for cigarette smoke-induced emphysema in mice. *Science* 277 (5334): 2002–2004.

Heinloth, A.N., Irwin, R.D., Boorman, G.A. et al. (2004). Gene expression profiling of rat livers reveals indicators of potential adverse effects. *Toxicol. Sci.* 80 (1): 193–202.

Huynh, R.A. and Mohan, C. (2017). Alzheimer's disease: Biomarkers in the genome, blood, and cerebrospinal fluid. *Front Neurol.* 8: 102.

IARC [International Agency for RTesearch on Cancer]. (1997). *Silica: Some Silicates, Coal Dust and Para-aramid Fibrils.* Lyon: International Agency for Research on Cancer.

Jablonski, K.A., Amici, S.A., Webb, L.M. et al. (2015). Novel markers to delineate murine M1 and M2 macrophages. *PLoS One* 10 (12): e0145342.

Jellinger, K.A., Janetzky, B., Attems, J., and Kienzl, E. (2008). Biomarkers for early diagnosis of Alzheimer disease: "Alzheimer Associated gene": A new blood biomarker? *J. Cell. Mol. Med.* 12 (4): 1094–1117.

Jf, W., Ra, J., Ciurlionis, R. et al. (2001). Clustering of hepatotoxins based on mechanism of toxicity using gene expression profiles. *Toxicol. Appl. Pharmacol.* 175 (1): 28–42.

Johnston, C.J., Driscoll, K.E., Finkelstein, J.N. et al. (2000). Pulmonary chemokine and mutagenic responses in rats after subchronic inhalation of amorphous and crystalline silica. *Toxicol. Sci.* 56 (2): 405–413.

Joseph, P. (2017). Transcriptomics in toxicology. *Food Chem. Toxicol.* 109 (Pt 1): 650–662.

Joseph, P., Umbright, C.M., Roberts, J.R. et al. (2021). Lung toxicity and gene expression changes in response to whole-body inhalation exposure to cellulose nanocrystal in rats. *Inhal. Toxicol.* 33 (2): 66–80.

Kany, S., Vollrath, J.T., and Relja, B. (2019). Cytokines in inflammatory disease. *Int. J. Mol. Sci.* 20 (23).

Kawasaki, H. (2015). A mechanistic review of silica-induced inhalation toxicity. *Inhal. Toxicol.* 27 (8): 363–377.

Kiddle, S.J., Sattlecker, M., Proitsi, P. et al. (2014). Candidate blood proteome markers of Alzheimer's disease onset and progression: A systematic review and replication study. *J. Alzheimers Dis.* 38 (3): 515–531.

Kim, D., Langmead, B., and Salzberg, S.L. (2015). HISAT: A fast spliced aligner with low memory requirements. *Nat. Methods* 12 (4): 357–360.

Kim, D., Pertea, G., Trapnell, C., Pimentel, H., Kelley, R., and Salzberg, S.L. (2013). TopHat2: Accurate alignment of transcriptomes in the presence of insertions, deletions and gene fusions. *Genome Biol.* 14 (4): R36.

Kim, H., Lee, S.G., and Rhie, J. (2017). Dermal and neural toxicity caused by acrylamide exposure in two Korean grouting workers: A case report. *Ann. Occup. Environ. Med.* 29: 50.

Lamm, S.H., Ferdosi, H., Dissen, E.K., Li, J., and Ahn, J. (2015). A systematic review and meta-regression analysis of lung cancer risk and inorganic arsenic in drinking water. *Int J Environ Res Public Health* 12 (12): 15498–15515.

Law, C.W., Chen, Y., Shi, W., and Smyth, G.K. (2014). voom: Precision weights unlock linear model analysis tools for RNA-seq read counts. *Genome Biol.* 15 (2): R29.

Li, B. and Dewey, C.N. (2011). RSEM: Accurate transcript quantification from RNA-Seq data with or without a reference genome. *BMC Bioinform.* 12: 323.

Lippmann, M. (2014). Toxicological and epidemiological studies of cardiovascular effects of ambient air fine particulate matter (PM2.5) and its chemical components: Coherence and public health implications. *Crit. Rev. Toxicol.* 44 (4): 299–347.

Liu, T., Jin, H., Ullenbruch, M. et al. (2004). Regulation of found in inflammatory zone 1 expression in bleomycin-induced lung fibrosis: Role of IL-4/IL-13 and mediation via STAT-6. *J. Immunol.* 173 (5): 3425–3431.

Market Analysis Report. (2020). Nanomaterials market size, share & trends report by product (carbon nanotubes, titanium dioxide) by application (medical, electronics, paints & coatings), by region, and segment forecasts, 2020–2027.https://www.grantviewresearch.com-analusis/nanotechnology-and-nanomaterials-market (accessed on May 25, 2021).

Mascarenhas, S., Mutnuri, S., and Ganguly, A. (2017). Deleterious role of trace elements: Silica and lead in the development of chronic kidney disease. *Chemosphere* 177: 239–249.

Mercer, P.F., Johns, R.H., Scotton, C.J. et al. (2009). Pulmonary epithelium is a prominent source of proteinase-activated receptor-1-inducible CCL2 in pulmonary fibrosis. *Am. J. Respir. Crit. Care Med.* 179 (5): 414–425.

Moon, J.Y., Eom, S.Y., Seo, J.W. et al. (2021). Effects of exposure to lead and cadmium on health of inhabitants of abandoned metal mine area in Korea. *Arch. Environ. Contam. Toxicol.* 80 (2): 490–498.

Moore, B.B. and Hogaboam, C.M. (2008). Murine models of pulmonary fibrosis. *Am. J. Physiol. Lung Cell Mol. Physiol.* 294 (2): L152–160.

Nakamura, A., Kaneko, N., Villemagne, V.L. et al. (2018). High performance plasma amyloid-beta biomarkers for Alzheimer's disease. *Nature* 554 (7691): 249–254.

Nau, G.J., Guilfoile, P., Chupp, G.L. et al. (1997). A chemoattractant cytokine associated with granulomas in tuberculosis and silicosis. *Proc. Natl. Acad. Sci. USA* 94 (12): 6414–6419.

NIOSH [National Institute for Occupational Safety and Health]. (1996). Preventing Silicosis and Deaths in Construction Workers. US Department of Health and Human Services (NIOSH). Publication Number 96–112. https://www.cdc.gov/niosh/docs/96-112/default.html (accessed on May 25, 2021).

NIOSH [National Institute for Occupational Safety and Health]. (2002). *Hazard Review: Health Effects of Occupational Exposures to Respirable Crystalline Silica*. US Department of Health and Human Services (NIOSH). Publication No. 2002–129. http://www.cdc.gov/niosh/docs/2002-129/02-129a.html (accessed May 25, 2021).

NRC [National Research Council]. (2007). *Applications of Toxicogenomic Technologies to Predictive Toxicology and Risk Assessment*. Washington, DC: National Academies Press.

OSHA [Occupational Safety and Health Administration]. (2013). OSHA'a proposed crystalline silica rule: Overview. OSHA Factsheet. https://www.osha.gov/silica/factsheets/OSHA_FS-3683_Silica_Overview.pdf. (accessed on May 25, 2021).

Page-McCaw, A., Ewald, A.J., and Werb, Z. (2007). Matrix metalloproteinases and the regulation of tissue remodelling. *Nat. Rev. Mol. Cell Biol.* 8 (3): 221–233.

Pang, J., Luo, Y., Wei, D. et al. (2021). Comparative transcriptome analyses reveal a transcriptional landscape of human silicosis lungs and provide potential strategies for silicosis treatment. *Front Genet.* 12: 652901.

Patel, H., Iniesta, R., Stahl, D., Dobson, R.J.B., and Newhouse, S.J. (2020). Working towards a blood-derived gene expression biomarker specific for Alzheimer's disease. *J. Alzheimers Dis.* 74 (2): 545–561.

Pellicano, M., Larbi, A., Goldeck, D. et al. (2012). Immune profiling of Alzheimer patients. *J. Neuroimmunol.* 242 (1-2): 52–59.

Peretz, A., Checkoway, H., Kaufman, J.D., Trajber, I., and Lerman, Y. (2006). Silica, silicosis, and lung cancer. *Isr. Med. Assoc. J.* 8 (2): 114–118.

Peretz, A., Peck, E.C., Bammler, T.K. et al. (2007). Diesel exhaust inhalation and assessment of peripheral blood mononuclear cell gene transcription effects: An exploratory study of healthy human volunteers. *Inhal. Toxicol.* 19 (14): 1107–1119.

Perkins, T.N., Shukla, A., Peeters, P.M. et al. (2012). Differences in gene expression and cytokine production by crystalline vs. amorphous silica in human lung epithelial cells. *Part Fibre Toxicol* 9 (1): 6.

Pollard, K.M. (2016). Silica, silicosis, and autoimmunity. *Front Immunol* 7: 97.

Prajapati, S.S., Mishra, R.A., Jhariya, B., and Dhatrak, S.V. (2021). Respirable dust and crystalline silica exposure among different mining sectors in India. *Arch. Environ. Occup. Health* 76 (7): 455–4614554.

Racette, B.A., Aschner, M., Guilarte, T.R., Dydak, U., Criswell, S.R., and Zheng, W. (2012). Pathophysiology of manganese-associated neurotoxicity. *Neurotoxicology* 33 (4): 881–886.

Radjendirane, V., Joseph, P., Lee, Y.H. et al. (1998). Disruption of the DT diaphorase (NQO1) gene in mice leads to increased menadione toxicity. *J. Biol. Chem.* 273 (13): 7382–7389.

Reynolds, H.Y. (2005). Lung inflammation and fibrosis: An alveolar macrophage-centered perspective from the 1970s to 1980s. *Am. J. Respir. Crit. Care Med.* 171 (2): 98–102.

Rimal, B., Greenberg, A.K., and Rom, W.N. (2005). Basic pathogenetic mechanisms in silicosis: Current understanding. *Curr. Opin. Pulm. Med.* 11 (2): 169–173.

Risso, D., Schwartz, K., Sherlock, G., and Dudoit, S. (2011). GC-content normalization for RNA-Seq data. *BMC Bioinform.* 12: 480.

Ritchie, M.E., Phipson, B., Wu, D. et al. (2015). limma powers differential expression analyses for RNA-sequencing and microarray studies. *Nucleic Acids Res.* 43 (7): e47.

Robinson, M.D., McCarthy, D.J., and Smyth, G.K. (2010). edgeR: A bioconductor package for differential expression analysis of digital gene expression data. *Bioinformatics* 26 (1): 139–140.

Ryckman, C., McColl, S.R., Vandal, K. et al. (2003). Role of S100A8 and S100A9 in neutrophil recruitment in response to monosodium urate monohydrate crystals in the air-pouch model of acute gouty arthritis. *Arthritis Rheum.* 48 (8): 2310–2320.

Rye, P.D., Booij, B.B., Grave, G. et al. (2011). A novel blood test for the early detection of Alzheimer's disease. *J. Alzheimers Dis.* 23 (1): 121–129.

Sager, T.M., Umbright, C.M., Mustafa, G.M. et al. (2020). Tobacco smoke exposure exacerbated crystalline silica-induced lung toxicity in rats. *Toxicol. Sci.* 178 (2): 375–390.

Salganik, M., Sergeyev, V.G., Shinde, V. et al. (2015). The loss of glucose-regulated protein 78 (GRP78) during normal aging or from siRNA knockdown augments human alpha-synuclein (alpha-syn) toxicity to rat nigral neurons. *Neurobiol. Aging* 36 (6): 2213–2223.

Scabilloni, J.F., Wang, L., Antonini, J.M., Roberts, J.R., Castranova, V., and Mercer, R.R. (2005). Matrix metalloproteinase induction in fibrosis and fibrotic nodule formation due to silica inhalation. *Am. J. Physiol. Lung Cell Mol. Physiol.* 288 (4): L709–717.

Sellamuthu, R., Umbright, C., Li, S., Kashon, M., and Joseph, P. (2011a). Mechanisms of crystalline silica-induced pulmonary toxicity revealed by global gene expression profiling. *Inhal. Toxicol.* 23 (14): 927–937.

Sellamuthu, R., Umbright, C., Roberts, J.R. et al. (2011b). Blood gene expression profiling detects silica exposure and toxicity. *Toxicol. Sci.* 122 (2): 253–264.

Sellamuthu, R., Umbright, C., Roberts, J.R. et al. (2012). Transcriptomic analysis of lungs and peripheral blood of crystalline silica-exposed rats. *Inhal. Tocxicol.* 24 (9): 570–579.

Sellamuthu, R., Umbright, C., Roberts, J.R. et al. (2013). Molecular insights into the progression of crystalline silica-induced pulmonary toxicity in rats. *J. Appl. Toxicol.* 33 (4): 301–312.

Sellamuthu, R., Umbright, C., Roberts, J.R., Young, S.H., Richardson, D., McKinney, W., Chen, B.T., Li, S., Kashon, M., and Joseph, P. 2017. Molecular mechanisms of pulmonary response progression in crystalline silica exposed rats. *Inhal Toxicol* 29: 53–64.

Sharma, S.K., Pande, J.N., and Verma, K. (1991). Effect of prednisolone treatment in chronic silicosis. *Am. Rev. Respir. Dis.* 143 (4 Pt 1): 814–821.

Silvestri, S., Ferrante, D., Giovannini, A. et al. (2020). Asbestos exposure of chrysotile miners and millers in Balangero, Italy. *Ann. Work Expo. Health* 64 (6): 636–644.

Stark, R., Grzelak, M., and Hadfield, J. (2019). RNA sequencing: The teenage years. *Nat. Rev. Genet.* 20 (11): 631–656.

Taulan, M., Paquet, F., Maubert, C., Delissen, O., Demaille, J., and Romey, M.C. (2004). Renal toxicogenomic response to chronic uranyl nitrate insult in mice. *Environ. Health Perspect.* 112 (16): 1628–1635.

Tomita, M., Nohno, T., Okuyama, T., Nishimatsu, S., and Adachi, J. (2002). Paraquat-induced gene expression in rat lung tissues using a differential display reverse transcription-polymerase chain reaction. *Arch. Toxicol.* 76 (9): 530–537.

Trapnell, C., Roberts, A., Goff, L. et al. (2012). Differential gene and transcript expression analysis of RNA-seq experiments with TopHat and Cufflinks. *Nat Protoc* 7 (3): 562–578.

Trojanek, J.B., Cobos-Correa, A., Diemer, S. et al. (2014). Airway mucus obstruction triggers macrophage activation and matrix metalloproteinase 12-dependent emphysema. *Am. J. Respir. Cell Mol. Biol.* 51 (5): 709–720.

Turro, E., Su, S.Y., Goncalves, A., Coin, L.J., Richardson, S., and Lewin, A. (2011). Haplotype and isoform specific expression estimation using multi-mapping RNA-seq reads. *Genome Biol.* 12 (2): R13.

Valentine, C.R., Valentine, J.L., Seng, J., Leakey, J., and Casciano, D. (1996). The use of transgenic cell lines for evaluating toxic metabolites of carbamazepine. *Cell Biol. Toxicol.* 12 (3): 155–165.

Voyle, N., Keohane, A., Newhouse, S. et al. (2016). A pathway-based classification method for analyzing gene expression for Alzheimer's disease diagnosis. *J. Alzheimers Dis.* 49 (3): 659–669.

Vupputuri, S., Parks, C.G., Nylander-French, L.A., Owen-Smith, A., Hogan, S.L., and Sandler, D.P. (2012). Occupational silica exposure and chronic kidney disease. *Ren Fail* 34 (1): 40–46.

Wagner, C.J., Schultz, C., and Mall, M.A. (2016). Neutrophil elastase and matrix metalloproteinase 12 in cystic fibrosis lung disease. *Mol. Cell. Pediatr.* 3 (1): 25.

Wagner, G.P., Kin, K., and Lynch, V.J. (2012). Measurement of mRNA abundance using RNA-seq data: RPKM measure is inconsistent among samples. *Theory Biosci.* 131 (4): 281–285.

Wang, C., Gong, B., Bushel, P.R. et al. (2014). The concordance between RNA-seq and microarray data depends on chemical treatment and transcript abundance. *Nat. Biotechnol.* 32 (9): 926–932.

Wang, S., Song, R., Wang, Z., Jing, Z., Wang, S., and Ma, J. (2018). S100A8/A9 in inflammation. *Front Immunol* 9: 1298.

Yamamoto, T. and Nishioka, K. (2003). Role of monocyte chemoattractant protein-1 and its receptor,CCR-2, in the pathogenesis of bleomycin-induced scleroderma. *J. Invest. Dermatol.* 121 (3): 510–516.

Zhang, J., Wu, Y., Zhang, Y., Leroith, D., Bernlohr, D.A., and Chen, X. (2008). The role of lipocalin 2 in the regulation of inflammation in adipocytes and macrophages. *Mol. Endocrinol.* 22 (6): 1416–1426.

Zhao, H., Jiang, Z., Lv, R. et al. (2020). Transcriptome profile analysis reveals a silica-induced immune response and fibrosis in a silicosis rat model. *Toxicol. Lett.* 333: 42–48.

Zheng, T., Zhu, Z., Wang, Z. et al. (2000). Inducible targeting of IL-13 to the adult lung causes matrix metalloproteinase- and cathepsin-dependent emphysema. *J. Clin. Invest.* 106 (9): 1081–1093.

8

Network Biology for Biomarker Discovery and Therapy in Cancer

Asim Bikas Das

Department of Biotechnology, National Institute of Technology Warangal, Warangal, 506004, Telangana, India.

Introduction

Recently "-omics" technologies have made significant progress in understanding of the etiology of cancer. The availability of a large amount of genomics, proteomics, and metabolomics data has allowed investigations into the cellular status of cancer cells at any given time point. Consequently cancer-specific treatments are successfully developed for a few specimens of this disease. Cancer is characterized by complex heterogeneity and uncontrolled cell growth. Cancer development and progression follow a classical evolutionary process, which consists in the step-wise accumulation of selected genomic or epigenetic changes. Some of these changes are chosen positively and lead to the growth of specific cell lineages; this growth is termed clonal evolution of cancer. Therefore cancer development is a dynamic process, which causes a continuous remodeling of cancer cells and leads to high heterogeneity (Zhang et al. 2018). Molecular heterogeneity is a common phenomenon among cancers that spring from different tissues, but it is also highly prominent in tumors derived from the same tissue. Moreover, the same genomic alteration in cancer does not show similar phenotypic changes, but involves instead various biological functions (Sumithra et al. 2019). For this reason genomic heterogeneity poses significant challenges when it comes to understanding the mechanism of disease progression and designing proper treatment strategies (Sawyers 2008).

Potential biological indicators or biomarkers are essential to improving cancer patient management in relation to diagnosis, stratification of cancer subtype, and monitoring of the patient's response during therapy. Owing to cancer's huge genomic heterogeneity, the discovery of reliable biomarkers remains unsuccessful for clinical application in many cancer types (Yan et al. 2016). Furthermore, despite the progress registered in understanding the molecular mechanism of cancer, the rate of new drug approvals has been prolonged. The failure of cancer drug discovery is mainly due to inaccurate drug targets and drug toxicity. This indicates that there is a need for an alternative approach to cancer biomarker discovery and drug development. The main limitation of the biomarker or potential target identification is that it focuses on single molecules as a target, which is to ignore the complex molecular interactions of disease cells. In cancer, genomic instability is the key driving force of tumorigenesis. The term "genomic instability" refers to an increased tendency to alter the genome during cell division and thereby to cause various genome-level alterations

Genomic and Epigenomic Biomarkers of Toxicology and Disease: Clinical and Therapeutic Actions,
First Edition. Edited by Saura C. Sahu.

(Shen 2011). Genome-wide analysis has proved that, because of genomic instability, a highly heterogeneous subpopulation of cells with a diverse mutation pattern evolves from the same tumor (Wang et al. 2014). As a result, small-scale measurement and analysis of molecular-level data may not be sufficient to identify a good target; holistic approaches could rather solve the problem. Since the invention of high-throughput technologies such as microarray and next-generation sequencing, researchers can perform large-scale or genome-wide measurements of genomics and epigenomics data. The compilation and analysis of such datasets—mRNA expression (transcriptome), DNA methylation (epigenome), single-nucleotide polymorphism (SNPs), copy-number variations (CNVs)—enhance our knowledge of biological systems. Complex and heterogeneous systems of cancer also evolved as a result of perturbations at multiple levels, for example gene, protein, and epigenome (Haukaas et al. 2016; Michailidou et al. 2017; Tyanova et al. 2016; Vareslija et al. 2019). Genes or proteins do not function in isolation; their interactions control the mechanisms that lead to healthy and diseased states in organisms. Furthermore, the genes involved in similar disorders are likely to interact with one another and have similar expression patterns. Any biological process inside a cell is regulated and coordinated at multiple levels by tightly controlled transcriptional, post-transcriptional, and post-translational molecular networks. In these circumstances, understanding the complex molecular interaction or network at each of the different molecular levels is crucial to understanding the mechanisms of disease formation and progression (Goh et al. 2007; Liu et al. 2020). Unlike single disease disorders or Mendelian diseases, complex and multifaceted diseases like cancer often need a system-level approach, because of the involvement of different molecular aberrations and environmental factors in their progression. Additionally, the interconnected nature of the biomolecules transmits the molecular abnormalities throughout the cellular system and intensifies the complexity, which poses significant challenges to the understanding of the disease mechanisms. So, representing the molecular interaction in the form of a network provides an opportunity to investigate, model, characterize, and understand complex cellular components in a holistic way, both in the healthy and in the diseased state of the cell (Zhang et al. 2014). Moreover, the construction of a cancer-specific molecular network from -omics data such as those yielded by genomics, proteomics, and metabolomics and from their integration and analysis could aid toward designing effective intervention strategies and discovering biomarkers. After constructing any molecular network, the next challenging task is analysis, which translates the network structure into meaningful and biologically relevant information. For example, the study of network structure and topology helps us understand the evolutionary mechanisms and the dynamical and functional behaviors of the network. For example, the scale-free structure of biological networks contains hub genes or proteins connected to many other genes or proteins (Jeong et al. 2001). Therefore hub genes or proteins are critically important in disease formation and in cellular dysfunction, and their identification is crucial for diagnosis and therapy. The architectural features of molecular interaction networks within a cell are to a great extent similar to architectural features in other complex systems, such as the internet, computer chips, and society. Consequently the existing mathematical frameworks (graph theory) or the quantifiable tools of network theory are applied to help us understand the architecture of biological networks in terms of organization and evolution (Barabasi et al. 2011; Barabasi and Oltvai 2004). Building the cancer-specific molecular network by considering all genetic alterations, then investigating the architecture of such a network via a graph theoretical approach can help us find the critical players in oncogenesis.

To provide a better understanding of this area of cancer research, the present chapter will introduce the basic concept of graphs and the different categories of graphs used to represent the biological network. The basic topological properties of the network will be discussed as well. Further, I will present various types of biological networks and their applications, in order to explain the mechanism of cancer and biomarker prediction. I will also offer an overview of disease networks. Finally, I will discuss how network theory can be applied to drug discovery and to precision medicine.

Categories of Graphs in Biology

In mathematics, networks are often termed graphs; the subject of "graph theory" has been developed precisely for the study and analysis of such networks. A network or graph is used to capture the informal relationship between a set of entities with connections between them. For example, a protein–protein interaction network (PPIN) consists of proteins (entities) and the interaction (relationship) between them. These entities are termed "nodes" or "vertices" in mathematics, and connections are called "edges" or "links." A graph G = (V, E) is composed of a set of vertices V and a set of edges E. An edge in the graph connects the vertices. Various categories of graph represent biomolecular interactions. The most commonly used graphs in network biology are undirected, directed, weighted, multi-edge, and bipartite graphs, as well as hypergraphs (Figure 8.1). In an *undirected graph*, the links between the vertices do not have a defined direction. Undirected networks are used for modeling the protein interaction network and the gene co-expression network. However, predicting the power of undirected networks is reduced, as the graph lacks the direction of flow of information. But the study of the architectural details

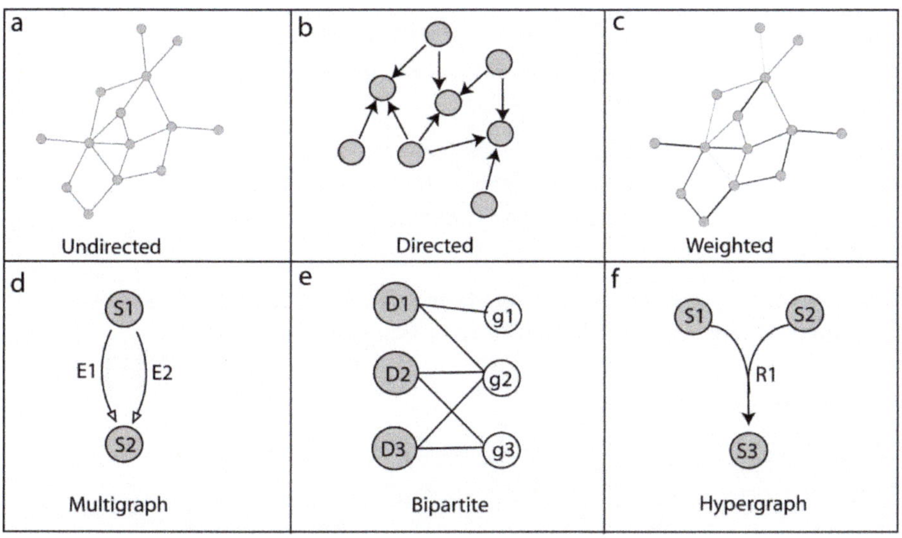

Figure 8.1 Categories of the graph. **a** Undirected graph, where edges are bidirectional. **b** Directed graph, where edges have specific directions. **c** Weighted graph, where edges have specific weights (the thickness of edges represents the strength of the relationship between two nodes). **d** Multigraph, where substance S1 is converted to S2 through two different reactions catalyzed by enzyme E1 and E2. **e** Bipartite graph, which shows the interaction between two types of nodes such as disease (D1, D2, and D3) and genes (g1, g2, and g3). **f** Hypergraph, where two substances, S1 and S2, connect to form substance S3 via reaction R1.

of such a network allows the understanding of the mechanism of biological processes. In a *directed graph*, links between the vertices have selected directions; for example, to represent the relationship between a transcription factor and gene, the directed graphs are used in the gene regulatory network. In a *weighted graph*, the links have specific weights. The values of the weights represent the strength of the relationship between the vertices. For example, to describe the correlation patterns between genes, a weighted gene co-expression network is used. The weighted gene co-expression network (WGCNA) is used for constructing gene co-expression networks in various cancers, in order to identify hub genes as potential biomarkers (Chen et al. 2016; Langfelder and Horvath 2008). In a *multigraph*, the same two nodes have multiple links or parallel links between them. Multigraphs are helpful to model the metabolic pathway where different reactions can transform the same substances. A *bipartite graph* (or bigraph) is an undirected graph whose vertices can be divided into two disjoint sets, V_1 and V_2, such that each link connects a V1 vertex to V2 vertex. The bipartite graph is often used to represent the disease–gene association network. Analysis of bipartite graphs revealed how a single gene is associated with several diseases. A *hypergraph* consists of a set of vertices V and a set of hyperedges E where an edge can join any number of vertices. Examples are metabolic networks where several substances often react to build other substances (Schreiber 2007; Koutrouli et al. 2020).

Network topology measures

Here we define network topology measures in order to compare and characterize various complex biological networks.

Degree and hub: A key measure of each node is the *degree*, which represents the number link it has to other nodes. The highest degree nodes in the network are called the *hub* (Barabasi 2016).

Betweenness centrality: Betweenness centrality measures the extent to which a node sits "between" a pair of other nodes, such that a path between those other nodes has to go through that node (Barabasi et al. 2011).

Module: Modules represent highly interlinked local regions in the network. A module is a group of closely related proteins or genes that perform specific biological functions through interaction with one another. The module is also known as the community (Barabasi et al. 2011).

Modularity: Modularity in the network refers to the pattern of connectedness in which nodes are grouped into highly connected subsets (Wagner et al. 2007).

Molecular Networks and Their Role in Cancer Biomarker Discovery

In biological networks, the "nodes" are biomolecules such as proteins, metabolites, or genes and the "edges" are the functional, physical, or chemical interactions between the corresponding biomolecules. Due to various nodes and edges, there are different forms of biological networks, for example protein–protein interaction networks, co-expression networks, gene transcriptional regulatory networks, or metabolic networks. The construction and analysis of networks at different molecular levels has offered an excellent opportunity to understand disease etiology and biomarker discovery. All the types of biological networks and their application in cancer research have been discussed here.

Protein–Protein Interaction Network

PPINs are widely used molecular networks aimed at studying human diseases, including cancer. PPINs are often represented as undirected networks. An undirected link connects two proteins if they interact physically. There are several approaches to determining protein–protein interaction (PPI), including by biophysical methods (such as X-ray crystallography, fluorescence microscopy) and by high-throughput methods (such as yeast two-hybrid). Proteins and their interactions play an essential role in biological processes and determine the molecular and cellular mechanisms that control healthy and diseased states in organisms (Safari-Alighiarloo et al. 2014). Therefore the construction and analysis of cancer-specific PPIN can be helpful for identifying molecular processes activated during cancer formation. There is an enormous amount of genetic and epigenetic modification in cancer, which can cause alterations in protein structure and expression and can lead to perturbations in coordinated PPI (Kuzmanov and Emili 2013). Again, the perturbations in PPI can cause changes in the overall architecture or topology of the network (Figure 8.2).

Therefore integrating all the genetic modification through the construction of PPIN and studying the effect of cancerous perturbations can provide valuable insight. For example, a cancer-specific mutation in p53 exhibits various structural changes that result in disturbances in PPIN

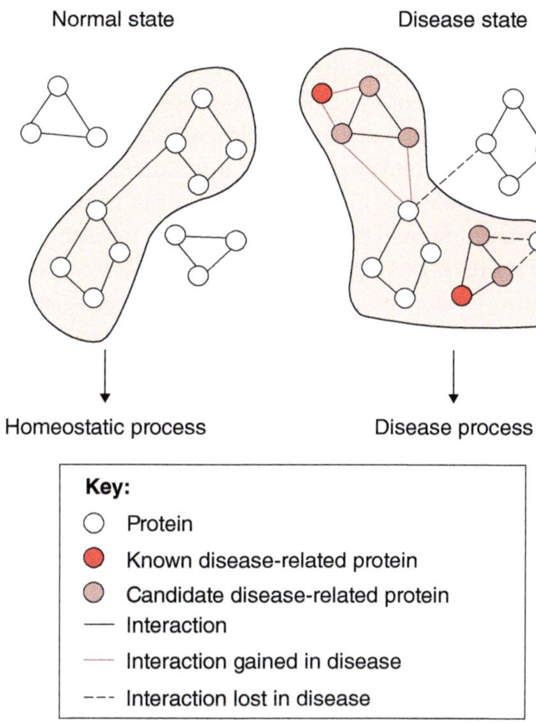

Figure 8.2 Alterations in protein-protein interactions in disease. A simplified view shows how the Protein-protein interaction alters in human disease. The normal patterns of PPIs are disturbed by genetic and epigenetic perturbations in cancer, which causes changes in PPIN topology and dynamics (Kuzmanov and Emili 2013).

(Joerger and Fersht 2007). Zhong et al. (2009) have shown that network perturbation in Mendelian disorders is due to the complete loss of gene products (node removal) or to alterations in interactions (edgetic perturbation). They concluded that edgetic perturbations probably play a vital role in the genotype-to-phenotype relationship in complex genetic disorders. The edgetic perturbations may occur as a result of mutations in the interacting interfaces of proteins. Because proteins interact through their interfaces, the cancerous mutation in the interfaces can change the topology of PPIN. Researchers have examined the effect of disease-causing mutations by using a structurally resolved PPI network. The mutations in the interacting surface of the proteins are specific to the disease and are associated with pathogenic processes (Wang et al. 2012). In consequence, investigating the alteration in the interacting interface of cancer-causing proteins will help describe their role in the PPIN. Kar et al. (2009) provided a detailed analysis of cancer-related human protein–protein interfaces and of the topological properties of the cancer network. Their study revealed that cancerous proteins have smaller, more planar, more charged, and less hydrophobic binding sites than non-cancer proteins, which results in low-affinity and high-specificity interactions in cancer. Cancerous proteins interact with their partners via distinct interfaces, and most of them are multi-interface and occupy the hub position in the network. Moreover, proteins known to be susceptible to cancerous mutations have more interaction partners than the non-cancer proteins. Jonsson and Bates (2006) observed that cancerous proteins tended to have, on average, twice as many interaction partners as non-cancer proteins. They also found that these proteins interact with one another and form bigger clusters or communities than the non-cancer proteins. Not only the mutation, but many oncogenes, too, are frequently upregulated in cancerous tissues. It has been observed that differentially upregulated genes are highly connected and occupy the hub position in the human PPIN (Wachi et al. 2005). Higher expression or upregulation of the gene is not always linked to tumorigenesis. Sumithra et al. (2019) have shown that an oncogenic splicing factor, Sam68, has cancer-specific interactions, and these interactions of Sam68 are directly linked with patient prognosis. Hence a study of the topological properties of PPINs (e.g., degree) (see the box "Network topology measures") of cancer cells that looks at genetic changes can provide crucial evidence for diagnostic and therapeutic targets that may have potential clinical applicability (Kuzmanov and Emili 2013). The integration of PPIs, data on differentially expressed genes, and protein structural and functional properties have shown promising results for identifying the cancer biomarker (Aragues et al. 2008). In this work, the authors built a cancer protein interaction network using the known cancer genes as seed genes. They defined the cancer linker degree (CLD) of a protein as the number of cancer genes to which that protein is connected, excluding the protein itself. They examined the relationship between the CLD of a protein and its likelihood of being a known cancer gene. Their findings suggest that the CLD of a protein is an excellent indicator when it comes to identifying a cancer-causing gene. They further added that proteins with a high CLD tend to be differentially expressed in multiple cancer types. Proteins involved in a specific disease tend to interact among themselves to form a disease-specific interaction network; on the basis of this understanding, the lung cancer-associated PPI network, termed OncoPPi was constructed using time-resolved Förster resonance energy transfer (TR-FRET) technology. OncoPPi reported 269 novel lung cancer-specific PPIs and identified novel mechanisms of action of hub proteins and cancer drivers such as MYC, STK11, RASSF1, and CDK4 (Figure 8.3). The degree and betweenness centrality (see the box "Network topology measures") of the nodes also revealed that MYC, AKT1, and STK11 are the three most critical hub protein in the OncoPPi network (Figure 8.3b). A cancer-specific PPIN is constructed mainly by integrating the PPI data from normal cells and the genomics data from a cancer cell. OncoPPi provides the experimentally determined cancer-associated PPI data of the selected set of lung cancer genes, which show potential significance

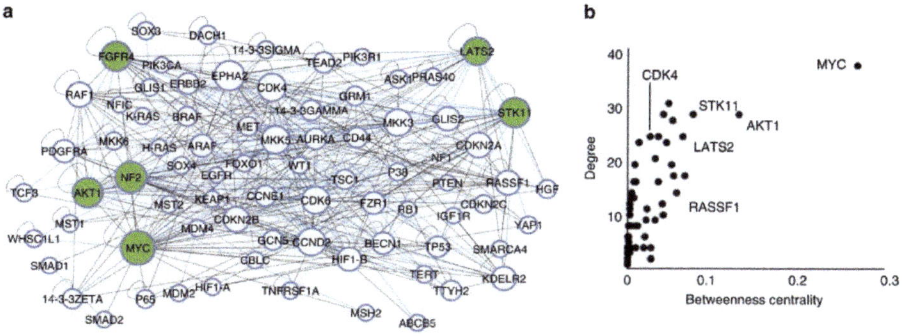

Figure 8.3 **a** OncoPPi network containing 83 lung cancer-associated proteins (nodes) connected via 397 interactions major hubs, which are highlighted in green. **b** Network topology analysis, which includes degree and betweenness centrality and identifies hub proteins in PPI (Li et al. 2017).

for cancer drug discovery. OncoPPi are freely available at the National Cancer Institute's Cancer Target Discovery and Development (CTD^2) Network Data Portal (https://ocg.cancer.gov/programs/ctd2/data-portal) Dashboard (http://ctd2-dashboard.nci.nih.gov/) (Ivanov et al. 2018; Li et al. 2017).

On the basis of these observations, scientific literature on cancer often recommends targeting hub gene for cancer treatment. Genome-wide studies illustrate that the deletion of a hub protein is more likely to be lethal than the deletion of a non-hub protein. Because hub proteins have more interactions, they are linked to more cellular processes, and the deletion of these hubs can result in cell death. This phenomenon is also known as the centrality–lethality rule (He and Zhang 2006). Targeting a network hub determined from cancer-associated PPIs (such as OncoPPi) could show an appropriate path for cancer-related therapeutic developments. But choosing the hubs in the network on the basis of genomics data may not be a valid target for controling cancer. However, the availability of cancer-associated PPIs is minimal. Therefore the best alternative approach is to find the cancer disease modules for drug targets (see the section "Network Medicine and Drug Design" in this chapter). In summary, the construction and analysis of PPI networks contribute significantly to identifying new cancer-causing proteins and to drug targets and biomarkers discovery.

Co-expression Network

Co-expression networks are also known as correlation networks; they are constructed by calculating the correlation coefficient between the concentration of genes or proteins. Two genes or proteins are connected if, in the set of samples, their concentrations are correlated. The magnitude of the correlation coefficient determines the relationship between two genes or proteins, such that a positive correlation coefficient between two genes indicates that they have a similar expression pattern across the set of samples (the expression of both genes increases or decreases simultaneously), which means that the genes in question are active in same biological processes. A negative correlation coefficient indicates that one expression increases, whereas another decreases. The most prominent methods of correlation measure are the Pearson and the Spearman correlation coefficients. The Pearson correlation describes the linear relationship between two continuous variables (such as two genes or proteins). Its value ranges from −1 to +1. By comparison, the Spearman correlation coefficient is estimated on the basis of ranked observations of two variables and ranges from −1 to +1. It can describe the linear as well as the non-linear relationship. The expression correlation can be defined categorically as a “strong” or a “weak” co-expression on the basis of the correlation coefficient value. If the correlation coefficient between genes is > 0.9, then it is generally said that the genes are strongly co-expressed; if it is 0.3 to 0.5, then their co-expression is weak. These correlation measures are used to construct the co-expression network, where each node represents a gene and each edge represents the strength of the co-expression relationship (correlation coefficient). Next, the clustering techniques are used to find the co-expression gene modules (groups of co-expressed genes) (see the box “Network topology measures”). Co-expression gene modules represent the group of genes with similar expression patterns across the samples. Clustering techniques such as k-means clustering and hierarchal clustering are used to find such co-expression modules. The biological significance of the co-expression modules can be identified using the biological processes and the functional enrichment analysis (Figure 8.4). Using a similar approach, gene co-expression networks in different cellular conditions, for instance in cancer, can be constructed so that we understand the regulatory mechanism of disease formation. However, to get an insight into a disease, the co-expression networks can be integrated with PPIs, methylome and transcription factors (TFs), and their target interactions data. This will help illustrate the mechanism of co-expression that is specific to a given biological context (van Dam et al. 2018).

Gene expression data were successfully used for the clinical diagnosis of cancer subtypes. The hierarchical clustering of gene expression data readily separated three subclasses of prostate cancer, indicating distinct gene expression patterns in each subtype (Lapointe et al. 2004). Gene co-expression-based analysis by Yu et al. (2020) has shown how genes'

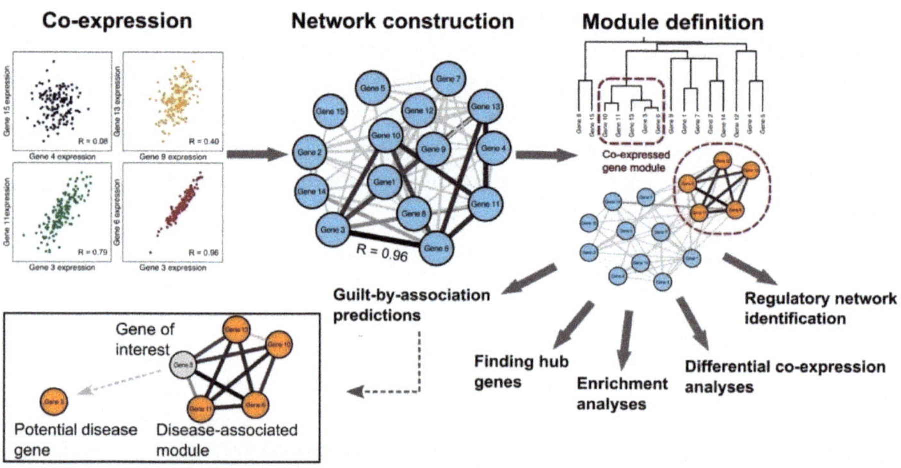

Figure 8.4 Construction and analysis of a co-expression network. The pairwise correlation coefficient is calculated for each possible gene pair in the mRNA expression data. Next, these pairwise correlations are represented as a co-expression network. Functional gene co-expression modules within these networks are identified using clustering analysis and functional enrichment analysis. Subsequently, these modules can be used to identify the hub gene, the differentially expressed modules, and the regulatory network. Potential disease genes can be identified using a guilt-by-association (GBA) approach that highlights genes co-expressed with multiple disease genes (van Dam et al. 2018).

functionalities change as cancer evolves from the early to the advanced stage. The authors have also constructed stage-specific co-expression networks for eight different human cancers with the help of gene expression data from TCGA. These stage-specific co-expression networks were then used to classify the cancer according to its stage, on the basis of the extent of network perturbation as a feature in a supervised machine-learning algorithm. The authors explain that, "for a given co-expression network of a specific stage, adding a new sample that intrinsically belongs to the stage should not alter significantly the structure of the co-expression network; in contrast, when a sample is added to the sample set of a different stage, it will affect the co-expression levels of some gene pairs, hence altering the structure of the co-expression network" (Yu et al. 2020). On the other hand, the analysis of sex- and stage-specific co-expression networks revealed the presence of conserved small-world networks across the stages of cancer in lung adenocarcinoma. Small-world networks across the stages of cancer present a synchronized system in the disordered environment of cancer and are probably involved in cancer evolution (Das 2020). These studies indicate that the construction and analysis of stage-specific co-expression networks create the opportunity to explore the progression of cancer. The identification of disease-specific differential co-expressed modules plays an important role in the improvement of cancer research. In general, in order to explain the molecular differences between multiple conditions (e.g., a diseased and a healthy state), one measures the difference in gene expression level. The most common method of comparing samples from different conditions is differential expression analysis. For example, gene expression data from cancerous and from healthy cells, obtained through microarray or RNAseq experiments, are compared with aview to finding the differentially expressed genes (DEGs) (i.e., the gene sets whose expression is altered in cancer) by differential expression analysis. But differential co-expression, or differential correlation, or differential network, or differentially co-expressed module identification and analysis provides the difference in gene–gene relationships between various conditions of interest. Differential co-expression analysis identifies the alteration between pairs of genes rather than between individual genes (Bhuva et al. 2019; McKenzie et al. 2016). The majority of differential co-expression analyses finds the differentially expressed gene modules or clusters. These clusters of genes or gene modules show a different pattern of expression under changing conditions or phenotypes. Several packages such as WGCNA (Langfelder and Horvath 2008), DiffCoEx (Tesson et al. 2010), and DICER (Amar et al. 2013) are used to identify the differential modules. However, these co-expression modules are often large, containing a large number of genes. Therefore it is important to identify the genes that can explain the clinical behavior of cancer. Most often, the researcher identifies hub genes in the modules because hub genes are highly connected and often associated with the clinical feature of diseases. Hubs can be intra-modular or inter-modular; intra-modular hubs are central to the specific module, inter-modular hubs are central to the entire network (van Dam et al. 2018). The weighted gene co-expression-based network analysis (WGCNA) of the gastric cancer transcriptome has identified clinically relevant differentially co-expressed gene modules. The expression of genes and the hub genes of the modules are specific to cancer subtypes; moreover, the gene expression patterns of the modules were linked with intestinal and diffuse subtypes and with the differentiation status of gastric cancer (Kalamohan et al. 2014). It has been observed that co-expressed modules identified through WGCNA are associated with breast cancer development. Modules that are highly correlated with breast cancer development are linked to cancer-specific biological processes such as cell migration, cell cycle, and cell–cell adhesion (Qiu et al. 2019). Therefore functional enrichment analysis (e.g., biological processes and pathways enrichment analysis) is the key step toward understanding the association of co-expressed modules with cancer development. Furthermore, survival analysis using Kaplan-Meier survival curves is another key step toward understanding the prognostic capability of the modules. If the higher expression of genes in the co-expressed modules is significantly associated with patient mortality, then those modules are linked to cancer pathogenesis or to the prognostic gene module. Differentially co-expressed prognostic gene modules in ovarian, bladder, blood, and breast cancer have shown important implications in the diagnosis and prognosis of cancer (Gov and Arga 2017; Shi and Tian 2019; Tang et al. 2018; Xiao et al. 2020). As a result, functional enrichment analysis and survival analysis are essential for establishing the differential co-expression modules or the hub genes within those modules as the potential therapeutic target in cancer. The data-driven approach in biology has revealed enormous opportunities to identify the molecular markers of disease. The correct use of tools and interpretation of results could lead to an effective biomarker discovery in cancer.

Gene Transcriptional Regulatory Networks

Gene regulatory networks (GRNs) have attracted a lot of interest in understanding the biology of human diseases. The regulation of gene expression is controlled by various regulatory elements in the genome—promoters, enhancers, silencers, and insulators (Figure 8.5a). The functioning of these regulatory elements depends on biochemical interactions that involve the DNA, the cellular proteins that makeup chromatin, and transcription factors (TF) (Doane and Elemento 2017). Therefore GRNs are commonly represented by many pairs of proteins and genes, in which a protein or a gene regulates the expression of another protein or gene. To adapt to dynamic cellular environments or to a diseased state, cells need to regulate the production of mRNA and of proteins; and, for that, the regulation of gene expression by TFs is one of the prevalent cellular processes. Deciphering TF–DNA interactions network or GRNs has potential applications in cancer biology and therapy. Elucidating the structures and dynamics of GRNs represents a significant challenge in systems biology, because a well-defined GRN contains a heterogeneous set of interactions. Although a GRN is often inferred on the basis of transcriptomic or gene expression data, expression data alone cannot differentiate direct from indirect regulation events; the integration of other omics data such as genomic and proteomic significantly increases network quality. GRNs are represented as directed networks; omics data—that is, mRNA expression data from microarray or RNAseq, DNA sequencing data of the transcription factor-binding site, and high-throughput chromatin immunoprecipitation assay (ChIP) data of DNA–protein interactions—are used precisely in order to construct the regulatory direction of GRNs in health and in various human diseases (Hecker et al. 2009) (Figure 8.5 b). The change of gene expression or the differentially expressed gene is a predominant phenomenon in a cancer cell. How are the expression changes regulated in cancer? There are several attempts to decode this regulatory mechanism using GRNs.

Aytes et al. (2014) have built genome-wide regulatory networks (interactomes) from gene expression data related to human and mouse prostate cancer, in order to identify the regulatory drivers of cancer. These researchers have generated a regulatory network using an in vivo perturbation study of mouse cancer models. The comparative analysis of mouse and human gene expression and perturbation data, followed by experimental and clinical validation, has found that FOXM1 and CENPF synergistically induce cancer growth by regulating target gene expression. The study suggests that FOXM1 and CENPF can be promising prognostic biomarkers, as their co-expression is associated with poor survival of

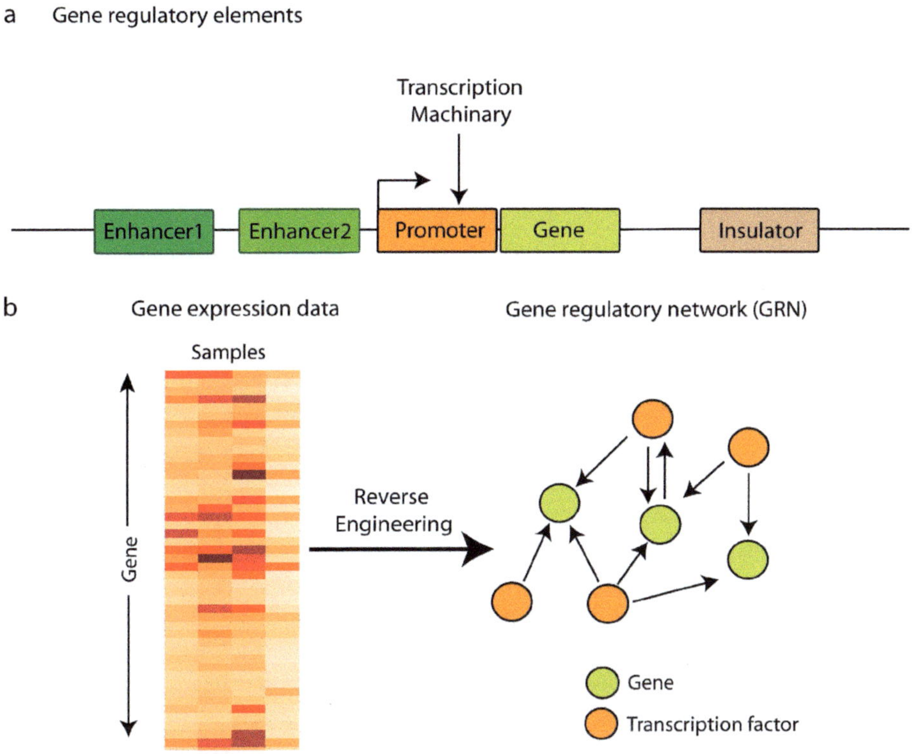

Figure 8.5 Gene regulatory elements and the GRN. **a** Multiple gene regulatory elements, including enhancer, promoter, and insulator, which are involved in controlling the transcription of a gene. **b** Reconstruction of GRN from gene expression data (heat map).

the patients. The epithelial–mesenchymal transition is the transdifferentiation process and a hallmark of tumor aggressiveness, through which epithelial cells develop the ability to invade (Fares et al. 2020). The mesenchymal phenotype is a prevalent characteristic of invasive glioma. The glioma-specific GRN reveals that two transcription factors, namely C/EBPβ and STAT3, are synergistic initiators and master regulators of mesenchymal transformation. The co-expression of C/EBPβ and STAT3 activates mesenchymal genes in malignant glioma and results in poor survival (Carro et al. 2010). Assembly GRN using gene expression profile and PPIN in B-cell revealed that MYB and FOXM1 act as synergistic *master regulators* of proliferation in the germinal center (GC) (Lefebvre et al. 2010). Researchers have also shown that the master regulator in the GRN can be identified as a potential target for cancer therapy (Madhamshettiwar et al. 2012). Therefore the GRN gives us some insight into the mechanistic processs of cancer formation. It also gives us insight into drug discovery, since the novel gene regulatory mechanisms and master regulators can be identified as the new drug targets.

Metabolic Network

The term "metabolic network" refers to the relationship between metabolites and the enzymes that interact with them to catalyze a biochemical reaction. Metabolic networks can be categorized into "metabolite-centric" and "enzyme-centric." Metabolite-centric networks are most commonly used; in them metabolites are represented as nodes, and two metabolites are connected if one metabolite is converted into the other through an enzymatic reaction. Enzyme-centric networks treat the enzymes as nodes and connect the enzymes involved in catalyzing reactions (Figure 8.6a, 8.6b; see Janga and Babu 2008). Metabolites are usually small molecules such as carbohydrates, fatty acids, nucleotides, amino acids, and lipids. Abnormal metabolism is associated with various human diseases, including cancer. Indeed, metabolic reprogramming is considered to be the hallmark of malignancy. Pioneering work by Otto Warburg in the 1920s revealed that cancer cells preferentially metabolize glucose through glycolysis, generating lactate as an end product, even in the presence of sufficient oxygen (Warburg 1956). Later it was shown that the processes of cell proliferation and growth need sufficient nutrients such as glucose, amino acids, nucleotides, and lipids to meet the extensive metabolic rewiring that happens in cancer cells (Zhu and Thompson 2019). Therefore identifying cancer-associated metabolic changes has potential applications in the discovery of a biomarker and in the development of novel therapeutics (Gatto et al. 2020; Mardinoglu et al. 2013). The global collection of all the metabolites present in the cell is known as a "metabolome." Giving a qualitative and a quantitative estimation of the metabolome is a great challenge for the biologist, because metabolomes are composed of various molecules and have multiple physical properties and chemical stability; therefore their isolation and their estimation are done using different analytical techniques. There is no single technique that can measure all metabolites in a cell, and metabolomics laboratories often use one or more methods to measure the metabolites. As a result, data are not uniform across laboratories; however, precision in measurement is required for biomarker discovery. Nuclear magnetic resonance and mass spectrometry are the most commonly used tools in metabolomics research (Lasky-Su and Clish 2017). The tumor-specific network of metabolites or metabolic network has shown promising results for an understanding of the deregulation of metabolic activity in the cancer cell. The genome-scale metabolic models (GEMs) are widely used to construct the tissue-specific metabolic network of various human diseases. Using a computational approach, GEMs predict the gene–protein–reaction associations for whole metabolic genes in an organism. These data are used in the simulation study to illustrate the metabolic fluxes for systems-level studies (Gu et al. 2019). The reconstructed metabolic networks from GEMs revealed that tumor metabolic networks are highly similar across cancers in terms of reactions, despite variations in the gene expression pattern. The GEM for 917 primary tumor samples across 13 cancer types illustrates that each network contains 4,721 reactions, of which 74% are core reactions (present in $>$ 95% cancer models). Out of this, 99.3% of the core reactions are present in healthy tissues. Thus the metabolic networks of cancerous and healthy cells have an almost similar structure. Still, the metabolic reprogramming of cancer occurs probably in order for metablism to adapt to the changing cellular environments and faster growth of cancer (Gatto et al. 2020). However, tissue-specific metabolic networks constructed using GEM and gene expression data in breast, lung, and kidney cancer revealed cancer-specific network modules. The clustering analysis of the kidney metabolic network revealed fifty functionally relevant communities, many of them well known to be involved in tumor metabolism (Figure 8.6 c; see Manipur et al. 2020).

Applying a similar approach, which integrates GEM and transcriptomic cancer data and then construction, WGCNA identified the metabolic alterations in subtypes of renal cell carcinoma (RCC). In that, amino acid metabolism and redox homeostasis are mostly altered (Pandey et al. 2020). Furthermore, integrating the cancer gene co-expression network with the metabolic network identified the crucial enzyme-coding metabolic genes in lung, leukemia, and breast cancer. Experimental evidence demonstrates that 70% of these metabolic

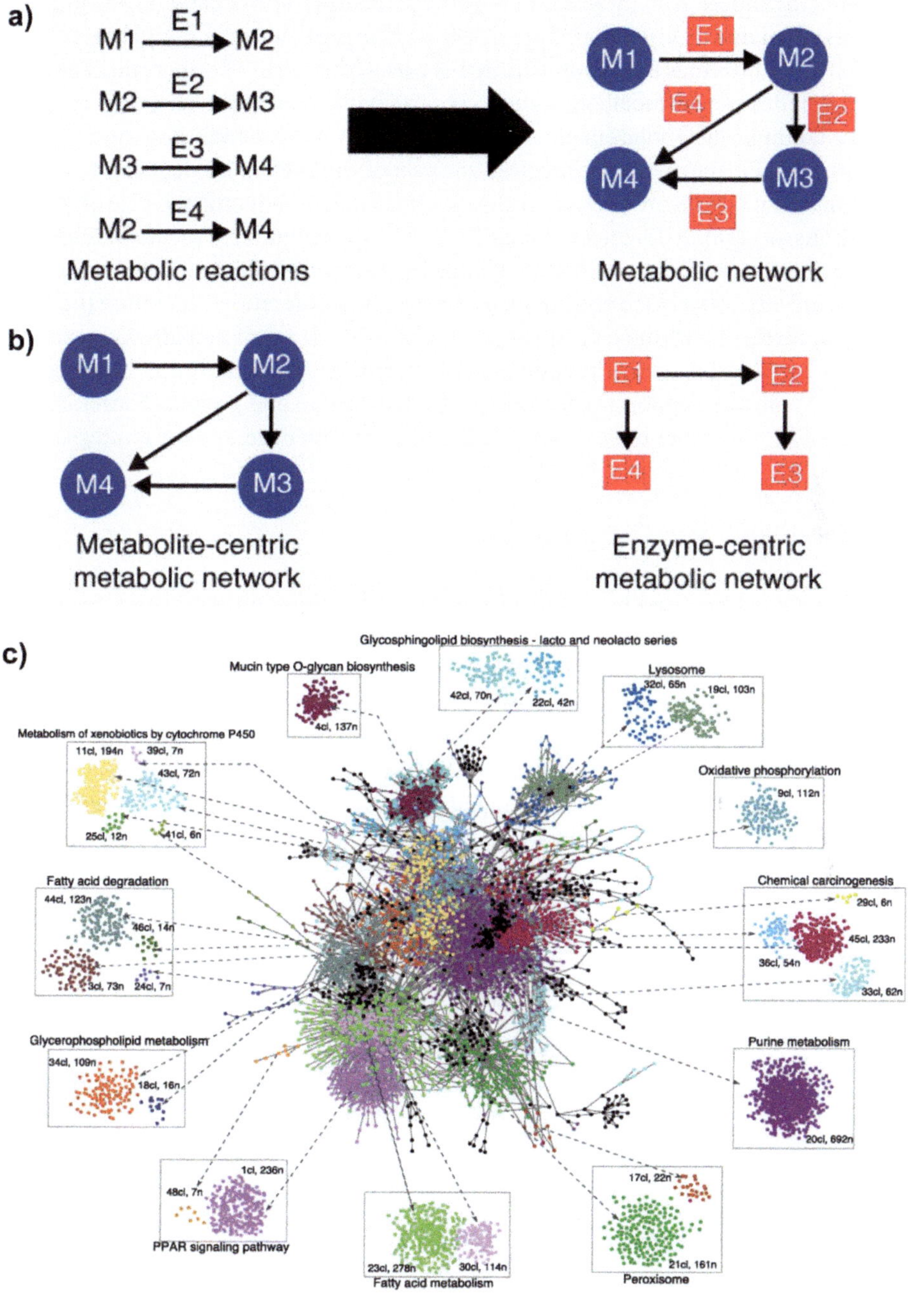

Figure 8.6 There are two categories of metabolic networks, illustrated respectively in **a** and **b**: metabolite-centric networks, where metabolites (M1, M2, M3, and M4) are represented as a node; and enzyme-centric networks, where enzyme (E1, E2, E3, and E4) are represented as node (Janga and Babu 2008). **c** presents the kidney cancer metabolic network (center) and the functionally relevant clusters (Manipur et al. 2020).

metabolites have anti-cancer properties (Chen et al. 2013). Targeting altered cancer metabolism is a promising therapeutic strategy for treatment. The construction and analysis of a genome-scale network model of cancer metabolism have predicted the genes involved in cancer cell proliferation, and those genes can be identified as potential cytostatic drug targets for reducing cancer growth. Many of these targets showed promising results in gene knockdown experiments (Folger et al. 2011). Several pieces of evidence demonstrate that the role of metabolic enzymes is a promising area of cancer research. The aberrant cell signaling in cancer induces non-metabolic functions of metabolic enzymes in many cellular functions. For example, metabolic enzymes phosphorylate proteins rather than their metabolites as substrates. Furthermore, topological analysis of the metabolic network revealed that most of the drug targets are present in the specific cluster of an enzyme-centric metabolic network (Asgari et al. 2013; Lu and Wang 2018). The deregulation of metabolic enzyme in glycolysis, the TCA cycle, the pentose phosphate pathway, lipid and amino acid metabolism pathways are also reported to contribute to cancer cell proliferation. Therefore the construction and analysis of enzyme-centric metabolic networks from these pathways can aid in finding new avenues for cancer treatment. The metabolic reprogramming in cancer is one of the key events that support the uncontrolled proliferation and growth. Therefore the integration of GEM with other omics data may provide the better therapeutic strategies.

Network Medicine and Drug Design

Cancer is not a disorder caused by a single gene, deregulation of multiple genes resulting in cancer formation. There is increasing evidence to suggest that the genes involved in the formation of this complex disease are not distributed randomly on the molecular network. Instead they work together, in biological modules or cell-signaling pathways. Indeed, genes associated with the same disease have a greater tendency to interact with one another. This causes the formation of a disease module (see the box "Network topology measures"), a connected subnetwork linked to a disease phenotype (Barabasi et al. 2011; Sharma et al. 2015) (Figure 8.7).

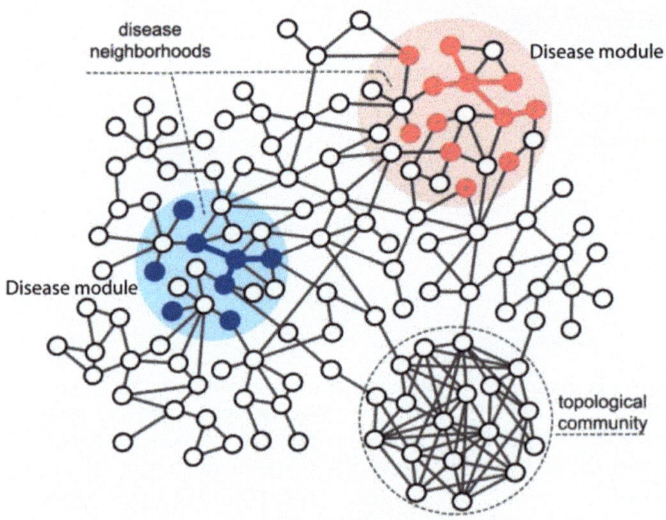

Figure 8.7 Proteins associated with the same disease or phenotype tend to localize in specific neighborhoods of the interactome, causing the formation of disease modules (marked with red and blue shades). Nodes in blue and red colors represent the disease protein, which forms the community structure in the interactome (Ghiassian et al. 2015).

Identifying disease modules of cancer can help the understanding of tumorigenesis, the discovery of new biomarkers, and drug designing. Several module-based studies have been published in the field of cancer. The process of modeling the network constructed using omic data identified the module relevant to cancer development in the breast. The results of network modeling identified the novel biomarker HMMR and its association with breast cancer. Previously, the role of HMMR had been unknown; the module-based study showed that HMMR is part of a BRCA1 functional module involved in breast cancer development (Pujana et al. 2007). It has been demonstrated that altered modularity of the PPI network may be a useful indicator of breast cancer prognosis (Taylor et al. 2009). Furthermore, the PPI sub-network efficiently discriminates against metastatic versus non-metastatic breast cancer (Chuang et al. 2007). A close relationship has been demonstrated between differentially expressed protein subnetwork and glioma subtype, that is, glioblastomas and oligodendrogliomas (Wuchty et al. 2010). This evidence firmly establishes the existence of disease modules in various human cancers. Network medicine-based drug-designing strategies attempt to target the disease module rather than the individual gene or protein or the hub in the network. An effective drug must target the protein within, or in the immediate vicinity of, the disease modules. To find a correct drug combination for a disease, a network-based drug-disease proximity measure was specifically developed. Using this proximity measure, one can quantify the therapeutic efficacy of a drug for a given disease (Barabasi et al. 2011; Guney et al. 2016). The highly interconnected nature of the human interactome and the multifactorial nature of cancer suggest that there are genetic connections of cancer with other human disorders. The disease association map of human tumor suppressor genes (TSGs) revealed that cancer is linked to various phenotypically unrelated human disorders such as neurological, developmental, immunological, hematological, metabolic, and cardiovascular diseases (Figure 8.8). Understanding the common genetic mechanisms that underlie diseases with distinct pathological symptoms will help us develop preventive and therapeutic strategies for cancer (Das 2019).

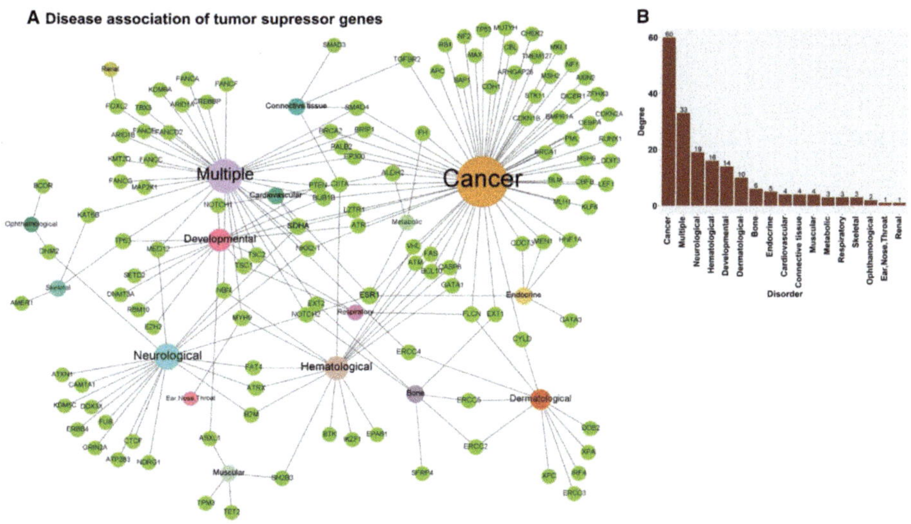

Figure 8.8 Disease association of TSGs. **a** Bi-partite network that shows that TSGs (green nodes) are connected to various human disorders. **b** Bar diagram that shows the total number of TSGs associated with the various diseases (the number at the top of each bar indicates the total number TSGs connected to a particular disease) (Das 2019).

The pioneering study by Goh et al. (2007) revealed the human disease network or diseasome, where a node represents disease and edges represent the various molecular relationships. Understanding such a disease–disease relationship will help us uncover the molecular basis of comorbidity. Cancer could be predicted from symptoms and signs of other human disorders, which in turn help us toward the early detection of cancer and its better clinical care.

Conclusion

Cancer is a highly complex, multifactorial disease, and the network-driven research approach will play a pivotal role in both improving our understanding of its complexity and helping us develop effective therapeutic interventions. The enormous amount of omic data, accumulated at different molecular levels—genomics, transcriptomics, proteomics, and metabolomics—is allowing us unprecedented insight into the molecular mechanisms of cancer. The integration of omics data with a network-based toolset can explain the regulatory programs and molecular organizational architectures of diseases, leading to the discovery of crucial biomarkers. In the present chapter I have discussed the molecular network at different levels and have summarized findings from published reports to the effect that these networks are essential to understanding cancer's etiology and pathogenesis. Despite demonstrated successes, network biology suffers from limitations that are due to the incompleteness of interactome maps of humans and to constraints in the existing tools. However, the rapid advancement of high-throughput technology and databases (such as TCGA) offers a way to overcome data limitations. This explains why network-driven approaches achieved overall success in cancer research. The fundamental problem of any research on a complex disease is to link all its molecular perturbations and to find the most disruptive point in the system, or the master regulator of the disease; and in this respect network biology offers the most comprehensive techniques to apply. In the future, the availability of more sophisticated tools and technology will lead to the development of personalized and precision medicine for cancer treatment.

References

Amar, D., Safer, H., and Shamir, R. (2013). Dissection of regulatory networks that are altered in disease via differential co-expression. *PLoS Comput. Biol.* 9: e1002955.

Aragues, R., Sander, C., and Oliva, B. (2008). Predicting cancer involvement of genes from heterogeneous data. *BMC Bioinform.* 9: 172.

Asgari, Y., Salehzadeh-Yazdi, A., Schreiber, F., and Masoudi-Nejad, A. (2013). Controllability in cancer metabolic networks according to drug targets as driver nodes. *PLoS One* 8: e79397.

Aytes, A., Mitrofanova, A., Lefebvre, C., Alvarez, M.J., Castillo-Martin, M., Zheng, T., Eastham, J.A., Gopalan, A., Pienta, K.J., Shen, M.M., Califano, A., and Abate-Shen, C. (2014). Cross-species regulatory network analysis identifies a synergistic interaction between FOXM1 and CENPF that drives prostate cancer malignancy. *Cancer Cell* 25: 638–651.

Barabasi, A.L. (2016). *Network Science*. Cambridge, MA: Cambridge University Press.

Barabasi, A.L., Gulbahce, N., and Loscalzo, J. (2011). Network medicine: A network-based approach to human disease. *Nat. Rev. Genet.* 12: 56–68.

Barabasi, A.L. and Oltvai, Z.N. (2004). Network biology: Understanding the cell's functional organization. *Nat. Rev. Genet.* 5: 101–113.

Bhuva, D.D., Cursons, J., Smyth, G.K., and Davis, M.J. (2019). Differential co-expression-based detection of conditional relationships in transcriptional data: Comparative analysis and application to breast cancer. *Genome Biol.* 20: 236.

Carro, M.S., Lim, W.K., Alvarez, M.J., Bollo, R.J., Zhao, X., Snyder, E.Y., Sulman, E.P., Anne, S.L., Doetsch, F., Colman, H., Lasorella, A., Aldape, K., Califano, A., and Iavarone, A. (2010). The transcriptional network for mesenchymal transformation of brain tumours. *Nature* 463: 318–325.

Chen, J., Ma, M., Shen, N., Xi, J.J., and Tian, W. (2013). Integration of cancer gene co-expression network and metabolic network to uncover potential cancer drug targets. *J. Proteome Res.* 12: 2354–2364.

Chen, J., Yu, L., Zhang, S., and Chen, X. (2016). Network analysis-based approach for exploring the potential diagnostic biomarkers of acute myocardial infarction. *Front. Physiol.* 7: 615.

Chuang, H.Y., Lee, E., Liu, Y.T., Lee, D., and Ideker, T. (2007). Network-based classification of breast cancer metastasis. *Mol. Syst. Biol.* 3: 140.

Das, A.B. (2019). Disease association of human tumor suppressor genes. *Mol. Genet. Genomics* 294: 931–940.

Das, A.B. (2020). Small-world networks of prognostic genes associated with lung adenocarcinoma development. *Genomics* 112: 4078–4088.

Doane, A.S. and Elemento, O. (2017). Regulatory elements in molecular networks. *Wiley Interdiscip. Rev. Syst. Biol. Med.* 9 (3): doi: 10.1002/wsbm.1374.

Fares, J., Fares, M.Y., Khachfe, H.H., Salhab, H.A., and Fares, Y. (2020). Molecular principles of metastasis: A hallmark of cancer revisited. *Signal Transduct. Target Ther.* 5: 28.

Folger, O., Jerby, L., Frezza, C., Gottlieb, E., Ruppin, E., and Shlomi, T. (2011). Predicting selective drug targets in cancer through metabolic networks. *Mol. Syst. Biol.* 7: 501.

Gatto, F., Ferreira, R., and Nielsen, J. (2020). Pan-cancer analysis of the metabolic reaction network. *Metab. Eng.* 57: 51–62.

Ghiassian, S.D., Menche, J., and Barabasi, A.L. (2015). A disease module detection (DIAMOnD) algorithm derived from a systematic analysis of connectivity patterns of disease proteins in the human interactome. *PLoS Comput. Biol.* 11: e1004120.

Goh, K.I., Cusick, M.E., Valle, D., Childs, B., Vidal, M., and Barabasi, A.L. (2007). The human disease network. *Proc. Natl. Acad. Sci. U.S.A.* 104: 8685–8690.

Gov, E. and Arga, K.Y. (2017). Differential co-expression analysis reveals a novel prognostic gene module in ovarian cancer. *Sci. Rep.* 7: 4996.

Gu, C., Kim, G.B., Kim, W.J., Kim, H.U., and Lee, S.Y. (2019). Current status and applications of genome-scale metabolic models. *Genome Biol.* 20: 121.

Guney, E., Menche, J., Vidal, M., and Barabasi, A.L. (2016). Network-based in silico drug efficacy screening. *Nat. Commun.* 7: 10331.

Haukaas, T.H., Euceda, L.R., Giskeodegard, G.F., Lamichhane, S., Krohn, M., Jernstrom, S., Aure, M.R., Lingjaerde, O.C., Schlichting, E., Garred, O., Due, E.U., Mills, G.B., Sahlberg, K.K., Borresen-Dale, A.L., and Bathen, T.F. (2016). Metabolic clusters of breast cancer in relation to gene- and protein-expression subtypes. *Cancer Metab.* 4: 12.

He, X. and Zhang, J. (2006). Why do hubs tend to be essential in protein networks? *PLoS Genet.* 2: e88.

Hecker, M., Lambeck, S., Toepfer, S., Van Someren, E., and Guthke, R. (2009). Gene regulatory network inference: Data integration in dynamic models: A review. *Biosystems* 96: 86–103.

Ivanov, A.A., Revennaugh, B., Rusnak, L., Gonzalez-Pecchi, V., Mo, X., Johns, M.A., Du, Y., Cooper, L.A.D., Moreno, C.S., Khuri, F.R., and Fu, H. (2018). The OncoPPi Portal: An integrative resource to explore and prioritize protein-protein interactions for cancer target discovery. *Bioinformatics* 34: 1183–1191.

Janga, S.C. and Babu, M.M. (2008). Network-based approaches for linking metabolism with environment. *Genome Biol.* 9: 239.

Jeong, H., Mason, S.P., Barabasi, A.L., and Oltvai, Z.N. (2001). Lethality and centrality in protein networks. *Nature* 411: 41–42.

Joerger, A.C. and Fersht, A.R. (2007). Structure-function-rescue: The diverse nature of common p53 cancer mutants. *Oncogene* 26: 2226–2242.

Jonsson, P.F. and Bates, P.A. (2006). Global topological features of cancer proteins in the human interactome. *Bioinformatics* 22: 2291–2297.

Kalamohan, K., Periasamy, J., Bhaskar Rao, D., Barnabas, G.D., Ponnaiyan, S., and Ganesan, K. (2014). Transcriptional co-expression network reveals the involvement of varying stem cell features with different dysregulations in different gastric cancer subtypes. *Mol. Oncol.* 8: 1306–1325.

Kar, G., Gursoy, A., and Keskin, O. (2009). Human cancer protein-protein interaction network: A structural perspective. *PLoS Comput. Biol.* 5: e1000601.

Koutrouli, M., Karatzas, E., Paez-Espino, D., and Pavlopoulos, G.A. (2020). A guide to conquer the biological network era using graph theory. *Front. Bioeng. Biotechnol.* 8: 34.

Kuzmanov, U. and Emili, A. (2013). Protein-protein interaction networks: Probing disease mechanisms using model systems. *Genome Med.* 5: 37.

Langfelder, P. and Horvath, S. (2008). WGCNA: An R package for weighted correlation network analysis. *BMC Bioinform.* 9: 559.

Lapointe, J., Li, C., Higgins, J.P., Van De Rijn, M., Bair, E., Montgomery, K., Ferrari, M., Egevad, L., Rayford, W., Bergerheim, U., Ekman, P., Demarzo, A.M., Tibshirani, R., Botstein, D., Brown, P.O., Brooks, J.D., and Pollack, J.R. (2004). Gene expression profiling identifies clinically relevant subtypes of prostate cancer. *Proc. Natl. Acad. Sci. USA* 101: 811–816.

Lasky-Su, J. and Clish, C.B. (2017). Metabolomics and network medicine. In: *Network Medicine*, J. Lasky-Su and C.B. Clish, 238–266. Cambridge, MA: Harvard University Press. https://doi.org/doi:10.4159/9780674545533-012.

Lefebvre, C., Rajbhandari, P., Alvarez, M.J., Bandaru, P., Lim, W.K., Sato, M., Wang, K., Sumazin, P., Kustagi, M., Bisikirska, B.C., Basso, K., Beltrao, P., Krogan, N., Gautier, J., Dalla-Favera, R., and Califano, A. (2010). A human B-cell interactome identifies MYB and FOXM1 as master regulators of proliferation in germinal centers. *Mol. Syst. Biol.* 6: 377.

Li, Z., Ivanov, A.A., Su, R., Gonzalez-Pecchi, V., Qi, Q., Liu, S., Webber, P., Mcmillan, E., Rusnak, L., Pham, C., Chen, X., Mo, X., Revennaugh, B., Zhou, W., Marcus, A., Harati, S., Johns, M.A., White, M.A., Moreno, C., Cooper, L.A., Du, Y., Khuri, F.R., and Fu, H. (2017). The OncoPPi network of cancer-focused protein-protein interactions to inform biological insights and therapeutic strategies. *Nat. Commun.* 8: 14356.

Liu, C., Ma, Y., Zhao, J., Nussinov, R., Zhang, Y., Cheng, F., and Zhang, Z. (2020). Computational network biology: Data, models, and applications. *Phys. Rep.* 846: 1–66.

Lu, S. and Wang, Y. (2018). Nonmetabolic functions of metabolic enzymes in cancer development. *Cancer Commun. (Lond.)* 38: 63.

Madhamshettiwar, P.B., Maetschke, S.R., Davis, M.J., Reverter, A., and Ragan, M.A. (2012). Gene regulatory network inference: Evaluation and application to ovarian cancer allows the prioritization of drug targets. *Genome Med.* 4: 41.

Manipur, I., Granata, I., Maddalena, L., and Guarracino, M.R. (2020). Clustering analysis of tumor metabolic networks. *BMC Bioinform.* 21: 349.

Mardinoglu, A., Gatto, F., and Nielsen, J. (2013). Genome-scale modeling of human metabolism: A systems biology approach. *Biotechnol. J.* 8: 985–996.

Mckenzie, A.T., Katsyv, I., Song, W.M., Wang, M., and Zhang, B. (2016). DGCA: A comprehensive R package for differential gene correlation analysis. *BMC Syst. Biol.* 10: 106.

Michailidou, K., Lindstrom, S., Dennis, J., Beesley, J., Hui, S., Kar, S., Lemacon, A., Soucy, P., Glubb, D., Rostamianfar, A., Bolla, M.K., Wang, Q., Tyrer, J., Dicks, E., Lee, A., Wang, Z., Allen, J., Keeman, R., Eilber, U., French, J.D., Qing Chen, X., Fachal, L., Mccue, K., Mccart Reed, A.E., Ghoussaini, M., Carroll, J.S., Jiang, X., Finucane, H., Adams, M., Adank, M.A., Ahsan, H., Aittomaki, K., Anton-Culver, H., Antonenkova, N.N., Arndt, V., Aronson, K.J., Arun, B., Auer, P.L., Bacot, F., Barrdahl, M., Baynes, C., Beckmann, M.W., Behrens, S., Benitez, J., Bermisheva, M., Bernstein, L., Blomqvist, C., Bogdanova, N.V., Bojesen, S.E., Bonanni, B., Borresen-Dale, A.L., Brand, J.S., Brauch, H., Brennan, P., Brenner, H., Brinton, L., Broberg, P., Brock, I.W., Broeks, A., Brooks-Wilson, A., Brucker, S.Y., Bruning, T., Burwinkel, B., Butterbach, K., Cai, Q., Cai, H., Caldes, T., Canzian, F., Carracedo, A., Carter, B.D., Castelao, J.E., Chan, T.L., David Cheng, T.Y., Seng Chia, K., Choi, J.Y., Christiansen, H., Clarke, C.L., Collee, M., Conroy, D.M., Cordina-Duverger, E., Cornelissen, S., Cox, D.G., Cox, A., Cross, S.S., Cunningham, J.M., Czene, K., Daly, M.B., Devilee, P., Doheny, K.F., Dork, T., Dos-Santos-Silva, I., Dumont, M., Durcan, L., Dwek, M., Eccles, D.M., Ekici, A.B., Eliassen, A.H., Ellberg, C., Elvira, M., Engel, C. et al. (2017). Association analysis identifies 65 new breast cancer risk loci. *Nature* 551: 92–94.

Pandey, N., Lanke, V., and Vinod, P.K. (2020). Network-based metabolic characterization of renal cell carcinoma. *Sci. Rep.* 10: 5955.

Pujana, M.A., Han, J.D., Starita, L.M., Stevens, K.N., Tewari, M., Ahn, J.S., Rennert, G., Moreno, V., Kirchhoff, T., Gold, B., Assmann, V., Elshamy, W.M., Rual, J.F., Levine, D., Rozek, L.S., Gelman, R.S., Gunsalus, K.C., Greenberg, R.A., Sobhian, B., Bertin, N., Venkatesan, K., Ayivi-Guedehoussou, N., Sole, X., Hernandez, P., Lazaro, C., Nathanson, K.L., Weber, B.L., Cusick, M.E., Hill, D.E., Offit, K., Livingston, D.M., Gruber, S.B., Parvin, J.D., and Vidal, M. (2007). Network modeling links breast cancer susceptibility and centrosome dysfunction. *Nat. Genet.* 39: 1338–1349.

Qiu, J., Du, Z., Wang, Y., Zhou, Y., Zhang, Y., Xie, Y., and Lv, Q. (2019). Weighted gene co-expression network analysis reveals modules and hub genes associated with the development of breast cancer. *Med. (Baltimore)* 98: e14345.

Safari-Alighiarloo, N., Taghizadeh, M., Rezaei-Tavirani, M., Goliaei, B., and Peyvandi, A.A. (2014). Protein–protein interaction networks (PPI) and complex diseases. *Gastroenterol. Hepatol. Bed. Bench.* 7: 17–31.

Sawyers, C.L. (2008). The cancer biomarker problem. *Nature* 452: 548–552.

Schreiber, F. (2007). Graph theory. In: *Analysis of Biological Networks*, (eds. B.H. Junker and F. Schreiber), 15–28. Hoboken, NJ: John Wiley & Sons, Inc. https://doi.org/10.1002/9780470253489.

Sharma, A., Menche, J., Huang, C.C., Ort, T., Zhou, X., Kitsak, M., Sahni, N., Thibault, D., Voung, L., Guo, F., Ghiassian, S.D., Gulbahce, N., Baribaud, F., Tocker, J., Dobrin, R., Barnathan, E., Liu, H., Panettieri, R.A., JR., Tantisira, K.G., Qiu, W., Raby, B.A., Silverman, E.K., Vidal, M., Weiss, S.T., and Barabasi, A.L. (2015). A disease module in the interactome explains disease heterogeneity, drug response and captures novel pathways and genes in asthma. *Hum. Mol. Genet.* 24: 3005–3020.

Shen, Z. (2011). Genomic instability and cancer: An introduction. *J. Mol. Cell. Biol.* 3: 1–3.

Shi, S. and Tian, B. (2019). Identification of biomarkers associated with progression and prognosis in bladder cancer via co-expression analysis. *Cancer Biomark.* 24: 183–193.

Sumithra, B., Saxena, U., and Das, A.B. (2019). A comprehensive study on genome-wide co-expression network of KHDRBS1/Sam68 reveals its cancer and patient-specific association. *Sci. Rep.* 9: 11083.

Tang, J., Kong, D., Cui, Q., Wang, K., Zhang, D., Gong, Y., and Wu, G. (2018). Prognostic genes of breast cancer identified by gene co-expression network analysis. *Front. Oncol.* 8: 374.

Taylor, I.W., Linding, R., Warde-Farley, D., Liu, Y., Pesquita, C., Faria, D., Bull, S., Pawson, T., Morris, Q., and Wrana, J.L. (2009). Dynamic modularity in protein interaction networks predicts breast cancer outcome. *Nat. Biotechnol.* 27: 199–204.

Tesson, B.M., Breitling, R., and Jansen, R.C. (2010). DiffCoEx: A simple and sensitive method to find differentially coexpressed gene modules. *BMC Bioinform.* 11: 497.

Tyanova, S., Albrechtsen, R., Kronqvist, P., Cox, J., Mann, M., and Geiger, T. (2016). Proteomic maps of breast cancer subtypes. *Nat. Commun.* 7: 10259.

Van Dam, S., Vosa, U., Van Der Graaf, A., Franke, L., and De Magalhaes, J.P. (2018). Gene co-expression analysis for functional classification and gene-disease predictions. *Brief Bioinform.* 19: 575–592.

Vareslija, D., Priedigkeit, N., Fagan, A., Purcell, S., Cosgrove, N., O'Halloran, P.J., Ward, E., Cocchiglia, S., Hartmaier, R., Castro, C.A., Zhu, L., Tseng, G.C., Lucas, P.C., Puhalla, S.L., Brufsky, A.M., Hamilton, R.L., Mathew, A., Leone, J.P., Basudan, A., Hudson, L., Dwyer, R., Das, S., O'Connor, D.P., Buckley, P.G., Farrell, M., Hill, A.D.K., Oesterreich, S., Lee, A.V., and Young, L.S. (2019). Transcriptome characterization of matched primary breast and brain metastatic tumors to detect novel actionable targets. *J. Natl. Cancer Inst.* 111: 388–398.

Wachi, S., Yoneda, K., and Wu, R. (2005). Interactome–transcriptome analysis reveals the high centrality of genes differentially expressed in lung cancer tissues. *Bioinformatics* 21: 4205–4208.

Wagner, G.P., Pavlicev, M., and Cheverud, J.M. (2007). The road to modularity. *Nat. Rev. Genet.* 8: 921–931.

Wang, X., Wei, X., Thijssen, B., Das, J., Lipkin, S.M., and Yu, H. (2012). Three-dimensional reconstruction of protein networks provides insight into human genetic disease. *Nat. Biotechnol.* 30: 159–164.

Wang, Y., Waters, J., Leung, M.L., Unruh, A., Roh, W., Shi, X., Chen, K., Scheet, P., Vattathil, S., Liang, H., Multani, A., Zhang, H., Zhao, R., Michor, F., Meric-Bernstam, F., and Navin, N.E. (2014). Clonal evolution in breast cancer revealed by single nucleus genome sequencing. *Nature* 512: 155–160.

Warburg, O. (1956). On the origin of cancer cells. *Science* 123: 309–314.

Wuchty, S., Zhang, A., Walling, J., Ahn, S., Li, A., Quezado, M., Oberholtzer, C., Zenklusen, J.C., and Fine, H.A. (2010). Gene pathways and subnetworks distinguish between major glioma subtypes and elucidate potential underlying biology. *J. Biomed. Inform.* 43: 945–952.

Xiao, J., Wang, X., and Bai, H. (2020). Clinical features and prognostic impact of coexpression modules constructed by WGCNA for diffuse large b-cell lymphoma. *Biomed. Res. Int.* 2020: 7947208.

Yan, W., Xue, W., Chen, J., and Hu, G. (2016). Biological networks for cancer candidate biomarkers Discovery. *Cancer Inform.* 15: 1–7.

Yu, X., Cao, S., Zhou, Y., Yu, Z., and Xu, Y. (2020). Co-expression based cancer staging and application. *Sci. Rep.* 10: 10624.

Zhang, B., Tian, Y., and Zhang, Z. (2014). Network biology in medicine and beyond. *Circ. Cardiovasc. Genet.* 7: 536–547.

Zhang, J., Spath, S.S., Marjani, S.L., Zhang, W., and Pan, X. (2018). Characterization of cancer genomic heterogeneity by next-generation sequencing advances precision medicine in cancer treatment. *Precis. Clin. Med.* 1: 29–48.

Zhong, Q., Simonis, N., Li, Q.R., Charloteaux, B., Heuze, F., Klitgord, N., Tam, S., Yu, H., Venkatesan, K., Mou, D., Swearingen, V., Yildirim, M.A., Yan, H., Dricot, A., Szeto, D., Lin, C., Hao, T., Fan, C., Milstein, S., Dupuy, D., Brasseur, R., Hill, D.E., Cusick, M.E., and Vidal, M. (2009). Edgetic perturbation models of human inherited disorders. *Mol. Syst. Biol.* 5: 321.

Zhu, J. and Thompson, C.B. (2019). Metabolic regulation of cell growth and proliferation. *Nat. Rev. Mol. Cell Biol.* 20: 436–450.

9

Epigenetic Biomarkers

Link to Maternal Exposure and Offspring Health Outcomes

Jairus Pulczinski[1], Moira Mccormick[1], Yuchen Sun[2], Musa Watfa[2], Robert YS Cheng[3], and Wan-Yee Tang[1,2]

[1] *Department of Environmental Health and Engineering, Johns Hopkins Bloomberg School of Public Health, Baltimore, MD*
[2] *Department of Environmental and Occupational Health, University of Pittsburgh Graduate School of Public Health, Pittsburgh, PA*
[3] *Molecular Mechanisms Section, Laboratory of Cancer and Immunometabolism, Center for Cancer Research, National Cancer Institute, Frederick, MD*
Corresponding author: Wan-Yee Tang Department of Environmental and Occupational Health, University of Pittsburgh School of Public Health, Pittsburgh, PA

Introduction

Gestation and early life represent a unique and remarkable biological time. During this period, a human zygote must grow and divide from single cell to roughly 20 trillion cells in the course of nine months. This growth is tied to distinct patterning events that form axes, tissue layers, and countless fate decisions and differentiation events that ultimately produce complex organ systems. Throughout this process, the fetus is dependent on the mother for its sustainment. This state is facilitated by fetal extra-embryonic tissue, which consists of the yolk sac, the amnion, the umbilical cord, and the placenta. These extra-fetal kinds of tissue are unique in that they are designed to be discarded after birth, yet are indispensable during gestation. The placenta in particular is responsible for facilitating gas exchanges between mother and fetus as well as nutrient absorption, waste excretion, some metabolism, immunomodulation, and hormonal signaling. Additionally, the placenta is not a passive organ but operates bidirectionally, signaling to both the mother and the fetus. Given the considerable amount of coordination and complexity that occurs during gestation, this process also opens a window of susceptibly for both the mother and the fetus. Depending of the stage of development and the duration of exposure, certain exposures may be deleterious or benign. It is abundantly clear, however, that maternal exposures can modulate fetal health and that these perturbations can have lasting health impacts into adulthood.

Dr. David Barker is credited with laying the groundwork for the developmental origins of health and disease (DoHaD) hypothesis. He observed a causal association of low birthweight (BW), premature birth (PTB), and intrauterine growth restriction (IUGR) with hypertension, coronary heart disease, and type II diabetes in adulthood (Barker 1997; Hales and Barker 2001, 1992). These findings prompted Barker to suggest that the maternal environment can signal to the fetus what environment to expect, allowing the fetus to prepare for and adapt to that kind of environment once it leaves the womb. If there is mismatch between the fetal phenotype and the environment it is born into, the offspring is more likely to

Genomic and Epigenomic Biomarkers of Toxicology and Disease: Clinical and Therapeutic Actions,
First Edition. Edited by Saura C. Sahu.

experience deleterious health effects. The six-month famine study of the Dutch Hunger Winter of 1944 to 1945 offered proof of the viability of this concept, as children who experienced the famine during gestation or in early life were at greater risk for glucose intolerance, obesity, and dyslipidemia than their siblings born after the famine (Roseboom et al. 2006). We now appreciate that maternal exposures mean much more than nutrition: they span diverse phenomena such as hypoxia, hypo- and hyper-nourishment, infection, hormones, drugs, and toxicants, each one of which is believed to alter the fetal epigenome. This is because epigenetics underpins developmental reprogramming by serving as a molecular link between the environment and gene expression (Tang and Ho 2007). Epigenetic modifications, including non-coding RNAs (ncRNAs), DNA methylation, and histone modifications, are mitotically and transgenerationally inheritable. These modifications work singularly or conjointly to affect DNA folding, nucleosome positioning, chromatin compaction, and ultimately nuclear organization. Environmental insults *in utero* appear to promote alterations in the germline epigenome (i.e., by DNA methylation) that escape epigenetic reprogramming after fertilization, just as an imprinted gene does. This germline epigenetic reprogramming alters the embryonic stem cell epigenome, and the derived somatic cell types that are sensitive to this altered epigenome are then more susceptible to promoting adult-onset disease (Martos et al. 2015). On the other hand, the maternal–fetal transfer is bidirectional. It is postulated that alterations in immunity, endocrine nexus, and trophoblast migration during pregnancy at the maternal–fetal interface may "reprogram" the pregnancy outcome (Cheng et al., 2018). How these signals contribute to the mother's health and whether epigenetic modifications are involved requires further investigation.

The existence of cutting-edge epigenetic techniques (e.g., microarrays and next-generation sequencing) and the collection of various biospecimens (placenta, cord, amniotic fluid, and maternal and cord blood) from either laboratory models or human cohorts allow us to identify epigenetic modifications at the single-cell, single-gene, and single-nucleus level. In this review, we summarize the evidence that demonstrates the relationship between maternal exposures and epigenetic alterations and their links to pregnancy outcomes and to later health problems in the offspring. We focus on the epigenetic effects of maternal insults on pregnancy, birth outcomes, allergic diseases, metabolic syndrome, and neurodevelopment. Findings suggest the use of epigenetic biomarkers for exposure assessment in order to predict the risk of disease later on. In addition, we discuss the challenges and future directions that may determine the success of understanding the mother's and the offspring's health.

Alterations in the Epigenetic Markers of Pregnancy or Birth Outcomes and Childhood Disease

The birth and childhood outcomes highlighted in this section are meant to provide an overview of some conditions in which there are substantial experimental data to suggest that epigenetic mechanisms may be useful as biomarkers of disease susceptibility or severity. Additionally, these outcomes represent common diseases that can have an immense impact on later life.

Preeclampsia

Preeclampsia is a condition characterized by hypertension that first presents during pregnancy, along with signs of maternal organ or placental dysfunction or proteinuria (Fox et al. 2019). In the United States, preeclampsia occurs in about 4% of all pregnancies and is linked to increased risk of mortality and morbidity for both mother and fetus. For the mother, preeclampsia can progress to eclampsia, which is characterized by the onset of seizures. For the fetus, preeclampsia increases the risk of PTB, low BW (LBW), placental abruption, and mortality, as well as the risk of disease in later life, for instance cardiovascular disease (CVD), metabolic disease, and neurocognitive disease (Lu and Hu 2019). The underlying pathophysiology of preeclampsia is poorly understood, but this condition is believed to be a product of placental dysfunction (Schneider 2017). Given the potential severity of preeclampsia, particularly if it is not diagnosed and addressed in a timely manner, there is considerable interest in the discovery of markers for its detection. These markers must strike a balance between being sensitive enough to capture the signs of worsening preeclampsia or the risk of stillbirth and avoiding the induction of pregnancy too early, as the only cure of preeclampsia is delivery (Carty et al. 2008).

ncRNAs, in particular long ncRNAs (lncRNAs) and microRNAs (miRNAs), have gained traction as possible biomarkers of preeclampsia, as diagnostic tools for its severity, and as potential mediators of its pathophysiology (Yin et al. 2020). For example, lncRNA *SH3PXD2A-AS1* expression in placenta is positively correlated with clinical severity in preeclampsia (Chen et al. 2020). Furthermore, two lncRNAs, *ENST00000527727* and *ENST00000415029*, were found to be associated with negative pregnancy outcomes, including fetal growth restriction (FGR) and placenta accreta (Dai et al. 2021; Wortman and Alexander 2013). *TERRA*, a lncRNA involved in telomere maintenance, together with the telomere length, was also associated with preeclampsia. TERRA expression and telomere length in placentae were significantly greater in preeclampsia than in healthy control pregnancies (Rahimi et al. 2020). The tissue-specific expression of lncRNAs and the co-expression of mRNAs have also been investigated in the context of preeclampsia. Fifty-three lncRNAs and 301 mRNAs were found to be differentially expressed in decidual tissue of preeclampsia, as compared to healthy controls. A co-expression network analysis of lncRNA-mRNA revealed genes associated with the progression of preeclampsia. Among them, five lncRNAs (*ANXA2P1*, *GTF2IP1*, *NACAP1*, *NCF1C*, and *OR7E37P*) were validated, which suggests that these lncRNAs may serve as biomarkers in the prediction and in the diagnosis of preeclampsia (He et al. 2021). Another study examined lncRNA profiles in early-onset severe preeclampsia by comparison to profiles in normal pregnancy. Among 527 differentially expressed lncRNAs, *MIR210HG* was chosen for an in vitro validation assay. The overexpression of *MIR210HG* was shown to inhibit trophoblast migration, in part via interactions with *miR-520a-3p*, which promote DKK-1 expression (Lei et al. 2021).

It has been suggested that miRNAs have an emerging role in preeclampsia; in consequence, they may serve as diagnostic markers for it. One study found elevated *miR-31-5p*, *miR-155-5p*, and *miR-214-3p* and reduced *miR-1290-3p* in maternal serum that was associated with preeclampsia and strongly correlated with hypertension and proteinuria symptoms (Kim et al. 2020). It is postulated that the induction of these miRNAs is due to NFKB activation, which leads to suppression of eNOS and PIGF. Strikingly, these selected miRNAs had a higher than 90% diagnostic accuracy in identifying preeclampsia (Kim et al. 2020). In addition, several miRNAs may be elevated postpartum, after pregnancy complications. One study identified *miR-98-5p*, *miR-222-3p*, *miR-210-3p*, *miR-155-5p*, *miR-296-3p*, *miR-181a-5p*, and *miR-29b-3p* that were upregulated in maternal plasma at delivery. Interestingly, At 1 year postpartum *miR-221-3p* was reduced in plasma from women with preeclampsia. This potentially indicates altered signaling induced by the condition (Murphy et al. 2015).

In addition to lncRNAs and miRNAs, DNA methylation has been investigated in the context of preeclampsia. *HLA-G* promoter methylation and increased *DNMT1* expression were found in preeclampsia placentae, unlike in the placentae of healthy controls. Knockdown of *DNMT1* by small interfering RNA (siRNA) increased *HLA-G* expression, potentially linking gene-specific DNA methylation and preeclampsia (Tang et al. 2015). Using genome-wide DNA methylation profiling, gene-specific DNA methylation changes in maternal peripheral blood were found to be associated with preeclampsia. Specifically, 207 CpGs were found to be differentially methylated in white blood cells (WBCs) from maternal peripheral blood of patients with preeclampsia during the first trimester (Anderson et al. 2014). In another study, global and gene-specific DNA methylation was assessed in maternal WBCs at the time of delivery. The global DNA hypermethylation of maternal blood was shown to be associated with preeclampsia, and methylation at *GRIN2B*, *GABRA1*, *PCDHB7*, and *BEX1* was associated with seizures, potentially providing a set of predictive biomarkers of these diseases (White et al. 2013). A methylation array analysis revealed 229 differentially methylated regions (DMRs) with 3,411 differentially methylated CpGs found in preterm preeclampsia placentae. The hypermethylation of *CDH11*, *COL5A1*, and *TNF* and the hypomethylation of *NCAM1* showed concordance in their expression in preterm preeclampsia placentas (Anton et al. 2014). A separate study looked at placental DNA methylation signatures in preeclampsia and severe preeclampsia by comparison to healthy controls, using an 850K methylation array. A total of 398 DMRs were identified, and fifteen loci were differentially methylated in both preeclampsia and severe preeclampsia groups. Among these, the methylation of *HIST1H3E*, *TICAM2*, and *ZNF417* has been previously shown to associate with preeclampsia severity (Lim et al. 2020). Taken together, these results suggest that either maternal blood or placental DNA methylation changes may improve our understanding of preeclampsia and our designs for a diagnostic tool for this negative health outcome during pregnancy.

Small for Gestational Age and Preterm Birth

Time of gestation and infant BW have long been known to influence the offspring's health. Barker's seminal work provided evidence that these factors were further linked to disease in adulthood. The terminology used for these outcomes varies. "Small for gestational age" (SGA) is a formula used when an infant's BW is below the tenth percentile for gestational age (Osuchukwu and Reed 2021). "Fetal growth restriction" (FGR or IUGR) is used to describe reduced fetal growth, defined as weight below 10% of the predicted fetal weight for gestational age. SGA and FGR are often used interchangeably (Osuchukwu and Reed 2021). "Low BW" (LBW) is used when an infant's weight is under 5 lbs and 8 oz. (roughly 2,500 grams), regardless of gestational age (Cutland et al. 2017). Finally, "preterm birth" (PTB) describes an infant born before the 37th week of pregnancy (Jain and Gyamfi-Bannerman 2018). The adverse events captured by these expressions affect a considerable number of cases, as over 8% of all births are deemed LBW and over 10% occur before term. The health risks for preterm and SGA children are considerable and may include mortality before the age of 1 and developmental delays. Given the negative health outcomes associated with these factors, research has investigated epigenetic biomarkers as predictive tools of LBW, SGA, PTB, and IUGR.

The expression of mRNA and miRNA in maternal peripheral blood collected at week 28 of pregnancy has shown utility in discriminating between term and preterm births. Multiple mRNA-miRNA pairs, specifically around the nitric oxide-signaling pathway, were found to be associated with PTB. By combining data about the expression of identified mRNAs and miRNAs—including *NCF1*, *MSRA*, *SDHA*, and *DDAH2* as well as *miR-342-3p*—with clinical data, researchers were able to predict which mothers would deliver before term and which ones would deliver in term more accurately than when using clinical information alone (Manuck et al. 2021). Similarly, miRNAs profiles were examined in maternal peripheral blood from a cohort of African American women with either term or preterm deliveries. From the profile, twelve out of forty-five miRNAs (*MIR-181A-3P*, *MIR-221-3P*, *MIR-33A-5P*, *MIR-6752-3P*, *MIR-1244*, *MIR-148A-3P*, *MIR-1-3P*, *MIR-1267*, *MIR-223-5P*, *MIR-199B-5P*, *MIR-133B* AND *MIR-144-3P*) were selected and used to develop a miRNA risk score. The score is able to identify the risk of spontaneous PTB by measuring these miRNAs in maternal blood collected early and late in the first trimester (Winger et al. 2020). Additionally, researchers have examined the utility of maternal plasma miRNAs for diagnosing fetal hypoxia; their goal was to identify pregnancies at risk for growth restriction and other complications. Maternal plasma was collected and compared between mothers with a preterm, fetal growth-restricted fetus and mothers matched in gestational age but with an appropriately sized fetus. Four miRNAs, *MIR-16-5P*, *MIR-103-3P*, *MIR-107-3P*, and *MIR-27B-3P*, were shown to be upregulated in pregnancies with FGR during weeks 32 to 37, which suggests that they may serve to identify complicated pregnancies (Tagliaferri et al. 2021). A similar study examined miRNA expression, specifically miRNAs associated with diabetes, CVD, and cerebrovascular disease in the context of common pregnancy complications like gestational hypertension, preeclampsia, gestational diabetes mellitus, preterm labor, FGR, or spontaneous PTB. A panel of twenty-nine miRNAs were assessed postpartum in the peripheral blood of mothers with complicated pregnancies and in their newborn. In these newborn, the blood was collected between the ages of three and eleven years. In peripheral blood leukocytes, twenty-three of the twenty-nine miRNAs were upregulated, *MIR-92A-3P* and *MIR-210-3P* being further elevated in children with overweight or obesity, while *MIR-342-3P* was further elevated in children with heart defects. Additionally, nineteen of these miRNAs could identify which children were born to complicated pregnancies (Hromadnikova et al. 2020) When the panel of twenty-nine miRNAs was examined postpartum in mothers with complicated pregnancies, twenty-six of them were upregulated. Findings suggest that the use of this set of miRNAs is able to discriminate between mothers with abnormal clinical findings and normal controls. Furthermore, several treatments for hypertension and infertility, as well as for overweight or obesity, resulted in a modulation of the aberrant expression of these miRNAs. Taken together, the peripheral blood miRNAs in this this panel may be useful for the identification of mothers at increased risk of developing cardiovascular complications and other long-term maternal health outcomes as a result of complicated pregnancies (Hromadnikova et al. 2021)

The cord blood DNA methylation of imprinted genes has been assessed in the context of LBW. A cohort of mothers were followed with the goal of identifying links such as epigenetic mechanisms between maternal antibiotic use and LBW. Five DMRs were found in the following imprinted genes that control fetal growth: *IGF2*, *H19*, *PLAGL1*, *MEG3*, and *PEG3*. These DMRs were associated with antibiotic use; however, only methylation at *PLAGL1* was associated with LBW. These data provide an intriguing link between maternal antibiotic exposure and BW outcomes, as well as a potential biomarker for LBW (Vidal et al. 2013). DNA methylation arrays have also allowed for the identification of blood DMRs in newborns and for their association with gestational age and BW. Using an EPIC methylation array on the umbilical cord blood, researchers identified decreased promoter methylation of *PTPRN2* and *HLADQB1*. The dysfunction of these genes may be linked to T cell dysregulation and increased reactive oxygen species production, suggesting that the aberrant DNA methylation of these genes could possibly mediate the consequence of IUGR (Krishna et al. 2020). Similarly, genome-wide DNA methylation analysis was performed on peripheral blood from preterm and term infants. DMRs were found in preterm infants (when compared to term infants) in twenty-five gene bodies and in fifty-eight promoters of protein-coding genes, ten candidate genes—*SLC7A5*, *SLC1A2*, *NPBWR1*, *APOL1*, *QPRT*, *LRG1*, *PRPH*, *GRIK5*, *TREM2*, and *MCHR1*—having been previously shown to have a neural function (Sparrow et al. 2016). A meta-analysis of genome-wide DNA methylation associations between cord blood DNA methylation and gestational age from seventeen cohorts provided evidence of differentially methylated CpGs enriched in embryonic development pathways in relation to gestational age at birth. Strikingly, not all the cord blood methylation changes persist into childhood and adolescence, which suggests that a follow-up longitudinal study is required to determine the relationship between DNA methylation changes and disease susceptibility in later life.

Placental DNA methylation has also been examined with the goal of identifying DMRs that may predict LBW or provide biological pathways that underlie LBW. A placental DNA methylation profiling study revealed fifteen CpGs that annotated genes involved in inflammation and oxidative stress. The methylation changes of these CpGs were associated with LBW. In particular, the methylation of a specific locus, cg06155341, was found to be linked to greater BW and to reduced *FOSL1* expression. *FOSL1* expression was negatively associated with BW (Tekola-Ayele et al. 2020). FOS proteins function in regulating cell proliferation, differentiation, and transformation, and an aberrant FOSL1 expression may alter the growth of the fetus, contributing to LBW. In addition to the placenta, the amnion represents an intriguing target for the epigenetic assessment of pregnancy or birth outcomes. Genome-wide methylation analysis was performed on the amnion—the inner layer of fetal extra-embryonic tissue in contact with the amniotic fluid—from both term and preterm births. Additionally, term vaginal births (deliveries with labor) were compared against term cesarean births (deliveries without labor). Surprisingly, term births share more epigenetic features with preterm births than with term births without labor. Differentially methylated CpGs were mapped onto sixty-four genes between term births with and without labor and onto sixty-one genes between term labor and preterm labor births (Kim et al. 2013). Although further research is needed to examine the functions of these genes in regulating labor or birth, this set of genes may be used for determining term and preterm births.

Allergic Diseases

Atopic diseases, specifically atopic dermatitis (eczema), allergic asthma, and allergic rhinitis, represent a constellation of allergic diseases often presenting in early childhood. The risk factors associated with these diseases are multifactorial; genetic background, environmental tobacco smoke (ETS) exposure, vitamin deficiency, FGR, early viral infection, allergen exposure, and stress are all linked to their development (Sly 2011). Given the diverse insults associated with atopic disease and the increasing prevalence of this class of diseases in the past few decades, the role of epigenetics in their regulation has gained traction. While the underlying pathophysiology of childhood asthma remains unclear, there is now a strong body of evidence indicating that epigenetic mechanisms are involved in allergy development and disease through maternal exposures (Gruzieva et al. 2021).

Maternal and cord blood DNA methylation is shown to have utility in predicting immunoglobulin E (IgE) sensitization, an important step in the development of allergic disease. Specifically, genome-wide DNA methylation analysis of peripheral mononuclear cells (PBMCs) was assessed in mother–infant or mother–child pairs, through measurements taken from maternal blood during pregnancy, from cord blood at delivery, and from peripheral blood from the child at the ages of two and five years. Thirty-four DMRs were associated with IgE sensitization to airborne allergens and another thirty-eight DMRs were found to be associated with IgE sensitization to food allergens (Acevedo et al. 2021). The DMRs of *TUBGCP5* and *AIF1* displayed the greatest increase in methylation while the DMRs of *DUSP22* and *CD300A* showed decreased methylation. Interestingly, *DUSP22* hypomethylation was induced through low-dose benzo[a]pyrene (BaP) exposure (Ouyang et al. 2012), and single-nucleotide polymorphisms (SNPs) in *DUSP22* were associated with asthma risk (Rathod et al. 2020), which suggests that genetic and epigenetic variation in *DUSP22* may serve as a marker for allergic asthma risk, if these findings are validated in other cohorts. Similarly, a study of the Taiwanese maternal and infant cohort demonstrated unique cord blood DNA methylation patterns at birth that were associated with the level of serum IgE at birth and at the ages of five, eight, and eleven years. DNA methylation changes at 273 CpG sites that tracked with IgE longitudinally were further validated in a separate cohort from the Isle of Wight. Among these differentially methylated CpG sites, sixteen were found to be aberrantly methylated and these methylation changes were associated with serum IgE, as seen in Taiwanese cohort (Han et al. 2021). Several of these annotated genes had been previously linked to allergic disease. For example, *MYO15B* and *BMI1* link respectively to hyper-IgE syndrome and to T helper 2 (TH2) differentiation.

In an epigenome-wide association study (EWAS), DNA methylation profiles were assessed prospectively in newborns across eight birth cohorts, to see whether DNA methylation signatures were associated with the development of allergic disease. Additionally, DNA methylation was assessed cross-sectionally in nine cohorts of children with and without asthma. In newborns, nine CpGs annotated to *CLNS1A*, *MIR-548*, *GPATCH2*, *SCOC*, *SUB1*, and *WDR20* were associated with asthmatic outcomes in later life, seven of the nine loci displaying increased methylation by comparison with healthy controls. Additionally, 179 CpGs and thirty-six DMRs were identified in school-age children with asthma, almost all displaying decreased methylation by comparison with healthy controls. These 179 CpGs were further validated in both nasal epithelial and eosinophilic samples; twenty of them achieved significance and direction of effect. These included *IL5RA*, for which small monoclonal antibodies currently exists for the treatment of severe eosinophilic asthma (Reese et al. 2019).

Placenta DNA methylation has been shown to be altered and associated with atopic dermatitis (AD) phenotypes. This effect was also most pronounced among children with high concurrent $PM_{2.5}$ exposure in the first trimester, and it was further enhanced by the presence of low cord blood vitamin D. Specifically, decreased placental CpG site-specific DNA methylation was seen in *AHRR*, *DPP10*, and *HLADRB1* (Yang et al. 2020). Interestingly, hypomethylation in a separate *AHRR* locus has been previously associated with an increased risk of eczema in mothers who smoked (Mukherjee et al. 2016). Seperately, the knockdown or mutation of *DPP10* has been shown to be associated with airway hyperresponsiveness, a hallmark of asthma, both in animal models and in vitro (Zhang et al. 2018). In another study, placental DNA methylation changes were shown to be associated with the likelihood of AD being diagnosed at the age of six months. In addition, several placenta DMRs were associated with AD diagnosis and stress or anxiety in mothers. Specifically, CpG loci at *MAD1L1*, *MMP27*, *LDLRAD4*, *RWDD2B*, *MICAL3*, and *TSPAN7* showed differential methylation in children with AD born to mothers with high stress. It has been postulated that the hypomethylation of *MMP27* may contribute to an increased MMP27 expression, which may increase inflammation and lead to placental remodeling (Kim et al. 2021).

While the vast majority of epigenetic studies related to allergic diseases have been carried out in the context of DNA methylation, there is a growing body of literature examining the relationship btween miRNAs in allergic disease risk (Rebane 2015). A study showed how a set of cord blood miRNAs that associated with cord blood IgE levels was in turn associated with allergic rhinitis. Specifically, eight miRNAs—*MIR-96*, *MIR-15A*, *MIR-197*, *MIR-191*, *MIR-21*, *MIR-20*, *MIR-126*, and *MIR-107*—were downregulated in relation to cord IgE level; and two of these (*MIR-21* and *MIR-126*) were found to be downregulated in monocytes collected from children with allergic rhinitis (Chen et al. 2010). *MIR-223* expression in maternal peripheral blood and in cord blood was associated with reduced T regulatory (Treg) cells in maternal blood and in cord blood. Children with lower Treg cells were known to be at an increased risk of developing AD within the first three years of their life. Maternal and cord blood MIR-233 levels were also associated with maternal cotinine levels, as well as with exposure to organic solvents (Herberth et al. 2014). All in all, the findings suggest that the assessment of blood MIR-233 levels in mothers and in newborns is a potential biomarker for not only AD but also environmental exposures such as to smoking (cotinine) and organic solvents.

Metabolic Syndrome

Metabolic syndrome is a collection of conditions related to disorders of the metabolism and subsequent pathologies, for example obesity, insulin resistance, dyslipidemia, and hypertension (Carson and Lawson 2018). These conditions are associated with an increased risk of CVD, cerebrovascular disease, and type II diabetes. Metabolic syndrome has been on the rise globally, as high caloric diets and sedentary behaviors contribute to obesity (Eckel et al. 2005; Ford et al. 2004). While lifestyle is a major factor in the development of metabolic syndrome, growing evidence indicates that early life exposures can increase this risk in the offspring (Li et al. 2011; Wang et al. 2017). For example, children exposed to famine in early life have been shown to be at increased risk of metabolic syndrome in adulthood. The effects of maternal metabolic syndrome on the developing fetus and its later health trajectory is not fully elucidated. In addition, women with metabolic syndrome are at greater risk of pregnancy complications, including gestational diabetes (Grieger et al. 2018). There is a strong body of evidence indicating that epigenetics may play a role in the pathogenesis of metabolic syndrome (Carson and Lawson 2018). For this reason, researchers sought to investigate epigenetic mechanisms in pregnancies that are complicated by metabolic syndrome and to identify early-life risk factors for the development of metabolic syndrome in adulthood. With this in mind, they investigated how gestational diabetes may alter the offspring's epigenome. Differential DNA methylation was found in the cord blood (99 CpG sites) and in the placenta (662 CpG sites) in mothers with gestational diabetes, as compared to mothers with uncomplicated pregnancies. Differential methylation patterns of *AHRR* and *PTPRN2* were identified in both the cord blood and the placenta (Awamleh et al. 2021). Interestingly, *AHRR* methylation is found in the cord blood of offspring born to obese mothers and of children born prematurely (Burris et al. 2015). These findings offer a link between maternal gestational diabetes on the one hand and altered placenta and cord blood DNA methylation on the other— a link that may affect the health of both mothers and children.

Altered DNA methylation has also been shown to be predictive of adiposity throughout life. *SLC6A4*, a gene that functions in regulating serotonin bioavailability and energy balance, was shown to be hypomethylated in cord blood. This methylation change was associated with increased fat mass at the age of four years and with increased total and percentual fat mass at the age of six–seven years. Decreased methylation of *SLC6A4* was also associated with increased weight gain in the mother during pregnancy. Additionally, *SLC6A4* was less likely to be methylated in the adipose tissue of obese adults and of adolescents with a higher BMI and a larger waist circumference. When all these data are taken together, the methylation status of *SLC6A4* represents a potential marker of adiposity from early life into adolescence and adulthood (Lillycrop et al. 2019). On the other hand, it is suggested that the DNA methylation patterns of imprinting genes are associated with metabolic syndrome in the offspring and serve as markers to predict childhood obesity or overweight. *IGF2*, a paternally imprinted gene, encodes a growth factor crucial for placental and fetal development. *H19*, a maternally imprinted lncRNA downstream of *IGF2*, can regulate cell growth, with an expression associated with *IGF2* repression (Pidsley et al. 2012). DNA methylation at *H19* DMR was shown to be associated with increased weight for one's age, despite the absence of change in intragenic DNA methylation at the *IGF2* DMR (Perkins et al. 2012). The reversible nature of epigenetic changes allows the modulation of disease phenotypes by altering the aberrant epigenetic changes. For example, epigenetic alterations have been observed after lifestyle interventions during pregnancy and have resulted in a modification of the offspring's metabolic syndrome. DNA methylation in the cord blood of children born to obese mothers was measured with and without lifestyle interventions, which consisted of physical activity with or without dietary advice. CpG site-specific methylation of *DISC1*, *GBX2*, *HERC2* and *HUWE1* was shown to mediate in part the effect of the lifestyle intervention on lean mass in the offspring (Jönsson et al. 2021).

In addition to DNA methylation, researchers examined whether exposures to pregnancy complications such as gestational diabetes, preterm labor, and FGR might alter miRNA expression, contributing to the increased risk of developing obesity or heart disease in later life. Peripheral blood miRNAs were assessed in children aged three to eleven, and twenty-three out of the twenty-nine miRNAs in the panel were found to be upregulated in children born to complicated pregnancies. Children in this category also presented with overweight or obesity or with heart defects. In particular, *MIR-92A-3P* and *MIR-210-3P* were upregulated in children with overweight or obesity, and *MIR-342-3P* elevated its expression in children with valve problems or heart defects. In addition, using a panel of nineteen miRNAs, the authors were able to identify 70% of the children at risk for obesity and overweight with 90% specificity, which indicated that a subset of miRNAs signatures may have utility as an early marker of obesity risk (Hromadnikova et al. 2020).

Neurodevelopmental Disorders

Neurodevelopmental disorders encompass a collection of impairments of the brain and neurological system: attention-deficit/hyperactivity disease (ADHD), autism spectrum disorder (ASD), intellectual impairments, motor control impairments, and cerebral palsy. In the United States roughly 15% of children are afflicted with neurodevelopmental disorders. In addition to genetics, a wide range of exposures, for example to environmental tobacco smoke (ETS), alcohol, mercury, lead, and pesticides, are linked to neurodevelopmental disorders. Given the gene–environment interactions seen in neurocognitive disorders, epigenetic mechanisms have been investigated in the context of development (Zahir and Brown 2011).

Cord DNA methylation patterns have been suggested as a diagnostic tool for ASD. Researchers measured cord blood DNA methylation with the goal of identifying epigenetic signatures of ASD. They found that, in both males and females, DMRs associated with ASD were more likely to be on the X chromosome, be expressed in the brain or in neural tissue, and be able to modify the binding with the methylation-sensitive transcription factors that contribute to brain development. Furthermore, many of these DMRs were mapped to genes previously associated with ASD, including *AFF2, CPXCR1, FAM155B, GRIA3, KLHL13, LINC00269, LOC101928201, LOC389906, MAGEB16, MIR3672, PCDH11X, SPANXN1, TDGF1P3, TMEM47, XRCC6P5*, and *ZIC3* (Mordaunt et al. 2020). In addition to cord blood, the differential DNA methylation of *CYP2E1* in placentae was associated with a high risk for ASD (Zhu et al. 2019).

Epigenetic mechanisms have also been suggested for assessing the risk for ADHD. DNA methylation patterns were assessed in the cord blood and in the peripheral blood of school-age children (seven to eleven years) in relation to ADHD symptoms. Nine CpGs were identified in cord blood that can predict later ADHD symptoms; they include *ERC2* and *CREB5*, which have been separately associated with ADHD. Surprisingly, no CpGs was associated with ADHD in the blood collected from school-age children. Nonetheless, these findings support the use of cord blood DNA methylation as an early diagnostic biomarker for ADHD and offer insights into the mechanisms that underlie the pathophysiology of ADHD (Neumann et al. 2020). Moreover, it has been postulated that the DNA methylation patterns examined in the placenta, specifically chorionic villus, and the matched maternal blood may be used to detect cases of Downs syndrome. In particular, thirteen DMRs were identified on chromosome 13: *TTC10, PROZ, KCTD12, PABPC3, LECT1, TEB1, FLJ46358, MGC48915, LOC283487, GPR12, DNAJC15*, and *ATP4B*. Taken together, these findings support the use of these DMRs as non-invasive biomarkers for Downs syndrome (Bunce et al. 2012).

Exposure to Prenatal Insults and Epigenetic Changes

While epigenetic mechanisms have shown to have utility as biomarkers of disease, there has been growing interest in the examination of epigenetics in the context of environmental exposures. These markers may have utility as biomarkers of exposure and may provide insight into how exposures can influence both the disease and its development. The exposures highlighted here are common, have been previously linked to negative outcomes in the offspring, and are accompanied by a growing body of evidence that indicates that epigenetic mechanisms may underlie their pathogenesis.

Air Pollution and Smoking

Early associations observed between the mother's exposure to air pollution and cord blood DNA methylation in the offspring have prompted researchers to investigate the potential of using epigenetic perturbations as biomarkers of such exposure. Seminal work by Perera et al. (2009) identified increased cord blood DNA methylation at the *ACSL3* promoter associated with maternal exposure to airborne polycyclic aromatic hydrocarbons (PAHs) and asthma symptoms in children prior to age five (Perera et al. 2009). This work provided early evidence that maternal exposure may alter epigenetic regulation in the offspring, potentially increasing the risk for later disease. Adding to the mechanistic background, it was demonstrated in the same cohort that maternal PAH exposure resulted in increased cord blood promoter methylation of *IFNγ*, a key type I cytokine, potentially indicative of the type II skewing seen in allergic disease (Tang et al. 2012). Just like exposure to ambient air pollution, exposure to ETS has been linked to a host of negative health consequences such as FGR, PTB, sudden infant death syndrome (SIDS), and lung injury, as well as to neurocognitive outcomes like ADHD (Cornelius and Day 2009; Delpisheh et al. 2008; Ion and Bernal 2015).

Air pollution exposure was modeled in a cohort of newborns and their mothers; cord blood was collected at the time of delivery, for genome-wide DNA methylation analysis. A group of 400 CpGs were found to have the best predictive value for $PM_{2.5}$ exposure across pregnancy. Furthermore, 140 CpGs were shown to have the strong predictive ability for NO_2 in the third trimester of pregnancy (Wang et al. 2022). Genetic variation in DNA methylation modification enzymes and epigenetic variation in LINE1 retrotransposons have also been shown to play important roles in the downstream cardiovascular consequences of exposure to prenatal air pollution. Exposure to multiple air pollutants ($PM_{2.5}$, PM_{10}, NO_2, and O_3) in the first trimester was associated with reduced blood LINE1 methylation in newborns, while exposure to ozone in the third trimester was associated with increased LINE1 methylation. Interestingly, SNPs in DNA methyltransferases, DNMT1, and DNMT3B, as well as in the DNA 5-methylcytosine dioxygenase TET2 and in the DNA repair enzyme TDG, both involved in DNA demethylation, were shown to mediate the magnitude of these associations, indicating that global DNA methylation can be perturbed by air pollution exposure (Breton et al. 2016). In addition to methylation changes seen in newborns, placental DNA methylation has been shown to mediate $PM_{2.5}$ exposure and the risk for allergic disease in later life. $PM_{2.5}$ exposure during the first trimester of pregnancy was associated with early-onset persistent AD, this effect being more pronounced for those with low cord blood vitamin D. Interestingly, cg16371648, mapped to *AHRR*, was hypomethylated in relation to air pollution. *AHRR* is known as the negative repressor of the aryl hydrocarbon receptor (AHR), which can be activated by ligands like dioxin and by components of ambient air pollution (with low molecular weight). Thus this finding suggests that cord blood *AHRR* methylation may be useful as a marker of exposures to air pollutants and may well contribute to our understanding of allergic diseases induced by airborne pollutants (Castañeda et al. 2018). Additionally, a meta-analysis examined the association between air pollution exposure throughout pregnancy and cord blood DNA methylation. The identified DMRs were further validated in separate cohorts of newborn and older children. In particular, the DMRs in *H19* and *MARCH11* identified in newborns and children were associated with prenatal PM_{10} and $PM_{2.5}$ exposure. Other DMRs, annotated to *FAM13A* and *NOTCH4*, were not replicated in cord blood from a separate newborn cohort but were associated with air pollution in a cohort of seven–nine-year-olds. Furthermore, in a cohort of adolescents, current PM_{10} exposure was associated with increased *NOTCH4* expression (Gruzieva et al. 2019). These results suggest that the methylation changes of *FAM13A* or *NOTCH4* (or both) may depend on the second insult. *FAM13A* and *NOTCH4* have been previously linked to lung function and asthma outcomes. Therefore it is worth following up the study by examining the relationship between the DNA methylation changes in these two genes and respiratory disease outcomes.

Mitochondrial DNA (mtDNA) methylation has been suggested to be a potential biomarker for exposures, given the fact that the mitochondria function to regulate cell fate in response to environmental exposures and that DNMT1 (Shock et al. 2011), DNMT3A (Devall et al. 2016), and TET1/2 proteins (Dzitoyeva et al. 2012) are also found in mitochondria. Covalent modifications by 5-methylcytosine or 5-hydroxymethylcytosine at mtDNA have been associated with mitochondrial function (Devall et al. 2014) and with environmental exposures like air pollutants (Breton et al. 2019; Byun et al. 2016, 2013). In a study, mitochondrial-derived peptide (MDP) levels in newborn cord blood were positively associated with prenatal traffic-related air pollution exposures. The changes in MDPs expression were shown to be associated with the DNA methylation of two regions of the mitochondrial displacement loop (D-loop) (Breton et al. 2019). Therefore the use of mtDNA provides new insights into the assessment of early-life exposures. In addition to DNA methylation, changes in miRNA expression have been shown to be perturbed by ambient air pollution. The ENVIRONAGE birth cohort study provided evidence that early-life exposure to particulate air pollution is associated with an altered expression of *miRnAs-16*, *miR-21*, *miR-146a*, and *miR-222* in placental tissue at birth (Tsamou et al. 2018). Many of these identified miRNAs target *PTEN*, whose expression is positively associated with third-trimester $PM_{2.5}$ exposure; thus they prove a potential mechanism for PM-induced effects on fetal programming.

Like air pollution, maternal self-reported smoking is associated with decreased methylation at cg05575921, annotated to *AHRR*. Furthermore, methylation at cg05575921 was associated with a decreased risk of asthma between the ages of eight and twenty-one years (Neophytou et al. 2019). In an EWAS of maternal smoking during pregnancy, 110 differentially methylated positions (DMPs) were identified in blood collected shortly after birth. One of these DMPs was annotated to *AHRR*, which is of note. Additionally, DNA methylation at three DMPs annotated to *GFI1*, *AHRR*, and *EXOC2* was shown to be a mediator between maternal smoking and low BW (Hannon et al. 2019). Other evidence of perturbations induced by exposure to maternal cigarette smoke include the altered methylation of *RUNX3*, which was associated with a decreased gestational age (Maccani et al. 2013).

Animal models are widely used to assess the epigenetic role of maternal exposures on offspring, as they allow identifications of tissue-specific, gene-specific, and exposure-specific epigenetic changes. Maternal inflammatory response plays an important role in maternal smoking-induced lung injuries in dams and their offspring. *Mir-1903* was shown to be a potential epigenetic biomarker of lung inflammation in the offspring of dams who were exposed to tobacco smoke during pregnancy (Ji et al. 2021). Prenatal exposure to nicotine has also been shown to selectively enhance *miR-181a* expression, which results in increased vascular tone, and this in turn can predispose the offspring to heart disease (Liu et al. 2019). Similarly, the mother's exposure to nicotine was associated with an increased expression of *miR-21* and *miR-124* as well as with a reduction in antioxidant enzyme levels and with an induction of *c-fos*, *c-myc*, *c-jun*, and *acta1* in the heart tissue of mice (Q. Huang et al. 2020). On the other hand, the effect of air pollution on offspring and on their epigenetic regulation is not limited to a single generation. Prenatal exposure to diesel exhaust particles in mice has been shown to increase asthma susceptibility transgenerationally; gene-specific DNA methylation changes were detected in F1, F2, and F3 mouse dendritic cells, with shared features associated with chromatin modifications (Gregory et al. 2017). Intergenerational effects of gestational secondhand smoke on offspring have also been observed. Specifically, the F1 and F2 offspring from dams exposed to sidestream cigarette smoke showed reduced expression of *miR-130a* and increased *miR-16* and *miR-221*. These miRNAs are believed to regulate HIF-1α response and were associated with increased allergic asthma and bronchopulmonary dysplasia in the progenies (Singh et al. 2017).

Metals

Certain metals, such as lead (Pb) and mercury (Hg), have long been known to be neurotoxicants for children when these are exposed to them pre- and perinatally. It is postulated that prenatal exposure to these metals may induce epigenetic changes that are linked to neurocognitive defects. Higher prenatal methylmercury (MeHg) exposure was associated with differential CpG site-specific DNA methylation in the peripheral blood at the age of seven, a process that may influence the offspring's neurodevelopment and mental health. Specifically, these CpG sites mapped to either *NR3C1*, a glucocorticoid receptor, or to *GRIN2B*, a glutamate receptor subunit involved in neurotransmission (Cediel Ulloa et al. 2021). In addition, total Hg and MeHg in the cord blood was shown to be associated with the methylation status of DMRs at *TCEANC2* and *ANGPT2* and near *PRPF18* and *FOXD2*. Three of the four CpGs at *TCEANC2* were further replicated in a cohort with matched cord blood and maternal blood (collected in the first trimester) Hg exposure, strengthening the associations between perturbed *TCEANC2* methylation and prenatal Hg exposure (Bakulski et al. 2015). In addition, arsenic (As) in food and drinking water is a major global health concern (Karagas et al. 2019; Mantha et al. 2017). It is estimated that over 200 million individuals are exposed to inorganic As (iAs) at levels above the WHO provisional guideline value of 10 μg/L (ppb) (Ahsan 2011; Shaji et al. 2021). Both iAs and its methylated metabolites can pass the placenta and be found in the fetus (Concha et al. 1998; Lindgren et al. 1984). It has been suggested that prenatal exposure to iAs may increase the risk of adverse health effects during early childhood as well as later on in life (Vahter 2008), perhaps through epigenetic modifications (Bailey and Fry 2014; Martin et al. 2017; Saintilnord and Fondufe-Mittendorf 2021). Prenatal As exposure, measured as maternal total urinary As, was associated with increased DNA methylation in the cord blood. Additionally, cord blood total As and maternal blood total As were associated with the methylation of transposon *Alu* in males only (Pilsner et al. 2012). Exposure to iAs was shown to influence placental functions, as well as to induce transcriptional and epigenetic changes in placentae (Cardenas et al. 2015; Green et al. 2016; Rahman et al. 2018; Winterbottom et al. 2019a, 2019b). In a study examining the association between exposure to twelve metals and global DNA methylation levels, associations were found between selenium (Se), cobalt (Co), copper (Cu), and Hg exposure and DNA methylation, Hg exposure being associated with reduced global DNA methylation and Se exposure being associated with increased DNA methylation (Weyde et al. 2021).

Organic Pollutants

Organic pollutants cover a broad range of toxicants, including persistent organic pollutants (POPs), endocrine disrupters, and pesticides. Exposure to these chemicals has been linked to developmental delays, for example to impaired neurological, immune, or reproductive function (Damstra 2002). Epigenetic changes such as DNA methylation may explain the mechanism of prenatal toxicity acquired from exposure to organic pollutants. For example, placental DNA methylation was found to be perturbed in relation to POPs exposure. Overall, POP levels—of organochloride pesticides, per- and polyfluoroalkyl substances (PFAS), polybrominated diphenyl ethers (PBDEs), polychlorinated biphenyls (PCBs)—in maternal plasma in early pregnancy were associated with 214 differentially methylated CpG sites in human placentae, collected from the fetal side within one hour from birth; twenty-four of these were correlated with the placental expression of twenty-one genes. In particular, high perfluoroundecanoic acid (PFUnDA) was associated with an increased CpG site-specific methylation of *TUSC3* and a decreased expression of *TUSC3* in the placenta. The changes in methylation and expression level of *TUSC3* were also correlated with shorter birth length. Notably, there were forty-four CpG sites and their methylation changes were correlated with at least one neonatal anthropometry measure; these measures included BW, birth length, and head circumference (Ouidir et al. 2020). Exposure to perfluorooctane sulfonic acid (PFOS) has been shown to alter *H19* expression in human trophoblast cells, an observation that offers mechanistic insight into this phenomenon. The increase in *H19* expression could be explained by its promoter hypomethylation. In addition, decreased *miR-19a* and *miR-19b* in the placenta of mice may contribute to the *h19* upregulation. These findings suggest that gestational exposure to POPs, just like to PFOS, can modulate the transcription of imprinting genes epigenetically, which results in alterations in placenta growth and fetus development (Li et al. 2020). In children exposed to polybrominated diethyl ethers (PBDEs) in utero, PBDE47 exposure was positively associated with decreased cord blood *TNFA* promoter methylation, in turn associated with increased TNFA expression, and this effectively linked PBDE exposure to increased inflammatory signaling (Dao et al. 2015). Similarly, women exposed to POPs (including PBDE47) had children with increased promoter methylation of thyroid development genes in the placenta, for instance *DIO3* in female infants and *MCT8* in male infants, which demonstrates sex-specific effects from exposure (Kim et al. 2019). It has been suggested that maternal exposure to POPs may induce the epigenetic reprogramming of germ cells. An animal study has demonstrated that maternal PCB exposure during the fetal primordial germ cell migration stage of pregnancy causes decreased expression of *Dnmt1*, *Dnmt3a*, *Dnmt3l*, and *Uhrf1* and increased expression of *Tet2* and *Tet3* in the mouse oocytes of the adult offspring. These changes in epigenetic enzyme expression were also accompanied by a decreased global DNA methylation and an increased global DNA hydroxymethylation, as well as by alterations in the methylation of imprinted genes, including *h19* and *Igfr2* (He et al. 2019). Epigenetic alterations in imprinting genes in germ cells may be transmitted to subsequent generations, although future study is required to prove this and to demonstrate the effect of epigenetic inheritance on progenies.

Exposure to organic pesticides in early life has been linked to a range of negative health outcomes in the offspring, and especially to neurocognitive disorders. For example, prenatal exposure to chlorpyrifos, an organophosphate pesticide, was associated with an increase in the cord blood DNA methylation of *PPARγ* and with poorer performance in the cognitive and language domains at the age of two years, especially in boys. In addition, histone (H) 3 lysine (K) 4 trimethylation of *PPARγ* in the cord blood was positively associated with gestational age, BW, birth length, and gross-motor performance (Chiu et al. 2021). Lindane, an organochlorine insecticide, has been demonstrated to alter the epigenetic regulation of cerebral CYP450s in rats, and prenatal exposure to lindane was followed by adult challenges associated with the altered H3 acetylation and the DNA methylation of promoters of cerebral *cyp450s* (Agrahari et al. 2019). Findings may provide not only new insights into the neurotoxicity of lindane but also new assessment tools for lindane exposure.

Bezophenone-3 (BP-3) (oxybezone) is an organic compound often used in sunscreen formulations. BP-3 has been shown to permeate the blood–brain barrier and has the potential to be an endocrine disruptor (Schlecht et al. 2004). Prenatal BP-3 exposure modulated the epigenetics of mouse neurons: there is reduced DNMT activity that results in global DNA hypomethylation, and hypomethylation of *gper1* and *bax* (Wnuk et al. 2018). In a separate study, prenatal exposure to BP-3 in mice attenuated histone deacetylase (HDAC) and sirtuin activities in the brain of the fetus (Wnuk et al. 2019). The measurement of HDAC and sirtuin in brain or other target tissues may allow the field to assess the BP-3 level and the neurotoxicity of BP-3, although further study is required.

In utero exposure to an endocrine disrupting chemical, bisphenol A (BPA) (Rubin 2011), was shown to be associated with sex-specific methylation changes, total BPA exposure being positively associated with *LINE1* global methylation for the human placenta (Nahar et al. 2015). Another study, which used human amniocytes sampled in the second trimester, demonstrated that thirty-six DMRs in males and fourteen DMRs in females were associated with the BPA levels in amniotic fluid (Bansal et al. 2020). In rats, BPA exposure during pregnancy significantly altered the expression of *miR-499-5p* in the hearts of dams and induced the expression of *miR-17-5p*, *miR-208-3p*, and *miR-210-3p* in the hearts of all fetuses (Rasdi et al. 2020). Although the role of these miRNAs requires further functional studies, the expression of patterns may be used for assessing maternal exposure to BPA and downstream effects on mother and offspring.

Maternal Diet

Maternal nutrition is crucial to fetal health, as the fetus is dependent on the mother for all its sustainment. Over- and undernutrition during pregnancy have been linked to negative health outcomes for the offspring, including the increased risk of a complicated pregnancy, CVD, and metabolic disease (Salam et al. 2013; Victora et al. 2008), with epigenetic mechanisms shown to be perturbed by these gestational conditions (Şanlı and Kabaran 2019). Additionally, deprivation of crucial micronutrients has been linked to birth defects; for example, folic acid deficiency is linked to the development of neural tube defects. Given that DNA methylation requires one-carbon metabolic substrates (including folic acid) for methylation reactions and given the links between low folate and neural tube defects, researchers investigated a range of one-carbon metabolism metabolites such as folic acid, betaine, 5′-methyltertrahydrofolate (5-MeTHF), and dimethylglycine in the maternal blood and paired these measures with transcriptomic changes and children's risk of ASD. In a study, differentially expressed gene (DMG) modules in maternal blood were found to associate with levels of maternal folic acid, betaine, 5-MeTHF, and dimethylglycine. Furthermore, DNA methylation-binding domain proteins, including MBD3L3, MBD3L4, and MBD3L5, showed a positive association with folic acid and 5-MeTHF, but a negative association with the risk of ASD, potentially linking methyl-binding proteins, one-carbon metabolites, and maternal blood DNA methylation to ASD outcomes (Y. Zhu et al. 2021).

Using an epigenome-wide DNA methylation profiling, cord blood DNA methylation patterns were compared with the maternal plasma folate levels. Forty-eight CpGs were identified that were significantly associated with maternal folate levels. Many of these were genes involved in development, but had not been previously associated with folate biology, for example *APC2*, *GRM8*, *SLC16A12*, OPCML, *PRPH*, *LHX1*, *KLK4*, and *PRSS21* (Joubert et al. 2016). In another birth cohort, seven DMRs were identified in the cord blood of infants whose mothers had been exposed to the highest serum folate levels (by comparison with mothers exposed to the lowest serum folate levels). In particular, high maternal folate was associated with the cord blood hypomethylation of *ZFP57*, a gene involved in regulating DNA methylation during development, and with increased H3/H4 acetylation at the *ZPF57* promoter (Amarasekera et al. 2014).

In addition to folate, combined micronutrient supplementation—for instance zinc and the vitamins A, B, C, and D—was shown to alter the epigenome of the offspring. In a study on pregnant women in the Gambia, before pregnancy, the women were provided either with placebo or with a micronutrient supplement that contained the vitamins A, B1, B2, B6, B12, C, D, and E, niacin, folic acid, iron, zinc, copper, selenium, and iodine; then, once the pregnancy was confirmed, supplementation was ceased and standard prenatal supplementation iron (60 mg) and folate (250 μg) were provided to both groups. The micronutrient supplementation was shown to alter the gene-specific DNA methylation of the offspring in a sex-specific manner, by reducing the methylation of *IGF2R* in girls and the methylation of *GTL2-2* in boys (Cooper et al. 2012). Conversely, maternal dietary vitamin A deficiency in rats was found to induce heart defects. Fetal hearts from mothers with vitamin A deficiency displayed upregulation of *Dnmt1* and downregulation of *Dnmt3a* and *Dnmt3b*. Additionally, in these same hearts, promoter hypermethylation of *GATA-4* was observed, along with reduced *GATA-4* mRNA level, which suggests that vitamin A deficiency in the mother can increase the risk of heart defects in the offspring, in part through the epigenetic regulation of *GATA4*, a transcription factor crucial for cardioblast survival and terminal cardiomyocyte differentiation (Feng et al. 2013). On the other hand, maternal nutrient restriction (MNR) was shown to regulate miRNA expression and to influence lipid metabolism in the offspring. Mice born in conditions of MNR displayed reduced *miR-181a* expression in the liver at postnatal day 1 when compared with the offspring of controls. Overexpression of *miR-181a* in BRL-3A rat liver cells reduced the lipid droplets after treatment with oleic acid for forty-eight hours, which also suppressed the expressions levels of *Sirt1*, *Foxo1*, *Klf6*, and *Pparγ*, indicating that *miR-181a* plays a role in the regulation of liver lipid metabolism (W. Zhu et al. 2021).

High-fat diets have been of interest, given their growing ubiquity and the increased prevalence of obesity. Maternal high-fat diet (HFD) in mice has been shown to disrupt one-carbon metabolism in the liver of the offsprings, leading to altered lipid homeostasis and DNA hypermethylation, particularly in CpG islands of specific genes involved in one-carbon metabolism (Peng et al. 2021). In mice, maternal HFD is also shown to increase renal global DNA methylation, albuminuria, glomerulosclerosis, and renal fibrosis. It is suggested that this process is in part mediated by DNA methylation, as the administration of hydralazine (an antihypertensive that induces DNA demethylation) to HFD mothers during gestation could reduce renal global DNA methylation and renal fibrotic markers at week 32 (Larkin et al. 2021). Maternal HFD has also been linked to heart disease in the offspring. In the hearts of rats exposed to maternal HFD, increased pro-fibrotic and pro-hypertrophic genes are observed, in part through the induction of EZH2 and the reduction in DNMT3B, which lead to reduced DNA methylation (Blin et al. 2020). In mice, maternal HFD has been shown to cause liver DNA hypermethylation in adulthood, as six loci identified in early life persist into adulthood (Seki et al. 2017). Maternal HFD has also been shown to alter epigenetic and gut microbiome pathways detrimentally, so as to favor the development of fatty liver disease and its progressive sequelae. Specifically, maternal HFD, followed by postnatal HFD, promoted DNA methylation changes in key genes of tissue development and metabolism—*Fgf21* and *Ppargc1β* (Wankhade et al. 2017). Maternal HFD has also been demonstrated to alter the methylation status of neonatal mice, in particular the hypermethylation of *Pgc-1α* (at a CpG site associated with metabolic disease in humans). Interestingly, this effect was suppressed by maternal exercise (Laker et al. 2014). And, also interestingly, in humans, increased maternal carbohydrate consumption was associated with a reduced methylation of the leptin gene, which implies that increased carbohydrate consumption increases leptin placental expression (Daniels et al. 2020). These findings demonstrate the dynamic epigenetic effects of maternal diet on the offspring's health outcomes.

New Challenges and Future Directions

Both laboratory and epidemiological studies suggested a mechanistic link between maternal exposures and disease outcomes in later life. Using experimental animal models, the investigation of the epigenetic effect of environmental exposures on maternal and fetal gene transcription can improve our understanding of pregnancy outcomes, fetal growth, and the etiology of fetal-based adult diseases. In addition, the assessment of epigenetic signatures can offer insights for future epigenetic epidemiology studies on childhood diseases to identify new risk factors for childhood diseases and to develop early intervention methods designed to reduce the risk of disease. In addition to easily accessible biospecimens (e.g., maternal and infant blood, cord blood, and amniotic fluid), epigenetic studies on placental tissues will provide a unique look into the role the placenta plays in mediating maternal and fetal outcomes. Nevertheless, there are challenges in assessing epigenetic changes in maternal–fetal health outcomes. For example, it has been suggested that epigenetic reprogramming occurs prenatally and postnatally, being dependent on the windows of exposure. The TaRGET II consortium aims to give a guideline to identifying the tissue-, exposure-, and time-specific epigenetic marks by using experimental animal models (Wang et al. 2018). Whether these epigenetic marks, identified in animal target tissues, can be validated in surrogate tissues (blood) from human birth cohorts and from prospective studies requires additional investigation. With the help of cutting-edge epigenomic approaches, we are now able to assess multi-omes (transcriptome, methylome, chromatin accessibility, and histone modifications) related to maternal exposures (Park et al. 2020). Nevertheless, it is critical to have a comprehensive analysis of molecular and pathophysiological changes (an integrative correlation analysis) in order to evaluate the epigenetic effect of environmental exposures in women and children. The inclusion of metabolic changes measured in urine and blood may offer additional information that links the effect of environmental toxins and nutrition on the epigenome, given that metabolites from one-carbon metabolism and the citric acid cycle serve as substrates for DNA and histone modifications. The development of tissue banks (larger sample sizes with proper exposure assessment) and of -omic data resources is believed to accelerate the understanding of epigenetics in women and children health. The number of publicly available epigenome resources such as the Gene Expression Omnibus (GEO) (Barrett et al. 2013; Edgar et al. 2002), the EWAS Atlas (Li et al. 2019), and the GWAS catalog (Buniello et al. 2019) is limited. It is urgent to establish consortiums and online resources that can integrate and compare data from both laboratory and human population studies. In this chapter we focused on the epigenetic effects of common environmental stressors on women's and children's health. But other environmental factors such as psychological stress, socioeconomic status, climate change, and new, emerging diseases (e.g. COVID-19) may affect pregnancy and parenting, ultimately modulating the epigenome and thus children's health. We also look forward to more evidence for the inheritance of epigenetic changes in subsequent generations. All in all, the analysis of epigenetic signatures in mother–infant pairs may allow us to make a better assessment of child development and of the risk for pregnancy-adverse outcomes, and to formulate strategies (e.g. lifestyle recommendations or therapeutics) designed to prevent or reduce the impact of environmental exposure on the communities affected by it.

References

Acevedo, N., Scala, G., Merid, S.K., Frumento, P., Bruhn, S., Andersson, A., Ogris, C., Bottai, M., Pershagen, G., Koppelman, G.H., Melén, E., Sonnhammer, E., Alm, J., Söderhäll, C., Kere, J., Greco, D., and Scheynius, A. (2021). Dna methylation levels in mononuclear leukocytes from the mother and her child are associated with ige sensitization to allergens in early life. *Int. J. Mol. Sci.* 22: 1–20.

Agrahari, A., Singh, A., Srivastava, A., Jha, R.R., Patel, D.K., Yadav, S., Srivastava, V., and Parmar, D. (2019). Overexpression of cerebral cytochrome P450s in prenatally exposed offspring modify the toxicity of lindane in rechallenged offspring. *Toxicol. Appl. Pharmacol.* 371: 20–37.

Ahsan, H. (2011). Arsenic in drinking-water background document for development of WHO guidelines for drinking-water quality. *Kaohsiung J. Med. Sci. 27 (9): 358–359.*

Amarasekera, M., Martino, D., Ashley, S., Harb, H., Kesper, D., Strickland, D., Saffery, R., and Prescott, S.L. (2014). Genome-wide DNA methylation profiling identifies a folate-sensitive region of differential methylation upstream of ZFP57-imprinting regulator in humans. *FASEB J.* 28: 4068–4076.

Anderson, C.M., Ralph, J.L., Wright, M.L., Linggi, B., and Ohm, J.E. (2014). DNA methylation as a biomarker for preeclampsia. *Biol. Res. Nurs.* 16: 409–420.

Anton, L., Brown, A.G., Bartolomei, M.S., and Elovitz, M.A. (2014). Differential methylation of genes associated with cell adhesion in preeclamptic placentas. *PLoS One* 9: e100148.

Awamleh, Z., Butcher, D.T., Hanley, A., Retnakaran, R., Haertle, L., Haaf, T., Hamilton, J., and Weksberg, R. (2021). Exposure to gestational diabetes mellitus (GDM) alters DNA methylation in placenta and fetal cord blood. *Diabetes Res. Clin. Pract.* 174: 108690.

Bailey, K.A. and Fry, R.C. (2014). Arsenic-associated changes to the epigenome: What are the functional consequences? *Curr. Environ. Heal. Reports* 1: 22–34.

Bakulski, K.M., Lee, H.J., Feinberg, J.I., Wells, E.M., Brown, S., Herbstman, J.B., Witter, F.R., Halden, R.U., Caldwell, K., Mortensen, M.E., Jaffe, A.E., Moye, J., Caulfield, L.E., Pan, Y., Goldman, L.R., Feinberg, A.P., and Daniele Fallin, M. (2015). Prenatal mercury concentration is associated with changes in DNA methylation at TCEANC2 in newborns. *Int. J. Epidemiol.* 44: 1249–1262.

Bansal, A., Robles-Matos, N., Wang, P.Z., Condon, D.E., Joshi, A., and Pinney, S.E. (2020). In utero bisphenol A exposure is linked with sex specific changes in the transcriptome and methylome of human amniocytes. *J. Clin. Endocrinol. Metab.* 105.

Barker, D.J.P. (1997). Maternal nutrition, fetal nutrition, and disease in later life. *Nutrition* 13 (9): 807–813.

Barrett, T., Wilhite, S.E., Ledoux, P., Evangelista, C., Kim, I.F., Tomashevsky, M., Marshall, K.A., Phillippy, K.H., Sherman, P.M., Holko, M., Yefanov, A., Lee, H., Zhang, N., Robertson, C.L., Serova, N., Davis, S., and Soboleva, A. (2013). NCBI GEO: Archive for functional genomics data sets: Update. *Nucleic Acids Res.* 41.

Cheng, S.B., Davis, S., and Sharma, S. (2018). Maternal–fetal cross talk through cell-free fetal DNA, telomere shortening, microchimerism, and inflammation. *American Journal of Reproductive Immunology. Am J Reprod Immunol. 79 (5).*

Blin, G., Liand, M., Mauduit, C., Chehade, H., Benahmed, M., Simeoni, U., and Siddeek, B. (2020). Maternal exposure to high-fat diet induces long-term derepressive chromatin marks in the heart. *Nutrients* 12: 181.

Breton, C.V., Song, A.Y., Xiao, J., Kim, S.J., Mehta, H.H., Wan, J., Yen, K., Sioutas, C., Lurmann, F., Xue, S., Morgan, T.E., Zhang, J., and Cohen, P. (2019). Effects of air pollution on mitochondrial function, mitochondrial DNA methylation, and mitochondrial peptide expression. *Mitochondrion* 46: 22–29.

Breton, C.V., Yao, J., Millstein, J., Gao, L., Siegmund, K.D., Mack, W., Whitfield-Maxwell, L., Lurmann, F., Hodis, H., Avol, E., and Gilliland, F.D. (2016). Prenatal air pollution exposures, DNA methyl transferase genotypes, and associations with newborn line1 and Alu methylation and childhood blood pressure and carotid intima-media thickness in the children's health study. *Environ. Health Perspect.* 124: 1905–1912.

Bunce, K., Chu, T., Surti, U., Hogge, W.A., and Peters, D.G. (2012). Discovery of epigenetic biomarkers for the noninvasive diagnosis of fetal disease. *Prenat. Diagn.* 32: 542–549.

Buniello, A., Macarthur, J.A.L., Cerezo, M., Harris, L.W., Hayhurst, J., Malangone, C., McMahon, A., Morales, J., Mountjoy, E., Sollis, E., Suveges, D., Vrousgou, O., Whetzel, P.L., Amode, R., Guillen, J.A., Riat, H.S., Trevanion, S.J., Hall, P., Junkins, H., Flicek, P., Burdett, T., Hindorff, L.A., Cunningham, F., and Parkinson, H. (2019). The NHGRI-EBI GWAS Catalog of published genome-wide association studies, targeted arrays and summary statistics 2019. *Nucleic Acids Res.* 47: D1005–D1012.

Burris, H.H., Baccarelli, A.A., Byun, H.M., Cantoral, A., Just, A.C., Pantic, I., Solano-Gonzalez, M., Svensson, K., Ortiz, M.T., Zhao, Y., Wright, R.O., and Téllez-Rojo, M.M. (2015). Offspring

DNA methylation of the aryl-hydrocarbon receptor repressor gene is associated with maternal BMI, gestational age, and birth weight. *Epigenetics* 10: 913–921.

Byun, H.M., Colicino, E., Trevisi, L., Fan, T., Christiani, D.C., and Baccarelli, A.A. (2016). Effects of air pollution and blood mitochondrial DNA methylation on markers of heart rate variability. *J. Am. Heart Assoc.* 5.

Byun, H.M., Panni, T., Motta, V., Hou, L., Nordio, F., Apostoli, P., Bertazzi, P.A., and Baccarelli, A.A. (2013). Effects of airborne pollutants on mitochondrial DNA methylation. *Part. Fibre Toxicol.* 10.

Cardenas, A., Houseman, E.A., Baccarelli, A.A., Quamruzzaman, Q., Rahman, M., Mostofa, G., Wright, R.O., Christiani, D.C., and Kile, M.L. (2015). In utero arsenic exposure and epigenome-wide associations in placenta, umbilical artery, and human umbilical vein endothelial cells. *Epigenetics* 10: 1054–1063.

Carson, C. and Lawson, H.A. (2018). Epigenetics of metabolic syndrome. *Physiol. Genomics. 50 (11): 947–955.*

Carty, D.M., Delles, C., and Dominiczak, A.F. (2008). Novel biomarkers for predicting preeclampsia. *Trends Cardiovasc. Med. 18 (6): 186–194.*

Castañeda, A.R., Pinkerton, K.E., Bein, K.J., Magaña-Méndez, A., Yang, H.T., Ashwood, P., and Vogel, C.F.A. (2018). Ambient particulate matter activates the aryl hydrocarbon receptor in dendritic cells and enhances Th17 polarization. *Toxicol. Lett.* 292: 85–96.

Cediel Ulloa, A., Gliga, A., Love, T.M., Pineda, D., Mruzek, D.W., Watson, G.E., Davidson, P.W., Shamlaye, C.F., Strain, J.J., Myers, G.J., Van Wijngaarden, E., Ruegg, J., and Broberg, K. (2021). Prenatal methylmercury exposure and DNA methylation in seven-year-old children in the Seychelles Child Development Study. *Environ. Int.* 147.

Chen, Q., Jiang, S., Liu, H., Gao, Y., Yang, X., Ren, Z., Gao, Y., Xiao, L., Hu, H., Yu, Y., Yang, X., and Zhong, M. (2020). Association of lncRNA SH3PXD2A-AS1 with preeclampsia and its function in invasion and migration of placental trophoblast cells. *Cell Death Dis* 11: 1–15.

Chen, R.F., Huang, H.C., Ou, C.Y., Hsu, T.Y., Chuang, H., Chang, J.C., Wang, L., Kuo, H.C., and Yang, K.D. (2010). MicroRNA-21 expression in neonatal blood associated with antenatal immunoglobulin e production and development of allergic rhinitis. *Clin. Exp. Allergy* 40: 1482–1490.

Chiu, K.C., Sisca, F., Ying, J.H., Tsai, W.J., Hsieh, W.S., Chen, P.C., and Liu, C.Y. (2021). Prenatal chlorpyrifos exposure in association with PPARγ H3K4me3 and DNA methylation levels and child development. *Environ. Pollut.* 274: 116511.

Concha, G., Vogler, G., Lezcano, D., Nermell, B., and Vahter, M. (1998). Exposure to inorganic arsenic metabolites during early human development. *Toxicol. Sci.* 44: 185–190.

Cooper, W.N., Khulan, B., Owens, S., Elks, C.E., Seidel, V., Prentice, A.M., Belteki, G., Ong, K.K., Affara, N.A., Constância, M., and Dunger, D.B. (2012). DNA methylation profiling at imprinted loci after periconceptional micronutrient supplementation in humans: Results of a pilot randomized controlled trial. *FASEB J.* 26: 1782–1790.

Cornelius, M.D. and Day, N.L. (2009). Developmental consequences of prenatal tobacco exposure. *Curr. Opin. Neurol. 22 (2): 121–125.*

Cutland, C.L., Lackritz, E.M., Mallett-Moore, T., Bardají, A., Chandrasekaran, R., Lahariya, C., Nisar, M.I., Tapia, M.D., Pathirana, J., Kochhar, S., and Muñoz, F.M. (2017). Low birth weight: Case definition & guidelines for data collection, analysis, and presentation of maternal immunization safety data. Vaccine.

Dai, C., Zhao, C., Xu, M., Sui, X., Sun, L., Liu, Y., Su, M., Wang, H., Yuan, Y., Zhang, S., Shi, J., Sun, J., and Li, Y. (2021). Serum lncRNAs in early pregnancy as potential biomarkers for the prediction of pregnancy-induced hypertension, including preeclampsia. *Mol. Ther.—Nucleic Acids* 24: 416–425.

Damstra, T. (2002). Potential effects of certain persistent organic pollutants and endocrine disrupting chemicals on the health of children. *J Toxicol. Clin. Toxicol.* 40 (4): 457–465.

Daniels, T.E., Sadovnikoff, A.I., Ridout, K.K., Lesseur, C., Marsit, C.J., and Tyrka, A.R. (2020). Associations of maternal diet and placenta leptin methylation. *Mol. Cell. Endocrinol.* 505.

Dao, T., Hong, X., Wang, X., and Tang, W.Y. (2015). Aberrant 5′-CpG methylation of cord blood TNFα associated with maternal exposure to polybrominated diphenyl ethers. *PLoS One* 10: e0138815.

Delpisheh, A., Brabin, L., Drummond, S., and Brabin, B.J. (2008). Prenatal smoking exposure and asymmetric fetal growth restriction. *Ann. Hum. Biol.* 35: 573–583.

Devall, M., Mill, J., and Lunnon, K. (2014). The mitochondrial epigenome: A role in Alzheimer's disease? *Epigenomics. 6 (6): 665–675.*

Devall, M., Roubroeks, J., Mill, J., Weedon, M., and Lunnon, K. (2016). Epigenetic regulation of mitochondrial function in neurodegenerative disease: New insights from advances in genomic technologies. *Neurosci. Lett. 625: 47–55.*

Dzitoyeva, S., Chen, H., and Manev, H. (2012). Effect of aging on 5-hydroxymethylcytosine in brain mitochondria. *Neurobiol. Aging* 33: 2881–2891.

Eckel, R.H., Grundy, S.M., and Zimmet, P.Z. (2005). The metabolic syndrome. *Lancet* 365: 1415–1428.

Edgar, R., Domrachev, M., and Lash, A.E. (2002). Gene Expression Omnibus: NCBI gene expression and hybridization array data repository. *Nucleic Acids Res.* 30: 207–210.

Feng, Y., Zhao, L.Z., Hong, L., Shan, C., Shi, W., and Cai, W. (2013). Alteration in methylation pattern of GATA-4 promoter region in vitamin A-deficient offspring's heart. *J. Nutr. Biochem.* 24: 1373–1380.

Ford, E.S., Giles, W.H., and Mokdad, A.H. (2004). Increasing prevalence of the metabolic syndrome among U.S. adults. *Diabetes Care* 27: 2444–2449.

Fox, R., Kitt, J., Leeson, P., Aye, C.Y.L., and Lewandowski, A.J. (2019). Preeclampsia: Risk Factors, Diagnosis, Management, and the Cardiovascular Impact on the Offspring. *J. Clin. Med.* 8: 1625.

Green, B.B., Karagas, M.R., Punshon, T., Jackson, B.P., Robbins, D.J., Houseman, E.A., and Marsit, C.J. (2016). Epigenome-wide assessment of DNA methylation in the placenta and arsenic exposure in the New Hampshire Birth Cohort Study (USA). *Environ. Health Perspect.* 124: 1253–1260.

Gregory, D.J., Kobzik, L., Yang, Z., McGuire, C.C., and Fedulov, A.V. (2017). Transgenerational transmission of asthma risk after exposure to environmental particles during pregnancy. *Am. J. Physiol.—Lung Cell. Mol. Physiol.* 313: L395–L405.

Grieger, J.A., Bianco-Miotto, T., Grzeskowiak, L.E., Leemaqz, S.Y., Poston, L., McCowan, L.M., Kenny, L.C., Myers, J.E., Walker, J.J., Dekker, G.A., and Roberts, C.T. (2018). Metabolic syndrome in pregnancy and risk for adverse pregnancy outcomes: A prospective cohort of nulliparous women. *PLoS Med.* 15.

Gruzieva, O., Merid, S.K., Koppelman, G.H., and Melén, E. (2021). An update on the epigenetics of asthma. *Curr. Opin. Allergy Clin. Immunol.* 21: 175–181.

Gruzieva, O., Xu, C.J., Yousefi, P., Relton, C., Merid, S.K., Breton, C.V., Gao, L., Volk, H.E., Feinberg, J.I., Ladd-Acosta, C., Bakulski, K., Auffray, C., Lemonnier, N., Plusquin, M., Ghantous, A., Herceg, Z., Nawrot, T.S., Pizzi, C., Richiardi, L., Rusconi, F., Vineis, P., Kogevinas, M., Felix, J.F., Duijts, L., Den Dekker, H.T., Jaddoe, V.W.V., Ruiz, J.L., Bustamante, M., Antó, J.M., Sunyer, J., Vrijheid, M., Gutzkow, K.B., Grazuleviciene, R., Hernandez-Ferrer, C., Annesi-Maesano, I., Lepeule, J., Bousquet, J., Bergström, A., Kull, I., Söderhäll, C., Kere, J., Gehring, U., Brunekreef, B., Just, A.C., Wright, R.J., Peng, C., Gold, D.R., Kloog, I., Demeo, D.L., Pershagen, G., Koppelman, G.H., London, S.J., Baccarelli, A.A., and Melén, E. (2019). Prenatal particulate air pollution and DNA methylation in newborns: An epigenome-wide meta-analysis. *Environ. Health Perspect.* 127.

Hales, C.N. and Barker, D.J.P. (1992). Type 2 (Non-insulin-dependent) diabetes mellitus: The thrifty phenotype hypothesis. *Diabetologia* 35 (7): 595–601.

Hales, C.N. and Barker, D.J.P. (2001). The thrifty phenotype hypothesis. *Br. Med. Bull.* 60: 5–20.

Han, L., Kaushal, A., Zhang, H., Kadalayil, L., Duan, J., Holloway, J.W., Karmaus, W., Banerjee, P., Tsai, S.F., Wen, H.J., Arshad, S.H., and Wang, S.L. (2021). DNA methylation at birth is associated with childhood serum immunoglobulin E levels. *Epigenetics Insights* 14.

Hannon, E., Schendel, D., Ladd-Acosta, C., Grove, J., Hansen, C.S., Hougaard, D.M., Bresnahan, M., Mors, O., Hollegaard, M.V., Bækvad-Hansen, M., Hornig, M., Mortensen, P.B., Børglum, A.D., Werge, T., Pedersen, M.G., Nordentoft, M., Buxbaum, J.D., Fallin, M.D., Bybjerg-Grauholm, J., Reichenberg, A., and Mill, J. (2019). Variable DNA methylation in neonates mediates the association between prenatal smoking and birth weight. *Philos. Trans. R. Soc. B Biol. Sci.* 374.

He, J., Liu, K., Hou, X., and Lu, J. (2021). Identification and validation of key non-coding RNAs and mRNAs using co-expression network analysis in pre-eclampsia. *Medicine (Baltimore)* 100: e25294.

He, Q.L., Wei, X.Y., Han, X.Y., Zhou, Q., Wang, H.Q., Ding, N.Z., Meng, X.Q., Schatten, H., Sun, Q.Y., and Liu, S.Z. (2019). Effects of 2,3′,4,4′5-pentachlorobiphenyl exposure during pregnancy on epigenetic imprinting and maturation of offspring's oocytes in mice. *Arch. Toxicol.* 93: 2575–2592.

Herberth, G., Bauer, M., Gasch, M., Hinz, D., Röder, S., Olek, S., Kohajda, T., Rolle-Kampczyk, U., Von Bergen, M., Sack, U., Borte, M., and Lehmann, I. (2014). Maternal and cord blood miR-223 expression associates with prenatal tobacco smoke exposure and low regulatory T-cell numbers. *J. Allergy Clin. Immunol.* 133 (543-550.e4).

Hromadnikova, I., Kotlabova, K., and Krofta, L. (2021). Association analysis in young and middle-aged mothers: Relation between expression of cardiovascular disease associated micrornas and abnormal clinical findings. *J. Pers. Med.* 11: 1–16.

Hromadnikova, I., Kotlabova, K., Krofta, L., and Sirc, J. (2020). Association analysis in children born from normal and complicated pregnancies—cardiovascular disease associated micrornas and the incidence of prehypertension/hypertension, overweight/ obesity,valve problems and heart defects. *Int. J. Mol. Sci.* 21: 1–20.

Huang, Q., Gang, H., Hu, Y., Tang, S., and Peng, D. (2020). Exposure of nicotine in neonatal rats induces activated inflammatory cytokines and altered biomarkers related to cardiac dysfunction. *Int. J. Pharmacol.* 16: 407–415.

Huang, Y., Zhu, J., Li, H., Wang, W., Li, Y., Yang, X., Zheng, N., Liu, Q., Zhang, Q., Zhang, W., and Liu, J. (2020). Cadmium exposure during prenatal development causes testosterone disruption in multigeneration via SF-1 signaling in rats. *Food Chem. Toxicol.* 135.

Ion, R. and Bernal, A.L. (2015). Smoking and preterm birth. *Reprod. Sci. 22 (8).*

Jain, J.A. and Gyamfi-Bannerman, C. (2018). Preterm labor. In *Evidence-Based Obstetrics and Gynecology* (eds. E. Norwitz, D. Miller, C. Zelop, and D. Keefe), 385–395. Oxford: Wiley Blackwell.

Ji, X., Yue, H., Li, G., and Sang, N. (2021). Maternal smoking-induced lung injuries in dams and offspring via inflammatory cytokines. *Environ. Int.* 156: 106618.

Jönsson, J., Renault, K.M., García-Calzón, S., Perfilyev, A., Estampador, A.C., Nørgaard, K., Lind, M.V., Vaag, A., Hjort, L., Michaelsen, K.F., Carlsen, E.M., Franks, P.W., and Ling, C. (2021). Lifestyle Intervention in pregnant women with obesity impacts cord blood DNA methylation, which associates with body composition in the offspring. *Diabetes* 70: 854–866.

Joubert, B.R., den Dekker, H.T., Felix, J.F., Bohlin, J., Ligthart, S., Beckett, E., Tiemeier, H., van Meurs, J.B., Uitterlinden, A.G., Hofman, A., Håberg, S.E., Reese, S.E., Peters, M.J., Kulle Andreassen, B., Steegers, E.A.P., Nilsen, R.M., Vollset, S.E., Midttun, Ø., Ueland, P.M., Franco, O.H., Dehghan, A., de Jongste, J.C., Wu, M.C., Wang, T., Peddada, S.D., Jaddoe, V.W.V., Nystad, W., Duijts, L., and London, S.J. (2016). Maternal plasma folate impacts differential DNA methylation in an epigenome-wide meta-analysis of newborns. *Nat. Commun.* 2016 71 (7): 1–8.

Karagas, M.R., Punshon, T., Davis, M., Bulka, C.M., Slaughter, F., Karalis, D., Argos, M., and Ahsan, H. (2019). Rice intake and emerging concerns on arsenic in rice: A review of the human evidence and methodologic challenges. *Curr. Environ. Heal. Reports. 6 (4): 361–372.*

Kim, H., Kang, M., Lee, S., Shin, Y., and Hong, S. (2021). Prenatal maternal anxiety promotes atopic dermatitis in offspring via placental DNA methylation changes. *Asian Pacific J. Allergy Immunol. 9 (3): 200–211.*

Kim, J., Pitlick, M.M., Christine, P.J., Schaefer, A.R., Saleme, C., Comas, B., Cosentino, V., Gadow, E., and Murray, J.C. (2013). Genome-wide analysis of DNA methylation in human amnion. *Sci. World J.* 2013 11.

Kim, S., Park, M., Kim, J.Y., Kim, T., Hwang, J.Y., Ha, K.S., Won, M.H., Ryoo, S., Kwon, Y.G., and Kim, Y.M. (2020). Circulating miRNAs associated with dysregulated vascular and trophoblast function as target-based diagnostic biomarkers for preeclampsia. *Cells* 9.

Kim, S., Cho, Y.H., Won, S., Ku, J.L., Moon, H.B., Park, J., Choi, G., Kim, S., and Choi, K. (2019). Maternal exposures to persistent organic pollutants are associated with DNA methylation of thyroid hormone-related genes in placenta differently by infant sex. *Environ. Int.* 130: 104956.

Krishna, R.G., Vishnu Bhat, B., Bobby, Z., Papa, D., Badhe, B., Kalidoss, V.K., and Karli, S. (2020). Identification of differentially methylated candidate genes and their biological significance in IUGR neonates by methylation EPIC array. *J. Matern. Neonatal. Med.* 1–9.

Laker, R.C., Lillard, T.S., Okutsu, M., Zhang, M., Hoehn, K.L., Connelly, J.J., and Yan, Z. (2014). Exercise prevents maternal high-fat diet-induced hypermethylation of the Pgc-1α gene and age-dependent metabolic dysfunction in the offspring. *Diabetes* 63: 1605–1611.

Larkin, B.P., Saad, S., Glastras, S.J., Nguyen, L.T., Hou, M., Chen, H., Wang, R., and Pollock, C.A. (2021). Low-dose hydralazine during gestation reduces renal fibrosis in rodent offspring exposed to maternal high fat diet. *PLoS One* 16.

Lei, D., Fang, C., Deng, N., Yao, B., and Fan, C. (2021). Long noncoding RNA expression profiling identifies MIR210HG as a novel molecule in severe preeclampsia. *Life Sci.* 270.

Li, J., Quan, X.J., Chen, G., Hong, J.W., Wang, Q., Xu, L.L., Wang, B.H., Yu, Z.H., and Yu, H.M. (2020). PFOS-induced placental cell growth inhibition is partially mediated by lncRNA H19 through interacting with miR-19a and miR-19b. *Chemosphere* 261.

Li, M., Zou, D., Li, Z., Gao, R., Sang, J., Zhang, Y., Li, R., Xia, L., Zhang, T., Niu, G., Bao, Y., and Zhang, Z. (2019). EWAS Atlas: A curated knowledgebase of epigenome-wide association studies. *Nucleic Acids Res.* 47: D983–D988.

Li, Y., Jaddoe, V.W., Qi, L., He, Y., Wang, D., Lai, J., Zhang, J., Fu, P., Yang, X., and Hu, F.B. (2011). Exposure to the Chinese famine in early life and the risk of metabolic syndrome in adulthood. *Diabetes Care* 34: 1014–1018.

Lillycrop, K.A., Garratt, E.S., Titcombe, P., Melton, P.E., Murray, R.J.S., Barton, S.J., Clarke-Harris, R., Costello, P.M., Holbrook, J.D., Hopkins, J.C., Childs, C.E., Paras-Chavez, C., Calder, P.C., Mori, T.A., Beilin, L., Burdge, G.C., Gluckman, P.D., Inskip, H.M., Harvey, N.C., Hanson, M.A., Huang, R.C., Cooper, C., and Godfrey, K.M. (2019). Differential SLC6A4 methylation: A predictive epigenetic marker of adiposity from birth to adulthood. *Int. J. Obes.* 43: 974–988.

Lim, J.H., Kang, Y.J., Bak, H.J., Kim, M.S., Lee, H.J., Kwak, D.W., Han, Y.J., Kim, M.Y., Boo, H., Kim, S.Y., and Ryu, H.M. (2020). Epigenome-wide DNA methylation profiling of preeclamptic placenta according to severe features. *Clin. Epigenetics* 12: 1–15.

Lindgren, A., Danielsson, B.R.G., Dencker, L., and Vahter, M. (1984). Embryotoxicity of arsenite and arsenate: Distribution in pregnant mice and monkeys and effects on embryonic cells in vitro. *Acta Pharmacol. Toxicol. (Copenh).* 54: 311–320.

Liu, B., Hu, X., Li, Y., Ke, J., Dasgupta, C., Huang, X., Walayat, A., Zhang, L., and Xiao, D. (2019). Epigenetic down-regulation of BK Ca channel by miR-181a contributes to the fetal and neonatal nicotine-mediated exaggerated coronary vascular tone in adult life. *Int. J. Cardiol.* 281: 82–89.

Lu, H.Q. and Hu, R. (2019). Lasting effects of intrauterine exposure to preeclampsia on offspring and the underlying mechanism. *AJP Rep.* 9: E275–E291.

Maccani, J.Z.J., Koestler, D.C., Houseman, E.A., Marsit, C.J., and Kelsey, K.T. (2013). Placental DNA methylation alterations associated with maternal tobacco smoking at the RUNX3 gene are also associated with gestational age. *Epigenomics* 5: 619–630.

Mantha, M., Yeary, E., Trent, J., Creed, P.A., Kubachka, K., Hanley, T., Shockey, N., Heitkemper, D., Caruso, J., Xue, J., Rice, G., Wymer, L., and Creed, J.T. (2017). Estimating inorganic arsenic exposure from U.S. rice and total water intakes. *Environ. Health Perspect* 125.

Manuck, T.A., Eaves, L.A., Rager, J.E., and Fry, R.C. (2021). Mid-pregnancy maternal blood nitric oxide-related gene and miRNA expression are associated with preterm birth. *Epigenomics* 13: 667–682.

Martin, E.M., Stýblo, M., and Fry, R.C. (2017). Genetic and epigenetic mechanisms underlying arsenic-associated diabetes mellitus: A perspective of the current evidence. *Epigenomics* 9 (5): 701–710.

Martos, S.N., Tang, W.Y., and Wang, Z. (2015). Elusive inheritance: Transgenerational effects and epigenetic inheritance in human environmental disease. *Prog. Biophys. Mol. Biol. 118: 44–54.*

Mordaunt, C.E., Jianu, J.M., Laufer, B.I., Zhu, Y., Hwang, H., Dunaway, K.W., Bakulski, K.M., Feinberg, J.I., Volk, H.E., Lyall, K., Croen, L.A., Newschaffer, C.J., Ozonoff, S., Hertz-Picciotto, I., Fallin, M.D., Schmidt, R.J., and LaSalle, J.M. (2020). Cord blood DNA methylome in newborns later diagnosed with autism spectrum disorder reflects early dysregulation of neurodevelopmental and X-linked genes. *Genome Med.* 121 (12): 1–25.

Mukherjee, N., Patil, V., Chen, S., Zhang, H., Arshad, S.H., Holloway, J.W., and Karmaus, W. (2016). Interaction of AHRR-methylation and gestational smoking influences adolescent eczema, but not asthma. *Eur. Respir. J.* 48: PA4594.

Murphy, M.S.Q., Casselman, R.C., Tayade, C., and Smith, G.N. (2015). Differential expression of plasma microRNA in preeclamptic patients at delivery and 1 year postpartum. *Am. J. Obstet. Gynecol.* 213: 367.e1–367.e9.

Nahar, M.S., Liao, C., Kannan, K., Harris, C., and Dolinoy, D.C. (2015). In utero bisphenol A concentration, metabolism, and global DNA methylation across matched placenta, kidney, and liver in the human fetus. *Chemosphere* 124: 54–60.

Neophytou, A.M., Oh, S.S., Hu, D., Huntsman, S., Eng, C., Rodríguez-Santana, J.R., Kumar, R., Balmes, J.R., Eisen, E.A., and Burchard, E.G. (2019). In utero tobacco smoke exposure, DNA methylation, and asthma in Latino children. *Environ. Epidemiol.* 3: e048.

Neumann, A., Walton, E., Alemany, S., Cecil, C., González, J.R., Jima, D.D., Lahti, J., Tuominen, S.T., Barker, E.D., Binder, E., Caramaschi, D., Carracedo, Á., Czamara, D., Evandt, J., Felix, J.F., Fuemmeler, B.F., Gutzkow, K.B., Hoyo, C., Julvez, J., Kajantie, E., Laivuori, H., Maguire, R., Maitre, L., Murphy, S.K., Murcia, M., Villa, P.M., Sharp, G., Sunyer, J., Raikkönen, K., Bakermans-Kranenburg, M., IJzendoorn, M., Van Guxens, M.,

Relton, C.L., and Tiemeier, H. (2020). Association between DNA methylation and ADHD symptoms from birth to school age: A prospective meta-analysis. *Transl. Psychiatry* 10: 1–11.
Osuchukwu, O.O. and Reed, D.J. (2021). *Small for Gestational Age*. StatPearls.
Ouidir, M., Mendola, P., Buck Louis, G.M., Kannan, K., Zhang, C., and Tekola-Ayele, F. (2020). Concentrations of persistent organic pollutants in maternal plasma and epigenome-wide placental DNA methylation. *Clin. Epigenetics* 12.
Ouyang, B., Baxter, C.S., Lam, H.M., Yeramaneni, S., Levin, L., Haynes, E., and Ho, S.M. (2012). Hypomethylation of dual specificity phosphatase 22 promoter correlates with duration of service in firefighters and is inducible by low-dose benzo[a]pyrene. *J. Occup. Environ. Med.* 54: 774–780.
Park, B., Khanam, R., Vinayachandran, V., Baqui, A.H., London, S.J., and Biswal, S. (2020). Epigenetic biomarkers and preterm birth. *Environ. Epigenetics* 6.
Peng, H., Xu, H., Wu, J., Li, J., Zhou, Y., Ding, Z., Siwko, S.K., Yuan, X., Schalinske, K.L., Alpini, G., Zhang, K.K., and Xie, L. (2021). Maternal high-fat diet disrupted one-carbon metabolism in offspring, contributing to nonalcoholic fatty liver disease. *Liver Int.* 41: 1305–1319.
Perera, F., Tang, W.Y., Herbstman, J., Tang, D., Levin, L., Miller, R., and Ho, S.M. (2009). Relation of DNA methylation of 5′-CpG island of ACSL3 to transplacental exposure to airborne polycyclic aromatic hydrocarbons and childhood asthma. *PLoS One* 4.
Perkins, E., Murphy, S.K., Murtha, A.P., Schildkraut, J., Jirtle, R.L., Demark-Wahnefried, W., Forman, M.R., Kurtzberg, J., Overcash, F., Huang, Z., and Hoyo, C. (2012). Insulin-like growth factor 2/H19 methylation at birth and risk of overweight and obesity in children. *J. Pediatr.* 161: 31–39.
Pidsley, R., Fernandes, C., Viana, J., Paya-Cano, J.L., Liu, L., Smith, R.G., Schalkwyk, L.C., and Mill, J. (2012). DNA methylation at the Igf2/H19 imprinting control region is associated with cerebellum mass in outbred mice. *Mol. Brain* 5: 1–9.
Pilsner, J.R., Hall, M.N., Liu, X., Ilievski, V., Slavkovich, V., Levy, D., Factor-Litvak, P., Yunus, M., Rahman, M., Graziano, J.H., and Gamble, M.V. (2012). Influence of prenatal arsenic exposure and newborn sex on global methylation of cord blood DNA. *PLoS One* 7: e37147–e37147.
Rahimi, G., Hosseini-Asl, S.S., Marghoub, M., Elahi, H.S., Amani, F., and Shahbazzadegan, S. (2020). Lncrna terra expression level and telomere length in placenta samples of healthy and preeclampsia women. *Iran. J. Obstet. Gynecol. Infertil.* 23: 9–18.
Rahman, M.L., Liang, L., Valeri, L., Su, L., Zhu, Z., Gao, S., Mostofa, G., Qamruzzaman, Q., Hauser, R., Baccarelli, A., and Christiani, D.C. (2018). Regulation of birthweight by placenta-derived miRNAs: Evidence from an arsenic-exposed birth cohort in Bangladesh. *Epigenetics* 13: 573–590.
Rasdi, Z., Kamaludin, R., Ab Rahim, S., Syed Ahmad Fuad, S.B., Othman, M.H.D., Siran, R., Mohd Nor, N.S., Abdul Hamid Hasani, N., and Sheikh Abdul Kadir, S.H. (2020). The impacts of intrauterine bisphenol A exposure on pregnancy and expression of miRNAs related to heart development and diseases in animal model. *Sci. Rep.* 10: 1–13.
Rathod, A., Duan, J., Zhang, H., Holloway, J.W., Ewart, S., Arshad, S.H., and Karmaus, W. (2020). Interweaving between genetic and epigenetic studies on childhoodasthma. *Epigenetics Insights* 13.
Rebane, A. (2015). microRNA and allergy. *Adv. Exp. Med. Biol.* 888: 331–352.
Reese, S.E., Xu, C.J., den Dekker, H.T., Lee, M.K., Sikdar, S., Ruiz-Arenas, C., Merid, S.K., Rezwan, F.I., Page, C.M., Ullemar, V., Melton, P.E., Oh, S.S., Yang, I.V., Burrows, K., Söderhäll, C., Jima, D.D., Gao, L., Arathimos, R., Küpers, L.K., Wielscher, M., Rzehak, P., Lahti, J., Laprise, C., Madore, A.M., Ward, J., Bennett, B.D., Wang, T., Bell, D.A., Vonk, J.M.,

Håberg, S.E., Zhao, S., Karlsson, R., Hollams, E., Hu, D., Richards, A.J., Bergström, A., Sharp, G.C., Felix, J.F., Bustamante, M., Gruzieva, O., Maguire, R.L., Gilliland, F., Baïz, N., Nohr, E.A., Corpeleijn, E., Sebert, S., Karmaus, W., Grote, V., Kajantie, E., Magnus, M.C., Örtqvist, A.K., Eng, C., Liu, A.H., Kull, I., Jaddoe, V.W.V., Sunyer, J., Kere, J., Hoyo, C., Annesi-Maesano, I., Arshad, S.H., Koletzko, B., Brunekreef, B., Binder, E.B., Räikkönen, K., Reischl, E., Holloway, J.W., Jarvelin, M.R., Snieder, H., Kazmi, N., Breton, C.V., Murphy, S.K., Pershagen, G., Anto, J.M., Relton, C.L., Schwartz, D.A., Burchard, E.G., Huang, R.C., Nystad, W., Almqvist, C., Henderson, A.J., Melén, E., Duijts, L., Koppelman, G.H., and London, S.J. (2019). Epigenome-wide meta-analysis of DNA methylation and childhood asthma. *J. Allergy Clin. Immunol.* 143: 2062–2074.

Roseboom, T., de Rooij, S., and Painter, R. (2006). The Dutch famine and its long-term consequences for adult health. *Early Hum. Dev. 82 (8): 485–491.*

Rubin, B.S. (2011). Bisphenol A: An endocrine disruptor with widespread exposure and multiple effects. *J. Steroid Biochem. Mol. Biol.* 127: 27–34.

Saintilnord, W.N. and Fondufe-Mittendorf, Y. (2021). Arsenic-induced epigenetic changes in cancer development. *Semin. Cancer Biol.* doi: 10.1016/j.semcancer.2021.03.019.

Salam, R.A., Das, J.K., Ali, A., Lassi, Z.S., and Bhutta, Z.A. (2013). Maternal undernutrition and intrauterine growth restriction. *Expert Rev. Obstet. Gynecol. 8 (6): 559–567.*

Şanlı, E. and Kabaran, S. (2019). Maternal obesity, maternal overnutrition and fetal programming: Effects of epigenetic mechanisms on the development of metabolic disorders. *Curr. Genomics* 20: 419–427.

Schlecht, C., Klammer, H., Jarry, H., and Wuttke, W. (2004). Effects of estradiol, benzophenone-2 and benzophenone-3 on the expression pattern of the estrogen receptors (ER) alpha and beta, the estrogen receptor-related receptor 1 (ERR1) and the aryl hydrocarbon receptor (AhR) in adult ovariectomized rats. *Toxicology* 205: 123–130.

Schneider, H. (2017). Placental dysfunction as a key element in the pathogenesis of preeclampsia. *Dev. Period Med.* 21: 309–316.

Seki, Y., Suzuki, M., Guo, X., Glenn, A.S., Vuguin, P.M., Fiallo, A., Du, Q., Ko, Y.A., Yu, Y., Susztak, K., Zheng, D., Greally, J.M., Katz, E.B., and Charron, M.J. (2017). In utero exposure to a high-fat diet programs hepatic hypermethylation and gene dysregulation and development of metabolic syndrome in male mice. *Endocrinology* 158: 2860–2872.

Shaji, E., Santosh, M., Sarath, K.V., Prakash, P., Deepchand, V., and Divya, B.V. (2021). Arsenic contamination of groundwater: A global synopsis with focus on the Indian Peninsula. *Geosci. Front.* 12: 101079.

Shock, L.S., Thakkar, P.V., Peterson, E.J., Moran, R.G., and Taylor, S.M. (2011). DNA methyltransferase 1, cytosine methylation, and cytosine hydroxymethylation in mammalian mitochondria. *Proc. Natl. Acad. Sci. USA* 108: 3630–3635.

Singh, S.P., Chand, H.S., Langley, R.J., Mishra, N., Barrett, T., Rudolph, K., Tellez, C., Filipczak, P.T., Belinsky, S., Saeed, A.I., Sheybani, A., Exil, V., Agarwal, H., Sidhaye, V.K., Sussan, T., Biswal, S., and Sopori, M. (2017). Gestational exposure to sidestream (secondhand) cigarette smoke promotes transgenerational epigenetic transmission of exacerbated allergic asthma and bronchopulmonary dysplasia. *J. Immunol.* 198: 3815–3822.

Sly, P.D. (2011). The early origins of asthma: Who is really at risk? *Curr. Opin. Allergy Clin. Immunol.* 11: 24–28.

Sparrow, S., Manning, J.R., Cartier, J., Anblagan, D., Bastin, M.E., Piyasena, C., Pataky, R., Moore, E.J., Semple, S.I., Wilkinson, A.G., Evans, M., Drake, A.J., and Boardman, J.P. (2016). Epigenomic profiling of preterm infants reveals DNA methylation differences at sites associated with neural function. *Transl. Psychiatry* 6: e716.

Tagliaferri, S., Cepparulo, P., Vinciguerra, A., Campanile, M., Esposito, G., Maruotti, G.M., Zullo, F., Annunziato, L., and Pignataro, G. (2021). miR-16-5p, miR-103-3p, and miR-27b-3p as early peripheral biomarkers of fetal growth restriction. *Front. Pediatr.* 9: 156.

Tang, W.Y. and Ho, S.M. (2007). Epigenetic reprogramming and imprinting in origins of disease. *Rev. Endocr. Metab. Disord.* 8 (2): 1743–182.

Tang, W.Y., Levin, L., Talaska, G., Cheung, Y.Y., Herbstman, J., Tang, D., Miller, R.L., Perera, F., and Ho, S.M. (2012). Maternal exposure to polycyclic aromatic hydrocarbons and 5′-CpG methylation of interferon-γ in cord white blood cells. *Environ. Health Perspect.* 120: 1195–1200.

Tang, Y., Liu, H., Li, H., Peng, T., Gu, W., and Li, X. (2015). Hypermethylation of the HLA-G promoter is associated with preeclampsia. *Mol. Hum. Reprod.* 21: 736–744.

Tekola-Ayele, F., Zeng, X., Ouidir, M., Workalemahu, T., Zhang, C., Delahaye, F., Delahaye, F., and Wapner, R. (2020). DNA methylation loci in placenta associated with birthweight and expression of genes relevant for early development and adult diseases. *Clin. Epigenetics* 12: 1–14.

Tsamou, M., Vrijens, K., Madhloum, N., Lefebvre, W., Vanpoucke, C., and Nawrot, T.S. (2018). Air pollution-induced placental epigenetic alterations in early life: A candidate miRNA approach. *Epigenetics* 13: 135–146.

Vahter, M. (2008). Health effects of early life exposure to arsenic. *Basic Clin. Pharmacol. Toxicol.* 102 (2): 204–211.

Victora, C.G., Adair, L., Fall, C., Hallal, P.C., Martorell, R., Richter, L., and Sachdev, H.S. (2008). *Maternal and Child Undernutrition: Consequences for Adult Health and Human Capital*. Lancet.

Vidal, A.C., Murphy, S.K., Murtha, A.P., Schildkraut, J.M., Soubry, A., Huang, Z., Neelon, S.E.B., Fuemmeler, B., Iversen, E., Wang, F., Kurtzberg, J., Jirtle, R.L., and Hoyo, C. (2013). Associations between antibiotic exposure during pregnancy, birth weight and aberrant methylation at imprinted genes among offspring. *Int. J. Obes.* 37: 907–913.

Wang, N., Wang, X., Li, Q., Han, B., Chen, Y., Zhu, C., Chen, Y., Lin, D., Wang, B., Jensen, M.D., and Lu, Y. (2017). The famine exposure in early life and metabolic syndrome in adulthood. *Clin. Nutr.* 36: 253–259.

Wang, T., Pehrsson, E.C., Purushotham, D., Li, D., Zhuo, X., Zhang, B., Lawson, H.A., Province, M.A., Krapp, C., Lan, Y., Coarfa, C., Katz, T.A., Tang, W.Y., Wang, Z., Biswal, S., Rajagopalan, S., Colacino, J.A., Tsai, Z.T.Y., Sartor, M.A., Neier, K., Dolinoy, D.C., Pinto, J., Hamanaka, R.B., Mutlu, G.M., Patisaul, H.B., Aylor, D.L., Crawford, G.E., Wiltshire, T., Chadwick, L.H., Duncan, C.G., Garton, A.E., McAllister, K.A., Bartolomei, M.S., Walker, C.L., and Tyson, F.L. (2018). The NIEHS TaRGET II Consortium and environmental epigenomics. *Nat. Biotechnol.* 36 (3): 225–227.

Wang, Y., Perera, F., Guo, J., Riley, K.W., Durham, T., Ross, Z., Ananth, C.V., Baccarelli, A., Wang, S., and Herbstman, J.B. (2022). A methodological pipeline to generate an epigenetic marker of prenatal exposure to air pollution indicators. *Epigenetics* 17 (1): 32–40.

Wankhade, U.D., Zhong, Y., Kang, P., Alfaro, M., Chintapalli, S.V., Thakali, K.M., and Shankar, K. (2017). Enhanced offspring predisposition to steatohepatitis with maternal high-fat diet is associated with epigenetic and microbiome alterations. *PLoS One* 12.

Weyde, K.V.F., Olsen, A.K., Duale, N., Kamstra, J.H., Skogheim, T.S., Caspersen, I.H., Engel, S.M., Biele, G., Xia, Y., Meltzer, H.M., Aase, H., and Villanger, G.D. (2021). Gestational blood levels of toxic metal and essential element mixtures and associations with global DNA methylation in pregnant women and their infants. *Sci. Total Environ.* 787: 147621.

White, W.M., Brost, B., Sun, Z., Rose, C., Craici, I., Wagner, S.J., Turner, S.T., and Garovic, V.D. (2013). Genome-wide methylation profiling demonstrates hypermethylation in maternal leukocyte DNA in preeclamptic compared to normotensive pregnancies. *Hypertens. Pregnancy* 32: 257–269.

Winger, E.E., Reed, J.L., Ji, X., Gomez-Lopez, N., Pacora, P., and Romero, R. (2020). MicroRNAs isolated from peripheral blood in the first trimester predict spontaneous preterm birth. *PLoS One* 15.

Winterbottom, E.F., Ban, Y., Sun, X., Capobianco, A.J., Marsit, C.J., Chen, X., Wang, L., Karagas, M.R., and Robbins, D.J. (2019a). Transcriptome-wide analysis of changes in the fetal placenta associated with prenatal arsenic exposure in the New Hampshire Birth Cohort Study. *Environ. Heal. A Glob. Access Sci. Source* 18.

Winterbottom, E.F., Moroishi, Y., Halchenko, Y., Armstrong, D.A., Beach, P.J., Nguyen, Q.P., Capobianco, A.J., Ayad, N.G., Marsit, C.J., Li, Z., Karagas, M.R., and Robbins, D.J. (2019b). Prenatal arsenic exposure alters the placental expression of multiple epigenetic regulators in a sex-dependent manner. *Environ. Heal. A Glob. Access Sci. Source* 18.

Wnuk, A., Rzemieniec, J., Litwa, E., Lasoń, W., and Kajta, M. (2018). Prenatal exposure to benzophenone-3 (BP-3) induces apoptosis, disrupts estrogen receptor expression and alters the epigenetic status of mouse neurons. *J. Steroid Biochem. Mol. Biol.* 182: 106–118.

Wnuk, A., Rzemieniec, J., Staroń, J., Litwa, E., Lasoń, W., Bojarski, A., and Kajta, M. (2019). Prenatal exposure to benzophenone-3 impairs autophagy, disrupts RXRs/PPARγ signaling, and alters epigenetic and post-translational statuses in brain neurons. *Mol. Neurobiol.* 56: 4820–4837.

Wortman, A.C. and Alexander, J.M. (2013). Placenta accreta, increta, and percreta. *Obstet. Gynecol. Clin. North Am. 40 (1): 137–154.*

Yang, S.I., Lee, S.H., Lee, S.Y., Kim, H.C., Bin, K.H., Kim, J.H., Lim, H., Park, M.J., Cho, H.J., Yoon, J., Jung, S., Yang, H.J., Ahn, K., Kim, K.W., Shin, Y.H., Suh, D.I., Won, H.S., Lee, M.Y., Kim, S.H., Choi, S.J., Kwon, J.Y., Jun, J.K., and Hong, S.J. (2020). Prenatal PM2.5 exposure and vitamin D–associated early persistent atopic dermatitis via placental methylation. *Ann. Allergy, Asthma Immunol.* 125: 665–673.e1.

Yin, Y., Liu, M., Yu, H., Zhang, J., and Zhou, R. (2020). Circulating microRNAs as biomarkers for diagnosis and prediction of preeclampsia: A systematic review and meta-analysis. *Eur. J. Obstet. Gynecol. Reprod. Biol. 253: 121–132.*

Zahir, F.R. and Brown, C.J. (2011). Epigenetic impacts on neurodevelopment: Pathophysiological mechanisms and genetic modes of action. *Pediatr. Res.* 69: 92–100.

Zhang, Y., Poobalasingam, T., Yates, L.L., Walker, S.A., Taylor, M.S., Chessum, L., Harrison, J., Tsaprouni, L., Adcock, I.M., Lloyd, C.M., Cookson, W.O., Moffatt, M.F., and Dean, C.H. (2018). Manipulation of dipeptidylpeptidase 10 in mouse and human in vivo and in vitro models indicates a protective role in asthma. *DMM Dis. Model. Mech* 11.

Zhu, W., Gui, W., Lin, X., Yin, X., Liang, L., and Li, H. (2021). Maternal undernutrition modulates hepatic microRNAs expression in the early life of offspring. *Exp. Cell Res.* 400.

Zhu, Y., Mordaunt, C.E., Durbin-Johnson, B.P., Caudill, M.A., Malysheva, O.V., Miller, J.W., Green, R., James, S.J., Melnyk, S.B., Fallin, M.D., Hertz-Picciotto, I., Schmidt, R.J., and LaSalle, J.M. (2021). Expression changes in epigenetic gene pathways associated with one-carbon nutritional metabolites in maternal blood from pregnancies resulting in autism and non-typical neurodevelopment. *Autism Res* 14: 11–28.

Zhu, Y., Mordaunt, C.E., Yasui, D.H., Marathe, R., Coulson, R.L., Dunaway, K.W., Jianu, J.M., Walker, C.K., Ozonoff, S., Hertz-Picciotto, I., Schmidt, R.J., and Lasalle, J.M. (2019). Placental DNA methylation levels at CYP2E1 and IRS2 are associated with child outcome in a prospective autism study. *Hum. Mol. Genet.* 28: 2659–2674.

10

The Role of Dynamic Epigenetic Changes in Modulating Homeostasis after Exposure to Low-dose Environmental Chemicals

Chongli Yuan, Jennifer L. Freeman, Junkai Xie, and Han Zhao

Purdue University, West Lafayette, Indiana

Introduction

The concept of epigenetics, which refers to heritable phenotype changes that do not involve genetic mutations, was first introduced in 1942 by Conrad Hal Waddington and gained increasing popularity in the past few decades. Unlike genetic mutations, which are almost permanently embedded in genome, epigenetic changes such as DNA methylation, histone post-translational modifications, and non-coding RNA processes are subject to dynamic changes at varying timescales. They can act independently and synergistically to respond to environmental challenges such as exposure to toxins; they can adjust cellular metabolomics and transcriptomics for fitness; and they can restore homeostasis. The acquired epigenetic changes, however, do not necessarily get fully restored after the cessation of exposure events, and thus can serve as potential biomarkers of past exposure events. Furthermore, the acquired epigenetic changes may reside on genes that regulate homeostasis and impact cellular fitness to adjust to future environmental challenges and to impose long-term disease risk. This chapter will thus evaluate the dynamic role of epigenetic changes in homeostasis-regulating gene sets and their potential as biomarkers of exposure.

Dynamic Epigenetic Changes after Environmental Exposure

Exposure to low-dose environmental chemicals such as heavy metals and pesticides does not always elicit immediate health outcomes of an adverse nature. Instead, exposure events of this kind can increase individuals' risk for various diseases years later, or even across generations. It has been suggested that epigenetic mechanisms constitute the major modifier that drives these adverse health outcomes later in life, especially when the envrionmental chemical exposures are at non-mutagenic doses. The hypothesis of developmental origins of health and disease (DOHaD) postulates that the window of development and growth presents a critical period of time in which exposure to certain environmental chemicals can have significant consequences for an individual's short-term and long-term health (Barker 2007). Part of this impact can be attributed to the adaptation of organisms to environmental challenges that are subsequently embedded in cellular "memory," and this

Genomic and Epigenomic Biomarkers of Toxicology and Disease: Clinical and Therapeutic Actions,
First Edition. Edited by Saura C. Sahu.

memory fosters different behaviors in response to similar survival challenges if and when such challenges are encountered later in life. It has long been postulated that the components that constitute cellular memory involve persistent changes in the transcriptome and in the epigenome.

DNA Methylation as a Potential Epigenetic Memory of Past Exposure Events

DNA methylation, one of the epigenetic modifications, refers to the covalent attachment of a methyl group to the fifth carbon on the cytosine and mainly on cytosine–phosphate–guanine (CpG) dinucleotides (Bird 2002; Miranda and Jones 2007). Conventionally, DNA hypermethylation at gene promotor regions has been associated with transcriptional suppression (Ng and Bird 1999). The mechanism of transcriptional suppression induced by DNA hypermethylation can be generally explained either through direct interference with the transcription factor binding or through the recruitment of transcriptional repression complexes that contain methyl CpG binding protein 2 (MeCP2) and histone deacetylases (HDACs) (Deaton and Bird 2011; Nan et al. 1998; Suzuki and Bird 2008).

Heavy metal contaminants in soil, air, and surface water are a common environmental concern in the United States. Early-life exposure to heavy metals such as lead (Pb) and arsenic (As) causes changes in gene expression in later life (Senut et al. 2012), and the altered gene expression pattern can be inherited cross-generationally (Lin et al. 2021; Mauro et al. 2016; Sen et al. 2015b; Yu et al. 2013). DNA methylation, as an inheritable epigenetic modification, has shown significant alteration in cases of similar exposure. These findings suggest the involvement of an epigenetic modification or the existence of inheritable changes embedded in the genome that can "record" histories of past exposure to environmental chemicals (Crews et al. 2007). In recent years, researchers have been using cell and animal models to study the impact of early-stage environmental toxins on the epigenome at later stages. For example, low-dose Pb is known to have low mutagenic activity or almost none (Tchounwou et al. 2012; Wise et al. 1993). Human neuroblastoma cell line SH-SY5Y, after being exposed to lead at non-mutagenic doses, showed an immediate decrease in DNA methylation; and the decreasing behavior persisted throughout the entire period of neuron differentiation (Lin et al. 2021). The exposure led to less neuron differentiation efficiency, and the resulting neurons showed shorter lengths of neurites. A similar finding, with persistent decrease in DNA methylation, was reported for arsenic exposure on V79 and HaCaT cells (Mauro et al. 2016). Other exposures to heavy metals such as cadmium, nickel, and chromium have also been reported to alter the DNA methylation level, either globally or at specific genetic loci (Govindarajan et al. 2002; Kondo et al. 2006; Takiguchi et al. 2003).

Aside from heavy metals, organic substances such as atrazine, heavily used as a pesticide in agricultural states like Indiana and Ohio, were detected in surface water owing to their good solubility (Cheremisinoff 2010; Raja et al. 2005). Particularly for agriculture workers, exposure to atrazine was associated with increased risk of various diseases and disorders, for example cancer (Simpkins et al. 2011), reproductive and endocrine disorder (Munger et al. 1997), and neurodegenerative diseases (Song et al. 2015). Persistent epigenome changes after the cessation of exposure have also been observed in animal and human cell line studies. HEK 293 cells exposed to atrazine showed a gradual decrease in global DNA methylation level over the exposing period (https://www.sciencedirect.com/science/article/pii/S0269749119357252). SH-SY5Y cells showed increased global DNA methylation immediately after exposure, but decreased DNA methylation after the period of neuron differentiation (Xie et al. 2021). The resulting neurons showed increased expression of alpha-synuclein and its aggregates A11, hallmarks for Parkinson's disease. Additionally, zebrafish exposed to atrazine at the developmental stage displayed an immediate decrease in global DNA methylation and DNA methyltransferase (DNMT) activities (Sanchez et al. 2016). Atrazine exposure affects not only the exposed generation but also its progeny. For example, the offspring of mice still exhibit significant epigenetic changes—such as decreases in H3K4me3 and DNA methylation changes in F3 (the third generation after exposure to atrazine) (Hao et al. 2016; Mcbirney et al. 2017). Collectively, all these findings suggest the involvement of epigenetic mechanisms in transducing atrazine effects to later stages in one's life. Such mechanisms act as a potential molecular memory of past exposure events.

Epigenetic modifications are dynamic in nature. A single epigenetic modification is typically co-regulated by "reader" and "eraser" proteins and is thus conceptually reversible. DNA methylation is synergistically modulated by DNMT and by methyl-cytosine dioxygenase (TET). In general, metals are known to increase the production of reactive oxygen species (ROS) via redox cycling (Fowler et al. 2004; Shen and Dryhurst 1998). The oxidative DNA damage can interfere with the interaction between methyltransferases and DNA (Valinluck et al. 2004), thus producing an alteration of cytosine methylation (Turk et al. 1995). The work by Lin et al. (2021) showed that the transcription level of DNMTs such as DNMT1, DNMT3A, and DNMT3B, and also that of demethylase TET1 in SH-SY5Y cells, can be altered after exposure to lead. *In vitro* studies have been carried out to show that Pb can inhibit DNMT1 activity (~33%) as a non-competitive inhibitor and can significantly lower the V_{max} and K_{cat} of corresponding DNMTs (Sanchez et al. 2016). Similarly, arsenic was reported to inhibit DNMT activity by 40% via similar mechanisms (Zhao et al. 1997). Apart from DNMT activity inhibition, arsenic induces the upregulation of arsenic methyltransferase (AS3MT), competing with DNMT for SAM as methyl donor and accumulating SAH, which further regulates SAM-dependent methyltransferase activity negatively, through feedback inhibition (Reichard and Puga 2010).

Exposure to organic chemicals such as atrazine can elicit immediate changes in the epigenome by affecting epigenetic enzymes as well. A previous study reported that atrazine exposure can inhibit DNMTs activity via non-competitive inhibition (Wirbisky-hershberger et al. 2017). Other studies, carried out on the carp, further suggested that the expressions of DNMT1 and of MBD2 were respectively up- and downregulated after atrazine exposure, which subsequently caused a global hypomethylation of DNA (Wang et al. 2014; Xing et al. 2015).

Histone Heterochromatin Markers and Their Modulators, Which Reshape the Epigenetic Landscape in Response to Past Exposure Events

Other than influencing DNA methylation, histone post-translational modifications dictate the transcriptional state of genes; they do this partly by modulating chromatin accessibility (Gibney and Nolan 2010; Klemm et al. 2019) and thus turning selected genes "on" or "off." The effects of these histone post-translational modifications on gene expression are diverse; they range from H3K4me3 and H3K27ac marked euchromatin for gene activation (Wysocka et al. 2006) and H3K27me3 marked facultative heterochromatin for gene repression (Wiles and Selker 2017) to H3K9me3 marked heterochromatin for permanent gene silencing (Mozzetta et al. 2015). In the previous section we have mentioned several studies that reported a DNA methylation change in response to environmental chemical exposure. Similarly, histone heterochromatin markers have shown significant alteration after exposure and tend to persist. When SH-SY5Y cells were exposed to lead and atrazine, the level of H3K27me3 and H3K9me3 changed immediately after exposure, and the change either persists or overcompensates through the differentiation state (Lin et al. 2021; Xie et al. 2021).

In an Alzheimer's disease (AD) mouse model, the global level of H3K27me3 was significantly elevated after developmental exposure to lead, and the level persisted throughout the lifespan of the animal (Eid et al. 2016). On the basis of their functions in gene regulation, H3K9me3 and H3K27me3 were commonly used as heterochromatin markers that represented densely compact regions in nucleus.

The level of histone lysine methylation, just like that of DNA methylation, is modulated by its corresponding "writers" and "erasers." An H3K9me3 eraser such as histone lysine demethylase 4A (KDM4A) was reported to be upregulated 1.52 times, and H3K27me3 erasers such as histone lysine demethylase 6A and 6B (KDM6A and KDM6B) showed more than a three-fold increase in SH-SY5Y cells after exposure to lead. When exposed to atrazine, these histone-modulating enzymes showed a different behavior from the one exhibited in Pb exposure. Exposure to atrazine at 3 ppb resulted in an approximately 11.5-fold decrease in KMT1A before differentiation; but after differentiation KDM4A showed a 1.4-fold increase in transcription, at 30 ppb (Lin et al. 2021). The dynamic balance of theses epigenetic modulators facilitates the establishment of the epigenetic landscape when cells encounter an environmental chemical challenge and helps them maintain and form epigenetic memories in the later stage.

Collectively these results suggest that epigenetic modifications are involved in the process of adaptation to environmental challenges. Transcriptomic changes tend to be set back to the original state after the removal of the exposure source, but changes in the epigenome may persist, and they can subsequently increase the susceptibility of cells to future challenges, as illustrated in Figure 10.1.

The Epigenetic Regulation of Homeostasis

The Epigenetic Regulation of Metabolic Homeostasis

Metabolic rewiring is an essential cellular mechanism of adaptation to external stimuli. The epigenetic mechanism integrates environmental stress with transcriptional regulation, which serves as another layer of cellular adaptation. These two layers of the adaptation mechanism are interconnected, having potentially a feedback loop. Metabolic activity

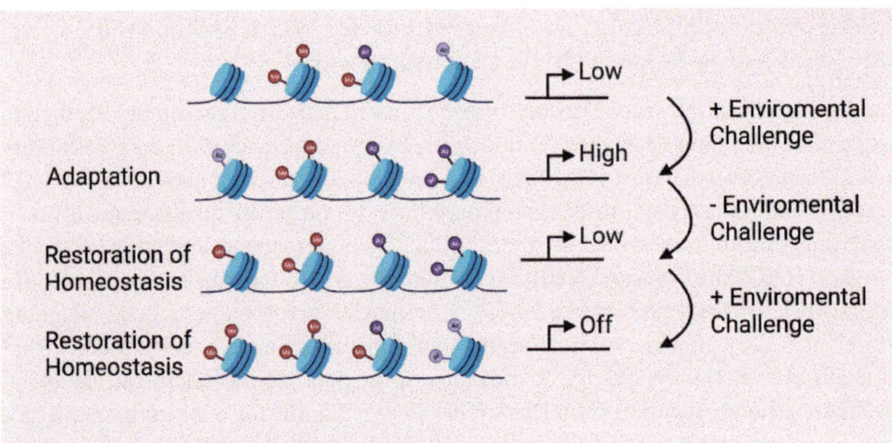

Figure 10.1 The potential roles of epigenetic modifications in adaptation to environmental challenges and in the restoration of homeostasis.

regulates the abundance of epigenetic enzyme co-factors, thus controlling the rate of epigenetic modification reaction. In turn, the epigenetic control of metabolic enzyme expression regulates metabolic homeostasis. In this section we will discuss how different types of epigenetic modifications regulate metabolic homeostasis; and we will do this by focusing on histone acetylation, histone methylation, and DNA methylation.

A Major Hypothesis in the Histone Acetylation Regulation of Metabolism

Histone acetylation, a highly dynamic post-translational marker with a turnover half-life of about two to forty minutes (Meagher 2014), is one of the best studied epigenetic modifications associated with metabolism. Histone acetyltransferases (HATs) add the acetyl group from acetyl-CoA to histone tails. Most of the acetyl-CoA is generated through the oxidation of glycolytic pyruvate, free fatty acid and, branched-chain amino acids (Feron 2019). It is well established that the intracellular level of acetyl-CoA, which is dependent on cell energy production, regulates histone acetylation and gene expression (Feron 2019). How histone acetylation regulates acetyl-CoA production, however, remains less understood. One of the prevailing hypotheses is that histone acetylation on metabolic enzyme encoding genes regulates transcription and further affects metabolic homeostasis, including lipid synthesis, glucose metabolism, and Ca^{2+} signaling. For example, decrease in the expression of acetyl-CoA carboxylase (ACC) was found to be regulated by histone acetylation and to induce a reduction in glycolysis activity in cancer cells (Corbet et al. 2016). P300 acetyltransferase regulates fatty acid synthase expression and increases lipid metabolism in prostate cancer cells (Gang et al. 2016). NAD^+-dependent deacetylase SIRT6 regulates glucose homeostasis by controlling the expression of key glucose metabolic genes (Zhong et al. 2010). In neurons, the calcium ion influx through NMDA receptors on the post-synapse results in a rapid increase of H3K9ac and decrease of H3K9me3 on the BDNF exon3 promotor and in a calcium-dependent phosphorylation of MeCP2 through calcium- or calmodulin-dependent protein kinase (CamK); and these processes lead to gene activation (Chen et al. 2003). In developmental and aging processes in mammals, the interplay between histone acetylation and metabolism is less defined. A decrease in both acetyl-CoA and histone acetylation levels was observed during the early aging of flies (Peleg et al. 2016). A perturbation of the acetyl-CoA synthesizing enzyme or an inhibition of HATs successfully alleviates aging-associated phenotypes and increases longevity in flies, indicating the essential role of histone acetylation and of the acetyl-CoA metabolism interaction during aging (Peleg et al. 2016).

Mitochondria, as major energy powerhouses of cells, are also subject to the regulation of histone acetylation. Each mitochondrion has its own membrane, its own DNA, and its own core protein machinery. ATP production and metabolic homeostasis are achieved through the electron transport chain (ETC) protein on the mitochondrial inner membrane, and they are encoded both by the mitochondrial DNA and by the nucleus DNA. It is not surprising that the nuclear epigenome affects mitochondrial activity and metabolic rewiring. Interestingly, the sirtuin-family protein Sirt3, HDAC, which is dependent on NAD^+, can translocate from the nucleus to mitochondria in response to cellular stress (Scher et al. 2007). Therefore it is also hypothesized that the translocation of HDACs in different cellular compartments, namely the nucleus and the mitochondria, maintains cellular homeostasis through a balance between the level of histone acetylation (i.e., the transcription level) and the level of mitochondrial protein acetylation (i.e., the mitochondrial activity level). Another competing hypothesis is that, instead of regulating the downstream gene expression, histone acetylation regulates cellular metabolism by maintaining the intracellular acetate pool through dynamic acetylation and deacetylation events. Specifically, extracellular pH decrease can induce a global decrease in histone acetylation as well as an increase of HDAC-derived acetate anion and proton excretion, which prevent the further acidification of intracellular content (Mcbrian et al. 2013).

Histone Methylation-Regulated Metabolic Enzymes are Cell Type-Dependent

Histone methylation is another well-characterized histone post-translational modification, which can activate or suppress transcription. The addition of the methyl group to histone lysine or arginine residues is catalyzed by histone methyltransferase (HMTs), using SAM as a methyl donor. SAM is generated through a methionine–homocysteine cycle in cytosol, which is dependent on a folate cycle and on ATP generated from mitochondria. The removal of the methyl group is catalyzed by histone demethylases (HDMs). HDMs consist of two major groups: lysine-specific demethylases (LSDs) and Jumonji-C domains (JmjCs). Both LSD and JmjC activities are TCA cycle-related and regulated by metabolites such as flavin adenine dinucleoride (FAD) and α-ketoglutarate (α-KG). Therefore mitochondria metabolism regulates histone methylation by controlling the activity of KDMs and HMTs via metabolite abundance. In turn, histone methylation can regulate mitochondrial activity. Dysregulation of HDMs and HMTs has been widely proved to affect mitochondrial metabolism. For example, the inhibition of SETD7 increases oxidative stress and triggers the mitochondria's antioxidant function through the upregulation of proinflammatory and antioxidant genes (He et al. 2015). The overexpression of LSD1 promotes oxidative metabolism in fat tissue (Duteil et al. 2014), while others report that the inhibition of LSD1 reduces glycolysis and promotes mitochondrial respiration in cancer cells (Sakamoto et al. 2015). These results indicate that histone methylation regulates metabolic enzymes distinctively in different cell types, depending on the basal energetic state of the cell.

DNA Methylation Connects the Epigenome in the Nucleus and the Epigenome in Mitochondria

DNA methylation, specifically 5-methylcytosine (5mC), is another important epigenetic marker that accumulates and reflects past exposure events and events that occur during aging. The process of accumulation and change of DNA methylation is generally known as "epigenetic memory" and "epigenetic clock." As with histone methylation, the addition of the methyl group to CpG sites is catalyzed by DNMTs using SAM as a methyl donor. Analogy from histone methylation can be made that the 5mC level is affected by the intracellular methyl donor pool, which is regulated by the methionine cycle in conformity with mitochondrial metabolism. In turn, several nucleus-encoded mitochondrial genes have been found to be regulated by DNA methylation: *ACCβ* (Oh et al. 2005), the low-density lipoprotein receptor (*LPL*) (Guay et al. 2013), and carnitine palmitoyltransferase 1A (*CPT1A*) (Irvin et al. 2014).

A mitochondrion contains its own DNA (mtDNA), which encodes part of the ETC complex proteins. mtDNA is also subject to methylation, which leaves its regulatory role largely unknown. Evidence has been shown of a correlation between mtDNA methylation and its encoded gene transcription and mtDNA replication (Sun et al. 2018; Tong et al. 2017). On the basis of this correlation, one can easily guess that the mitochondrial function can be partly regulated through mtDNA methylation. The link between mtDNA methylation and mitochondrial respiration, however, is still missing. It is also worth noticing that epigenetic enzyme-regulating DNA methylation, including DNMTs (DNMT1, DNMT3A and DNMT3B) and ten-eleven translocase 1/2 (TET1/2), is also found in mitochondria with low concentration (Bellizzi et al. 2013; Chen et al. 2012; Shock et al. 2011). There has been little evidence as to whether and how the redistribution of these epigenetic enzymes into different cellular compartments, specifically the nucleus and the mitochondria, would be affected by external stress and would cause changes in both the nuclear epigenome and mitochondrial metabolism.

In conclusion, the epigenome's regulation of metabolic homeostasis can be achieved through various mechanisms: (1) metabolic enzyme transcription change induced via histone acetylation, histone methylation, and DNA methylation; (2) maintenance of the intracellular pool of epigenetic enzyme co-factors by dynamic acetylation and deacetylation events; and (3) translocation of the epigenetic enzyme from the nucleus to mitochondria, a process that regulates mitochondrial respiration-related gene expression and activity.

Epigenetic Regulation of Autophagy

The maintenance of a healthy pool of organelles is essential for cell survival, especially in response to external stimuli. Decrease in protein turnover rates and the accumulation of damaged organelles are characteristic features of cell aging. The accumulation of DNA damage and misregulated protein expression are also common features of environmental exposure. As a quality control mechanism, cells adopt a catabolic pathway, autophagy, in order to degrade damaged cytoplasmic material and maintain organelle-level homeostasis. But, despite the role of autophagy in promoting cell survival, its excessive activation can also be considered a cell death pathway (Pesonen and Vähäkangas 2019). Therefore a balanced regulation of autophagy is essential for normal cell function at various stages, including development, differentiation, and aging. In this section we focus on the epigenetic regulation of autophagy.

The perturbation of HDACs can lead to the activation or suppression of autophagy in a cell-type dependent manner. For example, HDAC inhibitors promote autophagy in neuroblastoma (Körholz et al. 2021), while the opposite trend was observed in cardiac cells (Cao et al. 2011). The regulation of autophagy by HDAC6 is due to translocation into cytosol and binding to ubiquitinated autophagic machinery (Kawaguchi et al. 2003). Except for the cytosolic acetylation activity of HDACs toward specific autophagic pathways, the role of nuclear HDACs in regulating autophagy via histone acetylation is unclear.

Histone methylation and DNA methylation play different roles in autophagy by comparison to histone acetylation. Given the high specificity of HMTs and HDMs, histone methylation regulates autophagy via transcription regulation of the autophagy machinery. Drug inhibition and the knockdown of G9a induce LC3B expression via promoter binding; thus they promote autophagy (Artal-Martinez de Narvajas et al., 2013). The inhibition of LSD1 induces autophagy through the upregulation of mTORC in various cancer cell lines, including neuroblastoma and ovarian cancer (Ambrosio et al. 2017; Feng et al. 2016). Metabolic stress, specifically methionine deprivation, results in a decrease of intracellular SAM concentration and a reduction of H3K4me3 and DNA methylation; thus it triggers p53 signaling in human embryonic stem cell and the loss of pluripotency (Shiraki et al. 2014).

In conclusion autophagy, as a cellular organelle quality control mechanism, maintains its homeostasis during various cellular events such as differentiation, aging, and stress response. The epigenetic regulation of autophagy is achieved in multiple ways, for example (1) through translocation of an epigenetic enzyme into cytosol and direct interaction with the autophagy machinery; (2) through transcriptional regulation of the autophagy machinery; and (3) through transcriptional regulation of upstream stress-sensing pathways.

Epigenetic Regulation of Calcium Signaling

As an important intracellular secondary messenger, calcium regulates many biological processes and gene transcription (Naranjo and Mellström 2012; West et al. 2001). Cellular calcium level changes in response to factors such as extracellular environmental cues and epigenetic modifications, which further impact on the output of gene expression. In cardiomyocytes, an increased intracellular calcium level can lead to histone hyperacetylation on H3K9, K14, K27, and H4K16 by disrupting the class II HDAC to HAT (CBP and p300) balance in the nucleus (Sharma et al. 2014). Sarcoplasmic reticulum Ca^{2+}- ATPase (SERCA2a) is essential for Ca^{2+} homeostasis, regulates cardiac functions in cardiomyocytes, and is rich in CpG islands. Treated with hydrazine, which is a DNMT inhibitor, cardiomyocytes showed decreased promotor methylation in SERCA2, which increased its expression (Kao et al. 2011). The demethylation of SERCA showed increased activity and further increased intracellular calcium transiently—namely increased by more than 79% (Kao et al. 2011). In neurons, calcium ion influx through NMDA receptors on the post-synapse results in rapid increase of H3K9ac; and decrease of H3K9me3 on the BDNF exon3 promotor and calcium-dependent phosphorylation of MeCP2 through calcium- or calmodulin-dependent protein kinase (CamK), a process that lead to gene activation (Chen et al. 2003).

To summarize, the epigenetic regulation of homeostasis does not rely only on the direct transcriptional regulation of homeostasis-related genes. The translocation and the functional redirection of epigenetic enzymes also play an important role in various types of homeostasis.

Epigenetic Alterations Enriched in Homeostasis Pathways as a Potential Biomarker after Environmental Exposure

Homeostasis, or the state of plasticity to environmental challenges, is not necessarily a static event. Most homeostasis events are maintained by a balanced force, as detailed in Section 3, and thus retain the ability to restore normal cell functionality after exposure to low-dose environmental chemicals. The exposure events frequently leave a trace behind in the form of abnormal epigenetic modifications and make the system more susceptible to systematic failures when encountering the next challenge event, as illustrated in Figure 10.2. In this section, we will go through two examples to demonstrate epigenetic traces left on homeostasis-regulating genes.

Pb exposure is known to alter Ca^{2+} homeostasis, given the structural and valency similarity between the two substances. In addition to directly interfering with calcium transportation, binding, and storage, Pb exposure can alter the cell functions required for calcium homeostasis, thereby creating longer-term and potentially transgenerational health risks (Pounds 1984). The Ca^{2+} homeostatic system consists of a collection of Ca^{2+} pumps, channels, and downstream receptor systems that include calmodulin-binding proteins and protein kinase C (Pounds et al. 1991). Pb exposure can lead to

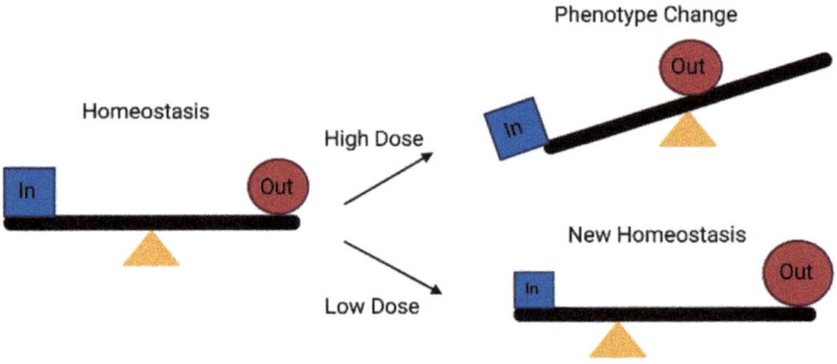

Figure 10.2 A schematic illustration of how low-dose environmental exposure may alter the state of homeostasis and the plasticity of the system. Example: in: Ca^{2+} transporter or ROS production; out: calmodulin proteins or ROS scavenger.

transcriptional changes in calmodulin-related genes, particularly kinases (Li et al. 2015; Schneider et al. 2012; Toscano et al. 2005) and NMDA receptors (Nihei and Guilarte 1999; Toscano et al. 2002; Zhang et al. 2002); and these, collectively, can cause abnormality in intracellular Ca^{2+} concentrations. The erythrocytes of lead-exposed workers have an intracellular concentration of Ca^{2+} of about 79 nM, which is more than twice the amount detected in the unexposed control group (Quintanar-escorza et al. 2007), although no immediate adverse health consequences were reported in these workers. Pb exposure is also a known trigger of aberrant DNA methylation changes, as discussed in Section 2. Whole-genome DNA methylation was performed on Pb-exposed mice and verified significant methylome alterations with prenatal or developmental exposure (Dosunmu et al. 2012; Dou et al. 2019). Similar observations were also made in human stem cells (Sen et al. 2015a) and population studies (Zhang et al. 2019). Among these alterations, calcium-signaling pathways and protein kinase C were commonly involved (Dosunmu et al. 2012; Zhang et al. 2019). Furthermore, Pb-induced alterations in Ca^{2+} homeostasis can be passed across generations. A recent work by Meyer et al. (2020) has revealed transcriptomic alterations in F2 after a developmental exposure to Pb at 10 μM using a zebrafish animal model (Meyer et al. 2020). Among these alterations, *camkk1a*, a calcium- or calmodulin-dependent protein kinase (kinase 1, alpha a) and a regulator of Ca^{2+}-signaling cascade, exhibits altered expression, along with *mapk1* (Meyer et al. 2020). Epigenetic regulating enzymes, including *dnmt3b*, *hdac4*, and *mecp2*, also exhibited significantly altered transcription, in a sex-dependent manner (Meyer et al. 2020). A controlled case study examined alterations in the DNA methylation profiles of neonatal blood cross generations with maternal exposure to Pb and found persistent DNA methylation alterations in selected genes, including *TRPV2*, a non-specific ion channel that is permeable to Ca^{2+} (Sen et al. 2015a). Collectively, these findings suggest that chronic and low-dose Pb exposure can alter the "equilibrium" point of calcium homeostasis by, or concurrently with, changes in the transcription or epigenetics of the genes that regulate calcium homeostasis. These alterations can persist across the lifespan and are potentially transmitted across generations, resulting in changed plasticity in adaptations to future challenges that involve Ca^{2+} homeostasis.

Energy production is fundamental to all living organisms, and is thus dynamically regulated to meet the varying metabolic demand during different developmental stages. Mitochondria are the main energy production center in eukaryotes; they use ETCs embedded in their inner membrane. Among these ETCs, Complex I is mainly responsible for the production of ATP. Complex I is also the major site of superoxide production. Superoxide is one of the ROSs that cause oxidative damages to various cellular components, including mitochondrial and nuclear DNA. ROS levels were co-regulated by an array of factors such as hypoxia-responsive proteins, growth factors (e.g., PI3K), and kinases (e.g., MAPK). These factors, along with the mitochondria, synergistically determine ROS homeostasis and shape cellular responses to oxidative stress. Many environmental chemicals are known to induce ROS, for example methyl-mercury, manganese, and rotenone (Ferrer et al. 2021; Neely et al. 2017; Reardon and Bhat 2007). At high doses, ROS can elicit acute cellular responses by causing DNA damage and can induce apoptosis, while at low doses it can alter the cellular response to oxidative stress in a persistent manner; hence it will be the focus of our discussion here.

Rotenone is a well-studied pesticide that targets specifically Complex I by impairing cellular respiration and by increasing superoxide production. Transcriptomic changes were characterized in rotenone-exposed rats, and significant upregulation was found in genes known to be induced in response to oxidative stress in the liver, the heart, and the brain stem (Heinz et al. 2017). Similar observations were made on a human brain spheroid model (Clapham 2007). Rotenone exposure has also been shown to alter epigenetics via altering chromatin organization, H3K27ac landscape, and DNA methylation (Freeman et al. 2020; Freeman and Wang 2020). A recent work by Lozoya et al. (2020) assessed the effects of developmental exposure to rotenone on the liver of mice, specifically on the DNA methylome and on tolerance to oxidative stress during the life span. The exposed group exhibited persistent mitochondria dysfunction along with lasting gene expression and DNA methylation alterations, although no changes in the expression or DNA methylation of oxidative phosphorylation (OXPHOS) was observed (Lozoya et al. 2020). Similar observations were made in drosophila, where paternal exposure to rotenone can result in metabolic alterations in the offspring via epigenetic-dependent inheritance (Seong et al. 2020). These collective findings seem to suggest that rotenone alters mitochondrial function, then the epigenome, potentially through the pathways detailed in Section 3, and cumulatively produces long-lasting impairments in anti-oxidant activity that are potentially inheritable.

It is also worth noting that various homeostasis pathways also correlate with one another. For example, Ca^{2+} homeostasis is known to regulate mitochondria fitness (Clapham 2007; Rizzuto et al. 2000). Interestingly, persistent changes in transcription or in epigenetics seem to occur primarily on genes that "consume" the target molecules, such as ROS-responsive genes in rotenone-exposed models and protein kinase C in Pb-exposed models. Further studies are thus warranted in the future to comprehensively assess the role of these downstream regulators of homeostasis and to help us understand the cross-talking between overlapping homeostatic pathways.

Conclusions

An epigenetic biomarker based on homeostasis-regulating genes can serve to assess the individual risk of developing diseases at a later stage in life or of transmitting disease risks across generations. Although the accumulating evidence supports this claim, more comprehensive and targeted studies are needed to identify highly selective biomarkers, which should be exposure- and disease-specific. This calls for synergestic efforts in the toxicolcoy and epigenetic community.

References

Ambrosio, S., Saccà, C.D., Amente, S., Paladino, S., Lania, L., and Majello, B. (2017). Lysine-specific demethylase LSD1 regulates autophagy in neuroblastoma through SESN2-dependent pathway. *Oncogene*. 36: 6701–6711.

Artal-Martinez De Narvajas, A., Gomez Timothy, S., Zhang, J.-S., Mann Alexander, O., Taoda, Y., Gorman Jacquelyn, A., Herreros-Villanueva, M., Gress Thomas, M., Ellenrieder, V., Bujanda, L., Kim, D.-H., Kozikowski Alan, P., Koenig, A., and Billadeau Daniel, D. (2013). Epigenetic regulation of autophagy by the methyltransferase G9a. *Mol. Cell. Biol.* 33: 3983–3993.

Barker, D.J. (2007). The origins of the developmental origins theory. *J. Intern. Med.* 261: 412–417.

Bellizzi, D., d'Aquila, P., Scafone, T., Giordano, M., Riso, V., Riccio, A., and Passarino, G. (2013). The control region of mitochondrial DNA shows an unusual CpG and non-CpG methylation pattern. *DNA Res.* 20: 537–547.

Bird, A. (2002). DNA methylation patterns and epigenetic memory. *Genes Dev.* 16: 6–21.

Cao, D.J., Wang, Z.V., Battiprolu, P.K., Jiang, N., Morales, C.R., Kong, Y., Rothermel, B.A., Gillette, T.G., and Hill, J.A. (2011). Histone deacetylase (HDAC) inhibitors attenuate cardiac hypertrophy by suppressing autophagy. *Proc. Natl. Acad. Sci.* 108: 4123.

Chen, H., Dzitoyeva, S., and Manev, H. (2012). Effect of aging on 5-hydroxymethylcytosine in the mouse hippocampus. *Restor. Neurol. Neurosci.* 30: 237–245.

Chen, W.G., Chang, Q., Lin, Y., Meissner, A., West, A.E., Griffith, E.C., Jaenisch, R., and Greenberg, M.E. (2003). Derepression of BDNF transcription involves calcium-dependent phosphorylation of MeCP2. *Science* 302: 885–889.

Cheremisinoff, N.P. (2010). *Handbook of Pollution Prevention and Cleaner Production, vol. 3: Best Practices in the Agrochemical Industry*. Amsterdam: William Andrew.

Clapham, D.E. (2007). Calcium signaling. *Cell* 131: 1047–1058.

Corbet, C., Pinto, A., Martherus, R., Santiago De Jesus, J.P., Polet, F., and Feron, O. (2016). Acidosis drives the reprogramming of fatty acid metabolism in cancer cells through changes in mitochondrial and histone acetylation. *Cell Metab.* 24: 311–323.

Crews, D., Gore, A.C., Hsu, T.S., Dangleben, N.L., Spinetta, M., Schallert, T., Anway, M.D., and Skinner, M.K. 2007. Transgenerational epigenetic imprints on mate preference. *Proc. Natl. Acad. Sci. USA* 104: 5942–5946.

Deaton, A.M. and Bird, A. (2011). CpG islands and the regulation of transcription. *Genes Dev.* 25: 1010–1022.

Dosunmu, R., Alashwal, H., and Zawia, N.H. (2012). Genome-wide expression and methylation profiling in the aged rodent brain due to early-life Pb exposure and its relevance to aging. *Mech. Ageing Dev.* 133: 435–443.

Dou, J.F., Farooqui, Z., Faulk, C.D., Barks, A.K., Jones, T., Dolinoy, D.C., and Bakulski, K.M. (2019). Perinatal lead (Pb) exposure and cortical neuron-specific DNA methylation in male mice. *Genes.* 10: 274.

Duteil, D., Metzger, E., Willmann, D., Karagianni, P., Friedrichs, N., Greschik, H., Günther, T., Buettner, R., Talianidis, I., Metzger, D., and Schüle, R. (2014). LSD1 promotes oxidative metabolism of white adipose tissue. *Nat. Commun.* 5: 4093.

Eid, A., Bihaqi, S.W., Renehan, W.E., and Zawia, N.H. (2016). Developmental lead exposure and lifespan alterations in epigenetic regulators and their correspondence to biomarkers of Alzheimer's disease. *Alzheimer's & Dementia (Amsterdam, Netherlands).* 2: 123–131.

Feng, S., Jin, Y., Cui, M., and Zheng, J. (2016). Lysine-specific demethylase 1 (LSD1) inhibitor S2101 induces autophagy via the AKT/mTOR pathway in SKOV3 ovarian cancer cells. *Med. Sci. Monit. Int. Med. J. Exp. Clin. Res.* 22: 4742–4748.

Feron, O. (2019). The many metabolic sources of acetyl-CoA to support histone acetylation and influence cancer progression. *Ann. Transl. Med.* 7: S277–S277.

Ferrer, B., Suresh, H., Santamaria, A., Rocha, J.B., Bowman, A.B., and Aschner, M. (2021). The antioxidant role of STAT3 in methylmercury-induced toxicity in mouse hypothalamic neuronal GT1-7 cell line. *Free Radic. Biol. Med.* 171: 245–259.

Fowler, B.A., Whittaker, M.H., Lipsky, M., Wang, G., and Chen, X.Q. (2004). Oxidative stress induced by lead, cadmium and arsenic mixtures: 30-day, 90-day, and 180-day drinking water studies in rats: An overview. *Biometals.* 17: 567–568.

Freeman, D.M., Lou, D., Li, Y., Martos, S.N., and Wang, Z. (2020). The conserved DNMT1-dependent methylation regions in human cells are vulnerable to neurotoxicant rotenone exposure. *Epigenet. Chromatin.* 13: 17.

Freeman, D.M. and Wang, Z. (2020). Epigenetic vulnerability of insulator CTCF motifs at Parkinson's disease-associated genes in response to neurotoxicant rotenone. *Front Genet.* 11: 627–627.

Gang, X., Yang, Y., Zhong, J., Jiang, K., Pan, Y., Karnes, R.J., Zhang, J., Xu, W., Wang, G., and Huang, H. (2016). P300 acetyltransferase regulates fatty acid synthase expression, lipid metabolism and prostate cancer growth. *Oncotarget*. 7: 15135–15149.

Gibney, E.R. and Nolan, C.M. (2010). Epigenetics and gene expression. *Heredity*. 105: 4–13.

Govindarajan, B., Klafter, R., Miller, M.S., Mansur, C., Mizesko, M., Bai, X., Lamontagne, K., and Arbiser, J.L. (2002). Reactive oxygen-induced carcinogenesis causes hypermethylation of p16 Ink4a and activation of MAP kinase. *Mol. Med.* 8: 1–8.

Guay, S.P., Brisson, D., Lamarche, B., Marceau, P., Vohl, M.C., Gaudet, D., and Bouchard, L. (2013). DNA methylation variations at CETP and LPL gene promoter loci: New molecular biomarkers associated with blood lipid profile variability. *Atherosclerosis*. 228: 413–420.

Hao, C., Gely-Pernot, A., Kervarrec, C., Boudjema, M., Becker, E., Khil, P., Tevosian, S., Jégou, B., and Smagulova, F. (2016). Exposure to the widely used herbicide atrazine results in deregulation of global tissue-specific RNA transcription in the third generation and is associated with a global decrease of histone trimethylation in mice. *Nucleic Acids Res.* 44: 9784–9802.

He, S., Owen, D.R., Jelinsky, S.A., and Lin, -L.-L. (2015). Lysine methyltransferase SETD7 (SET7/9) Regulates ROS Signaling through mitochondria and NFE2L2/ARE pathway. *Sci. Rep.* 5: 14368.

Heinz, S., Freyberger, A., Lawrenz, B., Schladt, L., Schmuck, G., and Ellinger-Ziegelbauer, H. (2017). Mechanistic investigations of the mitochondrial complex I inhibitor rotenone in the context of pharmacological and safety evaluation. *Sci. Rep.* 7: 45465-45465.

Irvin, M.R., Zhi, D., Joehanes, R., Mendelson, M., Aslibekyan, S., Claas, S.A., Thibeault, K.S., Patel, N., Day, K., Jones, L.W., Liang, L., Chen, B.H., Yao, C., Tiwari, H.K., Ordovas, J.M., Levy, D., Absher, D., and Arnett, D.K. (2014). Epigenome-wide association study of fasting blood lipids in the genetics of lipid-lowering drugs and diet network study. *Circulation*. 130: 565–572.

Kao, Y.-H., Cheng, -C.-C., Chen, Y.-C., Chung, -C.-C., Lee, T.-I., Chen, S.-A., and Chen, Y.-J. (2011). Hydralazine-induced promoter demethylation enhances sarcoplasmic reticulum Ca2+-ATPase and calcium homeostasis in cardiac myocytes. *Lab. Invest.* 91: 1291–1297.

Kawaguchi, Y., Kovacs, J.J., Mclaurin, A., Vance, J.M., Ito, A., and Yao, T.-P. (2003). The deacetylase HDAC6 regulates aggresome formation and cell viability in response to misfolded protein stress. *Cell*. 115: 727–738.

Klemm, S.L., Shipony, Z., and Greenleaf, W.J. (2019). Chromatin accessibility and the regulatory epigenome. *Nat. Rev. Genet.* 20: 207–220.

Kondo, K., Takahashi, Y., Hirose, Y., Nagao, T., Tsuyuguchi, M., Hashimoto, M., Ochiai, A., Monden, Y., and Tangoku, A. (2006). The reduced expression and aberrant methylation of p16INK4a in chromate workers with lung cancer. *Lung Cancer*. 53: 295–302.

Körholz, K., Ridinger, J., Krunic, D., Najafi, S., Gerloff, X.F., Frese, K., Meder, B., Peterziel, H., Vega-Rubin-de-Celis, S., Witt, O., and Oehme, I. (2021). Broad-spectrum HDAC inhibitors promote autophagy through FOXO transcription factors in neuroblastoma. *Cells* 10.

Li, S., Liu, X.L., Zhou, X.L., Jiang, S.J., and Yuan, H. (2015). Expression of calmodulin-related genes in lead-exposed mice. *Interdiscip. Toxicol.* 8: 155–158.

Lin, L.F., Xie, J., Sánchez, O.F., Bryan, C., Freeman, J.L., and Yuan, C. (2021). Low dose lead exposure induces alterations on heterochromatin hallmarks persisting through SH-SY5Y cell differentiation. *Chemosphere*. 264: 128486.

Lozoya, O.A., Xu, F., Grenet, D., Wang, T., Grimm, S.A., Godfrey, V., Waidyanatha, S., Woychik, R.P., and Santos, J.H. (2020). Single nucleotide resolution analysis reveals pervasive, long-lasting DNA methylation changes by developmental exposure to a mitochondrial toxicant. *Cell Rep.* 32: 108131.

Mauro, M., Caradonna, F., and Klein, C.B. (2016). Dysregulation of DNA methylation induced by past arsenic treatment causes persistent genomic instability in mammalian cells. *Environ. Mol. Mutagen.* 57: 137–150.

Mcbirney, M., King, S.E., Pappalardo, M., Houser, E., Unkefer, M., Nilsson, E., Sadler-Riggleman, I., Beck, D., Winchester, P., and Skinner, M.K. (2017). Atrazine induced epigenetic transgenerational inheritance of disease, lean phenotype and sperm epimutation pathology biomarkers. *PLOS ONE*. 12: e0184306.

Mcbrian, M.A.T.T.H.E.W.A., Behbahan, I.M.A.N.S., Ferrari, R., Su, T., Huang, T.-W., Li, K., Hong, C.A.N.D.I.C.E.S., Christofk, H.E.A.T.H.E.R.R., Vogelauer, M., Seligson, D.A.V.I.D.B., and Kurdistani, S.I.A.V.A.S.H.K. (2013). Histone acetylation regulates intracellular pH. *Mol. Cell.* 49: 310–321.

Meagher, R.B. (2014). "Memory and molecular turnover," 30 years after inception. *Epigenet. Chromatin.* 7: 37.

Mendonca, A., Sánchez, O.F., Xie, J., Carneiro, A., Lin, L., and Yuan, C. (2021). Identifying distinct heterochromatin regions using combinatorial epigenetic probes in live cells. *Biochimica Et Biophysica Acta (BBA) Gene Regul. Mech.* 1864: 194725.

Meyer, D.N., Crofts, E.J., Akemann, C., Gurdziel, K., Farr, R., Baker, B.B., Weber, D., and Baker, T.R. (2020). Developmental exposure to Pb2+ induces transgenerational changes to zebrafish brain transcriptome. *Chemosphere*. 244: 125527.

Miranda, T.B. and Jones, P.A. (2007). DNA methylation: The nuts and bolts of repression. *J. Cell. Physiol.* 213: 384–390.

Mozzetta, C., Boyarchuk, E., Pontis, J., and Ait-Si-Ali, S. (2015). Sound of silence: The properties and functions of repressive Lys methyltransferases. *Nat. Rev. Mol. Cell Biol.* 16: 499–513.

Munger, R., Isacson, P., Hu, S., Burns, T., Hanson, J., Lynch, C.F., Cherryholmes, K., Van Dorpe, P., and Hausler, W.J., JR (1997). Intrauterine growth retardation in Iowa communities with herbicide-contaminated drinking water supplies. *Environ. Health Perspect.* 105: 308–314.

Nan, X., Ng, H.H., Johnson, C.A., Laherty, C.D., Turner, B.M., Eisenman, R.N., and Bird, A. (1998). Transcriptional repression by the methyl-CpG-binding protein MeCP2 involves a histone deacetylase complex. *Nature*. 393: 386–389.

Naranjo, J.R. and Mellström, B. (2012). Ca2+-dependent transcriptional control of Ca2+ homeostasis *. *J. Biol. Chem.* 287: 31674–31680.

Neely, M.D., Davison, C.A., Aschner, M., and Bowman, A.B. (2017). From the cover: Manganese and rotenone-induced oxidative stress signatures differ in iPSC-derived human dopamine neurons. *Toxicol. Sci.* 159: 366–379.

Ng, H.H. and Bird, A. (1999). DNA methylation and chromatin modification. *Curr. Opin. Genet. Dev.* 9: 158–163.

Nihei, M.K. and Guilarte, T.R. (1999). NMDAR-2A subunit protein expression is reduced in the hippocampus of rats exposed to Pb2+ during development. *Mol. Brain. Res.* 66: 42–49.

Oh, S.-Y., Lee, M.-Y., Kim, J.-M., Yoon, S., Shin, S., Park, Y.N., Ahn, Y.-H., and Kim, K.-S. (2005). Alternative usages of multiple promoters of the acetyl-CoA carboxylase β gene are related to differential transcriptional regulation in human and rodent tissues*. *J. Biol. Chem.* 280: 5909–5916.

Peleg, S., Feller, C., Forne, I., Schiller, E., Sévin, D.C., Schauer, T., Regnard, C., Straub, T., Prestel, M., Klima, C., Schmitt Nogueira, M., Becker, L., Klopstock, T., Sauer, U., Becker, P.B., Imhof, A., and Ladurner, A.G. (2016). Life span extension by targeting a link between metabolism and histone acetylation in Drosophila. *EMBO Rep.* 17: 455–469.

Pesonen, M. and Vähäkangas, K. (2019). Autophagy in exposure to environmental chemicals. *Toxicol. Lett.* 305: 1–9.

Pounds, J.G. (1984). Effect of lead intoxication on calcium homeostasis and calcium-mediated cell function: A review. *Neurotoxicology* 5: 295–331.

Pounds, J.G., Long, G.J., and Rosen, J.F. (1991). Cellular and molecular toxicity of lead in bone. *Environ. Health Perspect.* 91: 17–32.

Quintanar-Escorza, M.A., González-Martínez, M.T., Navarro, L., Maldonado, M., Arévalo, B., and Calderón-Salinas, J.V. (2007). Intracellular free calcium concentration and calcium transport in human erythrocytes of lead-exposed workers. *Toxicol. Appl. Pharmacol.* 220: 1–8.

Raja, S., Ravikrishna, R., Kommalapati, R., and Valsaraj, K. (2005). Monitoring of fogwater chemistry in the gulf coast urban industrial corridor: Baton Rouge (Louisiana). *Environ. Monit. Assess.* 110: 99–120.

Reardon, A.M. and Bhat, H.K. (2007). Methylmercury neurotoxicity: Role of oxidative stress. *Toxicol. Environ. Chem.* 89: 535–554.

Reichard, J.F. and Puga, A. (2010). Effects of arsenic exposure on DNA methylation and epigenetic gene regulation. *Epigenomics*. 2: 87–104.

Rizzuto, R., Bernardi, P., and Pozzan, T. (2000). Mitochondria as all-round players of the calcium game. *J. Physiol. (Lond.)* 529 (Pt 1): 37–47.

Sakamoto, A., Hino, S., Nagaoka, K., Anan, K., Takase, R., Matsumori, H., Ojima, H., Kanai, Y., Arita, K., and Nakao, M. (2015). Lysine demethylase LSD1 coordinates glycolytic and mitochondrial metabolism in hepatocellular carcinoma cells. *Cancer Res.* 75: 1445.

Sanchez, O.F., Lee, J., Yu King Hing, N., Kim, S.-E., Freeman, J.L., and Yuan, C. (2016). Lead (Pb) exposure reduces global DNA methylation level by non-competitive inhibition and alteration of dnmt expression†. *Metallomics*. 9: 149–160.

Scher, M.B., Vaquero, A., and Reinberg, D. (2007). SirT3 is a nuclear NAD+-dependent histone deacetylase that translocates to the mitochondria upon cellular stress. *Genes Dev.* 21: 920–928.

Schneider, J.S., Mettil, W., and Anderson, D.W. (2012). Differential effect of postnatal lead exposure on gene expression in the hippocampus and frontal cortex. *J. Mol. Neurosci.* 47: 76–88.

Sen, A., Heredia, N., Senut, M.-C., Land, S., Hollocher, K., Lu, X., Dereski, M.O., and Ruden, D.M. (2015a). Multigenerational epigenetic inheritance in humans: DNA methylation changes associated with maternal exposure to lead can be transmitted to the grandchildren. *Sci. Rep.* 5: 14466.

Sen, A., Heredia, N., Senut, M.C., Land, S., Hollocher, K., Lu, X.Y., Dereski, M.O., and Ruden, D.M. (2015b). Multigenerational epigenetic inheritance in humans: DNA methylation changes associated with maternal exposure to lead can be transmitted to the grandchildren. *Sci. Rep.* 5.

Senut, M.-C., Cingolani, P., Sen, A., Kruger, A., Shaik, A., Hirsch, H., Suhr, S.T., and Ruden, D. 2012. Epigenetics of early-life lead exposure and effects on brain development.

Seong, K.-H., Ly, N.H., Katou, Y., Yokota, N., Nakato, R., Murakami, S., Hirayama, A., Fukuda, S., Kang, S., Soga, T., Shirahige, K., and Ishii, S. (2020). Paternal restraint stress affects offspring metabolism via ATF-2 dependent mechanisms in Drosophila melanogaster germ cells. *Commun. Biol.* 3: 208.

Sharma, A., Nguyen, H., Geng, C., Hinman, M.N., Luo, G., and Lou, H. (2014). Calcium-mediated histone modifications regulate alternative splicing in cardiomyocytes. *Proc. Natl. Acad. Sci.* 111: E4920–E4928.

Shen, X.M. and Dryhurst, G. (1998). Iron- and manganese-catalyzed autoxidation of dopamine in the presence of L-cysteine: Possible insights into iron- and manganese-mediated dopaminergic neurotoxicity. *Chem. Res. Toxicol.* 11: 824–837.

Shiraki, N., Shiraki, Y., Tsuyama, T., Obata, F., Miura, M., Nagae, G., Aburatani, H., Kume, K., Endo, F., and Kume, S. (2014). Methionine metabolism regulates maintenance and differentiation of human pluripotent stem cells. *Cell Metab.* 19: 780–794.

Shock, L.S., Thakkar, P.V., Peterson, E.J., Moran, R.G., and Taylor, S.M. (2011). DNA methyltransferase 1, cytosine methylation, and cytosine hydroxymethylation in mammalian mitochondria. *Proc. Natl. Acad. Sci.* 108: 3630.

Simpkins, J.W., Swenberg, J.A., Weiss, N., Brusick, D., Eldridge, J.C., Stevens, J.T., Handa, R.J., Hovey, R.C., Plant, T.M., and Pastoor, T.P. (2011). Atrazine and breast cancer: A framework assessment of the toxicological and epidemiological evidence. *Toxicol. Sci.* 123: 441–459.

Song, X.-Y., Li, J.-N., Wu, Y.-P., Zhang, B., and Li, B.-X. (2015). Atrazine causes autophagy-and apoptosis-related neurodegenerative effects in dopaminergic neurons in the rat nigrostriatal dopaminergic system. *Int. J. Mol. Sci.* 16: 13490–13506.

Sun, X., Vaghjiani, V., Jayasekara, W.S.N., Cain, J.E., and St. John, J.C. (2018). The degree of mitochondrial DNA methylation in tumor models of glioblastoma and osteosarcoma. *Clin. Epigenet.* 10: 157.

Suzuki, M.M. and Bird, A. (2008). DNA methylation landscapes: Provocative insights from epigenomics. *Nat. Rev. Genet.* 9: 465–476.

Takiguchi, M., Achanzar, W.E., Qu, W., Li, G., and Waalkes, M.P. (2003). Effects of cadmium on DNA-(cytosine-5) methyltransferase activity and DNA methylation status during cadmium-induced cellular transformation. *Exp. Cell Res.* 286: 355–365.

Tchounwou, P.B., Yedjou, C.G., Patlolla, A.K., and Sutton, D.J. (2012). Heavy metal toxicity and the environment. *Mol. Clin. Environ. Toxicol. 101: 133–164.* 101

Tong, H., Zhang, L., Gao, J., Wen, S., Zhou, H., and Feng, S. (2017). Methylation of mitochondrial DNA displacement loop region regulates mitochondrial copy number in colorectal cancer. *Mol. Med. Rep.* 16: 5347–5353.

Toscano, C.D., Hashemzadeh-Gargari, H., Mcglothan, J.L., and Guilarte, T.R. (2002). Developmental Pb2+ exposure alters NMDAR subtypes and reduces CREB phosphorylation in the rat brain. *Dev. Brain Res.* 139: 217–226.

Toscano, C.D., O'Callaghan, J.P., and Guilarte, T.R. (2005). Calcium/calmodulin-dependent protein kinase II activity and expression are altered in the hippocampus of Pb2+-exposed rats. *Brain Res.* 1044: 51–58.

Turk, P.W., Laayoun, A., Smith, S.S., and Weitzman, S.A. (1995). DNA adduct 8-hydroxyl-2′-deoxyguanosine (8-hydroxyguanine) affects function of human DNA methyltransferase. *Carcinogenesis.* 16: 1253–1255.

Valinluck, V., Tsai, H.H., Rogstad, D.K., Burdzy, A., Bird, A., and Sowers, L.C. (2004). Oxidative damage to methyl-CpG sequences inhibits the binding of the methyl-CpG binding domain (MBD) of methyl-CpG binding protein 2 (MeCP2). *Nucleic Acids Res.* 32: 4100–4108.

Wang, C., Zhang, Z., Yao, H., Zhao, F., Wang, L., Wang, X., Xing, H., and Xu, S. (2014). Effects of atrazine and chlorpyrifos on DNA methylation in the liver, kidney and gill of the common carp (*Cyprinus carpio L.*). *Ecotoxicol. Environ. Saf.* 108: 142–151.

West, A.E., Chen, W.G., Dalva, M.B., Dolmetsch, R.E., Kornhauser, J.M., Shaywitz, A.J., Takasu, M.A., Tao, X., and Greenberg, M.E. (2001). Calcium regulation of neuronal gene expression. *Proc. Natl. Acad. Sci. USA* 98: 11024–11031.

Wiles, E.T. and Selker, E.U. (2017). H3K27 methylation: A promiscuous repressive chromatin mark. *Curr. Opin. Genet. Dev.* 43: 31–37.

Wirbisky-Hershberger, S.E., Sanchez, O.F., Horzmann, K.A., Thanki, D., Yuan, C., and Freeman, J.L. (2017). Atrazine exposure decreases the activity of DNMTs, global DNA methylation levels, and dnmt expression. *Food Chem. Toxicol.* 109: 727–734.

Wise, J.P., Orenstein, J.M., and Patierno, S.R. (1993). Inhibition of lead chromate clastogenesis by ascorbate: Relationship to particle dissolution and uptake. *Carcinogenesis.* 14: 429–434.

Wysocka, J., Swigut, T., Xiao, H., Milne, T.A., Kwon, S.Y., Landry, J., Kauer, M., Tackett, A.J., Chait, B.T., and Badenhorst, P. (2006). A PHD finger of NURF couples histone H3 lysine 4 trimethylation with chromatin remodelling. *Nature.* 442: 86–90.

Xie, J., Lin, L., Sánchez, O.F., Bryan, C., Freeman, J.L., and Yuan, C. (2021). Pre-differentiation exposure to low-dose of atrazine results in persistent phenotypic changes in human neuronal cell lines. *Environ. Pollut.* 271: 116379.

Xing, H., Wang, C., Wu, H., Chen, D., Li, S., and Xu, S. (2015). Effects of atrazine and chlorpyrifos on DNA methylation in the brain and gonad of the common carp. *Comp. Biochem. Physiol. C: Toxicol. Pharmacol.* 168: 11–19.

Yu, Z., Chen, X., Zhang, J., Wang, R., and Yin, D. (2013). Transgenerational effects of heavy metals on L3 larva of Caenorhabditis elegans with greater behavior and growth inhibitions in the progeny. *Ecotoxicol. Environ. Saf.* 88: 178–184.

Zhang, X.X., He, Z., Feng, B., and Shao, H. (2019). An epigenome-wide DNA methylation study of workers with an occupational exposure to lead. *J. Appl. Toxicol.* 39: 1311–1319.

Zhang, X.-Y., Liu, A.-P., Ruan, D.-Y., and Liu, J. (2002). Effect of developmental lead exposure on the expression of specific NMDA receptor subunit mRNAs in the hippocampus of neonatal rats by digoxigenin-labeled in situ hybridization histochemistry. *Neurotoxicol. Teratol.* 24: 149–160.

Zhao, C.Q., Young, M.R., Diwan, B.A., Coogan, T.P., and Waalkes, M.P. (1997). Association of arsenic-induced malignant transformation with DNA hypomethylation and aberrant gene expression. *Proc. Natl. Acad. Sci. U.S.A.* 94: 10907–10912.

Zhong, L., D'Urso, A., Toiber, D., Sebastian, C., Henry, R.E., Vadysirisack, D.D., Guimaraes, A., Marinelli, B., Wikstrom, J.D., Nir, T., Clish, C.B., Vaitheesvaran, B., Iliopoulos, O., Kurland, I., Dor, Y., Weissleder, R., Shirihai, O.S., Ellisen, L.W., Espinosa, J.M., and Mostoslavsky, R. (2010). The histone deacetylase sirt6 regulates glucose homeostasis via Hif1α. *Cell.* 140: 280–293.

11

Emerging Non-invasive Molecular Biomarkers for Early Cancer Detection

Jacob Sobota, Yingxue Zhang, Eid Alshammari, and Zhe Yang

Department of Biochemistry, Microbiology, and Immunology, Wayne State University School of Medicine, Detroit, Michigan

Introduction

Cancer is the second leading cause of death worldwide, with an estimated total of almost 10 million deaths in 2020; and early detection is critical to patient survival. Across all cancers, detection in the early stages increases the patients' survival rate, for instance by as much as 70% between the localized stage and the metastatic stage in the case of breast cancer (Howlader et al. 2020). Pancreatic cancer, in particular, would benefit from an effective early detection, as its average five-year survival rate is below 10% as a result of most diagnoses being at stages III to IV (Young et al. 2018a). Gastric cancer makes up 1.5% of all cancer diagnoses, but it has a poor five-year survival rate of 32% owing to the lack of a general early screening option (Howlader et al. 2020; Leja et al. 2014). Currently a variety of methods are used across cancers for diagnosis; these methods can be costly, time-consuming, or potentially risky for the patient. Therefore the need to find effective and non-invasive biomarkers that should guide the early detection of cancer is critical for improving patient outcomes in cancer treatment. A biomarker is defined as "a characteristic that is measured as an indicator of normal biological processes, pathogenic processes, or responses to an exposure or intervention, including therapeutic interventions" (FDA–NIH Biomarker Working Group 2016). Non-invasive biomarkers are biomarkers where the measurement methods do not enter the body or are minimally intrusive. Non-invasive methods are generally preferred to invasive ones, on account of features such as reduced risk of complications, ease, and patient comfort.

The liquid biopsy, a capture of fluids of the body for biomarkers of interest, is an increasingly examined non-invasive alternative to tissue biopsies for cancer detection. Tissue biopsies are limited in the kinds of cancer they are suitable for in screening; for example, breast cancer tissue is comparatively easy to extract, but lung cancer tissue requires much more invasive procedures, including anesthesia and surgery. The appeal of liquid biopsies lies in their ease of use, minimal invasiveness, and high throughput. Usually the fluid being used is blood, but urine and tears are also possible sources of biomarkers for liquid biopsies. Within a fluid sample of circulating tumor DNA (ctDNA), circulating microRNAs (miRNAs), proteins, and exosomes are molecular characteristics currently being researched as promising biomarkers for early cancer detection.

Genomic and Epigenomic Biomarkers of Toxicology and Disease: Clinical and Therapeutic Actions, First Edition. Edited by Saura C. Sahu.

ctDNA originates in tumor cells and can be detected through mutations and methylation patterns from the tumor of origin. With ctDNA, the major challenge is detecting it against the rest of circulating cell-free DNA (cfDNA). This is because the fraction of ctDNA can vary widely (from 0.01% to over 90%) and can make up less than 0.1% of cfDNA, especially in the early stages of cancer (Diehl et al. 2005; Donaldson and Park 2018). cfDNA is extracellular DNA released by dying cells into circulation, in blood serum or other bodily fluids. It is typically a double-stranded fragment with a median length of 160–180 base pairs. cfDNA levels have been shown to be elevated in a number of conditions, including cancer, but also after physical exercise or in physical trauma. This makes cfDNA levels alone an unreliable biomarker for cancer diagnosis (Underhill et al. 2016; Wan et al. 2017). Compared to cfDNA, ctDNA is shorter in length and retains tumor-specific mutations and methylation patterns, which makes it possible for it to be explored as a cancer biomarker.

Circulating miRNA is also increasingly examined as an early marker of cancer. miRNAs are twenty-two nucleotide-long non-coding RNAs that regulate gene expression by inhibiting or degrading their target messenger RNAs (mRNAs). Circulating miRNA can be found outside cells, in a variety of bodily fluids such as blood and saliva. The dysregulation of miRNA in bodily fluids has been shown in multiple cancers (Brase et al. 2010; Mitchell et al. 2008; Tan et al. 2018). A change in miRNA types or levels provides information on the particular state of cancer and distinguishes cancer patients from healthy individuals. Practically speaking, extracellular miRNAs are stable in circulation, easy to store, and can be measured by multiple methods, including reverse transcription polymerase chain reaction (RT-PCR) and microarrays. Proteins circulating in bodily fluids are also examined as biomarkers for early cancer detection. The panels of proteins found circulating in the body differ from cancer to cancer. Some proteins are currently investigated in screening tests for certain cancers; for example, the carcinoembryonic antigen (CEA) text investigates the CEA family of glycoproteins in the case of gastric cancer (Feng et al. 2017). Exosomes are another entity in bodily fluids that can be developed into a non-invasive method for early cancer detection. Exosomes are a type of extracellular vesicle, generally 30–150 nm in size, that contain DNAs, mRNAs, non-coding RNAs such as miRNAs, and proteins. As tumor cells secrete exosomes more than non-malignant cells, the contents of exosomes can provide a picture of the environment inside tumor cells and may become useful for diagnosis (Rajagopal and Harikumar 2018).

Ongoing research on the molecular environment present in liquid biopsy samples in the early stages of cancer contributes to improving the early detection of cancers of all kinds. Here we will discuss emerging non-invasive molecular biomarkers such as ctDNA, miRNAs, and exosomes in the early detection of cancer. While every cancer is different, our discussion will address the cancer-specific application of each of these biomarkers for early detection. The goal of our discussion is to clarify current shortcomings in early cancer detection and to improve the use of non-invasive cancer biomarkers.

Conventional Methods of Early Cancer Diagnosis

A variety of diagnostic methods exist in current clinical practice for various cancer types. The leading diagnostic method for early lung cancer is low-dose computed tomography (CT). This method has high sensitivity for detecting lung nodules. However, because of this high sensitivity, many non-tumorous lung nodules are detected, which yields a false positive rate of 96% in some studies (Aberle et al. 2011). Often follow-up CTs or invasive surgery to obtain a sample of the nodule are required in order to confirm the malignancy of lung nodules. Conducting a tissue biopsy to aid in the diagnosis of lung cancer carries tangible risks of causing harm to the patient. Transthoracic needle biopsies are the leading method of obtaining tissue samples from lung nodules. Common complications from this tissue biopsy are pneumothorax and pulmonary hemorrhage. The incidence of these complications varies widely in studies, but averages around 20% for pneumothorax (Boskovic et al. 2014; Wiener et al. 2011). Thus, current methods of early diagnosis of lung cancer are capable of detecting the disease in localized stages, but the main problem is that the required follow-up procedures put cost, time, and health risk on the patients. An ideal early screening method for lung cancer would act to confirm the malignancy of lung nodules detected by CT in a quick manner, with little follow-up needed.

With breast cancer the usual method of diagnosis is imaging, for instance a mammogram or ultrasound; then, if an abnormality is found, confirmation is sought from a tissue sample obtained for this purpose. Given the location of the tumor, tissue biopsies for breast cancer are less likely than those for lung cancer to cause complications. However, with the prevalence of early screening efforts, the worry about overdiagnosis is commonly debated in relation to the usefulness of breast cancer screening (Løberg et al. 2015; Marmot et al. 2013). The X-rays used in mammograms have also been a source of concern, as the radiation from them creates the risk of developing a cancer that would have never occurred in the first place. New, non-invasive biomarkers should address the problems of current screening methods by matching or exceeding their diagnostic value, but should also address the problem of overdiagnosis by being able to evaluate quickly the likelihood of a given tumor's presenting symptoms or not in a case.

Gastric cancer is a heterogeneous disease, which presents a challenge in identifying reliable molecular biomarkers. In current gastric cancer screening, one method of diagnosis is *Helicobacter pylori* serology. The rationale behind this method is that *H. pylori* infection is one of the major risk factors for developing gastric cancer, which causes chronic inflammation of the gastric mucosa, damaging it over time. However, this method is insufficient on its own, as it cannot detect premalignant lesions. Another method of diagnosing gastric cancer is testing serum pepsinogen levels. Decreases in pepsinogen I (PGI) and the PGI/PGII ratio reflect gastric mucosa atrophy and can predict gastric cancer development (Leja et al. 2014). But the sensitivity of the PG test ranges from 37% to 62%, which means that it is unsuitable for a general population-based screening. Photofluorography via upper gastrointestinal series (UGIS) is a diagnostic method used in high-incidence populations such as the Japanese in gastric cancer screening. Studies conducted in Japan have shown a 40% to 60% reduction in mortality as a result of photofluorography screening (Hamashima et al. 2008). However, studies conducted in low-incidence populations have not shown reductions in mortality, which suggests that this method might be unsuitable for low-risk populations (Pisani et al. 1994). Upper gastrointestinal endoscopy is the gold standard for the confirmation of a gastric cancer diagnosis. This method is generally implemented in targeted high-risk populations, but it is not nearly as effective in low-incidence populations. Thus new biomarkers with sufficient sensitivity and specificity are needed for use in low-incidence populations. Any emerging biomarkers should be sensitive across multiple varieties of gastric cancer, if they are to address the heterogeneity of the disease.

Pancreatic ductal adenocarcinoma (PDAC) has one of the worst overall five-year survival rates. Although PDAC is the seventh leading cause of cancer death worldwide, its incidence in the population is very low. Current screening is limited to populations at high risk for pancreatic cancer, for instance individuals with germline mutations such as *STK11/LKB1*. The risk of pancreatic cancer is 132 times higher for individuals with the *STK11/LKB1* mutation than for individuals without it and nine times higher for first-degree relatives of any pancreatic cancer pantient than for other people (Klein et al. 2004; Matsubayashi 2011). The most common method of pancreatic cancer diagnosis is through multidetector CT, which is highly accurate in evaluating the extent of the cancer. However, CT cannot detect tumors under 1.5 cm in size, and its radioactive component makes it unsuitable for regular screening even in high-risk populations. Endoscopic ultrasound-guided fine needle aspiration is used for pancreatic lesions under 2 cm in size, but this technique is operationally challenging when it comes to obtaining quality imaging. Complicating matters, the majority of PDAC (around 90% of them) arise from microscopic intraductal lesions in the pancreas undetectable through available imaging methods (Morani et al. 2020; Pietryga and Morgan 2015). The current most widely used blood biomarker for PDAC is carbohydrate antigen 19–9 (CA19–9), a glycan that has utility in monitoring responses to treatment. However, the effectiveness of this maker in diagnosing asymptomatic patients is low; it has a median sensitivity of 79% and a specificity of 82% (Goonetilleke and Siriwardena 2007). Its usefulness is also limited by the fact that 5–10% of the general population lacks the enzyme needed to express it (Tsen et al. 2018). While the need for a biomarker for early PDAC detection is evident, researching biomarkers with greater sensitivity and specificity than CA19–9 for asymptomatic patients could shift the diagnosis to localized stages where treatment is more manageable.

Emerging Non-invasive Cancer Biomarkers

Diagnostic ctDNA Markers

ctDNA is a clinically accepted biomarker for genotyping known tumors and for guiding the treatment as an alternative to tissue biopsy. Using ctDNA avoids problems in tumor heterogeneity, as ctDNA is shed into circulation by all types of tumor cells. In addition, ctDNA can provide information on the tumor environment and prognosis through the mutations present in a sample. A multitude of detection techniques have been applied to the problem of detecting ctDNA in a liquid biopsy sample. The most sensitive methods used currently are predominantly PCR-based, for example digital droplet PCR (ddPCR) and BEAMing (beads, emulsion, amplification, and magnetics). These methods are rapid and have high sensitivity, which allows them to identify mutations reliably at a 0.01% concentration in a sample; but they are limited by the low number of species that can be detected in a single assay (Rodda et al. 2018). Next-generation sequencing techniques are also used in studies that can detect multiple sequence variations simultaneously in a sample.

In the context of searching for an early diagnosis for pancreatic cancer, ctDNA has mainly been investigated as a means of detecting somatic *KRAS* (Kirsten rat sarcoma viral oncogene homolog) mutations. Mutations in the *KRAS* gene have been reported in over 90% of PDAC cases as being among the earliest mutations to occur. These mutations are thought to be a driver in the progression of pancreatic intraepithelial neoplasia (PanIN) to invasive PDAC (Felsenstein et al. 2018). However, the actual utility of using *KRAS* mutations in ctDNA for the detection of early stages of PDAC has been contested on the grounds that results are inconsistent across studies. The false positive rate was found in several studies to be very low for distinguishing pancreatic cancer from healthy controls, with specificities greater than 99%, and also greater than CA19–9 (Dianxu et al. 2002; Sausen et al. 2015). In other studies, it has been found that there is a high concordance between *KRAS* mutations in ctDNA and matched tissue samples (Bettegowda et al. 2014; Cohen et al. 2017). However, a large case-control study using a next-generation sequencing (NGS) approach to detect mutations in the *KRAS* gene in patients with local, regional, and systemic stage PDAC detected only 10% of the localized stages, a rate with lower sensitivity than CA19–9 (Le Calvez-kelm et al. 2016). In this study, *KRAS* mutations were also found in 4.3% of chronic pancreatitis samples and in 3.7% of healthy control samples. The low sensitivity of these results is likely due to the detection limit of the NGS method used, which is an allelic fraction of 0.08%. Additionally, in other reserach there is high variance (20%–93%) in the reported sensitivity of using *KRAS* mutations to detect PDAC, which reflects the lack of standardized methodologies across different studies (Castells et al. 1999; Jiao et al. 2007; Le Calvez-kelm et al. 2016; Maire et al. 2002; Uemura et al. 2004). Until larger studies using consistent methodologies are conducted across wide populations, the "true" sensitivity of using the *KRAS* gene will remain uncertain.

In addition to mutational analysis, several studies have shown that ctDNA can be used as a means of detecting PDAC through its methylation states. The promoter regions of the genes *BNC1* and *ADAMTS1* have been shown to be methylated in pancreatic cancer. A combined analysis of the methylation states of these genes in ctDNA found that it has high sensitivity (95%) and specificity (92%) for localized stages of PDAC (Eissa et al. 2019). In a recent pan-cancer analysis with 6,689 participants both with and without cancer, whole-genome bisulfite sequencing (WGBS), a targeted methylation analysis technique, outperformed whole-genome sequencing and targeted sequencing (Liu et al. 2020). There are several reasons why methylation studies have shown greater sensitivity than mutational analysis-based methods. One is that methylation is more common than mutational sites, potentially enhancing the amount of ctDNA that can be detected in a sample. Another advantage is that epigenetic modifications can provide information on the tissue of origin of a sample. A third advantage is that methylation avoids the problem of clonal hematopoiesis (CH) and other highly prevalent non-malignant mutations in sequencing. CH represents a hurdle for mutational analysis methods, since somatic mutations occurring in non-malignant hematopoietic cells released into cfDNA by dying peripheral blood cells cause false positives. It has been seen that 25% of patients with non-hematologic cancers have CH, the CH frequency increasing with age (Coombs et al. 2017). In ctDNA analysis, CH can be identified and corrected for through matched cfDNA and white blood cell sequencing. This, however, adds a layer of complexity.

The use of ctDNA in lung cancer focuses on molecular profiling in diagnosed cases (Abbosh et al. 2017; Hanna et al. 2017; Santarpia et al. 2011). Lung cancer is typically classified into non-small-cell lung cancer (NSCLC) and small-cell lung cancer (SCLC), and these categories make up 85% and 15% of lung cancer cases respectively. A study compared the effectiveness of ctDNA to that of exosome-derived DNA in NSCLC diagnosis (Wan et al. 2018). Using a PCR assay method that targeted *EGFR* mutations, this study found ctDNA having 14% sensitivity for stage I and stage II lung cancer. Another study used a panel of fifty-five cancer driver genes and examined their ctDNA mutations in the plasma of patients with lung, breast, colorectal, or ovarian cancer (Phallen et al. 2017). This study found ctDNA in 45% of stage I lung cancer samples and in 62% of cases overall by using a capture and sequencing approach called TEC-Seq. In a recent study the ctDNA of the *RNF213* gene was found to have high specificity but low sensitivity in distinguishing benign lung nodules from lung cancer (Jiang et al. 2021). By comparison with the mutational analysis of ctDNA, methylation studies have received far more attention in NSCLC. Multiple genes have been found to be aberrantly methylated in lung cancer. Four genes, *BCAT1*, *CDO1*, *TRIM58*, and *ZNF177*, have been found to be hypermethylated in stage I lung cancer, and their use has been validated in brachial fluids (Diaz-Lagares et al. 2016). Methylation of the *SHOX2* locus has been identified as a diagnostic biomarker in lung cancer, more strongly associated with

SCLC and squamous cell lung cancer than with adenocarcinoma (Kneip et al. 2011). The ctDNA methylation of the *SHOX2* gene was found to be able to distinguish between malignant and non-malignant lung disease, as well as to suggest the stage of the cancer via the extent of methylation. Although *SHOX2* methylation has been seen to have low sensitivity in the early stages of lung cancer, a recent study showed that, in combination with *RASSF1A*, it displayed higher sensitivity than CEA in stage I lung cancers (Zhang et al. 2017). Another recent study using panels of *CDO1*, *TAC1*, *SOX17*, and *HOXA7* achieved up to 90% sensitivity with ctDNA methylation from plasma samples (Chen et al. 2020a). Overall, there is evidence that promoter methylation in ctDNA using panels of multiple genes is a promising biomarker in lung cancer diagnosis, although studies with large cohorts are needed before any clinical use can be seen.

ctDNA is also finding its value in the early detection of breast cancer. Targeted ddPCR has been used to detect ctDNA mutations of *PIK3CA* in early-stage breast cancer and has demonstrated good sensitivity and specificity (93% and 100%) (Beaver et al. 2014). A recent study using a NGS approach demonstrated the diagnostic value of ctDNA from a panel of 136 genes from the COSMIC database of somatic mutations (Zhang et al. 2019). When combined with predictive data from the Breast Imaging Reporting and Data System (BI-RADS) classification, the sensitivity of this gene panel could be raised from 74% to 92%, so that early-stage breast cancer could be categorized by pathological subtypes. Another study using the targeted NGS SafeSEQ approach with *PIK3CA* and *TP53* demonstrated that this approach could correlate ctDNA with age, tumor progression, and BI-RADS category (Rodriguez et al. 2019). With regard to methylation studies, single-target screening of promoter methylation has too low a sensitivity to be of clinical use in early breast cancer screening (Li et al. 2019a; Qian and Ruan 2018). However, as in the case of other cancers, the methylation-based screening performs best when a panel of promoter methylations is used. A recent study showed that a six-gene panel had respectively 80% and 72% sensitivity and specificity in the early detection of breast cancer when the putative cancer patients were compared to healthy controls (Shan et al. 2016).

Comparatively little work has been done in examining the use of mutational analysis of ctDNA in the early diagnosis of gastric cancer. Current research has focused on the quantification of cfDNA, although levels can be elevated in unrelated conditions, and this makes quantification unsuitable for patients who do not display symptoms of gastric cancer, or even for patients with related diseases that could be mistaken for gastric cancer. However, there has been interest in investigating gene promoter methylation for the detection of gastric cancer. *Reprimo*, *Rassf1A*, *CDH1*, *CDKN2A*, *MLH1*, *RUNX3*, *APC*, and *p16* are among the genes investigated for their methylation state in gastric cancer (Wen et al. 2017). Recently the *RPRML* gene was found to have tumor-suppressive properties, and its ctDNA has promising sensitivity and specificity when it comes to distinguishing gastric cancer patients from controls (Alarcón et al. 2020).

Using ctDNA also makes it possible to screen multiple kinds of cancer at once. A pan-cancer detection study using a method called CancerSEEK employed a panel of sixteen genetic alterations as well as protein biomarkers to evaluate the effectiveness of ctDNA in the simultaneous detection of multiple cancers (Cohen et al. 2018). Across 1,005 cancer patients with stage I–III cancer, this study achieved a specificity of over 99%, with varying sensitivities per each cancer; these ranged from 33% in breast cancers to 98% in ovarian cancers. The median sensitivity of stage I cancers was 48%, lower than the median sensitivities of 73% and 78% for stage II and stage III cancers. However, the approach taken in this study has been questioned. The case-control data were criticized for presumably not being representative of the screened populations, and it was objected that the numbers for each stage and histological type were too few to draw meaningful conclusions from (Young et al. 2018b). Another pan-cancer study used stored plasma samples from a longitudinal study and applied a methylation assay termed PanSeer to detect cancer in samples from ctDNA (Chen et al. 2020b). This study achieved a sensitivity and specificity of 88% and 96% in post-diagnosis patients and could detect 95% of the cases where asymptomatic individuals were later diagnosed with cancer. However, the study has several limitations related to its longitudinal the nature, namely uncertain factors involving sample preservation and patient follow-up.

Despite difficulties in using ctDNA for early detection, there are some recent studies that show alternative ways of doing it with higher sensitivity and specificity than in prior attempts. It has been shown that the difference in fragmentation from cfDNA to ctDNA can be used to improve detection in the sample without any need for a much deeper sequencing (Mouliere et al. 2018). An approach using a single-strand library preparation and hybrid capture-based cfDNA sequencing (SLHC-seq) enriched the damaged short fragments of cfDNA, which led to an increase in the sensitivity of ctDNA detection (Liu et al. 2019). The results obtained with this approach were consistent with tissue-based sequencing. Using a machine-learning approach called DELFI, the fragmentation profile of 236 cancer patients was found to have between 57% and 99% sensitivity and 98% specificity for the cancer types examined and, when this approach was combined with cfDNA mutation detection, the sensitivity rose to 91% (Cristiano et al. 2019).

MiRNAs as Biomarkers for Early Cancer Diagnosis

Another biomarker investigated for use in early cancer diagnosis is miRNAs, which act on multiple pathways in cancer development at a post-transcriptional level. Alterations in the expression level of miRNAs in serum have been reported in some studies years before the manifestation of clinical symptoms of cancer. This makes miRNAs ideal for an early diagnostic in a liquid biopsy. By comparison to ctDNA, miRNA is easier to detect in circulation. Serum and plasma have both been used to detect miRNA signatures in circulation, and neither has been conclusively proven to be a better source of unbiased miRNA expression. It has been reported that miRNA from red and white blood cells can bias circulating miRNA signatures, rendering whole blood samples unfavorable (Pritchard et al. 2012). Other bodily fluids such as saliva or urine have also been used to check miRNA signatures, although these are less common. The most frequently used methods for miRNA detection are quantitative real-time PCR (qRT-PCR), miRNA microarrays, and NGS methods, though ddPCR has been used in some studies as well. Microarrays target more sequences than qRT-PCR but have lower detection sensitivity. Like ctDNA analysis, NGS methods are capable of detecting previously unknown sequences. Although qRT-PCR is a common choice for the detection of miRNA, ddPCR is now being explored as an option for quantification, since it offers advantages such as not needing internal or external normalization.

Most current methods for miRNA detection require normalization in order to quantify miRNA expression. There is currently no commonly accepted optimal normalization strategy to use as a standard. This leads into the biggest challenge facing circulating miRNA as a diagnostic biomarker of cancer, which is that the lack of method standardization between studies raises problems with reproducibility. This is a fact that has been noticed for years, yet no widespread consensus standards have emerged to guide studies on cancer diagnosis using miRNA. A universal reference gene has been considered to be unlikely, and instead assessment on a case-by-case basis is suggested. The small non-coding RNA miR-16 and RNU6B are two common endogenous reference genes that are used across cancer types even though some studies have reported uncertainty about the stability of their expression (Kaija et al. 2020; Schaefer et al. 2010; Song et al. 2012). Future research should be focused on identifying suitable endogenous references that can be implemented to achieve reproducibility between findings.

Despite the lack of consistency in standardization, there have been a few miRNAs that are commonly used as diagnostic biomarkers in cancers. One of the most commonly studied miRNAs is miR-21, whose expression has been examined in a variety of cancers. In the case of pancreatic cancer, miR-21 has been included in multiple panels of miRNA signatures. miR-21 has been shown to be deregulated in precancerous PanIN lesions, and increased levels of miR-21 in plasma have been reported to be associated with worse survival rates after surgical resection (du Rieu et al. 2010; Karasek et al. 2018). One study found that miR-21 in serum, by itself, had a sensitivity and specificity of 83% and 78% in distinguishing patients from healthy controls (Alemar et al. 2016). A different study using a panel of seven miRNAs that included miR-21 had a sensitivity and specificity of 94% and 93% (Liu et al. 2012). More recently, a study identified miRNA signatures in PDAC cases. Sixteen miRNAs were significantly upregulated in PDAC, and ten of them were novel (Vila-Navarro et al. 2019). The same study showed that, after the inclusion of CA19–9, a panel of two miRNAs (miR-33a-3p and miR-320a) had an increased diagnostic performance, with 93% sensitivity and 85% specificity in distinguishing patients with PDAC or IPMNs from healthy controls. A very recent study found a panel of five miRNAs identified through NGS analysis that could distinguish PDAC from chronic pancreatitis and from healthy controls; the highest sensitivity and specificity for distinguishing from healthy controls were respectively 98% and 96%, though a larger sample size is needed to confirm the utility of these results for diagnosis (Khan et al. 2021). Another recent study identified eight miRNAs as potential diagnostic biomarkers for PDAC (Deng et al. 2020). However, a meta-analysis reported that combining miRNA signatures in panels made no significant difference from single miRNA signatures in pancreatic cancer studies; in the same analysis, miR-196b was found to have the highest sensitivity and specificity for early-stage PDAC (Peng et al. 2021).

The use of ddPCR and saliva for miRNA detection has also been examined in the context of pancreatic cancer diagnosis. A study using ddPCR validated the potential of miR-1290 as a biomarker for distinguishing pancreatic cancer patients from healthy subjects at risk of PDAC (Tavano et al. 2018). This study found that miR-1290 could not stratify healthy patients by risk level, and its sensitivity and specificity of 56% and 90% were outperformed by CA19–9. Although miR-1290 plasma levels were found to be uncorrelated to the stage of PDAC, the vast majority of PDAC cases in the study were at later or advanced stages. Overall, this study concluded that miR-1290 is unlikely to be a good diagnostic biomarker of PDAC, but ddPCR does show potential for quantifying miRNA in this type of cancer. Another study investigated the validity of using the whole saliva as a source of circulating miRNA for pancreatic cancer diagnosis (Humeau et al. 2015). Four miRNAs (hsa-miR-21, hsa-miR23a, hsa-miR-23b, and hsa-miR-29 c) were found, with 100% specificity, to be overexpressed in pancreatic cancer patients when compared to controls. However, the cohort size of the study was small (twenty-one participants in total), and thus its findings should be taken as a proof of the concept and validated with a larger multicenter cohort.

There are a multitude of studies on the use of miRNA signatures for the diagnosis of lung cancer. miR-21 has also been identified as an oncogene in NSCLC and has been seen to be significantly increased in early-stage NSCLC (Bica-Pop et al. 2018). A panel of serum miR-184 and serum miR-191 has been shown to have a sensitivity and a specificity of 87% and 86% for diagnosing NSCLC patients and separating them from healthy controls, and of 89%

and 85% for differentiating NSCLC from pneumonia (Ding et al. 2020). The expression levels of these miRNAs also had prognostic value in predicting the response to treatment. A recent study evaluated miR-25's diagnostic value for NSCLC from existing data and found 75% sensitivity and 81% specificity (Li et al. 2020). Another study constructed a panel of four upregulated miRNAs (miR-141, miR-200b, miR-193b, and miR-301) from miRNA expression profiles of NSCLC patients and healthy controls and reported a sensitivity and specificity of 96% and 95% for NSCLC (Nadal et al. 2015). A follow-up study validated this panel, obtaining 78% sensitivity and 91% specificity vis-à-vis healthy controls, or 81% sensitivity with unchanged specificity when only adenocarcinoma was included (Yang et al. 2019). The four miRNAs were also upregulated in SCLC and large-cell lung carcinoma, even though those two subtypes were not included in the data used to construct the panel.

One lung cancer study used a technology termed "FirePlex" to analyze miRNAs in plasma samples without the complex purification protocols typical of miRNA analysis (Leng et al. 2018). This study did not find any statistical difference from qRT-PCR methods when analyzing the same samples. Out of eleven miRNA biomarkers used in the study, a panel of two miRNAs (miR-205-5p and miR-210-3p) had 79% and 89% sensitivity and specificity for NSCLC, while combining all eleven miRNAs produced no significant enhancement. A significance of this study is that skipping most purification protocols through this technology does not affect the performance of miRNA quantification. By removing purification steps, this would cut down on the problems of reproducibility and inter-study variance. In addition to exploring new quantification methods, the use of miRNA as a source of biomarkers for lung cancer diagnosis has been examined with sputum. A recent study using a panel of two sputum miRNAs (miR-31-5p and miR-210-3p) and one plasma miRNA (miR-21-5p) showed a sensitivity of 86% and a specificity of 92% for NSCLC (Liao et al. 2020). These miRNAs performed independently of the disease's stage, and their performance was not improved by the addition of two other plasma miRNAs included in the study. Overall, the use of miRNA in lung cancer diagnosis has been demonstrated to make a promising diagnostic approach. There is, however, a distinct need for studies focusing on SCLC, which, despite accounting for a minority of lung cancer cases, has the worst prognosis.

As in most cancers, in breast cancer, too, miR-21 has been shown to be upregulated; this is detectable in serum at any stage (Anwar et al. 2019; Asaga et al. 2011; Chen and Wang 2014; Jinling et al. 2017; Yan et al. 2008). miR-21 has been consistently reported to have concordance between circulatory and tissue expression. One meta-analysis found a pooled 79% sensitivity and 85% specificity for miR-21 (Li et al. 2016). Two other miRNAs commonly identified in diagnostic studies of breast cancer are miR-10b and miR-155 (Grimaldi and Incoronato 2019; Roth et al. 2010). miR-155 is consistently reported as having concordance between circulation and tissue expression; both are upregulated (Liu et al. 2013; Roth et al. 2010). One study used the ratios between miRNAs to distinguish early-stage breast cancer patients from individuals with benign lesions and from healthy controls (Fang et al. 2018). The best combination of miRNA ratios gave 88% sensitivity and 78% specificity for breast cancer versus benign lesions and 89% sensitivity and 93% specificity versus healthy controls. This study showed the possibility of using the ratios of present miRNAs—rather than normalization with endogenous controls or spike-ins—to diagnose cancer. A recent study using a panel of four plasma-derived miRNAs (miR-1246, miR-206, miR-24, and miR-373) had a sensitivity and specificity of 98% and 96% for breast cancer (Jang et al. 2021). Another study constructed a panel of two miRNAs (miR-148b and miR-133a) from plasma samples of mainly stage I–II breast cancer patients and healthy controls and found an area under curve (AUC) of 0.86 (Shen et al. 2014).

Research into the diagnosis of gastric cancer has placed much more focus on miRNA biomarkers than on ctDNA. As in other cancers, in gastric cancer too a plethora of miRNAs have been identified as diagnostic biomarkers; and they include miR-21. Serum, plasma, and urine have been identified as sources of miRNAs for gastric cancer diagnosis. miRNAs in gastric juice have also been validated for diagnosing gastric cancer, although this method remains under-explored. One study found miR-21 in serum to have 88% sensitivity and 80% specificity for diagnosing gastric cancer (Wu et al. 2015). The serum miR-21 performed slightly better than peripheral blood miR-21, and both serum and peripheral blood miR-21 outperformed CA19–9 and CEA. A recent study focused on diagnosing early-gastric cancer found five miRNAs, four of which, when combined in a panel, could distinguish early gastric cancer from gastritis with an AUC of 0.83 (Zhu et al. 2019). Two of the miRNAs identified in the study could on their own differentiate early gastric cancer from advanced stages. Additionally, no significant difference was found between *H. pylori*-positive and *H. pylori*-negative gastritis. Another study identified serum miR-381 as a diagnostic biomarker for early gastric cancer, where it is downregulated (Li et al. 2019b). miR-381 outperformed CA19–9 and CEA in the study through higher sensitivity and specificity (84% and 98%). It was also found that miR-381 levels were lower in advanced cases than in early ones, and the lower levels correlated with gastric cancer progression. A study with a three-member panel of upregulated miRNAs (miR-627, miR-629, and miR-652) yielded 87% sensitivity and 86% specificity for differentiating gastric cancer from healthy controls (Shin et al. 2015). Notably, miR-652 was found to be downregulated in squamous cell lung cancer tissue and not upregulated, as in gastric cancer, which highlights the cancer-specific expression of this miRNA.

Less work has focused on urine as a source of miRNAs in gastric cancer, but a novel panel of miRNAs in urine was described in a recent study (Iwasaki et al. 2019). This study found two upregulated miRNAs, miR-6807-5p and miR-6856-5p, to be correlated with gastric cancer. When combined with the *H. pylori* infection status, the sensitivity and specificity of this panel were respectively 80% and 89%, and the panel could detect even stage I gastric cancer. Additionally, both serum and tissue miR-6807-5p and miR-6856-5p were found to be upregulated. The study shows promise for these miRNAs as biomarkers, and also for the potential of urine as an easy source of biomarkers for early cancer diagnosis. Another study, which used ddPCR, found four of the most reported miRNAs (miR-21, miR-93, miR-106a, and miR-106b) to be correlated with gastric cancer in tissue and plasma (Zhao et al. 2018a). This panel of miRNAs had a sensitivity of 85% and a specificity of 79% for differentiating gastric cancer from controls, and could distinguish stage I and II gastric cancer from III and IV stages. The study demonstrates the feasibility of using ddPCR to quantify miRNA and has yielded results in line with those of existing studies.

Exosomes as Cancer Biomarkers

Exosomes are another class of biomarker source with a potential for early cancer diagnosis. As this type of extracellular vesicle contains multiple molecular components inside it, its choice is attractive because they provide an idea of the intracellular environment. It has been shown that exosomes perform roles in intercellular signaling that are associated with tumor initiation and progression (Akoto and Saini 2021; Kogure et al. 2011; Melo et al. 2014; Rajagopal and Harikumar 2018). The molecular components of exosomes have been shown to have value as biomarkers for early cancer detection. DNA and miRNA contained in exosomes can be used as biomarkers, and exosomes themselves can also act as biomarkers through membrane content differences between cancer-derived and non-cancer-derived exosomes (Fan et al. 2018). The use of exosomes as biomarkers is considered a non-invasive method for cancer diagnosis, as they can be isolated from serum, plasma, saliva, and urine, among other sources (Lässer et al. 2011; Mitchell et al. 2009).

Exosome-derived *KRAS* gene mutations have been examined as pancreatic cancer biomarkers. One study achieved 75% sensitivity and 93% specificity in detecting PDAC by using *KRAS* mutations in exosomal DNA (Allenson et al. 2017). The PDAC detection rate of exosomal DNA in the study was found to be higher than that of ctDNA, but a significant percentage of healthy individuals also displayed mutations in exosomal *KRAS* genes. Another study examined exosomal miRNA for diagnosing localized PDAC (Xu et al. 2017). Two miRNAs found to be enriched in exosomes from pancreatic cancer cells, miR-196a and miR-1246, had AUCs of 0.81 and 0.73, respectively. A recent study examined a down-regulated exosomal miRNA, miR-1226-3p, and found that it shows promise as a diagnostic biomarker (Wang et al. 2021). In addition to DNA and miRNA, a protein enriched in the membranes of exosomes from cancer cells, glypican-1 (GPC1), has been investigated as a potential diagnostic biomarker for PDAC, but with inconsistent results. One study reported perfect sensitivity and specificity when examining whether exosomal membranes were GPC1-positive or GPC1-negative, while another study reported that they were not statistically significant (Lai et al. 2017; Melo et al. 2015). A recent study followed the methodology of the first to establish a reproducible, simple system of analyzing exosome GPC1 content (Xiao et al. 2020). This study found that GPC1 content had an AUC of 0.89 and, when combined with another exosomal protein, CD82, and with CA19–9, the AUC was improved to 0.94.

A few recent studies show promising exosomal biomarkers for NSCLC. One study combined the use of serum miRNA with exosomal miRNA in a panel of four serum and two exosomal miRNAs (Wu et al. 2020). The study found that this panel had 85% sensitivity and 93% specificity for diagnosing stage I and stage II NSCLC, and the miRNAs performed better in a panel than individually. Another study used a panel of four exosomal miRNAs to diagnose NSCLC and two other panels to differentiate adenocarcinoma from squamous cell lung cancer (Jin et al. 2017). This study found 80% sensitivity and 92% specificity for differentiating NSCLC patients from healthy controls, and comparable effectiveness for differentiating adenocarcinoma and squamous cell NSCLC subtypes. Other contents of exosomes have also been investigated for diagnostic utility in lung cancer. One study examined a circular RNA (circRNA), namely circRNA-002178, in relation to lung adenocarcinoma (Wang et al. 2020). This circRNA was found to be upregulated in lung adenocarcinoma tissue and detectable in exosomes even in the early stages. For differentiating lung cancer patients from healthy controls, this circRNA was found to have an excellent sensitivity and specificity, of 99% and 100% respectively. A recent study examined the lipid membrane content of exosomes as a means of diagnosing NSCLC (Fan et al. 2018). Two different types of multivariate statistical analysis were used in the study to determine important lipids after characterization of exosomes via mass spectrometry. Depending on the analysis used, the sensitivity and specificity for controls versus early NSCLC were 77% and 72%, or 65% and 77%.

Exosomal miRNAs have also been investigated as diagnostic biomarkers in breast cancer. The expression levels of three miRNAs between cell-free and exosomal fractions in serum were compared in breast cancer patients and in healthy individuals (Eichelser et al. 2014). The miRNAs were found to have higher expression in exosomes than in cell-free serum, and the upregulation of one miRNA, miR-373, was associated with aggressive breast cancers, indicating that it could be used as a diagnostic biomarker. Another study found exosomal miR-21 and miR-1246 having an AUC of 0.73 when combined to differentiate breast cancer patients from healthy subjects (Hannafon et al. 2016). In one study, exosomes were purified from urine samples derived from both breast cancer patients and healthy controls (Ando et al. 2019). Using the expression levels of miR-21 and matrix metalloproteinase-1 (MMP-1) in the exosomes, a 95% sensitivity and 79% specificity were achieved for screening breast cancer patients versus healthy controls. Another study examined urinary exosome miRNAs and found a four-miRNA panel with 99% sensitivity and 100% specificity (Hirschfeld et al. 2020). In one study, an exosomal long non-coding RNA (lncRNA), H19, was investigated as a biomarker for breast cancer (Zhong et al. 2020). The study found that upregulated H19 had 87% sensitivity and 71% specificity for differentiating breast cancer patients from healthy controls, which was higher than the score achieved by the CEA test for the same cohort.

Exosomes have also been examined in the early diagnosis of gastric cancer. A panel of two miRNAs (miR-19b-3p and miR-106a-5p) in serum exosomes had an AUC of 0.81 in validation testing for distinguishing gastric cancer from controls; this score was higher than that of CA19–9 (Wang et al. 2017). The panel had excellent sensitivity and specificity in the blind trial as well, though the sample size was small. Another study demonstrated the diagnostic value of exosomal miR-1246 for gastric cancer, where it was downregulated in tissue but overexpressed in serum exosomes (Shi et al. 2020). The sensitivity and specificity for differentiating stage I gastric cancer patients from healthy controls were respectively 86% and 74%, and exosomal miR-1246 overexpression could also differentiate gastric cancer from benign diseases. Recently, exosomal membrane-type 1 matrix metalloproteinase (MT1-MMP) mRNA expression was investigated as a biomarker for gastric cancer (Dong et al. 2019). MT1-MMP mRNA outperformed CEA expression in classifying gastric cancer patients and healthy controls. The diagnostic ability of MT1-MMP was increased to 76% sensitivity and 84% specificity when MT1-MMP was combined with CEA. One study investigated the exosomal lncRNA HOTTIP as a gastric cancer diagnostic biomarker and found a 70% sensitivity and 85% specificity, as well as a greater AUC than in CEA, CA19–9, and CA72–4 (Zhao et al. 2018b). A more recent study examined the lncRNA H19 and found 74% sensitivity and 84% specificity; it also found a greater AUC than in CEA, CA19–9, and CA72–4 (Zhou et al. 2020). Another recent study with a large cohort used lncRNA-GC1 and found a sensitivity of 87% and a specificity of 87% for distinguishing early gastric cancer from the state of healthy controls (Guo et al. 2020). LncRNA-GC1 could also distinguish precursor lesions from early gastric cancer, and it outperformed CEA, CA19–9, and CA72–4. Exosomal lncRNAs appear to be a promising class of biomarkers for gastric cancer with a potential for early diagnosis.

Concluding Remarks

Cancer continues to be the second leading cause of death across the world. There are many promising biomarkers for a non-invasive early detection of cancer. Non-invasive options for diagnosis are crucial: they could increase not only our effectiveness in detecting cancer early, but also access to early detection options through ease of implementation. Non-invasive diagnostic methods open the possibility of more frequent sampling for diagnosis, a method that can provide insight into changes in a patient over time. Different types of non-invasive biomarkers can also complement one another, as ctDNA, miRNA, and other components tend to be present in the same sources and can yield higher sensitivity and specificity in panels than individually. Non-invasive biomarkers can have different degrees of effectiveness, which are based not only on the stage of a cancer but on its subtype as well.

The foremost challenges standing between emergent non-invasive biomarkers and their clinical use are related to methodology. One factor causing discrepancies between studies is differences in their methodology, for instance differences between analytical technologies such as PCR-based methods and NGS methods. As normalization methods cause reproducibility issues between miRNA studies, this challenge must be overcome in order to ensure that miRNA is a reliable source of information; and this is done either through agreed standards or by using technology to circumvent the normalization process. Another challenge is the validation of biomarkers. Although many molecules and processes related to cancer have been explored as diagnostic biomarkers, there are many that are not yet validated, or the subject of a single study. Large cohorts between independent studies should be used to fully explore the diagnostic potential of promising biomarkers identified in smaller studies. While each kind of circulating biomarker has the potential to be an effective tool for the early diagnosis of cancer, the difficulty is in validating the ones that are effective and specifying in what contexts they work well and how we can measure them. Future investigations into emerging non-invasive biomarkers should be aware not only of the need to understand the biology surrounding their biomarker of interest, but also of how statistical differences and data processing can bias the results, even unintentionally.

References

Abbosh, C., Birkbak, N.J., Wilson, G.A., Jamal-Hanjani, M., Constantin, T., Salari, R., Le Quesne, J., Moore, D.A., Veeriah, S., Rosenthal, R., Marafioti, T., Kirkizlar, E., Watkins, T.B.K., Mcgranahan, N., Ward, S., Martinson, L., Riley, J., Fraioli, F., Bakir, A.L., Grönroos, M., Zambrana, E., Endozo, F., Bi, R., Fennessy, W.L., Sponer, F.M., Johnson, N., Laycock, D., Shafi, J., Czyzewska-Khan, J., Rowan, A., Chambers, T., Matthews, N., Turajlic, S., Hiley, C., Lee, S.M., Forster, M.D., Ahmad, T., Falzon, M., Borg, E., Lawrence, D., Hayward, M., Kolvekar, S., Panagiotopoulos, N., Janes, S.M., Thakrar, R., Ahmed, A., Blackhall, F., Summers, Y., Hafez, D., Naik, A., Ganguly, A., Kareht, S., Shah, R., Joseph, L., Marie Quinn, A., Crosbie, P.A., Naidu, B., Middleton, G., Langman, G., Trotter, S., Nicolson, M., Remmen, H., Kerr, K., Chetty, M., Gomersall, L., Fennell, D.A., Nakas, A., Rathinam, S., Anand, G., Khan, S., Russell, P., Ezhil, V., Ismail, B., Irvin-Sellers, M., Prakash, V., Lester, J.F., Kornaszewska, M., Attanoos, R., Adams, H., Davies, H., Oukrif, D., Akarca, A.U., Hartley, J.A., Lowe, H.L., Lock, S., Iles, N., Bell, H., Ngai, Y., Elgar, G., Szallasi, Z., Schwarz, R.F., Herrero, J., Stewart, A., Quezada, S.A., Peggs, K.S., Van Loo, P., Dive, C., Lin, C.J., Rabinowitz, M., Aerts, H. et al. (2017). Phylogenetic ctDNA analysis depicts early-stage lung cancer evolution. *Nature* 545: 446–451.

Aberle, D.R., Adams, A.M., Berg, C.D., Black, W.C., Clapp, J.D., Fagerstrom, R.M., Gareen, I.F., Gatsonis, C., Marcus, P.M., and Sicks, J.D. (2011). Reduced lung-cancer mortality with low-dose computed tomographic screening. *N. Engl. J. Med.* 365: 395–409.

Akoto, T. and Saini, S. (2021). Role of exosomes in prostate cancer metastasis. *Int. J. Mol. Sci.* 22.

Alarcón, M.A., Olivares, W., Córdova-Delgado, M., Muñoz-Medel, M., De Mayo, T., Carrasco-Aviño, G., Wichmann, I., Landeros, N., Amigo, J., Norero, E., Villarroel-Espíndola, F., Riquelme, A., Garrido, M., Owen, G.I., and Corvalán, A.H. (2020). The reprimo-like gene is an epigenetic-mediated tumor suppressor and a candidate biomarker for the non-invasive detection of gastric cancer. *Int. J. Mol. Sci.* 21.

Alemar, B., Izetti, P., Gregório, C., Macedo, G.S., Castro, M.A., Osvaldt, A.B., Matte, U., and Ashton-Prolla, P. (2016). miRNA-21 and miRNA-34a are potential minimally invasive biomarkers for the diagnosis of pancreatic ductal adenocarcinoma. *Pancreas* 45: 84–92.

Allenson, K., Castillo, J., San Lucas, F.A., Scelo, G., Kim, D.U., Bernard, V., Davis, G., Kumar, T., Katz, M., Overman, M.J., Foretova, L., Fabianova, E., Holcatova, I., Janout, V., Meric-Bernstam, F., Gascoyne, P., Wistuba, I., Varadhachary, G., Brennan, P., Hanash, S., Li, D., Maitra, A., and Alvarez, H. (2017). High prevalence of mutant KRAS in circulating exosome-derived DNA from early-stage pancreatic cancer patients. *Ann. Oncol.* 28: 741–747.

Ando, W., Kikuchi, K., Uematsu, T., Yokomori, H., Takaki, T., Sogabe, M., Kohgo, Y., Otori, K., Ishikawa, S., and Okazaki, I. (2019). Novel breast cancer screening: Combined expression of miR-21 and MMP-1 in urinary exosomes detects 95% of breast cancer without metastasis. *Sci. Rep.* 9: 13595.

Anwar, S.L., Sari, D.N.I., Kartika, A.I., Fitria, M.S., Tanjung, D.S., Rakhmina, D., Wardana, T., Astuti, I., Haryana, S.M., and Aryandono, T. (2019). Upregulation of circulating miR-21 expression as a potential biomarker for therapeutic monitoring and clinical outcome in breast cancer. *Asian Pac. J. Cancer Prev.* 20: 1223–1228.

Asaga, S., Kuo, C., Nguyen, T., Terpenning, M., Giuliano, A.E., and Hoon, D.S.B. (2011). Direct serum assay for microRNA-21 concentrations in early and advanced breast cancer. *Clin. Chem.* 57: 84–91.

Beaver, J.A., Jelovac, D., Balukrishna, S., Cochran, R., Croessmann, S., Zabransky, D.J., Wong, H.Y., Toro, P.V., Cidado, J., Blair, B.G., Chu, D., Burns, T., Higgins, M.J., Stearns, V., Jacobs, L., Habibi, M., Lange, J., Hurley, P.J., Lauring, J., Vandenberg, D., Kessler, J., Jeter, S., Samuels, M.L., Maar, D., Cope, L., Cimino-Mathews, A., Argani, P., Wolff, A.C., and Park, B.H. (2014). Detection of cancer DNA in plasma of patients with early-stage breast cancer. *Clin. Cancer Res.* 20: 2643–2650.

Bettegowda, C., Sausen, M., Leary, R.J., Kinde, I., Wang, Y., Agrawal, N., Bartlett, B.R., Wang, H., Luber, B., Alani, R.M., Antonarakis, E.S., Azad, N.S., Bardelli, A., Brem, H., Cameron, J.L., Lee, C.C., Fecher, L.A., Gallia, G.L., Gibbs, P., Le, D., Giuntoli, R.L., Goggins, M., Hogarty, M.D., Holdhoff, M., Hong, S.M., Jiao, Y., Juhl, H.H., Kim, J.J., Siravegna, G., Laheru, D.A., Lauricella, C., Lim, M., Lipson, E.J., Marie, S.K., Netto, G.J., Oliner, K.S., Olivi, A., Olsson, L., Riggins, G.J., Sartore-Bianchi, A., Schmidt, K., Shih, L.,.M., Oba-Shinjo, S.M., Siena, S., Theodorescu, D., Tie, J., Harkins, T.T., Veronese, S., Wang, T.L., Weingart, J.D., Wolfgang, C.L., Wood, L.D., Xing, D., Hruban, R.H., Wu, J., Allen, P.J., Schmidt, C.M., Choti, M.A., Velculescu, V.E., Kinzler, K.W., Vogelstein, B., Papadopoulos, N., and Diaz, L.A., JR. (2014). Detection of circulating tumor DNA in early- and late-stage human malignancies. *Sci. Transl. Med.* 6: 224ra24.

Bica-Pop, C., Cojocneanu-Petric, R., Magdo, L., Raduly, L., Gulei, D., and Berindan-Neagoe, I. (2018). Overview upon miR-21 in lung cancer: Focus on NSCLC. *Cell. Mol. Life Sci.* 75: 3539–3551.

Boskovic, T., Stanic, J., Pena-Karan, S., Zarogoulidis, P., Drevelegas, K., Katsikogiannis, N., Machairiotis, N., Mpakas, A., Tsakiridis, K., Kesisis, G., Tsiouda, T., Kougioumtzi, I., Arikas, S., and Zarogoulidis, K. (2014). Pneumothorax after transthoracic needle biopsy of lung lesions under CT guidance. *J. Thorac. Dis.* 6 (Suppl 1): S99–s107.

Brase, J.C., Wuttig, D., Kuner, R., and Sültmann, H. (2010). Serum microRNAs as non-invasive biomarkers for cancer. *Mol. Cancer* 9: 306.

Castells, A., Puig, P., Móra, J., Boadas, J., Boix, L., Urgell, E., Solé, M., Capellà, G., Lluís, F., Fernández-Cruz, L., Navarro, S., and Farré, A. (1999). K-ras mutations in DNA extracted from the plasma of patients with pancreatic carcinoma: Diagnostic utility and prognostic significance. *J. Clin. Oncol.* 17: 578–584.

Chen, C., Huang, X., Yin, W., Peng, M., Wu, F., Wu, X., Tang, J., Chen, M., Wang, X., Hulbert, A., Brock, M.V., Liu, W., Herman, J.G., and Yu, F. (2020a). Ultrasensitive DNA hypermethylation detection using plasma for early detection of NSCLC: A study in Chinese patients with very small nodules. *Clin. Epigenet.* 12: 39.

Chen, J. and Wang, X. (2014). MicroRNA-21 in breast cancer: Diagnostic and prognostic potential. *Clin. Transl. Oncol.* 16: 225–233.

Chen, X., Gole, J., Gore, A., He, Q., Lu, M., Min, J., Yuan, Z., Yang, X., Jiang, Y., Zhang, T., Suo, C., Li, X., Cheng, L., Zhang, Z., Niu, H., Li, Z., Xie, Z., Shi, H., Zhang, X., Fan, M., Wang, X., Yang, Y., Dang, J., Mcconnell, C., Zhang, J., Wang, J., Yu, S., Ye, W., Gao, Y., Zhang, K., Liu, R., and Jin, L. (2020b). Non-invasive early detection of cancer four years before conventional diagnosis using a blood test. *Nat. Commun.* 11: 3475.

Cohen, J.D., Javed, A.A., Thoburn, C., Wong, F., Tie, J., Gibbs, P., Schmidt, C.M., Yip-Schneider, M.T., Allen, P.J., Schattner, M., Brand, R.E., Singhi, A.D., Petersen, G.M., Hong, S.M., Kim, S.C., Falconi, M., Doglioni, C., Weiss, M.J., Ahuja, N., He, J., Makary, M.A., Maitra, A., Hanash, S.M., Dal Molin, M., Wang, Y., Li, L., Ptak, J., Dobbyn, L., Schaefer, J., Silliman, N., Popoli, M., Goggins, M.G., Hruban, R.H., Wolfgang, C.L., Klein, A.P., Tomasetti, C., Papadopoulos, N., Kinzler, K.W., Vogelstein, B., and Lennon, A.M. (2017). Combined circulating tumor DNA and protein biomarker-based liquid biopsy for the earlier detection of pancreatic cancers. *Proc. Natl. Acad. Sci. USA* 114: 10202–10207.

Cohen, J.D., Li, L., Wang, Y., Thoburn, C., Afsari, B., Danilova, L., Douville, C., Javed, A.A., Wong, F., Mattox, A., Hruban, R.H., Wolfgang, C.L., Goggins, M.G., Dal Molin, M., Wang, T.L., Roden, R., Klein, A.P., Ptak, J., Dobbyn, L., Schaefer, J., Silliman, N., Popoli, M., Vogelstein, J.T., Browne, J.D., Schoen, R.E., Brand, R.E., Tie, J., Gibbs, P., Wong, H.L., Mansfield, A.S., Jen, J., Hanash, S.M., Falconi, M., Allen, P.J., Zhou, S., Bettegowda, C., Diaz, L.A., JR., Tomasetti, C., Kinzler, K.W., Vogelstein, B., Lennon, A.M., and Papadopoulos, N. (2018). Detection and localization of surgically resectable cancers with a multi-analyte blood test. *Science* 359: 926–930.

Coombs, C.C., Zehir, A., Devlin, S.M., Kishtagari, A., Syed, A., Jonsson, P., Hyman, D.M., Solit, D.B., Robson, M.E., Baselga, J., Arcila, M.E., Ladanyi, M., Tallman, M.S., Levine, R.L., and Berger, M.F. (2017). Therapy-related clonal hematopoiesis in patients with non-hematologic cancers is common and associated with adverse clinical outcomes. *Cell Stem Cell* 21 (374–382.e4).

Cristiano, S., Leal, A., Phallen, J., Fiksel, J., Adleff, V., Bruhm, D.C., Jensen, S., Medina, J.E., Hruban, C., White, J.R., Palsgrove, D.N., Niknafs, N., Anagnostou, V., Forde, P., Naidoo, J., Marrone, K., Brahmer, J., Woodward, B.D., Husain, H., Van Rooijen, K.L., Ørntoft, M.W., Madsen, A.H., Van De Velde, C.J.H., Verheij, M., Cats, A., Punt, C.J.A., Vink, G.R., Van Grieken, N.C.T., Koopman, M., Fijneman, R.J.A., Johansen, J.S., Nielsen, H.J., Meijer, G.A., Andersen, C.L., Scharpf, R.B., and Velculescu, V.E. (2019). Genome-wide cell-free DNA fragmentation in patients with cancer. *Nature* 570: 385–389.

Deng, B., Wang, M., and Liu, Z. (2020). A panel of 8 miRNAs as a novel diagnostic biomarker in pancreatic cancer. *Med. (Baltimore)* 99: e22261.

Dianxu, F., Shengdao, Z., Tianquan, H., Yu, J., Ruoqing, L., Zurong, Y., and Xuezhi, W. (2002). A prospective study of detection of pancreatic carcinoma by combined plasma K-ras mutations and serum CA19-9 analysis. *Pancreas* 25: 336–341.

Diaz-Lagares, A., Mendez-Gonzalez, J., Hervas, D., Saigi, M., Pajares, M.J., Garcia, D., Crujerias, A.B., Pio, R., Montuenga, L.M., Zulueta, J., Nadal, E., Rosell, A., Esteller, M., and Sandoval, J. (2016). A novel epigenetic signature for early diagnosis in lung cancer. *Clin. Cancer Res.* 22: 3361–3371.

Diehl, F., Li, M., Dressman, D., He, Y., Shen, D., Szabo, S., Diaz, L.A., JR., Goodman, S.N., David, K.A., Juhl, H., Kinzler, K.W., and Vogelstein, B. (2005). Detection and quantification of mutations in the plasma of patients with colorectal tumors. *Proc. Natl. Acad. Sci. USA* 102: 16368–16373.

Ding, H., Wen, W., Ding, Q., and Zhao, X. (2020). Diagnostic valuation of serum miR-184 and miR-191 in patients with non-small-cell lung cancer. *Cancer Control* 27: 1073274820964783.

Donaldson, J. and Park, B.H. (2018). Circulating tumor DNA: Measurement and clinical utility. *Annu. Rev. Med.* 69: 223–234.

Dong, Z., Sun, X., Xu, J., Han, X., Xing, Z., Wang, D., Ge, J., Meng, L., and Xu, X. (2019). Serum membrane type 1-matrix metalloproteinase (MT1-MMP) mRNA protected by exosomes as a potential biomarker for gastric cancer. *Med. Sci. Monit.* 25: 7770–7783.

Du Rieu, M.L.C., Torrisani, J.R.M., Selves, J., Al Saati, T., Souque, A., Dufresne, M.N., Tsongalis, G.J., Suriawinata, A.A., Carrère, N., Buscail, L., and Cordelier, P. (2010). MicroRNA-21 is induced early in pancreatic ductal adenocarcinoma precursor lesions. *Clin. Chem.* 56: 603–612.

Eichelser, C., Stückrath, I., Müller, V., Milde-Langosch, K., Wikman, H., Pantel, K., and Schwarzenbach, H. (2014). Increased serum levels of circulating exosomal microRNA-373 in receptor-negative breast cancer patients. *Oncotarget* 5: 9650–9663.

Eissa, M.A.L., Lerner, L., Abdelfatah, E., Shankar, N., Canner, J.K., Hasan, N.M., Yaghoobi, V., Huang, B., Kerner, Z., Takaesu, F., Wolfgang, C., Kwak, R., Ruiz, M., Tam, M., Pisanic, T.R., 2ND, Iacobuzio-Donahue, C.A., Hruban, R.H., He, J., Wang, T.H., Wood, L.D., Sharma, A., and Ahuja, N. (2019). Promoter methylation of ADAMTS1 and BNC1 as potential biomarkers for early detection of pancreatic cancer in blood. *Clin. Epigenet.* 11: 59.

Fan, T.W.M., Zhang, X., Wang, C., Yang, Y., Kang, W.Y., Arnold, S., Higashi, R.M., Liu, J., and Lane, A.N. (2018). Exosomal lipids for classifying early and late stage non-small cell lung cancer. *Anal. Chim. Acta* 1037: 256–264.

Fang, R., Zhu, Y., Hu, L., Khadka, V.S., Ai, J., Zou, H., Ju, D., Jiang, B., Deng, Y., and Hu, X. (2018). Plasma microRNA pair panels as novel biomarkers for detection of early stage breast cancer. *Front. Physiol.* 9: 1879.

FDA–NIH Biomarker Working Group. (2016). *BEST: Biomarkers, EndpointS, and Other Tools.* Resource [Internet]. Silver Spring, MD: FDA.

Felsenstein, M., Hruban, R.H., and Wood, L.D. (2018). New developments in the molecular mechanisms of pancreatic tumorigenesis. *Adv. Anat. Pathol.* 25: 131–142.

Feng, F., Tian, Y., Xu, G., Liu, Z., Liu, S., Zheng, G., Guo, M., Lian, X., Fan, D., and Zhang, H. (2017). Diagnostic and prognostic value of CEA, CA19-9, AFP and CA125 for early gastric cancer. *BMC Cancer* 17: 737.

Goonetilleke, K.S. and Siriwardena, A.K. (2007). Systematic review of carbohydrate antigen (CA 19-9) as a biochemical marker in the diagnosis of pancreatic cancer. *Eur. J. Surg. Oncol.* 33: 266–270.

Grimaldi, A.M. and Incoronato, M. (2019). Clinical translatability of "identified" circulating mirnas for diagnosing breast cancer: Overview and update. *Cancers (Basel)* 11.

Guo, X., Lv, X., Ru, Y., Zhou, F., Wang, N., Xi, H., Zhang, K., Li, J., Chang, R., Xie, T., Wang, X., Li, B., Chen, Y., Yang, Y., Chen, L., and Chen, L. (2020). Circulating exosomal gastric cancer-associated long noncoding RNA1 as a biomarker for early detection and monitoring progression of gastric cancer: A multiphase study. *JAMA Surg.* 155: 572–579.

Hamashima, C., Shibuya, D., Yamazaki, H., Inoue, K., Fukao, A., Saito, H., and Sobue, T. (2008). The japanese guidelines for gastric cancer screening. *Jpn. J. Clin. Oncol.* 38: 259–267.

Hanna, N., Johnson, D., Temin, S., Baker, S.,.J.R., Brahmer, J., Ellis, P.M., Giaccone, G., Hesketh, P.J., Jaiyesimi, I., Leighl, N.B., Riely, G.J., Schiller, J.H., Schneider, B.J., Smith, T.J., Tashbar, J., Biermann, W.A., and Masters, G. (2017). Systemic therapy for stage IV non-small-cell lung cancer: American society of clinical oncology clinical practice guideline update. *J. Clin. Oncol.* 35: 3484–3515.

Hannafon, B.N., Trigoso, Y.D., Calloway, C.L., Zhao, Y.D., Lum, D.H., Welm, A.L., Zhao, Z.J., Blick, K.E., Dooley, W.C., and Ding, W.Q. (2016). Plasma exosome microRNAs are indicative of breast cancer. *Breast Cancer Res.* 18: 90.

Hirschfeld, M., Rücker, G., Weis, D., Berner, K., Ritter, A., Jäger, M., and Erbes, T. (2020). Urinary exosomal microRNAs as potential non-invasive biomarkers in breast cancer detection. *Mol. Diagn. Ther.* 24: 215–232.

Howlader, N., Noone, A., and Krapcho, M. 2020. SEER cancer statistics review, 1975–2017. Bethesda, MD: National Cancer Institute. Based on November 2019 SEER Data Submission posted to the SEER website, April 2020.

Humeau, M., Vignolle-Vidoni, A., Sicard, F., Martins, F., Bournet, B., Buscail, L., Torrisani, J., and Cordelier, P. (2015). Salivary microRNA in pancreatic cancer patients. *PLoS One* 10: e0130996.

Iwasaki, H., Shimura, T., Yamada, T., Okuda, Y., Natsume, M., Kitagawa, M., Horike, S.-I., and Kataoka, H. (2019). A novel urinary microRNA biomarker panel for detecting gastric cancer. *J. Gastroenterol.* 54: 1061–1069.

Jang, J.Y., Kim, Y.S., Kang, K.N., Kim, K.H., Park, Y.J., and Kim, C.W. (2021). Multiple microRNAs as biomarkers for early breast cancer diagnosis. *Mol. Clin. Oncol.* 14: 31.

Jiang, N., Zhou, J., Zhang, W., Li, P., Liu, Y., Shi, H., Zhang, C., Wang, Y., Zhou, C., Peng, C., Zhang, W., Hao, Y., Sun, Q., Li, Y., and Zhao, X. (2021). RNF213 gene mutation in circulating tumor DNA detected by targeted next-generation sequencing in the assisted discrimination of early-stage lung cancer from pulmonary nodules. *Thorac. Cancer* 12: 181–193.

Jiao, L., Zhu, J., Hassan, M.M., Evans, D.B., Abbruzzese, J.L., and Li, D. (2007). K-ras mutation and p16 and preproenkephalin promoter hypermethylation in plasma DNA of pancreatic cancer patients: In relation to cigarette smoking. *Pancreas* 34: 55–62.

Jin, X., Chen, Y., Chen, H., Fei, S., Chen, D., Cai, X., Liu, L., Lin, B., Su, H., Zhao, L., Su, M., Pan, H., Shen, L., Xie, D., and Xie, C. (2017). Evaluation of tumor-derived exosomal miRNA as potential diagnostic biomarkers for early-stage non-small cell lung cancer using next-generation sequencing. *Clin. Cancer Res.* 23: 5311.

Jinling, W., Sijing, S., Jie, Z., and Guinian, W. (2017). Prognostic value of circulating microRNA-21 for breast cancer: A systematic review and meta-analysis. *Artif. Cells Nanomed. Biotechnol.* 45: 1–6.

Kaija, H., Pakanen, L., and Porvari, K. (2020). RNU6B, a frequent reference in miRNA expression studies, differentiates between deaths caused by hypothermia and chronic cardiac ischemia. *Int. J. Legal Med.* 134: 159–162.

Karasek, P., Gablo, N., Hlavsa, J., Kiss, I., Vychytilova-Faltejskova, P., Hermanova, M., Kala, Z., Slaby, O., and Prochazka, V. (2018). Pre-operative plasma miR-21-5p is a sensitive biomarker and independent prognostic factor in patients with pancreatic ductal adenocarcinoma undergoing surgical resection. *Cancer Genomics Proteomics* 15: 321–327.

Khan, I.A., Rashid, S., Singh, N., Rashid, S., Singh, V., Gunjan, D., Das, P., Dash, N.R., Pandey, R.M., Chauhan, S.S., Gupta, S., and Saraya, A. (2021). Panel of serum miRNAs as potential non-invasive biomarkers for pancreatic ductal adenocarcinoma. *Sci. Rep.* 11: 2824.

Klein, A.P., Brune, K.A., Petersen, G.M., Goggins, M., Tersmette, A.C., Offerhaus, G.J., Griffin, C., Cameron, J.L., Yeo, C.J., Kern, S., and Hruban, R.H. (2004). Prospective risk of pancreatic cancer in familial pancreatic cancer kindreds. *Cancer Res.* 64: 2634–2638.

Kneip, C., Schmidt, B., Seegebarth, A., Weickmann, S., Fleischhacker, M., Liebenberg, V., Field, J.K., and Dietrich, D. (2011). SHOX2 DNA methylation is a biomarker for the diagnosis of lung cancer in plasma. *J. Thorac. Oncol.* 6: 1632–1638.

Kogure, T., Lin, W.L., Yan, I.K., Braconi, C., and Patel, T. (2011). Intercellular nanovesicle-mediated microRNA transfer: A mechanism of environmental modulation of hepatocellular cancer cell growth. *Hepatology* 54: 1237–1248.

Lai, X., Wang, M., Mcelyea, S.D., Sherman, S., House, M., and Korc, M. (2017). A microRNA signature in circulating exosomes is superior to exosomal glypican-1 levels for diagnosing pancreatic cancer. *Cancer Lett.* 393: 86–93.

Lässer, C., Seyed Alikhani, V., Ekström, K., Eldh, M., Torregrosa Paredes, P., Bossios, A., Sjöstrand, M., Gabrielsson, S., Lötvall, J., and Valadi, H. (2011). Human saliva, plasma and breast milk exosomes contain RNA: Uptake by macrophages. *J. Transl. Med.* 9: 9.

Le Calvez-Kelm, F., Foll, M., Wozniak, M.B., Delhomme, T.M., Durand, G., Chopard, P., Pertesi, M., Fabianova, E., Adamcakova, Z., Holcatova, I., Foretova, L., Janout, V., Vallee, M.P., Rinaldi, S., Brennan, P., Mckay, J.D., Byrnes, G.B., and Scelo, G. (2016). KRAS mutations in blood circulating cell-free DNA: A pancreatic cancer case-control. *Oncotarget* 7: 78827–78840.

Leja, M., You, W., Camargo, M.C., and Saito, H. (2014). Implementation of gastric cancer screening - the global experience. *Best Pract. Res. Clin. Gastroenterol.* 28: 1093–1106.

Leng, Q., Wang, Y., and Jiang, F. (2018). A direct plasma miRNA assay for early detection and histological classification of lung cancer. *Transl. Oncol.* 11: 883–889.

Li, C., Sun, L., Zhou, H., Yang, Y., Wang, Y., She, M., and Chen, J. (2020). Diagnostic value of microRNA-25 in patients with non-small cell lung cancer in Chinese population: A systematic review and meta-analysis. *Med. (Baltimore)* 99: e23425.

Li, M., Wang, C., Yu, B., Zhang, X., Shi, F., and Liu, X. (2019a). Diagnostic value of RASSF1A methylation for breast cancer: A meta-analysis. *Biosci. Rep.* 39.

Li, S., Yang, X., Yang, J., Zhen, J., and Zhang, D. (2016). Serum microRNA-21 as a potential diagnostic biomarker for breast cancer: A systematic review and meta-analysis. *Clin. Exp. Med.* 16: 29–35.

Li, Y., Sun, H., Guan, J., Ji, T., and Wang, X. (2019b). Serum microRNA-381: A potential marker for early diagnosis of gastric cancer. *Yonsei Med. J.* 60: 720–726.

Liao, J., Shen, J., Leng, Q., Qin, M., Zhan, M., and Jiang, F. (2020). MicroRNA-based biomarkers for diagnosis of non-small cell lung cancer (NSCLC). *Thorac. Cancer* 11: 762–768.

Liu, J., Mao, Q., Liu, Y., Hao, X., Zhang, S., and Zhang, J. (2013). Analysis of miR-205 and miR-155 expression in the blood of breast cancer patients. *Chin. J. Cancer Res.* 25: 46–54.

Liu, M.C., Oxnard, G.R., Klein, E.A., Swanton, C., Seiden, M.V., Liu, M.C., Oxnard, G.R., Klein, E.A., Smith, D., Richards, D., Yeatman, T.J., Cohn, A.L., Lapham, R., Clement, J., Parker, A.S., Tummala, M.K., Mcintyre, K., Sekeres, M.A., Bryce, A.H., Siegel, R., Wang, X., Cosgrove, D.P., Abu-Rustum, N.R., Trent, J., Thiel, D.D., Becerra, C., Agrawal, M., Garbo, L.E., Giguere, J.K., Michels, R.M., Harris, R.P., Richey, S.L., Mccarthy, T.A., Waterhouse, D.M., Couch, F.J., Wilks, S.T., Krie, A.K., Balaraman, R., Restrepo, A., Meshad, M.W., Rieger-Christ, K., Sullivan, T., Lee, C.M., Greenwald, D.R., Oh, W., Tsao, C.-K., Fleshner, N., Kennecke, H.F., Khalil, M.F., Spigel, D.R., Manhas, A.P., Ulrich, B.K., Kovoor, P.A., Stokoe, C., Courtright, J.G., Yimer, H.A., Larson, T.G., Swanton, C., Seiden, M.V., Cummings, S.R., Absalan, F., Alexander, G., Allen, B., Amini, H., Aravanis, A.M., Bagaria, S., Bazargan, L., Beausang, J.F., Berman, J., Betts, C., Blocker, A., Bredno, J., Calef, R., Cann, G., Carter, J., Chang, C., Chawla, H., Chen, X., Chien, T.C., Civello, D., Davydov, K., Demas, V., Desai, M., Dong, Z., Fayzullina, S., Fields, A.P., Filippova, D., Freese, P., Fung, E.T., Gnerre, S., Gross, S., Halks-Miller, M., Hall, M.P., Hartman, A.-R., Hou, C., Hubbell, E., Hunkapiller, N., Jagadeesh, K., Jamshidi, A., Jiang, R. et al. (2020). Sensitive and specific multi-cancer detection and localization using methylation signatures in cell-free DNA. *Ann. Oncol.* 31: 745–759.

Liu, R., Chen, X., Du, Y., Yao, W., Shen, L., Wang, C., Hu, Z., Zhuang, R., Ning, G., Zhang, C., Yuan, Y., Li, Z., Zen, K., Ba, Y., and Zhang, C.-Y. (2012). Serum microRNA expression profile as a biomarker in the diagnosis and prognosis of pancreatic cancer. *Clin. Chem.* 58: 610–618.

Liu, X., Liu, L., Ji, Y., Li, C., Wei, T., Yang, X., Zhang, Y., Cai, X., Gao, Y., Xu, W., Rao, S., Jin, D., Lou, W., Qiu, Z., and Wang, X. (2019). Enrichment of short mutant cell-free DNA fragments enhanced detection of pancreatic cancer. *EBioMedicine* 41: 345–356.

Løberg, M., Lousdal, M.L., Bretthauer, M., and Kalager, M. (2015). Benefits and harms of mammography screening. *Breast Cancer Res.* 17: 63.

Maire, F., Micard, S., Hammel, P., Voitot, H., Lévy, P., Cugnenc, P.H., Ruszniewski, P., and Puig, P.L. (2002). Differential diagnosis between chronic pancreatitis and pancreatic cancer: Value of the detection of KRAS2 mutations in circulating DNA. *Br. J. Cancer* 87: 551–554.

Marmot, M.G., Altman, D.G., Cameron, D.A., Dewar, J.A., Thompson, S.G., and Wilcox, M. (2013). The benefits and harms of breast cancer screening: An independent review. *Br. J. Cancer* 108: 2205–2240.

Matsubayashi, H. (2011). Familial pancreatic cancer and hereditary syndromes: Screening strategy for high-risk individuals. *J. Gastroenterol.* 46: 1249–1259.

Melo, S.A., Luecke, L.B., Kahlert, C., Fernandez, A.F., Gammon, S.T., Kaye, J., Lebleu, V.S., Mittendorf, E.A., Weitz, J., Rahbari, N., Reissfelder, C., Pilarsky, C., Fraga, M.F., Piwnica-Worms, D., and Kalluri, R. (2015). Glypican-1 identifies cancer exosomes and detects early pancreatic cancer. *Nature* 523: 177–182.

Melo, S.A., Sugimoto, H., O'Connell, J.T., Kato, N., Villanueva, A., Vidal, A., Qiu, L., Vitkin, E., Perelman, L.T., Melo, C.A., Lucci, A., Ivan, C., Calin, G.A., and Kalluri, R. (2014). Cancer exosomes perform cell-independent microRNA biogenesis and promote tumorigenesis. *Cancer Cell* 26: 707–721.

Mitchell, P.J., Welton, J., Staffurth, J., Court, J., Mason, M.D., Tabi, Z., and Clayton, A. (2009). Can urinary exosomes act as treatment response markers in prostate cancer? *J. Transl. Med.* 7: 4.

Mitchell, P.S., Parkin, R.K., Kroh, E.M., Fritz, B.R., Wyman, S.K., Pogosova-Agadjanyan, E.L., Peterson, A., Noteboom, J., O'Briant, K.C., Allen, A., Lin, D.W., Urban, N., Drescher, C.W., Knudsen, B.S., Stirewalt, D.L., Gentleman, R., Vessella, R.L., Nelson, P.S., Martin, D.B., and Tewari, M. (2008). Circulating microRNAs as stable blood-based markers for cancer detection. *Proc. Natl. Acad. Sci.* 105: 10513–10518.

Morani, A.C., Hanafy, A.K., Ramani, N.S., Katabathina, V.S., Yedururi, S., Dasyam, A.K., and Prasad, S.R. (2020). Hereditary and sporadic pancreatic ductal adenocarcinoma: Current update on genetics and imaging. *Radiol. Imaging Cancer* 2: e190020.

Mouliere, F., Chandrananda, D., Piskorz, A.M., Moore, E.K., Morris, J., Ahlborn, L.B., Mair, R., Goranova, T., Marass, F., Heider, K., Wan, J.C.M., Supernat, A., Hudecova, I., Gounaris, I., Ros, S., Jimenez-Linan, M., Garcia-Corbacho, J., Patel, K., Østrup, O., Murphy, S., Eldridge, M.D., Gale, D., Stewart, G.D., Burge, J., Cooper, W.N., Van Der Heijden, M.S., Massie, C.E., Watts, C., Corrie, P., Pacey, S., Brindle, K.M., Baird, R.D., Mau-Sørensen, M., Parkinson, C.A., Smith, C.G., Brenton, J.D., and Rosenfeld, N. (2018). Enhanced detection of circulating tumor DNA by fragment size analysis. *Sci. Transl. Med.* 10.

Nadal, E., Truini, A., Nakata, A., Lin, J., Reddy, R.M., Chang, A.C., Ramnath, N., Gotoh, N., Beer, D.G., and Chen, G. (2015). A novel serum 4-microRNA signature for lung cancer detection. *Sci. Rep.* 5: 12464.

Peng, C., Wang, J., Gao, W., Huang, L., Liu, Y., Li, X., Li, Z., and Yu, X. (2021). Meta-analysis of the diagnostic performance of circulating micrornas for pancreatic cancer. *Int. J. Med. Sci.* 18: 660–671.

Phallen, J., Sausen, M., Adleff, V., Leal, A., Hruban, C., White, J., Anagnostou, V., Fiksel, J., Cristiano, S., Papp, E., Speir, S., Reinert, T., Orntoft, M.W., Woodward, B.D., Murphy, D.,

Parpart-Li, S., Riley, D., Nesselbush, M., Sengamalay, N., Georgiadis, A., Li, Q.K., Madsen, M.R., Mortensen, F.V., Huiskens, J., Punt, C., Van Grieken, N., Fijneman, R., Meijer, G., Husain, H., Scharpf, R.B., Diaz, l.A., Jr., Jones, S., Angiuoli, S., Ørntoft, T., Nielsen, H.J., Andersen, C.L., and Velculescu, V.E. (2017). Direct detection of early-stage cancers using circulating tumor DNA. *Sci. Transl. Med.* 9.

Pietryga, J.A. and Morgan, D.E. (2015). Imaging preoperatively for pancreatic adenocarcinoma. *J. Gastrointest Oncol.* 6: 343–357.

Pisani, P., Oliver, W.E., Parkin, D.M., Alvarez, N., and Vivas, J. (1994). Case-control study of gastric cancer screening in Venezuela. *Br. J. Cancer* 69: 1102–1105.

Pritchard, C.C., Kroh, E., Wood, B., Arroyo, J.D., Dougherty, K.J., Miyaji, M.M., Tait, J.F., and Tewari, M. (2012). Blood cell origin of circulating microRNAs: A cautionary note for cancer biomarker studies. *Cancer Prev. Res. (Phila)* 5: 492–497.

Qian, X. and Ruan, L. (2018). APC gene promoter aberrant methylation in serum as a biomarker for breast cancer diagnosis: A meta-analysis. *Thorac. Cancer* 9: 284–290.

Rajagopal, C. and Harikumar, K.B. (2018). The origin and functions of exosomes in cancer. *Front. Oncol.* 8: 66.

Rodda, A.E., Parker, B.J., Spencer, A., and Corrie, S.R. (2018). Extending circulating tumor DNA analysis to ultralow abundance mutations: Techniques and challenges. *ACS Sens.* 3: 540–560.

Rodriguez, B.J., Córdoba, G.D., Aranda, A.G., Álvarez, M., Vicioso, L., Pérez, C.L., Hernando, C., Bermejo, B., Parreño, A.J., Lluch, A., Ryder, M.B., Jones, F.S., Fredebohm, J., Holtrup, F., Queipo-Ortuño, M.I., and Alba, E. (2019). Detection of TP53 and PIK3CA mutations in circulating tumor DNA using next-generation sequencing in the screening process for early breast cancer diagnosis. *J. Clin. Med.* 8.

Roth, C., Rack, B., Müller, V., Janni, W., Pantel, K., and Schwarzenbach, H. (2010). Circulating microRNAs as blood-based markers for patients with primary and metastatic breast cancer. *Breast Cancer Res.* 12: R90.

Santarpia, M., Altavilla, G., Salazar, M.F., Magri, I., Pettineo, G., Benecchi, S., and Rosell, R. (2011). Tyrosine kinase inhibitors for non-small-cell lung cancer: Finding patients who will be responsive. *Expert Rev. Respir. Med.* 5: 413–424.

Sausen, M., Phallen, J., Adleff, V., Jones, S., Leary, R.J., Barrett, M.T., Anagnostou, V., Parpart-Li, S., Murphy, D., Kay Li, Q., Hruban, C.A., Scharpf, R., White, J.R., O'Dwyer, P.J., Allen, P.J., Eshleman, J.R., Thompson, C.B., Klimstra, D.S., Linehan, D.C., Maitra, A., Hruban, R.H., Diaz, L.A., JR., Von Hoff, D.D., Johansen, J.S., Drebin, J.A., and Velculescu, V.E. (2015). Clinical implications of genomic alterations in the tumour and circulation of pancreatic cancer patients. *Nat. Commun.* 6: 7686.

Schaefer, A., Jung, M., Miller, K., Lein, M., Kristiansen, G., Erbersdobler, A., and Jung, K. (2010). Suitable reference genes for relative quantification of miRNA expression in prostate cancer. *Exp. Mol. Med.* 42: 749–758.

Shan, M., Yin, H., Li, J., Li, X., Wang, D., Su, Y., Niu, M., Zhong, Z., Wang, J., Zhang, X., Kang, W., and Pang, D. (2016). Detection of aberrant methylation of a six-gene panel in serum DNA for diagnosis of breast cancer. *Oncotarget* 7: 18485–18494.

Shen, J., Hu, Q., Schrauder, M., Yan, L., Wang, D., Medico, L., Guo, Y., Yao, S., Zhu, Q., Liu, B., Qin, M., Beckmann, M.W., Fasching, P.A., Strick, R., Johnson, C.S., Ambrosone, C.B., Zhao, H., and Liu, S. (2014). Circulating miR-148b and miR-133a as biomarkers for breast cancer detection. *Oncotarget* 5: 5284–5294.

Shi, Y., Wang, Z., Zhu, X., Chen, L., Ma, Y., Wang, J., Yang, X., and Liu, Z. (2020). Exosomal miR-1246 in serum as a potential biomarker for early diagnosis of gastric cancer. *Int. J. Clin. Oncol.* 25: 89–99.

Shin, V.Y., Ng, E.K., Chan, V.W., Kwong, A., and Chu, K.M. (2015). A three-miRNA signature as promising non-invasive diagnostic marker for gastric cancer. *Mol. Cancer* 14: 202.

Song, J., Bai, Z., Han, W., Zhang, J., Meng, H., Bi, J., Ma, X., Han, S., and Zhang, Z. (2012). Identification of suitable reference genes for qPCR analysis of serum microRNA in gastric cancer patients. *Dig. Dis. Sci.* 57: 897–904.

Tan, W., Liu, B., Qu, S., Liang, G., Luo, W., and Gong, C. (2018). MicroRNAs and cancer: Key paradigms in molecular therapy (Review). *Oncol. Lett.* 15: 2735–2742.

Tavano, F., Gioffreda, D., Valvano, M.R., Palmieri, O., Tardio, M., Latiano, T.P., Piepoli, A., Maiello, E., Pirozzi, F., and Andriulli, A. (2018). Droplet digital PCR quantification of miR-1290 as a circulating biomarker for pancreatic cancer. *Sci. Rep.* 8: 16389.

Tsen, A., Barbara, M., and Rosenkranz, L. (2018). Dilemma of elevated CA 19-9 in biliary pathology. *Pancreatology* 18: 862–867.

Uemura, T., Hibi, K., Kaneko, T., Takeda, S., Inoue, S., Okochi, O., Nagasaka, T., and Nakao, A. (2004). Detection of K-ras mutations in the plasma DNA of pancreatic cancer patients. *J. Gastroenterol.* 39: 56–60.

Underhill, H.R., Kitzman, J.O., Hellwig, S., Welker, N.C., Daza, R., Baker, D.N., Gligorich, K.M., Rostomily, R.C., Bronner, M.P., and Shendure, J. (2016). Fragment length of circulating tumor DNA. *PLoS Genet.* 12: e1006162.

Vila-Navarro, E., Duran-Sanchon, S., Vila-Casadesús, M., Moreira, L., Ginès, À., Cuatrecasas, M., Lozano, J.J., Bujanda, L., Castells, A., and Gironella, M. (2019). Novel circulating miRNA signatures for early detection of pancreatic neoplasia. *Clin. Transl. Gastroenterol.* 10: e00029.

Wan, J.C.M., Massie, C., Garcia-Corbacho, J., Mouliere, F., Brenton, J.D., Caldas, C., Pacey, S., Baird, R., and Rosenfeld, N. (2017). Liquid biopsies come of age: Towards implementation of circulating tumour DNA. *Nat. Rev. Cancer* 17: 223–238.

Wan, Y., Liu, B., Lei, H., Zhang, B., Wang, Y., Huang, H., Chen, S., Feng, Y., Zhu, L., Gu, Y., Zhang, Q., Ma, H., and Zheng, S.Y. (2018). Nanoscale extracellular vesicle-derived DNA is superior to circulating cell-free DNA for mutation detection in early-stage non-small-cell lung cancer. *Ann. Oncol.* 29: 2379–2383.

Wang, C., Wang, J., Cui, W., Liu, Y., Zhou, H., Wang, Y., Chen, X., Chen, X., and Wang, Z. (2021). Serum exosomal miRNA-1226 as potential biomarker of pancreatic ductal adenocarcinoma. *Onco Targets Ther* 14: 1441–1451.

Wang, J., Zhao, X., Wang, Y., Ren, F., Sun, D., Yan, Y., Kong, X., Bu, J., Liu, M., and Xu, S. (2020). circRNA-002178 act as a ceRNA to promote PDL1/PD1 expression in lung adenocarcinoma. *Cell Death Dis.* 11: 32.

Wang, N., Wang, L., Yang, Y., Gong, L., Xiao, B., and Liu, X. (2017). A serum exosomal microRNA panel as a potential biomarker test for gastric cancer. *Biochem. Biophys. Res. Commun.* 493: 1322–1328.

Wen, J., Zheng, T., Hu, K., Zhu, C., Guo, L., and Ye, G. (2017). Promoter methylation of tumor-related genes as a potential biomarker using blood samples for gastric cancer detection. *Oncotarget* 8: 77783–77793.

Wiener, R.S., Schwartz, L.M., Woloshin, S., and Welch, H.G. (2011). Population-based risk for complications after transthoracic needle lung biopsy of a pulmonary nodule: An analysis of discharge records. *Ann. Intern. Med.* 155: 137–144.

Wu, J., Li, G., Wang, Z., Yao, Y., Chen, R., Pu, X., and Wang, J. (2015). Circulating microRNA-21 is a potential diagnostic biomarker in gastric cancer. *Dis. Markers* 2015: 435656.

Wu, Q., Yu, L., Lin, X., Zheng, Q., Zhang, S., Chen, D., Pan, X., and Huang, Y. (2020). Combination of serum miRNAs with serum exosomal mirnas in early diagnosis for non-small-cell lung cancer. *Cancer Manag. Res.* 12: 485–495.

Xiao, D., Dong, Z., Zhen, L., Xia, G., Huang, X., Wang, T., Guo, H., Yang, B., Xu, C., Wu, W., Zhao, X., and Xu, H. (2020). Combined exosomal GPC1, CD82, and serum CA19-9 as multiplex targets: A specific, sensitive, and reproducible detection panel for the diagnosis of pancreatic cancer. *Mol. Cancer Res.* 18: 300.

Xu, Y.F., Hannafon, B.N., Zhao, Y.D., Postier, R.G., and Ding, W.Q. (2017). Plasma exosome miR-196a and miR-1246 are potential indicators of localized pancreatic cancer. *Oncotarget* 8: 77028–77040.

Yan, L.X., Huang, X.F., Shao, Q., Huang, M.Y., Deng, L., Wu, Q.L., Zeng, Y.X., and Shao, J.Y. (2008). MicroRNA miR-21 overexpression in human breast cancer is associated with advanced clinical stage, lymph node metastasis and patient poor prognosis. *Rna* 14: 2348–2360.

Yang, X., Su, W., Chen, X., Geng, Q., Zhai, J., Shan, H., Guo, C., Wang, Z., Fu, H., Jiang, H., Lin, J., Lagisetty, K.H., Zhang, J., Li, Y., Yang, S., Massion, P.P., Beer, D.G., Chang, A.C., Ramnath, N., and Chen, G. (2019). Validation of a serum 4-microRNA signature for the detection of lung cancer. *Transl. Lung Cancer Res.* 8: 636–648.

Young, M.R., Wagner, P.D., Ghosh, S., Rinaudo, J.A., Baker, S.G., Zaret, K.S., Goggins, M., and Srivastava, S. (2018a). Validation of biomarkers for early detection of pancreatic cancer: Summary of the alliance of pancreatic cancer consortia for biomarkers for early detection workshop. *Pancreas* 47: 135–141.

Young, R.P., Christmas, T., and Hopkins, R.J. (2018b). Multi-analyte assays and early detection of common cancers. *J. Thorac. Dis.* 10: S2165–s2167.

Zhang, C., Yu, W., Wang, L., Zhao, M., Guo, Q., Lv, S., Hu, X., and Lou, J. (2017). DNA methylation analysis of the SHOX2 and RASSF1A panel in bronchoalveolar lavage fluid for lung cancer diagnosis. *J. Cancer* 8: 3585–3591.

Zhang, X., Zhao, W., Wei, W., You, Z., Ou, X., Sun, M., Yin, Y., Tang, X., Zhao, Z., Hu, C., Liu, F., Deng, J., Mao, L., Zhou, D., Ren, Y., Li, X., Zhang, S., Liu, C., Geng, J., Yao, G., Song, B., Liu, Y., Li, D., Jiang, Y., Chen, Y., Zhao, Y., Yu, S., and Pang, D. (2019). Parallel analyses of somatic mutations in plasma circulating tumor DNA (ctDNA) and matched tumor tissues in early-stage breast cancer. *Clin.l Cancer Res.* 25: 6546.

Zhao, G., Jiang, T., Liu, Y., Huai, G., Lan, C., Li, G., Jia, G., Wang, K., and Yang, M. (2018a). Droplet digital PCR-based circulating microRNA detection serve as a promising diagnostic method for gastric cancer. *BMC Cancer* 18: 676.

Zhao, R., Zhang, Y., Zhang, X., Yang, Y., Zheng, X., Li, X., Liu, Y., and Zhang, Y. (2018b). Exosomal long noncoding RNA HOTTIP as potential novel diagnostic and prognostic biomarker test for gastric cancer. *Mol. Cancer* 17: 68.

Zhong, G., Wang, K., Li, J., Xiao, S., Wei, W., and Liu, J. (2020). Determination of serum exosomal H19 as a noninvasive biomarker for breast cancer diagnosis. *Onco Targets Ther* 13: 2563–2571.

Zhou, H., Shen, W., Zou, H., Lv, Q., and Shao, P. (2020). Circulating exosomal long non-coding RNA H19 as a potential novel diagnostic and prognostic biomarker for gastric cancer. *J. Int. Med. Res.* 48: 300060520934297.

Zhu, X.L., Ren, L.F., Wang, H.P., Bai, Z.T., Zhang, L., Meng, W.B., Zhu, K.X., Ding, F.H., Miao, L., Yan, J., Wang, Y.P., Liu, Y.Q., Zhou, W.C., and Li, X. (2019). Plasma microRNAs as potential new biomarkers for early detection of early gastric cancer. *World J. Gastroenterol.* 25: 1580–1591.

12

Aberrant DNA Methylation of Tumor Suppressor Genes and Oncogenes as Cancer Biomarkers

Eid Alshammari, Yingxue Zhang, Jacob Sobota, and Zhe Yang

Department of Biochemistry, Microbiology, and Immunology, Wayne State University School of Medicine, Detroit, Michigan, USA

Introduction

Cancer is, worldwide, an increasingly pressing health issue, with an estimate of 18.1 million new cases in 2018 (Bray et al. 2018). According to a recent population-based cohort study, cancer is the leading cause of death among middle-aged adults in developed countries (Dagenais et al. 2020). Unlike in the case of other diseases, for instance infectious, whose mortality rates have declined over the past few decades, cancer-related death rates have increased. This could be attributed to the complexity of cancer, which is characterized by uncontrolled cell growth, escape of immunosurveillance, and aberrant cellular regulation. In addition, the driving forces that contribute to carcinogenesis are multiple and diverse; they include genetic lesions, epigenome alterations, and environmental factors. Among those forces, epigenetic modifications have recently gained interest as an important player in cancer development. Unlike genetic mutations, whose impact on cancer is evident and has been extensively investigated, cancer epigenetics is a growing field of research, which could provide new insights into important tumorigenic pathways and potential targets for cancer therapy.

A link between genetics and developmental biology was proposed by the biologist Conrad Waddington, who was the first to coin the term "epigenetics": he described by it the mechanisms of phenotypic adaptation to the environment (Holliday 2006). Epigenetic modifications could be broadly defined as changes in gene expression as a result of some heritable or environmentally induced DNA alterations that do not affect the primary nucleotide sequence. Currently, epigenetics is known as an important mechanism by which an organism maintains normal development and adapts to environmental changes (Baylin and Jones 2011). The epigenetic status of the cell is under strict regulation, so that adequate cell differentiation be preserved. However, this regulation can be disrupted by various factors. Numerous studies pointed out the significance of epigenetic changes as an early event in cancer development and progression (Nebbioso et al. 2018). The epigenetic status is sensitive to genetic mutations and environmental factors such as harmful chemicals, toxins, or heavy metals (Teh et al. 2014). For this reason it is widely accepted that cancer is attributable to a mixture of genetic and epigenetic alternations effected or brought about by environmental change. The main epigenetic alterations that are associated with tumor development are aberrant DNA methylation, post-translational histone modification, chromatin remodeling, and non-coding

Genomic and Epigenomic Biomarkers of Toxicology and Disease: Clinical and Therapeutic Actions,
First Edition. Edited by Saura C. Sahu.

RNAs patterns (Margueron and Reinberg 2010). Not only have these modifications been observed at the initiation stage of carcinogenesis, but they are also associated with cancer progression and metastasis. Histone modifications, including acetylation, methylation, phosphorylation, and ubiquitination, which occur at the N-terminal tails of histones, can impact genome integrity and stability (Audia and Campbell 2016). DNA methylation, which takes place mostly at CpG dinucleotides in the human genome, can cause alterations in gene expression and dysregulation of tumor suppressor genes and proto-oncogenes (Das and Singal 2004; Kulis and Esteller 2010). Several signaling pathways that are involved in tumorigenesis are affected by these epigenetic modifications (Putiri and Robertson 2011). Thus, current data have established that epigenetic modifications can be used as biomarkers for the early detection of tumors, for drug discovery, and for the assessment of response to therapy.

DNA methylation is a central, well-studied epigenetic modification in health and disease conditions. It is achieved by covalent attachment of a methyl group to the carbon-5 atom of the cytosine within a CpG dinucleotide, through DNA methyltransferases (DNMTs), to form 5-methylcytosine (Esteller 2008). The main effect of DNA methylation is that it changes gene expression and generally results in gene silencing. In cancer, this change is associated with dysregulated signaling cascades and genes, as well as with the dysregulation of critical pathways that control normal cellular processes (Shen et al. 2012). The CpG dinucleotides aggregated at the gene promoters, called CpG islands, are usually unmethylated in normal cells (Deaton and Bird 2011). However, abnormal methylation of CpG islands is frequently observed in cancer; in fact it represents a hallmark of cancer. There are two forms of change in DNA methylation that contribute to cancer initiation and progression, namely *de novo* methylation of selected CpG islands and demethylation of particular sites of the genome (Kulis and Esteller 2010). The gain of DNA methylation, the "DNA hypermethylation" that occurs at the promoter CpG sequence, plays a critical role in cancer development by inhibiting key tumor suppressor genes involved in the regulation of cell cycle, DNA repair, apoptosis, and cell adhesion and motility (Putiri and Robertson 2011). The loss of DNA methylation, also known as "DNA hypomethylation," is another aberrant event that occurs early in human cancer, causing genomic instability through the impaired silencing of repetitive DNA (Ehrlich 2009; Negrini et al. 2010). Therefore malignant cells rely for their growth and metastasis on DNA hypermethylation to block the gene expression of significant tumor suppressor genes or on the activation of oncogenes through hypomethylation (or on both).

The use of DNA methylation as a biomarker could provide a powerful tool for cancer diagnosis, prognosis, and treatment. DNA methylation can be used to investigate how hypermethylated tumor suppressor genes or hypomethylated proto-oncogenes are involved in cancer initiation and development. More importantly, it will allow us not only to use the methylation status of these genes as epigenetic biomarkers for early cancer detection but also to monitor cancer progression and response to therapy, as well as to develop novel, epigenetic-based therapies. In cancer, potential DNA methylation biomarkers can be obtained from the genome-wide DNA methylation profiling of both healthy individuals and cancer patients. This initial approach is essential to identifying candidate genes or molecules as possible biomarkers that can be validated and further tested through multiple other epigenomic analyses. Here we review DNA methylation as an emerging and fast-developing field in cancer biomarker research. We focus on the role of aberrant DNA methylation as a major epigenetic alteration in the activation and inactivation of genes and on how their expression patterns are associated with the clinicopathological features of different cancers (Table 12.1). Our ultimate goal is to inform how methylation status of these genes could be used to identify novel diagnostic and prognostic biomarkers or therapeutic targets.

Table 12.1 DNA methylation-based cancer biomarkers.

Gene	Methylation status	Cancer type	Correlated clinicopathologic feature
P53	hypermethylated	Rectal cancer	Poor prognosis (Benard et al. 2014)
		Breast cancer	Shortened telomere length, progesterone receptor status, tumor cell proliferation, and breast cancer subtypes (Radpour et al. 2010, Zeng et al. 2020)
		Acute lymphoblastic leukemia	Decreased P53 expression level (Agirre et al. 2003)
RASSF1A	hypermethylated	Breast cancer	Increased BC risk, age, tumor stage, and poor survival (Euhus et al. 2008, Karray-Chouayekh et al. 2010)
		Esophageal squamous cell carcinoma	Regional lymph node involvement, histological differentiation, and tumor stage (Mao et al. 2011)
		Gastric cancer	Poor survival, metastasis, tumor late-stage, and tumor histology (Pimson et al. 2016)
		Non-small-cell lung carcinoma	Increased NSCL cancer risk, shorter disease-free survival DFS, shorter overall survival, and prediction of longer DFS with paclitaxel-carboplatin treatment (Li et al. 2012, De Fraipont et al. 2012)
		Colorectal cancer	Poor prognosis, tumor stage, and metastasis (Nilsson et al. 2013, Sinha et al. 2013)
CDH1	hypermethylated	Breast cancer	Lymphovascular invasion, infiltrating ductal histology, and lack of estrogen receptor (Shinozaki et al. 2005)
		Hepatocellular carcinoma	Poor overall survival (Wu et al. 2019)
		Gastric cancer	Disease risk, and *Helicobacter pylori* status (Zeng et al. 2015)
RAS	hypomethylated	Non-muscle invasive bladder cancer	Tumor stage, grade, type, and size (Tripathi et al. 2021)
		Hepatocellular carcinoma	Increased risk of tumor infiltration and metastasis, HCV3a-derived HCC risk (Shen et al. 1998, Maryam and Idrees 2018)
		Non-small-cell lung cancer	Induction of *Ras* allele loss and increased cancer risk (Vachtenheim et al. 1994)
ONECUT2	hypermethylated	Bladder cancer	Accurate prediction of bladder cancer in hematuria patients (Zhang et al. 2019, Wu et al. 2020)
		Upper tract urinary carcinoma	Prediction of upper tract urinary carcinoma risk in patients with hematuria (Xu et al. 2020)

(*Continued*)

Table 12.1 (Continued)

Gene	Methylation status	Cancer type	Correlated clinicopathologic feature
	hypomethylated	Gastric cancer	Pre-cancer lesions and early tumor stage (Seo et al. 2020)
BORIS	hypomethylated	Epithelial ovarian cancer	Advanced stage and poor prognosis (Woloszynska-Read et al. 2011)
		Hepatocellular carcinoma	Tumor size, TNM stage, shorter disease-free survival, microvascular involvement, postoperative recurrence, and worse overall survival (He et al. 2017, Chen et al. 2020b)
		Endometrial uterine cancer	Stages of cancer development and progression (Hoivik et al. 2014)

Aberrant DNA Methylation of Tumor Suppressor Genes

P53

Methylation of CpG islands at the promoter regions of tumor suppressor genes is a substantial epigenetic mechanism for the regulation of these genes (Jones 2012). Aberrant DNA methylation of those genes has been reported in many types of cancer as a potential critical driver of carcinogenesis (Heyn et al. 2014; Shimazu et al. 2015; Tessema et al. 2009). Methylation of CpG dinucleotides within the promoter region or CCWGG motifs of the tumor suppressor *p53* gene has been shown to repress the expression and function of this gene and its product, which may contribute to the onset and progression of cancer (Pogribny and James 2002). P53, a 53-kD nuclear phosphoprotein encoded by a 20-Kb gene located on chromosome 17, is the main component of the ARF-MDM2-p53 pathway (Lane 1994). As the "guardian of the genome," p53 is recognized as a master regulator of various critical cellular processes, including genomic stability, cell cycle control, senescence, and apoptotic cell death in response to a wide range of carcinogenic and damage signals (Chen 2016). *P53* is the most frequently inactivated tumor suppressor gene, which is mutated in approximately 50% of human cancers (Duffy et al. 2017). Although its inactivation frequency is largely due to the mutation or deletion of the gene, a growing body of evidence suggests that several other mechanisms, including amplification of p53 negative regulators and hypermethylation of p53 pathway genes, can serve as an alternative to control p53 activity and thus promote tumorigenesis (Herrero et al. 2016; Jones and Baylin 2007). Cytosine methylation at the promoter region of *p53* was found to lead to a reduction in its expression (Schroeder and Mass 1997). This result demonstrates that DNA methylation can transcriptionally regulate p53 and attenuate its critical function in the cell cycle and apoptosis, which may contribute to the development of cancers. Aberrant hypermethylation of the *p53* promoter has been reported in several human neoplasms (Amatya et al. 2005; Pogribny and James 2002; Zeng et al. 2020) (Table 12.1). For instance, 52% of a cohort of ovarian cancer patients exhibited DNA methylation in the *p53* promoter region (Chmelarova et al. 2013). Aberrant hypermethylation of *p53* has also been observed in 12%, 16%, and 32% of cervical cancer, breast cancer, and acute lymphoblastic leukemia patients, respectively (Agirre et al. 2003; Jha et al. 2012; Kang et al. 2001). These studies have revealed that methylation of the *p53* promoter is associated with the reduced expression of p53, which may be involved in cancer initiation and progression. However, one study showed that *p53* promoter methylation was not associated with the p53 mRNA and protein expression levels in glioblastoma samples (Jesionek-kupnicka et al. 2014). This suggests the existence of distinct, cancer-specific mechanisms that lead to perturbation of p53 expression. The perturbation may not depend only on *p53* promoter methylation but also on the hypermethylation of p53 pathway regulators such as *TBX2*, *TWIST1*, and *CDKN2A*. These gene products are known to inhibit ARF and MDM2, the other main components of the p53 pathway, and therefore disrupt p53 homeostasis (Nag et al. 2013). Hypermethylation of *PIK3R1* and *PIK3R5* also plays a role in the regulation of the *MDM2* gene in the p53 pathway by regulating the protein kinase B (*PKB*) gene, a key component of the AKT/PI3K signaling pathway (Lu et al. 2020). It has also been shown that *ARF* hypermethylation led to the deregulation of the ARF-MDM2-p53 pathway and decreased the tumor-suppressing activity of p53 (Nyiraneza et al. 2012).

Despite extensive investigations into the p53 pathway and its importance in tumor suppression, there is a lack of comprehensive analysis of the methylation statuses of *p53* and of their association with the expression levels we find across different types of human cancer. This leaves a gap in our knowledge and understanding of when, where, and why hypermethylation of *p53* is a significant event in human cancers. Whether methylated *p53* is a primary or a secondary driver of human malignancies remains largely unknown as well. But a few studies revealed the potential prognostic value of *p53* methylation, if combined with the methylation status of *APAF1* and *BCL2* genes. The methylation of these apoptotic genes was found to be correlated with the apoptotic status of rectal cancer cells and patient survival (Benard et al. 2014). The use of p53 as a diagnostic and prognostic biomarker has also been suggested in breast cancer (Radpour et al. 2010; Zeng et al. 2020), epithelial ovarian cancer (Chmelarova et al. 2013), and colorectal cancer (CRC) (Lam et al. 2016). These findings bring researchers a step closer to the discovery of clinically relevant methylation-based prognostic biomarkers for multiple invasive cancers.

RASSF1A

Ras-association domain family 1 isoform A (RASSF1A) is a well-defined tumor suppressor that gets frequently inactivated via aberrant promoter methylation in different types of cancer (Grawenda and O'Neill 2015). It consists of a microtubule-associated scaffold and a Ras-association domain and serves as a hub for the assembly of some tumor suppressor complexes. RASSF1A has a vital role in various biological processes, which range from participation in DNA repair and apoptosis to regulation of the cell cycle, microtubule network, and cell migration (Donninger et al. 2007). It is also involved in several cellular signaling pathways, including the RAS pathway, the Hippo pathway, and the estrogen receptor signaling pathway. Therefore it is not surprising that the inactivation of *RASSF1A* via aberrant promoter methylation has an adverse impact on these mechanisms and promotes the development and progression of cancer.

The methylation status of the *RASSF1A* gene in normal human tissues ranges from unmethylated patterns in the stomach and colon to slightly higher methylation levels in the liver and pancreas (Malpeli et al. 2019). The aberrant hypermethylation of the promoter CpG islands of *RASSF1A* has been shown to block its gene expression, to impact some other tumor suppressor genes, and to associate with the onset of different human cancers (Bai et al. 2016; Malpeli et al. 2019). The downregulation of *RASSF1A* resulted from the hypermethylation is frequently observed in various human cancers (Table 12.1). The correlation of *RASSF1A* methylation with cancer survival rates and disease risk suggests that this process has potential clinical value as a diagnostic and prognostic biomarker (Hesson et al. 2007). If we look at gastrointestinal (GI) cancers, the frequency of *RASSF1A* methylation ranges up to 31%, 36%, 50%, 75%, and 78% respectively in gastric cancer (GC), colorectal carcinoma, esophageal squamous cell carcinoma (ESCC), pancreatic endocrine tumor (PET), and hepatocellular carcinoma (HCC) (Malpeli et al. 2019). Hypermethylation of *RASSF1A* was found to associate with a significant risk of developing ESCC; a 5.9 odds ratio was signaled in a study with 112 ESCC patients and 235 normal controls (Zhou et al. 2013). Another study, with seventy-one esophageal cancer tissues and thirty-five adjacent normal tissues, showed that the promoter methylation of *RASSF1A* was associated with a loss of RASSF1A expression and a significant reduction in survival rates (Guo et al. 2016). Furthermore, it has been reported that, in a cohort of 124 ESCC samples, the hypermethylation of the *RASSF1A* promoter was significantly associated with several clinicopathological features such as regional lymph node involvement, histological differentiation, and tumor stage (Mao et al. 2011). A meta-analysis of 1,215 GC tissues showed a strong correlation between hypermethylation-induced RASSF1A inactivation and increased cancer susceptibility (Shi et al. 2014). Another study, which used blood samples from GC patients, revealed a high frequency of aberrant methylation of *RASSF1A* in 83% of patients, by comparison with 5.5% in the matched controls. This study also found a strong association between the high hypermethylation status of *RASSF1A* and metastasis, late stages of cancer, and lowered survival rates (Pimson et al. 2016). Together, these findings indicate that the hypermethylation status of *RASSF1A* can be used as a prognostic biomarker for gastrointestinal cancers.

The use of the hypermethylation status of *RASSF1A* as an epigenetic biomarker has also been suggested for lung cancer. A study using blood samples and lung cancer tissues from sixty-three lung cancer patients and thirty-six normal controls revealed RASSF1A as having potential diagnostic and prognostic values in combination with other tumor suppressor genes (Hsu et al. 2007). The methylation status of *RASSF1A* showed 84% concordance between the blood samples and the lung cancer tissues. The susceptibility to developing lung cancer was higher in lung cancer patients with *RASSF1A* methylation. This observation is supported by another study, where the aberrant promoter methylation of *RASSF1A* and *RARβ* was found respectively in 80% and 86% of lung cancer patients, while the control group lacked any methylation in these two genes (Li et al. 2012). In addition, the increased *RASSF1A* and *RARβ* methylation was correlated with a high risk for developing lung cancer. Methylated *RASSF1A* was also examined as a marker for neoadjuvant chemotherapy in lung cancer treatment. *RASSF1A* methylation was significantly associated with poor survival rates in 528 patients with non-small-cell lung carcinoma (NSCLC). However, differential *RASSF1A* methylation was associated with different outcomes of chemotherapy agents (de Fraipont et al. 2012). NSCLC patients who received a paclitaxel-carboplatin chemotherapy showed longer disease-free survival than patients who received a gemcitabine-based chemotherapy. This points to the potential clinical value of RASSF1A as a biomarker for guiding treatment decisions for NSCLC patients. *RASSF1A* and *RARβ* methylation is also useful for assessing the outcomes of chemotherapy agents with circulating cell-free DNA (cirDNA) from blood samples. The methylation status of *RASSF1A* and *RARβ* in cirDNA was up to three times higher in sixty patients with lung cancer than in thirty-two normal controls (de Fraipont et al. 2012; Ponomaryova et al. 2013). However, the index of methylation for *RASSF1A* and *RARβ* was significantly lowered when lung cancer patients received chemotherapy or underwent total tumor resection. This, again, indicates the value of using RASSF1A and RARβ as potential epigenetic biomarkers for lung cancer prognosis and treatment decisions (Ponomaryova et al. 2013).

The methylation frequency of *RASSF1A* has also been shown to be associated with a high susceptibility to breast cancer; this comes from a study of sixty-nine benign and malignant breast cancer patients and ninety-five healthy individuals (Euhus et al. 2008). Another study reported that the aberrant methylation of *RASSF1A* was detected in 87% of patients with breast tumors and correlated with tumor stage and poor survival rate (Karray-chouayekh et al. 2010). A methylation-sensitive high-resolution melting analysis revealed that 94% of breast cancer patients exhibited the methylation at the *RASSF1A* promoter. Interestingly, the tumor-adjacent regions, histologically intact, have significantly higher *RASSF1A* hypermethylation than the tumor-distant tissues. This finding indicates the role of the field cancerization that could be responsible for local cancer recurrence (Spitzwieser et al. 2015). A meta-analysis study of 1,795 breast cancer patients also suggested a strong correlation between *RASSF1A* hypermethylation and poor disease-free survival and overall survival rates (Jiang et al. 2012). Furthermore, it has been shown that the increased hypermethylation pattern in the *RASSF1A* promoter was significantly associated with tumor stage, metastasis, and poor prognosis in colorectal cancer (Nilsson et al. 2013; Sinha et al. 2013). Together, these pieces of evidence indicate that the epigenetic silencing of the *RASSF1A* gene through aberrant promoter hypermethylation has a pivotal role in the development and progression of breast cancer, which can be used as an epigenetic biomarker for the risk assessment, diagnosis, and prognosis of this disease.

Cadherin-1

Epithelial-cadherin (E-cadherin) is a master regulator of cell–cell adhesion, which is essential for the maintenance of normal epithelial architecture and tissue homeostasis (Kaszak et al. 2020; Oda and Takeichi 2011). Downregulation of the E-cadherin encoding gene (*CDH1*) by various mechanisms, including aberrant promoter hypermethylation, has been shown to stimulate cancer initiation, metastasis, and invasiveness (Kaszak et al. 2020). E-cadherin is a calcium-dependent transmembrane glycoprotein mainly expressed in epithelial tissues at the intercellular adherens junction (Berx et al. 1995; Oda and Takeichi 2011). It is composed of an extracellular domain with five cadherin repeats, a single transmembrane domain, and a cytoplasmic tail (Shapiro and Weis 2009). The calcium-binding sites within the extracellular domain mediate the E-cadherin interface between adjacent cells through calcium dimerization (Shapiro et al. 1995). The highly conserved cytoplasmic tail contains the juxtamembrane domain and the catenin-binding domain, which bind respectively to p120-catenin and β-catenin to recruit α-catenin (Ishiyama et al. 2010). The recruitment of α-catenin results in the formation of the cadherin–catenin complex, which links E-cadherin to the cellular actin cytoskeleton, thereby promoting cadherin-mediated cell adhesion (Takeichi 2018). It has been shown that the disruption of the cadherin–catenin complex contributes to the activation of Wnt oncogenic signaling cascades; it does so by promoting the release of β-catenin (Nelson and Nusse 2004). Upon induction by the binding of Wnt protein to the Frizzled receptor, the released β-catenin translocates to the nucleus and activates oncogenic genes, such as *c-myc*, to promote cancer cell migration, proliferation, and invasion (Nelson and Nusse 2004). Thus E-cadherin plays a vital role as a tumor suppressor gene through the sequestration of β-catenin, which is prevented from activating the Wnt pathway. E-cadherin also exerts its invasion suppressor role by maintaining epithelial cell polarization and differentiation through the formation of intercellular adhesion complexes (Bhagat et al. 2013).

Reduced E-cadherin expression enhances epithelial-to-mesenchymal transition (EMT), which is involved in the maintenance of cancer stem cells and in cancer drug resistance (Lombaerts et al. 2006; Song et al. 2019). A loss of E-cadherin expression through aberrant

DNA hypermethylation has been shown to downregulate the protein expression of E-cadherin, promoting cancer metastasis and invasiveness (Bhagat et al. 2013; Bruner and Derksen 2018; Machado et al. 2001; Wong et al. 2018). A study of twenty-two invasive lobular carcinomas (ILC) samples showed that about 77% of these samples exhibited DNA methylation at the *CDH1* promoter (Droufakou et al. 2001). The loss of E-cadherin expression has been found in 55% of the samples. In addition, 91% of the E-cadherin-negative samples were associated with *CDH1* promoter methylation. These findings indicate that the inactivation of the E-cadherin gene is due to DNA methylation. Another study investigated the DNA methylation status and protein expression of *CDH1* in fifty invasive ductal carcinomas (IDC) samples (Shargh et al. 2014). This study found that 94% of the samples had *CDH1* promoter hypermethylation, which was associated with a loss of E-cadherin protein expression in almost all the examined samples, by comparison with what was found in normal breast samples. This result suggests a strong correlation between *CDH1* promoter methylation and the depletion of E-cadherin expression in IDC. The methylation of the *CDH1* promoter was also found to increase with advancing breast tumor stages, the highest level being observed during the metastatic stage (Nass et al. 2000). In agreement with this finding, another investigation found that 53% of 151 primary breast tumors exhibited *CDH1* hypermethylation (Shinozaki et al. 2005). As *CDH1* promoter methylation increased to 90% in cases with sentinel lymph node metastasis, the potential clinical value of using hypermethylated *CDH1* as a predictive biomarker for metastatic breast cancers is clear.

A correlation between aberrant *CDH1* promoter hypermethylation and tumor progression and invasion has been reported in several other human cancers (Table 12.1). A meta-analysis study of 1,079 GC patients revealed a significantly higher level of DNA methylation at the *CDH1* promoter in the gastric mucosa samples from these patients than in the normal samples (Zeng et al. 2015). Another study proposed hypermethylated *CDH1* as a useful biomarker for the prediction of tumorigenesis and prognosis in individuals with a family history of GC and after *Helicobacter pylori* eradication (Kim et al. 2021). A meta-analysis of 375 HCC samples and 181 normal liver samples found a significant increase in the methylation level of the *CDH1* promoter in HCC samples (Wu et al. 2019). The samples with depleted *CDH1* gene expression showed poor overall survival by comparison with the samples with higher *CDH1* gene expression. This result suggests that *CDH1* can be used as a candidate prognostic biomarker for HCC patients. Another study revealed that 56% of colorectal cancer patients exhibited DNA methylation in the *CDH1* gene, while 48% of the samples showed a low level of *CDH1* mRNA, with a significant association with lymph node metastases and tumor stage (Garinis et al. 2002). Approximately 79% of the depleted CDH1 mRNA samples showing *CDH1* methylation suggest that aberrant DNA hypermethylation is a major cause for *CDH1* gene silencing, and therefore correlates with CRC metastasis. These observations show the important role of *CDH1*, as a tumor suppressor gene, in the regulation of cell–cell adhesion and how the loss of E-cadherin expression caused by aberrant DNA hypermethylation can promote tumor development and metastasis.

Aberrant DNA Methylation of Oncogenes

Ras

The epigenetic regulation of the *Ras* gene via DNA methylation and the existence of a negative correlation between its methylation status and gene expression point to the value of using Ras as a potential epigenetic biomarker for cancer. The Ras protein family is a class of small GTPases and it functions to convert extracellular signals into intracellular signaling cascades, so as to regulate diverse cellular processes (Chen et al. 2019). Ras, as one of the most mutated oncogenes in cancer, has been extensively studied as a significant cancer driver and as a potential target for cancer diagnosis and treatment. The Kirsten rat sarcoma viral oncogene homolog (*K-Ras*), the neuroblastoma Ras viral oncogene homolog (*N-Ras*), and the Harvey rat sarcoma viral oncogene homolog (*H-Ras*) are the three *Ras* genes that express four Ras proteins with two K-Ras isoforms that are due to alternative RNA splicing. Ras is a membrane-associated, GTP-regulated molecular switch cycling between the GDP-bound inactive state and the GTP-bound active state. The inactive state is induced by GTPase-activating proteins via GTP hydrolysis, while the active state is promoted by the activity of guanine nucleotide exchange factors (Hobbs et al. 2016). The activated form of Ras can bind to and upregulate various downstream target molecules that are involved in controlling cell growth, differentiation, and survival. However, mutations in *Ras* genes often result in constantly activated Ras proteins, and these in turn cause the overactivation of Ras-induced carcinogenic signaling pathways and the inactivation of multiple tumor-suppressing signaling pathways (Cox and Der 2010; Vigil et al. 2010).

DNA hypomethylation in the promoter region of *Ras* is considered to be an abnormal event in cancer that has been implicated in facilitating the aberrant expression and tumorigenic role of Ras. The aberrant hypomethylation patterns in *H-Ras* and, to a lesser extent, *K-Ras* were found in tumor samples from patients with colon or lung cancer (Feinberg and Vogelstein 1983). However, this early study did not examine the effect of these methylation alterations on gene expression. Another study showed hypomethylation in the promoter region of *H-Ras* and *K-Ras* in liver cancer tissues and pre-cancer lesions from rats that had been fed methyl-deficient and amino acid-defined diets (Bhave et al. 1988). The methyl-deficient diets were found to promote genomic hypomethylation as well as the hypomethylation of the *H-Ras* promoter when accompanied by sodium arsenite, an established human carcinogen (Okoji et al. 2002). Another study examined the methylation state of *H-Ras* and *K-Ras* in hepatomas and adjacent normal tissue from B6C3F1 mice (Vorce and Goodman 1989). This study revealed abnormal hypomethylation in *H-Ras* in all tumor samples, whereas *K-Ras* was hypomethylated in 50% of the samples. Importantly, aberrant *H-Ras* hypomethylation was associated with an increase in *H-Ras* gene expression in 75% of tumor samples. Another study with seventy-four NSCLCs and non-tumor tissue found that 18% of the samples have an allelic deletion at the *H-Ras* locus, and 71% of those exhibited hypomethylation at the 3′ region of the non-deleted allele (Vachtenheim et al. 1994). While the hypomethylation of *H-Ras* was found in 37% of heterozygous DNA samples without allele loss, 26% of the homozygous samples showed *H-Ras* hypomethylation with no mutation in the gene. This indicates that in lung cancer the DNA hypomethylation, not the mutation, is accompanied by a loss of heterozygosity. It also indicates that *H-Ras* hypomethylation is more frequent than allelic loss. This notion suggests that the hypomethylation may be an early event that predisposes to genetic instability and therefore leads to the allelic deletion of *H-Ras* observed in tumor samples. Thus the hypomethylation status of *H-Ras* can provide insights into tumorigenesis and how the hypomethylation of *H-Ras* can be used as a diagnostic biomarker and therapeutic target for cancer treatment.

The aberrant DNA methylation of *Ras* genes in several other types of cancer further highlights its potential clinical values in cancer (Table 12.1). *R-Ras* was found to be upregulated by DNA hypomethylation in a significant proportion of GCs (Nishigaki et al. 2005). In addition, a study of GC cells and colon cancer cells showed hypomethylation at the promoters of *H-Ras* and *c-myc* oncogenes. Treating those cancer cells with S Adenosylmethionine (SAM), a methyl donor, reversed the DNA hypomethylation effect and enhanced their promoter methylation levels. This SAM inhibition consequently led to a reduction in the transcript and protein levels of H-Ras and c-myc, thereby inhibiting tumor cell growth (Luo et al. 2010). Another study examined the methylation status of *H-Ras* in fifteen bladder cancer tissues and five control samples (Sun et al. 2012). This study revealed that the methylation pattern of the *H-Ras* promoter was significantly altered in the bladder tissue with cancer by comparison with normal bladder tissue. The hypomethylation of the *H-Ras* gene promoter was associated with the clinicopathological features of bladder cancer, suggesting that *H-Ras* hypomethylation could be an early epigenetic event in bladder carcinogenesis. A study that investigated the cross-generational effect of alcohol exposure has shown that ancestral alcohol exposure can cause an aberrant hypomethylation of the *H-Ras* promoter as well as a DNA hypermethylation of the *P53* gene, which subsequently altered their expression and therefore can trigger the onset of cancer (Hill et al. 2017). Using the methylation-specific PCR, one study with twenty-seven HCC patients and twenty-three non-HCC controls investigated the DNA methylation pattern of *Ras* genes in hepatitis C virus (HCV)-related HCC (Maryam and Idrees 2018). This study found a remarkable hypomethylation at the *N-Ras* promoter in the HCC patients by comparison with the controls. A significant correlation was found between the DNA hypomethylation of *H-Ras* and the increased risk of tumor infiltration and metastasis (Shen et al. 1998). Moreover, a recent study revealed that *H-Ras* and *K-Ras* were hypomethylated in muscle-invasive bladder cancer (MIBC) samples, whereas *N-Ras* showed a predominant hypomethylation pattern in non-muscle-invasive bladder cancer (NMIBC) samples (Tripathi et al. 2021). These abnormal hypomethylation patterns were associated with gene overexpression as well as with the tumor's stage, grade, type, and size. Together, the findings suggest that the promoter hypomethylation of *Ras* oncogenes can be used as a potential diagnostic biomarker for various types of cancer.

ONECUT2

Aberrant CpG methylation of the promoter of one cut homeobox 2 (ONECUT2) can serve as a promising epigenetic biomarker for the diagnosis and prognosis of several types of cancer (Seo et al. 2020) (Table 12.1). The *ONECUT2* gene is located on chromosome 18 and comprises two exons and one intron. The gene product consists of one cut domain and an atypical homologous domain. These domains bind to DNA sequences to regulate the expression of target genes. ONECUT2 is mainly expressed in the liver, gallbladder, and small intestines and less expressed in other tissues (Jacquemin et al. 1999). As a transcription factor, ONECUT2 promotes tissue development by regulating the expression of genes that are involved in cell differentiation, proliferation, and migration. *ONECUT2* is frequently dysregulated by DNA methylation in GC, upper tract urinary carcinoma (UTUC), and bladder cancer. It plays a crucial oncogenic role in angiogenesis and epithelial–mesenchymal transition and is associated with tumor progression, metastasis, and poor prognosis (Yu et al. 2020).

The methylation status of the *ONECUT2* promoter was proposed as diagnostic biomarkers for GC. A study of 160 primary GC patients found that the hypomethylation of the promoter of *ONECUT2* was negatively correlated with its gene expression, which indicates that aberrant DNA methylation leads to the upregulation of ONECUT2 (Seo et al. 2020). To confirm its oncogenic role, ONECUT2 was overexpressed in GC cells, which induced cell growth, proliferation, migration, and invasion (Chen et al. 2020a). Moreover, depletion of ONECUT2 resulted in tumor suppression (Seo et al. 2020; Wang et al. 2020). It has also been shown that upregulation of ONECUT2 led to the activation of acyl-CoA synthetase long-chain family member 5 (ACSL5), and both proteins contribute to cancer progression

(Seo et al. 2020). Another study demonstrated the potential diagnostic value of DNA methylation of *ONECUT2* in the detection of UTUC (Xu et al. 2020). The combined analysis of the methylation status of the *ONECUT2* promotor and high-throughput sequencing of other seventeen genes exhibited highly accurate diagnosis when using liquid biopsies from sixty-four hematuria patients and eighty-six non-UTUC samples. The prediction had a sensitivity value of 94% and a specificity value of 93%, while the positive and negative predictive values were respectively 92% and 95%. These results suggest that ONECUT2 can be used as a non-invasive epigenetic biomarker to predict the occurrence of UTUC.

The DNA methylation of *ONECUT2* can also be used to monitor the recurrence of tumors. Different marker combinations were analyzed for sensitivity and specificity in classifying recurrent bladder tumors using pre-trans-urethral resection urines (Kandimalla et al. 2012). A three-gene methylation panel of *ONECUT2*, *OTX1*, and *OSR1* achieved a sensitivity of 73% and a specificity of 90%, with an area under the curve (AUC) value of 0.86. This indicates the predictive value of this panel for detecting bladder tumor in urine samples and for monitoring bladder cancer patients. Another panel of methylation biomarkers comprising *ONECUT2* and three other genes (*HOXA9*, *PCDH17*, *POU4F2*) was proposed in order to help with differentiating bladder cancer from other benign lesions in hematuria patients (Wu et al. 2020). The panel showed high positive and negative predictive values (100% and 98%, respectively, with a ROC value of 0.87), with a sensitivity of 91% and a specificity of 73%. This indicates that it can have useful diagnostic applications and can eliminate unnecessarily invasive examinations for low-risk patients. Similar suggestions about using a combined urinary methylation assay that includes *ONECUT2* have been reported in relation to the accurate prediction of bladder cancer in hematuria patients (Zhang et al. 2019). These findings highlight that the aberrant DNA methylation of the *ONECUT2* promoter is a powerful clinical biomarker for the diagnosis and prognosis of several gastrointestinal and urinary tract cancers.

BORIS

The hypomethylation of the oncogenic promoter *BORIS*—brother of regulator of imprinted sites—is associated with its gene overexpression in numerous types of cancer (de Necochea-campion et al. 2011). BORIS, also known as CCCTC-binding factor-like (CTCFL), is an 11-zinc-finger DNA-binding protein that, in humans, is encoded by the *BORIS* gene located on chromosome 20 (Loukinov et al. 2002). It is a paralog of the transcription factor CCCTC-binding factor (CTCF), which is involved in chromatin architecture regulation, genomic imprinting, and genome organization (Marshall et al. 2014). BORIS is normally expressed in primary spermatocytes and spermatogonia in the testis, where it plays a vital role in development and early spermatogenesis (Sleutels et al. 2012; Suzuki et al. 2010). At the later stages of spermatogenesis, BORIS is suppressed through the DNA methylation of its promoter, to restrict its gene expression. Thus, normal somatic cells and tissues lack BORIS expression or have very low levels of it (Martin-kleiner 2012; Soltanian and Dehghani 2018). Several reports showed aberrant expression of BORIS in multiple human malignancies (D'Arcy et al. 2008; Kholmanskikh et al. 2008; Klenova et al. 2002). In HCC, BORIS was aberrantly expressed at a significantly higher level than in normal liver tissues and was associated with tumor size, grade, satellite lesion, and recurrence (Zhao et al. 2017). The prognostic outcome was the poorest in HCC patients with high BORIS expression and low suppressor of cytokine signaling 3 (SOCS3) expression. Another study demonstrated that aberrant BORIS expression was positively correlated with a high expression of progesterone receptor (PR) and estrogen receptor (ER) in breast tumor tissues and in different breast cancer cell lines (D'Arcy et al. 2008).

A demethylation of the *BORIS* promoter that leads to its reactivation underlies its aberrant expression in cancer (Hoffmann et al. 2006; Klenova et al. 2002). A positive correlation between the hypomethylation of *BORIS* promoter and its overexpression has been reported in various cancers, including head and neck cancer, lung cancer, epithelial ovarian cancer, prostate cancer, and HCC (Chen et al. 2020b; Hoffmann et al. 2006; Hong et al. 2005; Smith et al. 2009; Woloszynska-read et al. 2007). In epithelial ovarian cancer, hypomethylated and overexpressed *BORIS* was associated with advanced stage and poor survival (Woloszynska-read et al. 2011). In HCC, 60% of malignant plasma samples showed a significantly high level of hypomethylated *BORIS* at the promoter region, by comparison with 17% in benign lesion plasma samples (Chen et al. 2020b). This aberrant methylation status was significantly correlated with microvascular involvement and postoperative recurrence (Table 12.1). Also, the hypomethylated group showed worse overall survival than the hypermethylated group, which recommends the potential use of plasma BORIS as a non-invasive epigenetic biomarker for HCC prognosis (Chen et al. 2020b). A similar study found that the level of hypomethylated *BORIS* in HCC and in HCC cell lines was remarkably higher than in normal tissue controls (He et al. 2017). This hypomethylation was significantly correlated with tumor size, clinical TNM stage, and shorter disease-free survival and overall survival rates in HCC patients. BORIS also showed significant differential gene expression between endometrioid tumors and normal tissues (Hoivik et al. 2014). This aberrant expression, which was due to the demethylation of its promoter, increased gradually from the atypical hyperplasia, primary tumor, to metastatic lesions and was found to associate with poor survival. Therefore BORIS is a promising prognostic biomarker as well as a potential druggable target for drug invention.

Future Perspectives

The pressing need for novel strategies for cancer diagnosis and treatment has stimulated our interest in DNA methylation alterations as a promising field in which we could mine for, and develop, novel effective biomarkers. Numerous studies have sought to understand the function of DNA methylation and its association with clinicopathological parameters of human cancer. Conceptually, it is well perceived that aberrant DNA methylation is a promising cancer biomarker since it is an early and frequent event in cancer and can remain stable and be easily detected by different non-invasive methods. Nevertheless, only a small group of biomarkers have been approved and found their way into the clinic. This narrow application was due in part to technical limitations related to assessing samples and epigenetic data on a large scale and to the costly technologies involved. However, recent advances in molecular and epigenomic technologies, along with the continuous reduction in the cost of these technologies, have helped overcome these challenges and contributed, albeit at a slow rate, to a growing list of DNA methylation-based cancer biomarkers for early detection, diagnosis, prognosis, and drug response prediction (Koch et al. 2018; Locke et al. 2019). It is worth noticing that, because the investigations are very few, there is a significant knowledge gap concerning the mechanism and importance of promoter hypomethylation of oncogenes. Most of previous DNA methylation research has focused on DNA hypermethylation within the promoter region in tumor suppressor genes. Therefore more studies are needed to identify and examine clinically relevant methylated oncogenes. The incorporation of the oncogenes into the current DNA methylation-based cancer biomarker panels can improve the specificity of these biomarker panels and, eventually, patient outcomes.

DNA methylation studies sometimes lack a precise identification of DNA methylation sites within a promoter region, and this produces variation in the degrees of biomarker specificity for similar cancer tissues or cell lines. This is important, because the methylation-regulated transcriptional silencing or gene reactivation is greatly affected by the spread and exact location of DNA methylation. There are significant functional differences between the CpG islands even within the same promoter region (Homma et al. 2006). Thus it is critical to characterize DNA methylation sites with great precision in order to enhance the specificity profiles of candidate biomarkers for human cancers. There is also a paucity of comprehensive studies that one could mine for clinically relevant DNA methylation biomarkers in a large-scale genomic and epigenomic tumor analysis. However, multiple molecular and sequencing techniques such as next-generation sequencing (NGS) will enable the analysis of the whole methylomes for DNA methylation-based cancer biomarkers discovery. This will contribute to building up large-scale corpora of cancer epigenomics data that can be used to better predict an optimal correlation between DNA methylation sites and clinicopathological features. The bioinformatic interpretation of epigenetic information is also a critical process, especially given the rapid increase in available genomic and epigenomic datasets.

Whether an epigenetic modification is driving cancer initiation and progression and how its dynamic change determines disease outcomes remain largely outstanding questions. Future DNA methylation investigations should enhance our current knowledge about the role of DNA methylation under normal and pathological conditions, as well as about the broader effect of DNA methylation changes on gene expression, genome structure, and cell behavior. Understanding the mechanisms that underlie DNA methylation alteration events in each type of human cancer will enhance the current diagnosis and management of human cancers. However, the interpretation of methylomes alone cannot fully explain tumorigenesis. A combination of methylome-related information with genomic, transcriptomic, and other epigenetic information can improve current approaches to the early diagnosis and control of cancer, and thereby offer cancer patients better clinical outcomes.

References

Agirre, X., Vizmanos, J.L., Calasanz, M.J., Garcia-Delgado, M., Larrayoz, M.J., and Novo, F.J. (2003). Methylation of CpG dinucleotides and/or CCWGG motifs at the promoter of TP53 correlates with decreased gene expression in a subset of acute lymphoblastic leukemia patients. *Oncogene* 22: 1070–1072.

Amatya, V.J., Naumann, U., Weller, M., and Ohgaki, H. (2005). TP53 promoter methylation in human gliomas. *Acta Neuropathol.* 110: 178–184.

Audia, J.E. and Campbell, R.M. (2016). Histone modifications and cancer. *Cold Spring Harb. Perspect. Biol. 8 (4). doi: 10.1101/cshperspect.a019521.*

Bai, J., Zhang, X., Hu, K., Liu, B., Wang, H., Li, A., Lin, F., Zhang, L., Sun, X., Du, Z., and Song, J. (2016). Silencing DNA methyltransferase 1 (DNMT1) inhibits proliferation, metastasis and invasion in ESCC by suppressing methylation of RASSF1A and DAPK. *Oncotarget* 7: 44129–44141.

Baylin, S.B. and Jones, P.A. (2011). A decade of exploring the cancer epigenome: Biological and translational implications. *Nat. Rev. Cancer.* 11 (10): 726–734.

Benard, A., Zeestraten, E.C., Goossens-Beumer, I.J., Putter, H., Van De Velde, C.J., Hoon, D.S., and Kuppen, P.J. (2014). DNA methylation of apoptosis genes in rectal cancer predicts patient survival and tumor recurrence. *Apoptosis* 19: 1581–1593.

Berx, G., Staes, K., Van Hengel, J., Molemans, F., Bussemakers, M.J., Van Bokhoven, A., and Van Roy, F. (1995). Cloning and characterization of the human invasion suppressor gene E-cadherin (CDH1). *Genomics* 26: 281–289.

Bhagat, R., Premalata, C.S., Shilpa, V., Pallavi, V.R., Ramesh, G., Vijay, C.R., and Krishnamoorthy, L. (2013). Altered expression of beta-catenin, E-cadherin, and E-cadherin promoter methylation in epithelial ovarian carcinoma. *Tumour Biol.* 34: 2459–2468.

Bhave, M.R., Wilson, M.J., and Poirier, L.A. (1988). c-H-ras and c-K-ras gene hypomethylation in the livers and hepatomas of rats fed methyl-deficient, amino acid-defined diets. *Carcinogenesis* 9: 343–348.

Bray, F., Ferlay, J., Soerjomataram, I., Siegel, R.L., Torre, L.A., and Jemal, A. (2018). Global cancer statistics 2018: GLOBOCAN estimates of incidence and mortality worldwide for 36 cancers in 185 countries. *CA Cancer J. Clin.* 68: 394–424.

Bruner, H.C. and Derksen, P.W.B. (2018). Loss of E-cadherin-dependent cell-cell adhesion and the development and progression of cancer. *Cold Spring Harb. Perspect. Biol.* 10.

Chen, J. (2016). The cell-cycle arrest and apoptotic functions of p53 in tumor initiation and progression. *Cold Spring Harb. Perspect. Med.* 6: a026104.

Chen, J., Chen, J., Sun, B., Wu, J., and Du, C. (2020a). ONECUT2 accelerates tumor proliferation through activating ROCK1 expression in gastric cancer. *Cancer Manag. Res.* 12: 6113–6121.

Chen, M.M., Zhao, R.C., Chen, K.F., Huang, Y., Liu, Z.J., Wei, Y.G., Jian, Y., Sun, A.M., Qin, L., Li, B., and Qin, Y. (2020b). Hypomethylation of CTCFL promoters as a noninvasive biomarker in plasma from patients with hepatocellular carcinoma. *Neoplasma* 67: 909–915.

Chen, S., Li, F., Xu, D., Hou, K., Fang, W., and Li, Y. (2019). The function of RAS mutation in cancer and advances in its drug research. *Curr. Pharm. Des.* 25: 1105–1114.

Chmelarova, M., Krepinska, E., Spacek, J., Laco, J., Beranek, M., and Palicka, V. (2013). Methylation in the p53 promoter in epithelial ovarian cancer. *Clin. Transl. Oncol.* 15: 160–163.

Cox, A.D. and Der, C.J. (2010). Ras history: The saga continues. *Small GTPases* 1: 2–27.

D'Arcy, V., Pore, N., Docquier, F., Abdullaev, Z.K., Chernukhin, I., Kita, G.X., Rai, S., Smart, M., Farrar, D., Pack, S., Lobanenkov, V., and Klenova, E. (2008). BORIS, a paralogue of the transcription factor, CTCF, is aberrantly expressed in breast tumours. *Br. J. Cancer* 98: 571–579.

Dagenais, G.R., Leong, D.P., Rangarajan, S., Lanas, F., Lopez-Jaramillo, P., Gupta, R., Diaz, R., Avezum, A., Oliveira, G.B.F., Wielgosz, A., Parambath, S.R., Mony, P., Alhabib, K.F., Temizhan, A., Ismail, N., Chifamba, J., Yeates, K., Khatib, R., Rahman, O., Zatonska, K., Kazmi, K., Wei, L., Zhu, J., Rosengren, A., Vijayakumar, K., Kaur, M., Mohan, V., Yusufali, A., Kelishadi, R., Teo, K.K., Joseph, P., and Yusuf, S. (2020). Variations in common diseases, hospital admissions, and deaths in middle-aged adults in 21 countries from five continents (PURE): A prospective cohort study. *Lancet 395 (10226): 785–794 ???.*

Das, P.M. and Singal, R. (2004). DNA methylation and cancer. *J. Clin. Oncol.* 22: 4632–4642.

de Fraipont, F., Levallet, G., Creveuil, C., Bergot, E., Beau-Faller, M., Mounawar, M., Richard, N., Antoine, M., Rouquette, I., Favrot, M.C., Debieuvre, D., Braun, D., Westeel, V., Quoix, E., Brambilla, E., Hainaut, P., Moro-Sibilot, D., Morin, F., Milleron, B., Zalcman, G., and Intergroupe Francophone De Cancerologie, T. (2012). An apoptosis methylation prognostic signature for early lung cancer in the IFCT-0002 trial. *Clin. Cancer Res.* 18: 2976–2986.

De Necochea-Campion, R., Ghochikyan, A., Josephs, S.F., Zacharias, S., Woods, E., Karimi-Busheri, F., Alexandrescu, D.T., Chen, C.S., Agadjanyan, M.G., and Carrier, E. (2011). Expression of the epigenetic factor BORIS (CTCFL) in the human genome. *J. Transl. Med.* 9: 213.

Deaton, A.M. and Bird, A. (2011). CpG islands and the regulation of transcription. *Genes Dev. 25 (10): 1010–1022.*

Donninger, H., Vos, M.D., and Clark, G.J. (2007). The RASSF1A tumor suppressor. *J. Cell. Sci.* 120: 3163–3172.

Droufakou, S., Deshmane, V., Roylance, R., Hanby, A., Tomlinson, I., and Hart, I.R. (2001). Multiple ways of silencing E-cadherin gene expression in lobular carcinoma of the breast. *Int. J. Cancer* 92: 404–408.

Duffy, M.J., Synnott, N.C., and Crown, J. (2017). Mutant p53 as a target for cancer treatment. *Eur. J. Cancer* 83: 258–265.

Ehrlich, M. (2009). DNA hypomethylation in cancer cells. *Epigenomics* 1 (2): 239–259.

Esteller, M. (2008). Epigenetics in cancer. *N. Engl. J. Med. 358: 1248–1159.*

Euhus, D.M., Bu, D., Milchgrub, S., Xie, X.J., Bian, A., Leitch, A.M., and Lewis, C.M. (2008). DNA methylation in benign breast epithelium in relation to age and breast cancer risk. *Cancer Epidemiol. Biomarkers Prev.* 17: 1051–1059.

Feinberg, A.P. and Vogelstein, B. (1983). Hypomethylation of ras oncogenes in primary human cancers. *Biochem. Biophys. Res. Commun.* 111: 47–54.

Garinis, G.A., Menounos, P.G., Spanakis, N.E., Papadopoulos, K., Karavitis, G., Parassi, I., Christeli, E., Patrinos, G.P., Manolis, E.N., and Peros, G. (2002). Hypermethylation-associated transcriptional silencing of E-cadherin in primary sporadic colorectal carcinomas. *J. Pathol.* 198: 442–449.

Grawenda, A.M. and O'Neill, E. (2015). Clinical utility of RASSF1A methylation in human malignancies. *Br. J. Cancer* 113: 372–381.

Guo, Q., Wang, H.B., Li, Y.H., Li, H.F., Li, T.T., Zhang, W.X., Xiang, S.S., and Sun, Z.Q. (2016). Correlations of promoter methylation in WIF-1, RASSF1A, and CDH13 genes with the risk and prognosis of esophageal cancer. *Med. Sci. Monit.* 22: 2816–2824.

He, J., Huang, Y., Liu, Z., Zhao, R., Liu, Q., Wei, L., Yu, X., Li, B., and Qin, Y. (2017). Hypomethylation of BORIS is a promising prognostic biomarker in hepatocellular carcinoma. *Gene* 629: 29–34.

Herrero, A.B., Rojas, E.A., Misiewicz-Krzeminska, I., Krzeminski, P., and Gutierrez, N.C. (2016). Molecular mechanisms of p53 deregulation in cancer: An overview in multiple myeloma. *Int. J. Mol. Sci.* 17.

Hesson, L.B., Cooper, W.N., and Latif, F. (2007). The role of RASSF1A methylation in cancer. *Dis. Markers* 23: 73–87.

Heyn, H., Sayols, S., Moutinho, C., Vidal, E., Sanchez-Mut, J.V., Stefansson, O.A., Nadal, E., Moran, S., Eyfjord, J.E., Gonzalez-Suarez, E., Pujana, M.A., and Esteller, M. (2014). Linkage of DNA methylation quantitative trait loci to human cancer risk. *Cell Rep.* 7: 331–338.

Hill, S.Y., Rompala, G., Homanics, G.E., and Zezza, N. (2017). Cross-generational effects of alcohol dependence in humans on HRAS and TP53 methylation in offspring. *Epigenomics* 9: 1189–1203.

Hobbs, G.A., Der, C.J., and Rossman, K.L. (2016). RAS isoforms and mutations in cancer at a glance. *J. Cell. Sci.* 129: 1287–1292.

Hoffmann, M.J., Muller, M., Engers, R., and Schulz, W.A. (2006). Epigenetic control of CTCFL/BORIS and OCT4 expression in urogenital malignancies. *Biochem. Pharmacol.* 72: 1577–1588.

Hoivik, E.A., Kusonmano, K., Halle, M.K., Berg, A., Wik, E., Werner, H.M., Petersen, K., Oyan, A.M., Kalland, K.H., Krakstad, C., Trovik, J., Widschwendter, M., and Salvesen, H.B. (2014). Hypomethylation of the CTCFL/BORIS promoter and aberrant expression during endometrial cancer progression suggests a role as an Epi-driver gene. *Oncotarget* 5: 1052–1061.

Holliday, R. (2006). Epigenetics: A historical overview. *Epigenetics* 1 (2): 76–80.

Homma, N., Tamura, G., Honda, T., Matsumoto, Y., Nishizuka, S., Kawata, S., and Motoyama, T. (2006). Spreading of methylation within RUNX3 CpG island in gastric cancer. *Cancer Sci.* 97: 51–56.

Hong, J.A., Kang, Y., Abdullaev, Z., Flanagan, P.T., Pack, S.D., Fischette, M.R., Adnani, M.T., Loukinov, D.I., Vatolin, S., Risinger, J.I., Custer, M., Chen, G.A., Zhao, M., Nguyen, D.M., Barrett, J.C., Lobanenkov, V.V., and Schrump, D.S. (2005). Reciprocal binding of CTCF and BORIS to the NY-ESO-1 promoter coincides with derepression of this cancer-testis gene in lung cancer cells. *Cancer Res.* 65: 7763–7774.

Hsu, H.S., Chen, T.P., Hung, C.H., Wen, C.K., Lin, R.K., Lee, H.C., and Wang, Y.C. (2007). Characterization of a multiple epigenetic marker panel for lung cancer detection and risk assessment in plasma. *Cancer* 110: 2019–2026.

Ishiyama, N., Lee, S.H., Liu, S., Li, G.Y., Smith, M.J., Reichardt, L.F., and Ikura, M. (2010). Dynamic and static interactions between p120 catenin and E-cadherin regulate the stability of cell-cell adhesion. *Cell* 141: 117–128.

Jacquemin, P., Lannoy, V.J., Rousseau, G.G., and Lemaigre, F.P. (1999). OC-2, a novel mammalian member of the ONECUT class of homeodomain transcription factors whose function in liver partially overlaps with that of hepatocyte nuclear factor-6. *J. Biol. Chem.* 274: 2665–2671.

Jesionek-Kupnicka, D., Szybka, M., Malachowska, B., Fendler, W., Potemski, P., Piaskowski, S., Jaskolski, D., Papierz, W., Skowronski, W., Och, W., Kordek, R., and Zawlik, I. (2014). TP53 promoter methylation in primary glioblastoma: Relationship with TP53 mRNA and protein expression and mutation status. *DNA Cell Biol.* 33: 217–226.

Jha, A.K., Nikbakht, M., Jain, V., Sehgal, A., Capalash, N., and Kaur, J. (2012). Promoter hypermethylation of p73 and p53 genes in cervical cancer patients among north Indian population. *Mol. Biol. Rep.* 39: 9145–9157.

Jiang, Y., Cui, L., Chen, W.D., Shen, S.H., and Ding, L.D. (2012). The prognostic role of RASSF1A promoter methylation in breast cancer: A meta-analysis of published data. *PLoS One* 7: e36780.

Jones, P.A. (2012). Functions of DNA methylation: Islands, start sites, gene bodies and beyond. *Nat. Rev. Genet.* 13: 484–492.

Jones, P.A. and Baylin, S.B. (2007). The epigenomics of cancer. *Cell* 128: 683–692.

Kandimalla, R., Masius, R., Van Tilborg, A.A.G., Beukers, W., Orntoft, T., Dyrskjot, L., Bangma, C.H., and Zwarthoff, E.C. (2012). Abstract 4023: Genome-wide analysis of CpG island methylation identified OTX1, OSR1 and ONECUT2 as biomarkers for recurrent bladder cancer detection in voided urine. *Cancer Res.* 72: 4023–4023.

Kang, J.H., Kim, S.J., Noh, D.Y., Park, I.A., Choe, K.J., Yoo, O.J., and Kang, H.S. (2001). Methylation in the p53 promoter is a supplementary route to breast carcinogenesis: Correlation between CpG methylation in the p53 promoter and the mutation of the p53 gene in the progression from ductal carcinoma in situ to invasive ductal carcinoma. *Lab. Invest.* 81: 573–579.

Karray-Chouayekh, S., Trifa, F., Khabir, A., Boujelbane, N., Sellami-Boudawara, T., Daoud, J., Frikha, M., Jlidi, R., Gargouri, A., and Mokdad-Gargouri, R. (2010). Aberrant methylation of RASSF1A is associated with poor survival in Tunisian breast cancer patients. *J. Cancer Res. Clin. Oncol.* 136: 203–210.

Kaszak, I., Witkowska-Pilaszewicz, O., Niewiadomska, Z., Dworecka-Kaszak, B., Ngosa Toka, F., and Jurka, P. (2020). Role of cadherins in cancer-A review. *Int. J. Mol. Sci.* 21.

Kholmanskikh, O., Loriot, A., Brasseur, F., De Plaen, E., and De Smet, C. (2008). Expression of BORIS in melanoma: Lack of association with MAGE-A1 activation. *Int. J. Cancer* 122: 777–784.

Kim, H.J., Kim, N., Kim, H.W., Park, J.H., Shin, C.M., and Lee, D.H. (2021). Promising aberrant DNA methylation marker to predict gastric cancer development in individuals with family history and long-term effects of H. pylori eradication on DNA methylation. *Gastric Cancer* 24: 302–313.

Klenova, E.M., Morse, H.C., Ohlsson, R., 3RD, and Lobanenkov, V.V. (2002). The novel BORIS + CTCF gene family is uniquely involved in the epigenetics of normal biology and cancer. *Semin. Cancer Biol.* 12: 399–414.

Koch, A., Joosten, S.C., Feng, Z., De Ruijter, T.C., Draht, M.X., Melotte, V., Smits, K.M., Veeck, J., Herman, J.G., Van Neste, L., Van Criekinge, W., De Meyer, T., and Van Engeland, M. (2018). Analysis of DNA methylation in cancer: Location revisited. *Nat. Rev. Clin. Oncol.* 15: 459–466.

Kulis, M. and Esteller, M. (2010). DNA methylation and cancer. *Adv. Genet. 70: 27–56.*

Lam, K., Pan, K., Linnekamp, J.F., Medema, J.P., and Kandimalla, R. (2016). DNA methylation based biomarkers in colorectal cancer: A systematic review. *Biochim. Biophys. Acta* 1866: 106–120.

Lane, D.P. (1994). p53 and human cancers. *Br. Med. Bull.* 50: 582–599.

Li, W., Deng, J., Jiang, P., Zeng, X., Hu, S., and Tang, J. (2012). Methylation of the RASSF1A and RARbeta genes as a candidate biomarker for lung cancer. *Exp. Ther. Med.* 3: 1067–1071.

Locke, W.J., Guanzon, D., Ma, C., Liew, Y.J., Duesing, K.R., Fung, K.Y.C., and Ross, J.P. (2019). DNA methylation cancer biomarkers: Translation to the clinic. *Front. Genet.* 10: 1150.

Lombaerts, M., Van Wezel, T., Philippo, K., Dierssen, J.W., Zimmerman, R.M., Oosting, J., Van Eijk, R., Eilers, P.H., Van De Water, B., Cornelisse, C.J., and Cleton-Jansen, A.M. (2006). E-cadherin transcriptional downregulation by promoter methylation but not mutation is related to epithelial-to-mesenchymal transition in breast cancer cell lines. *Br. J. Cancer* 94: 661–671.

Loukinov, D.I., Pugacheva, E., Vatolin, S., Pack, S.D., Moon, H., Chernukhin, I., Mannan, P., Larsson, E., Kanduri, C., Vostrov, A.A., Cui, H., Niemitz, E.L., Rasko, J.E., Docquier, F.M., Kistler, M., Breen, J.J., Zhuang, Z., Quitschke, W.W., Renkawitz, R., Klenova, E.M., Feinberg, A.P., Ohlsson, R., Morse, H.C., 3RD, and Lobanenkov, V.V. (2002). BORIS, a novel male germ-line-specific protein associated with epigenetic reprogramming events, shares the same 11-zinc-finger domain with CTCF, the insulator protein involved in reading imprinting marks in the soma. *Proc. Natl. Acad. Sci. USA* 99: 6806–6811.

Lu, J., Wilfred, P., Korbie, D., and Trau, M. (2020). Regulation of canonical oncogenic signaling pathways in cancer via DNA methylation. *Cancers (Basel)* 12.

Luo, J., Li, Y.N., Wang, F., Zhang, W.M., and Geng, X. (2010). S-adenosylmethionine inhibits the growth of cancer cells by reversing the hypomethylation status of c-myc and H-ras in human gastric cancer and colon cancer. *Int. J. Biol. Sci.* 6: 784–795.

Machado, J.C., Oliveira, C., Carvalho, R., Soares, P., Berx, G., Caldas, C., Seruca, R., Carneiro, F., and Sobrinho-Simoes, M. (2001). E-cadherin gene (CDH1) promoter methylation as the second hit in sporadic diffuse gastric carcinoma. *Oncogene* 20: 1525–1528.

Malpeli, G., Innamorati, G., Decimo, I., Bencivenga, M., Nwabo Kamdje, A.H., Perris, R., and Bassi, C. (2019). Methylation dynamics of RASSF1A and its impact on cancer. *Cancers* 11.

Mao, W.M., Li, P., Zheng, Q.Q., Wang, C.C., Ge, M.H., Hu, F.J., Fang, X.H., Dong, Z.M., and Ling, Z.Q. (2011). Hypermethylation-modulated downregulation of RASSF1A expression is associated with the progression of esophageal cancer. *Arch. Med. Res.* 42: 182–188.

Margueron, R. and Reinberg, D. (2010). Chromatin structure and the inheritance of epigenetic information. *Nat. Rev. Genet. 11 (4): 285–296.*

Marshall, A.D., Bailey, C.G., and Rasko, J.E. (2014). CTCF and BORIS in genome regulation and cancer. *Curr. Opin. Genet. Dev.* 24: 8–15.

Martin-Kleiner, I. (2012). BORIS in human cancers – A review. *Eur. J. Cancer* 48: 929–935.

Maryam, M. and Idrees, M. (2018). Study of promoter hypomethylation profiles of RAS oncogenes in hepatocellular carcinoma derived from hepatitis C virus genotype 3a in Pakistani population. *J. Med. Virol.* 90: 1516–1523.

Nag, S., Qin, J., Srivenugopal, K.S., Wang, M., and Zhang, R. (2013). The MDM2-p53 pathway revisited. *J. Biomed. Res.* 27: 254–271.

Nass, S.J., Herman, J.G., Gabrielson, E., Iversen, P.W., Parl, F.F., Davidson, N.E., and Graff, J.R. (2000). Aberrant methylation of the estrogen receptor and E-cadherin 5′ CpG islands increases with malignant progression in human breast cancer. *Cancer Res.* 60: 4346–4348.

Nebbioso, A., Tambaro, F.P., Dell'Aversana, C., and Altucci, L. (2018). Cancer epigenetics: Moving forward. *PLoS Genet. 14 (6).*

Negrini, S., Gorgoulis, V.G., and Halazonetis, T.D. (2010). Genomic instability–an evolving hallmark of cancer. *Nat. Rev. Mol. Cell Biol. 11 (3): 220–228.*

Nelson, W.J. and Nusse, R. (2004). Convergence of Wnt, beta-catenin, and cadherin pathways. *Science* 303: 1483–1487.

Nilsson, T.K., Lof-Ohlin, Z.M., and Sun, X.F. (2013). DNA methylation of the p14ARF, RASSF1A and APC1A genes as an independent prognostic factor in colorectal cancer patients. *Int. J. Oncol.* 42: 127–133.

Nishigaki, M., Aoyagi, K., Danjoh, I., Fukaya, M., Yanagihara, K., Sakamoto, H., Yoshida, T., and Sasaki, H. (2005). Discovery of aberrant expression of R-RAS by cancer-linked DNA hypomethylation in gastric cancer using microarrays. *Cancer Res.* 65: 2115–2124.

Nyiraneza, C., Sempoux, C., Detry, R., Kartheuser, A., and Dahan, K. (2012). Hypermethylation of the 5′ CpG island of the p14ARF flanking exon 1beta in human colorectal cancer displaying a restricted pattern of p53 overexpression concomitant with increased MDM2 expression. *Clin. Epigenet.* 4: 9.

Oda, H. and Takeichi, M. (2011). Evolution: Structural and functional diversity of cadherin at the adherens junction. *J. Cell Biol.* 193: 1137–1146.

Okoji, R.S., Yu, R.C., Maronpot, R.R., and Froines, J.R. (2002). Sodium arsenite administration via drinking water increases genome-wide and Ha-ras DNA hypomethylation in methyl-deficient C57BL/6J mice. *Carcinogenesis* 23: 777–785.

Pimson, C., Ekalaksananan, T., Pientong, C., Promthet, S., Putthanachote, N., Suwanrungruang, K., and Wiangnon, S. (2016). Aberrant methylation of PCDH10 and RASSF1A genes in blood samples for non-invasive diagnosis and prognostic assessment of gastric cancer. *Peer J.* 4: e2112.

Pogribny, I.P. and James, S.J. (2002). Reduction of p53 gene expression in human primary hepatocellular carcinoma is associated with promoter region methylation without coding region mutation. *Cancer Lett.* 176: 169–174.

Ponomaryova, A.A., Rykova, E.Y., Cherdyntseva, N.V., Skvortsova, T.E., Dobrodeev, A.Y., Zav'yalov, A.A., Bryzgalov, L.O., Tuzikov, S.A., Vlassov, V.V., and Laktionov, P.P. (2013). Potentialities of aberrantly methylated circulating DNA for diagnostics and post-treatment follow-up of lung cancer patients. *Lung Cancer* 81: 397–403.

Putiri, E.L. and Robertson, K.D. (2011). Epigenetic mechanisms and genome stability. *Clin Epigenet. 2 (2): 299–314.*

Radpour, R., Barekati, Z., Haghighi, M.M., Kohler, C., Asadollahi, R., Torbati, P.M., Holzgreve, W., and Zhong, X.Y. (2010). Correlation of telomere length shortening with promoter methylation profile of p16/Rb and p53/p21 pathways in breast cancer. *Mod. Pathol.* 23: 763–772.

Schroeder, M. and Mass, M.J. (1997). CpG methylation inactivates the transcriptional activity of the promoter of the human p53 tumor suppressor gene. *Biochem. Biophys. Res. Commun.* 235: 403–406.

Seo, E.H., Kim, H.J., Kim, J.H., Lim, B., Park, J.L., Kim, S.Y., Lee, S.I., Jeong, H.Y., Song, K.S., and Kim, Y.S. (2020). ONECUT2 upregulation is associated with CpG hypomethylation at promoter-proximal DNA in gastric cancer and triggers ACSL5. *Int. J. Cancer* 146: 3354–3368.

Shapiro, L., Fannon, A.M., Kwong, P.D., Thompson, A., Lehmann, M.S., Grubel, G., Legrand, J.F., Als-Nielsen, J., Colman, D.R., and Hendrickson, W.A. (1995). Structural basis of cell-cell adhesion by cadherins. *Nature* 374: 327–337.

Shapiro, L. and Weis, W.I. (2009). Structure and biochemistry of cadherins and catenins. *Cold Spring Harb. Perspect. Biol.* 1: a003053.

Shargh, S.A., Sakizli, M., Khalaj, V., Movafagh, A., Yazdi, H., Hagigatjou, E., Sayad, A., Mansouri, N., Mortazavi-Tabatabaei, S.A., and Khorram Khorshid, H.R. (2014). Downregulation of E-cadherin expression in breast cancer by promoter hypermethylation and its relation with progression and prognosis of tumor. *Med. Oncol.* 31: 250.

Shen, L., Fang, J., Qiu, D., Zhang, T., Yang, J., Chen, S., and Xiao, S. (1998). Correlation between DNA methylation and pathological changes in human hepatocellular carcinoma. *Hepatogastroenterology* 45: 1753–1759.

Shen, X., He, Z., Li, H., Yao, C., Zhang, Y., He, L., Li, S., Huang, J., and Guo, Z. (2012). Distinct functional patterns of gene promoter hypomethylation and hypermethylation in cancer genomes. *PLoS One* 7: e44822.

Shi, D.T., Han, M., Gao, N., Tian, W., and Chen, W. (2014). Association of RASSF1A promoter methylation with gastric cancer risk: A meta-analysis. *Tumour Biol.* 35: 943–948.

Shimazu, T., Asada, K., Charvat, H., Kusano, C., Otake, Y., Kakugawa, Y., Watanabe, H., Gotoda, T., Ushijima, T., and Tsugane, S. (2015). Association of gastric cancer risk factors with DNA methylation levels in gastric mucosa of healthy Japanese: A cross-sectional study. *Carcinogenesis* 36: 1291–1298.

Shinozaki, M., Hoon, D.S., Giuliano, A.E., Hansen, N.M., Wang, H.J., Turner, R., and Taback, B. (2005). Distinct hypermethylation profile of primary breast cancer is associated with sentinel lymph node metastasis. *Clin. Cancer Res.* 11: 2156–2162.

Sinha, R., Hussain, S., Mehrotra, R., Kumar, R.S., Kumar, K., Pande, P., Doval, D.C., Basir, S.F., and Bharadwaj, M. (2013). Kras gene mutation and RASSF1A, FHIT and MGMT gene promoter hypermethylation: Indicators of tumor staging and metastasis in adenocarcinomatous sporadic colorectal cancer in Indian population. *PLoS One* 8: e60142.

Sleutels, F., Soochit, W., Bartkuhn, M., Heath, H., Dienstbach, S., Bergmaier, P., Franke, V., Rosa-Garrido, M., Van De Nobelen, S., Caesar, L., Van der Reijden, M., Bryne, J.C., Van Ijcken, W., Grootegoed, J.A., Delgado, M.D., Lenhard, B., Renkawitz, R., Grosveld, F., and Galjart, N. (2012). The male germ cell gene regulator CTCFL is functionally different from CTCF and binds CTCF-like consensus sites in a nucleosome composition-dependent manner. *Epigenet. Chromatin* 5: 8.

Smith, I.M., Glazer, C.A., Mithani, S.K., Ochs, M.F., Sun, W., Bhan, S., Vostrov, A., Abdullaev, Z., Lobanenkov, V., Gray, A., Liu, C., Chang, S.S., Ostrow, K.L., Westra, W.H., Begum, S., Dhara, M., and Califano, J. (2009). Coordinated activation of candidate proto-oncogenes and cancer testes antigens via promoter demethylation in head and neck cancer and lung cancer. *PLoS One* 4: e4961.

Soltanian, S. and Dehghani, H. (2018). BORIS: A key regulator of cancer stemness. *Cancer Cell Int.* 18: 154.

Song, Y., Ye, M., Zhou, J., Wang, Z., and Zhu, X. (2019). Targeting E-cadherin expression with small molecules for digestive cancer treatment. *Am. J. Transl. Res.* 11: 3932–3944.

Spitzwieser, M., Holzweber, E., Pfeiler, G., Hacker, S., and Cichna-Markl, M. (2015). Applicability of HIN-1, MGMT and RASSF1A promoter methylation as biomarkers for detecting field cancerization in breast cancer. *Breast Cancer Res.* 17: 125.

Sun, X.F., Li, L., Li, X.J., and Shen, W. (2012). Methylation pattern of oncogene HRAS gene promoter region and its clinical relevance to urocystic tumorigenesis. *Mol. Biol. Rep.* 39: 8431–8437.

Suzuki, T., Kosaka-Suzuki, N., Pack, S., Shin, D.M., Yoon, J., Abdullaev, Z., Pugacheva, E., Morse, H.C., Loukinov, D., 3RD, and Lobanenkov, V. (2010). Expression of a testis-specific form of Gal3st1 (CST), a gene essential for spermatogenesis, is regulated by the CTCF paralogous gene BORIS. *Mol. Cell. Biol.* 30: 2473–2484.

Takeichi, M. (2018). Multiple functions of alpha-catenin beyond cell adhesion regulation. *Curr. Opin. Cell Biol.* 54: 24–29.

Teh, A.L., Pan, H., Chen, L., Ong, M.-L., Dogra, S., Wong, J., Macisaac, J.L., Mah, S.M., Mcewen, L.M., Saw, S.-M., Godfrey, K.M., Chong, Y.-S., Kwek, K., Kwoh, C.-K., Soh, S.-E., Chong, M.F.F., Barton, S., Karnani, N., Cheong, C.Y., Buschdorf, J.P., Stünkel, W., Kobor, M.S., Meaney, M.J., Gluckman, P.D., and Holbrook, J.D. (2014). The effect of genotype and in utero environment on interindividual variation in neonate DNA methylomes. *Genome Res. 24 (7): 1064–1074.*

Tessema, M., Yu, Y.Y., Stidley, C.A., Machida, E.O., Schuebel, K.E., Baylin, S.B., and Belinsky, S.A. (2009). Concomitant promoter methylation of multiple genes in lung adenocarcinomas from current, former and never smokers. *Carcinogenesis* 30: 1132–1138.

Tripathi, K., Goel, A., Singhai, A., and Garg, M. (2021). Promoter hypomethylation as potential confounder of Ras gene overexpression and their clinical significance in subsets of urothelial carcinoma of bladder. *Mol. Biol. Rep.* 48: 2183–2199.

Vachtenheim, J., Horakova, I., and Novotna, H. (1994). Hypomethylation of CCGG sites in the 3′ region of H-ras protooncogene is frequent and is associated with H-ras allele loss in non-small cell lung cancer. *Cancer Res.* 54: 1145–1148.

Vigil, D., Cherfils, J., Rossman, K.L., and Der, C.J. (2010). Ras superfamily GEFs and GAPs: Validated and tractable targets for cancer therapy? *Nat. Rev. Cancer* 10: 842–857.

Vorce, R.L. and Goodman, J.I. (1989). Altered methylation of ras oncogenes in benzidine-induced B6C3F1 mouse liver tumors. *Toxicol. Appl. Pharmacol.* 100: 398–410.

Wang, G.H., Zhou, Y.M., Yu, Z., Deng, J.P., Liu, S.F., Wei, C.Z., Feng, Y., Mao, M., and Wang, Z. (2020). Up-regulated ONECUT2 and down-regulated SST promote gastric cell migration, invasion, epithelial-mesenchymal transition and tumor growth in gastric cancer. *Eur. Rev. Med. Pharmacol. Sci.* 24: 9378–9390.

Woloszynska-Read, A., James, S.R., Link, P.A., Yu, J., Odunsi, K., and Karpf, A.R. (2007). DNA methylation-dependent regulation of BORIS/CTCFL expression in ovarian cancer. *Cancer Immun.* 7: 21.

Woloszynska-Read, A., Zhang, W., Yu, J., Link, P.A., Mhawech-Fauceglia, P., Collamat, G., Akers, S.N., Ostler, K.R., Godley, L.A., Odunsi, K., and Karpf, A.R. (2011). Coordinated cancer germline antigen promoter and global DNA hypomethylation in ovarian cancer: Association with the BORIS/CTCF expression ratio and advanced stage. *Clin. Cancer Res.* 17: 2170–2180.

Wong, S.H.M., Fang, C.M., Chuah, L.H., Leong, C.O., and Ngai, S.C. (2018). E-cadherin: Its dysregulation in carcinogenesis and clinical implications. *Crit. Rev. Oncol. Hematol.* 121: 11–22.

Wu, X., Yao, X., Cao, Q., Wu, Z., Wang, Z., Liu, F., and Shen, L. (2019). Clinicopathological and prognostic significance of CDH1 hypermethylation in hepatocellular carcinoma: A meta-analysis. *Cancer Manag. Res.* 11: 857–864.

Wu, Y., Jiang, G., Zhang, N., Liu, S., Lin, X., Perschon, C., Zheng, S.L., Ding, Q., Wang, X., Na, R., Sun, J., and Xu, J. (2020). HOXA9, PCDH17, POU4F2, and ONECUT2 as a urinary biomarker combination for the detection of bladder cancer in chinese patients with hematuria. *Eur. Urol. Focus* 6: 284–291.

Xu, Y., Ma, X., Ai, X., Gao, J., Liang, Y., Zhang, Q., Ma, T., Mao, K., Zheng, Q., Wang, S., Jiao, Y., Zhang, X., and Li, H. (2020). A urine-based liquid biopsy method for detection of upper tract urinary carcinoma. *Front. Oncol.* 10: 597486.

Yu, J., Li, D., and Jiang, H. (2020). Emerging role of ONECUT2 in tumors. *Oncol. Lett.* 20: 328.

Zeng, F., Heng, J., Guo, X., Wang, Y., Wu, W., Tang, L., Chen, M., Wang, S., Deng, H., and Wang, J. (2020). The novel TP53 3′-end methylation pattern associated with its expression would be a potential biomarker for breast cancer detection. *Breast Cancer Res. Treat.* 180: 237–245.

Zeng, W., Zhu, J., Shan, L., Han, Z., Aerxiding, P., Quhai, A., Zeng, F., Wang, Z., and Li, H. (2015). The clinicopathological significance of CDH1 in gastric cancer: A meta-analysis and systematic review. *Drug Des. Devel. Ther.* 9: 2149–2157.

Zhang, N., Chen, S., Wu, L., Wu, Y., Jiang, G., Shao, J., Chen, L., Sun, J., Na, R., Wang, X., and Xu, J. (2019). Identification of cancer-specific methylation of gene combination for the diagnosis of bladder cancer. *J. Cancer* 10: 6761–6766.

Zhao, R., Chen, K., Zhou, J., He, J., Liu, J., Guan, P., Li, B., and Qin, Y. (2017). The prognostic role of BORIS and SOCS3 in human hepatocellular carcinoma. *Med. (Baltimore)* 96: e6420.

Zhou, S.L., Cui, J., Fan, Z.M., Li, X.M., Li, J.L., Liu, B.C., Zhang, D.Y., Liu, H.Y., Zhao, X.K., Song, X., Wang, R., Yan, Z.C., Yi, H.X., and Wang, L.D. (2013). Polymorphism of A133S and promoter hypermethylation in Ras association domain family 1A gene (RASSF1A) is associated with risk of esophageal and gastric cardia cancers in Chinese population from high incidence area in northern China. *BMC Cancer* 13: 259.

13

SMYD Protein Family as Promising Biomarkers for Cancer Diagnosis and Prognosis

Yingxue Zhang, Eid Alshammari, Jacob Sobota, and Zhe Yang

Department of Biochemistry, Microbiology, and Immunology, Wayne State University School of Medicine, Detroit, Michigan

Introduction

Cancer is a collection of diverse but related diseases, all characterized by an abnormal growth of cells. The key driver for cancer initiation and progression is the deregulation of the signaling pathways that safeguard normal cellular homeostasis (Hanahan and Weinberg 2011). Among the frequently altered pathways are the receptor tyrosine kinase (RTK)-RAS pathway, Wnt signaling, and cell cycle regulation (Anastas and Moon 2013; Sanchez-Vega et al. 2018). These pathways are typically affected by genetic alterations that fuel cancer growth either by promoting cell division or by inactivating cell cycle suppression. However, genetic alterations alone cannot account for the complexity of human carcinogenesis. It has been shown that epigenetic modifications are also frequently perturbed in many malignancies (Audia and Campbell 2016; Taby and Issa 2010). These changes support the transcriptional program crucial for the acquisition and maintenance of cancer hallmarks (Oleksiewicz and Machnik 2020). Epigenetics refers to the heritable changes in gene expression that do not involve alterations in the actual genetic sequence. Epigenetic regulation can occur through DNA methylation, histone modification, chromatin remodeling, and small non-coding RNA (ncRNA) regulation (Bannister and Kouzarides 2011; Esteller 2007). These alterations frequently play in concert, influencing the accessibility of chromosomal regions and thus affecting various activities that occur on DNA (Audia and Campbell 2016; Bannister and Kouzarides 2011). Given its roles in cancer, the epigenetic process has afforded a number of promising and effective opportunities for early detection, prognosis prediction, and therapeutic intervention in this disease.

SET and MYND domain-containing (SMYD) proteins are among the most important epigenetic regulators of cancer proliferation. The SMYD protein family is a special class of protein lysine methyltransferases (PKMTs) and is characterized by a conserved catalytic SET domain that is "split" by an MYND zinc finger (Figure 13.1; see Jiang et al. 2016, 2011, 2014; Sirinupong et al. 2011, 2010; Spellmon et al. 2015a, 2015b, 2017). There are five members in this protein family, SMYD1–5 (Spellmon et al. 2015a), and they are known to methylate numerous histone and non-histone proteins (Foreman et al. 2011; Kidder et al. 2017a; Yi et al. 2019). These methylations have linked SMYD proteins to the epigenetic regulation of signal transduction, cell cycle control, and gene expression in various cancers (Giakountis et al. 2017; Tracy et al. 2018; Yi et al. 2019; Zhang et al. 2021, 2019b). SMYD1, with a

Genomic and Epigenomic Biomarkers of Toxicology and Disease: Clinical and Therapeutic Actions,
First Edition. Edited by Saura C. Sahu.

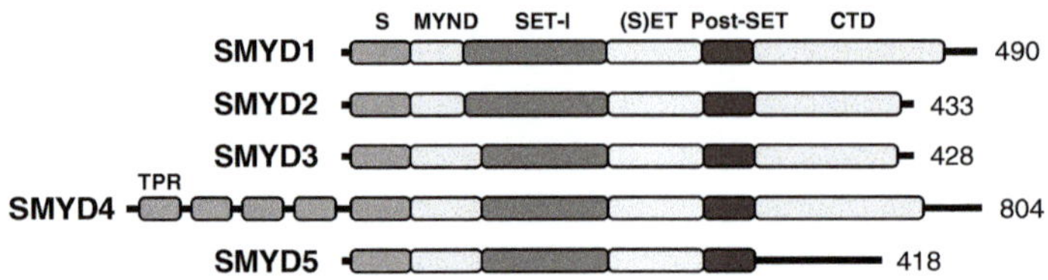

Figure 13.1 The SMYD protein family. S, S-sequence; MYND, myeloid-nervy-DEAF1; SET-I, insertion SET domain; (S)ET, core SET domain (suppressor of variegation-enhancer of zeste-trithorax); post-SET, SET C-terminal flanking domain; CTD, C-terminal domain.

cardiac-specific expression, was originally characterized as a critical epigenetic regulator in sarcomere assembly and in heart development (Liang et al. 2020; Rasmussen et al. 2015; Tan et al. 2006; Tracy et al. 2018). Recent studies have shown that SMYD1 might also be a novel mediator in medulloblastoma and prostate cancer with prognostic values (Wei et al. 2017; Zhang et al. 2019c). The overexpression of SMYD2 and SMYD3 has been demonstrated as a cancer biomarker in various types of cancers such as breast cancer, prostate cancer, and hepatocellular carcinoma (HCC) (Lobo et al. 2018; Song et al. 2019; Zuo et al. 2018a). It has been shown that the growth of cancer cell lines was impaired significantly by SMYD2 or SMYD3 knockdown (Kojima et al. 2020; Nakakido et al. 2015; Reynoird et al. 2016; Wang et al. 2008b). The oncogenic impact of SMYD2 and SMYD3 was mediated in part by their methyltransferase activity. SMYD2 methylated the tumor suppressor p53, thereby repressing its anti-tumor property (Huang et al. 2006). SMYD3 methylated MAP3K2, thus promoting the RAS-mediated cancer progression (Colon-Bolea and Crespo 2014; Mazur et al. 2014).

SMYD4 was first identified as a putative tumor suppressor gene, which participated in breast cancer development via inhibiting the expression of platelet-derived growth factor α (PDGFα) (Han et al. 2019; Hu et al. 2009). New evidence revealed that SMYD4 also played a role in the induction of the switch of non-cancer stem cells into cancer stem cells (CSCs) (Liu et al. 2020b). SMYD4 was shown to bind to the unmethylated Nanog promoter site, transactivating the expression of Nanog, an oncogenic reprogramming mediator, to promote carcinogenesis (Liu et al. 2020b). SMYD5 was known to maintain chromosome integrity via histone H4K20 methylation, which regulated heterochromatin structure and prevented spurious differentiation (Kidder et al. 2017a, 2017b). This methylation was also found to attenuate the expression of pro-inflammatory genes in LPS-induced inflammation and to associate with the transcriptional repressive activity of the NCoR co-repressor complex (Stender et al. 2012). SMYD5 was found to be overexpressed in multiple cancers, including gastric cancer and breast cancer (Meng et al. 2019; Song et al. 2019), but the underlying mechanism for its role in cancer initiation and development remains elusive.

Current data have clearly shown that SMYD members are promising biomarkers for cancer diagnosis and prognosis. The genetic and expression alterations of SMYD proteins were correlated with increased cancer morbidity and mortality (Komatsu et al. 2015; Song et al. 2019; Zhang et al. 2020a; Zuo et al. 2018a). SMYD protein-mediated lysine methylation was heavily associated with cancer progression (Huang et al. 2006; Li et al. 2018a; Yoshioka et al. 2017a). It has been shown that SMYD proteins acted as independent prognostic markers for patient overall survival in many cancers (Liu et al. 2015a, 2017a; Xu et al. 2018a; Zhang et al. 2020a). Here we review the genetic and epigenetic changes in SMYD proteins that are associated with cancer initiation and development. Our review focuses on the biomarker-oriented discussion of SMYD generic variants, copy number alterations, protein and mRNA expression changes, and histone and non-histone substrates. The far-reaching goal is to kindle the future development of SMYD proteins as clinical biomarkers for cancer diagnosis and prognosis.

SMYD Proteins as Cancer Biomarkers

Genetic Alterations as Biomarkers

The genomic alterations of *SMYD* genes have been found to associate with worse clinical outcomes in numerous types of cancer (Table 13.1). Genetic mutations in *SMYD1* and *SMYD2* were found to correlate with a reduced survival time (Galvan-Femenia et al. 2018; Wei et al. 2017). A frequent variable number tandem repeat (VNTR) polymorphism in *SMYD3* correlated with disease progression and poor prognosis (Li et al. 2018c; Liu et al. 2017b; Zhu et al. 2016). Copy number alterations have been found in *SMYD2*, *SMYD3*, and *SMYD4* genes that were associated with advanced disease stages and distant metastasis (Elsemman et al. 2016; Liu et al. 2015c; Noroxe et al. 2020). The acquired somatic mutations in the *SMYD1* gene were first identified in splenic marginal zone lymphoma (SMZL) (Peveling-Oberhag et al. 2015). The whole exome sequencing of the tumors revealed two recurrent mutations in three out of ten SMZL patients. Both these mutations are missense

Table 13.1 Genetic alterations as biomarkers.

Genetic alteration	Cancer type	Correlated clinicopathologic feature
SMYD1 mutation	Medulloblastoma group 3	Worse OS (Wei et al. 2017)
	Prostate cancer	Worse OS (Zhang et al. 2019c)
SMYD2 mutation	Non-small-cell lung cancer (advanced stage)	Worse OS; reduced time to progression; worse clinical outcome of first-line platinum-based treatment (Galvan-Femenia et al. 2018)
SMYD3 VNTR	Esophageal squamous cell carcinoma	Increased risk (Wang et al. 2008a)
	Ovarian cancer	Worse OS* (Liu et al. 2016a)
	Hepatocellular carcinoma	Worse OS; larger tumor size; advanced TNM staging; lymph node metastasis; drinking history; family history; HBV infection and cirrhosis (Li et al. 2018c)
SMYD2 CNA	Hepatocellular carcinoma (HCV-related)	Diagnosis of hepatocellular carcinoma# (Elsemman et al. 2016)
SMYD3 CNA	Breast cancer (basal-like)	Worse OS; larger tumor size; advanced LNM (Liu et al. 2015c)
	Hepatocellular carcinoma (HCV-related)	Diagnosis of hepatocellular carcinoma# (Elsemman et al. 2016)
SMYD4 CNA	Glioblastoma	Worse median OS (Noroxe et al. 2020)

*Identified as an independent prognostic biomarker.
#Identified as a diagnostic biomarker.
OS: overall survival; TNM: classification of malignant tumors; LNM: lymph node metastasis

mutations located within the catalytic SET domain. Because of this location, such mutations are potentially deleterious to the methyltransferase activity. This finding has revealed a potential for *SMYD1* as an alternative biomarker for SMZL diagnosis. Mutated *SMYD1* was also reported as a prognostic signature for group 3 medulloblastoma (G3MB) (Wei et al. 2017). The *SMYD1* gene harbors a non-silent somatic mutation in mouse medulloblastoma, where a single nucleotide variation results in R237Q mutation. The G3MB mice carrying this mutant have been shown to have a shortened life span.

Several germline genetic variations in *SMYD2* have been identified through genomic profiling of advanced-stage non-small-cell lung cancer (NSCLC) (Galvan-Femenia et al. 2018). These variations were found to correlate with poor clinical outcomes in patients who received the first-line platinum-based chemotherapy. Currently, the platinum-based chemotherapy is widely used for the treatment of most NSCLC patients at an advanced stage. In most cases, advanced NSCLC remains incurable, owing to a frequently acquired drug resistance (Galluzzi et al. 2012; Yin et al. 2016). *SMYD2* variants were shown to serve as genetic modifiers for therapy efficacy in NSCLC treatment. Three variants (rs6665343 G/A, rs11120295 C/T, and rs2291830 T/G) have been demonstrated to have an impact on the survival endpoints and to associate with decreased overall survival and with reduced time to progression (Galvan-Femenia et al. 2018). This finding indicates that *SMYD2* variants may have predictive and prognostic values that direct the choice of a combined therapy in advanced NSCLC treatment.

An *SMYD3* VNTR polymorphism has been found to correlate with disease progression in esophageal squamous cell carcinoma (ESCC), ovarian cancer, and HCC (Li et al. 2018c; Liu et al. 2016b; Zhu et al. 2016). A VNTR is a location in a genome where a short nucleotide sequence is organized as a tandem repeat. The length of such a repeat often shows variations among individuals and can affect gene expression. The polymorphism of a tandem repeat [(CCGCC)n] in the promoter region of *SMYD3* underlies its cancer association (Tsuge et al. 2005). This sequence is a binding site of transcription factor E2F1 (Tsuge et al. 2005). The wild-type three-repeat allele (3/3) has enhanced binding affinity by comparison with the two-repeat allele (3/2 or 2/2). The three-repeat allele was identified as a susceptibility factor associated with a significantly increased risk for the development of ESCC by comparison with the 3/2 or 2/2 genotype (Wang et al. 2008a). Interestingly, the association between the 3/3 genotype and increased risk of ESCC was more pronounced in males and elderly subjects from the age of fifty-five years on (Wang et al. 2008a). Among smokers, the association was even more significant and substantial, suggesting that the *SMYD3* VNTR polymorphism is a susceptibility factor for smoking-related ESCC (Wang et al. 2008a).

The *SMYD3* polymorphism was also identified as a risk factor for susceptibility and poor prognosis in ovarian cancer (Liu et al. 2016a). The frequencies of the genotype 3/3 and allele genotype 3 in patients with ovarian cancer were significantly higher than they were in the control group. The proportion of the genotype 3/3 in poorly differentiated tumors was significantly higher than the proportion of other genotypes. The genotype 3/3 was also associated with a worse disease stage and survival rate (Liu et al. 2016a). It was prevalent in patients with the International Federation of Gynecology and Obstetrics (FIGO) stage III–IV disease. The five-year survival rate was significantly lower for patients carrying this genotype. A COX regression analysis showed that the genotype 3/3 was an independent risk factor for ovarian cancer prognosis (Liu et al. 2016a). However, inconsistent results were found with respect to the association between the *SMYD3* VNTR genotypes and risk of HCC (Wang et al. 2007). An early study found no association between the *SMYD3* VNTR genotypes and risk of HCC or familial breast cancer (Binh et al. 2020; Frank et al. 2006;

Wang et al. 2007). However, a later study showed that the occurrence rate of the 3/3 genotype in HCC patients was significantly higher than the occurrence rate of the 2/2 or 2/3 genotypes (Li et al. 2018b). HCC patients with the 3/3 genotype had lower two- and three-year survival rates. The 3/3 genotype was also found to be associated with tumor–node–metastasis (TNM) staging, maximum tumor diameter, and lymph node metastases (Li et al. 2018b). Individuals were made more vulnerable to HCC by risk factors such as drinking history, HCC family history, cirrhosis, and hepatitis B virus (HBV) if they carried the 3/3 genotype (Binh et al. 2020; Li et al. 2018b). This study has raised the chances for the *SMYD3* VNTR polymorphism to become a diagnostic or prognostic biomarker for HCC patients. Existing discrepancies warrant additional studies with larger cohorts of cancer patients to confirm the clinical value of the *SMYD3* VNTR polymorphism.

Copy number alterations (CNAs) in *SMYD2*, *SMYD3*, and *SMYD4* genes may serve as promising diagnostic and prognostic biomarkers for HCC, breast cancer, and glioblastoma (Elsemman et al. 2016; Liu et al. 2015b; Noroxe et al. 2020). *SMYD2* and *SMYD3* genes, along with a number of other genes (*ASH1L*, *METTL13*, and *TARBP1*) on chromosome 1q, showed an increase in their copy number in hepatitis C virus (HCV)-related HCC (Elsemman et al. 2016). Amplification of chromosome 1q is a common genetic change in several malignancies (Bieche et al. 1995; Gratias et al. 2016; Skawran et al. 2008). Such a change has been associated with inferior outcomes in HCC, multiple myeloma, and brain tumors. In addition, genome rearrangements on chromosome 1q have been shown to affect S-Adenosylmethionine (SAM) metabolism (Elsemman et al. 2016). SAM is known to inhibit the growth of cancer cells by reversing the hypomethylation status of *c-MYC* and *H-RAS* genes (Luo et al. 2010). Since *SMYD2*, *SMYD3*, *ASH1L*, and *METTL13* genes encode SAM-dependent methyltransferases, their amplification may be involved in alternative uses of SAM that could provide another mechanism to promote HCC cell progression.

The copy number of *SMYD2* or *SMYD3* genes was significantly associated with basal-like breast cancer and shorter patient survival (Liu et al. 2015c). *SMYD2* and *SMYD3* genes were found to be concurrently or independently amplified in breast cancer along with two other histone methyltransferase genes (*ASH1L* and *SETDB1*). These genes showed different frequencies of CNAs in subtypes of breast cancer (Liu et al. 2015c). The frequencies of *SMYD3*, *ASH1L*, and *SETDB1* amplification were higher in the basal-like breast cancer: more than 25% of tumors exhibited high-level amplification by comparison with other subtypes. The expression level of *SMYD2* was significantly higher in the basal-like subtype, while the *SMYD3* gene was expressed at a significantly lower level than the non-basal subtypes (Liu et al. 2015c). These data indicate that the breast cancer, particularly the basal-like subtype, has a high frequency of CNAs in *SMYD2* and *SMYD3* genes.

The genomic profiling of glioblastoma patients revealed that the copy number alterations in the *SMYD4* gene were significantly correlated with worse median overall survival (Noroxe et al. 2020). Several clinical variables were age, performance status, steroid dose, and O6-methylguanine-DNA methyltransferase status; these were also found to be associated with *SMYD4* CNAs (Noroxe et al. 2020). This indicates that *SMYD4* could be a prognostic marker in glioblastoma patients and points to its potential for clinical utility. However, there was no correlation between *SMYD4* mutations and worse median overall survival. Thus, unlike *SMYD4* CNAs, *SMYD4* mutations may lack prognostic value for glioblastoma patients. Current data have provided evidence that genetic alterations in *SMYD* genes are associated with worse clinical outcomes in many cancers. The genomic profiling of *SMYD* genes has the potential to enable an early detection of cancer, a genomic-driven prognosis, and an evaluation of the resistance to certain anti-cancer therapies.

Abnormal Expression as Biomarkers

A growing number of studies have revealed the expression levels of SMYD proteins as promising diagnostic and prognostic biomarkers in cancer (Table 13.2). The abnormal expression of SMYD1 was associated with reduced progression-free survival (PFS) in patients with prostate cancer (Zhang et al. 2019c). SMYD2 overexpression has been reported as an independent prognostic indicator for seven cancers, including B lineage acute lymphoblastic leukemia (ALL), clear cell renal cell carcinoma, ESCC, gastric cancer, HCC, human papillomavirus (HPV)-unrelated head and neck squamous cell carcinoma, and papillary thyroid carcinoma (Komatsu et al. 2015, 2009; Ohtomo-Oda et al. 2016; Xu et al. 2018a; Yan et al. 2019; Zhang et al. 2020a; Zuo et al. 2018b). SMYD3 was also overexpressed in various types of cancer and has been suggested as an independent diagnostic and prognostic biomarker for colorectal cancer, ESCC, gastric cancer, HCC, NSCLC, prostate cancer, and ovarian cancer (Fei et al. 2017; Li et al. 2020; Liu et al. 2015a, 2015e, 2017a, 2016a; Lobo et al. 2018; Zhu et al. 2016). SMYD5 was recognized as a potential oncogene in breast cancer, gastric cancer, and liver cancer (Meng et al. 2019; Song et al. 2019). Its overexpression was correlated with poor clinical outcomes and prognosis (Meng et al. 2019).

Generally, highly expressed SMYD2 or SMYD3 was associated with worse overall survival (OS), aggressive progression, or resistance to anti-cancer therapies (Komatsu et al. 2009; Li et al. 2020; Shang and Wei 2019; Zhu and Huang 2020). These genes' expression level can also predict disease-free survival (DFS), disease-specific survival (DSS), event-free survival (EFS), or relapse-free survival (RFS) (Table 13.2). A univariate survival analysis indicated a significant association between SMYD2 overexpression and reduced OS in primary HCC patients (Zuo et al. 2018b). The expression of SMYD2 was found to be significantly upregulated at both mRNA and protein levels. A positive expression of SMYD2 was observed in 75% of HCC tissues, but only in 17% of adjacent non-tumorous tissues. A further multivariate analysis showed that positive SMYD2 expression, combined with vascular invasion, could serve as independent prognostic parameters for HCC patients (Zuo et al. 2018b). The overexpression of SMYD3 in HCC patients was associated not only with worse OS but also with reduced RFS (Wang et al. 2019). SMYD3 expression was found to be capable of independently predicting survival and recurrence after curative resection (Fei et al. 2017). SMYD2 and SMYD3 were also reported as independent prognostic indicators for other cancers (Liu et al. 2017a; Ohtomo-Oda et al. 2016; Xu et al. 2018b). The expression level of SMYD2 was correlated with worse RFS in breast cancer patients, while the overexpression of SMYD3 was correlated with both worse RFS and worse OS (Ma et al. 2018; Patani et al. 2011; Song et al. 2019). SMYD3 expression was positively associated with disease occurrence and with HBV infection history in HBV-related HCC (Binh et al. 2020). A low expression level of SMYD2 was associated with improved DSS and DFS in renal cancer, especially in patients with non-organ-confined tumors (Pires-luis et al. 2015).

The upregulation of SMYD2 or SMYD3 often correlated with aggressive cancer phenotypes. In gastric cancer, SMYD2 overexpression was associated with larger tumor size, enhanced lymphatic invasion, and higher rate of recurrence (Komatsu et al. 2015). In B lineage ALL patients, upregulated SMYD2 was associated with a high level of white blood cell (WBC) counts and a high percentage of peripheral blood blast cells (Zhang et al. 2020b). For SMYD3, overexpression was found to correlate with high WBC counts and a complex

Table 13.2 SMYD overexpression as biomarker.

	Cancer type	Correlated clinicopathologic feature
SMYD2	Acute lymphoblastic leukemia	Worse OS; unfavorable age; high WBC peripheral blood counts; resistance to chemotherapy (Brown et al. 2020, Sakamoto et al. 2014)
	Acute lymphoblastic leukemia – B lineage	Worse OS and EFS*; high WBC counts; high percentage of PB blast cells; high percentage of high-risk disease (Zhang et al. 2020b)
	Breast cancer	Worse RFS*#; more metastatic relapse (Song et al. 2019, Liu et al. 2015c)
	Breast cancer (triple-negative)	Worse OS (Li et al. 2018a)
	Cervical cancer	Worse OS; advanced FIGO stage; larger tumor size; poor prognosis (Sun et al. 2019)
	Chronic lymphocytic leukemia	High WBC counts; complex karyotype (Oliveira-Santos et al. 2016)
	Clear cell renal cell carcinoma	Worse OS and DFS*; high TNM stage; early tumor relapse (Yan et al. 2019)
	Colon cancer	Worse OS (Lai and Yang 2020); worse DFS (Ren et al. 2019); increased tumor growth; advanced vascular invasion; TNM state and LNM (Lai and Yang 2020); enhanced oxaliplatin resistance (Ren et al. 2019)
	Colorectal cancer	Worse OS and DFS; enhanced tumor diameter; advanced TNM stage; LNM; distant metastasis (Meng et al. 2020)
	Esophageal squamous cell carcinoma	Worse OS*; enhanced venous invasion; higher pT category and more recurrence (Komatsu et al. 2009)
	Gastric cancer	Worse OS*; larger tumor size; more aggressive lymphatic invasion; deeper tumor invasion; higher rates of LNM and recurrence (Komatsu et al. 2015)
	Gynecological cancer (OV, CESC and UCEC)	Worse OS (Zhang and Wang 2019)
	Hepatocellular carcinoma	Worse OS; larger tumor size; enhanced vascular invasion; differentiation and TNM stage (Zuo et al. 2018b)
	Human papillomavirus-unrelated non-multiple head and neck cancer	Worse OS*; enhanced venous invasion; advanced pT, pN and pStage (Ohtomo-Oda et al. 2016)
	Lower-grade gliomas	Worse OS (Yu et al. 2021)
	Non-small-cell lung cancer	Enhanced cisplatin resistance (Shang and Wei 2019)
	Ovarian clear cell carcinoma	Worse OS and DSS (Engqvist et al. 2020a)
	Pancreatic ductal adenocarcinoma	Worse OS; enhanced precancerous lesions (Reynoird et al. 2016)
	Papillary thyroid carcinoma	Worse OS*; larger tumor size; enhanced LNM and TNM stage (Xu et al. 2018a)
	Renal cell tumor	Improved DSS and DFS; patient age (Pires-Luis et al. 2015)

(Continued)

Table 13.2 (Continued)

	Cancer type	Correlated clinicopathologic feature
SMYD3	Bladder cancer	Worse OS and RFS; advanced histological grade and LNM (Wang et al. 2020b, Shen et al. 2016)
	Breast cancer	Worse OS and RFS (Song et al. 2019, Patani et al. 2011, Ma et al. 2018); reduced cisplatin sensitivity; advanced histological differentiation (Wang et al. 2020c); LNM, HER2 status and subtypes (Song et al. 2019, Ma et al. 2018)
	Breast cancer (claudin-low)	Worse RFS (Fenizia et al. 2019)
	Chronic lymphocytic leukemia	High WBC count; complex karyotype (Oliveira-Santos et al. 2016)
	Colon adenocarcinoma	Worse DFS; advanced tumor stage (Yue et al. 2020)
	Colorectal cancer	Worse OS*; advanced T stage (Liu et al. 2017a)
	Esophageal squamous cell carcinoma	Worse OS* (Liu et al. 2017b, Zhu et al. 2016); advanced lymph node metastasis (Zhu et al. 2016); more cancer occurrence (Dong et al. 2014)
	Gastric carcinoma	Worse OS* (Liu et al. 2015a, Liu et al. 2015e, Liu et al. 2015d, Liu et al. 2015f, Wang et al. 2018); metastasis-free survival (MFS) (Wang et al. 2018); larger tumor size, advanced lymph node metastasis (Liu et al. 2015a, Liu et al. 2015e, Liu et al. 2015d, Liu et al. 2015f); LNM, higher TNM stage (Wang et al. 2018)
	Glioma	Worse OS; higher pathologic grade (Dai et al. 2015)
	Hepatocellular carcinoma	Worse OS* (Zhou et al. 2019, Fei et al. 2017) and RFS (Zhou et al. 2019, Wang et al. 2019); unfavorable prognosis* (Fei et al. 2017); advanced clinical stage; larger tumor size; higher concentration of serum AFP; tumor relapse; HBV infection (Wang et al. 2019, Fei et al. 2017)
	Hepatocellular carcinoma (HBV related)	Occurrence; aggressive disease progression (Binh et al. 2020)
	Medullary thyroid cancer	Worse OS* (Khatami and Tavangar 2018) and DFS (Sponziello et al. 2014); frequent metastases; disease consistency (Khatami and Tavangar 2018, Sponziello et al. 2014)
	Non-small cell lung cancer	Worse OS* and DFS; advanced pathological stage; larger tumor size; presence of lymphovascular invasion; pleural invasion; distant metastasis (Li et al. 2020)
	Ovarian cancer	Enhanced invasion and adhesion of spheroids (Lyu et al. 2020)
	Ovarian cancer (epithelial)	Worse OS and DFS; frequent metastasis; advanced FIGO stage; more ascites (Zhang et al. 2019a)
	Pancreatic cancer	Worse OS; larger tumor size; advanced T stages; TNM stage; perineural invasion; lymph node metastasis (Zhu and Huang 2020)
	Prostate cancer	Worse DSS* (Lobo et al. 2018, Vieira et al. 2014); less differentiated; more aggressive tumors (Vieira et al. 2014)
	Renal cell carcinoma	Worse OS* and DFS; advanced tumor stage; histological and nuclear grade (Liu et al. 2020a)

(Continued)

Table 13.2 (Continued)

	Cancer type	Correlated clinicopathologic feature
SMYD5	Breast cancer	Worse OS; advanced SBR grade with poor clinical outcomes (Song et al. 2019)
	Gastric cancer	Worse OS (Meng et al. 2019)
	Liver cancer	Worse OS (Uhlen et al. 2017)

*Identified as an independent prognostic biomarker.
#Identified as a diagnostic biomarker.
OS: overall survival; RFS: relapse-free survival; DFS: disease-free survival; DSS: disease-specific survival; EFS: event free survival; LNM: lymph node metastasis; SBR grade: Scarff-Bloom-Richardson grade.
FIGO stage: The International Federation of Gynecology and Obstetrics staging system.
WBC: white blood cell

karyotype in chronic lymphocytic leukemia (CLL) patients (Oliveira-Santos et al. 2016). In HCC patients, upregulated SMYD3 was associated not only with an advanced clinical stage and with tumor relapse, but also with a higher concentration of serum alpha-fetoprotein (AFP) (Fei et al. 2017; Wang et al. 2019). AFP level is a commonly used biomarker for early cancer diagnosis. The underlying mechanism of the association between SMYD3 and AFP remains unknown, but their simultaneous overexpression could serve as a diagnostic biomarker for early cancer detection.

Patients with high expression levels of SMYD2 or SMYD3 have been shown to be more resistant to anti-cancer therapies. Increased oxaliplatin resistance was observed in colon cancer patients with SMYD2 overexpression (Ren et al. 2019). Cisplatin resistance in NSCLC patients was found to be associated with both SMYD2 and SMYD3 overexpression (Li et al. 2020; Shang and Wei 2019). SMYD2 and SMYD3 were involved in cisplatin resistance, likely through the regulation of p53 pathway activity and cell apoptosis. The knockdown of these proteins was found to inhibit NSCLC cell proliferation and tumorsphere formation (Shang and Wei 2019). Overexpressing these proteins in NSCLC cells made the cells become more resistant to apoptosis induced by cisplatin (Li et al. 2020). The overexpression of SMYD3 also reduced cisplatin sensitivity in breast cancer patients and correlated with advanced histological differentiation (Wang et al. 2020c). Consistently, ALL patients with low levels of SMYD2 expression were more sensitive to chemotherapy (Sakamoto et al. 2014). SMYD2 or SMYD3 depletion was found to increase cisplatin sensitivity in NSCLC cells (Li et al. 2020; Shang and Wei 2019). Thus the expression levels of SMYD2 and SMYD3 could be used as potential biomarkers for clinical decisions on more appropriate anti-cancer therapies.

SMYD5 overexpression correlated with worse OS and with poor clinical outcomes in several cancers. The increased SMYD5 mRNA level was associated with an advanced Scarff–Bloom–Richardson grade and with poor prognosis in breast cancer (Song et al. 2019). SMYD5 was upregulated in gastric cancer and associated with several clinicopathological features (Meng et al. 2019). High SMYD5 expression correlated negatively with the survival rate of liver cancer patients (Uhlen et al. 2017). However, the depletion of SMYD5 from human colon and lung cancer cells resulted in increased tumor growth and in the upregulation of genes that promote cell proliferation (Kidder et al. 2017a). This result suggests that SMYD5 could play a tumor-suppressive or tumor-promoting role in different contexts and cancers.

SMYD Protein-mediated Lysine Methylation as a Cancer Biomarker

Functional Impacts of Protein Lysine Methylation

SMYD proteins exert their diverse tumor-promoting roles through their ability to methylate numerous histone and non-histone proteins. Current data revealed that the functional impacts of SMYD-mediated lysine methylation into six categories. These are: effects on protein modification, protein–protein interaction, protein stability, enzyme activity, subcellular localization, and promoter binding (Hamamoto et al. 2015). For example, SMYD2 was found to methylate estrogen receptor α (ERα) at K266 in breast cancer cells (Zhang et al. 2013). This lysine methylation competitively inhibited acetylation of the same lysine residue by p300/cAMP. The crosstalk between these two modifications was found to regulate ERα target gene activation by affecting the chromatin recruitment of ERα (Zhang et al. 2013). The effect on protein–protein interaction was demonstrated through lysine methylation of heat shock protein 90 (HSP90). SMYD2 was found to methylate HSP90AB1 at K531 and K574 during cancer cell proliferation (Hamamoto et al. 2014). This lysine methylation promoted the dimerization and complex formation of the chaperone. Lysine methylation can also affect protein stability by inhibiting degradation through the ubiquitin-proteasome pathway. One example is the protein stability of the enhancer of zeste homolog 2 (EZH2) enhanced through SMYD2-mediated lysine methylation (Zeng et al. 2019). This lysine methylation played an essential role in fine-tuning EZH2 functions in chromatin recruitment and transcriptional repression, regulating breast cancer progression. Several enzymes in cancer-related signaling pathways, such as PARP1 (poly(ADP-ribose) polymerase 1), ALK, and AKT1, are substrates of SMYD proteins (Piao et al. 2014; Yi et al. 2019). Generally, lysine methylation of these enzymes enhanced their enzymatic activity, thereby facilitating tumor growth (Piao et al. 2014; Wang et al. 2017). The lysine 14 methylation of AKT1 by SMYD3 was essential for its kinase activity, which activated the AKT1 oncogenic activity in cancer cells (Yoshioka et al. 2016). As lysine residues are critical for the nuclear localization of proteins, one could speculate that the methylation of lysine may affect their subcellular localization (Hamamoto et al. 2015). SMYD2-mediated methylation of β-catenin was found to promote the translocation of β-catenin to the nucleus, which in turn activated Wnt signaling (Deng et al. 2017). Lysine methylation can also regulate the binding affinity of transcription factors to the promoters. Lysine methylation of p53 by SMYD2 markedly decreased its promoter binding ability and repressed its tumor suppressor activity (Huang et al. 2006). These diverse functional impacts have provided the molecular basis for SMYD-mediated lysine methylation as a cancer biomarker for diagnosis and outcome prediction.

Histone Lysine Methylation as a Biomarker

Numerous studies have demonstrated that histone lysine methylation is associated with the progression and clinical outcomes of cancer (Audia and Campbell 2016; Benard et al. 2014; Chervona and Costa 2012; Varier and Timmers 2011). Cancer prognosis can be predicted on the basis of histone lysine methylation both at the global level, across the genome, and at specific gene loci (Khan et al. 2015; Song et al. 2012; Vougiouklakis et al. 2020). Histone lysine methylation has also been identified in blood derived from cancer patients (Kaminska et al. 2019; Vougiouklakis et al. 2020). Circulating histones reflecting the abnormalities of tumor cells are promising non-invasive biomarkers for cancer diagnosis and prognosis (Liu et al. 2018). SMYD proteins have been found to methylate H2A.Z, H3, and

H4. These methylations have been reported to regulate tumor initiation and progression in various cancers (Table 13.3). SMYD3 di-methylated lysine 101 of H2A.Z1 at the promoter of *cyclin A1* in breast cancer cells (Tsai et al. 2016). H2A.Z1 is one of the two isoforms of H2A.Z. H2A.Z was considered an oncogenic histone variant overexpressed in breast, colorectal, liver, and bladder cancers (Giaimo et al. 2019; Monteiro et al. 2014). Because of the correlation between the levels of H2A.Z and short overall patient survival, H2A.Z can be

Table 13.3 SMYD3-mediated histone lysine methylation as biomarker.

Histone substrate	Gene locus	Function in cancer
H2A.Z1-K101 (me2)	*Cyclin A1* promoter	Regulate expression of cyclin A1 and G1-S progression in breast cancer (Tsai et al. 2016)
H3K4 (me1-3)	-	Independent predictor of poor survival for hepatocellular carcinoma (He et al. 2012)
	ASCL2 promoter	Involved in Wnt signaling; maintain cancer stem cell self-renewal and tumorigenicity (Wang et al. 2018)
	BCLAF1 promoter	Enhance BCLAF1 expression; activate autophagy in bladder cancer (Shen et al. 2016)
	CDK2 and *MMP2* promoters	Promote tumorigenicity and intrahepatic metastasis in hepatocellular carcinoma (Wang et al. 2019)
	ER target gene promoter	Regulate ER signaling in breast cancer (Kim et al. 2009)
	EZR promoter	Promote the mobility and invasiveness in esophageal squamous cell carcinoma (Zhang et al. 2018)
	hTERT promoter	Regulate TERT/telomerase expression (Liu et al. 2007)
	IGF-1R promoter	Activate IGF-1R expression; modulate cell proliferation, migration and invasion (Wang et al. 2020a)
	ITGB6 and *ITGAM* promoters	Promote implant metastasis in ovarian cancer (Lyu et al. 2020)
	MMP-9 promoter	Promote cancer progression (Cock-Rada et al. 2012)
	MYL9 gene promoter	Promote breast cancer cell migration (Luo et al. 2014)
	SLUG promoter	Regulate expression of ANKHD1 in hepatocellular carcinoma (Zhou et al. 2019)
	-	Promote homologous recombination (Chen et al. 2017)
H4K20 (me3)	*CCND2* promoter	Pro-oncogenic role in prostate carcinogenesis (Vieira et al. 2015)

useful as a tumor progression biomarker (Monteiro et al. 2014). SMYD3-mediated H2A.Z1 methylation was found to promote the malignant development of breast cancer by upregulating H2A.Z1 expression and by promoting G1-S progression (Tsai et al. 2016). Thus, an elevated H2A.Z1 K101me2 level could be a biomarker for breast cancer diagnosis and for prognosis prediction.

All SMYD proteins except SMYD5 were shown to methylate H3K4. Currently, only SMYD3-related H3K4 methylation was found to be involved in the development of cancers (Chen et al. 2017; Jiang et al. 2019; Lyu et al. 2020; Wu et al. 2019). H3K4 methylation was the only protein methylation biomarker approved for clinical use, and it has been used in diagnosis of colorectal cancer (Kaminska et al. 2019; Liu et al. 2018). However, numerous studies have demonstrated the prognostic value of H3K4 methylation in many types of cancer (Ellinger et al. 2010; Ye et al. 2020). Global H3K4 methylation levels were the independent predictors of PFS in renal cell carcinoma (Ellinger et al. 2010). Lower cellular levels of H3K4me1-3 were associated with an advanced histological grading and with renal tumor metastasis (Kumar et al. 2019). However, in HCC, a high level of H3K4me3, which was positively correlated with SMYD3 expression, was identified as a significant independent predictor of poor survival (He et al. 2012). The prognostic value of H3K4 methylation has also been demonstrated at specific gene loci. SMYD3 was found to methylate H3K4 at the *MMP-9* promoter (Cock-Rada et al. 2012; Wang et al. 2019). This methylation induced *MMP-9* expression and contributed to HCC tumorigenicity and metastasis. Other gene loci regulated by SMYD3-mediated H3K4 methylation are *IGF-1R*, *BCLAF1*, *MYL9*, *ER*, *integrin*, *CDK2*, *MMP2*, *SLUG*, *ASCL2*, *EZR*, and *hTER*, as well as homologous recombination genes involved in DNA repair (Table 13.3). Because most of these genes are oncogenic, these specific methylation sites could be an attractive prospect as prognostic markers and therapeutic targets.

H3K36 and H4K20 were two other sites in cancer that could be methylated by SMYD proteins. Deregulation of H3K36 methylation was associated with the pathogenesis of chondroblastoma and HPV-negative head and neck squamous cell carcinomas (Li et al. 2019; Papillon-cavanagh et al. 2017). SMYD2 was known to di-methylate H3K36 and to interact with the Sin3A complex, thus regulating histone deacetylase 1 (HDAC1)-mediated gene expression (Brown et al. 2006). SMYD2 was identified as an independent biomarker for the prognosis of HPV-negative head and neck cancer (Ohtomo-Oda et al. 2016). The dysregulation of H3K36 methylation might underlie its prognostic role in this cancer. H4K20 methylation was a key player in epigenetic regulation of genome maintenance (Jorgensen et al. 2013). Global H4K20 methylation level has been shown to be a prognostic parameter in several cancers such as bladder cancer, breast cancer, and prostate cancer (Behbahani et al. 2012; Schneider et al. 2011; Yokoyama et al. 2014). SMYD3 was shown to tri-methylate H4K20 and induce cyclin D2 expression, contributing to a more aggressive phenotype of prostate cancer (Vieira et al. 2015). These facts, together, demonstrated that SMYD protein-mediated histone methylation is a potential biomarker for cancer prognosis (Vieira et al. 2015).

Non-histone Lysine Methylation as a Biomarker

SMYD proteins are able to methylate numerous non-histone proteins that play important roles in human tumorigenesis. To date, a total of thirteen non-histone substrates were identified for SMYD2 and four for SMYD3 (Table 13.4). Most of these proteins were involved in the regulation of cancer development through modulating signaling pathways. These pathways include the receptor tyrosine kinase (RTK)-RAS pathway, the JAK/STAT3 pathway,

Table 13.4 Non-histone lysine methylation as biomarker.

	Substrate	**Methylation site**	**Function in cancer**
SMYD2	ALK	K1451, K1455, K1610	Enhance EML4-ALK phosphorylation and cell growth in NSCLC (Wang et al. 2017)
	β-catenin	K133me1	Promote β-catenin nuclear translocation; activate Wnt signaling (Deng et al. 2017)
	BMPR2	Kinase domain	Stimulate BMPR2 kinase activity; activate BMP pathway (Gao et al. 2017)
	ERα	K266me1	Repress ERα target gene expression (Zhang et al. 2013)
	EZH2	K307me1/me2	Enhance EZH2 stability, cell proliferation, EMT and invasion in breast cancer (Zeng et al. 2019)
	HSP90AB1	K531me, K574me1	Promote HSP90AB1 dimerization and chaperone complex formation; accelerate cancer cell proliferation (Hamamoto et al. 2014)
	MAPKAPK3	K355me1	Promote pancreatic ductal adenocarcinoma (Reynoird et al. 2016)
	p53	K370me1	Repress p53-mediated transcriptional regulation (Huang et al. 2006)
	PARP1	K528me1	Enhance poly(ADP-ribosyl)ation activity; promote cancer cell proliferation (Piao et al. 2014)
	PTEN	K313me2	Promote PTEN phosphorylation; negatively regulate PTEN tumor suppressor activity; activate phosphatidylinositol 3-kinase-AKT pathway; promote cell proliferation in breast cancer (Nakakido et al. 2015)
	RB	K810me1, K860me1	K810: enhance Ser 807/811 phosphorylation of RB1 (Cho et al. 2012) K860: provide a direct binding site for L3MBTL1; accelerate E2F transcriptional activity; promote cell cycle progression (Saddic et al. 2010)
	STAT3	K658	Activate STAT3; increase cell proliferation and survival in triple negative breast cancer (Li et al. 2018a)
	p65	K310	Activate p65; increase cell proliferation and survival in triple negative breast cancer (Li et al. 2018a)
SMYD3	AKT1	K14	Promote T308 phosphorylation, AKT1 activation and cancer progression (Yoshioka et al. 2016)
	HER2	K175	Enhance HER2 homodimerization, autophosphorylation and activation; promote cancer progression (Yoshioka et al. 2017b)
	MAP3K2	K260me1/2/3	Activate Ras/Raf/MEK/ERK signaling; prevent MAP3K2 from binding to PP2A; lead to the formation of lung and pancreatic adenocarcinomas (Mazur et al. 2014, Colon-Bolea and Crespo 2014)
	VEGFR1	K831me2	Enhance VEGFR1 kinase activity and carcinogenesis (Kunizaki et al. 2007)

the Wnt signaling pathway, the bone morphogenic protein (BMP) pathway, and the cell cycle/checkpoint pathway (Figure 13.2). Lysine methylation of these non-histone substrates has extended the value of SMYD proteins for cancer subtype recognition, early detection, and prognosis prediction.

The RTK-RAS Signaling Pathway

The tumorigenic roles of SMYD2 and SMYD3 were closely associated with the receptor tyrosine kinase (RTK)-RAS pathway. Several of their non-histone substrates were key receptors and downstream regulators in this pathway. These were three SMYD2 substrates, namely anaplastic lymphoma kinase (ALK), MAPKAPK3, and phosphatase and tensin homologue (PTEN) and four SMYD3 substrates, namely human epidermal growth factor receptor 2 (HER2), vascular endothelial growth factor receptor 1 (VEGFR1), protein kinase B (AKT1), and MAP3K2. Among these proteins, ALK, HER2, and VEGFR1 belong to the receptor tyrosine kinase family, a family of transmembrane protein receptors that can be activated through ligand binding (Hubbard and Till 2000). Ligand binding induced the homodimerization and autophosphorylation of these RTKs that led to the activation of several downstream signaling pathways (Hubbard and Till 2000). SMYD proteins were associated with the two activated pathways: the mitogen-activated protein kinase (MAPK) pathway and the phosphatidylinositol 3-kinase (PI3K)-AKT pathway. The substrates MAP3K2 and

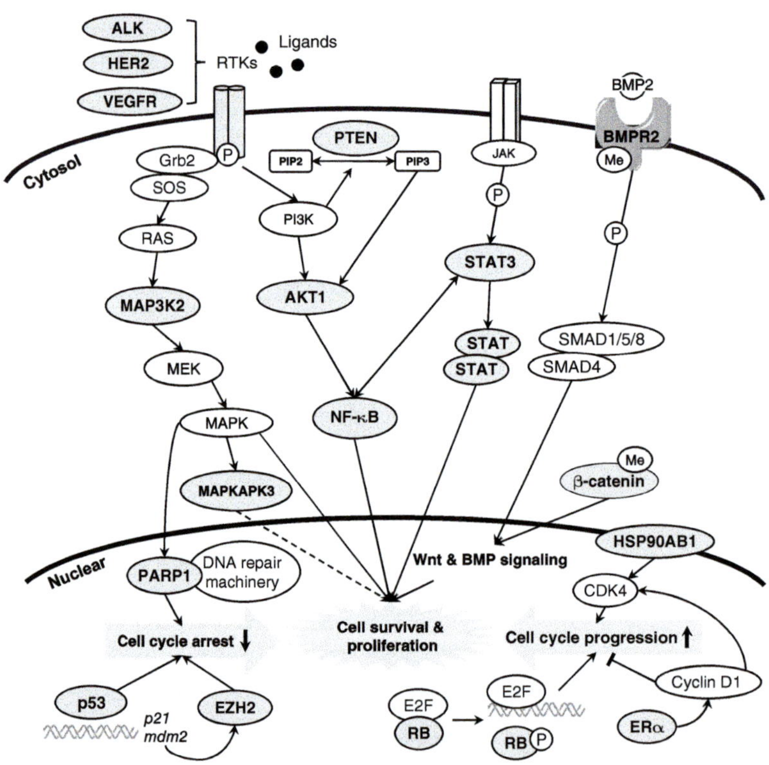

Figure 13.2 SMYD proteins involved in cancer-promoting signaling pathways. SMYD substrates are shown in grey. These substrates are part of several cancer signaling pathways, which include the RTK-RAS pathway, the PI3K-AKT pathway, the RAS-MAPK pathway, the JAK-STAT3 pathway, Wnt signaling, the BMP pathway, and cell cycle regulation. The crosstalks between these pathways contribute to cancer cell survival and proliferation.

MAPKAPK3 were the downstream regulators in the MAPK pathway that modulated signal transduction and amplification. The substrates AKT1 and PTEN were the crucial mediators in the PI3K-AKT pathway, and PTEN played a regulatory role in AKT1 activation. The methylation of these proteins either enhanced their oncogenic activity or repressed the tumor suppressor function (Nakakido et al. 2015; Yi et al. 2019). Because these proteins were heavily associated with the carcinogenic process of various cancers, their methylation marks could be a predictive biomarker for cancer progression.

SMYD2 was found to methylate ALK at lysine 1451, 1455, and 1610 in *EML4-ALK* fusion-driven NSCLC (Wang et al. 2017). The *EML4-ALK* fusion characterized a specific subtype of NSCLC and was found in 4–6% of lung adenocarcinomas. This fusion drove a constitutive oncogenic activation of ALK, which caused tumorigenesis (Soda et al. 2007). ALK methylation was found to promote the phosphorylation of EML4-ALK and facilitate cancer cell growth (Wang et al. 2017). The substitution of lysine 1610 negatively affected ALK phosphorylation and cell proliferation. The combination of an SMYD2 inhibitor and an ALK inhibitor was found to additively suppress cancer cell growth (Wang et al. 2017). These findings revealed that SMYD2 might be a promising target in the development of a novel treatment for EML4-ALK-driven NSCLC. The other two RTKs, HER2 and VEGFR1, were methylated by SMYD3. HER2 was amplified in various human breast cancer cell lines (Feng et al. 2018). HER2 overexpression was linked to breast cancer cell proliferation and tumor progression (Schwartz and Erban 2017). Breast cancer cells that overexpressed HER2 were more likely to progress to metastasis (Bredin et al. 2020; Schwartz and Erban 2017). SMYD3 was found to tri-methylate HER2 at lysine 175 (Yoshioka et al. 2017b). This methylation enhanced HER2 homodimerization and autophosphorylation, subsequently leading to an increased phosphorylation of the downstream effectors AKT and PLCγ1 and promoting breast cancer progression. HER2 overexpression, observed in 18–25% of breast cancers, was identified as an independent marker of poor prognosis (Menard et al. 2001). The upregulation of SMYD3 was also correlated with poor OS and RFS in breast cancer patients (Ma et al. 2018; Song et al. 2019). Thus, SMYD3-mediated HER2 methylation could serve as an alternative maker to improve the prognosis of breast carcinomas. SMYD3 methylated VEGFR1 at lysine 831 (Kunizaki et al. 2007). This methylation enhanced VEGFR1 kinase activity and ligand-dependent autophosphorylation. VEGFR1 played a role in pathologic angiogenesis in cancer (Freire Valls et al. 2019). Upon activation, it induced the recruitment of endothelial progenitor cells, thereby promoting the formation of new blood vessels (Huss et al. 2001). VEGFR1 overexpression promoted breast cancer cell survival and was associated with worse OS independently of anti-cancer treatment (Lee et al. 2007; Schneider et al. 2008; Spanheimer et al. 2015). Thus, tumors overexpressing SMYD3 could show enhanced signal transduction by VEGFR1 and enhanced angiogenesis.

The SMYD-mediated methylation of MAP3K2 and of MAPKAPK3 enhanced the MAPK pathway activation, thus promoting cancer cell proliferation. MAP3K2 was methylated by SMYD3 at lysine 260 (Mazur et al. 2014). This methylation not only enhanced the MAPK signaling and the ERK activation, but also prevented MAP3K2 from binding to the phosphatase PP2A. PP2A is a key negative regulator of the MAPK pathway; it acts by dephosphorylating activated signaling molecules. Reduced PP2A binding resulted in a reduced dephosphorylation of MAP3K2 (Mazur et al. 2014). This subsequently led to prolonged RAS-ERK signaling and promoted the formation of lung and pancreatic adenocarcinomas (Mazur et al. 2014). Depletion of SMYD3 was found to synergize with a MEK inhibitor and repress the pancreatic neoplasia (Mazur et al. 2014). Another MAPK pathway molecule, MAPKAPK3, can be methylated by SMYD2. SMYD2 mono-methylated MAPKAPK3 at lysine 355 in pancreatic ductal adenocarcinoma (PDAC) cells (Reynoird et al. 2016). Currently the molecular consequences of this methylation remain unclear. One possible model is that this methylation might regulate MAPKAPK3 activation and then contribute to the promotion of RAS-driven PDAC development. This model is supported by the fact that MAPKAPK3 is a stress response kinase, activated by ERK-mediated phosphorylation and involved in hyperinflammation linked to PDAC development (Reynoird et al. 2016). Also, the expression level of SMYD2 was elevated in PDAC, and the pharmacological inhibition of SMYD2 or MAPKAPK3 restricted PDAC growth (Reynoird et al. 2016). Because the transcriptional MAPK Pathway Activity Score (MPAS) has been demonstrated to be a clinically relevant biomarker in multiple cancer types (Wagle et al. 2018), the involvement of SMYD2 and SMYD3 in the MAPK pathway suggests possible clinical utilities for the related methylation marks.

A SMYD-mediated methylation also regulates the PI3K-AKT pathway and was involved in promoting cell proliferative processes. The hyperactivation of the PI3K-AKT pathway was found in a variety of malignancies (Fresno Vara et al. 2004; Tokunaga et al. 2006). SMYD proteins were involved in this pathway by methylating the central mediator AKT1 and the major antagonist PTEN (Nakakido et al. 2015; Yoshioka et al. 2016). AKT1, a serine/threonine-protein kinase, was overexpressed in various types of cancer (Etemadmoghadam and Bowtell 2014). When activated, AKT phosphorylated more than 100 substrates, for example NF-κB, GSK3β, mTOR, and MDM2, and this produced a wide range of biological effects (Fresno Vara et al. 2004). AKT1 was methylated by SMYD3 at lysine 14 within the pleckstrin homology (PH) domain (Yoshioka et al. 2016). The PH domain is a site for binding to phosphatidylinositol-3,4,5-triphosphate (PIP3). This binding resulted in PH membrane localization, triggered phosphorylation of the regulatory threonine residue (Thr308), and led to ATK1 activation (Yoshioka et al. 2016). SMYD3 overexpression enhanced Thr308 phosphorylation and the activation of downstream pathways (Yoshioka et al. 2016). When lysine 14 was substituted, the amount of AKT1 proteins at the plasma membrane was significantly reduced, and the phosphorylation level of Thr308 was attenuated. Cancer cells overexpressing lysine 14-substituted AKT1 showed a decreased growth rate in comparison with those overexpressing wild-type AKT1. These facts, together, indicate that the SMYD3-mediated methylation is essential for ATK1 activation and cancer cell proliferation. PTEN, another SMYD substrate in PI3K-AKT signaling, was methylated by SMYD2 (Nakakido et al. 2015). PTEN, a well-known tumor suppressor, is often mutated or lost in cancer cells (Keniry and Parsons 2008; Milella et al. 2015). PTEN converts PIP3 into phosphatidylinositol-4,5-bisphosphate (PIP2), acting as a major antagonist of the PI3K-AKT pathway (Myers et al. 1998). SMYD2 mono-methylated PTEN at lysine 313 (Nakakido et al. 2015). This methylation was found to attenuate PTEN tumor suppressor activity by enhancing the phosphorylation of serine 380. Lysine 313 substitution resulted in a diminished Ser380 phosphorylation accompanied by an increased PTEN activity (Nakakido et al. 2015). The knockdown of SMYD2 was found to suppress the cell growth of breast cancer cells and to attenuate the phosphorylation levels of AKT1 (Nakakido et al. 2015). Numerous studies have highlighted AKT1 and PTEN as prognostic and predictive biomarkers in cancer (Etemadmoghadam and Bowtell 2014; Lopez-Cortes et al. 2018; Ngeow et al. 2017). Thus, the regulatory roles of SMYD-mediated methylation hold promise for exploring the PI3K-AKT pathway as a cancer biomarker.

The identification of NF-κB and STAT3 as SMYD2 substrates has suggested that SMYD2 might be an important coordinator that mediated cross talks between different cancer signaling pathways (Li et al. 2018a). NF-κB, one of the downstream substrates of AKT, is known to regulate various cytokine-dependent cellular activities related to inflammation

(Bai et al. 2009; Haffner et al. 2006; Singh et al. 2007). The interaction between NF-κB and STAT3 played vital roles in controlling the communication between cancer cells and inflammatory cells (Li et al. 2018a). SMYD2 methylated p65 subunit of NF-κB at lysine 310. This methylation was found to contribute to the hyper-phosphorylation and activation of p65 in breast cancer cells (Li et al. 2018a). The activated p65 promoted cell proliferation and survival, thereby enhancing tumor growth. SMYD2 methylated STAT3 at lysine 685. This methylation led to increased STAT3 phosphorylation, dimerization, and nuclear accumulation (Li et al. 2018a). STAT3, a key transcription factor in JAK/STAT signaling, promoted cancer growth by activating the transcription of genes involved in cell division (Yu et al. 2014, 2009). However, SMYD2 affected only a subset of STAT3 target genes, and this implies a specific regulatory mechanism. There existed a positive feedback loop between SMYD2, STAT3, and NF-κB that conferred a synergistic effect on cancer growth (Li et al. 2018a). The interaction between NF-κB and STAT3 facilitated NF-κB recruitment to the promoter of STAT3 and vice versa. STAT3 enhanced p65 methylation and activation by recruiting SMYD2. The interaction between p65 and SMYD2 contributed to STAT3 methylation and activation (Li et al. 2018a). These findings demonstrated the role of SMYD2 in integrating the RTK-RAS and JAK/STAT signaling pathways and its potential prognostic value in cancer.

Wnt Signaling Pathway

SMYD2 was involved in the Wnt signaling pathway by methylating β-catenin (Deng et al. 2017). The aberrant activation of Wnt signaling played important roles in the tumorigenesis of many different types of cancer, most notably colorectal cancer (CRC) (Meng et al. 2020).Upon activation, β-catenin, a key modulator of this pathway, translocated from the cytoplasm to the nucleus, where it activated the gene expression of target proteins. SMYD2 methylated β-catenin at lysine 133 (Deng et al. 2017). This methylation promoted β-catenin nuclear translocation, which resulted in an increased expression of *cMYC* and *CCND1* genes in colon cancer cells. The substitution of lysine 133 almost completely abolished β-catenin nuclear localization. Treatment with a SMYD2 inhibitor caused a significant reduction of β-catenin nuclear localization and a subsequent induction of colon cancer cell death (Deng et al. 2017). SMYD2 was also found to activate Wnt signaling by suppressing the expression of adenomatous polyposis coli 2 (APC2) (Meng et al. 2020). APC2, a negative regulator of the Wnt/β-catenin pathway, inhibited the signaling by promoting a rapid degradation of β-catenin (Mohamed et al. 2019). APC2 downregulation by SMYD2 was found to induce the epithelial–mesenchymal transition program that increased the metastatic ability of CRC (Meng et al. 2020). These data support the involvement of SMYD2 in CRC via the regulation of the Wnt signaling pathway and its potential use as a biomarker for CDC diagnosis.

The interaction between Wnt and BMP pathways has been extensively studied in the context of regulating the apoptosis and proliferation of cancer cells (Huang and Klein 2004). In addition to Wnt signaling, SMYD2 was also involved in BMP signaling and was essential for a robust BMP-induced gene expression (Gao et al. 2017). BMP proteins belong to the transforming growth factor-β (TGF-β) family and are a diverse class of receptors that contain over twenty growth factors. They are essential for bone formation and are involved in the development of several types of cancer (Bach et al. 2018). SMYD2 promoted BMP signaling by methylating BMP type 2 receptor (BMPR2) (Gao et al. 2017). This methylation was found to stimulate BMPR2 kinase activity and the activation of the BMP pathway. A high expression of BMPR2 was correlated with a poor OS in ovarian cancer patients (Fukuda et al. 2020). BMPR2 enhanced ovarian cancer cell proliferation, whereas the BMP receptor kinase inhibitors inhibited the growth of these cells (Fukuda et al. 2020). Notably, a high expression of SMYD2 was also associated with worse OS and DSS in ovarian cancer patients (Engqvist et al. 2020b). These findings have validated the prognostic value of SMYD2 and BMPR2 in ovarian cancer. The combination of SMYD2 and BMP inhibitors would be worth exploring in the treatment of ovarian cancer.

Cell Cycle Regulation

Dysregulations of signaling pathways that control DNA repair, cell cycle checkpoint, and cell proliferation are common hallmarks of cancer (Sanchez-Vega et al. 2018). SMYD2 was implicated in these cellular processes by methylating PARP1, p53, EZH2, retinoblastoma protein (RB), HSP90, and ERα. PARP1 methylation by SMYD2 enhanced the repair of DNA double-strand breaks generated by radiation and chemotherapy (Piao et al. 2014). The methylation of p53 and EZH2 by SMYD2 suppressed cell cycle arrest, thus facilitating the growth and invasion of cancer cells (Huang et al. 2006; Zeng et al. 2019). The SMYD2-mediated methylation of RB and HSP90 promoted cell cycle progression, while ERα methylation repressed ERα target gene activation upon estrogen depletion (Cho et al. 2012; Hamamoto et al. 2014; Zhang et al. 2013). These findings have not only linked SMYD2-mediated methylation to different aspects of cell cycle regulation but also paved the way for an investigation of the marks of methylation as cancer biomarkers.

SMYD2 mono-methylated lysine 528 of PARP1 within the auto-modification region (Piao et al. 2014). Methylated PARP1 showed an increased poly(ADP-ribosyl)ation activity upon oxidative stress. PARP1, a molecular sensor of DNA breaks, was involved in cell cycle arrest and repair of DNA single- and double-strand breaks through the organization of DNA repair machinery (Ray Chaudhuri and Nussenzweig 2017). Increased PARP1 activity was known as a mechanism by which cancer cells escaped from apoptosis signals induced by DNA damage (Pietrzak et al. 2018). Several PARP1 inhibitors have been taken into clinical trials (Luo et al. 2020; Zhu et al. 2020). The inhibition of PARP1 enzyme activity was proven not only to enhance the cytotoxicity of DNA damaging agents but also to induce G0/G1 cell cycle arrest in cancer cells (Weaver and Yang 2013; Yang et al. 2013). SMYD2 depletion resulted in the reduction of PARP1 enzymatic activity, thereby increasing the susceptibility of cancer cells to DNA-damaging chemotherapy (Piao et al. 2014). This was in agreement with SMYD2 overexpression observed in a variety of cancer tissues and associated with drug resistance (Sakamoto et al. 2014; Shang and Wei 2019). Targeting these enzymes would represent a valuable strategy for improving the outcomes of cancer chemotherapy or radiotherapy.

SMYD2 was involved in cell cycle arrest also by methylating p53 (Fan et al. 2015; Huang et al. 2006). *p53* is one of the most important tumor suppressor genes, and is involved in various human tumorigenesis. p53 can induce cell cycle arrest, apoptosis, or senescence when cells are exposed to various forms of stress from the environment or within the cells (Boutelle and Attardi 2021; Meek 2015). SMYD2 mono-methylated p53 at K370 within the regulatory domain, at the extreme carboxyl terminus (Huang et al. 2006). This methylation not only impaired the expression of p21, a downstream target of p53, but also induced p53 dissociation from the p21 protein. p21, as an essential mediator of p53-dependent cell cycle arrest, was known to inhibit G1 progression and induce cellular senescence by blocking the activity of cyclin-dependent kinases (CDKs) (Karimian et al. 2016). The reduced interaction between p53 and p21 was associated with an increased cellular proliferation and with reduced apoptosis (Huang et al. 2006). Consistently, the knockdown of SMYD2 enhanced p53-mediated apoptosis in cancer cells under a DNA damage condition (Huang et al. 2006). p53 methylation also affected the interaction with the MDM2 protein that led to reduced protein level. MDM2 contributed to p53-dependent cell cycle arrest by regulating p53 stability and p21 inhibitory activity (Vassilev et al. 2004). Diminished MDM2 levels underlay the resistant phenotype in tumors with p53 mutation (Cao et al. 2019).

A recent study unveiled an EZH2-mediated epigenetic mechanism behind p53-dependent tissue sensitivity to DNA damage (Kuser-Abali et al. 2018). SMYD2 might play a role in this process by regulating EZH2 stability (Zeng et al. 2019). EZH2 was involved in tumorigenesis by compacting chromatin and suppressing gene expression (Duan et al. 2020; Xu et al. 2019). The compacted chromatin was also found to decrease sensitivity to DNA damage (Margueron et al. 2008). It was shown that p53 and MDM2 modulated DNA damage sensitivity by targeting EZH2 for ubiquitination and degradation (Wienken et al. 2016). Inhibiting EZH2 might be able to sensitize p53-mutant cancer cells to chemotherapy. SMYD2 methylated EZH2 at lysine 307, and this enhanced the stability of EZH2 (Zeng et al. 2019). This methylation inhibited EZH2 degradation through the ubiquitin-proteasome pathway. SMYD2 and EZH2 activities were found to collectively modulate the invasion of breast cancer cells by repressing the expression of RASSF1 (Zeng et al. 2019). RASSF1 was known to inhibit the accumulation of cyclin D1 and thereby to induce cell cycle arrest (Shivakumar et al. 2002; Thaler et al. 2009). These findings demonstrated the role of SMYD2 in boosting cancer cell resistance to DNA damage by enhancing EZH2 stability through methylation. A multivariate survival analysis showed that, independently of tumor proliferation and clinical factors, the expression levels of EZH2 constituted the strongest prognostic factor in mantle-cell lymphoma pathogenesis (Zeng et al. 2019). Kaplan–Meier survival analysis showed that a high expression of SMYD2 or EZH2 was associated with poor OS in patients with breast cancer, lung cancer, and gastric cancer (Zeng et al. 2019). This opened the possibility of a combined use of SMYD2 and EZH2 in order to achieve a more accurate prognosis in cancer.

In addition to the negative regulation of cell cycle arrest, SMYD2 can directly promote cell cycle progression by methylating RB and HSP90. RB normally inhibited the G1/S transition of the cell cycle and was often mutated in cancer (Angus et al. 2002; Giacinti and Giordano 2006). RB interacted with E2F transcription factors and regulated genes related to S-phase entry (Johnson et al. 2016). In a hypo-phosphorylated state, RB bound to E2F and repressed the expression of E2F target genes. When RB was hyper-phosphorylated, E2F was dissociated from RB and transactivated genes that drove cell cycle progression. SMYD2 mono-methylated RB at lysine 810 (Cho et al. 2012). This methylation enhanced RB phosphorylation at serine 807 and 811, leading to augmented E2F transcriptional activity and promoting cell cycle progression. SMYD2 was also reported to mono-methylate RB at lysine 860, in response to DNA damage (Saddic et al. 2010). This methylation regulated RB binding to the transcriptional repressor L3MBTL1. L3MBTL1 regulated chromatin condensation and gene repression via its methyllysine-binding activity (Gao et al. 2011; Trojer et al. 2007). The association between L3MBTL1 and methylated RB suggested that SMYD2-mediated methylation might serve as a docking site for recruiting L3MBTL1 to the promoters of specific E2F target genes; these repress transcription, thus promoting cell cycle progression.

HSP90 methylation by SMYD2 was found to promote cell proliferation by regulating HSP90 chaperone activities (Hamamoto et al. 2014). HSP90, an evolutionarily conserved molecular chaperone, participated in the stabilization and activation of more than 200 client proteins (Schopf et al. 2017). Among these proteins are key mediators of various cell signaling pathways and cell cycle control. Cancer cells used HSP90 chaperone machinery to protect an array of mutated and overexpressed oncoproteins from misfolding and degradation (Gupta et al. 2019; Trepel et al. 2010). Thus HSP90 was recognized as a critical facilitator of oncogene addiction and cancer cell survival (Gupta et al. 2019; Miyata et al. 2013). SMYD2 methylated HSP90AB1 at lysine 531 and 574 (Hamamoto et al. 2014). These methylations enhanced cell cycle progression by maintaining the stability of HER2 and CDK4. These two oncoproteins were HSP90 clients. HER2 is known to promote G1 to S progression in breast cancer cells by regulating RB phosphorylation (Le et al. 2006). CDK4 was identified as a key player in cell cycle control, as it mediated progression through the G1 phase, when the cell was ready to initiate DNA synthesis (Baker and Reddy 2012). The enhanced stability of HER2 and CDK4 induced through HSP90AB1 methylation was found to facilitate cell cycle development and tumorigenesis (Hamamoto et al. 2014). HSP90 also played a reciprocal role in regulating SMYD2 activity. The H3K4 methyltransferase activity of SMYD2 can be enhanced through interaction with HSP90 (Abu-Farha et al. 2008). This enhancement led to an increased induction of genes involved in the cell cycle through epigenetic transcription regulation.

The SMYD2-mediated ERα methylation revealed that SMYD2 plays a complex role in controlling cell cycle progression. ERα was mono-methylated by SMYD2 at lysine 266 (Zhang et al. 2013). Under an estrogen-depleted condition, this methylation attenuated the recruitment of ERα to the chromatin. As a result, the activation of ERα target genes—including growth regulation by estrogen in breast cancer 1 (GREB1), progesterone receptor (PR), and trefoil factor 1 (TFF1)—was prevented. However, upon estrogen stimulation, ERα methylation was diminished, which enabled the p300/CREB-binding protein-mediated ERα acetylation and activated the expression of its target genes (Zhang et al. 2013). These findings suggested that SMYD2 could negatively regulate cell cycle development by attenuating ERα signaling. Given the opposing functions of SMYD2 in cell cycle progression, care must be taken when using SMYD2 proteins simply as drug targets or biomarkers. A comprehensive understanding of SMYD2-mediated methylation could help develop substrate-specific methylation marks or their combination for use in favorable and unfavorable cancer prognosis.

Concluding Remarks

SMYD proteins are promising biomarkers for cancer diagnosis and prognosis on account of their involvement in multiple aspects of cancer initiation and progression. The clinical values of SMYD proteins are clearly evident, since not only the aberrant changes in their genes and expression levels but also the methylation statuses of their various histone and non-histone substrates are associated with patient survival. Such multifaceted associations offer diverse opportunities to explore the clinical values of SMYD proteins at multiple levels, both individually and in combination. Ubiquitous overexpression in various types of cancer also raises possibilities of exploring SMYD2 or SMYD3 as a universal biomarker for cancer detection and prediction. The pharmacological inhibition of SMYD proteins that repress tumor progression and promote cancer cell apoptosis demonstrates the therapeutic values of these proteins in cancer treatment. But the main challenge we are currently facing is the lack of a complete picture of SMYD substrates. Another challenge is their multifunctional nature, as they have both tumor-promoting and tumor-suppressing roles. Further investigation should be directed toward defining the molecular mechanisms that govern the broad substrate specificity of these proteins. Understanding their spatiotemporal methylation patterns will allow the development of specific SMYD-based strategies for more precise types of cancer diagnosis and prognosis.

References

Abu-Farha, M., Lambert, J.P., Al-Madhoun, A.S., Elisma, F., Skerjanc, I.S., and Figeys, D. (2008). The tale of two domains: Proteomics and genomics analysis of SMYD2, a new histone methyltransferase. *Mol. Cell Proteomics* 7: 560–572.

Anastas, J.N. and Moon, R.T. (2013). WNT signalling pathways as therapeutic targets in cancer. *Nat. Rev. Cancer* 13: 11–26.

Angus, S.P., Fribourg, A.F., Markey, M.P., Williams, S.L., Horn, H.F., Degregori, J., Kowalik, T.F., Fukasawa, K., and Knudsen, E.S. (2002). Active RB elicits late G1/S inhibition. *Exp. Cell Res.* 276: 201–213.

Audia, J.E. and Campbell, R.M. (2016). Histone modifications and cancer. *Cold Spring Harb Perspect Biol.* 8: a019521.

Bach, D.H., Park, H.J., and Lee, S.K. (2018). The dual role of bone morphogenetic proteins in cancer. *Mol. Ther. Oncolytics* 8: 1–13.

Bai, D., Ueno, L., and Vogt, P.K. (2009). Akt-mediated regulation of NFkappaB and the essentialness of NFkappaB for the oncogenicity of PI3K and Akt. *Int. J. Cancer* 125: 2863–2870.

Baker, S.J. and Reddy, E.P. (2012). CDK4: A key player in the cell cycle, development, and cancer. *Genes Cancer* 3: 658–669.

Bannister, A.J. and Kouzarides, T. (2011). Regulation of chromatin by histone modifications. *Cell Res.* 21: 381–395.

Behbahani, T.E., Kahl, P., Von Der Gathen, J., Heukamp, L.C., Baumann, C., Gutgemann, I., Walter, B., Hofstadter, F., Bastian, P.J., Von Ruecker, A., Muller, S.C., Rogenhofer, S., and Ellinger, J. (2012). Alterations of global histone H4K20 methylation during prostate carcinogenesis. *BMC Urol.* 12 (5).

Benard, A., Goossens-Beumer, I.J., Van Hoesel, A.Q., De Graaf, W., Horati, H., Putter, H., Zeestraten, E.C.M., Van De Velde, C.J.H., and Kuppen, P.J.K. (2014). Histone trimethylation at H3K4, H3K9 and H4K20 correlates with patient survival and tumor recurrence in early-stage colon cancer. *Bmc Cancer* 14.

Bieche, I., Champeme, M.H., and Lidereau, R. (1995). Loss and gain of distinct regions of chromosome 1q in primary breast cancer. *Clin. Cancer Res.* 1: 123–127.

Binh, M.T., Hoan, N.X., Giang, D.P., Tong, H.V., Bock, C.T., Wedemeyer, H., Toan, N.L., Bang, M.H., Kremsner, P.G., Meyer, C.G., Song, L.H., and Velavan, T.P. (2020). Upregulation of SMYD3 and SMYD3 VNTR 3/3 polymorphism increase the risk of hepatocellular carcinoma. *Sci. Rep.* 10: 2797.

Boutelle, A.M. and Attardi, L.D. (2021). p53 and tumor suppression: It takes a network. *Trends Cell Biol.* 31: 298–310.

Bredin, P., Walshe, J.M., and Denduluri, N. (2020). Systemic therapy for metastatic HER2-positive breast cancer. *Semin. Oncol.* 47: 259–269.

Brown, M.A., Edwards, M.A., Alshiraihi, I., Geng, H.M., Dekker, J.D., and Tucker, H.O. (2020). The lysine methyltransferase SMYD2 is required for normal lymphocyte development and survival of hematopoietic leukemias. *Genes Immun.* 21: 119–130.

Brown, M.A., Sims, R.J., 3RD, Gottlieb, P.D., and Tucker, P.W. (2006). Identification and characterization of Smyd2: A split SET/MYND domain-containing histone H3 lysine 36-specific methyltransferase that interacts with the Sin3 histone deacetylase complex. *Mol. Cancer* 5: 26.

Cao, Z., Xue, J., Cheng, Y., Wang, J., Liu, Y., Li, H., Jiang, W., Li, G., Gui, Y., and Zhang, X. (2019). MDM2 promotes genome instability by ubiquitinating the transcription factor HBP1. *Oncogene* 38: 4835–4855.

Chen, Y.J., Tsai, C.H., Wang, P.Y., and Teng, S.C. (2017). SMYD3 promotes homologous recombination via regulation of H3K4-mediated gene expression. *Sci. Rep.* 7: 3842.

Chervona, Y. and Costa, M. (2012). Histone modifications and cancer: Biomarkers of prognosis? *Am. J. Cancer. Res.* 2: 589–597.

Cho, H.S., Hayami, S., Toyokawa, G., Maejima, K., Yamane, Y., Suzuki, T., Dohmae, N., Kogure, M., Kang, D., Neal, D.E., Ponder, B.A., Yamaue, H., Nakamura, Y., and Hamamoto, R. (2012). RB1 methylation by SMYD2 enhances cell cycle progression through an increase of RB1 phosphorylation. *Neoplasia* 14: 476–486.

Cock-Rada, A.M., Medjkane, S., Janski, N., Yousfi, N., Perichon, M., Chaussepied, M., Chluba, J., Langsley, G., and Weitzman, J.B. (2012). SMYD3 promotes cancer invasion by epigenetic upregulation of the metalloproteinase MMP-9. *Cancer Res.* 72: 810–820.

Colon-Bolea, P. and Crespo, P. (2014). Lysine methylation in cancer: SMYD3-MAP3K2 teaches us new lessons in the Ras-ERK pathway. *Bioessays* 36: 1162–1169.

Dai, B., Wan, W.Q., Zhang, P., Zhang, Y.S., Pan, C.C., Meng, G.L., Xiao, X.R., Wu, Z., Jia, W., Zhang, J.T., and Zhang, L.W. (2015). SET and MYND domain-containing protein 3 is overexpressed in human glioma and contributes to tumorigenicity. *Oncol. Rep.* 34: 2722–2730.

Deng, X., Hamamoto, R., Vougiouklakis, T., Wang, R., Yoshioka, Y., Suzuki, T., Dohmae, N., Matsuo, Y., Park, J.H., and Nakamura, Y. (2017). Critical roles of SMYD2-mediated beta-catenin methylation for nuclear translocation and activation of Wnt signaling. *Oncotarget* 8: 55837–55847.

Dong, S.W., Zhang, H., Wang, B.L., Sun, P., Wang, Y.G., and Zhang, P. (2014). Effect of the downregulation of SMYD3 expression by RNAi on RIZ1 expression and proliferation of esophageal squamous cell carcinoma. *Oncol. Rep.* 32: 1064–1070.

Duan, R., Du, W., and Guo, W. (2020). EZH2: A novel target for cancer treatment. *J. Hematol. Oncol.* 13: 104.

Ellinger, J., Kahl, P., Mertens, C., Rogenhofer, S., Hauser, S., Hartmann, W., Bastian, P.J., Buttner, R., Muller, S.C., and Von Ruecker, A. (2010). Prognostic relevance of global histone H3 lysine 4 (H3K4) methylation in renal cell carcinoma. *Int. J. Cancer* 127: 2360–2366.

Elsemman, I.E., Mardinoglu, A., Shoaie, S., Soliman, T.H., and Nielsen, J. (2016). Systems biology analysis of hepatitis C virus infection reveals the role of copy number increases in regions of chromosome 1q in hepatocellular carcinoma metabolism. *Mol. Biosyst.* 12: 1496–1506.

Engqvist, H., Parris, T.Z., Kovacs, A., Ronnerman, E.W., Sundfeldt, K., Karlsson, P., and Helou, K. (2020a). Validation of novel prognostic biomarkers for early-stage clear-cell, endometrioid and mucinous ovarian carcinomas using immunohistochemistry. *Front Oncol.* 10.

Engqvist, H., Parris, T.Z., Kovacs, A., Ronnerman, E.W., Sundfeldt, K., Karlsson, P., and Helou, K. (2020b). Validation of novel prognostic biomarkers for early-stage clear-cell, endometrioid and mucinous ovarian carcinomas using immunohistochemistry. *Front Oncol.* 10: 162.

Esteller, M. (2007). Cancer epigenomics: DNA methylomes and histone-modification maps. *Nat. Rev. Genet.* 8: 286–298.

Etemadmoghadam, D. and Bowtell, D. (2014). AKT1 gene amplification as a biomarker of treatment response in ovarian cancer: Mounting evidence of a therapeutic target. *Gynecol. Oncol.* 135: 409–410.

Fan, J.D., Lei, P.J., Zheng, J.Y., Wang, X., Li, S., Liu, H., He, Y.L., Wang, Z.N., Wei, G., Zhang, X., Li, L.Y., and Wu, M. (2015). The selective activation of p53 target genes regulated by SMYD2 in BIX-01294 induced autophagy-related cell death. *PLoS One* 10: e0116782.

Fei, X., Ma, Y., Liu, X.W., and Meng, Z.X. (2017). Overexpression of SMYD3 is predictive of unfavorable prognosis in hepatocellular carcinoma. *Tohoku J. Exp. Med.* 243: 219–226.

Feng, Y., Spezia, M., Huang, S., Yuan, C., Zeng, Z., Zhang, L., Ji, X., Liu, W., Huang, B., Luo, W., Liu, B., Lei, Y., Du, S., Vuppalapati, A., Luu, H.H., Haydon, R.C., He, T.C., and Ren, G. (2018). Breast cancer development and progression: Risk factors, cancer stem cells, signaling pathways, genomics, and molecular pathogenesis. *Genes Dis.* 5: 77–106.

Fenizia, C., Bottino, C., Corbetta, S., Fittipaldi, R., Floris, P., Gaudenzi, G., Carra, S., Cotelli, F., Vitale, G., and Caretti, G. (2019). SMYD3 promotes the epithelial-mesenchymal transition in breast cancer. *Nucleic Acids Res.* 47: 1278–1293.

Foreman, K.W., Brown, M., Park, F., Emtage, S., Harriss, J., Das, C., Zhu, L., Crew, A., Arnold, L., Shaaban, S., and Tucker, P. (2011). Structural and functional profiling of the human histone methyltransferase SMYD3. *PLoS One* 6: e22290.

Frank, B., Hemminki, K., Wappenschmidt, B., Klaes, R., Meindl, A., Schmutzler, R.K., Bugert, P., Untch, M., Bartram, C.R., and Burwinkel, B. (2006). Variable number of tandem repeats polymorphism in the SMYD3 promoter region and the risk of familial breast cancer. *Int. J. Cancer* 118: 2917–2918.

Freire Valls, A., Knipper, K., Giannakouri, E., Sarachaga, V., Hinterkopf, S., Wuehrl, M., Shen, Y., Radhakrishnan, P., Klose, J., Ulrich, A., Schneider, M., Augustin, H.G., Ruiz De Almodovar, C., and Schmidt, T. (2019). VEGFR1(+) metastasis-associated macrophages contribute to metastatic angiogenesis and influence colorectal cancer patient outcome. *Clin. Cancer Res.* 25: 5674–5685.

Fresno Vara, J.A., Casado, E., De Castro, J., Cejas, P., Belda-Iniesta, C., and Gonzalez-Baron, M. (2004). PI3K/Akt signalling pathway and cancer. *Cancer Treat. Rev.* 30: 193–204.

Fukuda, T., Fukuda, R., Tanabe, R., Koinuma, D., Koyama, H., Hashizume, Y., Moustakas, A., Miyazono, K., and Heldin, C.H. (2020). BMP signaling is a therapeutic target in ovarian cancer. *Cell Death Discov.* 6.

Galluzzi, L., Senovilla, L., Vitale, I., Michels, J., Martins, I., Kepp, O., Castedo, M., and Kroemer, G. (2012). Molecular mechanisms of cisplatin resistance. *Oncogene* 31: 1869–1883.

Galvan-Femenia, I., Guindo, M., Duran, X., Calabuig-Farinas, S., Mercader, J.M., Ramirez, J.L., Rosell, R., Torrents, D., Carreras, A., Kohno, T., Jantus-Lewintre, E., Camps, C., Perucho, M., Sumoy, L., Yokota, J., and De Cid, R. (2018). Genomic profiling in advanced stage non-small-cell lung cancer patients with platinum-based chemotherapy identifies germline variants with prognostic value in SMYD2. *Cancer Treat Res. Commun.* 15: 21–31.

Gao, C., Herold, J.M., Kireev, D., Wigle, T., Norris, J.L., and Frye, S. (2011). Biophysical probes reveal a "compromise" nature of the methyl-lysine binding pocket in L3MBTL1. *J. Am. Chem. Soc.* 133: 5357–5362.

Gao, S., Wang, Z., Wang, W., Hu, X., Chen, P., Li, J., Feng, X., Wong, J., and Du, J.X. (2017). The lysine methyltransferase SMYD2 methylates the kinase domain of type II receptor BMPR2 and stimulates bone morphogenetic protein signaling. *J. Biol. Chem.* 292: 12702–12712.

Giacinti, C. and Giordano, A. (2006). RB and cell cycle progression. *Oncogene* 25: 5220–5227.

Giaimo, B.D., Ferrante, F., Herchenrother, A., Hake, S.B., and Borggrefe, T. (2019). The histone variant H2A.Z in gene regulation. *Epigenetics Chromatin* 12: 37.

Giakountis, A., Moulos, P., Sarris, M.E., Hatzis, P., and Talianidis, I. (2017). Smyd3-associated regulatory pathways in cancer. *Semin. Cancer Biol.* 42: 70–80.

Gratias, E.J., Dome, J.S., Jennings, L.J., Chi, Y.Y., Tian, J., Anderson, J., Grundy, P., Mullen, E.A., Geller, J.I., Fernandez, C.V., and Perlman, E.J. (2016). Association of chromosome 1q gain with inferior survival in favorable-histology wilms tumor: A report from the children's oncology group. *J. Clin. Oncol.* 34: 3189–3194.

Gupta, S.D., Bommaka, M.K., and Banerjee, A. (2019). Inhibiting protein-protein interactions of Hsp90 as a novel approach for targeting cancer. *Eur. J. Med. Chem.* 178: 48–63.

Haffner, M.C., Berlato, C., and Doppler, W. (2006). Exploiting our knowledge of NF-kappaB signaling for the treatment of mammary cancer. *J. Mammary Gland Biol. Neoplasia* 11: 63–73.

Hamamoto, R., Saloura, V., and Nakamura, Y. (2015). Critical roles of non-histone protein lysine methylation in human tumorigenesis. *Nat. Rev. Cancer* 15: 110–124.

Hamamoto, R., Toyokawa, G., Nakakido, M., Ueda, K., and Nakamura, Y. (2014). SMYD2-dependent HSP90 methylation promotes cancer cell proliferation by regulating the chaperone complex formation. *Cancer Lett.* 351: 126–133.

Han, S., Zou, H., Lee, J.W., Han, J., Kim, H.C., Cheol, J.J., Kim, L.S., and Kim, H. (2019). miR-1307-3p stimulates breast cancer development and progression by targeting SMYD4. *J. Cancer* 10: 441–448.

Hanahan, D. and Weinberg, R.A. (2011). Hallmarks of cancer: The next generation. *Cell* 144: 646–674.

He, C.C., Xu, J.Y., Zhang, J.L., Xie, D., Ye, H., Xiao, Z.Y., Cai, M.Y., Xu, K., Zeng, Y.J., Li, H.G., and Wang, J. (2012). High expression of trimethylated histone H3 lysine 4 is associated with poor prognosis in hepatocellular carcinoma. *Hum. Pathol.* 43: 1425–1435.

Hu, L., Zhu, Y.T., Qi, C., and Zhu, Y.J. (2009). Identification of Smyd4 as a potential tumor suppressor gene involved in breast cancer development. *Cancer Res.* 69: 4067–4072.

Huang, H.C. and Klein, P.S. (2004). Interactions between BMP and Wnt signaling pathways in mammalian cancers. *Cancer Biol. Ther.* 3: 676–678.

Huang, J., Perez-Burgos, L., Placek, B.J., Sengupta, R., Richter, M., Dorsey, J.A., Kubicek, S., Opravil, S., Jenuwein, T., and Berger, S.L. (2006). Repression of p53 activity by Smyd2-mediated methylation. *Nature* 444: 629–632.

Hubbard, S.R. and Till, J.H. (2000). Protein tyrosine kinase structure and function. *Annu. Rev. Biochem.* 69: 373–398.

Huss, W.J., Hanrahan, C.F., Barrios, R.J., Simons, J.W., and Greenberg, N.M. (2001). Angiogenesis and prostate cancer: Identification of a molecular progression switch. *Cancer Res.* 61: 2736–2743.

Jiang, Y., Holcomb, J., Spellmon, N., and Yang, Z. (2016). Purification of histone lysine methyltransferase SMYD2 and co-crystallization with a target peptide from estrogen receptor α. *Methods Mol. Biol.* 1366: 207–217.

Jiang, Y., Lyu, T., Che, X., Jia, N., Li, Q., and Feng, W. (2019). Overexpression of SMYD3 in ovarian cancer is associated with ovarian cancer proliferation and apoptosis via methylating H3K4 and H4K20. *J. Cancer* 10: 4072–4084.

Jiang, Y., Sirinupong, N., Brunzelle, J., and Yang, Z. (2011). Crystal structures of histone and p53 methyltransferase SmyD2 reveal a conformational flexibility of the autoinhibitory C-terminal domain. *PLoS One* 6: e21640.

Jiang, Y., Trescott, L., Holcomb, J., Zhang, X., Brunzelle, J., Sirinupong, N., Shi, X., and Yang, Z. (2014). Structural insights into estrogen receptor α methylation by histone methyltransferase SMYD2, a cellular event implicated in estrogen signaling regulation. *J. Mol. Biol.* 426: 3413–3425.

Johnson, J., Thijssen, B., Mcdermott, U., Garnett, M., Wessels, L.F., and Bernards, R. (2016). Targeting the RB-E2F pathway in breast cancer. *Oncogene* 35: 4829–4835.

Jorgensen, S., Schotta, G., and Sorensen, C.S. (2013). Histone H4 lysine 20 methylation: Key player in epigenetic regulation of genomic integrity. *Nucleic Acids Res.* 41: 2797–2806.

Kaminska, K., Nalejska, E., Kubiak, M., Wojtysiak, J., Zolna, L., Kowalewski, J., and Lewandowska, M.A. (2019). Prognostic and predictive epigenetic biomarkers in oncology. *Mol Diagn Ther* 23: 83–95.

Karimian, A., Ahmadi, Y., and Yousefi, B. (2016). Multiple functions of p21 in cell cycle, apoptosis and transcriptional regulation after DNA damage. *DNA Repair (Amst.)* 42: 63–71.

Keniry, M. and Parsons, R. (2008). The role of PTEN signaling perturbations in cancer and in targeted therapy. *Oncogene* 27: 5477–5485.

Khan, S.A., Reddy, D., and Gupta, S. (2015). Global histone post-translational modifications and cancer: Biomarkers for diagnosis, prognosis and treatment? *World J. Biol. Chem.* 6: 333–345.

Khatami, F. and Tavangar, S.M. (2018). Genetic and epigenetic of medullary thyroid cancer. *Iran. Biomed. J.* 22: 142–150.

Kidder, B.L., He, R., Wangsa, D., Padilla-Nash, H.M., Bernardo, M.M., Sheng, S., Ried, T., and Zhao, K. (2017a). SMYD5 controls heterochromatin and chromosome integrity during embryonic stem cell differentiation. *Cancer Res.* 77: 6729–6745.

Kidder, B.L., Hu, G., Cui, K., and Zhao, K. (2017b). SMYD5 regulates H4K20me3-marked heterochromatin to safeguard ES cell self-renewal and prevent spurious differentiation. *Epigenetics Chromatin* 10: 8.

Kim, H., Heo, K., Kim, J.H., Kim, K., Choi, J., and An, W. (2009). Requirement of histone methyltransferase SMYD3 for estrogen receptor-mediated transcription. *J. Biol. Chem.* 284: 19867–19877.

Kojima, M., Sone, K., Oda, K., Hamamoto, R., Kaneko, S., Oki, S., Kukita, A., Kawata, A., Honjoh, H., Kawata, Y., Kashiyama, T., Sato, M., Taguchi, A., Miyamoto, Y., Tanikawa, M., Tsuruga, T., Nagasaka, K., Wada-Hiraike, O., Osuga, Y., and Fujii, T. (2020). The histone methyltransferase SMYD2 is a novel therapeutic target for the induction of apoptosis in ovarian clear cell carcinoma cells. *Oncol. Lett.* 20.

Komatsu, S., Ichikawa, D., Hirajima, S., Nagata, H., Nishimura, Y., Kawaguchi, T., Miyamae, M., Okajima, W., Ohashi, T., Konishi, H., Shiozaki, A., Fujiwara, H., Okamoto, K., Tsuda, H., Imoto, I., Inazawa, J., and Otsuji, E. (2015). Overexpression of SMYD2 contributes to malignant outcome in gastric cancer. *Br. J. Cancer* 112: 357–364.

Komatsu, S., Imoto, I., Tsuda, H., Kozaki, K.I., Muramatsu, T., Shimada, Y., Aiko, S., Yoshizumi, Y., Ichikawa, D., Otsuji, E., and Inazawa, J. (2009). Overexpression of SMYD2 relates to tumor cell proliferation and malignant outcome of esophageal squamous cell carcinoma. *Carcinogenesis* 30: 1139–1146.

Kumar, A., Kumari, N., Sharma, U., Ram, S., Singh, S.K., Kakkar, N., Kaushal, K., and Prasad, R. (2019). Reduction in H3K4me patterns due to aberrant expression of methyltransferases and demethylases in renal cell carcinoma: Prognostic and therapeutic implications. *Sci. Rep.* 9.

Kunizaki, M., Hamamoto, R., Silva, F.P., Yamaguchi, K., Nagayasu, T., Shibuya, M., Nakamura, Y., and Furukawa, Y. (2007). The lysine 831 of vascular endothelial growth factor receptor 1 is a novel target of methylation by SMYD3. *Cancer Res.* 67: 10759–10765.

Kuser-Abali, G., Gong, L., Yan, J., Liu, Q., Zeng, W., Williamson, A., Lim, C.B., Molloy, M.E., Little, J.B., Huang, L., and Yuan, Z.M. (2018). An EZH2-mediated epigenetic mechanism behind p53-dependent tissue sensitivity to DNA damage. *Proc. Natl. Acad. Sci. U.S.A.* 115: 3452–3457.

Lai, Y.Z. and Yang, Y. (2020). SMYD2 facilitates cancer cell malignancy and xenograft tumor development through ERBB2-mediated FUT4 expression in colon cancer. *Mol. Cell. Biochem.* doi: 10.1007/s11010-020-03738-2.

Le, X.F., Bedrosian, I., Mao, W., Murray, M., Lu, Z., Keyomarsi, K., Lee, M.H., Zhao, J., and Bast, R.C., JR. (2006). Anti-HER2 antibody trastuzumab inhibits CDK2-mediated NPAT and histone H4 expression via the PI3K pathway. *Cell Cycle* 5: 1654–1661.

Lee, T.H., Seng, S., Sekine, M., Hinton, C., Fu, Y., Avraham, H.K., and Avraham, S. (2007). Vascular endothelial growth factor mediates intracrine survival in human breast carcinoma cells through internally expressed VEGFR1/FLT1. *PLoS Med.* 4: 1101–1116.

Li, J., Ahn, J.H., and Wang, G.G. (2019). Understanding histone H3 lysine 36 methylation and its deregulation in disease. *Cell. Mol. Life Sci.* 76: 2899–2916.

Li, J., Zhao, L., Pan, Y., Ma, X., Liu, L., Wang, W., and You, W. (2020). SMYD3 overexpression indicates poor prognosis and promotes cell proliferation, migration and invasion in nonsmall cell lung cancer. *Int. J. Oncol.* 57: 756–766.

Li, L.X., Zhou, J.X., Calvet, J.P., Godwin, A.K., Jensen, R.A., and Li, X. (2018a). Lysine methyltransferase SMYD2 promotes triple negative breast cancer progression. *Cell Death Dis* 9: 326.

Li, R.D., Tang, Y.H., Wang, H.L., Yang, D., Sun, L.J., and Li, W. (2018b). The SMYD3 VNTR 3/3 polymorphism confers an increased risk and poor prognosis of hepatocellular carcinoma in a Chinese population. *Pathol. Res. Pract.* 214: 625–630.

Li, R.D., Tang, Y.H., Wang, H.L., Yang, D., Sun, L.J., and Li, W. (2018c). The SMYD3 VNTR 3/3 polymorphism confers an increased risk and poor prognosis of hepatocellular carcinoma in a Chinese population. *Pathol. Res. Pract.* 214: 625–630.

Liang, Q., Cai, M., Zhang, J., Song, W., Zhu, W., Xi, L., and Tian, Z. (2020). Role of muscle-specific histone methyltransferase (Smyd1) in exercise-induced cardioprotection against pathological remodeling after myocardial infarction. *Int J Mol Sci* 21.

Liu, C., Fang, X.L., Ge, Z., Jalink, M., Kyo, S., Bjorkhom, M., Gruber, A., Sjoberg, J., and Xu, D.W. (2007). The telomerase reverse transcriptase (hTERT) gene is a direct target of the histone methyltransferase SMYD3. *Cancer Res.* 67: 2626–2631.

Liu, C., Liu, L., Wang, K., Li, X.F., Ge, L.Y., Ma, R.Z., Fan, Y.D., Li, L.C., Liu, Z.F., Qiu, M., Hao, Y.C., Shi, Z.F., Xia, C.Y., Straat, K., Huang, Y., Ma, L.L., and Xu, D. (2020a). VHL-HIF-2alpha axis-induced SMYD3 upregulation drives renal cell carcinoma progression via direct trans-activation of EGFR. *Oncogene* 39: 4286–4298.

Liu, H., Li, Y., Li, J., Liu, Y., and Cui, B. (2018). H3K4me3 and Wdr82 are associated with tumor progression and a favorable prognosis in human colorectal cancer. *Oncol. Lett.* 16: 2125–2134.

Liu, H., Liu, Y., Kong, F., Xin, W., Li, X., Liang, H., and Jia, Y. (2015a). Elevated levels of SET and MYND domain-containing Protein 3 are correlated with overexpression of transforming growth factor-beta1 in gastric cancer. *J. Am. Coll. Surg.* 221: 579–590.

Liu, L., Kimball, S., Liu, H., Holowatyj, A., and Yang, Z.Q. (2015b). Genetic alterations of histone lysine methyltransferases and their significance in breast cancer. *Oncotarget* 6: 2466–2482.

Liu, L.X., Kimball, S., Liu, H., Holowatyj, A., and Yang, Z.Q. (2015c). Genetic alterations of histone lysine methyltransferases and their significance in breast cancer. *Oncotarget* 6: 2466–2482.

Liu, N., Sun, S., and Yang, X. (2017a). Prognostic significance of stromal SMYD3 expression in colorectal cancer of TNM stage I-III. *Int. J. Clin. Exp. Pathol.* 10: 8901–8907.

Liu, S., Cheng, K., Zhang, H., Kong, R., Wang, S., Mao, C., and Liu, S. (2020b). Methylation status of the nanog promoter determines the switch between cancer cells and cancer stem cells. *Adv. Sci. (Weinh)* 7: 1903035.

Liu, T.T., Xu, H., Gao, W.P., Zhang, S.X., Zhou, X.H., Tang, J., and Liu, Q.N. (2016a). SET and MYND domain-containing protein 3 (SMYD3) polymorphism as a risk factor for susceptibility and poor prognosis in ovarian cancer. *Med. Sci. Monitor* 22: 5131–5140.

Liu, T.T., Xu, H., Gao, W.P., Zhang, S.X., Zhou, X.H., Tang, J., and Liu, Q.N. (2016b). SET and MYND domain-containing Protein 3 (SMYD3) polymorphism as a risk factor for susceptibility and poor prognosis in ovarian cancer. *Med. Sci. Monit.* 22: 5131–5140.

Liu, X., Zheng, Z., Chen, C., Guo, S., Liao, Z., Li, Y., Zhu, Y., Zou, H., Wu, J., Xie, W., Zhang, P., Xu, L., Wu, B., and Li, E. (2017b). Network analyses elucidate the role of SMYD3 in esophageal squamous cell carcinoma. *FEBS Open Bio.* 7: 1111–1125.

Liu, Y., Deng, J.Y., Luo, X.G., Pan, Y., Zhang, L., Zhang, R.P., and Liang, H. (2015d). Overexpression of SMYD3 was associated with increased STAT3 activation in gastric cancer. *Med. Oncol.* 32.

Liu, Y., Liu, H., Luo, X., Deng, J., Pan, Y., and Liang, H. (2015e). Overexpression of SMYD3 and matrix metalloproteinase-9 are associated with poor prognosis of patients with gastric cancer. *Tumour Biol.* 36: 4377–4386.

Liu, Y., Luo, X.G., Deng, J.Y., Pan, Y., Zhang, L., and Liang, H. (2015f). SMYD3 overexpression was a risk factor in the biological behavior and prognosis of gastric carcinoma. *Tumor Biol.* 36: 2685–2694.

Lobo, J., Rodrigues, A., Antunes, L., Graca, I., Ramalho-Carvalho, J., Vieira, F.Q., Martins, A.T., Oliveira, J., Jeronimo, C., and Henrique, R. (2018). High immunoexpression of Ki67, EZH2, and SMYD3 in diagnostic prostate biopsies independently predicts outcome in patients with prostate cancer. *Urol. Oncol.* 36: 161 e7–161 e17.

Lopez-Cortes, A., Leone, P.E., Freire-Paspuel, B., Arcos-Villacis, N., Guevara-Ramirez, P., Rosales, F., and Paz, Y.M.C. (2018). Mutational analysis of oncogenic AKT1 gene associated with breast cancer risk in the high altitude Ecuadorian mestizo population. *Biomed. Res. Int.* 2018: 7463832.

Luo, J., Dai, X.C., Hu, H., Chen, J., Zhao, L.J., Yang, C.Y., Sun, J.F., Zhang, L.M., Wang, Q., Xu, S.L., Xu, Y., Liu, N.B., Ying, G.G., and Wang, P. (2020). Fluzoparib increases radiation sensitivity of non-small cell lung cancer (NSCLC) cells without BRCA1/2 mutation, a novel PARP1 inhibitor undergoing clinical trials. *J. Cancer Res. Clin. Oncol.* 146: 721–737.

Luo, J., Li, Y.N., Wang, F., Zhang, W.M., and Geng, X. (2010). S-adenosylmethionine inhibits the growth of cancer cells by reversing the hypomethylation status of c-myc and H-ras in human gastric cancer and colon cancer. *Int. J. Biol. Sci.* 6: 784–795.

Luo, X.G., Zhang, C.L., Zhao, W.W., Liu, Z.P., Liu, L., Mu, A., Guo, S., Wang, N., Zhou, H., and Zhang, T.C. (2014). Histone methyltransferase SMYD3 promotes MRTF-A-mediated transactivation of MYL9 and migration of MCF-7 breast cancer cells. *Cancer Lett.* 344: 129–137.

Lyu, T., Jiang, Y., Jia, N., Che, X., Li, Q., Yu, Y., Hua, K., Bast, R.C., JR., and Feng, W. (2020). SMYD3 promotes implant metastasis of ovarian cancer via H3K4 trimethylation of integrin promoters. *Int. J. Cancer* 146: 1553–1567.

Ma, S.J., Liu, Y.M., Zhang, Y.L., Chen, M.W., and Cao, W. (2018). Correlations of EZH2 and SMYD3 gene polymorphisms with breast cancer susceptibility and prognosis. *Biosci. Rep.* 38.

Margueron, R., Li, G.H., Sarma, K., Blais, A., Zavadil, J., Woodcock, C.L., Dyniacht, B.D., and Reinberg, D. (2008). Ezh1 and Ezh2 maintain repressive chromatin through different mechanisms. *Mol. Cell* 32: 503–518.

Mazur, P.K., Reynoird, N., Khatri, P., Jansen, P.W., Wilkinson, A.W., Liu, S., Barbash, O., Van Aller, G.S., Huddleston, M., Dhanak, D., Tummino, P.J., Kruger, R.G., Garcia, B.A., Butte, A.J., Vermeulen, M., Sage, J., and Gozani, O. (2014). SMYD3 links lysine methylation of MAP3K2 to Ras-driven cancer. *Nature* 510: 283–287.

Meek, D.W. (2015). Regulation of the p53 response and its relationship to cancer. *Biochem. J.* 469: 325–346.

Menard, S., Fortis, S., Castiglioni, F., Agresti, R., and Balsari, A. (2001). HER2 as a prognostic factor in breast cancer. *Oncology* 61: 67–72.

Meng, F., Liu, X., Lin, C., Xu, L., Liu, J., Zhang, P., Zhang, X., Song, J., Yan, Y., Ren, Z., and Zhang, Y. (2020). SMYD2 suppresses APC2 expression to activate the Wnt/beta-catenin pathway and promotes epithelial-mesenchymal transition in colorectal cancer. *Am. J. Cancer. Res.* 10: 997–1011.

Meng, X., Zhao, Y., Liu, J., Wang, L., Dong, Z., Zhang, T., Gu, X., and Zheng, Z. (2019). Comprehensive analysis of histone modification-associated genes on differential gene expression and prognosis in gastric cancer. *Exp. Ther. Med.* 18: 2219–2230.

Milella, M., Falcone, I., Conciatori, F., Cesta Incani, U., Del Curatolo, A., Inzerilli, N., Nuzzo, C.M., Vaccaro, V., Vari, S., Cognetti, F., and Ciuffreda, L. (2015). PTEN: Multiple functions in human malignant tumors. *Front Oncol.* 5: 24.

Miyata, Y., Nakamoto, H., and Neckers, L. (2013). The therapeutic target Hsp90 and cancer hallmarks. *Curr. Pharm. Des.* 19: 347–365.

Mohamed, N.E., Hay, T., Reed, K.R., Smalley, M.J., and Clarke, A.R. (2019). APC2 is critical for ovarian WNT signalling control, fertility and tumour suppression. *Bmc Cancer* 19.

Monteiro, F.L., Baptista, T., Amado, F., Vitorino, R., Jeronimo, C., and Helguero, L.A. (2014). Expression and functionality of histone H2A variants in cancer. *Oncotarget* 5: 3428–3443.

Myers, M.P., Pass, I., Batty, I.H., Van Der Kaay, J., Stolarov, J.P., Hemmings, B.A., Wigler, M.H., Downes, C.P., and Tonks, N.K. (1998). The lipid phosphatase activity of PTEN is critical for its tumor suppressor function. *Proc. Natl. Acad. Sci. U.S.A.* 95: 13513–13518.

Nakakido, M., Deng, Z., Suzuki, T., Dohmae, N., Nakamura, Y., and Hamamoto, R. (2015). Dysregulation of AKT pathway by SMYD2-mediated lysine methylation on PTEN. *Neoplasia* 17: 367–373.

Ngeow, J., Sesock, K., and Eng, C. (2017). Breast cancer risk and clinical implications for germline PTEN mutation carriers. *Breast Cancer Res. Treat.* 165: 1–8.

Noroxe, D.S., Yde, C.W., Ostrup, O., Michaelsen, S.R., Schmidt, A.Y., Kinalis, S., Torp, M.H., Skjoth-Rasmussen, J., Brennum, J., Hamerlik, P., Poulsen, H.S., Nielsen, F.C., and Lassen, U. (2020). Genomic profiling of newly diagnosed glioblastoma patients and its potential for clinical utility: A prospective, translational study. *Mol. Oncol.* 14: 2727–2743.

Ohtomo-Oda, R., Komatsu, S., Mori, T., Sekine, S., Hirajima, S., Yoshimoto, S., Kanai, Y., Otsuji, E., Ikeda, E., and Tsuda, H. (2016). SMYD2 overexpression is associated with tumor cell proliferation and a worse outcome in human papillomavirus-unrelated nonmultiple head and neck carcinomas. *Hum. Pathol.* 49: 145–155.

Oleksiewicz, U. and Machnik, M. (2020). Causes, effects, and clinical implications of perturbed patterns within the cancer epigenome. *Semin. Cancer Biol.* doi: 10.1016/j.semcancer.2020.12.014.

Oliveira-Santos, W., Rabello, D.A., Lucena-Araujo, A.R., De Oliveira, F.M., Rego, E.M., Silva, F.P., and Saldanha-Araujo, F. (2016). Residual expression of SMYD2 and SMYD3 is associated with the acquisition of complex karyotype in chronic lymphocytic leukemia. *Tumor Biol.* 37: 9473–9481.

Papillon-Cavanagh, S., Lu, C., Gayden, T., Mikael, L.G., Bechet, D., Karamboulas, C., Ailles, L., Karamchandani, J., Marchione, D.M., Garcia, B.A., Weinreb, I., Goldstein, D., Lewis, P.W., Dancu, O.M., Dhaliwal, S., Stecho, W., Howlett, C.J., Mymryk, J.S., Barrett, J.W., Nichols, A.C., Allis, C.D., Majewski, J., and Jabado, N. (2017). Impaired H3K36 methylation defines a subset of head and neck squamous cell carcinomas. *Nat. Genet.* 49: 180–185.

Patani, N., Jiang, W.G., Newbold, R.F., and Mokbel, K. (2011). Histone-modifier gene expression profiles are associated with pathological and clinical outcomes in human breast cancer. *Anticancer Res.* 31: 4115–4125.

Peveling-Oberhag, J., Wolters, F., Doring, C., Walter, D., Sellmann, L., Scholtysik, R., Lucioni, M., Schubach, M., Paulli, M., Biskup, S., Zeuzem, S., Kuppers, R., and Hansmann, M.L. (2015). Whole exome sequencing of microdissected splenic marginal zone lymphoma: A study to discover novel tumor-specific mutations. *BMC Cancer* 15: 773.

Piao, L., Kang, D., Suzuki, T., Masuda, A., Dohmae, N., Nakamura, Y., and Hamamoto, R. (2014). The histone methyltransferase SMYD2 methylates PARP1 and promotes poly(ADP-ribosyl)ation activity in cancer cells. *Neoplasia* 16: 257–64, 264 e2.

Pietrzak, J., Spickett, C.M., Ploszaj, T., Virag, L., and Robaszkiewicz, A. (2018). PARP1 promoter links cell cycle progression with adaptation to oxidative environment. *Redox Biol.* 18: 1–5.

Pires-Luis, A.S., Vieira-Coimbra, M., Vieira, F.Q., Costa-Pinheiro, P., Silva-Santos, R., Dias, P.C., Antunes, L., Lobo, F., Oliveira, J., Goncalves, C.S., Costa, B.M., Henrique, R., and Jeronimo, C. (2015). Expression of histone methyltransferases as novel biomarkers for renal cell tumor diagnosis and prognostication. *Epigenetics* 10: 1033–1043.

Rasmussen, T.L., Ma, Y., Park, C.Y., Harriss, J., Pierce, S.A., Dekker, J.D., Valenzuela, N., Srivastava, D., Schwartz, R.J., Stewart, M.D., and Tucker, H.O. (2015). Smyd1 facilitates heart development by antagonizing oxidative and ER stress responses. *PLoS One* 10: e0121765.

Ray Chaudhuri, A. and Nussenzweig, A. (2017). The multifaceted roles of PARP1 in DNA repair and chromatin remodelling. *Nat. Rev. Mol. Cell Biol.* 18: 610–621.

Ren, H.L., Wang, Z., Chen, Y., Liu, Y.J., Zhang, S., Zhang, T.T., and Li, Y.T. (2019). SMYD2-OE promotes oxaliplatin resistance in colon cancer through MDR1/P-glycoprotein via MEK/ERK/AP1 pathway. *Onco Targets Ther.* 12: 2585–2594.

Reynoird, N., Mazur, P.K., Stellfeld, T., Flores, N.M., Lofgren, S.M., Carlson, S.M., Brambilla, E., Hainaut, P., Kaznowska, E.B., Arrowsmith, C.H., Khatri, P., Stresemann, C., Gozani, O., and Sage, J. (2016). Coordination of stress signals by the lysine methyltransferase SMYD2 promotes pancreatic cancer. *Genes Dev.* 30: 772–785.

Saddic, L.A., West, L.E., Aslanian, A., Yates, J.R., 3RD, Rubin, S.M., Gozani, O., and Sage, J. (2010). Methylation of the retinoblastoma tumor suppressor by SMYD2. *J. Biol. Chem.* 285: 37733–37740.

Sakamoto, L.H.T., De Andrade, R.V., Felipe, M.S.S., Motoyama, A.B., and Silva, F.P. (2014). SMYD2 is highly expressed in pediatric acute lymphoblastic leukemia and constitutes a bad prognostic factor. *Leuk. Res.* 38: 496–502.

Sanchez-Vega, F., Mina, M., Armenia, J., Chatila, W.K., Luna, A., La, K.C., Dimitriadoy, S., Liu, D.L., Kantheti, H.S., Saghafinia, S., Chakravarty, D., Daian, F., Gao, Q., Bailey, M.H., Liang, W.W., Foltz, S.M., Shmulevich, I., Ding, L., Heins, Z., Ochoa, A., Gross, B., Gao, J., Zhang, H., Kundra, R., Kandoth, C., Bahceci, I., Dervishi, L., Dogrusoz, U., Zhou, W., Shen, H., Laird, P.W., Way, G.P., Greene, C.S., Liang, H., Xiao, Y., Wang, C., Iavarone, A., Berger, A.H., Bivona, T.G., Lazar, A.J., Hammer, G.D., Giordano, T., Kwong, L.N., Mcarthur, G., Huang, C., Tward, A.D., Frederick, M.J., Mccormick, F., Meyerson, M., Cancer Genome Atlas Research, N., Van Allen, E.M., Cherniack, A.D., Ciriello, G., Sander, C., and Schultz, N. (2018). Oncogenic signaling pathways in the cancer genome atlas. *Cell* 173: 321–337 e10.

Schneider, A.C., Heukamp, L.C., Rogenhofer, S., Fechner, G., Bastian, P.J., Von Ruecker, A., Muller, S.C., and Ellinger, J. (2011). Global histone H4K20 trimethylation predicts cancer-specific survival in patients with muscle-invasive bladder cancer. *BJU Int.* 108: E290–6.

Schneider, B.P., Radovich, M., Sledge, G.W., Robarge, J.D., Li, L., Storniolo, A.M., Lemler, S., Nguyen, A.T., Hancock, B.A., Stout, M., Skaar, T., and Flockhart, D.A. (2008). Association of polymorphisms of angiogenesis genes with breast cancer. *Breast Cancer Res. Treat.* 111: 157–163.

Schopf, F.H., Biebl, M.M., and Buchner, J. (2017). The HSP90 chaperone machinery. *Nat. Rev. Mol. Cell Biol.* 18: 345–360.

Schwartz, R.S. and Erban, J.K. (2017). Timing of metastasis in breast cancer. *N. Engl. J. Med.* 376: 2486–2488.

Shang, L. and Wei, M.J. (2019). Inhibition of SMYD2 sensitized cisplatin to resistant cells in NSCLC through activating p53 pathway. *Front Oncol.* 9.

Shen, B., Tan, M., Mu, X., Qin, Y., Zhang, F., Liu, Y., and Fan, Y. (2016). Upregulated SMYD3 promotes bladder cancer progression by targeting BCLAF1 and activating autophagy. *Tumour Biol.* 37: 7371–7381.

Shivakumar, L., Minna, J., Sakamaki, T., Pestell, R., and White, M.A. (2002). The RASSF1A tumor suppressor blocks cell cycle progression and inhibits cyclin D1 accumulation. *Mol. Cell. Biol.* 22: 4309–4318.

Singh, S., Shi, Q., Bailey, S.T., Palczewski, M.J., Pardee, A.B., Iglehart, J.D., and Biswas, D.K. (2007). Nuclear factor-kappaB activation: A molecular therapeutic target for estrogen receptor-negative and epidermal growth factor receptor family receptor-positive human breast cancer. *Mol. Cancer Ther.* 6: 1973–1982.

Sirinupong, N., Brunzelle, J., Doko, E., and Yang, Z. (2011). Structural insights into the autoinhibition and posttranslational activation of histone methyltransferase SmyD3. *J. Mol. Biol.* 406: 149–159.

Sirinupong, N., Brunzelle, J., Ye, J., Pirzada, A., Nico, L., and Yang, Z. (2010). Crystal structure of cardiac-specific histone methyltransferase SmyD1 reveals unusual active site architecture. *J. Biol. Chem.* 285: 40635–40644.

Skawran, B., Steinemann, D., Weigmann, A., Flemming, P., Becker, T., Flik, J., Kreipe, H., Schlegelberger, B., and Wilkens, L. (2008). Gene expression profiling in hepatocellular carcinoma: Upregulation of genes in amplified chromosome regions. *Mod. Pathol.* 21: 505–516.

Soda, M., Choi, Y.L., Enomoto, M., Takada, S., Yamashita, Y., Ishikawa, S., Fujiwara, S., Watanabe, H., Kurashina, K., Hatanaka, H., Bando, M., Ohno, S., Ishikawa, Y., Aburatani, H., Niki, T., Sohara, Y., Sugiyama, Y., and Mano, H. (2007). Identification of the transforming EML4-ALK fusion gene in non-small-cell lung cancer. *Nature* 448: 561–566.

Song, J., Liu, Y., Chen, Q., Yang, J., Jiang, Z., Zhang, H., Liu, Z., and Jin, B. (2019). Expression patterns and the prognostic value of the SMYD family members in human breast carcinoma using integrative bioinformatics analysis. *Oncol. Lett.* 17: 3851–3861.

Song, J.S., Kim, Y.S., Kim, D.K., Park, S.I., and Jang, S.J. (2012). Global histone modification pattern associated with recurrence and disease-free survival in non-small cell lung cancer patients. *Pathol. Int.* 62: 182–190.

Spanheimer, P.M., Lorenzen, A.W., De Andrade, J.P., Kulak, M.V., Carr, J.C., Woodfield, G.W., Sugg, S.L., and Weigel, R.J. (2015). Receptor tyrosine kinase expression predicts response to sunitinib in breast cancer. *Ann. Surg. Oncol.* 22: 4287–4294.

Spellmon, N., Holcomb, J., Trescott, L., Sirinupong, N., and Yang, Z. (2015a). Structure and function of SET and MYND domain-containing proteins. *Int J Mol Sci* 16: 1406–1428.

Spellmon, N., Sun, X., Sirinupong, N., Edwards, B., Li, C., and Yang, Z. (2015b). Molecular dynamics simulation reveals correlated inter-lobe motion in protein lysine methyltransferase SMYD2. *PLoS One* 10: e0145758.

Spellmon, N., Sun, X., Xue, W., Holcomb, J., Chakravarthy, S., Shang, W., Edwards, B., Sirinupong, N., Li, C., and Yang, Z. (2017). New open conformation of SMYD3 implicates conformational selection and allostery. *AIMS Biophys.* 4: 1–18.

Sponziello, M., Durante, C., Boichard, A., Dima, M., Puppin, C., Verrienti, A., Tamburrano, G., Di Rocco, G., Redler, A., Lacroix, L., Bidart, J.M., Schlumberger, M., Damante, G., Russo, D., and Filetti, S. (2014). Epigenetic-related gene expression profile in medullary thyroid cancer revealed the overexpression of the histone methyltransferases EZH2 and SMYD3 in aggressive tumours. *Mol. Cell. Endocrinol.* 392: 8–13.

Stender, J.D., Pascual, G., Liu, W., Kaikkonen, M.U., Do, K., Spann, N.J., Boutros, M., Perrimon, N., Rosenfeld, M.G., and Glass, C.K. (2012). Control of proinflammatory gene programs by regulated trimethylation and demethylation of histone H4K20. *Mol. Cell* 48: 28–38.

Sun, J.J., Li, H.L., Ma, H., Shi, Y., Yin, L.R., and Guo, S.J. (2019). SMYD2 promotes cervical cancer growth by stimulating cell proliferation. *Cell Biosci.* 9 (75).

Taby, R. and Issa, J.P. (2010). Cancer epigenetics. *CA Cancer J. Clin.* 60: 376–392.

Tan, X., Rotllant, J., Li, H., De Deyne, P., and Du, S.J. (2006). SmyD1, a histone methyltransferase, is required for myofibril organization and muscle contraction in zebrafish embryos. *Proc. Natl. Acad. Sci. U.S.A.* 103: 2713–2718.

Thaler, S., Schad, A., and Dammann, R. (2009). RASSF1A mediates p21Cip1/Waf1-dependent cell cycle arrest and senescence through modulation of the Raf-MEK-ERK pathway and inhibition of Akt in vitro and in a double-conditional xenograft model. *Cancer Res.* 69.

Tokunaga, E., Kimura, Y., Mashino, K., Oki, E., Kataoka, A., Ohno, S., Morita, M., Kakeji, Y., Baba, H., and Maehara, Y. (2006). Activation of PI3K/Akt signaling and hormone resistance in breast cancer. *Breast Cancer* 13: 137–144.

Tracy, C., Warren, J.S., Szulik, M., Wang, L., Garcia, J., Makaju, A., Russell, K., Miller, M., and Franklin, S. (2018). The Smyd family of methyltransferases: Role in cardiac and skeletal muscle physiology and pathology. *Curr. Opin. Physiol.* 1: 140–152.

Trepel, J., Mollapour, M., Giaccone, G., and Neckers, L. (2010). Targeting the dynamic HSP90 complex in cancer. *Nat. Rev. Cancer* 10: 537–549.

Trojer, P., Li, G., Sims, R.J., 3RD, Vaquero, A., Kalakonda, N., Boccuni, P., Lee, D., Erdjument-Bromage, H., Tempst, P., Nimer, S.D., Wang, Y.H., and Reinberg, D. (2007). L3MBTL1, a histone-methylation-dependent chromatin lock. *Cell* 129: 915–928.

Tsai, C.H., Chen, Y.J., Yu, C.J., Tzeng, S.R., Wu, I.C., Kuo, W.H., Lin, M.C., Chan, N.L., Wu, K.J., and Teng, S.C. (2016). SMYD3-mediated H2A.Z.1 methylation promotes cell cycle and cancer proliferation. *Cancer Res.* 76: 6043–6053.

Tsuge, M., Hamamoto, R., Silva, F.P., Ohnishi, Y., Chayama, K., Kamatani, N., Furukawa, Y., and Nakamura, Y. (2005). A variable number of tandem repeats polymorphism in an E2F-1 binding element in the 5 flanking region of SMYD3 is a risk factor for human cancers. *Nat. Genet.* 37: 1104–1107.

Uhlen, M., Zhang, C., Lee, S., Sjöstedt, E., Fagerberg, L., Bidkhori, G., Benfeitas, R., Arif, M., Liu, Z., Edfors, F., Sanli, K., Von Feilitzen, K., Oksvold, P., Lundberg, E., Hober, S., Nilsson, P., Mattsson, J., Schwenk, J.M., Brunnström, H., Glimelius, B., Sjöblom, T., Edqvist, P.H., Djureinovic, D., Micke, P., Lindskog, C., Mardinoglu, A., and Ponten, F. (2017). A pathology atlas of the human cancer transcriptome. *Science* 357.

Varier, R.A. and Timmers, H.T. (2011). Histone lysine methylation and demethylation pathways in cancer. *Biochim. Biophys. Acta* 1815: 75–89.

Vassilev, L.T., Vu, B.T., Graves, B., Carvajal, D., Podlaski, F., Filipovic, Z., Kong, N., Kammlott, U., Lukacs, C., Klein, C., Fotouhi, N., and Liu, E.A. (2004). In vivo activation of the p53 pathway by small-molecule antagonists of MDM2. *Science* 303: 844–848.

Vieira, F.Q., Costa-Pinheiro, P., Almeida-Rios, D., Graca, I., Monteiro-Reis, S., Simoes-Sousa, S., Carneiro, I., Sousa, E.J., Godinho, M.I., Baltazar, F., Henrique, R., and Jeronimo, C.

(2015). SMYD3 contributes to a more aggressive phenotype of prostate cancer and targets Cyclin D2 through H4K20me3. *Oncotarget* 6: 13644–13657.

Vieira, F.Q., Costa-Pinheiro, P., Ramalho-Carvalho, J., Pereira, A., Menezes, F.D., Antunes, L., Carneiro, I., Oliveira, J., Henrique, R., and Jeronimo, C. (2014). Deregulated expression of selected histone methylases and demethylases in prostate carcinoma. *Endocr. Relat. Cancer* 21: 51–61.

Vougiouklakis, T., Bernard, B.J., Nigam, N., Burkitt, K., Nakamura, Y., and Saloura, V. (2020). Clinicopathologic significance of protein lysine methyltransferases in cancer. *Clin. Epigenetics* 12: 146.

Wagle, M.C., Kirouac, D., Klijn, C., Liu, B., Mahajan, S., Junttila, M., Moffat, J., Merchant, M., Huw, L., Wongchenko, M., Okrah, K., Srinivasan, S., Mounir, Z., Sumiyoshi, T., Haverty, P.M., Yauch, R.L., Yan, Y., Kabbarah, O., Hampton, G., Amler, L., Ramanujan, S., Lackner, M.R., and Huang, S.M.A. (2018). A transcriptional MAPK pathway activity score (MPAS) is a clinically relevant biomarker in multiple cancer types. *Npj Pre. Oncol.* 2.

Wang, G., Huang, Y., Yang, F., Tian, X., Wang, K., Liu, L., Fan, Y., Li, X., Li, L., Shi, B., Hao, Y., Xia, C., Nie, Q., Xin, Y., Shi, Z., Ma, L., Xu, D., and Liu, C. (2020a). High expression of SMYD3 indicates poor survival outcome and promotes tumour progression through an IGF-1R/AKT/E2F-1 positive feedback loop in bladder cancer. *Aging (Albany NY)* 12: 2030–2048.

Wang, G.L., Huang, Y., Yang, F.L., Tian, X.J., Wang, K., Liu, L., Fan, Y.D., Li, X.F., Li, L.C., Shi, B.K., Hao, Y.C., Xia, C.Y., Nie, Q.S., Xin, Y., Shi, Z.F., Ma, L.L., Xu, D.W., and Liu, C. (2020b). High expression of SMYD3 indicates poor survival outcome and promotes tumour progression through an IGF-1R/AKT/E2F-1 positive feedback loop in bladder cancer. *Aging-Us* 12: 2030–2048.

Wang, H., Liu, Y., Tan, W., Zhang, Y., Zhao, N., Jiang, Y., Lin, C., Hao, B., Zhao, D., Qian, J., Lu, D., Jin, L., Wei, Q., Lin, D., and He, F. (2008a). Association of the variable number of tandem repeats polymorphism in the promoter region of the SMYD3 gene with risk of esophageal squamous cell carcinoma in relation to tobacco smoking. *Cancer Sci.* 99: 787–791.

Wang, L., Xu, M.L., Wang, C., Dong, Q.Q., Miao, Z., Chen, X.Y., Wang, N., He, H.P., Zhang, T.C., and Luo, X.G. (2020c). SET and MYND domain-containing protein 3 inhibits tumor cell sensitivity to cisplatin. *Oncol. Lett.* 19: 3469–3476.

Wang, R., Deng, X., Yoshioka, Y., Vougiouklakis, T., Park, J.H., Suzuki, T., Dohmae, N., Ueda, K., Hamamoto, R., and Nakamura, Y. (2017). Effects of SMYD2-mediated EML4-ALK methylation on the signaling pathway and growth in non-small-cell lung cancer cells. *Cancer Sci.* 108: 1203–1209.

Wang, S.Z., Luo, X.G., Shen, J., Zou, J.N., Lu, Y.H., and Xi, T. (2008b). Knockdown of SMYD3 by RNA interference inhibits cervical carcinoma cell growth and invasion in vitro. *BMB Rep.* 41: 294–299.

Wang, T., Wu, H., Liu, S., Lei, Z., Qin, Z., Wen, L., Liu, K., Wang, X., Guo, Y., Liu, Q., Liu, L., Wang, J., Lin, L., Mao, C., Zhu, X., Xiao, H., Bian, X., Chen, D., Xu, C., and Wang, B. (2018). SMYD3 controls a Wnt-responsive epigenetic switch for ASCL2 activation and cancer stem cell maintenance. *Cancer Lett.* 430: 11–24.

Wang, X.Q., Miao, X., Cai, Q., Garcia-Barcelo, M.M., and Fan, S.T. (2007). SMYD3 tandem repeats polymorphism is not associated with the occurrence and metastasis of hepatocellular carcinoma in a Chinese population. *Exp. Oncol.* 29: 71–73.

Wang, Y., Xie, B.H., Lin, W.H., Huang, Y.H., Ni, J.Y., Hu, J., Cui, W., Zhou, J., Shen, L., Xu, L.F., Lian, F., and Li, H.P. (2019). Amplification of SMYD3 promotes tumorigenicity and intrahepatic metastasis of hepatocellular carcinoma via upregulation of CDK2 and MMP2. *Oncogene* 38: 4948–4961.

Weaver, A.N. and Yang, E.S. (2013). Beyond DNA repair: Additional functions of PARP-1 in cancer. *Front Oncol.* 3: 290.

Wei, L., Murphy, B.L., Wu, G., Parker, M., Easton, J., Gilbertson, R.J., Zhang, J., and Roussel, M.F. (2017). Exome sequencing analysis of murine medulloblastoma models identifies WDR11 as a potential tumor suppressor in Group 3 tumors. *Oncotarget* 8: 64685–64697.

Wienken, M., Dickmanns, A., Nemajerova, A., Kramer, D., Najafova, Z., Weiss, M., Karpiuk, O., Kassem, M., Zhang, Y., Lozano, G., Johnsen, S.A., Moll, U.M., Zhang, X., and Dobbelstein, M. (2016). MDM2 associates with polycomb repressor complex 2 and enhances stemness-promoting chromatin modifications independent of p53. *Mol. Cell* 61: 68–83.

Wu, X., Xu, Q., Chen, P., Yu, C., Ye, L., Huang, C., and Li, T. (2019). Effect of SMYD3 on biological behavior and H3K4 methylation in bladder cancer. *Cancer Manag. Res.* 11: 8125–8133.

Xu, J., Wang, Z., Lu, W., Jiang, H., Lu, J., Qiu, J., and Ye, G. (2019). EZH2 promotes gastric cancer cells proliferation by repressing p21 expression. *Pathol. Res. Pract.* 215: 152374.

Xu, W.C., Chen, F.L., Fei, X., Yang, X.Q., and Lu, X.F. (2018b). Overexpression of SET and MYND domain-containing Protein 2 (SMYD2) is associated with tumor progression and poor prognosis in patients with papillary thyroid carcinoma. *Med. Sci. Monitor* 24: 7357–7365.

Yan, L., Ding, B., Liu, H., Zhang, Y., Zeng, J., Hu, J., Yao, W., Yu, G., An, R., Chen, Z., Ye, Z., Xing, J., Xiao, K., Wu, L., and Xu, H. (2019). Inhibition of SMYD2 suppresses tumor progression by down-regulating microRNA-125b and attenuates multi-drug resistance in renal cell carcinoma. *Theranostics* 9: 8377–8391.

Yang, L., Huang, K., Li, X.R., Du, M., Kang, X., Luo, X., Gao, L., Wang, C., Zhang, Y.Q., Zhang, C., Tong, Q.S., Huang, K., Zhang, F.X., and Huang, D. (2013). Identification of Poly(ADP-Ribose) polymerase-1 as a cell cycle regulator through modulating Sp1 mediated transcription in human hepatoma cells. *Plos One* 8.

Ye, X.D., Qiu, B.Q., Xiong, D., Pei, X., Jie, N., Xu, H., Zhu, S.Q., Long, X., Xu, Z., Wu, H.B., Xu, J.J., Huang, Y.S., and Wu, Y.B. (2020). High level of H3K4 tri-methylation modification predicts poor prognosis in esophageal cancer. *J. Cancer* 11: 3256–3263.

Yi, X., Jiang, X.J., and Fang, Z.M. (2019). Histone methyltransferase SMYD2: Ubiquitous regulator of disease. *Clin. Epigenetics* 11: 112.

Yin, J.Y., Li, X., Zhou, H.H., and Liu, Z.Q. (2016). Pharmacogenomics of platinum-based chemotherapy sensitivity in NSCLC: Toward precision medicine. *Pharmacogenomics* 17: 1365–1378.

Yokoyama, Y., Matsumoto, A., Hieda, M., Shinchi, Y., Ogihara, E., Hamada, M., Nishioka, Y., Kimura, H., Yoshidome, K., Tsujimoto, M., and Matsuura, N. (2014). Loss of histone H4K20 trimethylation predicts poor prognosis in breast cancer and is associated with invasive activity. *Breast Cancer Res.* 16: R66.

Yoshioka, Y., Suzuki, T., Matsuo, Y., Nakakido, M., Tsurita, G., Simone, C., Watanabe, T., Dohmae, N., Nakamura, Y., and Hamamoto, R. (2016). SMYD3-mediated lysine methylation in the PH domain is critical for activation of AKT1. *Oncotarget* 7: 75023–75037.

Yoshioka, Y., Suzuki, T., Matsuo, Y., Tsurita, G., Watanabe, T., Dohmae, N., Nakamura, Y., and Hamamoto, R. (2017a). Protein lysine methyltransferase SMYD3 is involved in tumorigenesis through regulation of HER2 homodimerization. *Cancer Med* 6: 1665–1672.

Yoshioka, Y., Suzuki, T., Matsuo, Y., Tsurita, G., Watanabe, T., Dohmae, N., Nakamura, Y., and Hamamoto, R. (2017b). Protein lysine methyltransferase SMYD3 is involved in tumorigenesis through regulation of HER2 homodimerization. *Cancer Med.* 6: 1665–1672.

Yu, H., Lee, H., Herrmann, A., Buettner, R., and Jove, R. (2014). Revisiting STAT3 signalling in cancer: New and unexpected biological functions. *Nat. Rev. Cancer* 14: 736–746.

Yu, H., Pardoll, D., and Jove, R. (2009). STATs in cancer inflammation and immunity: A leading role for STAT3. *Nat. Rev. Cancer* 9: 798–809.

Yu, H., Zhang, D., and Lian, M. (2021). Identification of an epigenetic prognostic signature for patients with lower-grade gliomas. *CNS Neurosci Ther* 27: 470–483.

Yue, F.R., Wei, Z.B., Yan, R.Z., Guo, Q.H., Liu, B., Zhang, J.H., and Li, Z. (2020). SMYD3 promotes colon adenocarcinoma (COAD) progression by mediating cell proliferation and apoptosis. *Exp. Ther. Med.* 20: 11.

Zeng, Y., Qiu, R., Yang, Y., Gao, T., Zheng, Y., Huang, W., Gao, J., Zhang, K., Liu, R., Wang, S., Hou, Y., Yu, W., Leng, S., Feng, D., Liu, W., Zhang, X., and Wang, Y. (2019). Regulation of EZH2 by SMYD2-mediated lysine methylation is implicated in tumorigenesis. *Cell Rep.* 29: 1482–1498 e4.

Zhang, L., Jin, Y., Yang, H., Li, Y., Wang, C., Shi, Y., and Wang, Y. (2019a). SMYD3 promotes epithelial ovarian cancer metastasis by downregulating p53 protein stability and promoting p53 ubiquitination. *Carcinogenesis* 40: 1492–1503.

Zhang, P., Ruan, J., Weng, W., and Tang, Y. (2020a). Overexpression of SET and MYND domain-containing protein 2 (SMYD2) is associated with poor prognosis in pediatric B lineage acute lymphoblastic leukemia. *Leuk. Lymphoma* 61: 437–444.

Zhang, P., Ruan, J.F., Weng, W.W., and Tang, Y.M. (2020b). Overexpression of SET and MYND domain-containing protein 2 (SMYD2) is associated with poor prognosis in pediatric B lineage acute lymphoblastic leukemia. *Leuk. Lymphoma* 61: 437–444.

Zhang, X., Tanaka, K., Yan, J., Li, J., Peng, D., Jiang, Y., Yang, Z., Barton, M.C., Wen, H., and Shi, X. (2013). Regulation of estrogen receptor alpha by histone methyltransferase SMYD2-mediated protein methylation. *Proc. Natl. Acad. Sci. U.S.A.* 110: 17284–17289.

Zhang, X. and Wang, Y. (2019). Identification of hub genes and key pathways associated with the progression of gynecological cancer. *Oncol. Lett.* 18: 6516–6524.

Zhang, X.D., Huang, G.W., Xie, Y.H., He, J.Z., Guo, J.C., Xu, X.E., Liao, L.D., Xie, Y.M., Song, Y.M., Li, E.M., and Xu, L.Y. (2018). The interaction of lncRNA EZR-AS1 with SMYD3 maintains overexpression of EZR in ESCC cells. *Nucleic Acids Res.* 46: 1793–1809.

Zhang, Y., Hayden, S., Spellmon, N., Xue, W., Martin, K., Muzzarelli, K., Kovari, L., and Yang, Z. (2021). Sperm chromatin-condensing protamine enhances SMYD5 thermal stability. *Biochem. Biophys. Res. Commun.* 550: 1–7.

Zhang, Y., Li, C., and Yang, Z. (2019b). Is MYND domain-mediated assembly of SMYD3 complexes involved in calcium dependent signaling? *Front Mol. Biosci.* 6: 121.

Zhang, Y., Yan, L., Yao, W., Chen, K., Xu, H., and Ye, Z. (2019c). Integrated analysis of genetic abnormalities of the histone lysine methyltransferases in prostate cancer. *Med. Sci. Monit.* 25: 193–239.

Zhou, Z.Y., Jiang, H., Tu, K.S., Yu, W., Zhang, J.L., Hu, Z.G., Zhang, H.Y., Hao, D.K., Huang, P.B., Wang, J., Wang, A.J., Xiao, Z.Y., and He, C.C. (2019). ANKHD1 is required for SMYD3 to promote tumor metastasis in hepatocellular carcinoma. *J. Exp. Cli. Cancer Res.* 38.

Zhu, C.L. and Huang, Q. (2020). Overexpression of the SMYD3 promotes proliferation, migration, and invasion of pancreatic cancer. *Dig. Dis. Sci.* 65: 489–499.

Zhu, H., Wei, M., Xu, J., Hua, J., Liang, C., Meng, Q., Zhang, Y., Liu, J., Zhang, B., Yu, X., and Shi, S. (2020). PARP inhibitors in pancreatic cancer: Molecular mechanisms and clinical applications. *Mol. Cancer* 19: 49.

Zhu, Y., Zhu, M.X., Zhang, X.D., Xu, X.E., Wu, Z.Y., Liao, L.D., Li, L.Y., Xie, Y.M., Wu, J.Y., Zou, H.Y., Xie, J.J., Li, E.M., and Xu, L.Y. (2016). SMYD3 stimulates EZR and LOXL2 transcription to enhance proliferation, migration, and invasion in esophageal squamous cell carcinoma. *Hum. Pathol.* 52: 153–163.

Zuo, S.R., Zuo, X.C., He, Y., Fang, W.J., Wang, C.J., Zou, H., Chen, P., Huang, L.F., Huang, L.H., Xiang, H., and Liu, S.K. (2018a). Positive expression of SMYD2 is associated with poor prognosis in patients with primary hepatocellular carcinoma. *J. Cancer* 9: 321–330.

Zuo, S.R., Zuo, X.C., He, Y., Fang, W.J., Wang, C.J., Zou, H., Chen, P., Huang, L.F., Huang, L.H., Xiang, H., and Liu, S.K. (2018b). Positive expression of SMYD2 is associated with poor prognosis in patients with primary hepatocellular carcinoma. *J. Cancer* 9: 321–330.

14

Toward Precision Medicine

Epigenetic Alterations in Human Melanoma

Carmen Elena Condrat[1,2], Elena Codruta Dobrica[3*], Sanda Maria Cretoiu[4], and Dragos Cretoiu[2,4]*

[1] *Carol Davila University of Medicine and Pharmacy, Bucharest, Romania*
[2] *Alessandrescu-Rusescu National Institute for Mother and Child Health, Fetal Medicine Excellence Research Center Bucharest, Romania*
[3] *Dermatology Department, Elias University Emergency Hospital, Bucharest, Romania*
[4] *Department of Cell and Molecular Biology and Histology, Carol Davila University of Medicine and Pharmacy, Bucharest, Romania*
[*] *Authors have contributed equally*

Introduction

Malignant melanoma (MM) is a serious type of skin cancer that arises from the malignant transformation of neural crest-derived melanocytes. While certainly not the most common form of skin cancer, as it accounts for around 5% of cutaneous tumors, MM is the most lethal one, being responsible for over 80% of deaths related to skin cancer [1]. Its prognosis can be favorable: five-year survival rates are 90–100% in cases of early diagnosis—that is, when lesions are in stages 0 or I, which necessitate only surgical removal. But these rates fall to around 18–28% if patients reach stage IV [2]. Moreover, the incidence of cutaneous MM has seen a steadily increasing trend over the years, as a result of increased life expectancy and improved detection (among other factors) [3,4]. In Europe, MM is the sixth most common type of cancer, having reached an age-standardized incidence rate of 22.9 per 100.000 people in 2020 (Figure 14.1).

Melanomas may develop either in previously existing melanocytic nevi or *de novo*, in apparently normal tegument areas. Table 14.1 indicates the most significant risk factors, although there is no consensus as to the most frequent etiologic pathways involved: these seem differ according to the location of MM.

While more frequent in sun-exposed areas, MM may also occur in concealed regions of the body such as the perineum, the palms and the soles of the feet, or mucosal surfaces of the gastrointestinal tract [15,16]. Among the lesions considered to precede MM, few are of particular interest, for example the common acquired nevus, the dysplastic nevus, the congenital nevus, and the cellular blue nevus. Cutaneous melanoma typically has two phases of growth. Initially the growth is radial, in other words tumor cells occupy the surrounding epidermis; afterwards the growth is vertical, in depth, invading the dermis. Depending on their depth, lesions are classified into thin, which measure 1 mm or less, moderate, which range between 1 mm and 4 mm, and thick lesions, which are larger than 4 mm. MMs are also divided into four major histopathological subtypes according to their pattern: superficial spreading, nodular, lentigo maligna, and acral lentiginous melanoma—but other forms, albeit rare, may occur too [17].

Genomic and Epigenomic Biomarkers of Toxicology and Disease: Clinical and Therapeutic Actions, First Edition. Edited by Saura C. Sahu.

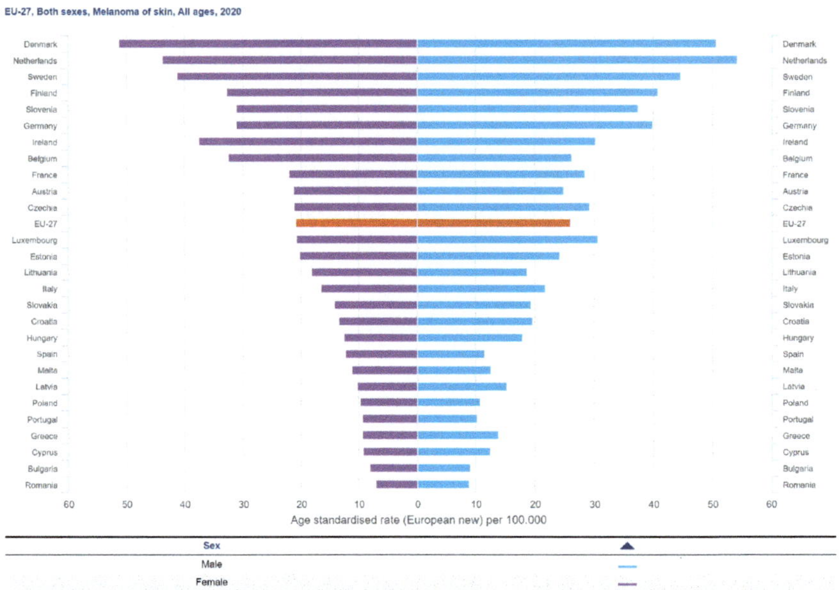

Figure 14.1 Estimated age-standardized incidence rates of skin melanoma in Europe by country, comparison by sex, using data from the European Cancer Information System (ECIS). EU-27 orange bar depicts the average age standardised rate of MM per 100.000 inhabitants in the 27 countries of the EU [5]. Generated by accession of https://ecis.jrc.ec.europa.eu website, March 1st, 2021.

Table 14.1 Summary of the main risk factors involved in the etiology of melanoma. Source: Sanda Maria Cretoiu.

Risk factor	Elevated risk	References
Family history	Affirmative history in 5% to 10% of patients Increased risk, 2.2-fold higher in patients with at least one affected relative Familial atypical multiple mole–melanoma (FAMMM) syndrome in more than two family members (formerly termed B-K mole syndrome, dysplastic nevus syndrome), 100% risk to develop MM	[6,7]
Socioeconomic status	Advanced stages are detected in patients with lower socioeconomic status Lack of information about the disease and decreased perception of the risk	[8,9]
Lifetime exposure to sunlight	Low latitude and high UVB and UVA radiation exposure Time and number of severe and painful sun blisters Use of tanning appliances or extensive exposure on tanning beds Use of inappropriate sunscreen blockers: using only UVB blockers creates the premises for a higher exposure to UVA by comparison with people who do not use sun blockers and are protected by shade or by non-exposure to the sun	[10–12]
Personal characteristics	Blue eyes, blond or red hair, freckles, skin phototype (pale or white skin) and presence of benign and/or dysplastic melanocytic nevi Sex: males have an increased risk of developing melanoma on the head, neck, or trunk, while females are at risk of developing melanoma on the lower limb Immunosuppressive status: chemotherapy, patients who suffered transplantation, patients with hematologic malignancies	[13,14]

The diagnosis of MM generally starts with a clinical evaluation and a dermoscopy assessment, while the histopathological examination is typically used for confirmation. However, owing to the heterogeneity of the condition and the occurrence of histological mimics of melanoma, molecular biomarkers and immunohistochemistry have been increasingly employed in its diagnosis [18,19]. Currently MM staging follows four systems, namely the Breslow scale, the Clark classification, the number staging, and the tumor, nodes, metastases (TNM) staging system [20]. The last one is the cornerstone of the widely adopted American Joint Committee on Cancer (AJCC) melanoma staging system, which estimates prognosis and guides the therapeutic process on the basis of lesion thickness, lymph node involvement, and distant metastasis [21]. Broadly speaking, the current therapeutic approaches to melanoma consist of surgical excision, chemotherapy, radiation therapy, photodynamic treatment, immunotherapy, and targeted therapy. Depending on the patient's overall health and on the stage and location of the primary tumor, the therapeutic plan can be limited to a single therapy or can adopt combined therapies [22]. In this respect, investigating the genetic profile of melanocytes as well as the molecular features indicative of malignant progression continues to ensure the development of novel therapeutic targets and potential agents. However, complex epigenetic mechanisms of histone modifications, chromatin remodeling, and differential expression of long non-coding RNAs (lncRNAs) that are thought to be responsible for the transformation of stem cells into cancerous ones should be exploited in the search of novel therapies.

The detailed examination of the human genome allowed the deciphering of the role of epigenetic changes in gene regulation and expression. In consequence, the main aim of this chapter is to discuss the relationships between melanoma epigenetics and cancer tumorigenesis and progression, and also the possibility of developing new pharmacological agents.

Epigenetic Alterations in Melanoma

Recent advances in elucidating the development and progression of melanoma have provided valuable insight into the role of epigenetic events, at the same time highlighting the need for further exploration in the future. Epigenetic alterations in MM play one of the most critical roles in melanoma progression. Gene expression is finely regulated through the cooperation between chromatin structures that are dependent on epigenetic factors, transcription factors, and translational mechanisms [23]. In the human genome, chromatin consists of a tightly packaged DNA molecule associated with proteins, most of which are histones—H2A, H2B, H3, and H4. Change in the structure of chromatin is a dynamic process that begins post-transcriptionally and is addressed to the histones N-terminal region (methylation, acetylation, sumoylation, glycosylation, ubiquitination, and phosphorylation) or to the incorporation of some histone variants; this leads to chromatin remodeling, under the control of specific enzymes, for each process, for example histone methyltransferases/demethylases and histone acetyltransferases/deacetylases for the first two processes (Figure 14.2) [24,25].

The term "epigenetic alterations" refers to heritable changes that occur in the DNA structure and do not involve its coding sequence, but still lead to modifications in gene expression. These broadly include the addition or removal of methyl groups from DNA molecules, changes in histone proteins that provide chromosomes with structural support, and the rearrangement of chromatin to a state that is more accessible for transcription [26,27]. These alterations typically take place over time and in the presence of certain environmental factors, and have increasingly been shown to be partly responsible for various phases of tumorigenesis.

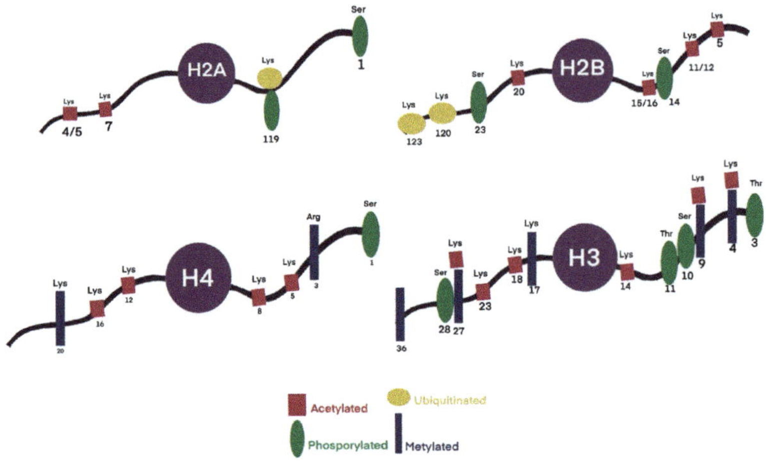

Figure 14.2 Post-translational changes in histones N-terminal "tail" regions projecting outward from nucleosomes. These modifications are able to affect the interactions between nucleosomes and are responsible for chromatin accessibility and for the epigenetic regulation of gene expression. Lys = lysine, Ser = serine, Arg = arginine, Thr = threonine. Source: Elena Codruta Dobrica.

DNA Methylation

DNA methylation patterns are essential for the normal functioning of human cells and are related to the regulation of gene expression, the silencing of transposable elements (retroviral elements), X-chromosome inactivation, genomic imprinting, aging, and carcinogenesis [28]. These patterns, which are present during development and differentiation, are prone to turning into a form of epigenetic memory that gets inherited through successive cell cycles. The DNA methylation of C-5 of cytosines is restricted to CpG sites [29]. Tumors are characterized by reduced DNA methylation and contain aberrantly hypermethylated CpG islands.

In melanoma, the term "aberrant DNA methylation" may refer to focal hypermethylation or to global hypomethylation. The exact trigger of these events still remains unknown, but it has been speculated that the overexpression or increased activity of enzymes in the DNA methyltransferase (DNMT) family, as well as an abnormal targeting of these enzymes, could play a causal role [30].

Perhaps the most notable effect of focal DNA hypermethylation is the silencing of tumor suppressor gene promoters. Among the most thoroughly investigated DNA hypermethylation tumor suppressor genes are PTEN, a known inhibitor of the PI3K pathway [31]; p14ARF, an inhibitor of the mouse double minute 2 homolog (MDM2) [32]; Ras association domain family 1 isoform A (RASSF1A), a regulator of apoptosis and cell cycle [33]; p16INK4A, an inhibitor of the cyclinD1/CDK4 complex and a cell cycle regulator [34]; and O-6-methylguanine-DNA methyltransferase (MGMT), which is involved in DNA repair [35].

As PTEN is the main negative regulator of the PI3K-AKT signaling pathway, PTEN promoter hypermethylation results in the loss of cell proliferation blockade, and this loss in turn leaves room for cell cycle progression and subsequent cell growth [36]. In melanoma serum samples, several studies have reported identifying the promoter methylation of PTEN in over 60% of cases [31,37]. Moreover, PTEN methylation has been shown to be a significant prognostic factor in patients with melanoma, being associated with overall poor survival rates [37,38]. Further on, p14ARF acts as a tumor suppressor as a result of its ability to both hinder ribosome biogenesis and induce cell cycle arrest followed by apoptosis by inhibiting MDM2, which, in turn activates p53 [39]. In a recent study using methylation-specific PCR, hypermethylation of p14ARF was identified in over half of the metastatic melanoma serum samples, while promoter methylation of both p14ARF and p16INK4A could be observed in 12% of cases [32]; similar findings were later reported by Venza and colleagues [40]. The latter study also looked at the dominant mechanism of tumor suppressor silencing and found that, while p16INK4A primarily undergoes promoter methylation, the silencing of p14ARF more often occurs because of histone hypoacetylation. Under normal circumstances, p16INK4A induces the activation of the retinoblastoma protein (Rb) by inhibiting CDK4/6, a process that leads to G1/S cell cycle arrest [41]. In the serum of metastatic cutaneous melanoma, promoter methylation of p16INK4A has been reported in around 25% of cases [34,41].

RASSF1A is a known cell cycle regulator whose hypermethylation, leading to its loss of expression, has repeatedly been reported as correlated with advanced stage melanoma. To this extent, the methylation of two areas of the RASSF1A CpG promoter region has been identified in 55% of the melanoma tumors examined by Spugnardi et al. using methylation-specific PCR [33]. Their study further demonstrated that RASSF1A re-expression was possible after a treatment with 5′-aza-2′-deoxycytidine. Similarly, Hoon et al. later reported the hypermethylation of RASSF1A CpG islands in 57% of metastatic melanoma samples and in 15% of primary melanoma tumors, highlighting its potential role in melanoma progression [42]. In a later study, Tanemura et al. reported RASSF1A methylation in almost half of the metastatic melanoma tissues analyzed, whereas early-stage melanoma tissues (I, II) did not exhibit this epigenetic phenomenon [43].

On the other hand, given that active DNA demethylation in mammals has been shown to result in 5-hydroxymethylcytosine (5-hmc) as either an intermediate or an end product, studies have pointed toward its role in melanoma. Since isocitrate dehydrogenase type 2 (IDH2) and the ten-eleven translocation (TET) enzymes play significant roles in transforming 5-methylcytosine (5-mc) into 5-hmc, it has been shown that loss of TET activity and underexpression of IDH2 lead to decreased 5-hmc levels, which have been correlated with melanoma progression [44,45]. Furthermore, in melanoma models, restoring the expression of IDH2 and TET2 has been shown to bring back 5-hmc levels, thereby halting tumor growth and invasion [46]. Several studies have suggested the use of 5-hmc levels as a distinguishing marker between melanocytic nevi and MMs [47,48]. Moreover, in a recent study on melanoma cell lines, Chatterjee et al. have identified a global DNA hypomethylated state that could be linked to an increased expression of programmed death-ligand 1 (PD-L1). PD-L1, when binding to the PD-1 receptor found on activated T cells, manages to inhibit their activity, thus facilitating immune evasion. Herein lies the reason for an interest in modulating the expression of PD-L1 on melanoma cells [49].

Histone Modifications

The basic structural unit of eukaryotic chromatin, the nucleosome, contains a DNA strand wrapped around a set of eight histones. While generally very much organized, chromatin is dynamic, and changes in its structure are responsible for crucial cellular processes [50]. In this regard, shifting between the two distinct varieties of chromatin—namely euchromatin, which is genetically active owing to its uncoiled form, and heterochromatin, which prevents DNA transcription owing to its firmly packed form—is primarily mediated by particular enzymes that drive histone modifications, such as histone deacetylases (HDACs), histone demethylases (HDMs), histone acetyltransferases (HATs), and histone methyltransferases

(HMTs); these are also referred to as writer, reader, and eraser enzymes (Figure 14.2) [51,52]. In melanoma progression, the role of such enzymes has been highlighted by several studies. For instance, the SET domain bifurcated 1 (SETDV1) protein, an HMT that mediates trimethylation at histone H3 lysine 9 (H3K9me3), has been shown to promote gene expression silencing in an epigenetic manner [53]. Later, Ceol and colleagues investigated SETDB1 in a zebrafish BRAF V600E melanoma model and found that not only was it highly expressed in melanoma, but it also promoted its progression, thus confirming its oncogenic potential [54]. Afterwards, Orouji and colleagues further confirmed SETDB1 as a potential therapeutic target when they demonstrated that it also promotes melanoma invasion and metastasis by catalyzing mono-methylation at lysine 4 of histone H3 (H3K4me1), in proximity to the regulatory sites of the thrombospondin 1 (THBS1) protein, which is known to mediate cell-to-cell adhesion and promote an aggressive tumor behavior [55].

Further acting as a chromatin writer, the enhancer of zeste homolog 2 (EZH2) HMT mediates the trimethylation at histone H3 lysine 27 (H3K27me3), and, in doing so, manages to activate the non-canonical nuclear factor-κB (NF-κB) signaling cascade, thus bypassing cellular senescence and promoting tumorigenic transformation in melanoma cells [56]. Moreover, using murine melanoma models, Zingg and colleagues have recently discovered that, in combination with BRAF V600E and/or NRAS Q61K mutations, EZH2 stimulates melanoma progression by silencing the primary cilium genes, thus promoting the canonical Wnt pathway, which is critical in the initial steps of melanomagenesis [57].

The roles of effector or chromatin reader proteins, capable of identifying chromatin modifications and subsequently implementing specific regulatory processes, have also been observed in cases of melanoma. In this regard, the overexpression of certain members of the bromodomain and extra-terminal (BET) protein family (BRD2, BRD3, BRD4, and BRDT) of epigenetic readers has been shown to favor tumor cell survival in melanoma [58]. By binding to acetylated lysine residues on histones and to sequence-specific DNA-binding factors, bromodomains mediate gene transcription and expression [59]. In melanoma tissues, BRD2 and BRD4 may be overexpressed, thus stimulating the NF-κB signaling pathway, which is involved in tumor initiation and progression through its ability to regulate key cellular processes such as cell proliferation, differentiation, and death, along with cytokine and chemokine production [60]. In a similar manner, Gallagher and colleagues later found that BET protein inhibitor I-BET151 managed to induce the downregulation of the NF-κB transcription factor in cultured melanoma cells, thus reducing tumor cell growth, cytokine production, and resistance to vemurafenib [61].

Eraser enzymes, known to remove histone post-translational modifications, are most commonly represented as consisting of the four classes of HDACs, which eliminate acetyl groups from histone tails, and HDMs. Because limiting histone deacetylation with the aid of HDAC inhibitors, either broad-spectrum or selective, has proven effective in counteracting the proliferative activity of melanoma cells, more attention has been paid to the roles of HDACs in melanoma tumorigenesis [62]. For instance, HDAC6, which belongs to the second class of HDACs, has been shown to play a key role in immune surveillance as a result of its ability to regulate the expression of PD-L1 by promoting the activation of signal transducer and activator of transcription 3 (STAT3) and, to a lesser degree, STAT1, in melanoma cells [63]. Additionally, it has been revealed that, by genetically or pharmacologically inhibiting HDAC6, melanoma cells not only display an increased immunogenicity—because of an enhanced expression of major histocompatibility complex (MHC) class I molecules and tumor-associated antigens—but also exhibit lower proliferation rates and G1 cell cycle arrest [64]. On the other hand, certain HDACs, such as HDAC3 and HDAC8, when elevated in metastatic melanoma, can be linked to better survival rates [65]. However, when exposed to stress, for example through the use of a combination therapy that consists of BRAF and MEK inhibitors, the higher HDAC8 expression promotes drug resistance, since it can regulate MAPK through its target, the activator protein 1 (AP-1) transcription factor, by reducing the acetylation of one of its components, c-Jun [66].

When we approach the study of melanoma progression, lysine-specific demethylase 6B (KDM6B), an HDM that catalyzes the demethylation of histone 3 in position 27 (H3K27), has also emerged as a valuable player. Specifically, Park and colleagues have recently managed to demonstrate that tumor growth and overall melanoma progression and metastasis are supported by KDM6B, as it can increase macrophage infiltration, trans-endothelial migration, and angiogenesis in melanoma by activating the bone morphogenetic protein (BMP) signaling pathway [67]. Similarly, increased expression of KDM5B, which mediates the demethylation of histone 3 in position 4 (H3K4), is known to promote drug resistance in melanoma cells by favoring the survival of certain melanoma-propagating cells that possess stem or progenitor cell qualities (or both) [68]. Vogel and colleagues later demonstrated that selectively targeting KDM5B reduced colony formation and cellular invasion by reprogramming the metabolome and the behavior of melanoma cells so that they regain sensitivity to drugs such as BRAF and proteasome inhibitors [69].

Non-coding RNAs as Epigenetic Regulators

Non-coding RNA is the RNA that cannot be translated into proteins, and it accounts for most of the transcribed genetic material [70].

This type of RNA is classified into two broad categories: small non-coding RNA (non-coding RNA fragments containing less than 200 bp nucleotides) and lncRNA (non-coding RNA fragments containing more than 200 bp nucleotides) [71]. These are further classified into multiple subcategories; the most important ones are shown in Figure 14.3.

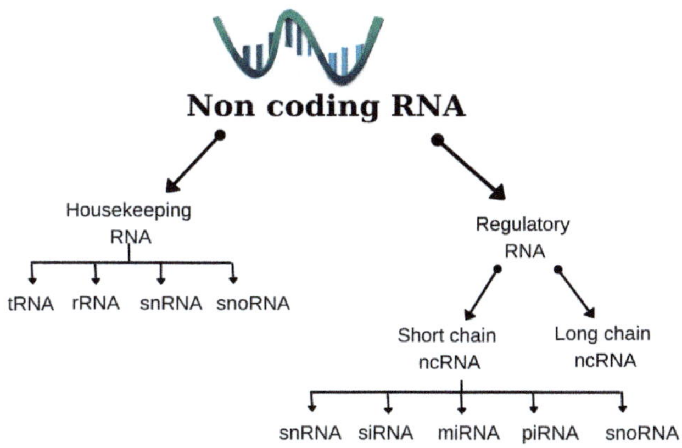

Figure 14.3 Classification of main types of non-coding RNA (ncRNAs). Two large categories of noncoding RNAs are distinguished: housekeeping and regulatory noncoding RNAs. Housekeeping ncRNAs are divided into ribosomal (rRNA), transfer (tRNA), small nuclear (snRNA), and small nucleolar RNAs (snoRNAs) depending on their structure and function. Regulatory noncoding RNAs are classified depending on the number of nucleotides into short chain ncRNAs (<200 nucleotides) or long chain ncRNAs (>200 nucleotides). Short ncRNAs are subdivided into small nuclear RNA (snRNA), small interfering RNAs (siRNAs), microRNAs (miRNAs), Piwi-associated RNAs (piRNAs) and small nucleolar RNA (snoRNA). Source: Elena Codruta Dobrica.

Multiple functions of certain non-coding RNAs have been identified in recent decades; for example, non-coding RNAs have an essential role in regulating DNA expression [72]. This function can be achieved either directly, through the action of the non-coding RNA on the protein structures involved in transcription processes, or indirectly, by regulating the two mechanisms mentioned earlier: DNA methylation by dimethyltransferases (DMTs) and histone changes [73]. Moreover, recent studies show an increase in the expression of these non-coding RNAs in tumor cells, which raises the idea of a possible role of non-coding RNAs in the carcinogenic processes in these cells: initiation, progression, and proliferation [71].

The following lncRNAs were shown to have an altered expression in MM and to play a role in melanoma development and progression: ANRIL, BANCR, CASC15, GAS5, HOTAIR, Llme23, MALAT1, PAUPAR, RMEL3, SAMMSON, SNHG5, SPRY4-IT1, and UCA1 [74].

Ten of these long-chain non-coding RNAs have been found to be involved in processes of proliferation, migration, invasion, metastasis by degradation of the extracellular matrix, and cellular transition to aggressive forms with highly invasive potential. These are SPRY4-IT1, BANCR, HOTAIR, MALAT 1, UCA1, CASC15, ANRIL, SLNCR1, SAMMSON, and LLME23 [75,76].

Other two of the thirteen lncRNAs listed here, namely RMEL3 and GAS5, appear to have beneficial effects, which support the body's fight against melanoma neoplastic cells: they inhibit migration, invasion, and proliferation and they decrease the survival of tumor cells.

Most of these RNAs are overexpressed in tumor cells to perform their functions; the only RNA known to be downregulated is GAS5 [77,78]. Their blocking seems to be an important starting point in the treatment of melanoma, including that of treatment-resistant forms, because it would stop the aforementioned processes [79].

One of the most studied lncRNAs is HOTAIR, which is overexpressed in tumor cells of many cancers, including metastatic melanoma (especially lymph node metastases). The presence of this non-coding RNA seems to be a useful biomarker in the differential diagnosis of melanoma and dysplastic lesions, since HOTAIR is expressed only in the cells of the former [80]. Moreover, MM is up regulated in metastases comparable with the primary tumor, and this leads to the conclusion that it is responsible for the metastatic development of secondary tumors [74]. lncRNAs are molecules difficult to preserve; moreover, they are difficult to study in vivo. Therefore there is a great need to develop in the future new study techniques and new experimental models, human or animals, with the aim of achieving new therapies that can target the non-coding RNAs.

Pathogenesis and Molecular Biology of Melanoma

MM is generally considered a condition more often encountered in fair-skinned populations, especially when they exhibit sun-seeking behavior. Moreover, it has been proven time and again that pale complexions are more predisposed to developing various types of skin cancer, including MM, primarily because of their lower production of melanin [81]. Ultraviolet (UV) radiation exposure poses a considerable risk, owing to the mutagenic effects of UV, which include the formation of pyrimidine dimers through photochemical reactions and cytosine deamination to uracil within DNA [82,83]. Further host risk factors that are associated with an increased sensitivity to UV rays, thus potentially also increasing the risk of melanoma development, are a positive family history, genetic susceptibility, expressed not only as skin types I and II but also as genodermatoses such as xeroderma pigmentosum [84], and familial atypical multiple mole melanoma (FAMMM) syndrome [7], along with a high number of melanocytic nevi [85–87].

Melanocytes originate from undifferentiated precursors that migrate from the neural crest to their destination, namely the hair and the skin, the inner ear cochlea, the iris, and the choroid coat [88]. In embryogenesis, their survival and established course are facilitated by complex biochemical cascades, such as the endothelin-3/endothelin receptor type B (EDNRB) signaling pathway, the Wingless signaling (Wnt)/β-catenin transduction cascade, and the cell-fate regulating Notch signaling pathway [89,90]. Additionally, multiple transcription factors and target genes are further involved in this process, such as the microphthalmia-associated transcription factor (MITF), the hairy/enhancer of split homologue-1 (Hes-1), the sex-determining region Y (SRY)-box10 (SOX10), the paired box gene 3 (PAX 3), and the B-cell lymphoma 2 (Bcl-2) family of regulator proteins responsible for apoptosis mediating [91–95]. Consequently, the aberrant protein expression that is due to mutations in these genes (among others) contributes to abnormal cell growth paired with halting cell death (Figure 14.4). As the development of cutaneous melanoma takes place after an aberrant chain of events involving numerous signaling pathways that are decisive for the dynamic state between cell proliferation and death [96], the gradually improved understanding of the cellular and molecular mechanisms fundamental to the genesis and progression of melanoma has to some extent facilitated progress in the treatment of advanced stages of cutaneous melanoma. In this regard, novel therapeutic strategies achieved thanks to the evolving field of high-throughput genomics involve the targeting of specific pathways, such as the mitogen-activated protein kinase (MAPK) pathway, and of checkpoint proteins [97]—all of which substantially increase survival rates in metastatic melanoma patients [98]. However, as a result of certain individual characteristics, these therapies do not apply to all patients and, what is more, results tend to be transient, mainly because of acquired drug resistance [58]. It has been demonstrated that genetic and epigenetic modifications are involved in both innate and acquired resistance to these regimens [20], and this conclusion encourages large-scale efforts to better understand the molecular mechanisms of melanoma in order to develop more efficient therapies.

It is currently surmised that the first step in the cascade of events involved in the etiology of melanoma is the inflammatory response that occurs because of the UV-induced keratinocyte damage, which in turn promotes both angiogenesis and melanoma cell migration [99].

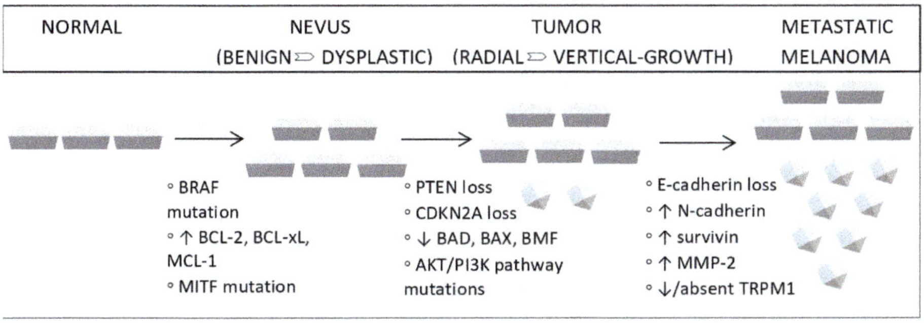

Figure 14.4 Main molecular events in melanoma progression. BRAF = B-Raf proto-oncogene, serine/threonine kinase; BCL-2 = B-cell lymphoma 2; BCL-xL = B-cell lymphoma-extra-large; MCL-1 = Myeloid-cell leukemia 1; MITF = microphthalmia-associated transcription factor; PTEN = phosphatase and tensin homolog; CDKN2A = cyclin-dependent kinase inhibitor 2A; BAD = BCL-2-associated agonist of cell death; BAX = BCL-2-associated X protein; BMF = Bcl-2-modifying factor; AKT/PI3K = protein kinase B/phosphoinositide 3-kinase signaling pathway; MMP-2 = matrix metalloproteinase-2; TRPM-1 = transient receptor potential melastatin-1. Source: Carmen Elena Condrat.

Further on, UV rays can trigger the mutation of the TP53 tumor suppressor gene in melanocytes that express the BRAF V600E mutation, thus terminating the anti-oncogene response of the p53 protein [100]. As the BRAF V600E missense mutation—which changes valine into glutamic acid—is present in around 40–60% of melanoma cases [101] and even in greater proportion in the vertical growth and metastatic phases [102], this mutation has been increasingly regarded as a promising therapeutic target. The BRAF mutation is the most frequent somatic mutation that leads to abnormal signaling through the MAPK pathway and, in doing so, promotes both apoptosis suppression and the deregulation of cell cycle [103]. In a similar manner, aberrant MAPK signaling is also caused by mutations in the neuroblastoma RAS viral oncogene homolog (NRAS), which occurs in up to 30% of melanoma cases [104]. However, the NRAS mutation is also responsible for the aberrant activation of the PI3K pathway, which it achieves by extending the NRAS-active (GTP-bound) state [105]. While mutations in these oncogenes very rarely coexist, several studies have noticed that BRAF-positive melanoma patients who did not respond to BRAF inhibitor therapy were likely to present NRAS mutations [20,106]. The third most commonly mutated gene in melanoma is the tumor suppressor neurofibromatosis type 1 (NF1) gene, encountered in up to 15% of melanoma cases [107]. Under normal circumstances, NF1 converts active, GTP-bound RAS, to its inactive form, GDP-bound, thus inhibiting the NRAS protein, while its mutation consequently leads to the hyperactivation of the MAPK and PI3K signaling cascades [108,109]. On the other hand, although rather rare, the PI3K/AKT pathway can also be abnormally activated through the inactivation of the PTEN tumor suppressor gene [110,111], recent studies suggesting that PTEN loss be used as a predictor of melanoma survival [112].

Further studies have provided insight into the involvement of the cell death-controlling BCL-2 family of proteins, which consists of both pro-apoptotic proteins such as BAD, BAX, BAK, and BMF and anti-apoptotic proteins such as BCL-xL, BCL-2, and MCL-1 [91,113]. In this regard, it has been suggested that a dysfunction in the dynamics of the BCL-2 family that consists either in a decreased activity of the pro-apoptotic proteins or in an increased expression of the anti-apoptotic proteins may result in failed BAK/BAX activation and in subsequent exaggerated cellular proliferation [114]. However, further investigation into this matter is no doubt required, as the findings so far are contradictory: recent studies suggest that lowly expressed pro-apoptotic proteins may actually predict a favorable prognosis. This is expressed in higher progression-free survival rates [115].

Additional studies have examined the role of epithelial–mesenchymal transition (EMT) in melanocytes, as this is the first step in cell migration and metastasis [116]. Kreizenbeck et al. have analyzed the prognostic value of adhesion molecules E-cadherin, N-cadherin, and P-cadherin, known mediators of EMT. They found that loss of E-cadherin and of P-cadherin were characteristic of metastatic lesions, while an increased expression of N-cadherin could be associated either with EMT-positive tumors prone to metastasis or with well-differentiated, non-invasive tumors [117]. Later studies, however, suggested that a switch from E-cadherin to N-cadherin may predict progression from primary to metastatic melanoma [118,119].

In the case of familial melanoma, germline mutations in the tumor suppressor cyclin-dependent kinase inhibitor 2A (CDKN2A) gene and in the cyclin-dependent kinase 4 (CDK4) oncogene have been implicated time and again [120,121]. CDKN2A encodes tumor suppressors p14 alternate reading frame (p14ARF) and p16 inhibitor of cyclin-dependent kinase 4A (p16INK4a), the latter being involved in cell cycle arrest on account of its ability to inhibit the cyclinD1/CDK4 complex [122]. A mutated CDKN2A gene cannot successfully exert its inhibitory effect over this complex, and this incapacity leads to cell cycle progression and abnormal proliferation through the E2 factor (E2F) transcriptional network [123].

Current Therapeutic Approaches

Depending on the stage and location of the tumor as well as on the patient's overall health, several therapeutic strategies have been developed over time. Initial stages generally benefit from classical surgical excision or, less commonly, from Mohs surgery—especially when the esthetic result is important [124]. Should the lesion be thicker (stage IB) or ulcerated, sentinel lymph node biopsy (SLNB) is typically performed, to assess the extent to which cancer cells have spread [125]. If lymph nodes are free from cancer cells, follow-up is recommended, while a positive SLNB result requires either lymph node dissection or regular ultrasound monitoring. Moreover, some studies advise additional treatment in the form of immunotherapy or targeted therapy. As the tumor advances, wide excision coupled with lymph node dissection is typically followed by adjuvant immunotherapy, targeted therapy, radiation therapy, chemotherapy, or local injections with interleukin-2 (IL-2), Bacillus Calmette–Guérin (BCG) vaccine, or Talimogene laherparepvec (T-VEC) vaccine—depending on tumor location, patient characteristics, and drug accessibility [126,127].

Immunotherapy is a valuable player in melanoma therapy. It has several systemic and local treatment modalities. While the response might initially be slow, this type of treatment can lead to patient reassessment for residual metastasectomy [128]. Immunotherapy strategies currently approved for use in melanoma are briefly described in Table 14.2.

Table 14.2 Types of immunotherapies currently approved for use in melanoma patients.

Treatment class	Treatment type	Drugs	Action
Checkpoint inhibitors	PD-1 inhibitor	Pembrolizumab, nivolumab	Modulates the immune response encouraging T cells to identify and attack cancer cells [129].
	CTLA4-inhibitor	Ipilimumab	
Cytokines	Cytokine	Interferons including pegylated interferon alfa-2b, interleukins, hematopoietic growth factors	Stimulates malignant-specific immune cells of the host, overall increasing the immune system's response [130].
Vaccines	Vaccine	BCG vaccine	Induces an immune response by activating both arms of the immune system, innate and acquired [131].
Oncolytic viruses	Oncolytic virus therapy	T-VEC vaccine	Modified viral agents are used to infect and kill cancer cells, thereby releasing tumor antigens that subsequently stimulate the host immune response [132].
Non-specific immune stimulation	TLR agonists	Imiquimod	Apoptosis induction by activating the caspase-cascade system and boosting the immune system by mediating the release of multiple cytokines [133].

PD-1 = programmed cell death protein 1; CTLA4 = cytotoxic T-lymphocyte-associated protein 4; BCG = Bacillus Calmette–Guérin; T-VEC = Talimogene laherparepvec; TLR = Toll-like receptors. Source: Carmen Elena Condrat.

Targeted therapy relies mostly on targeting the BRAF protein and the mitogen-activated protein kinase (MEK), a downstream kinase of both BRAF and NRAS in the MAPK pathway [134]. BRAF inhibitors such as vemurafenib and dabrafenib directly attack the BRAF protein, generally producing rapid tumor shrinkage in BRAF V600E-positive melanomas. However, due to the occurrence of BRAF inhibitor resistance, single-agent therapy has gradually been replaced by combinatorial treatment. When MEK inhibitors such as trametinib and cobimetinib are added to the therapeutic scheme, the simultaneous inhibition of both BRAF and MEK ensue superior response and survival rates, especially among patients with unresectable melanoma or with metastatic melanoma (or with both) [135]. But, for a maintained response, continued combined therapy is necessary, as it has been noticed that treatment discontinuation leads to relapse within a year, in about half of the patients who respond well to treatment [136]. Another somatic mutation less often encountered in melanoma, specifically the activating mutation in the receptor tyrosine kinase (C-KIT) gene found in around 15% of acral melanoma cases [137], has been targeted through therapeutic agents such as imatinib and nilotinib, and with promising results [138].

Radiation therapy, although not commonly used in melanoma, finds its utility either when surgery cannot be performed or after lymph node dissection, and also in a more palliative setting. Furthermore, stereotactic radiosurgery using radiation in high doses can be applied to tumors that spread to the brain [139].

Chemotherapy may also be employed in advanced cases, but generally not as a first treatment, since the superior effectiveness of immunotherapy and targeted therapy has been proven on numerous occasions. While chemotherapy is not as useful here as in other types of cancer, it can benefit some patients, especially when combined with other therapeutic agents for palliation. Chemo agents used in the treatment of melanoma include alkylating agents such as dacarbazine, temozolomide, and nitrosoureas such as carmustine, platinum agents, vinca alkaloids, and taxanes [140]. Isolated limb infusion with cytotoxic agents may also be selected for surgically unresectable tumors confined to the leg or arm, to limit the side effects [141].

Therapeutic Implications of Epigenetic Changes in Melanoma

The abovementioned epigenetic changes play an important role in the initiation and progression of melanoma, especially through transcriptional alterations caused by histone changes, DNA methylation, and chromatin remodeling [58]. The methylation of tumor suppressor genes such as PTEN, LOX, COL1A2 (their epigenetic changes occur in about 50–80% of melanoma cases) leads to their inactivation, thus playing a key role in the appearance of melanoma; it is among the first pieces of evidence of the involvement of epigenetics in pathogenesis. Subsequently, advanced genomics studies have identified epigenetic methylation changes in the entire DNA of melanoma cells [142].

These hypotheses are supported by the fact that cell populations with an identical BRAF V600E mutation develop melanoma in different proportions, which suggests the possible influence of environmental factors [143]. Thus these mutations are necessary, but not sufficient to initiate the process of tumorigenesis just by themselves [142].

In addition to the involvement of epigenetics in the initiation and progression of a tumor, increasing attention has been paid to how epigenetics influences resistance to a targeted treatment. This is one of the most important clinical and practical implications of epigenetics, so that understanding these mechanisms may open up new perspectives on patients' response to treatment.

Thus the main therapies used in melanoma are BRAF inhibitors or common BRAF-MEK inhibitors, the latter being the most common mutations found in melanoma patients. Although the response to these therapies is impressive in the short term, it is associated with a high degree of recurrence at a short interval (fewer than twelve months) through resistance to treatment. Epigenetics seems to play an increasingly important role in the emergence of this resistance: a significant percentage of the patients who develop it do not have any of the mechanisms known so far to be responsible for it, namely RAS, MEK, and BRAF mutations [144].

The main epigenetic mechanisms that appear to lead to resistance to vemurafenib or vemurafenib-trametinib are the histone 3 demethylation of lysines by specific demethylases and the activation of genes involved in deacetylation processes. These processes were found to be present in cells with an increased resistance to treatment and an aggressive character [54,55,143,145].

On the other hand, at the opposite pole of these mechanisms are new perspectives on the use of epigenetics to modulate the expression of growth factors involved in resistance to treatment. Epigenetics seems to play a dual role, both as a resistance-inducing factor and as a tool in the mechanisms to fight it [146]. There are currently numerous epigenetic modulators, for example decitabine, azacitabine, vorinostat, panobinostat, or valproic acid, that are used in trials in combination with anti-cancer therapy (targeted therapy or immunotherapy). They modulate by introducing changes in cellular metabolism and thereby causing apoptosis (decreasing oxidative phosphorylation and electron transport in the mitochondrial chain II), or by controlling the expression of growth factors or their receptors (decreasing HER2/EGFR, increasing estrogen receptors), or by increasing the cells' sensitivity to the receptors involved in the activation of apoptosis pathways (tumor necrosis factor (TNF)-related apoptosis-induced ligand and its receptor), or by increasing the immune response (in order to counteract the escape from the immune response of cancer cells) [146–151].

The abovementioned therapies have shown promising results, especially in hematologic cancers resistant to conventional chemotherapy. Apart from the fact that these epigenetic modulators are used in combating resistance to treatment, encouraging results seem to be observed in their use for the treatment of aggressive forms of melanoma with non-cutaneous localizations, such as uveal melanoma. The latter, along with other kinds of solid tumors, represents therapeutic challenges in terms of epigenetic modulators that are due to the increased differentiation of the cells from which the tumor cell derives. Thus, multiple phase I and II clinical trials in various stages of research appear to be effective in decreasing migration and invasion alongside cell growth and cell proliferation, and in increasing p53 gene expression and apoptosis [152].

Another important role of epigenetics appears to be in the diagnosis of melanoma and dysplastic lesions. Recent studies raise the possibility of using specific DNA methylation patterns as biomarkers to assess malignancy. Conway et al. highlight the presence of hypermethylation regions (for KCNK4, GSTM2, TRIP6, FRZB, COL1A2, NPR2 genes) and hypomethylation (for a set of sixteen genes, including CARD15/NOD2, MPO, ITK, and CD2) in cancer cells, but not at the level of cells that belong to benign nevi [153]. Another study, by Gao et al., uses a different set of genes (CLDN11, CDH11, PPP1R3C, MAPK13, and GNMT), in which it has identified specific methylation patterns that allow the differentiation of malignant lesions from benign skin lesions [154].

Also, the hypomethylation of certain genes (MEOX2, OLIG3, PON3) seems to be a prognostic factor for survival (positive correlation). At the same time, decreased PTEN expression appears to be associated with poor prognosis in terms of survival, the presence of this change in patients with BRAF V600E positive melanoma being associated with faster brain metastasis [155,156].

Conclusion

Melanoma is a type of tumor that, although it has a low incidence, is responsible for the highest percentage of skin-related mortality. The complexity of the pathophysiological mechanisms of initiation and progression, as well as the variable response to treatment caused by the appearance of chemoresistance to significant degrees, make the diagnosis of melanoma a grim diagnosis and a therapeutic challenge for both the patient and the doctor. The impossibility of explaining these mechanisms through pre-existing, stable genetic changes has thus placed the focus of attention on epigenetics. The changes in DNA methylation, the histones modifications that lead to chromatin remodeling, and also the action of non-coding RNA as a regulator of these processes all seem to explain the chemoresistance to treatment and the variable aggressiveness of this pathology. Moreover, new scientific data propose the use of epigenetic changes both as biomarkers of the disease and as key tools in combating resistance; the advance new modulatory drugs associated with classical chemotherapy. Although data from clinical trials are still scarce, as most of these trials are in the I–II stages and the long-term effects on maintaining the therapeutic response are unknown, the current results are promising. We can consider epigenetics to be a new step toward a personalized therapy for pathologies with poor prognosis, with high therapeutic resistance, and with high mortality in the short term; and advanced melanoma ia one of the leading representatives of diseases in this category.

References

1 Bertolotto, C. (2013). Melanoma: From melanocyte to genetic alterations and clinical options. *Scientifica (Cairo)* 2013: 1–22.

2 Svedman, F.C., Pillas, D., Taylor, A., Kaur, M., Linder, R., and Hansson, J. (2016). Stage-specific survival and recurrence in patients with cutaneous malignant melanoma in Europe: A systematic review of the literature. *Clin. Epidemiol.* 8: 109–122.

3 Bay, C., Kejs, A.M.T., Storm, H.H., and Engholm, G. (2015). Incidence and survival in patients with cutaneous melanoma by morphology, anatomical site and TNM stage: A Danish population-based register study 1989–2011. *Cancer Epidemiol.* 39 (1): 1–7.

4 Rigel, D.S., Russak, J., and Friedman, R. (2010). The evolution of melanoma diagnosis: 25 years beyond the ABCDs. *CA Cancer J. Clin.* 60 (5): 301–316.

5 European Cancer Information System https://ecis.jrc.ec.europa.eu (accessed March 1st, 2021).

6 Eckerle Mize, D., Bishop, M., Resse, E., and Sluzevich, J. (2011). Familial atypical multiple mole melanoma syndrome. In Cancer Syndromes, ed. D.L. Riegert-Johnson, L.A. Boardman, T. Hefferon, and M. Roberts, n.p. Bethesda, MD: National Center for Biotechnology Information.

7 Lynch, H.T. and Shaw, T.G. (2016). Familial atypical multiple mole melanoma (FAMMM) syndrome: History, genetics, and heterogeneity. *Fam. Cancer* 15 (3): 487–491.

8 Mandalà, M., Imberti, G.L., Piazzalunga, D., Belfiglio, M., Lucisano, G., Labianca, R., Marchesi, L., Merelli, B., Robone, S., Poletti, P., Milesi, L., and Tondini, C. (2011). Association of socioeconomic status with Breslow thickness and disease-free and overall survival in stage I–II primary cutaneous melanoma. *Mayo Clin. Proc.* 86 (2): 113–119.

9 Harvey, V.M., Patel, H., Sandhu, S., Wallington, S.F., and Hinds, G. (2014). Social determinants of racial and ethnic disparities in cutaneous melanoma outcomes. *Cancer Control* 21 (4): 343–349.

10 Dale Wilson, B., Moon, S., and Armstrong, F. (2012). Comprehensive review of ultraviolet radiation and the current status on sunscreens. *J. Clin. Aesthet. Dermatol.* 5 (9): 18–23.

11 Watson, M., Holman, D.M., and Maguire-Eisen, M. (2016). Ultraviolet radiation exposure and its impact on skin cancer risk. *Semin. Oncol. Nurs.* 32 (3): 241–254.

12 Gandini, S., Sera, F., Cattaruzza, M.S., Pasquini, P., Zanetti, R., Masini, C., Boyle, P., and Melchi, C.F. (2005). Meta-analysis of risk factors for cutaneous melanoma: III. Family history, actinic damage and phenotypic factors. *Eur. J. Cancer* 41 (14): 2040–2059.

13 MacKie, R.M., Hauschild, A., and Eggermont, A.M.M. (2009). Epidemiology of invasive cutaneous melanoma. *Ann. Oncol.* 20 (Suppl. 6): vi1–vi7.

14 Williams, L.H., Shors, A.R., Barlow, W.E., Solomon, C., and White, E. (2011). Identifying persons at highest risk of melanoma using self-assessed risk factors. *J. Clin. Exp. Dermatol. Res.* 2 (6). https://doi.org/10.4172/2155-9554.1000129.

15 Mun, G.-H. (2012). Management of malignant melanoma. *Arch. Plast. Surg.* 39 (5): 565–574.

16 Khalid, U., Saleem, T., Imam, A.M., and Khan, M.R. (2011). Pathogenesis, diagnosis and management of primary melanoma of the colon. *World J. Surg. Oncol.* 9 (1).

17 Perera, E., Gnaneswaran, N., Jennens, R., and Sinclair, R. (2013). Malignant melanoma. *Healthcare (Basel)* 2 (1): 1–19.

18 Kim, R.H. and Meehan, S.A. (2017). Immunostain use in the diagnosis of melanomas referred to a tertiary medical center: A 15-year retrospective review (2001–2015): Immunostain use in referred melanomas. *J. Cutan. Pathol.* 44 (3): 221–227.

19 Hoon, D.S., Bostick, P., Kuo, C., Okamoto, T., Wang, H.J., Elashoff, R., and Morton, D.L. (2000). Molecular markers in blood as surrogate prognostic indicators of melanoma recurrence. *Cancer Res.* 60 (8): 2253–2257.

20 Liu, Y. and Sheikh, M.S. (2014). Melanoma: Molecular pathogenesis and therapeutic management. *Mol. Cell. Pharmacol.* 6 (3): 228.

21 Gershenwald, J.E., Scolyer, R.A., Hess, K.R., Sondak, V.K., Long, G.V., Ross, M.I., Lazar, A.J., Faries, M.B., Kirkwood, J.M., McArthur, G.A., Haydu, L.E., Eggermont, A.M.M., Flaherty, K.T., Balch, C.M., and Thompson, J.F. (2017). Melanoma staging: Evidence-based changes in the American Joint Committee on Cancer, eighth edition cancer staging manual. *CA Cancer J. Clin.* 67 (6): 472–492.

22 Shah, D.J. and Dronca, R.S. (2014). Latest advances in chemotherapeutic, targeted, and immune approaches in the treatment of metastatic melanoma. *Mayo Clin. Proc.* 89 (4): 504–519.

23 Goldberg, A.D., Allis, C.D., and Bernstein, E. (2007). Epigenetics: A landscape takes shape. *Cell* 128 (4): 635–638.

24 Allis, C.D. and Jenuwein, T. (2016). The molecular hallmarks of epigenetic control. *Nat. Rev. Genet.* 17 (8): 487–500.

25 Lawrence, M., Daujat, S., and Schneider, R. (2016). Lateral thinking: How histone modifications regulate gene expression. *Trends Genet.* 32 (1): 42–56.

26 Karpf, A.R. (2013). Epigenetic alterations in oncogenesis. *Preface. Adv. Exp. Med. Biol.* 754: v–vii.

27 Ilango, S., Paital, B., Jayachandran, P., Padma, P.R., and Nirmaladevi, R. (2020). Epigenetic alterations in cancer. *Front. Biosci. (Landmark Ed.)* 25: 1058–1109.

28 Kim, M. and Costello, J. (2017). DNA methylation: An epigenetic mark of cellular memory. *Exp. Mol. Med.* 49 (4): e322–e322.

29 Howell, P.M., Jr, Liu, S., Ren, S., Behlen, C., Fodstad, O., and Riker, A.I. (2009). Epigenetics in human melanoma. *Cancer Control* 16 (3): 200–218.

30 Micevic, G., Theodosakis, N., and Bosenberg, M. (2017). Aberrant DNA methylation in melanoma: Biomarker and therapeutic opportunities. *Clin. Epigenet.* 9 (1). https://doi.org/10.1186/s13148-017-0332-8.

31 Mirmohammadsadegh, A., Marini, A., Nambiar, S., Hassan, M., Tannapfel, A., Ruzicka, T., and Hengge, U.R. (2006). Epigenetic silencing of the PTEN gene in melanoma. *Cancer Res.* 66 (13): 6546–6552.

32 Freedberg, D.E., Rigas, S.H., Russak, J., Gai, W., Kaplow, M., Osman, I., Turner, F., Randerson-Moor, J.A., Houghton, A., Busam, K., Timothy Bishop, D., Bastian, B.C., Newton-Bishop, J.A., and Polsky, D. (2008). Frequent P16-independent inactivation of P14ARF in human melanoma. *J. Natl. Cancer Inst.* 100 (11): 784–795.

33 Spugnardi, M., Tommasi, S., Dammann, R., Pfeifer, G.P., and Hoon, D.S.B. (2003). Epigenetic inactivation of RAS association domain family protein 1 (RASSF1A) in malignant cutaneous melanoma. *Cancer Res.* 63 (7): 1639–1643.

34 Jonsson, A., Tuominen, R., Grafström, E., Hansson, J., and Egyhazi, S. (2010). High frequency of P16(INK4A) promoter methylation in NRAS-mutated cutaneous melanoma. *J. Invest. Dermatol.* 130 (12): 2809–2817.

35 Tuominen, R., Jewell, R., Van Den Oord, J.J., Wolter, P., Stierner, U., Lindholm, C., Hertzman Johansson, C., Lindén, D., Johansson, H., Frostvik Stolt, M., Walker, C., Snowden, H., Newton-Bishop, J., Hansson, J., and Egyházi Brage, S. (2015). MGMT promoter methylation is associated with temozolomide response and prolonged progression-free survival in disseminated cutaneous melanoma: MGMT associated with TMZ response and PFS. *Int. J. Cancer* 136 (12): 2844–2853.

36 Tian, F., Ding, D., and Li, D. (2015). Fangchinoline targets PI3K and suppresses PI3K/AKT signaling pathway in SGC7901 cells. *Int. J. Oncol.* 46 (6): 2355–2363.

37 Lahtz, C., Stranzenbach, R., Fiedler, E., Helmbold, P., and Dammann, R.H. (2010). Methylation of PTEN as a prognostic factor in malignant melanoma of the skin. *J. Invest. Dermatol.* 130 (2): 620–622.

38 Roh, M.R., Gupta, S., Park, K.-H., Chung, K.Y., Lauss, M., Flaherty, K.T., Jönsson, G., Rha, S.Y., and Tsao, H. (2016). Promoter methylation of PTEN is a significant prognostic factor in melanoma survival. *J. Invest. Dermatol.* 136 (5): 1002–1011.

39 Sherr, C.J. (2001). The INK4a/ARF network in tumour suppression. *Nat. Rev. Mol. Cell Biol.* 2 (10): 731–737.

40 Venza, M., Visalli, M., Biondo, C., Lentini, M., Catalano, T., Teti, D., and Venza, I. (2015). Epigenetic regulation of P14ARF and P16INK4A expression in cutaneous and uveal melanoma. *Biochim. Biophys. Acta* 1849 (3): 247–256.

41 Serrano, M., Lee, H., Chin, L., Cordon-Cardo, C., Beach, D., and DePinho, R.A. (1996). Role of the INK4a locus in tumor suppression and cell mortality. *Cell* 85 (1): 27–37.

42 Hoon, D.S.B., Spugnardi, M., Kuo, C., Huang, S.K., Morton, D.L., and Taback, B. (2004). Profiling epigenetic inactivation of tumor suppressor genes in tumors and plasma from cutaneous melanoma patients. *Oncogene* 23 (22): 4014–4022.

43 Tanemura, A., Terando, A.M., Sim, M.-S., Van Hoesel, A.Q., De Maat, M.F.G., Morton, D.L., and Hoon, D.S.B. (2009). CpG island methylator phenotype predicts progression of malignant melanoma. *Clin. Cancer Res.* 15 (5): 1801–1807.

44 Gambichler, T., Sand, M., and Skrygan, M. (2013). Loss of 5-hydroxymethylcytosine and ten-eleven translocation 2 protein expression in malignant melanoma. *Melanoma Res.* 23 (3): 218–220.

45 Gustafson, C.B., Yang, C., Dickson, K.M., Shao, H., Van Booven, D., Harbour, J.W., Liu, Z.-J., and Wang, G. (2015). Epigenetic reprogramming of melanoma cells by vitamin C treatment. *Clin. Epigenet.* 7 (1): 51.

46 Lian, C.G., Xu, Y., Ceol, C., Wu, F., Larson, A., Dresser, K., Xu, W., Tan, L., Hu, Y., Zhan, Q., Lee, C.-W., Hu, D., Lian, B.Q., Kleffel, S., Yang, Y., Neiswender, J., Khorasani, A.J., Fang, R., Lezcano, C., Duncan, L.M., Scolyer, R.A., Thompson, J.F., Kakavand, H., Houvras, Y., Zon, L.I., Mihm, M.C., Jr, Kaiser, U.B., Schatton, T., Woda, B.A., Murphy, G.F., and Shi, Y.G. (2012). Loss of 5-hydroxymethylcytosine is an epigenetic hallmark of melanoma. *Cell* 150 (6): 1135–1146.

47 Lee, J.J., Granter, S.R., Laga, A.C., Saavedra, A.P., Zhan, Q., Guo, W., Xu, S., Murphy, G.F., and Lian, C.G. (2015). 5-hydroxymethylcytosine expression in metastatic melanoma versus nodal nevus in sentinel lymph node biopsies. *Mod. Pathol.* 28 (2): 218–229.

48 Rodić, N., Zampella, J., Sharma, R., Burns, K.H., and Taube, J.M. (2015). Diagnostic utility of 5-hydroxymethylcytosine immunohistochemistry in melanocytic proliferations: Diagnostic utility of 5-hydroxymethylcytosine immunohistochemistry. *J. Cutan. Pathol.* 42 (11): 807–814.

49 Chatterjee, A., Rodger, E.J., Ahn, A., Stockwell, P.A., Parry, M., Motwani, J., Gallagher, S.J., Shklovskaya, E., Tiffen, J., Eccles, M.R., and Hersey, P. (2018). Marked global DNA hypomethylation is associated with constitutive PD-L1 expression in melanoma. *iScience* 4: 312–325.

50 Bernstein, E. and Hake, S.B. (2006). The nucleosome: A little variation goes a long way. *Biochem. Cell Biol.* 84 (4): 505–517.

51 Zentner, G.E. and Henikoff, S. (2013). Regulation of nucleosome dynamics by histone modifications. *Nat. Struct. Mol. Biol.* 20 (3): 259–266.

52 Marmorstein, R. and Zhou, -M.-M. (2014). Writers and readers of histone acetylation: Structure, mechanism, and inhibition. *Cold Spring Harb. Perspect. Biol.* 6 (7): a018762.

53 Schultz, D.C., Ayyanathan, K., Negorev, D., Maul, G.G., and Rauscher, F.J., 3rd. (2002). SETDB1: A novel KAP-1-associated histone H3, lysine 9-specific methyltransferase that contributes to HP1-mediated silencing of euchromatic genes by KRAB zinc-finger proteins. *Genes Dev.* 16 (8): 919–932.

54 Ceol, C.J., Houvras, Y., Jane-Valbuena, J., Bilodeau, S., Orlando, D.A., Battisti, V., Fritsch, L., Lin, W.M., Hollmann, T.J., Ferré, F., Bourque, C., Burke, C.J., Turner, L., Uong, A., Johnson, L.A., Beroukhim, R., Mermel, C.H., Loda, M., Ait-Si-Ali, S., Garraway, L.A., Young, R.A., and Zon, L.I. (2011). The histone methyltransferase SETDB1 is recurrently amplified in melanoma and accelerates its onset. *Nature* 471 (7339): 513–517.

55 Orouji, E., Federico, A., Larribère, L., Novak, D., Lipka, D.B., Assenov, Y., Sachindra, S., Hüser, L., Granados, K., Gebhardt, C., Plass, C., Umansky, V., and Utikal, J. (2019). Histone methyltransferase SETDB1 contributes to melanoma tumorigenesis and serves as a new potential therapeutic target. *Int. J. Cancer* 145 (12): 3462–3477.

56 De Donatis, G.M., Pape, E.L., Pierron, A., Cheli, Y., Hofman, V., Hofman, P., Allegra, M., Zahaf, K., Bahadoran, P., Rocchi, S., Bertolotto, C., Ballotti, R., and Passeron, T. (2016). NF-KB2 induces senescence bypass in melanoma via a direct transcriptional activation of EZH2. *Oncogene* 35 (21): 2735–2745.

57 Zingg, D., Debbache, J., Peña-Hernández, R., Antunes, A.T., Schaefer, S.M., Cheng, P.F., Zimmerli, D., Haeusel, J., Calçada, R.R., Tuncer, E., Zhang, Y., Bossart, R., Wong, -K.-K., Basler, K., Dummer, R., Santoro, R., Levesque, M.P., and Sommer, L. (2018). EZH2-mediated primary cilium deconstruction drives metastatic melanoma formation. *Cancer Cell* 34 (1): 69–84.e14.

58 Strub, T., Ballotti, R., and Bertolotto, C. (2020). The "ART" of epigenetics in melanoma: From histone "alterations, to resistance and therapies." *Theranostics* 10 (4): 1777–1797.

59 Morgado-Pascual, J.L., Rayego-Mateos, S., Tejedor, L., Suarez-Alvarez, B., and Ruiz-Ortega, M. (2019). Bromodomain and extraterminal proteins as novel epigenetic targets for renal diseases. *Front. Pharmacol.* 10: 1315.

60 Segura, M.F., Fontanals-Cirera, B., Gaziel-Sovran, A., Guijarro, M.V., Hanniford, D., Zhang, G., González-Gomez, P., Morante, M., Jubierre, L., Zhang, W., Darvishian, F., Ohlmeyer, M., Osman, I., Zhou, -M.-M., and Hernando, E. (2013). BRD4 sustains melanoma proliferation and represents a new target for epigenetic therapy. *Cancer Res.* 73 (20): 6264–6276.

61 Gallagher, S.J., Mijatov, B., Gunatilake, D., Gowrishankar, K., Tiffen, J., James, W., Jin, L., Pupo, G., Cullinane, C., McArthur, G.A., Tummino, P.J., Rizos, H., and Hersey, P. (2014). Control of NF-KB activity in human melanoma by bromodomain and extra-terminal protein inhibitor I-BET151. *Pigment Cell Melanoma Res.* 27 (6): 1126–1137.

62 Fiziev, P., Akdemir, K.C., Miller, J.P., Keung, E.Z., Samant, N.S., Sharma, S., Natale, C.A., Terranova, C.J., Maitituoheti, M., Amin, S.B., Martinez-Ledesma, E., Dhamdhere, M., Axelrad, J.B., Shah, A., Cheng, C.S., Mahadeshwar, H., Seth, S., Barton, M.C., Protopopov, A., Tsai, K.Y., Davies, M.A., Garcia, B.A., Amit, I., Chin, L., Ernst, J., and Rai, K. (2017). Systematic epigenomic analysis reveals chromatin states associated with melanoma progression. *Cell Rep.* 19 (4): 875–889.

63 Lienlaf, M., Perez-Villarroel, P., Knox, T., Pabon, M., Sahakian, E., Powers, J., Woan, K., Lee, V., Cheng, C., Deng, F., Smalley, S., K S, M., Montecino, M., Kozikowski, A., Pinilla-Ibarz, J., Sarnaik, A., Seto, E., Weber, J., Sotomayor, E.M., and Villagra, A. (2016). Essential role of HDAC6 in the regulation of PD-L1 in melanoma. *Mol. Oncol.* 10 (5): 735–750.

64 Woan, K.V., Lienlaf, M., Perez-Villaroel, P., Lee, C., Cheng, F., Knox, T., Woods, D.M., Barrios, K., Powers, J., Sahakian, E., Wang, H.W., Canales, J., Marante, D., Smalley, K.S.M., Bergman, J., Seto, E., Kozikowski, A., Pinilla-Ibarz, J., Sarnaik, A., Celis, E., Weber, J., Sotomayor, E.M., and Villagra, A. (2015). Targeting histone deacetylase 6 mediates a dual anti-melanoma effect: Enhanced antitumor immunity and impaired cell proliferation. *Mol. Oncol.* 9 (7): 1447–1457.

65 Wilmott, J.S., Colebatch, A.J., Kakavand, H., Shang, P., Carlino, M.S., Thompson, J.F., Long, G.V., Scolyer, R.A., and Hersey, P. (2015). Expression of the class 1 histone deacetylases HDAC8 and 3 are associated with improved survival of patients with metastatic melanoma. *Mod. Pathol.* 28 (7): 884–894.

66 Emmons, M.F., Faião-Flores, F., Sharma, R., Thapa, R., Messina, J.L., Becker, J.C., Schadendorf, D., Seto, E., Sondak, V.K., Koomen, J.M., Chen, Y.A., Lau, E.K., Wan, L., Licht, J.D., and Smalley, K.S.M. (2019). HDAC8 regulates a stress response pathway in melanoma to mediate escape from BRAF inhibitor therapy. *Cancer Res.* 79 (11): 2947–2961.

67 Park, W.-Y., Hong, B.-J., Lee, J., Choi, C., and Kim, M.-Y. (2016). H3K27 demethylase JMJD3 employs the NF-KB and BMP signaling pathways to modulate the tumor microenvironment and promote melanoma progression and metastasis. *Cancer Res.* 76 (1): 161–170.

68 Liu, X., Zhang, S.-M., McGeary, M.K., Krykbaeva, I., Lai, L., Jansen, D.J., Kales, S.C., Simeonov, A., Hall, M.D., Kelly, D.P., Bosenberg, M.W., and Yan, Q. (2019). KDM5B promotes drug resistance by regulating melanoma-propagating cell subpopulations. *Mol. Cancer Ther.* 18 (3): 706–717.

69 Vogel, F.C.E., Bordag, N., Zügner, E., Trajkovic-Arsic, M., Chauvistré, H., Shannan, B., Váraljai, R., Horn, S., Magnes, C., Thomas Siveke, J., Schadendorf, D., and Roesch, A.

(2019). Targeting the H3K4 demethylase KDM5B reprograms the metabolome and phenotype of melanoma cells. *J. Invest. Dermatol.* 139 (12): 2506–2516.e10.

70 Joshi, A. and Romanowska, J. (2020). Recent advances in computational-based approaches in epigenetics studies. In Epigenetics Methods, vol. 18, ed. T. Tollefsbol, 569–590. San Diego, CA: Elsevier.

71 Peng, Y., Li, J., and Zhu, L. (2019). Cancer and non-coding RNAs. In Nutritional Epigenomics, vol. 14, ed. B.S. Ferguson, 119–132. San Diego, CA: Elsevier.

72 Latronico, A.C. and Silveira, L.F. (2019). Genetic and epigenetic control of puberty. In Encyclopedia of Endocrine Diseases, ed. I. Huhtaniemi and L. Martini, 126–136. San Diego, CA: Elsevier.

73 Duffy, H.B.D. and Roth, T.L. (2020). Epigenetics and development. In Reference Module in Neuroscience and Biobehavioral Psychology, 386–399. San Diego, CA: Elsevier.

74 Richtig, G., Ehall, B., Richtig, E., Aigelsreiter, A., Gutschner, T., and Pichler, M. (2017). Function and clinical implications of long non-coding RNAs in melanoma. *Int. J. Mol. Sci.* 18 (4): 715.

75 Hulstaert, E., Brochez, L., Volders, P.-J., Vandesompele, J., and Mestdagh, P. (2017). Long non-coding RNAs in cutaneous melanoma: Clinical perspectives. *Oncotarget* 8 (26): 43470–43480.

76 Mazar, J., Zhao, W., Khalil, A.M., Lee, B., Shelley, J., Govindarajan, S.S., Yamamoto, F., Ratnam, M., Aftab, M.N., Collins, S., Finck, B.N., Han, X., Mattick, J.S., Dinger, M.E., and Perera, R.J. (2014). The functional characterization of long noncoding RNA SPRY4-IT1 in human melanoma cells. *Oncotarget* 5 (19): 8959–8969.

77 Chen, L., Yang, H., Xiao, Y., Tang, X., Li, Y., Han, Q., Fu, J., Yang, Y., and Zhu, Y. (2016). LncRNA GAS5 is a critical regulator of metastasis phenotype of melanoma cells and inhibits tumor growth in vivo. *Onco. Targets. Ther.* 9: 4075–4087.

78 Chen, L., Yang, H., Xiao, Y., Tang, X., Li, Y., Han, Q., Fu, J., Yang, Y., and Zhu, Y. (2016). Lentiviral-mediated overexpression of long non-coding RNA GAS5 reduces invasion by mediating MMP2 expression and activity in human melanoma cells. *Int. J. Oncol.* 48 (4): 1509–1518.

79 Yu, X., Zheng, H., Tse, G., Chan, M.T.V., and Wu, W.K.K. (2018). Long non-coding RNAs in melanoma. *Cell Prolif.* 51 (4): e12457.

80 Tang, L., Zhang, W., Su, B., and Yu, B. (2013). Long noncoding RNA HOTAIR is associated with motility, invasion, and metastatic potential of metastatic melanoma. *Biomed Res. Int.* 2013: 251098.

81 Psaty, E.L., Scope, A., Halpern, A.C., and Marghoob, A.A. (2010). Defining the patient at high risk for melanoma. *Int. J. Dermatol.* 49 (4): 362–376.

82 Hodis, E., Watson, I.R., Kryukov, G.V., Arold, S.T., Imielinski, M., Theurillat, J.-P., Nickerson, E., Auclair, D., Li, L., Place, C., Dicara, D., Ramos, A.H., Lawrence, M.S., Cibulskis, K., Sivachenko, A., Voet, D., Saksena, G., Stransky, N., Onofrio, R.C., Winckler, W., Ardlie, K., Wagle, N., Wargo, J., Chong, K., Morton, D.L., Stemke-Hale, K., Chen, G., Noble, M., Meyerson, M., Ladbury, J.E., Davies, M.A., Gershenwald, J.E., Wagner, S.N., Hoon, D.S.B., Schadendorf, D., Lander, E.S., Gabriel, S.B., Getz, G., Garraway, L.A., and Chin, L. (2012). A landscape of driver mutations in melanoma. *Cell* 150 (2): 251–263.

83 Ikehata, H. and Ono, T. (2011). The mechanisms of UV mutagenesis. *J. Radiat. Res.* 52 (2): 115–125.

84 Paszkowska-Szczur, K., Scott, R.J., Serrano-Fernandez, P., Mirecka, A., Gapska, P., Górski, B., Cybulski, C., Maleszka, R., Sulikowski, M., Nagay, L., Lubinski, J., and Dębniak, T. (2013). xeroderma pigmentosumgenes and melanoma risk: XPgenes and melanoma risk. *Int. J. Cancer* 133 (5): 1094–1100.

85 Bauer, J. and Garbe, C. (2003). Acquired melanocytic nevi as risk factor for melanoma development: A comprehensive review of epidemiological data. *Pigment Cell Res.* 16 (3): 297–306.

86 Hawkes, J.E., Truong, A., and Meyer, L.J. (2016). Genetic predisposition to melanoma. *Semin. Oncol.* 43 (5): 591–597.

87 Olsen, C.M., Zens, M.S., Stukel, T.A., Sacerdote, C., Chang, Y.-M., Armstrong, B.K., Bataille, V., Berwick, M., Elwood, J.M., Holly, E.A., Kirkpatrick, C., Mack, T., Bishop, J.N., Østerlind, A., Swerdlow, A.J., Zanetti, R., Green, A.C., Karagas, M.R., and Whiteman, D.C. (2009). Nevus density and melanoma risk in women: A pooled analysis to test the divergent pathway hypothesis. *Int. J. Cancer* 124 (4): 937–944.

88 Elsevier (2010). Encyclopedia of Movement Disorders, 1st. San Diego, CA: Academic Press.

89 Schouwey, K., Delmas, V., Larue, L., Zimber-Strobl, U., Strobl, L.J., Radtke, F., and Beermann, F. (2007). Notch1 and Notch2 receptors influence progressive hair graying in a dose-dependent manner. *Dev. Dyn.* 236 (1): 282–289.

90 Osawa, M., Egawa, G., Mak, -S.-S., Moriyama, M., Freter, R., Yonetani, S., Beermann, F., and Nishikawa, S.-I. (2005). Molecular characterization of melanocyte stem cells in their niche. *Development* 132 (24): 5589–5599.

91 Anvekar, R.A., Asciolla, J.J., Missert, D.J., and Chipuk, J.E. (2011). Born to be alive: A role for the BCL-2 family in melanoma tumor cell survival, apoptosis, and treatment. *Front. Oncol.* 1 (34). https://doi.org/10.3389/fonc.2011.00034.

92 Aoki, Y., Saint-Germain, N., Gyda, M., Magner-Fink, E., Lee, Y.-H., Credidio, C., and Saint-Jeannet, J.-P. (2003). Sox10 regulates the development of neural crest-derived melanocytes in xenopus. *Dev. Biol.* 259 (1): 19–33.

93 Takeda, K., Yasumoto, K., Takada, R., Takada, S., Watanabe, K., Udono, T., Saito, H., Takahashi, K., and Shibahara, S. (2000). Induction of melanocyte-specific microphthalmia-associated transcription factor by Wnt-3a. *J. Biol. Chem.* 275 (19): 14013–14016.

94 Lang, D., Lu, M.M., Huang, L., Engleka, K.A., Zhang, M., Chu, E.Y., Lipner, S., Skoultchi, A., Millar, S.E., and Epstein, J.A. (2005). Pax3 functions at a nodal point in melanocyte stem cell differentiation. *Nature* 433 (7028): 884–887.

95 Moriyama, M., Osawa, M., Mak, -S.-S., Ohtsuka, T., Yamamoto, N., Han, H., Delmas, V., Kageyama, R., Beermann, F., Larue, L., and Nishikawa, S.-I. (2006). Notch signaling via Hes1 transcription factor maintains survival of melanoblasts and melanocyte stem cells. *J. Cell Biol.* 173 (3): 333–339.

96 Sarkar, D., Leung, E.Y., Baguley, B.C., Finlay, G.J., and Askarian-Amiri, M.E. (2015). Epigenetic regulation in human melanoma: Past and future. *Epigenetics* 10 (2): 103–121.

97 Shin, M.H., Kim, J., Lim, S.A., Kim, J., and Lee, K.-M. (2020). Current insights into combination therapies with MAPK inhibitors and immune checkpoint blockade. *Int. J. Mol. Sci.* 21 (7): 2531.

98 Wilmott, J.S., Hersey, P., Long, G.V., and Scolyer, R.A. (2015). Synergistic effects of MAPK and immune checkpoint inhibitors in melanoma: What is the best combination strategy? *Melanoma Manag.* 2 (1): 15–19.

99 Bald, T., Quast, T., Landsberg, J., Rogava, M., Glodde, N., Lopez-Ramos, D., Kohlmeyer, J., Riesenberg, S., Van Den Boorn-Konijnenberg, D., Hömig-Hölzel, C., Reuten, R., Schadow, B., Weighardt, H., Wenzel, D., Helfrich, I., Schadendorf, D., Bloch, W., Bianchi, M.E., Lugassy, C., Barnhill, R.L., Koch, M., Fleischmann, B.K., Förster, I., Kastenmüller, W., Kolanus, W., Hölzel, M., Gaffal, E., and Tüting, T. (2014). Ultraviolet-radiation-induced inflammation promotes angiotropism and metastasis in melanoma. *Nature* 507 (7490): 109–113.

100 Viros, A., Sanchez-Laorden, B., Pedersen, M., Furney, S.J., Rae, J., Hogan, K., Ejiama, S., Girotti, M.R., Cook, M., Dhomen, N., and Marais, R. (2014). Ultraviolet radiation accelerates BRAF-driven melanomagenesis by targeting TP53. *Nature* 511 (7510): 478–482.

101 Sullivan, R.J. and Flaherty, K.T. (2011). BRAF in melanoma: Pathogenesis, diagnosis, inhibition, and resistance. *J. Skin Cancer* 2011: 423239.

102 Dong, J., Phelps, R.G., Qiao, R., Yao, S., Benard, O., Ronai, Z., and Aaronson, S.A. (2003). BRAF Oncogenic Mutations Correlate with Progression Rather than Initiation of Human Melanoma. *Cancer Res.* 63 (14): 3883–3885.

103 Cohen, C., Zavala-Pompa, A., Sequeira, J.H., Shoji, M., Sexton, D.G., Cotsonis, G., Cerimele, F., Govindarajan, B., Macaron, N., and Arbiser, J.L. (2002). Mitogen-actived protein kinase activation is an early event in melanoma progression. *Clin. Cancer Res.* 8 (12): 3728–3733.

104 Jakob, J.A., Bassett, R.L., Jr, Ng, C.S., Curry, J.L., Joseph, R.W., Alvarado, G.C., Rohlfs, M.L., Richard, J., Gershenwald, J.E., Kim, K.B., Lazar, A.J., Hwu, P., and Davies, M.A. (2012). NRAS mutation status is an independent prognostic factor in metastatic melanoma. *Cancer* 118 (16): 4014–4023.

105 Giehl, K. (2005). Oncogenic Ras in tumour progression and metastasis. *Biol. Chem.* 386 (3). https://doi.org/10.1515/bc.2005.025.

106 Muñoz-Couselo, E., Adelantado, E.Z., Ortiz, C., García, J.S., and Perez-Garcia, J. (2017). NRAS-mutant melanoma: Current challenges and future prospects. *Onco. Targets. Ther.* 10: 3941–3947.

107 Whittaker, S.R., Theurillat, J.-P., Van Allen, E., Wagle, N., Hsiao, J., Cowley, G.S., Schadendorf, D., Root, D.E., and Garraway, L.A. (2013). A genome-scale RNA interference screen implicates NF1 loss in resistance to RAF inhibition. *Cancer Discov.* 3 (3): 350–362.

108 Nissan, M.H., Pratilas, C.A., Jones, A.M., Ramirez, R., Won, H., Liu, C., Tiwari, S., Kong, L., Hanrahan, A.J., Yao, Z., Merghoub, T., Ribas, A., Chapman, P.B., Yaeger, R., Taylor, B.S., Schultz, N., Berger, M.F., Rosen, N., and Solit, D.B. (2014). Loss of NF1 in cutaneous melanoma is associated with RAS activation and MEK dependence. *Cancer Res.* 74 (8): 2340–2350.

109 Krauthammer, M., Kong, Y., Bacchiocchi, A., Evans, P., Pornputtapong, N., Wu, C., McCusker, J.P., Ma, S., Cheng, E., Straub, R., Serin, M., Bosenberg, M., Ariyan, S., Narayan, D., Sznol, M., Kluger, H.M., Mane, S., Schlessinger, J., Lifton, R.P., and Halaban, R. (2015). Exome sequencing identifies recurrent mutations in NF1 and RASopathy genes in sun-exposed melanomas. *Nat. Genet.* 47 (9): 996–1002.

110 Pollock, P.M., Walker, G.J., Glendening, J.M., Que Noy, T., Bloch, N.C., Fountain, J.W., and Hayward, N.K. (2002). PTEN inactivation is rare in melanoma tumours but occurs frequently in melanoma cell lines. *Melanoma Res.* 12 (6): 565–575.

111 Davies, M.A. (2012). The role of the PI3K-AKT pathway in melanoma. *Cancer J.* 18 (2): 142–147.

112 Cabrita, R., Mitra, S., Sanna, A., Ekedahl, H., Lövgren, K., Olsson, H., Ingvar, C., Isaksson, K., Lauss, M., Carneiro, A., and Jönsson, G. (2020). The role of PTEN loss in immune escape, melanoma prognosis and therapy response. *Cancers (Basel)* 12 (3): 742.

113 Eberle, J. and Hossini, A.M. (2008). Expression and function of Bcl-2 proteins in melanoma. *Curr. Genomics* 9 (6): 409–419.

114 Mukherjee, N., Schwan, J.V., Fujita, M., Norris, D.A., and Shellman, Y.G. (2015). Alternative treatments for melanoma: Targeting BCL-2 family members to DE-Bulk and kill cancer stem cells. *J. Invest. Dermatol.* 135 (9): 2155–2161.

115 Guttà, C., Rahman, A., Aura, C., Dynoodt, P., Charles, E.M., Hirschenhahn, E., Joseph, J., Wouters, J., De Chaumont, C., Rafferty, M., Warren, M., Van Den Oord, J.J., Gallagher, W.M., and Rehm, M. (2020). Low expression of pro-apoptotic proteins Bax, Bak and Smac indicates prolonged progression-free survival in chemotherapy-treated metastatic melanoma. *Cell Death Dis* 11 (2): 124.

116 Guarino, M., Rubino, B., and Ballabio, G. (2007). The role of epithelial–mesenchymal transition in cancer pathology. *Pathology* 39 (3): 305–318.

117 Kreizenbeck, G.M., Berger, A.J., Subtil, A., Rimm, D.L., and Gould Rothberg, B.E. (2008). Prognostic significance of cadherin-based adhesion molecules in cutaneous malignant melanoma. *Cancer Epidemiol. Biomarkers Prev.* 17 (4): 949–958.

118 Hao, L., Ha, J.R., Kuzel, P., Garcia, E., and Persad, S. (2012). Cadherin switch from E- to N-cadherin in melanoma progression is regulated by the PI3K/PTEN pathway through twist and snail: Cadherin switch is regulated by PI3K/PTEN Pathway. *Br. J. Dermatol.* 166 (6): 1184–1197.

119 Yan, S., Holderness, B.M., Li, Z., Seidel, G.D., Gui, J., Fisher, J.L., and Ernstoff, M.S. (2016). Epithelial–mesenchymal expression phenotype of primary melanoma and matched metastases and relationship with overall survival. *Anticancer Res.* 36 (12): 6449–6456.

120 Liu, L., Dilworth, D., Gao, L., Monzon, J., Summers, A., Lassam, N., and Hogg, D. (1999). Mutation of the CDKN2A 5′ UTR creates an aberrant initiation codon and predisposes to melanoma. *Nat. Genet.* 21 (1): 128–132.

121 FitzGerald, M.G., Harkin, D.P., Silva-Arrieta, S., MacDonald, D.J., Lucchina, L.C., Unsal, H., O'Neill, E., Koh, J., Finkelstein, D.M., Isselbacher, K.J., Sober, A.J., and Haber, D.A. (1996). Prevalence of germ-line mutations in P16, P19ARF, and CDK4 in familial melanoma: Analysis of a clinic-based population. *Proc. Natl. Acad. Sci. USA* 93 (16): 8541–8545.

122 Potrony, M., Badenas, C., Aguilera, P., Puig-Butille, J.A., Carrera, C., Malvehy, J., and Puig, S. (2015). Update in genetic susceptibility in melanoma. *Ann. Transl. Med.* 3 (15): 210.

123 Zeng, H., Jorapur, A., Shain, A.H., Lang, U.E., Torres, R., Zhang, Y., McNeal, A.S., Botton, T., Lin, J., Donne, M., Bastian, I.N., Yu, R., North, J.P., Pincus, L., Ruben, B.S., Joseph, N.M., Yeh, I., Bastian, B.C., and Judson, R.L. (2018). Bi-allelic loss of CDKN2A initiates melanoma invasion via BRN2 activation. *Cancer Cell* 34 (1): 56–68.e9.

124 Beaulieu, D., Fathi, R., Srivastava, D., and Nijhawan, R. (2018). Current perspectives on Mohs micrographic surgery for melanoma. *Clin. Cosmet. Investig. Dermatol.* 11: 309–320.

125 Nakamura, Y. (2019). The role and necessity of sentinel lymph node biopsy for invasive melanoma. *Front. Med. (Lausanne)* 6: 231.

126 Coventry, B.J. (2019). Therapeutic Vaccination Immunomodulation: Forming the basis of all cancer immunotherapy. *Ther. Adv. Vaccines Immunother.* 7: 251513551986223.

127 Kibbi, N., Ariyan, S., Faries, M., and Choi, J.N. (2015). Treatment of in-transit melanoma with intralesional bacillus Calmette-Guérin (BCG) and topical imiquimod 5% cream: A report of 3 cases. *J. Immunother.* 38 (9): 371–375.

128 Tyrell, R., Antia, C., Stanley, S., and Deutsch, G.B. (2017). Surgical resection of metastatic melanoma in the era of immunotherapy and targeted therapy. *Melanoma Manag.* 4 (1): 61–68.

129 Berraondo, P. (2019). Mechanisms of action for different checkpoint inhibitors. *HemaSphere* 3 (S2): 28–30.

130 Nicholas, C. and Lesinski, G.B. (2011). Immunomodulatory cytokines as therapeutic agents for melanoma. *Immunotherapy* 3 (5): 673–690.

131 Kremenovic, M., Schenk, M., and Lee, D.J. (2020). Clinical and Molecular Insights into BCG immunotherapy for melanoma. *J. Intern. Med.* 288 (6): 625–640.

132 Grigg, C., Blake, Z., Gartrell, R., Sacher, A., Taback, B., and Saenger, Y. (2016). Talimogene laherparepvec (T-Vec) for the treatment of melanoma and other cancers. *Semin. Oncol.* 43 (6): 638–646.

133 Bubna, A.K. (2015). Imiquimod: Its role in the treatment of cutaneous malignancies. *Indian J. Pharmacol.* 47 (4): 354–359.

134 Dossett, L.A., Kudchadkar, R.R., and Zager, J.S. (2015). BRAF and MEK inhibition in melanoma. *Expert Opin. Drug Saf.* 14 (4): 559–570.

135 Broman, K.K., Dossett, L.A., Sun, J., Eroglu, Z., and Zager, J.S. (2019). Update on BRAF and MEK inhibition for treatment of melanoma in metastatic, unresectable, and adjuvant settings. *Expert Opin. Drug Saf.* 18 (5): 381–392.

136 Robert, C., Grob, J.J., Stroyakovskiy, D., Karaszewska, B., Hauschild, A., Levchenko, E., Chiarion Sileni, V., Schachter, J., Garbe, C., Bondarenko, I., Gogas, H., Mandalá, M., Haanen, J.B.A.G., Lebbé, C., Mackiewicz, A., Rutkowski, P., Nathan, P.D., Ribas, A., Davies, M.A., Flaherty, K.T., Burgess, P., Tan, M., Gasal, E., Voi, M., Schadendorf, D., and Long, G.V. (2019). Five-year outcomes with dabrafenib plus trametinib in metastatic melanoma. *N. Engl. J. Med.* 381 (7): 626–636.

137 Tzen, C.-Y., Wu, Y.-H., and Tzen, C.-Y. (2014). Characterization of KIT mutation in melanoma. *Dermatol. Sin.* 32 (1): 7–12.

138 Wei, X., Mao, L., Chi, Z., Sheng, X., Cui, C., Kong, Y., Dai, J., Wang, X., Li, S., Tang, B., Lian, B., Yan, X., Bai, X., Zhou, L., Guo, J., and Si, L. (2019). Efficacy evaluation of imatinib for the treatment of melanoma: Evidence from a retrospective study. *Oncol. Res.* 27 (4): 495–501.

139 Hadi, I., Roengvoraphoj, O., Bodensohn, R., Hofmaier, J., Niyazi, M., Belka, C., and Nachbichler, S.B. (2020). Stereotactic radiosurgery combined with targeted/ immunotherapy in patients with melanoma brain metastasis. *Radiat. Oncol.* 15 (1): 37.

140 Luke, J.J. and Schwartz, G.K. (2013). Chemotherapy in the management of advanced cutaneous malignant melanoma. *Clin. Dermatol.* 31 (3): 290–297.

141 Giles, M.H. and Coventry, B.J. (2013). Isolated limb infusion chemotherapy for melanoma: An overview of early experience at the Adelaide melanoma unit. *Cancer Manag. Res.* 5: 243–249.

142 Martinez-Cardús, A., Vizoso, M., Moran, S., and Manzano, J.L. (2015). Epigenetic mechanisms involved in melanoma pathogenesis and chemoresistance. *Ann. Transl. Med.* 3 (15): 209.

143 Patton, E.E., Widlund, H.R., Kutok, J.L., Kopani, K.R., Amatruda, J.F., Murphey, R.D., Berghmans, S., Mayhall, E.A., Traver, D., Fletcher, C.D.M., Aster, J.C., Granter, S.R., Look, A.T., Lee, C., Fisher, D.E., and Zon, L.I. (2005). BRAF mutations are sufficient to promote nevi formation and cooperate with P53 in the genesis of melanoma. *Curr. Biol.* 15 (3): 249–254.

144 Hammerlindl, H. and Schaider, H. (2018). Epigenetics in melanoma development and drug resistance. In: Human Skin Cancers: Pathways, Mechanisms, Targets and Treatments, ed. M. Blumenberg, n.p. London: InTech.

145 Menon, D.R., Das, S., Krepler, C., Vultur, A., Rinner, B., Schauer, S., Kashofer, K., Wagner, K., Zhang, G., Rad, E.B., Haass, N.K., Soyer, H.P., Gabrielli, B., Somasundaram, R., Hoefler, G., Herlyn, M., and Schaider, H. (2015). A stress-induced early innate response causes multidrug tolerance in melanoma. *Oncogene* 34 (34): 4545.

146 Quagliano, A., Gopalakrishnapillai, A., and Barwe, S.P. (2020). Understanding the mechanisms by which epigenetic modifiers avert therapy resistance in cancer. *Front. Oncol.* 10: 992.

147 Yang, J., Nie, J., Ma, X., Wei, Y., Peng, Y., and Wei, X. (2019). Targeting PI3K in cancer: Mechanisms and advances in clinical trials. *Mol. Cancer* 18 (1).

148 Marengo, B., Garbarino, O., Speciale, A., Monteleone, L., Traverso, N., and Domenicotti, C. (2019). MYC expression and metabolic redox changes in cancer cells: A synergy able to induce chemoresistance. *Oxid. Med. Cell. Longev.* doi: 10.1155/2019/7346492.

149 Desbats, M.A., Giacomini, I., Prayer-Galetti, T., and Montopoli, M. (2020). Metabolic plasticity in chemotherapy resistance. *Front. Oncol.* 10.

150 Chang, F., Lee, J.T., Navolanic, P.M., Steelman, L.S., Shelton, J.G., Blalock, W.L., Franklin, R.A., and McCubrey, J.A. (2003). Involvement of PI3K/Akt pathway in cell cycle progression, apoptosis, and neoplastic transformation: A target for cancer chemotherapy. *Leukemia* 17 (3): 590–603.

151 Gonzalez, H., Hagerling, C., and Werb, Z. (2018). Roles of the immune system in cancer: From tumor initiation to metastatic progression. *Genes Dev.* 32 (19–20): 1267–1284.

152 Chokhachi Baradaran, P., Kozovska, Z., Furdova, A., and Smolkova, B. (2020). Targeting epigenetic modifications in uveal melanoma. *Int. J. Mol. Sci.* 21 (15).

153 Conway, K., Edmiston, S.N., Khondker, Z.S., Groben, P.A., Zhou, X., Chu, H., Kuan, P.F., Hao, H., Carson, C., Berwick, M., Olilla, D.W., and Thomas, N.E. (2011). DNA-methylation profiling distinguishes malignant melanomas from benign nevi: Methylation profiling of melanomas and benign nevi. *Pigment Cell Melanoma Res.* 24 (2): 352–360.

154 Gao, L., Van Den Hurk, K., Moerkerk, P.T.M., Goeman, J.J., Beck, S., Gruis, N.A., Van Den Oord, J.J., Winnepenninckx, V.J., Van Engeland, M., and Van Doorn, R. (2014). Promoter CpG island hypermethylation in dysplastic nevus and melanoma: CLDN11 as an epigenetic biomarker for malignancy. *J. Invest. Dermatol.* 134 (12): 2957–2966.

155 Bucheit, A.D., Chen, G., Siroy, A., Tetzlaff, M., Broaddus, R., Milton, D., Fox, P., Bassett, R., Hwu, P., Gershenwald, J.E., Lazar, A.J., and Davies, M.A. (2014). Complete loss of PTEN protein expression correlates with shorter time to brain metastasis and survival in stage IIIB/C melanoma patients with BRAFV600 mutations. *Clin. Cancer Res.* 20 (21): 5527–5536.

156 Wouters, J., Vizoso, M., Martinez-Cardus, A., Carmona, F.J., Govaere, O., Laguna, T., Joseph, J., Dynoodt, P., Aura, C., Foth, M., Cloots, R., Van Den Hurk, K., Balint, B., Murphy, I.G., McDermott, E.W., Sheahan, K., Jirström, K., Nodin, B., Mallya-Udupi, G., Van Den Oord, J.J., Gallagher, W.M., and Esteller, M. (2017). Comprehensive DNA methylation study identifies novel progression-related and prognostic markers for cutaneous melanoma. *BMC Med.* 15 (1): 101.

15

Currents Trends and Future Perspectives in Our Epigenetic Signatures

What a Diet Can Trigger

Elena-Codruța Dobrică[1,2], Mihnea-Alexandru Găman[3,4], Matei-Alexandru Cozma[5], and Sanda Maria Cretoiu[6]

[1] Department of Pathophysiology, University of Medicine and Pharmacy of Craiova, Craiova, Romania
[2] Department of Dermatology, "Elias" University Emergency Hospital, 011461, Bucharest, Romania
[3] Faculty of Medicine, "Carol Davila" University of Medicine and Pharmacy, Bucharest, Romania
[4] Department of Hematology, Center of Hematology and Bone Marrow Transplantation, Fundeni Clinical Institute, Bucharest, Romania
[5] Department of Gastroenterology, Colentina Clinical Hospital, Bucharest, Romania
[6] Department of Cell and Molecular Biology and Histology, Carol Davila University of Medicine and Pharmacy, Bucharest, Romania

How Epigenetics Influences Life: Short Overview

Epigenetics is defined as the totality of changes in gene expression, but not in the succession of nucleobases from the structure of DNA that can be transmitted from one cell to another during cell division, or even from one generation to another [1]. The first mention of the term "epigenetics" was recorded in the late 1940s by Conrad Waddington, but attempts to demonstrate this concept in the scientific world failed [2]. Almost fifteen years later, Waddington manages to show that changes induced by environmental factors in the phenotype of *Drosophila melanogaster* can be passed from one generation to another, confirming his previously noted ideas [3]. At the heart of the concept of epigenetics was one of the important statements made by the pioneer of phylogenetics, Valentin Haecker, who laid the foundation for theories that try to explain how the genotype leads to the phenotype. Waddington clearly mentioned the importance of Haecker's 1918 paper *Phänogenetik*, in his own reference work [4,5]. Half a century later, in 1975, Robin Holliday and his team took an important step in advancing the evolution and understanding of epigenetics. They proposed the existence of methyltransferases to explain how the gene switch is inherited in order to preserve different phenotypes of cells in different tissues, under the conditions of an identical DNA [6,7].

From a linguistic and conceptual point of view, the term "epigenetics" comes from the composition of two other nouns, "epigenesis" and "genetics"; the first emphasizes the process by which living organisms differentiate and develop gradually, starting from a basic undifferentiated structure, while the second highlights the heritable nature of the process, through the ability to transmit features from one generation to another [8]. In modern times, one of the first researchers to promote the term "epigenesis" was William Harvey in the mid-seventeenth century. A century later, the embryologist Caspar Friedrich Wolff advanced the concept of epigenesis in a theory that rebutted the idea of preformation, very popular at that time. After two more centuries, in the early 1930, the experiments

Genomic and Epigenomic Biomarkers of Toxicology and Disease: Clinical and Therapeutic Actions, First Edition. Edited by Saura C. Sahu.

performed by Conrad Hall Waddington juxtaposed the two notions of epigenesis and genetics, giving rise to a third: epigenetics [9]. Although the initial definition of epigenetics as "the branch of biology which studies the causal interactions between genes and their products which bring the phenotype into being" is still correct, over the past two centuries the definitions have progressively narrowed [10].

The epigenetic changes induced by environmental factors are hardly reversible; they are durable changes. This explains their persistence from one division to another, and even from one generation to another, in contrast to transcriptional control [11].

The epigenetic mechanisms are multiple, the most important ones being changes in the DNA through processes of methylation, phosphorylation, acetylation, and so on, changes in histones and in the regulation of gene expression through the action of non-coding RNA. All these processes lead to chromatin remodeling, with changes in gene expression but without changes in the nucleotide sequence [11,12].

Speaking of heredity, one cannot overlook the fact that what constitutes it is the sum of all the factors that surround the human body and that, apart from heritable characters, a lot of epigenetic variations make us who we are. Diet, including malnutrition, hypernutrition, and special micronutrient deficiencies, is one of the most important factors directly involved in epigenetic modifications, although it is frequently played down, overlooked, or ignored when discussing the origin of the diseases. Figure 15.1 synthesizes the main links between nutrition and epigenetics and how they are reflected in one's phenotype.

DNA methylation is one of the most studied epigenetic phenomena. It begins in the embryo, being involved in important processes in both intrauterine and postnatal life. Notable among these processes are the implantation and development of the embryo, the remodeling of the chromatin structure, the inactivation of the X chromosome, the silencing of genes, the modulation of regulatory T lymphocytes, and the initiation of carcinogenesis [13]. Methylation occurs under the influence of enzymes called DNA methyltransferases (DMT) (1, 2, 3A, 3B, 3 C, 3 L), which cause the transfer of a methyl group from the amino acid S-adenosyl-L-methionine to position 5 of the nucleobase cytosine [14]. The main functions and characteristics of DMTs are presented in Table 15.1.

DNA methylation is one of the most stable epigenetic changes, which can be transmitted from the mother cell to the two daughter cells during cell division, but also from one generation to the next; these changes are specific ones, characteristic of a certain type of tissue and responsible for activating certain genes and inactivating others, in each cell, depending on the functions it has to perform [15].

Multiple studies show the influence of nutritional factors in changing the methylation pattern, the restrictive diet in obese patients being associated with alterations of many methylation sites located in genetic regions with an important role in carcinogenesis [20].

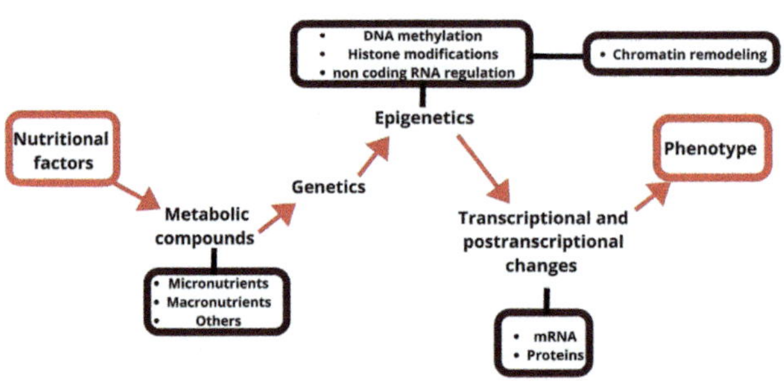

Figure 15.1 From nutritional factors to phenotype. The mechanism through which the diet influences the phenotype and the main epigenetic mechanisms involved. *Source:* Elena-Codruta Dobrica.

Table 15.1 DMTs subtypes and their function.

DMT subtype	Function	Characteristics	Reference
DMT1	Global DNA methylation	DNA that is hemimethylated	[15]
DMT 3A, DMT 3B	De novo DNA methylation	DNA that is unmethylated	[16]
DMT 3 C	DNA methylation	Young transposons in male germ cells	[17]
DMT3L	DNA-methyltransferase-like protein	Control the activity of DMT3A	[18]
DMT2	RNA methyltransferase	Does not methylate DNA in mammals	[19]

Another type of intensely studied epigenetic process is that of histone changes. These are octamer structures made up of four types of protein chains (H2A, H2B, H3 and H4), and they have an essential role in DNA packaging [21]. Histone changes are classified into acetylation, methylation, phosphorylation, and ubiquitination processes, which take place at the amino acid level at the N-terminal sites. These changes affect the way chromatin is packaged, leading to a different availability of DNA fragments to be transcribed. They thus control the activity of genetic expression in both directions, that is, though activation and through suppression [22,23].

Histone changes have been associated with carcinogenesis and aging; they impact life expectancy in both directions, depending on the type of histone at which acetylation, methylation, phosphorylation, or ubiquitination takes place and on the amino acids involved [24,25]. Moreover, it has been observed that these changes are influenced by diet, either through caloric restrictions that determine changes in the H3 and H4 chains or through the consumption of foods with an inhibitory role of histone deacetylases (grapes, broccoli, soybeans, etc.) [23,26]. The diagram in Figure 15.2 summarizes the main environmental factors that determine epigenetic changes through histone modification, DNA methylation, and non-coding RNA.

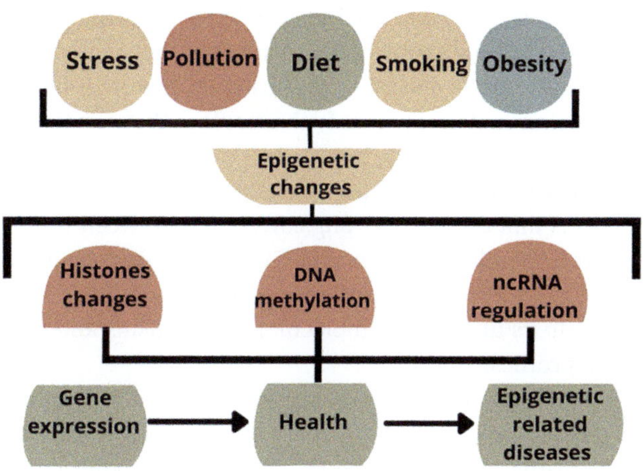

Figure 15.2 The complex interrelationship between environmental factors, epigenetics, and chronic diseases. Lifelong interactions with various environmental factors lead to epigenetic changes such as DNA methylation, histone changes, and non-coding RNA with a role in altering gene expression and in the initiation and progression of cancer, cardiovascular disease, diabetes, neurological diseases, and so on. *Source*: Elena-Codruta Dobrica.

Special attention has been paid in recent years to non-coding RNA as one of the essential epigenetic processes. Initially non-coding RNA was considered an intermediate molecule in the transformation of gene structures from DNA into effector proteins, but many of its functions have now been discovered and found to play a role in the direct and indirect regulation of gene expression [27]. Non-coding RNA is classified according to the number of nucleotide base pairs involved in long-chain non-coding RNA (lncRNA) (over 200 nucleotide base pairs) and in short-chain non-coding RNA (fewer than 200 nucleotide base pairs) [28]. Attention has recently shifted from the study of microRNA (miRNA) to the study of lncRNA, which is thought to play a key role in the development of many chronic diseases such as cancer and cardiovascular disease and in neurological diseases such as Alzheimer's and amyotrophic lateral sclerosis [29]. Moreover, lncRNA expression is influenced by nutritional factors, having different patterns of expression in the obese population, in those suffering from type 2 diabetes, and in those who have impaired insulin secretion [30]. LncRNA also plays a direct role in regulating carbohydrate and lipid metabolism through its roles in pancreatic beta cell development and insulin secretion, in the repression of certain liver enzymes (liver glucokinase), and in the regulation of metabolic pathways involved in the synthesis and transport of cholesterol and triglycerides [31-33].

Although epigenetic processes are essential for adaptation to external changes and for evolution, recent studies show the involvement of these processes in the initiation, proliferation, and progression of cancers. This aspect opens new perspectives on the study of epigenetic molecular mechanisms and on how they can be used in understanding carcinogenesis and other chronic diseases [34].

Diet-induced Epigenetic Alterations: Transgenerational Effects and Early Life Nutrition

In the past two decades we have witnessed a worldwide increase in human longevity, up to an average age of eighty in the developed countries [35]. Along with this higher life expectancy, one also observes an increase in chronic diseases and their impact on the quality of life. We are witnessing an obesity and type 2 diabetes pandemic, has resulted from significant changes in lifestyle, especially in relation to diet, physical activity, and sleep [36]. In addition, some other pathologies included in the category of non-communicable diseases (NCD) are on the rise. Cardiovascular diseases (17.9 million people a year), cancers (9.3 million a year) chronic respiratory diseases (4.1 million a year) and diabetes (1.5 million a year) burden our society with high socioeconomic costs and increased morbidity and mortality: each year 41 million people die from these pathologies, a number that represents 71% of worldwide deaths [37]. All these metabolic diseases are caused by genetic inheritance only to a small degree, while increasing evidence points to the fact that they represent an epigenetically mediated effect.

Epigenetic mechanisms at work during prenatal and early postnatal life are attracting greater interest, as it is now quite obvious that there is a strong correlation between nutritional and environmental factors and the spiralling incidence of NCDs that affect modern society (Figure 15.3); and this correlation is relevant. It is certain that prognostic epigenetic analyses will be used in the future to help us understand molecular mechanisms and disease-specific DNA methylation pattern profiles.

At the beginning of the 1980s several studies pointed out that there might be a close relationship between living conditions during childhood, nutritional status, height, weight, and mortality rate [38-40]. In 1990, the British epidemiologist David Barker introduced the

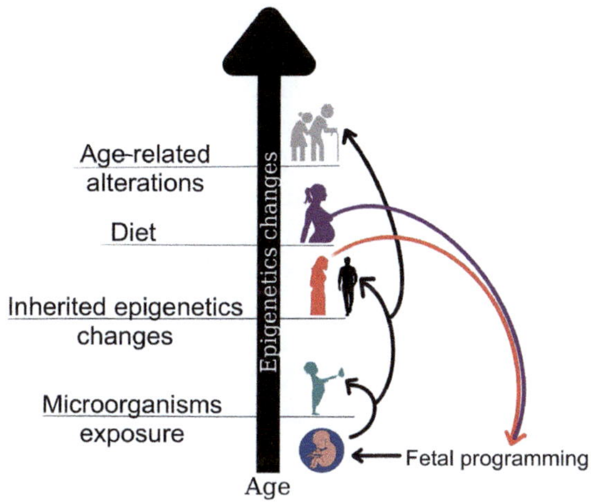

Figure 15.3 The dynamics of epigenetics over generations. Epigenetic changes increase as life progresses, bringing on exposure to many key factors. Changes during crucial stages of life (e.g. before conception and during pregnancy) will have effects on the fetus by contributing to pre- and postnatal programming. These changes will influence the development of the fetus into a baby and all the transitions thereafter (child, youth, adult, older person). *Source*: Elena-Codruta Dobrica.

idea that "the womb may be more important than the home," which is the leading hypothesis in the theory of prenatal programming and points out that factors such as maternal environment, placental and birth weight, and the composition of infant food are all responsible for long-term health outcomes; many diseases can stem from them [41]. The theory was further developed by Baker's team into what is known today as "the thrifty phenotype hypothesis," to demonstrate that the temporary or permanent malfunctioning of an organism is due to factors that are active during a specific time frame in the fetus' life [42]. Lately the developmental origins of health and disease (DOHaD) hypothesis generated the concept of developmental programming, according to which early adaptive processes take place at the very beginning of our existence to adjust our homeostatic systems to the environment; these processes shape the predictive adaptive response necessary to improve our chances of survival [43]. But they are also prone to creating a disadvantage for the offspring in the form of maladaptations, if the offspring lands in an unexpected or adverse postnatal environment [44].

On this view, low-weight newborn babies are at risk of developing some metabolic syndrome (hypertension, diabetes, and dyslipidemia), often in correlation with maternal malnutrition during pregnancy. The scientific community criticized this kind of approach; however, the development of molecular techniques and the existence of a genome-wide association platform showed that only a small percentage (10%) of cases of type 2 diabetes is of genetic origin [45].

Even if the primary etiological factors in diabetes remain unexplained, there is no doubt that early life determinants occupy a central position among them [46]. Moreover, the concept of "metabolic memory" or "legacy effect" has been developed for diabetic and cardiovascular patients [47,48]. Chronic maternal malnutrition results in an epigenetic programming of the fetus towards developing resource-saving mechanisms in order to survive. Studies conducted on children born during and after major periods of famine—such as the Great Chinese Famine, the Austrian hunger crisis, and the Dutch famine—revealed that these subjects were at high risk of metabolic syndrome [49-51]. After sixty years, the individuals exposed to the Dutch famine showed lower levels of methylation in their DNA of the IGF2 gene than their siblings of same sex who had not been subjected to the famine. This result supports the idea that epigenetic changes are acquired very early on and persist throughout life [52]. Aside from undernutrition, several other maternal factors—such as low-protein and high-fat diets—should be taken into account.

Epigenetic alterations occur in relation to environmental factors, parents' dietary patterns playing a significant role (Figure 15.4).

Epigenetic modifications are transmissible from parent to offspring, either intergenerationally or transgenerationally. The term "intergenerational epigenetic inheritance" refers to the impact of *in utero* exposure on the developing embryo and fetus and its germline, while the term "transgenerational inheritance" refers to the inheritance transmitted from one generation to another. Nowadays it has become increasingly evident that lifestyle nutritional modifications can affect epigenetic processes in the development of the fetus, because the genome is unmethylated in the blastocyst.

So a maternal diet that is either restricted or excessive can provide methyl donors, for example folate, methionine, or choline, which influence the DNA methylation patterns [53,54]. The major pathways involving folate and methionine metabolism are shown in Figure 15.5.

Maternal hyperglycemia and amino acid depletion are considered to be metabolic stressors for the embryonic stem cell lineages, which will become adapted to stabilizing the growth of the conceptus and to ensuring postnatal adaptability [55,56].

Interesting animal studies performed in recent years come to support Barker's hypothesis. In a study on female mice exposed to cigarette smoke as a factor of fetal underdevelopment, Huang et al. showed that, if the offspring are also subjected to a postnatal high-fat diet (HFD), they will develop increased adiposity and various metabolic disorders. Moreover,

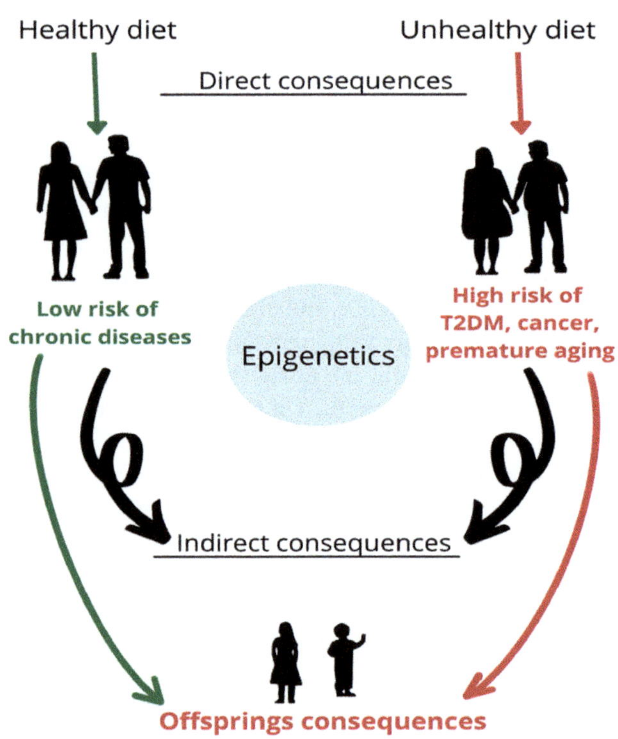

Figure 15.4 The direct and indirect consequences of nutrition. Parental exposure to inappropriate nutritional factors has not only direct effects (i.e. effects on them, in the form of obesity, cardiovascular disease, or type 2 diabetes) but also indirect effects (i.e. effects on the fetus, in the form of an increased risk of chronic diseases in adulthood). T2DM: Type 2 Diabetes mellitus. *Source*: Elena-Codruta Dobrica.

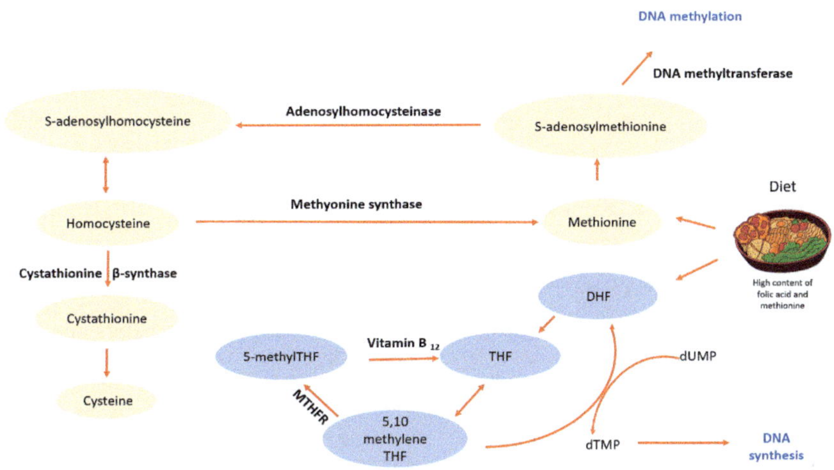

Figure 15.5 The role of diet in folic acid and methionine cycles. Tetrahydrofolate can be obtained by two mechanisms: conversion of dihydrofolate to tetrahydrofolate, a reaction catalyzed by dihydrofolate reductase (DHFR); and conversion of 5-methyltetrahydrofolate from liver stores to tetrahydrofolate, with the aid of vitamin B12. THF = tetrahydrofolate; DHF = dihydrofolate; dTMP = deoxythymidine monophosphate; dUMP = 2′-deoxyuridine-5′-monophosphate; MTHFR = methylenetetrahydrofolate reductase 5, 10-methylene-THF reductase. *Source*: Sanda-Maria Cretoiu.

the male offspring from mothers simultaneously exposed to cigarette smoke and a HFD will develop exaggerated metabolic disorders on account of a disruption of metabolic regulators [57]. The adipocytokine, adiponectin, and leptin genes are downregulated through DNA hypermethylation in murine models of offspring exposed to HFD *in utero* [58]. These epigenetic changes lead to the development of a metabolic syndrome-like condition and are transmitted for three generations [59].

Vanhees et al. (2014) showed that the future of the offspring can also be influenced by HFDs and maternal low consumption of protein [60]. Zhang et al. demonstrated that male offspring of female mice exposed to a HFD four weeks before mating and throughout gestation and lactation developed hyperglycemia and insulin resistance, especially when the pups were also fed with a HFD [61]. Maternal overnutrition impacted the fetal liver and produced *Irs2* and *Map2k4* gene methylation as long-term modifications, which predisposed the offspring to diabetes [61]. Another recent study focusing on the intergenerational effect of a HFD used wild-type C57BL6/J mating pairs, fed with high saturated fats from the period before conception through lactation. Their results were sex-specific and consistent with those of previous studies. Female offspring developed obesity under the influence of the maternal HFD, which amplified the effect of the HFD in adult life, while males were affected only by their own HFD in adulthood [62].

The future reproductive potential of the offspring can also be affected by a HFD during gestation (Figure 15.6), through a transgenerational effect.

Tsoulis et al. studied the rat offspring of mothers exposed to a HFD during gestation and lactation and demonstrated that fetuses develop significantly fewer oocytes during pregnancy and later in life suffer increased follicular atresia [63]. Aiken et al. demonstrated that “developmental programming effects on the female reproductive tract persist in a second-generation of offspring following a grand-maternal nutritional stimulus” (they were referring to the ovarian reserve influenced by low-protein diet in pregnancy) [64]. Another

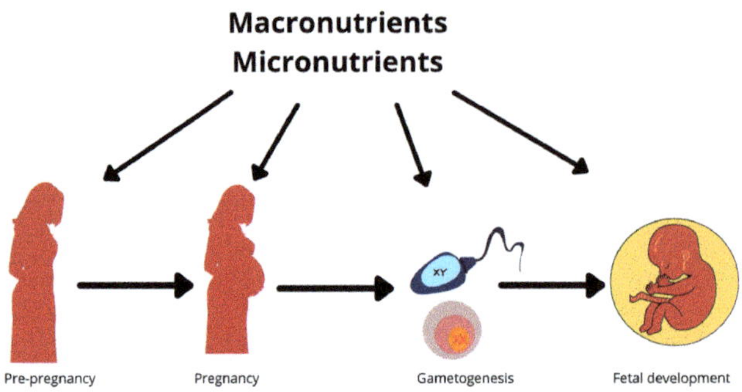

Figure 15.6 The transgenerational effect of nutrients. The mother's exposure, both before and during pregnancy, to micronutrients and macronutrients causes epigenetic changes that will be transmitted over generations, with subsequent influences in adulthood. Parental dietary intake will impact three generations: (1) the parental generation, represented by the mother responsible for the intake of nutrients; (2) the first generation, represented by the fetus exposed to nutrients during its intrauterine development; and (3) the second generation, represented by the fetal gametes exposed to nutrients during gametogenesis. *Source*: Elena-Codruta Dobrica and Sanda-Maria Cretoiu.

study showed that a low-protein diet during preconception, pregnancy, and lactation significantly depleted primordial follicles in rat offspring and affected their follicular health later in life [65]

Maternal micronutrient concentrations are also very important in the regulation of DNA methylation, one good example being vitamin B12 concentrations. An elevated level of B12 during pregnancy is accompanied by a decrease in global DNA methylation in the fetus, while high concentrations of B12 in the plasma of newborns are associated with a decreased methylation of IGFBP3, which is responsible for intrauterine growth [66]. The offspring can develop increased adiposity and insulin resistance if a methyl-deficient maternal diet was adopted during the periconceptional period; moreover, if folic acid and methionine were also deficient, the offspring will develop a dysregulation of the immune function and high blood pressure [67]. Maternal prenatal folic acid supplementation exacerbated the adverse effects of a HFD in adulthood, on lipid metabolism in offspring [68]. Folic acid supplementation may have a different impact, depending on the gestational age: during early gestation, it is followed by wheezing; in late pregnancy, it is followed by asthma in the offspring [69]. Another study showed that prenatal or early folic acid supplementation prevented wheezing in the offspring of atopic women [70]. A systematic review and meta-analysis performed by Criedr et al. found no evidence of a correlation between the periconceptional administration of folic acid and the development of asthma in offspring [70].

Flavonoids which are important antioxidants and are represented by genistein in soybeans and by quercetin in onions—as well as by other antioxidants such as lycopene in tomatoes, resveratrol in grapes and berries, sulforaphane in broccoli, epigallocatechin-3-gallate (EGCG) in green tea, or curcumin and allyl sulfur compounds in spices—are less studied for their effect on the offspring, although they are present in many dietary supplements on the market [71,72]. Tea and coffee polyphenols consumed by mothers can inhibit DNMT1 (DNA cytosine-5-methyltransferase 1)-catalyzed DNA methylation [73].

Quercetin was shown to play an important role in the liver of mice fetuses pre-exposed to it: it upregulated Nrf2 (nuclear factor erythroid 2-related factor 2), an important regulator of cellular resistance to oxidants. Thus, offspring exposed to low concentrations of pro-oxidants will not experience cell damage. Moreover, fetal exposure to genistein is followed by an increased gene expression of Nrf2. Simultaneous exposure to quercetin and genistein will protect the liver of adult mice [74].

The role of iron in fetal programming should not be neglected, iron being responsible for oxidative stress and the formation of ROS, and also being frequently prescribed for the prophylaxis and treatment of anemia during pregnancy. Excess of heme iron intake from diet has been shown to be associated with gestational diabetes, as it appears from a prospective cohort study of 3,158 pregnant women [75]. Iron and zinc were shown to interact with folate-mediated one-carbon metabolism (FOCM) and DNA methyltransferases, influencing DNA methylation [76]. Table 15.2 lists a few macro-, micro-, and phytonutrients with their epigenetic interference, as it is presented in the current literature, without being exhaustive.

Paternal genetic and epigenetic information is transmitted to the offspring through sperm. Different factors such as smoking, social stress, or the father's diet cause the transmission of

Table 15.2 Epigenetic interference of food and nutrients in DNA methylation, histone modification, and miRNAs.

	Epigenetic interference			**References**
Food/*Nutrient*	**Histone modification**	**DNA methyltransferase pathway**	**MiRNAs**	
Beef liver, eggs, fish/*Choline*	*	*	*	[77,78]
Seeds, nuts, animal fats, and fish oils/*Fatty acids*		*		[79,80]
Wheat bran, wheat bread spinach, shrimps/*Betaine*	*	*	*	[81,82]
Honey, agave, molasses, dried fruit, fruits, fruit juices, and sweet corn/*High-glucose*	*	*		[83,84]
Dairy, cold rolled oats, legumes, cooled potatoes, cooled rice, underripe bananas/ *Butyrate source*	*	*		[85,86]
Vegetables, cereales, yeast/*low in folate and methionine*	*	*	*	[87,88]
Fruits and vegetables/*Vitamin C*		*		[89–91]
Orange and yellow fruits and vegetables/*Vitamin E*	*	*	*	[92,93]
Grape, blueberries, cranberries, peanuts, pistachios/*Resveratrol*	*	*		[94,95]
Tomatoes, papaya, watermelon/*Lycopene*	*			[96]
Soya/*Genistein*	*	*	*	[97,98]
Broccoli/*Sulforphane*	*	*		[99]
Broccoli/*Isothyociante*	*	*	*	[100,101]
Green tea/*Epigallocatechin-3-gallate*	*	*		[102,103]
Curcuma/*Curcumin*	*	*	*	[104]
Citrus, onion, apple, berries/*Quercetin*	*	*		[105–107]

paternally acquired phenotypes via the RNA present in sperm (for detailed reviews, see [108,109]). Paternal obesity seems to affect the status of methylation in the offspring, especially for the developing sperm [110]. Soubry et al. 2015 found an increase in DNA methylation at the IGF2 and H19 differentially methylated regions (DMRs) among newborns from obese mothers. By contrast, there is a permanent negative association between DNA methylation in the offspring and paternal obesity [111]. The Newborn Epigenetics Study (NEST), which links preconceptional obesity and DNA methylation profiles in the offspring, aimed to find out whether the mother's intake of B group vitamins is associated with aberrant methylation and loss of imprinting in insulin-like growth factor-2 (IGF2) and Delta, drosophila homolog-like-1 (DLK1) and with subsequent weight gain during the first three years of the offspring's life [111].The transgenerational effect was described by Kati et al., who demonstrated that the grandparents' diet during puberty (between the ages of eight and twelve years) affects the longevity of their grandchildren through epigenetic inheritance [112].

Genomic imprinting in one generation is transmitted to the other. In general, low weight at birth indicates an adverse growth trajectory during fetal life and is frequently followed by the development of metabolic diseases during adult life (this is reviewed in detail in [113]). After birth, the neonate is also susceptible to interventions in the developmental programming, for example alterations in milk composition. The "lactocrine hypothesis" is addressed by numerous studies that have shown that nutritional programming is related to the energy provided by the mother's milk [114,115]. Research on rat pups demonstrated that, when fed milk that is rich in carbohydrates by comparison with high-fat milk, the pups immediately develop hyperinsulinemia, which leads to adult obesity [116]. Breastfeeding is of tremendous importance to the growth and development of a child; the main influences are summarized in Figure 15.7.

Rat studies also demonstrated that increased consumption of milk can "malprogram" the mechanism of energy homeostasis [117]. Overnourishment during the suckling period

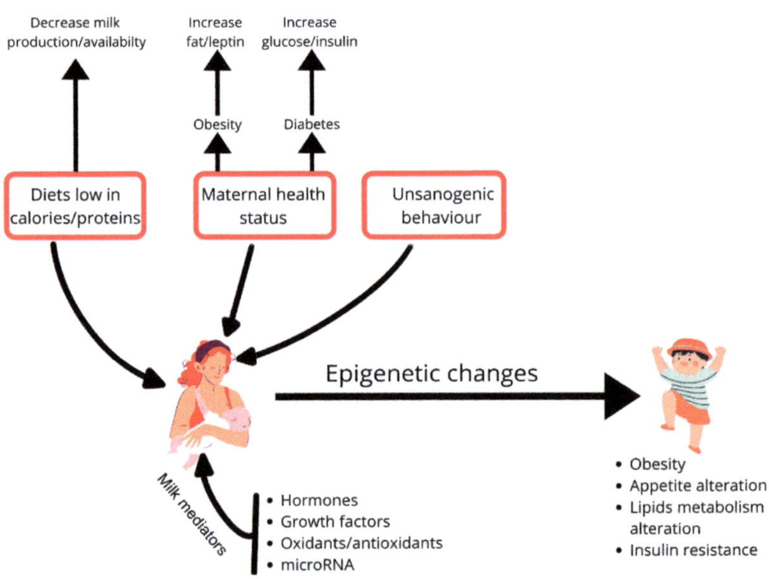

Figure 15.7 The role of breastfeeding in the growth and development of the child. Factors that quantitatively and qualitatively influence the breastfeeding process and their effects on the pathologies developed by the child in the future. *Source*: Elena-Codruta Dobrica.

is followed by hyperleptinemia in the adipose tissue, hyperinsulinemia in the pancreas, and decreased thermogenesis in the brown adipose tissue. Hyperinsulinemia determines an increased orexigenic response, which is followed in the postweaning period by hyperphagia and increased body weight, then reflected in adult life as metabolic syndrome [118,119].

With the development of milk-feeding formulas, a lot of possibilities were created for overfeeding babies. Also, the dietary patterns for the infants were changed with the introduction of fruits, fruit juices, and vegetables after six months of life. The Women, Infants and Children (WIC) Infant Feeding Practices Study revealed that one third of the babies between one and four months of life received sugar water, fruit-flavored drinks, sodas, tea, and coffee [120,121].

In the future, epigenetic epidemiologic studies will be needed to show the impact of genotype and environmental exposure on disease risk, although there are many obstacles to the investigation of the associations between a phenotype and thousands of biologic variables, and the subject could be extended in many directions.

Exposure to Chemicals through Food Chain

Endocrine-disrupting chemicals, also known as endocrine disruptors, are chemical substances found in our food or in the environment that are involved in the perturbation of hormonal homeostasis. In 2009 the Endocrine Society published a scientific statement about the effects of endocrine disruptors on humans of both sex [122]. The US Environmental Protection Agency (EPA) defines an endocrine-disrupting chemical as "an exogenous agent that interferes with synthesis, secretion, transport, metabolism, binding action, or elimination of natural blood-borne hormones that are present in the body and are responsible for homeostasis, reproduction, and developmental process." The action of such chemicals starts after binding to nuclear or membrane receptors and can not only have an impact on the exposed individual but also have epigenetic effects on the offspring and on subsequent generations (this is reviewed in detail by [123] and [124]).

One can classify endocrine disruptors largely into two major groups: chemicals and plant hormones (phytoestrogens). Chemical compounds include pesticides such as methoxychlor, chlorpyrifos, dichlorodiphenyltrichloroethane (DDT), fungicides (vinclozolin), herbicides (atrazine), industrial chemicals (PCBs, dioxins), and plastics (phthalates, BPA (bisphenol A), alkylphenols). However, a lot of other substances can be added, for instance pharmaceuticals (ethinyl estradiol, diethylstilbestrol), nutraceuticals, personal care products (parabens, phthalates, triclosan), and cookware with perfluorooctanoic acid used for teflon lining.

Endocrine disruptors are capable of altering DNA methylation patterns, the most significant modifications consisting in epimutations in the germline that are transgenerationally transmitted to progeny [125]. These epigenetic modifications in the developing germ cells within the fetus are irreversible.

Maternal exposure to BPA in animal experiments can interfere with the development of carcinogenic processes in the offspring and can cause altered methylation in the fetal forebrain and behavioral changes [126]. The hypomethylation of the promoter of nucleosome binding protein-1 (Nsbp1) is a permanent epigenetic mark of neonatal exposure to BPA or estrogens [127]. In addition, BPA can directly influence the genes responsible for DNA methylation [128]. BPA exposure during gestation results in alterations in the ductal morphology of the mammary glands, as animals studied revealed [129,130]

After birth, BPA mimics the 17 β-estradiol hormone and affects babies fed with formula: BPA is broken down by heat, acids, or bases from formula packages or polycarbonate bottles, prematurely activating primordial follicles and altering the levels of sex-steroid hormones. The effects on adult life consist in early puberty and infertility in both sexes, polycystic ovary syndrome (PCOS), and the development of breast and prostate cancer [131,132].

Ethinylestradiol, another endocrine disruptor, can impair brain estrogen signaling in the offsprings of mothers exposed to oral contraceptives [133,134]. Rat model studies also showed that maternal exposure to ethinylestradiol increased the risk of breast cancer for two generations [135]. Exposure to diethylstilbestrol (DES) before birth produces several developmental abnormalities in the reproductive tract and increases tumor incidence [136].

Organotins—a class of chemical compounds represented mainly by tributyltin, a substance used in antifouling paint for marine vessels—have been banned. However, some organotins are found in vinyl (PVC) products, including toys, school supplies, rain jackets, backpacks wallpaper, biocides, VCT floor tiles, surface disinfectants including hospital and veterinary ones, shower curtains, scintillation detectors for γ- and X-rays, and many other industry applications [137]. In experimental animals, organotins induced thymic atrophy, decreasing the number of lymphocytes [138]. In vivo experiments on *Sebastiscus marmoratus* demonstrated that tributyltin can cause abnormal testicular development, affecting Sertoli cells and spermatogenesis [139]. Moreover, a recent study shows that tributyltin causes epigenetic changes such as hypermethylation leading, to decreased NRF-1 expression [140].

Future studies are needed to underline the role of chemical endocrine disruptors and organotins and the pathways by which they interfere in, and alter, DNA methylation.

Nutritional Genomic Approaches in Cancers

The interest in exchanges between nutrition and cancer has steadily grown over the years. To date, early 66,000 studies have been published in PubMed/MEDLINE (https://www.ncbi.nlm.nih.gov/pubmed), the earliest research papers dating back to the 1910s. The past five years have witnessed an explosion of publications tackling this topic (Figure 15.8).

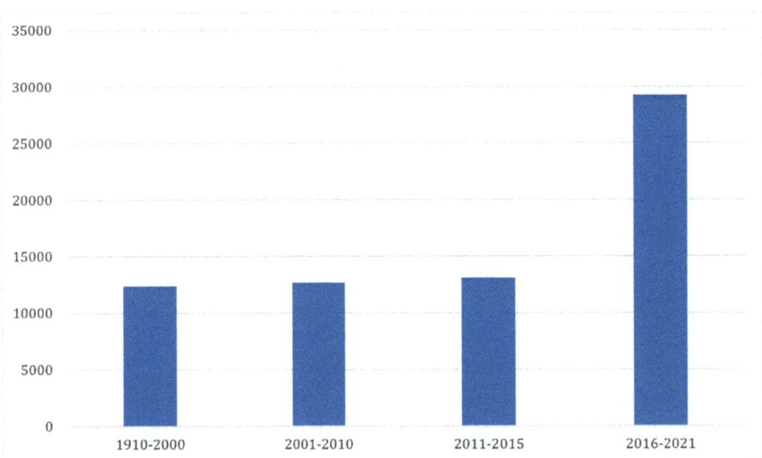

Figure 15.8 The number of studies focused on nutrition and cancer identified in the PubMed/MEDLINE database has increased steadily over the years. *Source*: Mihnea-Alexandru Gaman.

However, it is important to mention that the benefits of a healthy diet and its impact on human health have been hypothesized since time immemorial; in the classical world, Hippocrates is credited to have said: "Let food be thy medicine and medicine be thy food." Thus it comes as no surprise that nutrigenetics and nutrigenomics, and in particular the use of nutraceuticals, phytochemicals, and functional foods, have grasped the attention of the global scientific community [141,142]

According to recent statistics on deaths and mortality in the United States published by the National Center for Health Statistics of the Centers for Disease Control and Prevention (CDC), malignancies stand out as the second cause of death and are outperformed only by cardiovascular disorders [143]. Hence novel strategies of chemoprevention are warranted.

Nowadays it is obvious that inactivation of certain tumor-suppressor genes arises because of hypermethylation within the promoter regions, while the hypomethylation process induces genomic instability. Both hyper- and hypomethylation, which are altered pathways, have been associated with human tumors, hypomethylation being frequent in neoplastic tissues where vitamin levels are low [144].

DNA methylation is carried out via a pathway that involves folate and vitamin B-12; thus deficiencies in these substances cause global DNA hypomethylation [145]. Folic acid is needed in the synthesis of the amino acid methionine, which in turn is required for the process of DNA methylation [146,147].

Dietary factors and patterns have been shown to impact one's risk of cancer. For example, subjects who follow the Mediterranean diet, known to lower oxidative stress and inflammation levels through its properties, seem to be less prone to developing malignancies [148]. But a diet rich in fats and proteins has been linked to the development of colorectal or prostate cancer [149]. Several dietary factors, for example salt intake and beer or liquor consumption, elevate the risk of gastric cancer. Others, such as the consumption of white vegetables or fruit, display a protective effect against this malignancy [150]. In this respect, nutrigenomics or nutritional genomics has emerged as a science that evaluates the impact of diet-derived bioactive compounds on the expression of genes [147,151]

Alcohol is one of the most consumed beverages worldwide, despite playing a putative role in the development of a myriad of malignancies, for example breast, prostate, and gastrointestinal cancers [152]. Early exposure to alcohol has been shown to result in epigenetic alterations in the offspring of murine models—for example in an acetylation of histones, which influences cognition and behaviour in the alcohol-exposed mice [153]. In particular, alcohol consumption disrupts FOCM and causes folate deficiency.

According to a recent meta-analysis, an elevated intake of folate has been demonstrated to exhibit a protective effect against the development of pancreatic cancer [154]. Another meta-analysis revealed a protective action of folate and vitamin B6 against the development of renal cell carcinoma [155]. In addition, breast cancer risk seems to be reduced via an elevated consumption of vitamin B2, vitamin B6 and folate, as the latter two reduce the risk of estrogen and progesterone receptor-negative breast cancer specifically [156]. Moreover, vitamin B6 has been identified as a suppressor of the programmed death-ligand 1 and might serve in the near future as a lead in drug discovery and development in the field as immunotherapy, as programmed death-ligand 1 inhibitors are already in use or tested in clinical trials for several malignancies: non-small-cell lung cancer, hepatocellular carcinoma, melanoma, bladder, and breast or cervical cancers [157–161]. Moreover, vitamin B12 has emerged as an epigenetic regulator in colorectal cancer (CRC), as it has been proven that serum vitamin B12 levels are negatively associated with the methylation of DNA in subjects diagnosed with this malignancy [162]. Moreover, vitamin B12 and folic acid were found to be at very low levels in tissue samples of squamous cell carcinoma [163] and previous studies claimed that even the atypisms found in the bronchial squamous metaplasia of smokers decreased after folate and vitamin B-12 supplementation [164].

Vitamin C is another epigenetic regulator—on account of its properties to reduce the quantity of dysfunctional DNA, to preserve the stability of the human genome, and to reduce the hypermethylation of histones—and has been employed in the treatment of blood cancers in combination with DNA methyltransferase inhibitors [165–167]. In addition, vitamin C is involved in the remodelling of the epigenome by stimulating the function of the ten-eleven translocation proteins and Jumonji-C domain-containing histone demethylases, which are involved in the generation of induced pluripotent stem cells via the reprogramming of somatic cells [168,169].

Vitamin D levels during pregnancy are linked to histone modifications, mainly acetylation, and also to DNA methylation [170]. Low levels of vitamin D show negative associations with a tumor suppressor that is often inactive in CRC [171]. During pregnancy studies have shown that supplementation with vitamin D can protect against multiple sclerosis, and also against osteoporosis [172,173].

Several phytochemicals have demonstrated promising effects in the prevention and management of cancer, yet most of the data about these molecules are derived from research on cell lines or murine models. One of the most globally investigated phytochemicals are flavonoids. Quercetin is a flavone that has shown antineoplastic effects in thyroid and colon cancer, respectively through upregulation of pro-NAG-1 cytokine and through pro-apoptotic and antiproliferative action [174,175]. Luteolin is a flavonol that has been formulated into nanoethosomes and has been tested in the management of hepatocellular carcinoma, where it has displayed antioxidative stress, decreasing the expression of the GPC3 gene and the levels of malondialdehyde and nitric oxide, as well as elevating the concentrations of superoxide dismutase and glutathione [176]. L-asparaginase administered in combination with genistein, which is an isoflavone, resulted in the enhancement of programmed cell death, mitochondrial dysfunction, and reduction in anti-apoptosis molecules; and it elevated the cytotoxicity of the anticancer treatment in the HL-60 cell line [177]. Hesperidin, a flavanone, has exhibited pro-apoptotic actions in the HeLa cancer cell line used in the study of cervix malignancies, particularly via ROS generation; it has also decreased the expression of Jab1 mRNA and the activation of caspase-3 [178]. Epigallocatechin-3-gallate, a flavan-3-ol, led to mitochondrial dysfunction, generated oxidative stress, and suppressed the growth of the acute lymphoblastic leukemia Jurkat cell line [179]. Delphinidin, an anthocyanin, demonstrated antineoplastic actions in colon cancer cell lines as a result of its ability to modulate two signaling pathways, namely MAPKinase and JAK/STAT-3, through the production of oxygen free radicals, reduction in the levels of antioxidants, peroxidation of lipids, mitochondrial dysfunction, and DNA disruption in HCT116 colon cancer cells [180]. As flavonoids are widely present in the diet of individuals who consume vegetables, fruits, and cocoa or drink wine or green tea, this class of phytochemicals might play important roles in the nutritional approach to cancer therapy in the future [181].

Epigenetics and Prebiotics, Probiotics, and Postbiotics

The definition of probiotics has undergone many changes in recent years, and there is still no single version that is unanimously accepted. There is nevertheless a definition that the most important international health agencies—for example the Food and Agriculture Organization, the World Health Organization, Codex, the World Gastroenterology Organization, and the

European Food Safety Authority—agree on. According to it, probiotics are "live microorganisms which can provide health benefits on the host when administered in adequate amounts" [182–185] (this definition is found in the same words in all these sources). Until a beneficial effect on health is demonstrated, no bacteria can be classified in this way, and therefore the use of the term "probiotics" is restricted in some countries of the European Union, to prevent any confusion in the absence of a real health benefit [184].

Although there are many bacterial species that can be used as probiotics, the most common are four classes of lactic acid-producing microorganisms such as *Bifidobacterium* spp., *Lactococcus* spp., *Escherichia coli* strain Nissle 1917, and *Bacillus coagulans* [182,186]. Many of the bacteria already mentioned, especially those of the genus *Lactobacillus* spp., are already present in the human commensal flora, for instance in the oral cavity and in the gastrointestinal or urogenital tract [186]. Different species can be administered in different amounts to the same patient, and for this reason many different beneficial effects can be obtained using just a handful of microorganisms from one patient to another [182]

As for prebiotics, they were first defined by Gibson and Roberfroid in the *Journal of Nutrition* in 1995 as a "nondigestible dietary ingredient that beneficially affects the host by selectively stimulating the growth or activity of a limited number of bacteria in the colon" [182]. Since then, the definition has been revised numerous times, last time in 2017 and by the same authors. Prebiotics are currently defined as a substrate that is selectively used by host microorganisms because it confers a health benefit [182,185].

Also, a series of criteria have been established for a substance to satisfy in order to be considered a prebiotic: (1) it must have resistance to gastric acidity, enzymatic hydrolysis, or gastrointestinal absorption; (2) it must be subjected to fermentation by intestinal microflora; and (3) it must produce a selective stimulation of the growth and activity of intestinal bacteria associated with health benefits [182]. On the basis of these criteria, the most appropriate probiotics—and also the ones most commonly used as such—are lactulose, inulin, fructo-oligosaccharides, galacto-oligosaccharides, and human milk oligosaccharides [182]. Moreover, it has been shown in recent years that some bacteria used as probiotics have a certain specificity for prebiotics through very specific enzymes, for instance β-Galactosidase from *Bifidobacterium adolescentis* and β-fructofuranosidase from some *Bacteroidetes* species [182].

These new data, together with the concomitant administration of trace elements and bacterial species with an affinity for these elements, have led to the emergence of a new principle. Their combination has been called "symbiotic" and uses the benefits of each element to achieve a new effect [182,187]. A suitable example of this new principle is the concomitant administration of *Bifidobacterium animalis* and polydextrose. This operation improved the action of metformin in mice, boosting the glycemic response and lowering the overall plasma glucose concentration [188].

The human microbiome represents the totality of microorganisms that colonize the gastrointestinal tract. It is acquired during the first three years of life, is composed of 500 to 1,000 species of bacteria, and changes constantly throughout life, being influenced by diet, lifestyle, stress, antibiotics, and host genetics [189]. Because in recent years it has been shown that the microbiome is involved in the onset and progression of countless diseases such as allergies, inflammatory bowel disease, infections, cancers, multiple sclerosis, and autoimmune diseases, the positive manipulation or restoration of an altered gut microbiome through microbiota-targeted drugs, through bacteriophages or the use of engineered bacteria, through the transplantation of a healthy gut microbiome, and through probiotics or prebiotics has come to arouse great interest [189,190]. But we should point out that this area is still in its infancy and the main mechanisms involved are not yet fully understood [190].

The administration of prebiotics and probiotics has been shown to have multiple benefits, both for healthy individuals and for individuals who suffer from dysbiosis to a degree, for example from antibiotic-associated diarrhoea, necrotizing enterocolitis, and recurrent pouchitis [182,186,187]. The additional benefits of prebiotics and probiotics occur either through their direct metabolic output or through their interaction with resident flora and host immunity. Among these broad, clinically validated mechanisms are the following: altering the composition of an already existing normal microbial ecosystem through a competitive exclusion of pathogens (this is achieved via adherence to epithelial cells with pili and via antimicrobial resistance or production, for the sake of competing with other microorganisms); maintaining gut lumen epithelial barrier integrity and permeability; downregulating hypersensitivity reactions; and improving mineral and nutrient absorption [182,186,187]. Moreover, probiotics use various vitamins, proteins, and carbohydrates not absorbed or digested by the host to produce important secondary metabolites such as short-chain fatty acids (SCFAs), for instance acetate, propionate, and butyrate. These can provide energy for the host and may modulate immune responses [189].

SCFAs are the most important products resulting from the carbohydrate metabolism of colonic bacteria [183]. These are bioactive molecules with metabolic and immunological activity that have been shown to lower the risk of cardiovascular disease, diabetes mellitus, and CRC [182,183]. Their most important effects are: protection from infections, dose-dependent decrease in glucagon levels and increase in insulin levels (this is achieved by increasing glucagon-like peptide-1 activity), improvement of glucose tolerance, and overall anti-inflammatory and immunological modulation activity in the colonic mucosa; the latter can offer some protection against allergies and autoimmune diseases [182,183,190]. CRC is an increasingly serious public health problem, especially in developed countries, where its incidence is increasing, especially among young individuals [186]. In recent years, it has become clear that lifestyle plays a special role in the occurrence of this type of neoplasm—especially eating habits and environmental factors [186]. The microbiome has begun to be seen more and more as a central factor in carcinogenesis and, as a result, it is in everyone's interest to find different methods of regenerating or modifying it through prebiotics, probiotics, or symbiotics, in the hope of influencing treatment in some way, or even preventing diseases like CRC [183].

A particularly important role is played by the products that result from the metabolism of various food compounds performed by the colonic bacterial flora. They cause genetic or epigenetic changes at the cellular level, for instance aberrant methylation of various nucleotides, various patterns of DNA methylation, and various microRNAs, which have been shown to be markers of the initiation and progression of CRC [183,186]. All these changes underlie the process of carcinogenesis and, because they are partially reversible, many efforts are put into their early detection and prevention, and even in their use as therapeutic targets [186].

"Type C" promoter genes are specific to neoplastic cells. The aberrant methylation of genes belonging to this group represents epigenetic events that underlie subsequent changes associated with the occurrence and progression of CRC [187]. Of these genes, MINT2 is a specific marker for CRC; and, in a study that included sixteen promoter genes from this group, Worthley et al. demonstrated that additional symbiosis was associated with the lowest methylation level of the MINT2 gene [187]. Moreover, the same authors demonstrated the existence of beneficial effects in patients with a history of CRC after concomitant supplementation of the diet with resistant starch and *B. lactis*. These effects include decreased colorectal epithelium proliferation, promotion of beneficial bacterial communities, decreased DNA damage at the cellular level, and modulation of surface cytokines such as interleukin-2 and interferon-1 [187].

There are countless other bacterial species that have shown promising results in the prevention of CRC. *L. acidophilus* increased the activity of genes with tumor suppressor activity among NCK2025 cells, and *L. reuteri* caused a decrease in cancer cell proliferation, an effect most likely achieved by generating SCFA [191]. Moreover, *L. casei* has shown a number of promising results when used in various combinations with other bacterial species: along with *B. lactis*, it stimulates increased NK cell activity; and it causes increased matrix metalloproteinase activity −9 when co-administered with *L. rhamnosus* [183,191].

Another class that has attracted much interest in recent years is that of postbiotics. Although their study began in the 1970s, the last three years have been characterized by a spectacular increase in scientific work on this topic. Thus, out of the total of 242 articles on postbiotics on the PubMed/MEDLINE platform [191], 193 have been published between 2019 and 2021. In 2019, after several years in which it did not have a clear notion of postbiotics, the International Scientific Association for Probiotics and Prebiotics came to define them as a "preparation of inanimate microorganisms and/or their components that confers a health benefit on the host" [192]. These are products of the microbial metabolism of the gut's microbiota, substances secreted by external microorganisms in controlled environments (bacteria, yeasts), enzymes (with an antioxidant role), cell debris, SCFAs that contribute to homeostasis and modulate the activity of resident or supplemented microorganisms [192,193]. They can promote wound healing by decreasing the risk of infection, neoplasms, and atherosclerosis through their immunomodulatory, antitumoral, and autophagy functions and by modulating the lipids metabolism [194].

Conclusion and Future Perspectives

Although the concept of epigenetics has been proposed as early as the beginning of the twentieth century, the mechanisms underlying epigenetic processes have begun to be truly understood only toward the end of the past decade, which has seen an explosion in the number of publications dedicated to epigenetics. Many risk factors increasingly difficult to avoid—the modern lifestyle, the western diet rich in refined carbohydrates and saturated fats and poor in unprocessed fruits and vegetables, stress, and a polluted environment—highlight the need to study the relationship between exposure to these factors and gene activity. New findings in the field of epigenetics emphasize that the activity of genes is regulated by the abovementioned factors through the influence they have on DNA methylation, histone modification, and the regulation of non-coding RNA. All these processes control gene expression in both directions: some activate them, others silence them. Moreover, the changes caused by environmental factors appear to be stable and transgenerationally transmissible. Their role in fetal programming, which has a major impact on the metabolic pathways of the future child and adult, emphasizes once again the importance of these factors in the development of NCDs such as cardiovascular and neurological disease, obesity, or diabetes.

The diet thus becomes not only a tool of hygienic–dietary treatment that precedes the drug, but an essential tool in the prevention of diseases with the highest mortality and financial load worldwide. Through what we eat we become responsible not only for our health, but also for that of our descendants, who inherit our DNA with all the epigenetic changes we have gained through interaction with environmental and dietary factors.

References

1 Dupont, C., Armant, D., and Brenner, C. (2009). Epigenetics: Definition, mechanisms and clinical perspective. *Semin. Reprod. Med.* 27 (05): 351–357.
2 Waddington, C.H. (1942). The epigenotype. *Endeavour* 1: 18–20.
3 Waddington, C.H. (1956). Genetic assimilation of the Bithorax phenotype. *Evolution* 10 (1): 257–293.
4 Hoßfeld, U., Watts, E., and Levit, G.S. (2017). Valentin Haecker (1864–1927) as a pioneer of phenogenetics: Building the bridge between genotype and phenotype. *Epigenetics* 12 (4): 247–253.
5 Haecker, V. (1918). Entwicklungsgeschichtliche Eigenschaftsanalyse (Phänogenetik). Jena: Gustav Fischer.
6 Jeggo, P. and Gounari, F. (2014). Robin holliday 1932–2014. *Nat. Struct. Mol. Biol.* 21 (6): 501–502.
7 Holliday, R. and Pugh, J.E. (1975). DNA modification mechanisms and gene activity during development. *Science* 187: 226–232.
8 Bard, J.B.L. (2008). Waddington's legacy to developmental and theoretical biology. *Biol. Theory* 3 (3): 188–197.
9 Maienschein, J. (2017). Epigenesis and preformationism. In The Stanford Encyclopedia of Philosophy, ed. E.N. Zalta. Metaphysics Research Lab, Stanford University. https://plato.stanford.edu/archives/spr2017/entries/epigenesist.
10 Waddington, C.H. (1968). Towards a theoretical biology. *Nature* 218 (5141): 525–527.
11 Notterman, D.A. and Mitchell, C. (2015). Epigenetics and understanding the impact of social determinants of health. *Pediatr. Clin. North Am.* 62 (5): 1227–1240.
12 Kanherkar, R.R., Bhatia-Dey, N., and Csoka, A.B. (2014). Epigenetics across the human lifespan. *Front. Cell Dev. Biol.* 2: 49.
13 Moore, L.D., Le, T., and Fan, G. (2013). DNA Methylation and its basic function. *Neuropsychopharmacology* 38 (1): 23–38.
14 Jin, B., Li, Y., and Robertson, K.D. (2011). DNA methylation: Superior or subordinate in the epigenetic hierarchy? *Genes Cancer* 2 (6): 607–617.
15 Kim, M. and Costello, J. (2017). DNA Methylation: An epigenetic mark of cellular memory. *Exp. Mol. Med.* 49 (4): e322–e322.
16 Okano, M., Bell, D.W., Haber, D.A., and Li, E. (1999). DNA methyltransferases Dnmt3a and Dnmt3b are essential for de novo methylation and mammalian development. *Cell* 99 (3): 247–257.
17 Barau, J., Teissandier, A., Zamudio, N., Roy, S., Nalesso, V., Hérault, Y., Guillou, F., and Bourc'His, D. (2016). The DNA methyltransferase DNMT3C protects male germ cells from transposon activity. *Science* 354 (6314): 909–912.
18 Chedin, F., Lieber, M.R., and Hsieh, C.-L. (2002). The DNA methyltransferase-like protein DNMT3L stimulates de novo methylation by Dnmt3a. *Proc. Natl. Acad. Sci. USA* 99 (26): 16916–16921.
19 Ashapkin, V.V., Kutueva, L.I., and Vanyushin, B.F. (2016). Dnmt2 is the most evolutionary conserved and enigmatic cytosine DNA methyltransferase in eukaryotes. *Russ. J. Genet.* 52 (3): 237–248.
20 Nicoletti, C.F., Cortes-Oliveira, C., Noronha, N.Y., Pinhel, M.A.S., Dantas, W.S., Jácome, A., Marchini, J.S., Gualano, B., Crujeiras, A.B., and Nonino, C.B. (2020). DNA methylation pattern changes following a short-term hypocaloric diet in women with obesity. *Eur. J. Clin. Nutr* 74 (9): 1345–1353.

21 Mariño-Ramírez, L., Kann, M.G., Shoemaker, B.A., and Landsman, D. (2005). Histone structure and nucleosome stability. *Expert Rev. Proteomics* 2 (5): 719–729.

22 Swygert, S.G. and Peterson, C.L. (2014). Chromatin dynamics: Interplay between remodeling enzymes and histone modifications. *Biochim. Biophys. Acta* 1839 (8): 728–736.

23 Alaskhar Alhamwe, B., Khalaila, R., Wolf, J., Von Bülow, V., Harb, H., Alhamdan, F., Hii, C.S., Prescott, S.L., Ferrante, A., Renz, H., Garn, H., and Potaczek, D.P. (2018). Histone modifications and their role in epigenetics of atopy and allergic diseases. *Allergy Asthma Clin. Immunol.* 14 (1): 39.

24 Molina-Serrano, D., Kyriakou, D., and Kirmizis, A. (2019). Histone modifications as an intersection between diet and longevity. *Front. Genet.* 10: 192.

25 Tsurumi, A. and Li, W.X. (2012). Global heterochromatin loss: A unifying theory of aging? *Epigenetics* 7 (7): 680–688.

26 Daniel, M. and Tollefsbol, T.O. (2015). Epigenetic linkage of aging, cancer and nutrition. *J. Exp. Biol.* 218 (Pt 1): 59–70.

27 Dey, B.K., Mueller, A.C., and Dutta, A. (2014). Long non-coding RNAs as emerging regulators of differentiation, development, and disease. *Transcription* 5 (4): e944014.

28 Kung, J.T.Y., Colognori, D., and Lee, J.T. (2013). Long noncoding RNAs: Past, present, and future. *Genetics* 193 (3): 651–669.

29 Lekka, E. and Hall, J. (2018). Noncoding RNAs in disease. *FEBS Lett.* 592 (17): 2884–2900.

30 Green, C.D., Huang, Y., Dou, X., Yang, L., Liu, Y., and Han, J.-D.J. (2017). Impact of dietary interventions on noncoding RNA networks and MRNAs encoding chromatin-related factors. *Cell Rep.* 18 (12): 2957–2968.

31 Zhang, T.-N., Wang, W., Yang, N., Huang, X.-M., and Liu, C.-F. (2020). Regulation of glucose and lipid metabolism by long non-coding RNAs: Facts and research progress. *Front. Endocrinol. (Lausanne)* 11: 457.

32 Ekine-Afolabi, B.A., Njan, A.A., Rotimi, S.O., R I, A., Elbehi, A.M., Cash, E., and Adeyeye, A. (2020). The Impact of diet on the involvement of non-coding RNAs, extracellular vesicles, and gut microbiome-virome in colorectal cancer initiation and progression. *Front. Oncol.* 10: 583372.

33 Wijesinghe, S.N., Nicholson, T., Tsintzas, K., and Jones, S.W. (2021). Involvements of long noncoding rnas in obesity-associated inflammatory diseases. *Obes. Rev.* 22 (4): e13156.

34 Deans, C. and Maggert, K.A. (2015). What do you mean, "Epigenetic"? *Genetics* 199 (4): 887–896.

35 WHO (2021). Global health estimates: Life expectancy and leading causes of death and disability. WHO. https://www.who.int/data/gho/data/themes/mortality-and-global-health-estimates.

36 Stefan, N., Birkenfeld, A.L., and Schulze, M.B. (2021). Global pandemics interconnected: Obesity, impaired metabolic health and COVID-19. *Nat. Rev. Endocrinol.* 17 (3): 135–149.

37 WHO (2021). Noncommunicable diseases. WHO, April 13. https://www.who.int/news-room/fact-sheets/detail/noncommunicable-diseases.

38 Waaler, H.T. (2009). Height. weight and mortality the norwegian experience. *Acta Med. Scand.* 215 (S679): 1–56.

39 Barker, D. (1986). Infant mortality, childhood nutrition, and ischaemic heart disease in england and wales. *Lancet* 327 (8489): 1077–1081.

40 Barker, D.J., Osmond, C., and Law, C.M. (1989). The Intrauterine and early postnatal origins of cardiovascular disease and chronic bronchitis. *J. Epidemiol. Community Health* 43 (3): 237–240.

41 Barker, D.J. (1990). The fetal and infant origins of adult disease. *BMJ* 301 (6761): 1111.

42 Hales, C.N. and Barker, D.J. (1992). Type 2 (Non-Insulin-Dependent) diabetes mellitus: The thrifty phenotype hypothesis. *Diabetologia* 35 (7): 595–601.

43 Gluckman, P.D., Hanson, M.A., Beedle, A.S., and Spencer, H.G. (2008). Predictive adaptive responses in perspective. *Trends Endocrinol. Metab.* 19 (4): 109–110. author reply 112.

44 Wadhwa, P.D., Buss, C., Entringer, S., and Swanson, J.M. (2009). Developmental origins of health and disease: Brief history of the approach and current focus on epigenetic mechanisms. *Semin. Reprod. Med.* 27 (5): 358–368.

45 Ahlqvist, E., Ahluwalia, T.S., and Groop, L. (2011). Genetics of Type 2 diabetes. *Clin. Chem.* 57 (2): 241–254.

46 Vaag, A.A., Grunnet, L.G., Arora, G.P., and Brøns, C. (2012). The thrifty phenotype hypothesis revisited. *Diabetologia* 55 (8): 2085–2088.

47 Khunti, K., Kosiborod, M., and Ray, K.K. (2018). Legacy benefits of blood glucose, blood pressure and lipid control in individuals with diabetes and cardiovascular disease: Time to overcome multifactorial therapeutic inertia? *Diabetes Obes. Metab.* 20 (6): 1337–1341.

48 Ranjit Unnikrishnan, I., Anjana, R., and Mohan, V. (2011). Importance of controlling diabetes early: The concept of metabolic memory, legacy effect and the case for early insulinisation. *JAPI* 59: 8–12.

49 Li, Y., Jaddoe, V.W., Qi, L., He, Y., Wang, D., Lai, J., Zhang, J., Fu, P., Yang, X., and Hu, F.B. (2011). Exposure to the Chinese famine in early life and the risk of metabolic syndrome in adulthood. *Diabetes Care* 34 (4): 1014–1018.

50 Thurner, S., Klimek, P., Szell, M., Duftschmid, G., Endel, G., Kautzky-Willer, A., and Kasper, D.C. (2013). Quantification of excess risk for diabetes for those born in times of hunger, in an entire population of a nation, across a century. *Proc. Natl. Acad. Sci. USA* 110 (12): 4703–4707.

51 Schulz, L.C. (2010). The dutch hunger winter and the developmental origins of health and disease. *Proc. Natl. Acad. Sci. USA* 107 (39): 16757–16758.

52 Heijmans, B.T., Tobi, E.W., Stein, A.D., Putter, H., Blauw, G.J., Susser, E.S., Slagboom, P.E., and Lumey, L.H. (2008). Persistent epigenetic differences associated with prenatal exposure to famine in humans. *Proc. Natl. Acad. Sci. USA* 105 (44): 17046–17049.

53 Jang, H. and Serra, C. (2014). Nutrition, epigenetics, and diseases. *Clin. Nutr. Res.* 3 (1): 1–8.

54 Eckert, J.J., Porter, R., Watkins, A.J., Burt, E., Brooks, S., Leese, H.J., Humpherson, P.G., Cameron, I.T., and Fleming, T.P. (2012). Metabolic induction and early responses of mouse blastocyst developmental programming following maternal low protein diet affecting life-long health. *PLoS One* 7 (12): e52791.

55 Kwong, W.Y., Wild, A.E., Roberts, P., Willis, A.C., and Fleming, T.P. (2000). Maternal undernutrition during the preimplantation period of rat development causes blastocyst abnormalities and programming of postnatal hypertension. *Development* 127 (19): 4195–4202.

56 Watkins, A.J., Ursell, E., Panton, R., Papenbrock, T., Hollis, L., Cunningham, C., Wilkins, A., Perry, V.H., Sheth, B., Kwong, W.Y., Eckert, J.J., Wild, A.E., Hanson, M.A., Osmond, C., and Fleming, T.P. (2008). Adaptive responses by mouse early embryos to maternal diet protect fetal growth but predispose to adult onset disease. *Biol. Reprod.* 78 (2): 299–306.

57 Huang, T., Yang, M., Zeng, Y., Huang, X., Wang, N., Chen, Y., Li, P., Yuan, J., Chen, C., Oliver, B.G., and Yi, C. (2021). Maternal high fat diet consumption exaggerates metabolic disorders in mice with cigarette-smoking induced intrauterine undernutrition. *Front. Nutr.* 8: 638576.

58 Masuyama, H. and Hiramatsu, Y. (2012). Effects of a high-fat diet exposure in utero on the metabolic syndrome-like phenomenon in mouse offspring through epigenetic changes in adipocytokine gene expression. *Endocrinology* 153 (6): 2823–2830.

59 Masuyama, H., Mitsui, T., Nobumoto, E., and Hiramatsu, Y. (2015). The effects of high-fat diet exposure in utero on the obesogenic and diabetogenic traits through epigenetic changes in adiponectin and leptin gene expression for multiple generations in female mice. *Endocrinology* 156 (7): 2482–2491.

60 Vanhees, K., Vonhögen, I.G.C., Van Schooten, F.J., and Godschalk, R.W.L. (2014). You are what you eat, and so are your children: The impact of micronutrients on the epigenetic programming of offspring. *Cell. Mol. Life Sci.* 71 (2): 271–285.

61 Zhang, Q., Xiao, X., Zheng, J., Li, M., Yu, M., Ping, F., Wang, T., and Wang, X. (2019). A maternal high-fat diet induces DNA methylation changes that contribute to glucose intolerance in offspring. *Front. Endocrinol. (Lausanne)* 10: 871.

62 Mamounis, K.J., Shvedov, N.R., Margolies, N., Yasrebi, A., and Roepke, T.A. (2020). The effects of dietary fatty acids in the physiological outcomes of maternal high-fat diet on offspring energy homeostasis in mice. *J. Dev. Orig. Health Dis.* 11 (3): 273–284.

63 Tsoulis, M.W., Chang, P.E., Moore, C.J., Chan, K.A., Gohir, W., Petrik, J.J., Vickers, M.H., Connor, K.L., and Sloboda, D.M. (2016). Maternal high-fat diet-induced loss of fetal oocytes is associated with compromised follicle growth in adult rat offspring. *Biol. Reprod.* 94 (4): 94.

64 Aiken, C.E., Tarry-Adkins, J.L., and Ozanne, S.E. (2015). Transgenerational developmental programming of ovarian reserve. *Sci. Rep.* 5 (1): 16175.

65 Winship, A.L., Gazzard, S.E., Cullen Mcewen, L.A., Bertram, J.F., and Hutt, K.J. (2018). Maternal low-protein diet programmes low ovarian reserve in offspring. *J. Reprod. Fertil.* 156 (4): 299–311.

66 McKay, J.A., Groom, A., Potter, C., Coneyworth, L.J., Ford, D., Mathers, J.C., and Relton, C.L. (2012). Genetic and non-genetic influences during pregnancy on infant global and site specific DNA methylation: Role for folate gene variants and vitamin B12. *PLoS One* 7 (3): e33290.

67 Sinclair, K.D., Allegrucci, C., Singh, R., Gardner, D.S., Sebastian, S., Bispham, J., Thurston, A., Huntley, J.F., Rees, W.D., Maloney, C.A., Lea, R.G., Craigon, J., McEvoy, T.G., and Young, L.E. (2007). DNA methylation, insulin resistance, and blood pressure in offspring determined by maternal periconceptional B vitamin and methionine status. *Proc. Natl. Acad. Sci. USA* 104 (49): 19351–19356.

68 Yang, X., Huang, Y., Sun, C., and Li, J. (2017). Maternal prenatal folic acid supplementation programs offspring lipid metabolism by aberrant DNA methylation in hepatic ATGL and adipose LPL in rats. *Nutrients* 9 (9). https://doi.org/10.3390/nu9090935.

69 Sharland, E., Montgomery, B., and Granell, R. (2011). Folic acid in pregnancy: Is there a link with childhood asthma or wheeze? *Aust. Fam. Physician* 40 (6): 421–424.

70 Alfonso, V.H., Bandoli, G., Von Ehrenstein, O., and Ritz, B. (2018). Early folic acid supplement initiation and risk of adverse early childhood respiratory health: A population-based study. *Matern. Child Health J.* 22 (1): 111–119.

71 Vanhees, K., Coort, S., Ruijters, E.J.B., Godschalk, R.W.L., Van Schooten, F.J., and Barjesteh Van Waalwijk Van Doorn-khosrovani, S. (2011). Epigenetics: Prenatal exposure to genistein leaves a permanent signature on the hematopoietic lineage. *FASEB J.* 25 (2): 797–807.

72 Ong, T.P., Moreno, F.S., and Ross, S.A. (2011). Targeting the epigenome with bioactive food components for cancer prevention. *J. Nutrigenet. Nutrigenomics* 4 (5): 275–292.

73 Lee, W.J. and Zhu, B.T. (2006). Inhibition of DNA methylation by caffeic acid and chlorogenic acid, two common catechol-containing coffee polyphenols. *Carcinogenesis* 27 (2): 269–277.

74 Vanhees, K., Van Schooten, F.J., Van Waalwijk Van Doorn-Khosrovani, S.B., Van Helden, S., Munnia, A., Peluso, M., Briedé, J.J., Haenen, G.R.M.M., and Godschalk, R.W.L. (2013). Intrauterine exposure to flavonoids modifies antioxidant status at adulthood and decreases oxidative stress-induced DNA damage. *Free Radic. Biol. Med.* 57: 154–161.

75 Qiu, C., Zhang, C., Gelaye, B., Enquobahrie, D.A., Frederick, I.O., and Williams, M.A. (2011). Gestational diabetes mellitus in relation to maternal dietary heme iron and nonheme iron intake. *Diabetes Care* 34 (7): 1564–1569.

76 Chango, A. and Pogribny, I.P. (2015). Considering maternal dietary modulators for epigenetic regulation and programming of the fetal epigenome. *Nutrients* 7 (4): 2748–2770.

77 Zeisel, S. (2017). Choline, other methyl-donors and epigenetics. *Nutrients* 9 (5). https://doi.org/10.3390/nu9050445.

78 Tiffon, C. (2018). The impact of nutrition and environmental epigenetics on human health and disease. *Int. J. Mol. Sci.* 19 (11): 3425.

79 Sarabi, M.M. and Naghibalhossaini, F. (2018). The impact of polyunsaturated fatty acids on DNA methylation and expression of DNMTs in human colorectal cancer cells. *Biomed. Pharmacother.* 101: 94–99.

80 Moradi Sarabi, M., Mohammadrezaei Khorramabadi, R., Zare, Z., and Eftekhar, E. (2019). Polyunsaturated fatty acids and DNA methylation in colorectal cancer. *World J. Clin. Cases* 7 (24): 4172–4185.

81 Kalani, A., Kamat, P.K., Tyagi, S.C., and Tyagi, N. (2013). Synergy of homocysteine, microRNA, and epigenetics: A novel therapeutic approach for stroke. *Mol. Neurobiol.* 48 (1): 157–168.

82 Cai, D., Jia, Y., Song, H., Sui, S., Lu, J., Jiang, Z., and Zhao, R. (2014). Betaine supplementation in maternal diet modulates the epigenetic regulation of hepatic gluconeogenic genes in neonatal piglets. *PLoS One* 9 (8): e105504.

83 El-Osta, A., Brasacchio, D., Yao, D., Pocai, A., Jones, P.L., Roeder, R.G., Cooper, M.E., and Brownlee, M. (2008). Transient high glucose causes persistent epigenetic changes and altered gene expression during subsequent normoglycemia. *J. Exp. Med.* 205 (10): 2409–2417.

84 Siebel, A.L., Fernandez, A.Z., and El-Osta, A. (2010). Glycemic memory associated epigenetic changes. *Biochem. Pharmacol* 80 (12): 1853–1859.

85 Mali, P., Chou, B.-K., Yen, J., Ye, Z., Zou, J., Dowey, S., Brodsky, R.A., Ohm, J.E., Yu, W., Baylin, S.B., Yusa, K., Bradley, A., Meyers, D.J., Mukherjee, C., Cole, P.A., and Cheng, L. (2010). Butyrate greatly enhances derivation of human induced pluripotent stem cells by promoting epigenetic remodeling and the expression of pluripotency-associated genes. *Stem Cells* 28 (4): 713–720.

86 Martin, S., Hardy, T., and Tollefsbol, T. (2013). Medicinal chemistry of the epigenetic diet and caloric restriction. *Curr. Med. Chem.* 20 (32): 4050–4059.

87 Crider, K.S., Yang, T.P., Berry, R.J., and Bailey, L.B. (2012). Folate and DNA methylation: A review of molecular mechanisms and the evidence for folate's role. *Adv. Nutr.* 3 (1): 21–38.

88 Kim, Y.-I. (2005). Nutritional epigenetics: Impact of folate deficiency on DNA methylation and colon cancer susceptibility. *J. Nutr.* 135 (11): 2703–2709.

89 Eryılmaz Pehlivan, F. (2018). Vitamin C: An Epigenetic Regulator. In Vitamin C, ed. J.G. LeBlanc, 1-12. London, England: IntechOpen.

90 Beckett, E.L., Yates, Z., Veysey, M., Duesing, K., and Lucock, M. (2014). The role of vitamins and minerals in modulating the expression of microRNA. *Nutr. Res. Rev* 27 (1): 94–106.

91 Kim, Y.J., Ku, S.Y., Rosenwaks, Z., Liu, H.C., Chi, S.W., Kang, J.S., Lee, W.J., Jung, K.C., Kim, S.H., Choi, Y.M., Kim, J.G., and Moon, S.Y. (2010). MicroRNA expression profiles are altered by gonadotropins and vitamin C Status during in vitro follicular growth. *Reprod. Sci.* 17 (12): 1081–1089.

92 Remely, M., Ferk, F., Sterneder, S., Setayesh, T., Kepcija, T., Roth, S., Noorizadeh, R., Greunz, M., Rebhan, I., Wagner, K.-H., Knasmüller, S., and Haslberger, A. (2017). Vitamin E modifies high-fat diet-induced increase of DNA strand breaks, and changes in expression and DNA methylation of Dnmt1 and MLH1 in C57BL/6J male mice. *Nutrients* 9 (6). https://doi.org/10.3390/nu9060607.

93 Rimbach, G., Moehring, J., Huebbe, P., and Lodge, J.K. (2010). Gene-regulatory activity of alpha-tocopherol. *Molecules* 15 (3): 1746–1761.

94 Fernandes, G.F.S., Silva, G.D.B., Pavan, A.R., Chiba, D.E., Chin, C.M., and Dos Santos, J.L. (2017). Epigenetic regulatory mechanisms induced by resveratrol. *Nutrients* 9 (11). https://doi.org/10.3390/nu9111201.

95 Kala, R., Shah, H.N., Martin, S.L., and Tollefsbol, T.O. (2015). Epigenetic-based combinatorial resveratrol and pterostilbene alters DNA damage response by affecting SIRT1 and DNMT enzyme expression, including SIRT1-dependent γ-H2AX and telomerase regulation in triple-negative breast cancer. *BMC Cancer* 15 (1): 672.

96 Fu, L.-J., Ding, Y.-B., Wu, L.-X., Wen, C.-J., Qu, Q., Zhang, X., and Zhou, -H.-H. (2014). The effects of lycopene on the methylation of the GSTP1 promoter and global methylation in prostatic cancer cell lines PC3 and LNCaP. *Int. J. Endocrinol* 2014: 620165.

97 Zhang, Y. and Chen, H. (2011). Genistein, an epigenome modifier during cancer prevention. *Epigenetics* 6 (7): 888–891.

98 Matsuda, S., Minami, A., Ono, Y., and Kitagishi, Y. (2015). Neuroprotection of Genistein in Alzheimer's Disease. In Diet and Nutrition in Dementia and Cognitive Decline, ed. C.R. Martin and V.R. Preedy, 1003–1010. San Diego, CA: Elsevier.

99 Kaufman-Szymczyk, A., Majewski, G., Lubecka-Pietruszewska, K., and Fabianowska-Majewska, K. (2015). The role of sulforaphane in epigenetic mechanisms, including interdependence between histone modification and DNA methylation. *Int. J. Mol. Sci.* 16 (12): 29732–29743.

100 Gerhauser, C. (2013). Epigenetic impact of dietary isothiocyanates in cancer chemoprevention. *Curr. Opin. Clin. Nutr. Metab. Care* 16 (4): 405–410.

101 Sundaram, M.K., R, P., Haque, S., Akhter, N., Khan, S., Ahmed, S., and Hussain, A. (2021). Dietary isothiocyanates inhibit cancer progression by modulation of epigenome. *Semin. Cancer Biol.* https://doi.org/10.1016/j.semcancer.2020.12.021.

102 Nandakumar, V., Vaid, M., and Katiyar, S.K. (2011). (-)-epigallocatechin-3-gallate reactivates silenced tumor suppressor genes, cip1/P21 and P16INK4a, by reducing DNA methylation and increasing histones acetylation in human skin cancer cells. *Carcinogenesis* 32 (4): 537–544.

103 Yiannakopoulou, E.C. (2015). Targeting DNA methylation with green tea catechins. *Pharmacology* 95: 111–116. doi: 10.1159/000375503.

104 Maugeri, A., Mazzone, M.G., Giuliano, F., Vinciguerra, M., Basile, G., Barchitta, M., and Agodi, A. (2018). Curcumin modulates DNA methyltransferase functions in a cellular model of diabetic retinopathy. *Oxid. Med. Cell. Longev.* 2018: 5407482.

105 Kedhari Sundaram, M., Hussain, A., Haque, S., Raina, R., and Afroze, N. (2019). Quercetin modifies 5′CpG promoter methylation and reactivates various tumor

suppressor genes by modulating epigenetic marks in human cervical cancer cells. *J. Cell. Biochem.* 120 (10): 18357–18369.
106 Burdeos, G.C., Blank, R., and Wolffram, S. (2020). Influence of quercetin on the global DNA methylation pattern in pigs. *Food Funct.* 11 (9): 7421–7426.
107 Stepani, V., Novak, R., and Gall, K. (2014). Epigenome, Cancer Prevention and Flavonoids and Curcumin. In Epigenetics and Epigenomics, ed. C.J. Payne, 173–177. London, England: InTech.
108 Xu, X., Miao, Z., Sun, M., and Wan, B. (2021). Epigenetic mechanisms of paternal stress in offspring development and diseases. *Int. J. Genomics* 2021: 6632719.
109 Cunningham, A.M., Walker, D.M., and Nestler, E.J. (2021). Paternal transgenerational epigenetic mechanisms mediating stress phenotypes of offspring. *Eur. J. Neurosci.* 53 (1): 271–280.
110 Soubry, A., Murphy, S.K., Wang, F., Huang, Z., Vidal, A.C., Fuemmeler, B.F., Kurtzberg, J., Murtha, A., Jirtle, R.L., Schildkraut, J.M., and Hoyo, C. (2015). Newborns of obese parents have altered DNA methylation patterns at imprinted genes. *Int. J. Obes.* 39 (4): 650–657.
111 Soubry, A., Schildkraut, J.M., Murtha, A., Wang, F., Huang, Z., Bernal, A., Kurtzberg, J., Jirtle, R.L., Murphy, S.K., and Hoyo, C. (2013). Paternal obesity is associated with IGF2 hypomethylation in newborns: Results from a newborn epigenetics study (NEST) cohort. *BMC Med.* 11 (1): 29.
112 Kaati, G., Bygren, L.O., Pembrey, M., and Sjöström, M. (2007). Transgenerational response to nutrition, early life circumstances and longevity. *Eur. J. Hum. Genet.* 15 (7): 784–790.
113 Ismail, H. and Chang, Y.-L. (2012). Sequelae of fetal growth restriction. *J. Med. Ultrasound* 20 (4): 191–200.
114 Ellsworth, L., Harman, E., Padmanabhan, V., and Gregg, B. (2018). Lactational programming of glucose homeostasis: A window of opportunity. *J. Reprod. Fertil.* 156 (2): R23–R42.
115 Quinn, E.A. (2021). Centering human milk composition as normal human biological variation. *Am. J. Hum. Biol.* 33 (1): e23564.
116 Hiremagalur, B.K., Vadlamudi, S., Johanning, G.L., and Patel, M.S. (1993). Long-term effects of feeding high carbohydrate diet in pre-weaning period by gastrostomy: A new rat model for obesity. *Int. J. Obes. Relat. Metab. Disord.* 17 (9): 495–502.
117 Patel, M.S. and Srinivasan, M. (2011). Metabolic programming in the immediate postnatal life. *Ann. Nutr. Metab.* 58 (s2): 18–28.
118 Glavas, M.M., Kirigiti, M.A., Xiao, X.Q., Enriori, P.J., Fisher, S.K., Evans, A.E., Grayson, B.E., Cowley, M.A., Smith, M.S., and Grove, K.L. (2010). Early overnutrition results in early-onset arcuate leptin resistance and increased sensitivity to high-fat diet. *Endocrinology* 151 (4): 1598–1610.
119 Liu, Z., Lim, C.Y., Su, M.Y.-F., Soh, S.L.Y., Shui, G., Wenk, M.R., Grove, K.L., Radda, G.K., Han, W., and Xiao, X. (2013). Neonatal overnutrition in mice exacerbates high-fat diet-induced metabolic perturbations. *J. Endocrinol.* 219 (2): 131–143.
120 McCann, M.F., Baydar, N., and Williams, R.L. (2007). Breastfeeding attitudes and reported problems in a national sample of WIC participants. *J. Hum. Lact.* 23 (4): 314–324.
121 McCann, M.F., Baydar, N., and Williams, R.L. (2008). Consumption of soft drinks and other sweet drinks by WIC infants. *Am. J. Public Health* 98 (10): 1735.

122 Diamanti-Kandarakis, E., Bourguignon, J.-P., Giudice, L.C., Hauser, R., Prins, G.S., Soto, A.M., Zoeller, R.T., and Gore, A.C. (2009). Endocrine-disrupting chemicals: An endocrine society scientific statement. *Endocr. Rev* 30 (4): 293–342.

123 Anway, M.D. and Skinner, M.K. (2006). Epigenetic transgenerational actions of endocrine disruptors. *Endocrinology* 147 (6): s43–s49.

124 Skinner, M.K., Manikkam, M., and Guerrero-Bosagna, C. (2011). Epigenetic transgenerational actions of endocrine disruptors. *Reprod. Toxicol.* 31 (3): 337–343.

125 Skinner, M.K., Manikkam, M., and Guerrero-Bosagna, C. (2010). Epigenetic transgenerational actions of environmental factors in disease etiology. *Trends Endocrinol. Metab.* 21 (4): 214–222.

126 Palanza, P., Gioiosa, L., Vom Saal, F.S., and Parmigiani, S. (2008). Effects of developmental exposure to bisphenol A on brain and behavior in mice. *Environ. Res.* 108 (2): 150–157.

127 Tang, W.-Y., Morey, L.M., Cheung, Y.Y., Birch, L., Prins, G.S., and Ho, S.-M. (2012). Neonatal exposure to estradiol/bisphenol A alters promoter methylation and expression of Nsbp1 and Hpcal1 genes and transcriptional programs of Dnmt3a/b and Mbd2/4 in the rat prostate gland throughout life. *Endocrinology* 153 (1): 42–55.

128 Song, X., Miao, M., Zhou, X., Li, D., Tian, Y., Liang, H., Li, R., and Yuan, W. (2019). Bisphenol A exposure and sperm ACHE hydroxymethylation in men. *Int. J. Environ. Res. Public Health* 16 (1): 152.

129 Vandenberg, L.N., Maffini, M.V., Wadia, P.R., Sonnenschein, C., Rubin, B.S., and Soto, A.M. (2007). Exposure to environmentally relevant doses of the xenoestrogen bisphenol-A alters development of the fetal mouse mammary gland. *Endocrinology* 148 (1): 116–127.

130 Wadia, P.R., Cabaton, N.J., Borrero, M.D., Rubin, B.S., Sonnenschein, C., Shioda, T., and Soto, A.M. (2013). Low-dose BPA exposure alters the mesenchymal and epithelial transcriptomes of the mouse fetal mammary gland. *PLoS One* 8 (5): e63902.

131 Patel, S., Brehm, E., Gao, L., Rattan, S., Ziv-Gal, A., and Flaws, J.A. (2017). Bisphenol A exposure, ovarian follicle numbers, and female sex steroid hormone levels: Results from a CLARITY-BPA study. *Endocrinology* 158 (6): 1727–1738.

132 Prins, G.S., Patisaul, H.B., Belcher, S.M., and Vandenberg, L.N. (2019). CLARITY-BPA academic laboratory studies identify consistent low-dose bisphenol A effects on multiple organ systems. *Basic Clin. Pharmacol. Toxicol.* 125 (S3): 14–31.

133 Strifert, K. (2015). An epigenetic basis for autism spectrum disorder risk and oral contraceptive use. *Med. Hypotheses* 85 (6): 1006–1011.

134 National Toxicology Program (2011). Diethylstilbestrol. *Rep. Carcinog* 12: 159–161.

135 De Assis, S., Warri, A., Cruz, M.I., Laja, O., Tian, Y., Zhang, B., Wang, Y., Huang, T.H.-M., and Hilakivi-Clarke, L. (2012). High-fat or ethinyl-oestradiol intake during pregnancy increases mammary cancer risk in several generations of offspring. *Nat. Commun* 3 (1): 1053.

136 Soloway, A.H. (2014). Potential carcinogens from steroid hormones and diethyl stilbesterol (DES): Chemical relationships between breast, ovarian and prostate cancers. *J. Drug Metab. Toxicol.* 05 (02). https://doi.org/10.4172/2157-7609.1000170.

137 Champ, M.A.A. (2000). Review of organotin regulatory strategies, pending actions, related costs and benefits. *Sci. Total Environ* 258 (1–2): 21–71.

138 Schmeits, P.C.J., Van Kol, S., Van Loveren, H., Peijnenburg, A.A.C.M., and Hendriksen, P.J.M. (2014). The effects of tributyltin oxide and deoxynivalenol on the transcriptome of the mouse thymoma cell line EL-4. *Toxicol. Res. (Camb.)* 3 (4): 254–265.

139 Zhang, J., Zuo, Z., He, C., Cai, J., Wang, Y., Chen, Y., and Wang, C. (2009). Effect of tributyltin on testicular development in sebastiscus marmoratus and the mechanism involved. *Environ. Toxicol. Chem.* 28 (7): 1528–1535.

140 Hanaoka, S., Ishida, K., Tanaka, S., Sakamoto, S., Okuda, K., Sanoh, S., Ohta, S., and Kotake, Y. (2018). Tributyltin induces epigenetic changes and decreases the expression of nuclear respiratory factor-1. *Metallomics* 10 (2): 337–345.

141 Braicu, C., Mehterov, N., Vladimirov, B., Sarafian, V., Nabavi, S.M., Atanasov, A.G., and Berindan-Neagoe, I. (2017). Nutrigenomics in cancer: Revisiting the effects of natural compounds. *Semin. Cancer Biol.* 46: 84–106.

142 Yeung, A.W.K., Mocan, A., and Atanasov, A.G. (2018). Let food be thy medicine and medicine be thy food: A bibliometric analysis of the most cited papers focusing on nutraceuticals and functional foods. *Food Chem.* 269: 455–465.

143 CDC (2021). Deaths and mortality. CDC, October 19. https://www.cdc.gov/nchs/fastats/deaths.htm.

144 Nam, A.-R., Lee, K.-H., Hwang, H.-J., Schabort, J.J., An, J.-H., Won, S.-H., and Cho, J.-Y. (2020). Alternative methylation of intron motifs is associated with cancer-related gene expression in both canine mammary tumor and human breast cancer. *Clin. Epigenetics* 12 (1): 110.

145 Johanning, G.L., Heimburger, D.C., and Piyathilake, C.J. (2002). DNA methylation and diet in cancer. *J. Nutr.* 132 (12): 3814S–3818S.

146 Dumitrescu, R.G. (2018). Alcohol-induced epigenetic changes in cancer. *Methods Mol. Biol.* 1856: 157–172.

147 Sales, N.M.R., Pelegrini, P.B., and Goersch, M.C. (2014). Nutrigenomics: Definitions and advances of this new science. *J. Nutr. Metab.* 2014: 202759.

148 Mentella, M.C., Scaldaferri, F., Ricci, C., Gasbarrini, A., and Miggiano, G.A.D. (2019). Cancer and mediterranean diet: A review. *Nutrients* 11 (9): 2059.

149 Narita, S., Nara, T., Sato, H., Koizumi, A., Huang, M., Inoue, T., and Habuchi, T. (2019). Research evidence on high-fat diet-induced prostate cancer development and progression. *J. Clin. Med.* 8 (5): 597.

150 Fang, X., Wei, J., He, X., An, P., Wang, H., Jiang, L., Shao, D., Liang, H., Li, Y., Wang, F., and Min, J. (2015). Landscape of dietary factors associated with risk of gastric cancer: A systematic review and dose-response meta-analysis of prospective cohort studies. *Eur. J. Cancer* 51 (18): 2820–2832.

151 Nasir, A., Bullo, M.M.H., Ahmed, Z., Imtiaz, A., Yaqoob, E., Jadoon, M., Ahmed, H., Afreen, A., and Yaqoob, S. (2020). Nutrigenomics: Epigenetics and cancer prevention: A comprehensive review. *Crit. Rev. Food Sci. Nutr* 60 (8): 1375–1387.

152 Alattas, M., Ross, C.S., Henehan, E.R., and Naimi, T.S. (2020). Alcohol policies and alcohol-attributable cancer mortality in U.S. States. *Chem. Biol. Interact.* 315 (108885): 108885.

153 Cantacorps, L., Alfonso-Loeches, S., Guerri, C., and Valverde, O. (2019). Long-term epigenetic changes in offspring mice exposed to alcohol during gestation and lactation. *J. Psychopharmacol.* 33 (12): 1562–1572.

154 Fu, H., Zeng, J., Liu, C., Gu, Y., Zou, Y., and Chang, H. (2021). Folate intake and risk of pancreatic cancer: A systematic review and updated meta-analysis of epidemiological studies. *Dig. Dis. Sci.* 66 (7): 2368–2379.

155 Clasen, J.L., Heath, A.K., Scelo, G., and Muller, D.C. (2020). Components of one-carbon metabolism and renal cell carcinoma: A systematic review and meta-analysis. *Eur. J. Nutr.* 59 (8): 3801–3813.

156 Zeng, J., Gu, Y., Fu, H., Liu, C., Zou, Y., and Chang, H. (2020). Association between one-carbon metabolism-related vitamins and risk of breast cancer: A systematic review and meta-analysis of prospective studies. *Clin. Breast Cancer* 20 (4): e469–e480.

157 Yuan, J., Li, J., Shang, M., Fu, Y., and Wang, T. (2021). Identification of vitamin B6 as a PD-L1 suppressor and an adjuvant for cancer immunotherapy. *Biochem. Biophys. Res. Commun* 561: 187–194.

158 Kaiser, M., Semeraro, M.D., Herrmann, M., Absenger, G., Gerger, A., and Renner, W. (2021). Immune aging and immunotherapy in cancer. *Int. J. Mol. Sci.* 22 (13): 7016.

159 Callejo, A., Frigola, J., Iranzo, P., Carbonell, C., Diaz, N., Marmolejo, D., Assaf, J.D., Cedrés, S., Martinez-Marti, A., Navarro, A., Pardo, N., Amat, R., and Felip, E. (2021). Interrelations between patients' clinicopathological characteristics and their association with response to immunotherapy in a real-world cohort of NSCLC patients. *Cancers (Basel)* 13 (13): 3249.

160 Dyhl-Polk, A., Mikkelsen, M.K., Ladekarl, M., and Nielsen, D.L. (2021). Clinical trials of immune checkpoint inhibitors in hepatocellular carcinoma. *J. Clin. Med.* 10 (12): 2662.

161 Dimitriou, F., Staeger, R., Ak, M., Maissen, M., Kudura, K., Barysch, M.J., Levesque, M.P., Cheng, P.F., Dummer, R., and Mangana, J. (2021). Frequency, treatment and outcome of immune-related toxicities in patients with immune-checkpoint inhibitors for advanced melanoma: Results from an institutional database analysis. *Cancers (Basel)* 13 (12): 2931.

162 Boughanem, H., Hernandez-Alonso, P., Tinahones, A., Babio, N., Salas-Salvadó, J., Tinahones, F.J., and Macias-Gonzalez, M. (2020). Association between serum vitamin B12 and Global DNA methylation in colorectal cancer patients. *Nutrients* 12 (11): 3567.

163 Piyathilake, C.J., Johanning, G.L., Macaluso, M., Whiteside, M., Oelschlager, D.K., Heimburger, D.C., and Grizzle, W.E. (2000). Localized folate and vitamin B-12 deficiency in squamous cell lung cancer is associated with global DNA hypomethylation. *Nutr. Cancer* 37 (1): 99–107.

164 Heimburger, D.C., Alexander, C.B., Birch, R., Butterworth, C.E., Jr, Bailey, W.C., and Krumdieck, C.L. (1988). Improvement in bronchial squamous metaplasia in smokers treated with folate and vitamin B12. Report of a preliminary randomized, double-blind intervention trial. *JAMA* 259 (10): 1525–1530.

165 Kovina, A.P., Petrova, N.V., Razin, S.V., and Kantidze, O.L. (2020). L-Ascorbic acid in the epigenetic regulation of cancer development and stem cell reprogramming. *Acta Naturae* 12 (4): 5–14.

166 Mikkelsen, S.U., Gillberg, L., Lykkesfeldt, J., and Grønbæk, K. (2021). The role of vitamin C in epigenetic cancer therapy. *Free Radic. Biol. Med.* 170: 179–193.

167 Brabson, J.P., Leesang, T., Mohammad, S., and Cimmino, L. (2021). Epigenetic regulation of genomic stability by vitamin C. *Front. Genet.* 12: 675780.

168 Lee Chong, T., Ahearn, E.L., and Cimmino, L. (2019). Reprogramming the epigenome with vitamin C. *Front. Cell Dev. Biol.* 7: 128.

169 Wang, T., Chen, K., Zeng, X., Yang, J., Wu, Y., Shi, X., Qin, B., Zeng, L., Esteban, M.A., Pan, G., and Pei, D. (2011). The histone demethylases Jhdm1a/1b enhance somatic cell reprogramming in a vitamin-C-Dependent manner. *Cell Stem Cell* 9 (6): 575–587.

170 Karlic, H. and Varga, F. (2011). Impact of vitamin D metabolism on clinical epigenetics. *Clin. Epigenetics* 2 (1): 55–61.

171 Tapp, H.S., Commane, D.M., Bradburn, D.M., Arasaradnam, R., Mathers, J.C., Johnson, I.T., and Belshaw, N.J. (2013). Nutritional factors and gender influence age-related dna methylation in the human rectal mucosa. *Aging Cell* 12 (1): 148–155.

172 Hanwell, H.E.C. and Banwell, B. (2011). Assessment of evidence for a protective role of vitamin D in multiple sclerosis. *Biochim. Biophys. Acta* 1812 (2): 202–212.

173 Javaid, M.K., Crozier, S.R., Harvey, N.C., Gale, C.R., Dennison, E.M., Boucher, B.J., Arden, N.K., Godfrey, K.M., and Cooper, C. (2006). Princess anne hospital study group. Maternal vitamin D status during pregnancy and childhood bone mass at age 9 years: A longitudinal study. *Lancet* 367 (9504): 36–43.

174 Hong, Y., Lee, J., Moon, H., Ryu, C.H., Seok, J., Jung, Y.-S., Ryu, J., and Baek, S.J. (2021). Quercetin induces anticancer activity by upregulating pro-NAG-1/GDF15 in differentiated thyroid cancer cells. *Cancers (Basel)* 13 (12): 3022.

175 Özgöçmen, M., Bayram, D., Armağan, İ., Türel, G.Y., Sevimli, M., and Şenol, N. (2021). Is quercetin beneficial for colon cancer? A cell culture study, using the apoptosis pathways. *Anticancer Agents Med. Chem.* 21. https://doi.org/10.2174/1871520621666210624110547.

176 Elsayed, M.M.A., Okda, T.M., Atwa, G.M.K., Omran, G.A., Abd Elbaky, A.E., and Ramadan, A.E.H. (2021). Design and optimization of orally administered luteolin nanoethosomes to enhance its anti-tumor activity against hepatocellular carcinoma. *Pharmaceutics* 13 (5). https://doi.org/10.3390/pharmaceutics13050648.

177 Vodnik, V.V., Mojić, M., Stamenović, U., Otoničar, M., Ajdžanović, V., Maksimović-Ivanić, D., Mijatović, S., Marković, M.M., Barudžija, T., Filipović, B., Milošević, V., and Šošić-Jurjević, B. (2021). Development of genistein-loaded gold nanoparticles and their antitumor potential against prostate cancer cell lines. *Mater. Sci. Eng. C Mater. Biol. Appl.* 124 (112078): 112078.

178 Pandey, P., Khan, F., and Maurya, P. (2021). Targeting Jab1 using hesperidin (dietary phytocompound) for inducing apoptosis in HeLa cervical cancer cells. *J. Food Biochem.* 45 (7): e13800.

179 Tofolean, I.T., Ganea, C., Ionescu, D., Filippi, A., Garaiman, A., Goicea, A., Gaman, M.-A., Dimancea, A., and Baran, I. (2016). Cellular determinants involving mitochondrial dysfunction, oxidative stress and apoptosis correlate with the synergic cytotoxicity of epigallocatechin-3-Gallate and menadione in human leukemia jurkat T cells. *Pharmacol. Res.* 103: 300–317.

180 Zhang, Z., Pan, Y., Zhao, Y., Ren, M., Li, Y., Lu, G., Wu, K., and He, S. (2021). Delphinidin modulates JAK/STAT3 and MAPKinase signaling to induce apoptosis in HCT116 cells. *Environ. Toxicol.* 36 (8): 1557–1566.

181 Kopustinskiene, D.M., Jakstas, V., Savickas, A., and Bernatoniene, J. (2020). Flavonoids as anticancer agents. *Nutrients* 12 (2): 457.

182 Al Theyab, A., Almutairi, T., Al-Suwaidi, A.M., Bendriss, G., McVeigh, C., and Chaari, A. (2020). Epigenetic effects of gut metabolites: Exploring the path of dietary prevention of type 1 diabetes. *Front. Nutr.* 7: 563605.

183 Salek Farrokhi, A., Mohammadlou, M., Abdollahi, M., Eslami, M., and Yousefi, B. (2020). Histone deacetylase modifications by probiotics in colorectal cancer. *J. Gastrointest. Cancer* 51 (3): 754–764.

184 Hill, C., Guarner, F., Reid, G., Gibson, G.R., Merenstein, D.J., Pot, B., Morelli, L., Canani, R.B., Flint, H.J., Salminen, S., Calder, P.C., and Sanders, M.E. (2014). Expert consensus document. The International scientific association for probiotics and prebiotics consensus statement on the scope and appropriate use of the term probiotic: Expert consensus document. *Nat. Rev. Gastroenterol. Hepatol.* 11 (8): 506–514.

185 Floch, M.H. (2014). Probiotics and Prebiotics. *Gastroenterol. Hepatol. (N. Y* 10 (10): 680–681.

186 Kumar, M., Nagpal, R., Verma, V., Kumar, A., Kaur, N., Hemalatha, R., Gautam, S.K., and Singh, B. (2013). Probiotic metabolites as epigenetic targets in the prevention of colon cancer. *Nutr. Rev.* 71 (1): 23–34.

187 Worthley, D.L., Le Leu, R.K., Whitehall, V.L., Conlon, M., Christophersen, C., Belobrajdic, D., Mallitt, K.-A., Hu, Y., Irahara, N., Ogino, S., Leggett, B.A., and Young, G.P. (2009). A human, double-blind, placebo-controlled, crossover trial of prebiotic, probiotic, and synbiotic supplementation: Effects on luminal, inflammatory, epigenetic, and epithelial biomarkers of colorectal cancer. *Am. J. Clin. Nutr* 90 (3): 578–586.

188 Stenman, L.K., Waget, A., Garret, C., Briand, F., Burcelin, R., Sulpice, T., and Lahtinen, S. (2015). Probiotic B420 and prebiotic polydextrose improve efficacy of antidiabetic drugs in mice. *Diabetol. Metab. Syndr* 7 (1): 75.

189 Morovic, W. and Budinoff, C.R. (2021). Epigenetics: A new frontier in probiotic research. *Trends Microbiol.* 29 (2): 117–126.

190 Miro-Blanch, J. and Yanes, O. (2019). Epigenetic regulation at the interplay between gut microbiota and host metabolism. *Front. Genet.* 10: 638.

191 Dasari, S., Kathera, C., Janardhan, A., Praveen Kumar, A., and Viswanath, B. (2017). Surfacing role of probiotics in cancer prophylaxis and therapy: A systematic review. *Clin. Nutr.* 36 (6): 1465–1472.

192 Wegh, C.A.M., Geerlings, S.Y., Knol, J., Roeselers, G., and Belzer, C. (2019). Postbiotics and their potential applications in early life nutrition and beyond. *Int. J. Mol. Sci.* 20 (19): 4673.

193 Paul, B., Barnes, S., Demark-Wahnefried, W., Morrow, C., Salvador, C., Skibola, C., and Tollefsbol, T.O. (2015). Influences of diet and the gut microbiome on epigenetic modulation in cancer and other diseases. *Clin. Epigenetics* 7 (1): 112.

16

Genetic and Epigenetic Biomarkers of Organophosphate Compounds, Dialkyl Phosphate Exposure, and Their Relation to Biological Effects

David S. Hernández-Toledano and Libia Vega

Department of Toxicology, Center for Research and Advanced Studies of the National Polytechnic Institute, México

Introduction

Organic synthetic phosphorus compounds are a vast group of chemicals that have several applications. These compounds consist of a phosphate group, usually from phosphoric acid, modified with diverse chemical groups that include alkyl, aryl, halogenated, ester, or amine substituents; these potentially change the properties of the original phosphate group (Timperley and Tattersall 2014). Within these compounds, organophosphates (OPs) are a phosphate group esterified to its substituents; they have different applications, the most common ones being in pesticides, warfare agents, flame retardants, and plasticizers. The OP compounds are ubiquitously distributed in the environment, including in such diverse matrices as air, water, foods, and soils as a result of industrial processes, agricultural and urban applications, and parent compound degradation. This scenario brings the possibility of chronic exposure, both occupational and non-occupational, which in turn increases the risk that the populations affected would develop various diseases, for example several types of cancer, neurological conditions, infertility, and immunotoxic effects.

OP compounds are not harmless. High-dose exposure, whether occupational, accidental, or intentional, causes high toxicity, mostly related to neurotoxic effects. To date, numerous reports have described other toxicological effects that can also be produced by low doses and chronic exposure patterns.

Among these harmful effects, diverse studies have pointed to the effects of OPs on the DNA. The outcome can be cell death, infertility, cancer development, and more. The exact mechanisms and molecular targets are still unclear; however, the effects have been reported in several models, thanks to the identification of genotoxic biomarkers. In addition to genotoxicity, more recent studies are now showing that OPs also modify gene expression and contribute to disease development. This gene expression may be related to different molecular targets, which still remain to be identified; but the recognition of potential biomarkers for these gene changes is important, as it contributes to our knowledge of the toxicity of OP compounds.

Genomic and Epigenomic Biomarkers of Toxicology and Disease: Clinical and Therapeutic Actions,
First Edition. Edited by Saura C. Sahu.

Organophosphate Compounds

OP compounds are synthetic chemicals, first introduced as pesticides and warfare agents during World War II. These substances, esters derived from phosphoric or thio-phosphoric acids, possess a central phosphorus atom bonded to sulfur or oxygen atoms and diverse esterified substituents, such as alkyl, aryl, amine, cyanide, or halogenated substituents (Figure 16.1).

During the nineteenth century, the first OP compounds were synthesized, but it was not until their biological effects were registered that OPs were used as potential warfare agents and insecticides. In addition to insecticides and warfare agents, other products have been developed from OP compounds for different applications, depending on the chemical or physical properties and biological effects of each compound. For example, pesticides and warfare agents are used mainly because of OP's affinity for cholinesterase, which causes neurotoxic effects (Casida and Quistad 2004). Ophthalmic products take advantage of these molecular targets; thus OPs are used to induce dilation of the pupil (miosis) in the treatment of some eye diseases, such as cataracts. Flame retardants are mainly trialkyl-modified OP compounds with a high boiling point and a low volatility that can react with H^+ and OH^- ions to slow combustion (van der Veen and de Boer 2012). Additionally, OPs can also be used as lubricants, plasticizers, and pore size regulators, which probably explains their presence in a range of products as wide as floor polish, hydraulic fluids, textiles, rubber, polyurethane foam, cutting oils, resins, glues, thermoplastics, epoxies, and phenolic resins (Andresen et al. 2004; Casida 2017; Terry 2012; van der Veen and de Boer 2012).

The global production of OP esters was 500,000 tons worldwide in 2015 (Wang et al. 2015). The increased production of OP-derived substances resulted in augmented exposure, both occupational and non-occupational. Exposure can occur even via OP degradation products or metabolites, which are found in agricultural, industrial, and domestic environments and are present in the air, in water, in dust, and in foods (Chen et al. 2013; Hernández et al. 2019; Tamaro et al. 2018). The main concern raised by the distribution of these substances regards their biological effects. Initially it was thought that these OP substances have only acute neurotoxic effects, but evidence has now increasingly related them to much wider array—from oxidant damage, alkylating damage, and signaling pathway disruption to infertility, endocrine disruption, immunotoxicity, embryo defects, and cancer development.

OP pesticides are one of the most studied groups of OP substances. They were originally produced as alternatives to organochloride pesticides, by reason of their lower toxicity and lower environmental persistence. Like many other pesticide compounds, OP pesticides react with the active catalytic site of acetylcholinesterase (AChE) (Casida and Quistad 2004), causing acute neurotoxic effects. However, ecotoxicological and epidemiological studies of occupational and non-occupational exposures have shown a wide range of effects, suggesting different mechanisms and targets that are not fully explained by AChE

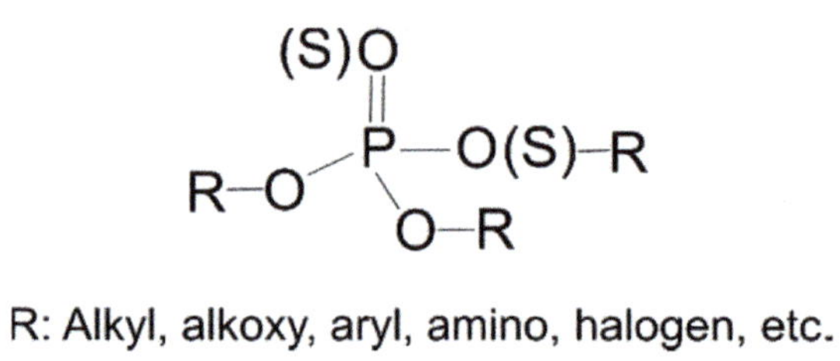

Figure 16.1 General structure of an OP. R substituents are different chemical groups that may be the same or different in the molecule.

or hydrolase inhibition. One of these effects is genotoxicity, which has been explained as resulting from the oxidative stress caused by OP metabolism and which can be associated with the development of cancer from chronic or low-dose OP exposures. Additionally, *in vitro* models have shown that OP pesticides modify gene expression, which may contribute to understanding OP toxicity. OP genotoxicity and its role in gene expression will be further discussed in the following sections.

OP flame retardants have different toxic effects, and possibly non-genotoxic or mutagenic effects. They can be classified into two categories: halogenated and non-halogenated. The halogenated group has the greatest toxicological relevance, because retardants of this type produce effects such as carcinogenicity, immunotoxicity, and reproductive problems. This may be related to the presence of the halogen atom, frequently chloride, which makes them more reactive and more persistent (Li et al. 2019).

As mentioned earlier, OPs can exert toxicity for long periods after exposure. This presupposes the participation of their metabolites, since OP pesticides are quickly metabolized once they enter an organism: they result in hydrolyzed, oxidized, or conjugated products. Among oxidized products, the oxon form is described as one of the most reactive; and it is also mainly responsible for the neurotoxic effects. Also, hydrolysis of the leaving group can form specific metabolites. These metabolites possess a specific chemical structure, which varies depending on the pesticide and confers different biological properties. During OP pesticide hydrolysis, non-specific metabolites are formed; they are called dialkyl phosphates (DAPs). Traditionally, DAPs have been used to determine the presence of these compounds, both clinically and in the environment. This usage is due to the fact that DAPs have longer half-life times that their parent compounds. They can be less reactive than oxon or halogenated substituents, which have been accepted as not exerting biological effects; and they can be easily used as exposure biomarkers. However, recent reports have found *in vitro* and *in vivo* effects of these non-specific metabolites, which may also be implicated in OP toxicity.

Dialkyl Phosphates

DAPs are classified as phosphoryl, thiophosphoryl, or di-thiophosphoryl esters, which all contain a phosphorus atom bonded to oxygen or sulfur atoms, just like other OP compounds, and esterified to alkyl substituents (methyl, ethyl, isopropyl, butyl, etc.; see Figure 16.2A). These are small hydrophilic molecules with acidic properties that are modified according to the alkyl substituents. An increased size of the alkyl substituent results in a decrease in the acidity of the molecule. Also, an increase in the number of sulfur atoms decreases the acidity in water but can increase it in alcoholic solutions. In the case of phosphorothioates (one sulfur atom), an oxo tautomer is present in a solution that can be stabilized according to the polarity of the solvent; however, a thiono tautomer is more stable (Figure 16.2B; see Fest and Schmidt 1982).

Alkyl phosphates are used for the synthesis of several products, principally pesticides, flame retardants, emulsifiers, and plasticizers, increasing the presence of OP compounds in the environment. Specifically, DAPs are produced for the manufacture of several OP pesticides, since they are esters or amide derivatives. From OP pesticides, we can identify two major groups of DAPs, ethyl and methyl DAPs, which differ in their alkyl esterified substituents (Figure 16.3). The ethyl group consists of diethyldithiophosphate (DEDTP), diethylthiophosphate (DETP), and diethylphosphate (DEP), the while methyl DAPs are dimethyldithiophosphate (DMDTP), dimethylthiophosphate (DMTP), and dimethylphosphate (DMP) (Kavvalakis and Tsatsakis 2012). The dithiophosphates DEDTP and DMDTP

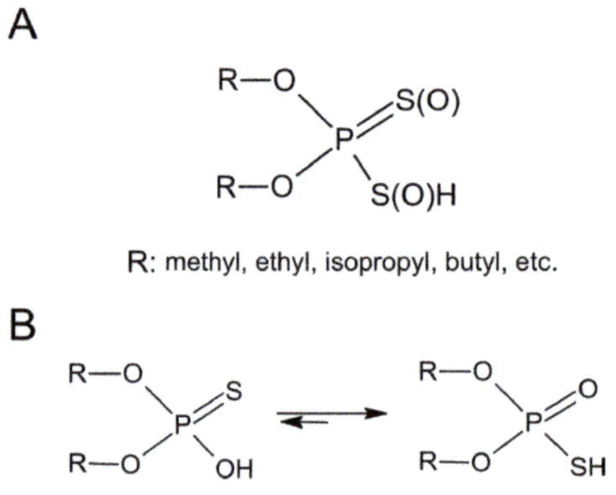

Figure 16.2 (A) Dialkyl phosphate general formula; (B) thiono-thiol equilibrium in OP thionic acids (modified from Fest and Schmidt 1982).

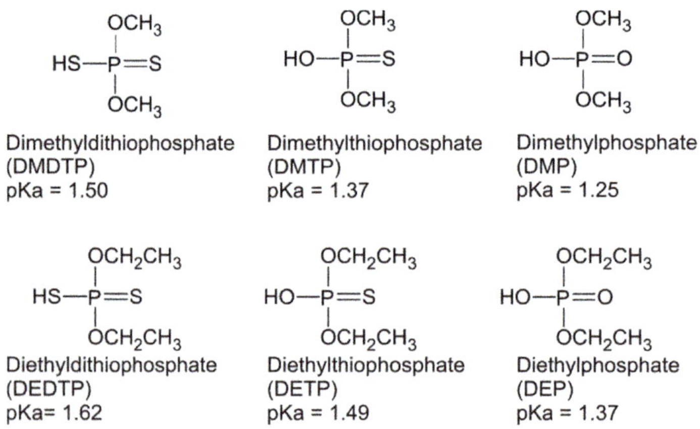

Figure 16.3 DAP structures and pKa. Upper row: methyl DAPs. Lower row: ethyl DAPs.

are synthesized from phosphorus pentasulfide and the corresponding alcohol (ethylic or methyl alcohols), while the thiophosphates DETP and DMTP are produced from such diverse precursors as dialkyl phosphorochloridothionates, dialkyl H-thiophosphonates, or trialkyl phosphites. In any case, dithiophosphates and thiophosphates can be used to produce OP pesticides or trialkyl OP compounds such as flame retardants, or they can be used for synthetic oligo-deoxyribonucleotides (Timperley et al. 2014).

DAPs can also be produced from OP pesticide metabolism or hydrolysis. This process results in their release from the parent compound. In addition, the combustion and degradation of flame retardants produce the most oxidized forms, such as DMP (Korobeinichev et al. 2007), which are released directly into the environment. Besides, microorganisms and physical conditions such as acidity, radiation, and heat can degrade OP compounds in the environment as a result of pesticide application or pollution from industries; and this increases the presence of DAPs. OPs and DAPs have been reported in several matrices such as soils, lakes, collected rainwater, air, dust, and fruits (Andresen et al. 2004; Lasagna et al. 2020).

DAPs are best known for their use as exposure biomarkers of OPs, specifically OP pesticides. Recently, some investigations have provided evidence that these metabolites may not be harmless. DAPs can inhibit enzymes such as AChE, can induce genotoxicity by inducing DNA strand breaks and micronucleus formation both *in vivo* and *in vitro*, and can generate immunotoxicity by modifying cytokines and macrophage responses, altering microbiota, causing inflammatory responses, and impairing neurotransmitter transport in glial cells (Imaizumi et al. 1993; Vega et al. 2009; Esquivel-Senties et al. 2010; Medina-Buelvas et al. 2019a; Medina-Buelvas et al. 2019b; Olivares-Bañuelos et al. 2019; Yang et al. 2019, Yang et al. 2020; Hernández-Toledano et al. 2020). Taken together, these reports show that DAPs can affect various organs and systems that suffer general toxic effects often associated only with parent compounds. Hence research should be extended into DAPs as well.

Toxicokinetics

OP compounds can be absorbed through respiratory, oral, and dermal pathways. Dermal absorption is slower; however, it could speed up, depending on the presence of lipophilic solvents in the pesticide formulation. Respiratory and dermal exposures usually occur in occupational populations during the preparation, manipulation, or application of pesticide formulations. Acute oral exposure is mainly due to accidental or intentional self-poisoning events. Additionally, OPs can be easily absorbed orally, through the ingestion of contaminated food or water. Once in the organism, they reach systemic circulation, where they can be distributed to different organs and tissues, for example adipose tissue, the liver, and the kidneys (Kwong 2002; Timchalk 2010).

OPs are quickly metabolized in the first couple of hours after exposure, producing various metabolites. From these metabolites, the most quantified biomarkers are DAPs and a specific metabolite that is composed of the pesticide-leaving group. DAPs produced or studied *in vivo*, both in accidents and in cases of self-poisoning, can reach peak concentration in the blood within three hours from exposure and are slowly eliminated in a biphasic manner through urine, with post-exposure half-life times around 50 to 110 hours (Drevenkar et al. 1993; Timchalk et al. 2007; Vasilić et al. 1999).

In the particular case of direct exposure to DAPs, the latter exhibit high bioavailability and are easily absorbed orally during the ingestion of contaminated food (Chen et al. 2013). However, direct ingestion of these metabolites is followed by a quicker elimination phase than in the case of DAPs produced through OP pesticide exposure (Timchalk et al. 2007).

OP Exposure

Exposure to OP compounds can occur from direct contact, accidental poisoning, and self-poisoning, produced either occupationally or non-occupationally. Occupational OP exposures are mainly related to pesticide application, agricultural work, floriculture, and the chemical manufacture of OP compounds (Bolognesi and Holland 2016; Silva Pinto et al. 2020). In some cases, the greater exposure levels of OP found in the relevant populations are due to the lack of protective equipment and to low levels of knowledge about the correct use of these products. Additionally, a deficiency in security measures can cause the distribution of OP residues to non-occupationally exposed populations, such as the families of occupationally exposed workers.

Non-occupational exposures occur as a result of the high distribution of OPs in the environment. OPs and DAPs can be found in soil, urban dust, and rainwater (Quirós-Alcalá et al. 2011; Sidhu et al. 2019; Tamaro et al. 2018; Weerasekera et al. 2009); additionally, OP residues and DAPs may be ingested in fruits, juices, or contaminated water (Chen et al. 2013; Kumar et al. 2016; Zhang et al. 2008). Although exposure levels are lower than in the case of occupationally exposed populations, exposure can be chronic. In addition, such exposures have been problematic, as their real levels may be misinterpreted because of the occupational sources associated with various toxic effects.

Regulations

OP pesticides came to be regulated a few decades ago, initially on account of their toxic effects, as found in several studies. The first step was a risk assessment of OP pesticides; their acute effects, such as AChE inhibition and neurotoxicity, were looked into. Such assessments also considered exposure to OP mixtures, but the studies treated AChE inhibition as the only mechanism for facilitating the estimation of safety levels. But the practice of examining an additive effect during mixed exposures without considering the different toxicity potentials of OP compounds is a cause for concern. What is more, this approach does not look at OP effects that are reported at concentrations lower than that which causes AChE inhibition.

Many developed countries, including the United Kingdom, the European Union, and the United States, banned the agricultural use of several OP pesticides on the strength of risk assessments. Sometimes there is restricted use in special cases, such as that of vector control, or use is permitted for special considerations (Kozawa et al. 2009). Maximum residue levels were determined—in addition to safety levels and tolerance doses—primarily in order to evaluate the correct application of OP pesticides (Jara and Winter 2019). In the absence of regulation, many developing countries are still producing, applying, or selling these pesticides despite the ban policies, especially as the cost of pesticides is low. Organizations such as the Pesticide Action Network (PAN) have classified OP pesticides according to their diverse toxic effects, in an attempt to provide more information about OP risk and to contribute to pesticide control policies.

Apart from OP nerve agents, which were banned during the Chemical Weapons Convention, not all OP compounds have received the same attention with regard to safety. There are few or no existing regulations concerning safety levels, maximum residue levels, or maximum doses of daily ingestion for other OP compounds such as flame retardants, their metabolites, or even DAPs, although they can be found in food, water, and other consumed products. In the particular case of flame retardants, the first policies recently put in place are still focused on gathering evidence about their toxic effects (Blum et al. 2019).

OP Biomarkers

As previously mentioned, DAPs are mostly employed as exposure biomarkers for OP pesticides; however, many other molecular targets can also be used for the evaluation of OP compounds.

DAPs

DAPs could be the most frequent biomarkers of exposure to OP pesticides, given their longer lifetime in the organism by comparison with the lifetime of parent compounds; and this can be quantified in several biological matrices—for instance blood, plasma, urine, and hair (Kavvalakis and Tsatsakis 2012)—using analytical methods such as high-performance liquid chromatography (HPLC). These biomarkers have been extensively used in epidemiologic studies and risk assessments for OP compounds.

However, the determination of DAPs is non-specific. Given that the same DAP could be used for the synthesis of several pesticides, or can even be oxidized into another DAP, the quantification of these molecules reflects only the exposure to OPs, without being able to identity the specific parent compound. Additionally, these quantifications could be misleading, because DAPs can also be present in soil, dust, and some food products. The result would then be an overestimation when determining occupational exposures to a specific source, as happens in epidemiologic studies. This issue has raised concerns about the veracity of quantifications and associations with OP exposure outcomes. Even so, these quantifications are still used in evaluating OP exposures.

Cholinesterases

Different OP compounds can participate in nucleophilic attacks on different targets, causing the formation of adducts. The most common and most studied OP adduct is the phosphorylation of the serine 203 residue at the catalytic site of AChE, a process that results in the inhibition of enzyme activity and in the generation of neurotoxic effects related to acute OP exposure (Fukuto 1990). The activity of erythrocytes or serum AChE is used in many studies as an effect biomarker, to evaluate the exposure level, which may be relevant in clinical assessments during acute intoxication. AChE activity can be determined through different methods, but the colorimetric Ellman assay has become the standard method for OP exposure detection because it is easy to operate, reliable, and of low cost (Marsillach et al. 2013).

However, this biomarker is not OP-specific, since other substances such as carbamate pesticides can also produce AChE inhibition and the consequent neurotoxicity. Another disadvantage that should be considered is that the activity of AChE may not fully reflect the toxicity of OP compounds. A recent study proved that the OP pesticide chlorpyrifos (CPF) can induce genotoxicity during a low-dose chronic scheme without affecting AChE activity or antioxidant responses (Kopjar et al. 2018), which may not be useful in non-occupational and low-dose exposures.

AChE is a serine hydrolase containing the motif GXSXG on the catalytic site, like many other hydrolases that may also be affected by OP exposure, for example butyrylcholinesterase (BChE), fatty acid amide (FAAT), neuropathy target esterase (NTE), acyl peptide hydrolase (APH), and arylformamidase (AFMID) (Casida and Quistad 2004). BChE, just like AChE, is used to evaluate OP toxicity with the Ellman assay.

As previously discussed, AChE can be inhibited by several OP compounds thanks to the reactivity of the phosphate group, which will not provide enough information about a specific OP. This may be relevant when evaluating certain exposures, because many OP compounds—pesticides, warfare agents, flame retardants—and their metabolites, including DAPs, can inhibit AChE to a greater or lesser extent (Imaizumi et al. 1993).

Adducts

Recently, some studies have found that OP compounds can also form adducts with other proteins, which may be useful as exposure and effect biomarkers. As part of these efforts, an *in vitro* study showed that about seventy proteins can be identified as targets for the formation of a di-alkoxylphosphate adduct that is mostly present in tyrosine residues (Onder et al. 2018). This represents a stable modification, which can be evaluated over longer periods of time than in the case of the AChE or DAP evaluations. The new molecular targets may serve as potential exposure biomarkers, but further studies should clarify their utility. In this group, we point out two targets: albumin and tubulin.

Albumin is an abundant protein present in plasma that participates in the modulation of oncotic pressure and the transport of xenobiotics and endogenous molecules (Moman et al. 2020). *In vitro* and *in vivo* studies have proved that albumin can be the target of several OPs, including pesticides, warfare agents, and OP probes that form an adduct at tyrosine residue 411 of human albumin. This adduct has also been detected in human serum in accidental exposures, which confirms its potential as an OP biomarker. Unlike AChE adducts, albumin adducts are more stable and avoid hydrolysis of alkyl substituents, which make the adduct available for detection for a longer period of time (Li et al. 2010; Peeples et al. 2005).

The disadvantages of albumin adducts are related to the methodology, which requires gas chromatography or HPLC techniques coupled with mass spectrometry to identify the adducts (Crow et al. 2014; Peeples et al. 2005).

Tubulin is a cytoskeletal protein that forms microtubules that participate in several cellular functions. Exposure to OP substances *in vitro* and *in vivo* has revealed the formation of adducts on different amino acid residues, mostly tyrosines and lysine residues, in which OPs formed di-alkoxyphosphate adducts (Chu et al. 2018; Grigoryan et al. 2009; Jiang et al. 2010; Schopfer and Lockridge 2018). Because tubulin is a ubiquitous and abundant protein, its modification could represent a good exposure biomarker.

The identification of this type of adduct can be done with mass spectrometry techniques. More recently, Onder and collaborators (2018) developed the anti-depY, an antibody capable of detecting the di-alkoxyphosphate tyrosine modification, which may facilitate the determination of exposure to OPs. However, as with DAPs and AChE, this approach will not identify the specific OP, since the detectable chemical modification belongs to the common phosphate group in the molecule. Until now, this modification has been detected only with *in vitro* and *in vivo* models; hence it should be verified in human exposures, if we want to evaluate its potential as a biomarker.

OP Toxic Effects

As we mentioned in the beginning, originally OP pesticides were produced because they had lower toxicity and less environmental persistence than their organochloride counterparts. Like many other pesticide compounds, OP pesticides react with the active catalytic site of AChE, causing acute neurotoxic effects through the promotion of nuclear substitution reactions between AChE and OP (Casida and Quistad 2004). This, in turn, results in enzyme inhibition, increasing the acetylcholine stimulation of nerves and muscles and causing hyperactivity, tremors, convulsions, paralysis, and death. AChE is also present in different non-target species—mammals, birds, fish, and reptiles (Fukuto 1990)—in which the increased concentrations of this neurotransmitter heighten processes of salivation, lacrimation, bronchial secretions, bronchoconstriction, sweating, nausea, vomiting, bradycardia, headache, tremor, weakness, ataxia, convulsion, coma, and death (Sinha and Sharma 2003). In addition to cholinergic effects, OP compounds cause toxic effects in several organs and systems in human populations (Figure 16.4) and *in vivo* models (Table 16.1).

Genotoxicity of OP Compounds

Several *in vitro* and *in vivo* studies, including epidemiological ones, have evidenced the potential effects of OP compounds on DNA (Figure 16.5). This is important, since DNA damage has been used as a possible explanation for the association between chronic

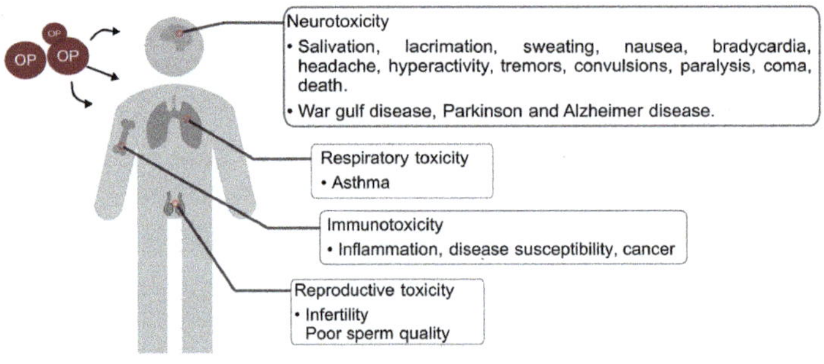

Figure 16.4 Toxic effects observed in human populations exposed to OP compounds.

Table 16.1 Toxic effects associated to OP exposures on *in vivo* models.

OP Compound	Biological effect	Model	Reference
Chlorpyrifos, Mipafox, DFP	AChE and neuropathy target esterase	mouse hen	(Makhaeva et al. 2014)
Chlorpyrifos	Neurodegeneration and cognitive impairment	rats	(Voorhees et al. 2019)
Parathion, Diazinon	Damage in lymphoid organs	mouse	(Galloway and Handy 2003)
Chlorpyrifos, Diazinon, Malathion, Methyl parathion	↓ Immune cells	fish, rat	(Galloway and Handy 2003)
Methyl parathion	↑ Bacterial infection	rabbit	(Galloway and Handy 2003)
Chlorpyrifos Dimethoate	↑ Sperm DNA damage ↓ Sperm quality Testicular damage	rat	(Sayim 2007; Babazadeh and Najafi 2017)
DEDTP	↑ Acute parasite infection	mouse	(D. M. Medina-Buelvas et al. 2019b)
DEDTP	Micronucleus in polychromatic erythrocytes Immunomodulatory effects ↑ Tumor volume	mouse	(D. Medina-Buelvas et al. 2019a)
DEP	↑ Opportunistic pathogens in microbiota	rat	(Yang et al. 2019)

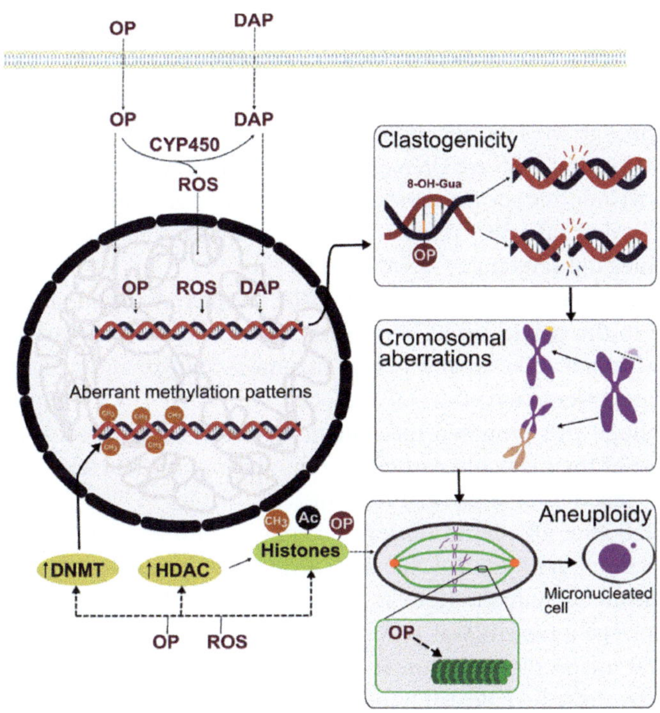

Figure 16.5 Genetic and epigenetic effects of OP compounds.

exposure to these substances and the development of cancer. In particular, OP pesticides have induced several genotoxic events, for example chromosomal aberrations, micronucleus, single and double DNA strand breaks, and DNA adduct formation (including DNA mono adducts, DNA intra-strand crosslinks, and DNA–protein crosslinks) (Ojha and Gupta 2015, 2017). Nevertheless, the exact molecular mechanisms are still unknown. Some hypotheses have pointed out that this could be a secondary effect caused, for example, by the reactive oxidant species produced by OP metabolism, or could even be a consequence of triggering apoptosis signaling. Exposure to DAP has also shown genotoxicity induction with *in vitro* and *in vivo* models, mainly presented by DNA strand breaks and micronucleus formation at low doses. This suggests the participation of DAP in the general genotoxicity associated with its parent compounds. However, more research is needed to clarify its effects. Whatever the mechanisms of OP and DAP, we have attempted to review some of the evidence concerning their possible effects.

Suggested Mechanisms

Clastogenicity

Clastogenicity consists in a break of DNA strands that can occur in one or both chains of these strands. The break could be produced by several mechanisms. OP pesticides can induce clastogenicity, which has been reported indirectly, through the identification of chromosome fragmentation, which leads to micronucleus formation. It has also been detected directly, through single-cell electrophoresis, also known as the comet assay, which identifies breaks in single- and double-stranded DNA. The exact mechanism is still to be elucidated, but the most accepted explanations revolve around the oxidant stress produced by OP exposure, since both *in vitro* and *in vivo* evidence indicates that OP compounds produce reactive oxygen species (ROS) such as superoxide ions and hydrogen peroxide. The formation of these ROSs can be related to the metabolism of the pesticides, which is catalyzed in part by P450 cytochrome isoforms. In addition, it has been reported that OPs inhibit the function of antioxidant enzymes, creating a misbalance of ROS in the cells. The increasing ROS leads to oxidant damage in several macromolecules, including in the DNA. This damage can be detected through the identification of oxidized nucleotides that cause the formation of alkali-labile sites (ALS). ALS are DNA sites characterized by the loss of the DNA base. These apurinic or apyrimidinic sites are susceptible to alkaline solutions, which results in DNA strand breaks. In normal cells, ALS can be repaired by base or nucleotide excision pathways, which produce the severing of DNA strands.

Another possibility is the direct interaction of OPs with DNA strands. A recent study showed that chronic exposure to the pesticide CPF in Wistar rats resulted in increasing DNA strand breaks at doses that do not affect cholinesterase enzyme activity or antioxidant responses, which suggests a direct interaction of the OP molecule with some DNA component (Kopjar et al. 2018). Although this proposed mechanism needs to be clarified, it is possible that the phosphate part of the molecule participates in nucleophilic reactions, as described for some proteins, for instance albumin and tubulin.

Like OP pesticides, DAP metabolites also cause clastogenic effects. With adult and fetal human hepatic cell lines HepG2 and WRL-68, DEDTP and DETP generated DNA strand breaks in a concentration-dependent manner. Here it is probable that the ethyl DAPs affected the DNA indirectly, since genotoxicity was prevented through the co-treatment of sulconazole, a cytochrome P450 unspecific inhibitor. It is also suggested that, under the metabolic conditions provided by these hepatic cell lines, the OP metabolites may be activated during further metabolism, generating ROS, or even producing possibly highly reactive unstable intermediates during the sulphur oxidation process, which may react directly with DNA (Vega et al. 2009). However, the exact mechanism is still to be clarified.

Aneuploidy

Aneuploidy refers to the gain or loss of chromosomes that results from a missegregation event during cell division. It has considerable biological importance, because the modification of genetic material favors cell transformations that may participate in cancer development. This event is also highly present in the transformed cell, since its mitotic machinery and cell checkpoint may be altered.

OP molecules can generate this kind of genotoxicity. *In vitro* studies have revealed that OP pesticides such as dichlorvos cause micronucleus induction by altering the mitotic spindle; this generates multipolar and monopolar spindles, which affect the chromosome segregation clearly observed in immunofluorescence images of mitotic spindles. This effect is related to the mislocalization of the Kif2a (Fiore et al. 2013) kinase, which plays an important role in microtubule dynamics (Yi et al. 2016). Another possible explanation for OP-mediated aneuploidy is the direct alteration of cytoskeleton elements. Recent reports have shown tubulin to be a molecular target of different OP compounds, which may interact with amino acid residues such as tyrosines and lysines, generating organophosphate adducts (Lockridge and Schopfer 2010). The consequences of these modifications include microtubule polymerization disruption, the alteration of microtubule integrity (Grigoryan and Lockridge 2009), and the reduction of interactions with microtubule-associated proteins (MAPs) (Jiang et al. 2010), which are crucial for microtubule dynamics during cell division. This effect has also been reported in spermatic cells of populations occupationally exposed to a mixture of OP pesticides. The result is in an increase of null-sex chromosome cells (Recio et al. 2001).

In human hepatic HepG2 and WRL-68 cells, the methylated DAP DMTP induced a micronucleus formation similar to DEDTP and DETP. This effect depends on metabolic conditions because, in peripheral blood mononuclear cells (PBMCs), it occurred only at high concentrations. In one study DMTP did not induce nucleoplasmic bridges (NPB), which represents a genotoxic event caused principally by clastogenic events. This suggests that the micronucleus formation may be due to aneuploidogenic events (Hernández-Toledano et al. 2020).

Biomarkers of Genotoxicity

The evaluation of the genotoxicity of OP compounds and their DAPs is mostly done through the identification of the micronucleus, the measurement of DNA strand breaks via comet assay, and DNA fragmentation using fluorescence probes.

The easiest study applicable to *in vitro, in vivo*, and epidemiologic approaches is the micronucleus test. Briefly, the method consists in identifying the extranuclear genetic material that resembles a small cell nucleus located in the cytoplasm of cells; this is done with the help of an optical microscope and a differential stain, such as the Giemsa–Wright stain. The simplest micronucleus assay permits the identification of a genotoxic event. Further steps and determination can provide more information about the genotoxic mechanism, since a micronucleus can be constituted either of entire chromosomes or of a part of them; and these are generated respectively by aneuploidy or by clastogenic events. In addition to its low cost, the method can be performed in different biological matrices, such as blood and buccal cells; it is a non-invasive method. Therefore micronucleus determination is suitable for epidemiological studies as an exposure biomarker, but mainly as an effect biomarker.

Epigenetic Effects of OP Compounds

The controversial relationship between OP exposure and toxic outcomes without a clear role led to the hypothesis of new toxic mechanisms of OP compounds, such as epigenetic alterations (Figure 16.4). These modifications are now considered and studied in different models.

Recent research has shown that OP pesticides could modify gene expression during embryo development and adulthood through occupational exposures that are related, for example, to cancer development and neurological diseases (Slotkin and Seidler 2011). The mechanisms are still unclear, but it is possible that OP compounds act through several modes of action, including several epigenetic events such as DNA methylation, histone modification, and non-coding RNA expression (Cavalli and Heard 2019). For example, some OP pesticides—terbufos, diazinon, parathion, fenofos, dichlorvos, and others similar to these—induce a modification of the methylation patterns of several genes, some of which are related to cancer development (Zhang et al. 2012; Zhang and Wallace et al. 2012). Additionally, other OP pesticides, such as CPF, form adducts with several proteins, including histones, modifying histone methylation and increasing non-coding RNAs such as miRNA-132 and miRNA-212 (Burke et al. 2017; Onder et al. 2018), which may contribute to the modification of gene expression. Interestingly, many of these reports from *in vivo* models used low doses with biological relevance but without acute cholinergic effects or AChE inhibition, which means that it is important to study the effects of chronic exposure to OP compounds.

Until now, the epigenetic effects of OP compounds have been limited to OP pesticides and, to a lesser extent, OP flame retardants in different models, mainly with *in vitro* approaches. Because no epigenetic modifications have been reported for DAPs, the next sections are focused only on OP pesticides.

Suggested Mechanisms

DNA Methylation

Studies have shown that OP pesticides and OP flame retardants have induced aberrations in methylation patterns *in vitro* and *in vivo*. The hypermethylation and hypomethylation of several genes, including oncogenic genes, are among these modifications and may be caused by the increased expression of DNA methyl transferases DNMT1 and DNMT3b (Li et al. 2020; Zhang and Andrew et al. 2012; Zhang and Wallace et al. 2012).

Methylation also plays an important role during chromosome segregation. The centromeric regions of chromosomes are normally methylated, which in turn confers low recombination in this region; and it is hypothesized that hypomethylation can modify the interaction and attachment of chromosomes to the mitotic spindle and can in consequence produce a micronucleus through aneuploidogenic events (Luzhna et al. 2013).

OP pesticides also affect gene expression through the hypermethylation and hypomethylation of gametic cells, and this occurs even during embryo development. CPF has shown increased methylation in the XIST promoter in bovine spermatic cells. XIST is responsible for silencing the X chromosome, which, it has been suggested, affects embryo development, producing infertility (Pallotta et al. 2019).

DNA methylation effects occur with *in vitro* and *in vivo* models, as well as in human populations. In occupationally exposed populations, specific gene promoters (Herrera-Moreno et al. 2019; Paredes-Céspedes et al. 2019) and global DNA methylation levels have been evaluated through the assay of the long interspersed nucleotide element 1 (LINE-1) (Benitez-Trinidad et al. 2018). A recent study in elderly populations found that low-level chronic exposure to OPs induced modifications in DNA methylation patterns in blood cells and saliva samples. These modifications occurred in genes related to the nicotinic and muscarinic acetylcholine receptor pathways (Paul et al. 2018), which could favor neurotoxic OP effects.

The exact mechanism of DNA methylation changes after OP exposure is still unclear, but the alkylating properties of some OPs and ROS may be implicated in this effect. Specifically, ROS can trigger aberrations in methylation patterns, both hypomethylation and hypermethylation, through the oxidation of guanine nucleotides, through the hydroxylation of cytosines (the catalyst of DNA methylation through deprotonate cytosine molecules), or through the upregulation of DNMT expression (Wu and Ni 2015).

Histone Modification

In addition to altering methylation patterns, OP compounds can modify histones. Some studies have shown that pesticides such as CPF induce the methylation of lysine 4 of histone 3 (H3K4), which is associated with the expression of genes related to neural development (Burke et al. 2017). Additionally, hypermethylation of H3K4 is also related to chromosome missegregation by disrupting mitotic spindles (Luzhna et al. 2013). Also, CPF induced the overexpression of histone deacetylases (HDAC) in an *in vivo* model with a chemically induced tumor (Ventura et al. 2019). The overexpression of HDAC is also related to an increase in ROS (Wu and Ni 2015). This increase in HDAC may be partly responsible for the changes in oncogenic gene expression during occupational exposure to OP compounds.

Adduct Formation

As previously mentioned, OP pesticides can form adducts with different proteins. In a recent approach that tests an antibody for tyrosine-diethoxyphosphate, an adduct produced by CPF exposure, several proteins were identified, and the histones H1.4, H2B, H1x, H1.3, and H4 were detected among them (Onder et al. 2018). In the case of H4, CPF formed an adduct in four tyrosine residues: Y51, Y72, Y88, and Y98. Interestingly, Y51 and Y72 are sites of histone phosphorylation, which is related to DNA repair and cell cycle progression. The particular phosphorylation of Y72 is required to recruit histone methyltransferases (Chou et al. 2014; Hossain et al. 2016). The blockade of these tyrosine sites in histones may perturb cell responses and methylation patterns. The Y88 and Y98 residues are located at the zone of interaction with H3, which makes them less accessible to OP compounds under certain conditions (Singh and Gunjan 2011).

Previous reports also found that OP pesticides and metabolites form adducts in different amino acid residues such as serine, threonine, and lysine as well as in tyrosine (Lockridge and Schopfer 2010). These residues have important post-translational implications as modification sites. However, research is still needed to determine the participation of these residues in histone alteration by OP compounds.

Potential Epigenetic Biomarkers

The identification of new epigenetic biomarkers can be a useful tool for evaluating the development of diseases and conditions related to OP exposure—for example cancer, neurological disorders, neurodegenerative diseases, and neurodevelopmental effects.

Additionally, it may help to evaluate the levels of OP exposure, given that many reports have shown that DNA methylation occurs at low doses of OP exposure, and even at doses with non-acute cholinergic effects, which can be relevant for biomonitoring populations chronically exposed to low doses. According to a study of long-term exposure in a human population (Paul et al. 2018), DNA methylation changes occurred in acetylcholine receptor-related genes. This can result in finding useful targets for assaying OP exposure. In addition, the identification of these methylation changes can be performed in a biological matrix such as blood and saliva samples, which makes this type of changes a suitable candidate for a biomarker of effect.

More candidate genes are possible, for example oncogenic and tumor suppressor genes, which can be used to evaluate the cancer risk in occupationally exposed populations. However, these modifications are not specific to OP compounds, and for this reason such genes are less reliable as biomarkers.

Despite the *in vitro, in vivo*, and epidemiological studies showing changes in DNA methylation patterns, their potential use as a biomarker needs to be further studied and improved.

In addition to DNA methylation, existing reports have found histone modification, mainly by post-translational modifications; but this modification evidence needs to be confirmed in human populations, and further research could identify its role in inducing OP toxic effects.

Concluding Remarks

The evaluation of OP compound exposure is often limited to acute neurotoxic effects, which have been extensively studied. However, new studies have found increased risks of non-neurological diseases and other non-neurological effects occurring after chronic and low-level exposures.

Current reports demonstrate that genotoxicity is a toxic effect induced by several OP compounds such as pesticides, flame retardants, and their DAP metabolites, which makes this phenomenon a potential biomarker of effect: it could be relevant in the risk assessment of these compounds. Although such effects have been studied, the exact mechanisms by which OP compounds induce genotoxicity are unclear. Fortunately, recent investigations are now considering epigenetic alterations related to xenobiotic exposure. Here OP pesticides showed the induction of aberrant DNA methylation patterns in specific genes and histone modifications. This information will be useful for proposing new effect biomarkers that could complement genotoxic ones in risk assessment studies. However, epigenetic modifications still need to be further researched with more OP pesticides, OP flame retardants, and their metabolites in order to be as specific as possible.

Interestingly, recent investigations of DAPs also suggest that non-specific metabolites can contribute to overall OP toxicity, at least in genotoxic outcomes, while epigenetic effects are still to be investigated. The identification of new effect biomarkers will be useful not only for assessing the risk of disease development but also for making progress in OP toxicity research, because the mechanisms of the found biological effects are still unclear. The inclusion of novel research into epigenetic changes that happen after exposure to OP compounds also contributes to the investigation of non-cholinergic effects, which may help to fill the gaps in OP compound toxicological profiles.

References

Andresen, J.A., Grundmann, A., and Bester, K. (2004). Organophosphorus flame retardants and plasticisers in surface waters. *Sci. Total Environ.* 332 (1–3): 155–166. doi: 10.1016/j.scitotenv.2004.04.021.

Babazadeh, M. and Najafi, G. (2017) Effect of chlorpyrifos on sperm characteristics and testicular tissue changes in adult male rats. Vet. Res. Forum 8 (4): 319–326. http://www.ncbi.nlm.nih.gov/pubmed/2932679.

Benitez-Trinidad, A.B. et al. (2018). Relationship between LINE-1 methylation pattern and pesticide exposure in urban sprayers. *Food and Chem. Toxicol.* 113: 125–133. doi: 10.1016/j.fct.2018.01.035.

Blum, A. et al. (2019). Organophosphate ester flame retardants: Are they a regrettable substitution for polybrominated diphenyl ethers? *Environm. Sci. Technol. Lett.* 6 (11): 638–649. doi: 10.1021/acs.estlett.9b00582.

Bolognesi, C. and Holland, N. (2016). The use of the lymphocyte cytokinesis-block micronucleus assay for monitoring pesticide-exposed populations. *Mutat. Res. Rev. Mutat. Res.* 770: 183–203. doi: 10.1016/j.mrrev.2016.04.006.

Burke, R.D. et al. (2017). Developmental neurotoxicity of the organophosphorus insecticide chlorpyrifos: From clinical findings to preclinical models and potential mechanisms. *J. Neurochem.* 142: 162–177. doi: 10.1111/jnc.14077.

Casida, J.E. (2017). Organophosphorus xenobiotic toxicology. *Annu. Rev. Pharmacol. Toxicol.* 57: 309–327. doi: 10.1146/annurev-pharmtox-010716-104926.

Casida, J.E. and Quistad, G.B. (2004). Organophosphate toxicology: Safety aspects of nonacetylcholinesterase secondary targets. *Chem. Res. Toxicol.* 17 (8): 983–998. doi: 10.1021/tx0499259.

Cavalli, G. and Heard, E. (2019). Advances in epigenetics link genetics to the environment and disease. *Nature* 571 (7766): 489–499. doi: 10.1038/s41586-019-1411-0.

Chen, L. et al. (2013). Absorption and excretion of organophosphorous insecticide biomarkers of malathion in the rat: Implications for overestimation bias and exposure misclassification from environmental biomonitoring. *Regul. Toxicol. Pharmacol.* 65 (3): 287–293. doi: 10.1016/J.YRTPH.2012.12.010.

Chou, R.H. et al. (2014). EGFR modulates DNA synthesis and repair through tyr phosphorylation of histone H4. *Dev. Cell* 30 (2): 224–237. doi: 10.1016/j.devcel.2014.06.008.

Chu, S. et al. (2018). Covalent binding of the organophosphate insecticide profenofos to tyrosine on A- and B-tubulin proteins. *Chemosphere* 199: 154–159. doi: 10.1016/j.chemosphere.2018.02.003.

Crow, B.S. et al. (2014). Simultaneous measurement of tabun, sarin, soman, cyclosarin, VR, VX, and VM adducts to tyrosine in blood products by isotope dilution UHPLC-MS/MS. *Anal. Chem.* 86 (20): 10397–10405. doi: 10.1021/ac502886c.

Drevenkar, V. et al. (1993). Chlorpyrifos metabolites in serum and urine of poisoned persons. *Chem. Biol. Interact.* 87 (1–3): 315–322. doi: 10.1016/0009-2797(93)90059-8.

Esquivel-Senties, M.S. et al. (2010). Organophosphorous pesticide metabolite (DEDTP) induces changes in the activation status of human lymphocytes by modulating the interleukin 2 receptor signal transduction pathway. *Toxicol. Appl. Pharmacol.* 248 (2): 122–133. doi: 10.1016/j.taap.2010.07.017.

Fest, C. and Schmidt, K.-J. (1982). Chemical section. In: *The Chemistry of Organophosphorus Pesticides*, C. Fest and K.-J. Schmidt, 50–182. Berlin: Springer. https://doi.org/10.1007/978-3-642-68441-8_3.

Fiore, M. et al. (2013). The pesticide dichlorvos disrupts mitotic division by delocalizing the kinesin Kif2a from centrosomes. *Environ. Mol. Mutagen.* John Wiley & Sons, Ltd. 54 (4): 250–260. 10.1002/em.21769.

Fukuto, T.R. (1990). Mechanism of action of organophosphorus and carbamate insecticides. *Environ. Health Perspect.* 87: 245–254. doi: 10.1289/ehp.9087245.

Galloway, T. and Handy, R. (2003). Immunotoxicity of organophosphorous pesticides. *Ecotoxicology* 12 (1–4): 345–363. doi: 10.1023/a:1022579416322.

Grigoryan, H. et al. (2009). Mass spectrometry identifies multiple organophosphorylated sites on tubulin. *Toxicol. Appl. Pharmacol.* Academic Press. 240 (2): 149–158. doi: 10.1016/J.TAAP.2009.07.020.

Grigoryan, H. and Lockridge, O. (2009). Nanoimages show disruption of tubulin polymerization by chlorpyrifos oxon: Implications for neurotoxicity. *Toxicol. Appl. Pharmacol.* 240 (2): 143–148. doi: 10.1016/j.taap.2009.07.015.

Hernández, A.F. et al. (2019). Biomonitoring of common organophosphate metabolites in hair and urine of children from an agricultural community. *Environ. Int.* 131: 104997. doi: 10.1016/j.envint.2019.104997.

Hernández-Toledano, D.S., Estrada-Muñiz, E., and Vega, L. (2020). Genotoxicity of the organophosphate pesticide malathion and its metabolite dimethylthiophosphate in human cells in vitro. *Mutat. Res. Genet Toxicol. Environ. Mutagen.* 856–857. doi: 10.1016/j.mrgentox.2020.503233.

Herrera-Moreno, J.F. et al. (2019). Modified CDKN2B (p15) and CDKN2A (p16) DNA methylation profiles in urban pesticide applicators. *Environ. Sci. Pollut. Res.* 15124–15135. doi: 10.1007/s11356-019-04658-5.

Hossain, M.B. et al. (2016). TIE2-mediated tyrosine phosphorylation of H4 regulates DNA damage response by recruiting ABL1. *Sci. Adv.* 2 (4). doi: 10.1126/sciadv.1501290.

Imaizumi, H. et al. (1993). Metabolism and toxicity of acid phosphate esters, metabolites of organophosphorous insecticides, in rat. *Eisei Kagaku* (公益社団法人 日本薬学会). 39 (6): 566–571. doi: 10.1248/jhs1956.39.6_566.

Jara, E.A. and Winter, C.K. (2019). Safety levels for organophosphate pesticide residues on fruits, vegetables, and nuts. *Int. J. Food Contam.* 6: 1. doi: 10.1186/s40550-019-0076-7.

Jiang, W. et al. (2010). Mice treated with chlorpyrifos or chlorpyrifos oxon have organophosphorylated tubulin in the brain and disrupted microtubule structures, suggesting a role for tubulin in neurotoxicity associated with exposure to organophosphorus agents. *Toxicol. Sci.* 115 (1): 183–193. doi: 10.1093/toxsci/kfq032.

Kavvalakis, M.P. and Tsatsakis, A.M. (2012). The atlas of dialkylphosphates: Assessment of cumulative human organophosphorus pesticides' exposure. *Forensic Sci. Int.* 218 (1–3): 111–122. doi: 10.1016/j.forsciint.2011.10.019.

Kopjar, N. et al. (2018). Evaluation of chlorpyrifos toxicity through a 28-day study: Cholinesterase activity, oxidative stress responses, parent compound/metabolite levels, and primary DNA damage in blood and brain tissue of adult male Wistar rats. *Chem. Biol. Interact.* 279: 51–63. doi: 10.1016/j.cbi.2017.10.029.

Korobeinichev, O.P., Shvartsberg, V.M., and Shmakov, A.G. (2007). The chemistry of combustion of organophosphorus compounds. *Russ. Chem. Rev.* 76 (11): 1094–1121. 10.1070/rc2007v076n11abeh003713.

Kozawa, K. et al. (2009). Toxicity and actual regulation of organophosphate pesticides. *Toxin. Rev.* 28 (4): 245–254. doi: 10.3109/15569540903297808.

Kumar, S., Kaushik, G., and Villarreal-Chiu, J.F. (2016). Scenario of organophosphate pollution and toxicity in India: A review. *Environ. Sci. Pollut. Res.* 23 (10): 9480–9491. doi: 10.1007/s11356-016-6294-0.

Kwong, T.C. (2002). Organophosphate pesticides: biochemistry and clinical toxicology. *Ther. Drug Monit.* 24 (1): 144–149. doi: 10.1097/00007691-200202000-00022.

Lasagna, M. et al. (2020). Chlorpyrifos subthreshold exposure induces epithelial–mesenchymal transition in breast cancer cells. *Ecotoxicol. Environ. Saf.* 205: 111312. doi: 10.1016/j.ecoenv.2020.111312.

Li, B. et al. (2010). Detection of adduct on tyrosine 411 of albumin in humans poisoned by dichlorvos. *Toxicol. Sci.* 116 (1): 23–31. doi: 10.1093/toxsci/kfq117.

Li, J. et al. (2019). A review on organophosphate ester (OPE) flame retardants and plasticizers in foodstuffs: Levels, distribution, human dietary exposure, and future directions. *Environ. Int.* 127: 35–51. doi: 10.1016/j.envint.2019.03.009.

Li, R. et al. (2020). Early-life exposure to the organophosphorus flame-retardant tris (1,3-dichloro-2-propyl) phosphate induces delayed neurotoxicity associated with DNA methylation in adult zebrafish. *Environ. Int.* 134: 105293. doi: 10.1016/j.envint.2019.105293.

Lockridge, O. and Schopfer, L.M. (2010). Review of tyrosine and lysine as new motifs for organophosphate binding to proteins that have no active site serine. *Chem. Biol. Interact.* 187 (1–3): 344–348. doi: 10.1016/j.cbi.2010.03.002.

Luzhna, L., Kathiria, P., and Kovalchuk, O. (2013). Micronuclei in genotoxicity assessment: From genetics to epigenetics and beyond. *Front Genet.* 4: 1–17. doi: 10.3389/fgene.2013.00131.

Makhaeva, G.F. et al. (2014). Further studies toward a mouse model for biochemical assessment of neuropathic potential of organophosphorus compounds. *J. Applied Toxicol.* 34 (12): 1426–1435. doi: 10.1002/jat.2977.

Marsillach, J., Costa, L.G., and Furlong, C.E. (2013). Protein adducts as biomarkers of exposure to organophosphorus compounds. *Toxicology* 307: 46–54. doi: 10.1016/j.tox.2012.12.007.

Medina-Buelvas, D. et al. (2019a). Genotoxic and immunotoxic effects of the organophosphate metabolite diethyldithiophosphate (DEDTP) in vivo. *Toxicol. Appl. Pharmacol.* 366: 96–103. 10.1016/j.taap.2019.01.023.

Medina-Buelvas, D.M. et al. (2019b). Increased heart fibrosis and acute infection in a murine Chagas disease model associated with organophosphorus pesticide metabolite exposure. *Sci. Rep.* 9 (1): 17539. doi: 10.1038/s41598-019-54218-7.

Moman, R., Gupta, N., and Varacallo, M. (2020). Physiology, albumin. In NCBI Bookshelf, StatPearls [Internet]. Treasure Island, FL: StatPearls Publishing. https://www.ncbi.nlm.nih.gov/books/NBK459198.

Ojha, A. and Gupta, Y.K. (2015). Evaluation of genotoxic potential of commonly used organophosphate pesticides in peripheral blood lymphocytes of rats. *Hum. Exp. Toxicol.* 34 (4): 390–400. doi: 10.1177/0960327114537534.

Ojha, A. and Gupta, Y.K. (2017). Study of commonly used organophosphate pesticides that induced oxidative stress and apoptosis in peripheral blood lymphocytes of rats. *Hum. Exp. Toxicol.* 36 (11): 1158–1168. doi: 10.1177/0960327116680273.

Olivares-Bañuelos, T.N. et al. (2019). The neurotoxin diethyl dithiophosphate impairs glutamate transport in cultured Bergmann glia cells. *Neurochem. Int.* 123: 77–84. doi: 10.1016/j.neuint.2018.06.004.

Onder, S. et al. (2018). Mass spectral detection of diethoxyphospho-tyrosine adducts on proteins from HEK293 cells using monoclonal antibody depy for enrichment. *Chem. Res. Toxicol.* 31 (6): 520–530. doi: 10.1021/acs.chemrestox.8b00083.

Pallotta, M.M. et al. (2019). In vitro exposure to CPF affects bovine sperm epigenetic gene methylation pattern and the ability of sperm to support fertilization and embryo development. *Environ. Mol. Mutagen. 60 (1):* 85–95. doi: 10.1002/em.22242.

Paredes-Céspedes, D.M. et al. (2019). Pesticide exposure modifies DNA methylation of coding region of WRAP53α, an antisense sequence of p53, in a Mexican population. *Chem. Res. Toxicol.* 32 (7): 1441–1448. doi: 10.1021/acs.chemrestox.9b00153.

Paul, K.C. et al. (2018). Organophosphate pesticide exposure and differential genome-wide DNA methylation. *Sci. Total Environ.* 645: 1135–1143. doi: 10.1016/j.scitotenv.2018.07.143.

Peeples, E.S. et al. (2005). Albumin, a new biomarker of organophosphorus toxicant exposure, identified by mass spectrometry. *Toxicol. Sci.* 83 (2): 303–312. doi: 10.1093/toxsci/kfi023.

Quirós-Alcalá, L. et al. (2011). Pesticides in house dust from urban and farmworker households in California: An observational measurement study. *Environ. Health: A Glob. Access Sci. Source* 10 (1): 19. doi: 10.1186/1476-069X-10-19.

Recio, R. et al. (2001). Organophosphorous pesticide exposure increases the frequency of sperm sex null aneuploidy. *Environ. Health Perspect.* 109 (12): 1237–1240. http://www.ncbi.nlm.nih.gov/pubmed/11748030.

Sayim, F. (2007). Histopathological effects of dimethoate on testes of rats. *Bull Environ. Contam. Toxicol.* 78 (6): 479–484. doi: 10.1007/s00128-007-9196-5.

Schopfer, L.M. and Lockridge, O. (2018). Chlorpyrifos oxon promotes tubulin aggregation via isopeptide cross-linking between diethoxyphospho-Lys and Glu or ASP: Implications for neurotoxicity. *J. Biol. Chem.* 293 (35): 13566–13577. doi: 10.1074/jbc.RA118.004172.

Sidhu, G.K. et al. (2019). Toxicity, monitoring and biodegradation of organophosphate pesticides: A review. *Crit. Rev. Environ. Sci. Technol.* 49 (13): 1135–1187. doi: 10.1080/10643389.2019.1565554.

Silva Pinto, B.G. et al. (2020). Occupational exposure to pesticides: Genetic danger to farmworkers and manufacturing workers: A meta-analytical review. *Sci. Total Environ.* 748: 141382. doi: 10.1016/j.scitotenv.2020.141382.

Singh, R.K. and Gunjan, A. (2011). Histone tyrosine phosphorylation comes of age. *Epigenetics* 6 (2): 153–160. doi: 10.4161/epi.6.2.13589.

Sinha, P.K. and Sharma, A. (2003). Organophosphate poisoning: A review. *Med. J. Indonesia* 120–126. doi: 10.13181/mji.v12i2.100.

Slotkin, T.A. and Seidler, F.J. (2011). Developmental exposure to organophosphates triggers transcriptional changes in genes associated with Parkinson's disease in vitro and in vivo. *Brain Res. Bull.* 86 (5–6): 340–347. doi: 10.1016/j.brainresbull.2011.09.017.

Tamaro, C.M. et al. (2018). Characterization of organophosphate pesticides in urine and home environment dust in an agricultural community. *Biomarkers* 23 (2): 174–187. doi: 10.1080/1354750X.2017.1395080.

Terry, A.V.V. (2012). Functional consequences of repeated organophosphate exposure: Potential non-cholinergic mechanisms. *Pharmacol. Ther.* 134 (3): 355–365. doi: 10.1016/j.pharmthera.2012.03.001.

Timchalk, C. et al. (2007). Comparative pharmacokinetics of the organophosphorus insecticide chlorpyrifos and its major metabolites diethylphosphate, diethylthiophosphate and 3,5,6-trichloro-2-pyridinol in the rat. *Toxicology* 237 (1–3): 145–157. doi: 10.1016/J.TOX.2007.05.007.

Timchalk, C. (2010). Organophosphorus insecticide pharmacokinetics. In *Hayes' Handbook of Pesticide Toxicology*, ed. R. Krieger, 1409–1433. London: Academic Press. doi: 10.1016/B978-0-12-374367-1.00066-5.

Timperley, C.M., Renard, P.Y., and Gomez, C. (2014). Phosphoryl compounds. In *Best Synthetic Methods: Organophosphorus (V) Chemistry*, ed. C.M. Timperley, 365–562. London: Academic Press. doi: 10.1016/B978-0-08-098212-0.00004-2.

Timperley, C.M. and Tattersall, J. (2014). General overview. In *Best Synthetic Methods: Organophosphorus (V) Chemistry*, ed. C.M. Timperley, 1–89. London: Academic Press. doi: 10.1016/B978-0-08-098212-0.00001-7.

van der Veen, I. and de Boer, J. (2012). Phosphorus flame retardants: Properties, production, environmental occurrence, toxicity and analysis. *Chemosphere* 88 (10): 1119–1153. doi: 10.1016/j.chemosphere.2012.03.067.

Vasilić, Ž., Štengl, B., and Drevenkar, V. (1999). Dimethylphosphorus metabolites in serum and urine of persons poisoned by malathion or thiometon. *Chem. Biol. Interact.* 119–120: 479–487. doi: 10.1016/S0009-2797(99)00061-7.

Vega, L. et al. (2009). Diethylthiophosphate and diethyldithiophosphate induce genotoxicity in hepatic cell lines when activated by further biotransformation via Cytochrome P450. *Mutat. Res. Genet. Toxicol. Environ. Mutagen.* 679 (1–2): 39–43. doi: 10.1016/j.mrgentox.2009.07.009.

Ventura, C. et al. (2019). Effects of the pesticide chlorpyrifos on breast cancer disease. Implication of epigenetic mechanisms. *J. Steroid Biochem. Mol. Biol.* 186: 96–104. doi: 10.1016/j.jsbmb.2018.09.021.

Voorhees, J.R. et al. (2019). Occupational-like organophosphate exposure disrupts microglia and accelerates deficits in a rat model of Alzheimer's disease. *Npj Aging Mech. Dis.* 5 (1): 1–14. doi: 10.1038/s41514-018-0033-3.

Wang, R. et al. (2015). Occurrence and spatial distribution of organophosphate ester flame retardants and plasticizers in 40 rivers draining into the Bohai Sea, north China. *Environ. Pollut.* 198: 172–178. doi: 10.1016/j.envpol.2014.12.037.

Weerasekera, G. et al. (2009). A mass spectrometry-based method to measure dialkylphosphate degradation products of organophosphorous insecticides in dust and orange juice. *J. Environ. Monit.* 11 (7): 1345–1351. doi: 10.1039/b821841b.

Wu, Q. and Ni, X. (2015). ROS-mediated DNA methylation pattern alterations in carcinogenesis. *Curr. Drug Targets* 16: 13–19. https://doi.org/10.2174/1389450116666150113121054.

Yang, F. et al. (2019). Effects of diethyl phosphate, a non-specific metabolite of organophosphorus pesticides, on serum lipid, hormones, inflammation, and gut microbiota. *Molecules* 24 (10): 21–23. doi: 10.3390/molecules24102003.

Yang, F.W. et al. (2020). Assessment of the endocrine-disrupting effects of diethyl phosphate, a nonspecific metabolite of organophosphorus pesticides, by in vivo and in silico approaches. *Environ. Int.* 135: 105383. doi: 10.1016/j.envint.2019.105383.

Yi, Z.Y. et al. (2016). Kif2a regulates spindle organization and cell cycle progression in meiotic oocytes. *Sci. Rep.* 6 (1): 1–12. doi: 10.1038/srep38574.

Zhang, X. et al. (2008). Dialkylphosphates (DAPs) in fruits and vegetables may confound biomonitoring in organophosphorus insecticide exposure and risk assessment. *J. Agric. Food Chem.* 56 (22): 10638–10645. doi: 10.1021/jf8018084.

Zhang, X., Andrew, D.,.W. et al. (2012). DNA methylation alterations in response to pesticide exposure. *Environ. Mol. Mutagen.* 53: 542–549. doi: 10.1002/em.21718.

Zhang, X., Wallace, A.D. et al. (2012). Genome-wide study of DNA methylation alterations in response to diazinon exposure in vitro. *Environ. Toxicol. Pharmacol.* 34 (3): 959–968. doi: 10.1016/j.etap.2012.07.012.

17

Genetic, Epigenetic, and Anatomical Factors in Agenesis and Development of Female Reproductive Tract

Tadaaki Nakajima[1], *Tomomi Sato*[2], *and Taisen Iguchi*[2]

[1] *Department of Science, Yokohama City University, Yokohama, Japan*
[2] *Graduate School of Nanobioscience, Yokohama City University, Yokohama, Japan*

Introduction

Female reproductive organs, including the oviducts, the uterus, and the upper part of the vagina, are developed from the same primordium: the Müllerian duct. Many developmental studies, mainly on mice and chickens, have revealed the involvement of genes and proteins in the development of the Müllerian duct (Roly et al. 2018; Santana Gonzalez et al. 2021). A developmental failure of the Müllerian duct is considered to provide the etiology for agenesis in the female reproductive tract, the phenomenon known as the Mayer–Rokitansky–Küster–Hauser (MRKH) syndrome. The second most common etiology of primary amenorrhea is MRKH syndrome, with an estimated incidence of about one in 4,500 women (Choussein et al. 2017). In patients with MRKH syndrome, most cases are sporadic; however, familial clustering has been detected as well, and many candidates for the responsible genes have been proposed (Ledig and Wieacker 2018; Watanabe et al. 2017). The search for the genes responsible for this syndrome is complicated, because many factors that are essential to the development of the Müllerian duct are potential candidates, and these factors are interrelated. Even a comprehensive understanding of the many aspects of developmental biology that are involved in the onset of MRKH syndrome is insufficient. In this chapter we attempt to summarize not only the involvement of genetic and epigenetic factors, but also that of anatomical structures, detailing the development of the Müllerian duct. Moreover, we advance a research proposal for biomarkers of gene mutations, polymorphisms, gene expression, and anatomy related to MRKH syndrome.

Estrogen signaling is one of the most important factors related to the functions of the uterus in adult women. In mammals, estrogen binds to and activates estrogen receptors (ESR) 1 and 2, also known as ERα and ERβ (Hewitt and Korach 2018). But ESR1 and ESR2 are expressed even in the developmental stages, in both mouse and human uteri (Cunha et al. 2021; Nakajima et al. 2015), and the inhibition of ESRs in neonatal mice causes permanently stimulated cell proliferation in the uterine epithelium throughout life (Nakajima et al. 2015). Here we describe the role of ESRs in the developing uterus and provide a chronology for the availability of biomarkers of the uterine function and of endometrial cancer.

Genomic and Epigenomic Biomarkers of Toxicology and Disease: Clinical and Therapeutic Actions,
First Edition. Edited by Saura C. Sahu.

The Dysregulation of Genomic Factors in the Development of the Müllerian Duct Causes MRKH Syndrome

The Wolffian duct, which is the primordium for the male reproductive tract, develops from the urogenital ridge of the intermediate mesoderm and forms the mesonephros (Shaw and Renfree 2014). After the formation of the Wolffian duct, the Müllerian duct develops from the mesonephros. The initiation of the Müllerian duct occurs at the coelomic epithelium of the mesonephros, and the specialized epithelium invaginates to the mesenchyme and elongates to the urogenital sinus at the caudal region along the Wolffian duct (Orvis and Behringer 2007). After the elongation of the Müllerian duct, the regions of the oviducts, uterus, and vagina are determined. This developmental pattern of the female reproductive tract has been investigated in mice, which have a bicornuate unicollis, and in chickens. Even in humans, which have a simplex uterus, the Müllerian duct develops in a similar manner (Cunha et al. 2018). However, some genes in these developmental stages show differences by species. For example, doublesex and mab-3-related transcription factor 1 (*Dmrt1*) gene are required for the elongation of the Müllerian duct of chickens (Ayers et al. 2015), but *Dmrt1* is not expressed in the Müllerian duct of mice (Raymond et al. 1999). For this reason we summarized the responsible genes at each developmental step, but it is prudent to recognize that some of these genes may have different roles in the development of the human Müllerian duct.

The transcription factors, paired box 2 (*Pax2*) and LIM homeobox protein 1 (*Lhx1*), are expressed at the initiation stage of the Müllerian duct, and knockout mice for those genes display uterine agenesis, which indicates that they are essential for development of the Müllerian duct in the mouse (Kobayashi et al. 2004; Torres et al. 1995). The *Pax2* and *Lhx1* genes are regulated differently because *Lhx1* is expressed in the remnant of the Müllerian duct in *Pax2* knockout mice (Kobayashi et al. 2004). *Pax8*, belonging to same family of *Pax2* gene, is also expressed in the Müllerian duct, and *Pax8* knockout mice have only residual uterine tissue that consists of smooth muscle cells (Mittag et al. 2007). *Pax2/8* double knockout mice do not develop Wolffian or Müllerian ducts (Bouchard et al. 2002). The regulatory system of *Pax2* and *Lhx1* expression has been revealed in chicken development. Bone morphogenetic protein (BMP) and fibroblast growth factor (FGF) signaling induce respectively *Pax2* and *Lhx1* expression in the coelomic epithelium of mesonephros (Atsuta and Takahashi 2016). Further, the Wolffian duct is necessary for FGF receptor gene expression to induce *Lhx1* expression (Atsuta and Takahashi 2016). Wingless-type MMTV integration site family member 4 (*Wnt4*) knockout in mice blocks the invagination of the specified coelomic epithelium (Roly et al. 2018; Vainio et al. 1999). At the next step, *Wnt9B* expression in the Wolffian duct is required for the elongation of Müllerian ducts (Carroll et al. 2005). Other Wnt family members participate mainly in the formation of uterine glands (Allison Stewart et al. 2013; Cooke et al. 2013) rather than in the development of the Müllerian duct. A double knockout of murine retinoic acid receptors (*RAR*s) and retinoid X receptors (*RXR*s) disrupts RA signaling and prevents the formation of the Müllerian duct (Kastner et al. 1997; Lohnes et al. 1994), indicating the importance of RA signaling for development. BMP, FGF, Wnt, and RA signaling are required to induce intermediate mesodermal cells from murine and human pluripotent stem cells (Lam et al. 2014; Taguchi and Nishinakamura 2017); therefore resources for those signals for the induction of embryonic intermediate mesoderm may be reused for the development of the Müllerian duct. RA signaling is also reused and regulates the fate determination of the uterus and vagina after the development of the Müllerian ducts (Nakajima et al. 2016, 2019).

The description "Type I MRKH syndrome" designates the majority of cases, in which agenesis of the female reproductive tract occurs without other defects. Agenesis with additional anomalies is labelled Type II MRKH (Backhouse et al. 2019). The congenital nature and familial clustering of MRKH syndrome raise the possibility that this syndrome is encoded by specific genes (Backhouse et al. 2019). Candidate mutations of genes and chromosomes responsible for MRKH syndrome have been variously reported in patients (Ledig and Wieacker 2018; Watanabe et al. 2017). In much of the western world, in a large cohort of families, single-allele mutations have been identified, suggesting autosomal dominant inheritance for MRKH syndrome (Williams et al. 2017). In this cohort study, mutations in the two well-known candidate genes, *WNT4* and HNF1 homeobox B (*HNF1B*), occur at exceedingly low frequencies, and mutations of other candidate genes, such as *LHX1*, *TBX6*, and *WNT9B* are rarely detected (Williams et al. 2017). In Turkish patients, regions of homozygosity may manifest autosomal, recessively inherited causes (Demir Eksi et al. 2018). In a large, Chinese cohort study, gene-disrupting variants in *PAX8*, *BMP4*, *BMP7*, *TBX6*, *HOXA10*, *EMX2*, and *WNT9B* have been identified (Chen et al. 2021). A whole exome-sequencing study of 111 unrelated people with MRKH syndrome has identified variants in many genes, including *HOXA10*, *PAX2*, *RARA*, *WNT4*, *WNT9B*, and *TBX6* (Mikhael et al. 2021). Transcription factors HNF1B and EMX2 activate the expression of *Lhx1*, *Pax2*, and *Wnt9B* in the development of mouse ureteric buds derived from the Wolffian duct (Lokmane et al. 2010). TBX6 and HOXA10 are essential respectively for the development of the mesoderm and uterus in mice (Benson et al. 1996; Muraoka et al. 2018; Wehn et al. 2020). Therefore the responsible genes may be partially dependent on ethnic differences, and almost all cases are rare, many candidate mutations (Table 17.1) being detected in relation to the development of the Müllerian duct. Recently a new candidate gene, GREB1-like retinoic acid receptor coactivator, which is important for kidney development, was reported (Herlin et al. 2019; Jacquinet et al. 2020). In Type I and Type II MRKH syndrome, various candidates for responsible genes have been identified, which suggests that multiple genes may cause the disease and that a consistent genetic biomarker for MRKH syndrome may be difficult to develop (Figure 17.1).

Table 17.1 List of candidate genes involved in the development of the Müllerian duct and MRKH syndrome.

Gene	Reference	Loss of function in animal models	Reference	Expression in animal models	Reference
BMP4	Chen et al. 2021	Loss of *Pax2* and *Lhx1* expression	Atsuta and Takahashi 2016	-	-
BMP7	Chen et al. 2021	Loss of *Pax2* and *Lhx1* expression	Atsuta and Takahashi 2016	-	-
EMX2	Demir Eksi et al. 2018 Chen et al. 2021	Agenesis of Müllerian duct	Miyamoto et al., 1997	From initiation stage	Miyamoto et al. 1997
GREB1l	Herlin et al. 2019 Jacquinet et al. 2020	Agenesis of kidney and genital tract	De Tomasi et al., 2017	-	-

(*Continued*)

Table 17.1 (Continued)

Gene	Reference	Loss of function in animal models	Reference	Expression in animal models	Reference
HNF1B	Williams et al. 2017 Ledig and Wieacker 2018	Loss of *Lhx1*, *Pax2*, and *Wnt9B* expression	Lokmane et al. 2010	Wolffian ductal epithelium	Lokmane et al. 2010
HOXA10	Chen et al. 2021 Mikhael et al. 2021	Proximal uterus into oviduct	Benson et al. 1996	Mesenchyme after elongation	Ma et al. 1998
LAMC1	Mikhael et al. 2021	Agenesis of kidney and genital tract	Willem et al. 2002	-	-
LHX1	Williams et al. 2017 Ledig and Wieacker 2018	Agenesis of Müllerian duct	Kobayashi et al. 2004	Epithelium from initiation stage	Kobayashi et al. 2004 Atsuta and Takahashi 2016
PAX2	Mikhael et al. 2021	Agenesis of Müllerian duct	Torres et al. 1995	Epithelium from initiation stage	Torres et al. 1995 Atsuta and Takahashi 2016
PAX8	Williams et al. 2017 Smol et al. 2020 Chen et al. 2021	Loss of endometrium	Mittag et al. 2007	Müllerian ductal epithelium	Vainio et al. 1999
RARA	Mikhael et al. 2021	Agenesis of Müllerian duct	Lohnes et al. 1994 Kastner et al. 1997	Müllerian duct	Nakajima et al. 2016
TBX6	Ledig and Wieacker 2018 Chen et al. 2021 Mikhael et al. 2021	Essential for development of mesoderm	Muraoka et al. 2018 Wehn et al. 2020	-	-
WNT4	Williams et al. 2017 Ledig and Wieacker 2018 Mikhael et al. 2021	Arrest of Müllerian ductal invagination	Vainio et al. 1999 Roly et al. 2018	Müllerian ductal mesenchyme	Vainio et al. 1999
WNT9B	Williams et al. 2017 Ledig and Wieacker 2018 Chen et al. 2021 Mikhael et al. 2021	Arrest of Müllerian ductal elongation	Carroll et al. 2005	Wolffian ductal epithelium	Carroll et al. 2005

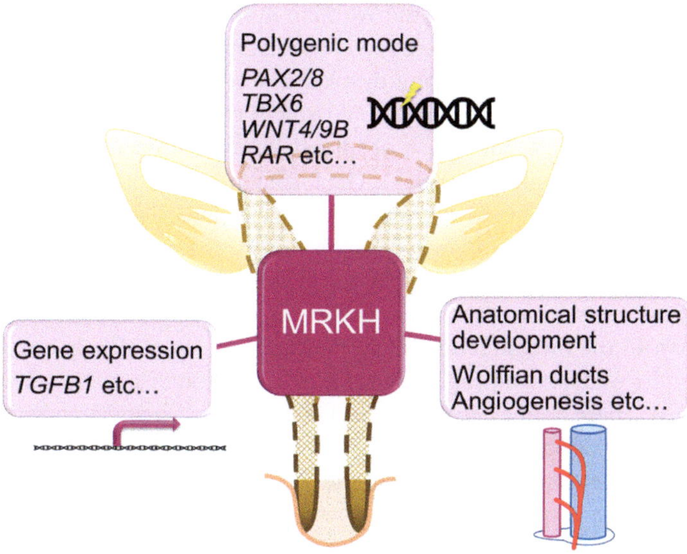

Figure 17.1 Genetic, epigenetic, and anatomical factors involved in the Mayer–Rokitansky–Küster–Hauser (MRKH) syndrome.

The Relationship of Epigenomic and Anatomical Factors with MRKH Syndrome

To predict the potential significance of variants, an Italian cohort study applied protein–protein interaction network analysis to DNA copy number profiling (Pontecorvi et al. 2021). In that study, some new genes encoding the hub protein, such as mitogen-activated protein kinase 3, have been hypothesized (Pontecorvi et al. 2021); however, this gene mutation has not been reported in patients with MRKH syndrome. Therefore we need other approaches as well, including approaches related to epigenetic and protein expression levels, for a study of biomarkers for MRKH syndrome. A combination of microarray and methylation analyses reveals differences in the expression of many genes and the methylation of DNA in MRKH syndrome (Rall et al. 2011).

To give us a grasp of the comprehensive gene activity in MRKH syndrome, the transcription landscape in the endometrial tissue of uterine rudiments was profiled in a large cohort of MRKH patients using RNA-sequencing (Hentrich et al. 2020). In that study of gene expression, downregulation of transforming growth factor beta 1 (*TGFB1*) was found to be the central gene in a differentially expressed gene network of transcription factors and cytokines (Hentrich et al. 2020) (Figure 17.1). Their work accelerated research on the biology of MRKH syndrome; however, gene expression profiles may differ between the uterine remnant and the developing primordium of the uterus in patients with MRKH syndrome. Since *TGFB1* is expressed in various types of tissues (Fagerberg et al. 2014), it is hard to use the expression of this gene as a MRKH biomarker.

The morphology of neighboring tissues also influences the development of Müllerian ducts. The Wolffian duct exhibits *Wnt9B* expression that promotes the elongation of Müllerian ducts (Carroll et al. 2005); however, mutant mice expressing diphtheria toxin A subunit in the Wolffian duct to eliminate it show truncated Müllerian ducts, without effects on *Wnt9b* or *Lhx1* expression in the Wolffian duct (Chiga et al. 2014). These results led us to consider a hypothesis according to which the Wolffian duct itself is required for the elongation of the Müllerian ducts. In fact the deletion of a sonic hedgehog gene that is expressed only in the Wolffian duct induces uterine dysplasia, with a reduction of the mesenchyme and with agenesis of the vascular vessels (Haraguchi et al. 2021). Remnant cells of the Wolffian duct contribute to the formation of smooth muscle, stroma, and vascular vessels in the adult uterus (Haraguchi et al. 2021). The formation of vascular vessels may also be involved in the development of the uterus. Expression of EGF-like domain 7, which is a marker of active angiogenesis, is detected in the uterus of the adult mouse, indicating vascular growth and remodeling throughout the estrus cycle (Bambino et al. 2014). In the uterus of adult mice and humans, angiogenesis is regulated by estrogen (Ma et al. 2001; Taylor et al. 2001). In the human fetal uterus, marker proteins of the tip cells and pericytes, which are employed for sprouting and for the stabilization of new vessels, are detected (Rusu et al. 2013). These studies give certain evidence of active angiogenesis in the uterus of mice and humans during developmental and post-developmental stages. The vasculature and blood flow may be essential for the maintenance of the uterus because, in uterine transplantation in primates, uterine vessels are anastomosed and a lack of uterine blood flow induces irreversible rejection (Kisu et al. 2018, 2019). Further, a non-surgically created neovagina of patients with MRKH shows weak vaginal blood flow (Both et al. 2018). Taken together, the development of the Wolffian duct and the blood vessels of the female reproductive tract are essential for uterine development and maintenance, and blood flow in the developing uterus may offer candidate biomarkers for MRKH syndrome (Figure 17.1).

Estrogen Receptors in the Developing Uterus Are Irreversibly Involved in Uterine Function and Disease

Estrogen and its receptors, ESR1 and ESR2, regulate functions, including cell proliferation, protein secretion, and implantation in the adult uterus of mice (Hewitt and Korach 2018). Even during the developmental stage, ESRs have a role in the uterus. Exposure to synthetic estrogen, diethylstilbestrol (DES), *in utero* induces DES syndrome in humans, and model mice show atrophic thread-like uteri and permanent stimulation of cell proliferation and cornification in the vaginal epithelium (Iguchi et al. 2021). In neonatally DES-treated mice, vascular density in the developing uterine stroma is low and maturation of vessels is delayed; therefore angiogenic suppression is primarily responsible for uterine atrophy derived from DES treatment (Yamashita et al. 2013). Transcriptome analysis in MRKH syndrome suggests that deficient ESR signals may lead to an abnormal development of the female reproductive tract (Rall et al. 2014). Further, the children of women and men prenatally exposed to DES show a higher prevalence of MRKH syndrome (Wautier et al. 2019). Thus, polymorphisms of the *ESR1* and *ESR2* genes relating to expression and functions may affect sensitivity to external estrogen, inducing malformation in the developing female reproductive tract (Figure 17.2).

The inhibition of ESR functions in neonatal mice induces the irreversible stimulation of cell proliferation in the uterine epithelium throughout adulthood, and the phenotype is similar to that of *Esr2* knockout mice (Nakajima et al. 2015). Cell proliferation in the uterine epithelium of neonatal *Esr1* knockout mice is comparable (Nanjappa et al. 2015). These studies indicate that native ESR2 without external estrogen acts as an

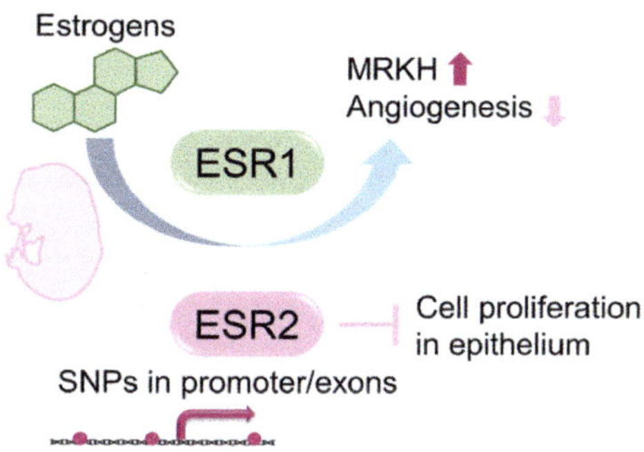

Figure 17.2 The role of estrogen receptors (ESRs) in the developing uterus, and single nucleotide polymorphisms (SNPs) in the ESR2 gene.

inhibitor of basal cell proliferation of the uterine epithelium. Single-nucleotide polymorphisms (SNPs) of the *ESR2* gene are common among patients with endometriosis and ovarian cancer (Kyriakidis and Papaioannidou 2016; Smolarz and Romanowicz 2021). Although SNPs in the promoter region of the *ESR2* gene are not associated with the development of endometrial cancer (Lattrich et al. 2014), SNPs of the *ESR2* gene regulate *ESR2* mRNA expression, both with activation and inhibition (Hamdi et al. 2019; Philips et al. 2012). Taken together, SNPs of *ESR1* and *ESR2* genes may be useful as biomarkers in several types of disease of the female reproductive tract, such as cancer and dysplasia (Figure 17.2).

Conclusion

Diagnostically, MRKH syndrome is characterized by agenesis of the uterus and vagina; therefore the disease can be detected through simple observation with magnetic resonance imaging. However, no treatment can be administered to patients at this stage without regenerative therapy, because the uterus has not been formed. Diagnosis using some biomarkers at an early developmental stage may allow us to treat MRKH syndrome, in which polygenic, epigenetic, and anatomical factors may be pleiotropically involved in the induction of agenesis. Thus it is challenging to find a consistent genetic or epigenetic biomarker for MRKH syndrome. Development of integrated biomarkers among polygenic, epigenetic, and anatomical factors involved in the development and maintenance of Müllerian ducts is needed. Angiogenesis and blood flow in the developing uterus may be new anatomical biomarkers for MRKH syndrome. ESRs in the developing uterus are also involved in the development and functions of the uterus, and SNPs of *ESR1* and *ESR2* relating to expression and functions may be biomarkers for sensitivity to exogenous estrogen during the developing stage and during cell proliferation of the adult uterine epithelium. In the near future we hope that the availability of SNPs of *ESR1* and *ESR2* as candidate biomarkers for diseases of the female reproductive tract such as MRKH syndrome will be investigated.

Acknowledgment

This work was supported by JSPS KAKENHI Grant Numbers JP21K06244 (T.N.) and 18K06315, 15K07132, and 24570074 (T.S.).

References

Allison Stewart, C. et al. (2013). CTNNB1 in mesenchyme regulates epithelial cell differentiation during müllerian duct and postnatal uterine development. *Mol. Endocrinol.* 27 (9): 1442–1454. doi:10.1210/me.2012-1126.

Atsuta, Y. and Takahashi, Y. (2016). Early formation of the Müllerian duct is regulated by sequential actions of BMP/Pax2 and FGF/Lim1 signaling. *Development* 143 (19): 3549–3559. doi:10.1242/dev.137067.

Ayers, K.L. et al. (2015).DMRT1 is required for Müllerian duct formation in the chicken embryo. *Dev. Biol.* 400 (2): 224–236. doi: 10.1016/j.ydbio.2015.02.001.

Backhouse, B. et al. (2019). Identification of candidate genes for Mayer–Rokitansky–Küster–Hauser syndrome using genomic approaches. *Sexual Development* 13 (1): 26–34. doi:10.1159/000494896.

Bambino, K. et al. (2014). Epidermal growth factor-like domain 7 is a marker of the endothelial lineage and active angiogenesis. *Genesis* 52 (7): 657–670. doi:10.1002/dvg.22781.

Benson, G.V. et al. (1996). Mechanisms of reduced fertility in Hoxa-10 mutant mice: Uterine homeosis and loss of maternal Hoxa-10 expression. *Development* 122 (9): 2687–2696. doi:10.1242/dev.122.9.2687.

Both, S. et al. (2018). Sexual response in women with Mayer–Rokitansky–Küster–Hauser syndrome with a nonsurgical neovagina. *Am. J. Obstet. Gynecol.* 219 (3): 283.e1–283.e8. doi:10.1016/j.ajog.2018.07.012.

Bouchard, M. et al. (2002). Nephric lineage specification by Pax2 and Pax8. *Genes Dev.* 16 (22): 2958–2970. doi:10.1101/gad.240102.

Carroll, T.J. et al. (2005). Wnt9b plays a central role in the regulation of mesenchymal to epithelial transitions underlying organogenesis of the mammalian urogenital system. *Dev. Cell* 9 (2): 283–292. doi:10.1016/j.devcel.2005.05.016.

Chen, N. et al. (2021). Perturbations of genes essential for Müllerian duct and Wölffian duct development in Mayer–Rokitansky–Küster–Hauser syndrome. *Am. J. Hum. Genet.* 1–9. doi:10.1016/j.ajhg.2020.12.014.

Chiga, M. et al. (2014). Preformed wolffian duct regulates Müllerian duct elongation independently of canonical Wnt signaling or Lhx1 expression. *Int. J. Dev. Biol.* 58 (9): 663–668. doi:10.1387/ijdb.140261rn.

Choussein, S. et al. (2017). Mullerian dysgenesis: A critical review of the literature. *Arch. Gynecol. Obstet.* 295 (6): 1369–1381. doi:10.1007/s00404-017-4372-2.

Cooke, P.S. et al. (2013). Uterine glands: Development, function and experimental model systems. *Mol. Hum. Reprod.* 19 (9): 547–558. doi:10.1093/molehr/gat031.

Cunha, G.R. et al. (2018). Development of the human female reproductive tract. *Differentiation* 103: 46–65. doi: 10.1016/j.diff.2018.09.001.

Cunha, G.R. et al. (2021). Ontogeny of estrogen receptors in human male and female fetal reproductive tracts. *Differentiation* 118 (October 2020): 107–131. doi:10.1016/j.diff.2020.10.001.

De Tomasi, L. et al. (2017). Mutations in GREB1L cause bilateral kidney agenesis in humans and mice. *Am. J. Hum. Genet.* 101 (5): 803–814. doi:10.1016/j.ajhg.2017.09.026.

Demir Eksi, D. et al. (2018). Copy number variation and regions of homozygosity analysis in patients with Müllerian aplasia. *Mol. Cytogenet.* 11 (1): 1–8. doi:10.1186/s13039-018-0359-3.

Fagerberg, L. et al. (2014). Analysis of the human tissue-specific expression by genome-wide integration of transcriptomics and antibody-based proteomics. *Mol. Cell. Proteom.* 13 (2): 397–406. doi:10.1074/mcp.M113.035600.

Hamdi, Y. et al. (2019). Functional analysis of promoter variants in genes involved in sex steroid action, DNA repair and cell cycle control. *Genes* 10 (3): 1–22. doi:10.3390/genes10030186.

Haraguchi, R. et al. (2021). New insights into development of female reproductive tract-hedgehog-signal response in wolffian tissues directly contributes to uterus development. *Int. J. Mol. Sci.* 22 (3): 1–13. doi:10.3390/ijms22031211.

Hentrich, T. et al. (2020). The endometrial transcription landscape of MRKH syndrome. *Front. Cell Devel. Biol.* 8 (September): 1–13. doi:10.3389/fcell.2020.572281.

Herlin, M.K. et al. (2019). Whole-exome sequencing identifies a GREB1L variant in a three-generation family with Müllerian and renal agenesis: A novel candidate gene in Mayer–Rokitansky–Küster–Hauser (MRKH) syndrome. A case report. *Hum. Reprod.* 34 (9): 1838–1846. doi:10.1093/humrep/dez126.

Hewitt, S.C. and Korach, K.S. (2018). Estrogen receptors: New directions in the new millennium. *Endocr. Rev.* 39 (5): 664–675. doi: 10.1210/er.2018-00087.

Iguchi, T. et al. (2021). New frontiers of developmental endocrinology opened by researchers connecting irreversible effects of sex hormones on developing organs. *Differentiation* 118: 4–23. doi:10.1016/j.diff.2020.10.003.

Jacquinet, A. et al. (2020). GREB1L variants in familial and sporadic hereditary urogenital adysplasia and Mayer–Rokitansky–Kuster–Hauser syndrome. *Clin. Genet.* 98 (2): 126–137. doi:10.1111/cge.13769.

Kastner, P. et al. (1997). Genetic evidence that the retinoid signal is transduced by heterodimeric RXR/RAR functional units during mouse development. *Development* 124: 313–326.

Kisu. et al. (2019). Long-term outcome and rejection after allogeneic uterus transplantation in cynomolgus macaques. *J. Clin. Med.* 8 (10): 1572. doi:10.3390/jcm8101572.

Kisu, I. et al. (2018). Basic research on uterus transplantation in nonhuman primates in Japan. *J. Obstet. Gynaecol. Res.* 44 (10): 1871–1881. doi:10.1111/jog.13724.

Kobayashi, A. et al. (2004). Requirement of Lim1 for female reproductive tract development. *Development* 131 (3): 539–549. doi:10.1242/dev.00951.

Kyriakidis, I. and Papaioannidou, P. (2016). Estrogen receptor beta and ovarian cancer: A key to pathogenesis and response to therapy. *Arch. Gynecol. Obstet.* Springer Berlin Heidelberg. 293 (6): 1161–1168. doi:10.1007/s00404-016-4027-8.

Lam, A.Q. et al. (2014). Rapid and efficient differentiation of human pluripotent stem cells into intermediate mesoderm that forms tubules expressing kidney proximal tubular markers. *J. Am. Soc. Nephrol.* 25 (6): 1211–1225. doi:10.1681/ASN.2013080831.

Lattrich, C. et al. (2014). Polymorphisms in the promoter region of estrogen receptor β gene in endometrial cancer. *Arch. Gynecol. Obstet.* 289 (3): 631–635. doi:10.1007/s00404-013-3012-8.

Ledig, S. and Wieacker, P. (2018). Clinical and genetic aspects of Mayer–Rokitansky–Küster–Hauser syndrome. *Medizinische Genetik* 30 (1): 3–11. doi:10.1007/s11825-018-0173-7.

Lohnes, D. et al. (1994). Function of the retinoic acid receptors (RARs) during development. (I) Craniofacial and skeletal abnormalities in RAR double mutants. *Development* 120 (10): 2723–2748. doi:10.1242/dev.120.10.2723.

Lokmane, L. et al. (2010). vHNF1 functions in distinct regulatory circuits to control ureteric bud branching and early nephrogenesis. *Development* 137 (2): 347–357. doi:10.1242/dev.042226.

Ma, L. et al. (1998). Abdominal B (AbdB) Hoxa genes: Regulation in adult uterus by estrogen and progesterone and repression in Müllerian duct by the synthetic estrogen diethylstilbestrol (DES). *Dev. Biol.* 197 (2): 141–154. doi:10.1006/dbio.1998.8907.

Ma, W. et al. (2001). Adult tissue angiogenesis: Evidence for negative regulation by estrogen in the uterus. *Mol. Endocrinol.* 15 (11): 1983–1992. doi:10.1210/mend.15.11.0734.

Mikhael, S. et al. (2021). Genetics of agenesis/hypoplasia of the uterus and vagina: Narrowing down the number of candidate genes for Mayer–Rokitansky–Küster–Hauser Syndrome. *Hum. Genet.* 140 (4): 667–680. doi:10.1007/s00439-020-02239-y.

Mittag, J. et al. (2007). Congenital hypothyroid female Pax8-deficient mice are infertile despite thyroid hormone replacement therapy. *Endocrinology* 148 (2): 719–725. doi:10.1210/en.2006-1054.

Miyamoto, N. et al. (1997). Defects of urogenital development in mice lacking Emx2. *Development* 124 (9): 1653–1664. doi:10.1242/dev.124.9.1653.

Muraoka, N. et al. (2018). Tbx6 induces nascent mesoderm from pluripotent stem cells and temporally controls cardiac versus somite lineage diversification. *Stem Cell* 23 (3): 382–395. doi:10.1016/j.stem.2018.07.001.

Nakajima, T. et al. (2015). Neonatal estrogen receptor β is important in the permanent inhibition of epithelial cell proliferation in the mouse uterus. *Endocrinology* 156 (9). doi:10.1210/en.2015-1012.

Nakajima, T., Iguchi, T., and Sato, T. (2016). Retinoic acid signaling determines the fate of uterine stroma in the mouse Müllerian duct. *Proc. Natl. Acad. Sci. USA* 113 (50): 10.1073/pnas.1608808113.

Nakajima, T., Yamanaka, R., and Tomooka, Y. (2019). Elongation of Müllerian ducts and connection to urogenital sinus determine the borderline of uterine and vaginal development. *Biochem. Biophs. Rep.* 17: 44–50. doi:10.1016/j.bbrep.2018.10.013.

Nanjappa, M.K. et al. (2015). Neonatal uterine and vaginal cell proliferation and adenogenesis are independent of estrogen receptor 1 (ESR1) in the mouse. *Biol. Reprod.* 92 (3): 1–10. doi:10.1095/biolreprod.114.125724.

Orvis, G.D. and Behringer, R.R. (2007). Cellular mechanisms of Müllerian duct formation in the mouse. *Dev. Biol.* 306 (2): 493–504. doi: 10.1016/j.ydbio.2007.03.027.

Philips, S. et al. (2012). Functional characterization of a genetic polymorphism in the promoter of the ESR2 gene. *Horm. Cancer.* 3 (1–2): 37–43. doi:10.1007/s12672-011-0086-2.

Pontecorvi, P. et al. (2021). Protein–protein interaction network analysis applied to DNA copy number profiling suggests new perspectives on the aetiology of Mayer–Rokitansky–Küster–Hauser syndrome. *Sci. Rep.* 11 (1): 1–11. doi:10.1038/s41598-020-79827-5.

Rall, K. et al. (2011). A combination of transcriptome and methylation analyses reveals embryologically relevant candidate genes in MRKH patients. *Orphanet. J. Rare Dis.* 6 (32): 1–12. doi:10.1186/1750-1172-6-32.

Raymond, C.S. et al. (1999). Expression of Dmrt1 in the genital ridge of mouse and chicken embryos suggests a role in vertebrate sexual development. *Dev. Biol.* 215 (2): 208–220. doi:10.1006/dbio.1999.9461.

Roly, Z.Y. et al. (2018). The cell biology and molecular genetics of Müllerian duct development. *Wiley Interdiscip. Rev. Dev. Biol.* 7 (3): 1–13. doi:10.1002/wdev.310.

Rusu, M.C. et al. (2013). Sprouting angiogenesis in human midterm uterus and fallopian tube is guided by endothelial tip cells. *Anat. Sci. Int.* 88 (1): 25–30. doi:10.1007/s12565-012-0154-x.

Santana Gonzalez, L. et al. (2021). Mechanistic drivers of Müllerian duct development and differentiation into the oviduct. *Front. Cell Devel. Biol.* 9: 1–19. doi:10.3389/fcell.2021.605301.

Shaw, G. and Renfree, M.B. (2014). Wolffian duct development. *Sexual Development* 8 (5): 273–280. doi: 10.1159/000363432.

Smol, T. et al. (2020). Mayer–Rokitansky–Künster–Hauser syndrome due to 2q12.1q14.1 deletion: PAX8 the causing gene? *Eur. J. Med. Genet.* 63 (4): 103812. doi:10.1016/j.ejmg.2019.103812.

Smolarz, B. and Romanowicz, H. (2021). Association between single nucleotide polymorphism of the CYP19A1 and ESR2 genes and endometriosis. *Arch. Gynecol. Obstet.* 304 (2): 439–445. doi:10.1007/s00404-021-06051-5.

Taguchi, A. and Nishinakamura, R. (2017). Higher-order kidney organogenesis from pluripotent stem cells. *Cell Stem Cell* 21 (6): 730–746.e6. doi:10.1016/j.stem.2017.10.011.

Taylor, R.N. et al. (2001). Endocrine and paracrine regulation of endometrial angiogenesis. *Ann. NY Acad. Sci.* 943: 109–121. doi:10.2217/17460816.3.2.117.

Torres, M. et al. (1995). Pax-2 controls multiple steps of urogenital development. *Development* 121 (12): 4057–4065. doi:10.1242/dev.121.12.4057.

Vainio, S. et al. (1999). Female development in mammals is regulated by Wnt-4 signalling. *Nature* 397 (6718): 405–409. doi:10.1038/17068.

Watanabe, K. et al. (2017). Recent advances in the molecular mechanisms of Mayer–Rokitansky–Küster–Hauser syndrome (Review). *Biomed. Rep.* 7 (2): 123–127. doi:10.3892/br.2017.929.

Wautier, A. et al. (2019). Genital tract and reproductive characteristics in daughters of women and men prenatally exposed to diethylstilbestrol (DES). *Therapie* 75 (5): 2020–2021. doi:10.1016/j.therap.2019.10.004.

Wehn, A.K. et al. (2020). Functionally distinct roles for T and Tbx6 during mouse development. *Biol Open* 9 (8): doi:10.1242/BIO.054692.

Willem, M. et al. (2002). Specific ablation of the nidogen-binding site in the laminin γ1 chain interferes with kidney and lung development. *Development* 129 (11): 2711–2722. doi:10.1242/dev.129.11.2711.

Williams, L.S. et al. (2017). Genetic analysis of Mayer–Rokitansky–Kuster–Hauser syndrome in a large cohort of families. *Fertil. Steril.* 108 (1): 145–151.e2. doi:10.1016/j.fertnstert.2017.05.017.

Yamashita, S. et al. (2013). Mechanisms of angiogenic suppression in uteri exposed to diethylstilbestrol neonatally in the mouse. *Biol. Reprod.* 88 (5): 1–13. 10.1095/biolreprod.112.106443.

18

Cause or Consequence

Epigenomic DNA Methylation Changes in Arsenic-Mediated in Vitro Transformation of Human Prostate Cells

B. Alex Merrick[1], Dhiral P. Phadke[2], Ruchir R. Shah[2], Deepak Mav[2], and Erik J. Tokar[1,ǂ]

[1] *Mechanistic Toxicology Branch, Division of the National Toxicology Program, National Institute of Environmental Health Sciences, NC*
[2] *Sciome, LLC, NC*
[ǂ] *Contributed equally to this work*

Introduction

Inorganic arsenic is an environmental human carcinogen: lung, bladder, skin, liver, and urinary tract organs are recognized sites of tumor formation after prolonged exposure to arsenic-contaminated drinking water supplies (Mandal 2017; Nordstrom 2002). The mammalian metabolism of inorganic arsenic occurs in a sequence of oxidation and reduction reactions, with the participation of methyltransferases and the cofactor S-adenosylmethionine (SAM), to form monomethylarsonous acid (MMA^{III}), monomethylarsonic acid (MMA^{V}), dimethylated metabolites such dimethylarsinous acid (DMA^{III}), and dimethylarsinic acid (DMA^{V}) for urinary excretion (Sattar et al. 2016; Vahter 2009). However, exceeding the metabolic methylation capacity of the liver and other organs can lead to arsenic accumulation, toxicity, and tumor formation (Arslan et al. 2017; Yu et al. 2000). In the lab, arsenic accumulates in experimental animals, under paradigms of increasing levels of arsenic exposure provided in the animal's drinking water; those increasing arsenic levels vary by tissue and age (Markowski et al. 2011). Epidemiological studies have similarly shown that increasing levels of arsenic in drinking water are associated with an increased incidence of arsenic-related diseases in the skin, the lung, and other tissues (Hasanvand et al. 2020; Kapaj et al. 2006; Shao et al. 2019).

Studies of arsenic carcinogenesis have revealed that it is an extremely complex process, incompletely understood. Although inorganic arsenic is not directly mutagenic in point-mutation detection assays (Rossman et al. 1980), its ability to cause cytogenetic damage, including chromosomal aberrations, sister chromatid exchange, and micronuclei formation, is well documented (Roy et al. 2018). It has been proposed that DNA damage occurs indirectly, through arsenic-induced formation of reactive oxygen species from the disruption of cellular redox reactions and mitochondrial functions (Hu et al. 2020; Soza-Ried et al. 2019). Additional contributors to DNA damage may come from arsenic impairment of DNA repair enzymes (Kocadal et al. 2020; Zhou et al. 2021). Epigenetic effects of arsenic have been proposed as major contributors to malignant transformation (Bustaffa et al. 2014; Pogribny and Rusyn 2013). These effects range from altered DNA methylation (Roy et al. 2020) to histone modifications (Howe and Gamble 2016), altered miRNA expression (Ferragut Cardoso et al. 2020), and signaling pathway disruption (Soza-Ried et al. 2019).

Genomic and Epigenomic Biomarkers of Toxicology and Disease: Clinical and Therapeutic Actions, First Edition. Edited by Saura C. Sahu.

The exact mechanism for how arsenic produces DNA damage and its role in epigenetic dysfunction in cancer formation are unknown. It is especially unclear whether DNA damage is required for arsenic carcinogenesis or whether epigenetic changes alone can provide sufficient changes in cellular signaling, metabolism, and cell proliferation. Epidemiological studies are just now making great strides in describing arsenic-related DNA methylation changes associated with gene expression, to benchmark arsenic's potential for the development of disease (Bozack et al. 2020), though there is a great need for specific molecular signatures of arsenic-derived tumors. Advances in mechanistic understanding should spur the development of predictive molecular indicators of tumors caused by arsenic exposure.

Arsenic affects many organs of the reproductive and urinary tracts, including the prostate gland (Zimta et al. 2019). Chronic in vitro exposure to arsenic can malignantly transform many different types of immortalized cells (Chen and Costa 2017). When immortalized human prostate epithelial RWPE-1 cells were continually exposed to sodium arsenite in vitro over a twenty-nine-week period, a fully transformed cell line emerged. These cells were termed chronic arsenic exposed prostate epithelial (CAsE-PE) cells (Achanzar et al. 2002). CAsE-PE cells display many malignant characteristics, including loss of cell contact inhibition in vitro, anchorage-independent growth in soft agar, malignant tumor growth upon injection into nude mice, resistance to apoptosis, secretion of metalloproteinases such as MMP-9, and androgen-independent growth (Achanzar et al. 2002; Benbrahim-Tallaa et al. 2007; Brambila et al. 2002). KRAS transcript and protein overexpression were well documented in CAsE-PE cells early on (Benbrahim-Tallaa et al. 2005) and, more recently, oncogenic mutations in KRAS in CAsE-PE cells have been demonstrated (Merrick et al. 2019).

However, the potential for epigenetic contributions to transformation needs to be further explored, through work that includes a more global assessment of arsenic's effects upon DNA methylation. Prior work from our lab has shown a differential methylation of cancer-relevant transcripts in CAsE-PE cells in relation to control cells, with inverse correlation of methylation to altered gene expression (Pelch et al. 2015). For example, an increased expression of HYAL1 and S100P was correlated with hypomethylation near the transcriptional start site of these genes, while a decreased expression of NES and NTM transcripts in CasE-PE cells was correlated with hypermethylation near the transcriptional start site. In the current study, our hypothesis was that genome-wide analysis of DNA methylation marks in CAsE-PE cells would show demethylation events associated with gene activation and increased methylation for gene suppression that, in aggregate, would contribute to malignant transformation.

Materials and Methods

Cells and Cell Culture

Parent RWPE-1 cells were originally isolated from normal human prostate epithelial cells, and immortalized with human papillomavirus 18 (HPV 18); they are diploid, anchorage-dependent and non-tumorigenic, showing no tumors in nude mice (Webber et al. 1997). RWPE-1 cells were grown in keratinocyte serum-free medium at 37°C, 5% CO_2 with 50 μg/ml bovine pituitary extract, 5 ng/ml epidermal growth factor, containing an antibiotic–antimycotic mixture. CAsE-PE cells were developed through continual exposure of parental RWPE-1 cells to 5 μM sodium arsenite for twenty-nine to thirty weeks, with media changes at three times per week (Achanzar et al. 2002). RWPE-1 (control) and

CAsE-PE (arsenic-transformed) cells were harvested from three independent cultures and frozen at −80°C, until RNA extraction and two independent cultures for DNA extraction of each cell line were made.

RNA-seq and Differential Expression

Gene expression by RNA-seq was performed as previously described (Merrick et al. 2019). Briefly, RNA was isolated from three separate cultures of time-matched RWPE-1 and CAsE-PE cells at twenty-nine weeks using RNeasy spin columns (Qiagen, Valencia, CA). PolyA-tailed mRNA was isolated and random hexamer primed to create a cDNA library from which 100 base-pair (bp) paired-end reads were generated and sequenced on an Illumina HiSeq2000 instrument (Illumina, San Diego, California). Fastq files were analyzed for quality control using FastQC. Reads were aligned to hg19 of the human genome using TopHat, and then entered into UCSC genome browser tracks for the visual inspection of the normalized signal at any genomic location. The UCSC browser tracks contain RPKM (reads per kilobase per million reads) normalized read counts. The deep sequencing fastq data files for expression analysis and alignment are stored in the Sequence Read Archive (SRA) under Study Accession Nos. SUB4913016 and PRJNA514436. Cuffdiff version 2.2.0 was used to determine differential gene expressed transcripts between RWPE-1 and CAsE-PE cells on the basis of the human RefSeq transcriptome.

BisDNA-seq

DNA was isolated from time-matched RWPE-1 and CAsE-PE cells at thirty weeks using DNeasy spin columns (Qiagen in Valencia, California) and fragmented by adaptive focused acoustic energy (Covaris in Massachusetts). DNA was primed with random hexamers, and libraries were created at 400 bp median fragment size and prepared for 100 bp paired-end sequencing on a HiSeq2000 Illumina instrument. Average Phred scores of ≥ 20 per position were used for alignment. Eluted methylated DNA fragments were equally apportioned (a) for DNA-seq sequencing for improved alignment of RNA-seq data to the genome; and (b) for MBD2 enrichment and bisulfite reduction before DNA sequencing (BisDNA-seq), in order to determine the exact sites of methylation. Briefly, fragmented DNA was affinity-enriched for 5′-methyl cytosines (5mC) after DNA fragment binding to MBD2 (methyl-CpG-binding domain protein) paramagnetic beads (Invitrogen, Life Technologies, Carlsbad, California) and subsequent elution with 1 M NaCl. Methylation-enriched DNA fractions were treated with bisulfite (Zymo Research, Irvine, California), in order to convert unmethylated cytosines into uracil (detected as thymidine after PCR), while methylated cytosines remained unchanged. Libraries were constructed for NextGen sequencing analysis (Illumina HiSeq 2000) from either untreated methylation-enriched DNA (DNA-seq) or bisulfite-treated methylation-enriched DNA (BisDNA-seq). Reads were Bowtie 2.2.3-aligned to the reference genome (hg19) for RNA-seq, while the Bismark algorithm was used for BisDNA-seq 0.12.4 alignment. Methylated and non-methylated counts were used to call significantly differentially methylated bases (DMBs called at p value < 0.01) using the Rao-Scott likelihood ratio test (Rao and Scott 1987; Zhang et al. 2018). A BedGraph file was created and calls were made for differentially methylated bases and clusters. A differentially methylated cluster (DMC) was defined as having at least five differentially methylated bases (DMB) within a 300 bp DNA sequence length. Only DMCs that presented an average ≥ 50% methylation difference between CAsE-PE and RWPE-1 control were reported. The analysis strategy has been briefly outlined in Figures 18.1 and 18.2.

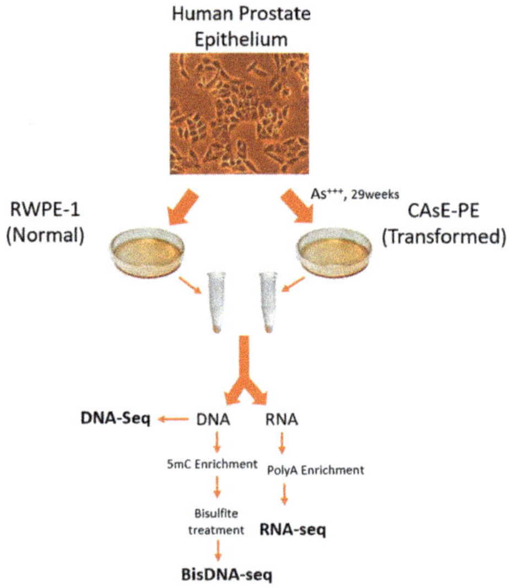

Figure 18.1 Experimental approach. DNA and RNA were extracted separately from RWPE-1 prostate cells and CAsE-PE transformed cells for NextGen sequencing analysis (Illumina). Sequencing was done on genomic DNA (DNA-seq) and methylation-enriched DNA after further bisulfite reduction (BisDNA-seq). RNA-seq was performed for gene expression. A Cufflinks pipeline was used to assemble transcripts and estimate their abundance, and CuffDiff was used for differential expression using RefSeq annotation. DNA methylation within regulatory regions reduces or silences transcript expression and, conversely, less methylation can increase transcription and expression.

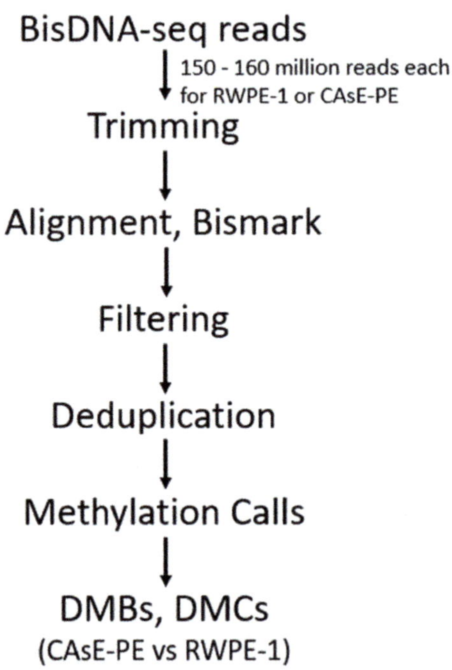

Figure 18.2 Differentially methylated bases (DMBs) and clusters (DMCs). Isolated DNA from RWPE and CAsE-PE (arsenic) cells were bisulfite-treated, libraries were constructed, and BisDNA-seq analysis was performed to generate 150–160 million reads from 100 bp paired-end fragments. Reads were trimmed for quality, filtered, and after the removal of duplicated regions sequences were aligned to hg19 using Bismark software. After filtering, methylation calls were made from 66–89 million reads. Methylated cytosines (DMBs) were organized into differentially methylated clusters (DMCs), defined as a minimum of 5 CpG sites within a 300bp length. In downstream analysis, we considered only the most significantly differentially methylated clusters with an average methylation difference $\geq$ 50%.

Validation of Methylated Sequences by Roche454 Sequencing

Annotated gene regions with substantial changes in gene expression by RNA-seq and cytosine methylation by BisDNA-seq were examined for quantitative validation through pyrosequencing on a 454 FLX sequencer (Roche, Branford, Connecticut). Target regions are provided later in the text (see "Results"). DNA from RWPE-1 and CAsE-PE cells was bisulfite-converted using the EpiTect Bisulfite Kit at Qiagen. PCR primers were designed to target bisulfite-converted regions using the online program MethPrimer (http://www.urogene.org/methprimer). Bisulfite-converted DNA was amplified, and specificity was checked for single amplicon products through visualization on a 1% agarose gel before purification. PCR products of the gene targets were pooled in equimolar amounts and titrated for optimal library input concentration for bulk emulsion PCR. Pooled libraries were amplified by bulk emPCR to generate sufficient numbers of beads for a full picotiter plate run. Image processing and base calls were performed, and bisulfite-treated DNA reads were aligned against corresponding untreated DNA sequences by Bismark (Krueger and Andrews 2011). The relative numbers of reads of CAsE-PE and RWPE-1 (control) reads were compared for each site.

Results

Gene Expression by RNA-seq

In this study, our genomic approach to DNA methylation involved MBD2 bead-capture of DNA fragments containing CpG base pairs. Our bioinformatic objective was to identify CpG changes in DNA methylation in CAsE-PE cells, then to relate methylation changes to differential gene expression from RWPE-1 cells and determine the relevance of such changes to malignant transformations. We had previously performed RNA-seq on three separate RWPE-1 and CAsE-PE cell cultures, and gene expression data were reported as described previously (Merrick et al. 2019). Briefly, RNA-seq reads for each sample accounted for > 200X coverage of the transcriptome. Reads were aligned according to RefSeq-annotated sequences for a total alignment of 46,090 transcripts and isoforms. CuffDiff showed 7,265 transcript differences, for which 3,261 transcripts were upregulated and 4,004 transcripts were downregulated in CAsE-PE cell by comparison with RWPE-1 controls. The pronounced overexpression of KRAS (> 400 fold) and its downstream transcripts in CAsE-PE cells was a notable finding. Pathway analysis of differentially expressed transcripts demonstrated increases in transcripts representing cell growth, proliferation, morphology, and death or survival pathways that supported transcriptional changes highly associated with cancer and with reproductive and developmental diseases.

DNA Base-pair Methylation and Methylation Clusters

The determination of DNA methylation was made from BisDNA-seq of two separate RWPE-1 and CAsE-PE DNA isolations after enrichment for 5mC (5′-methyl cytosine), favoring a concentration of CpG sites by MBD2 affinity capture (Figure 18.1). From the Bismark analysis pipeline (Figure 18.2), 150–160 million combined reads were accumulated for each cell type. The alignment strategy accounted for > 90% of reads. Figure 18.3 shows the apportionment of methylated C/G bases (e.g. CpGs) that accounted for the 56,434,896 CpG sites (both strands) predicted for the human genome (hg19 build). For both cell types, BisDNA-seq data showed that 60.09% of CpG genomic sites were accounted for in 33,913,012 reads by at least one or more reads in captured 5mC DNA, while the remaining 22,521,884 CpG sites were not measured under the current MBD2 capture and BisDNA-seq procedures. A histogram of total CpG reads (methylated or unmethylated) is shown in Figure 18.3. As expected, the histogram shows a concentration of CpG reads on a discrete number of CpG sites. In other words, some DNA regions contained more CpG sites than others. These data suggest that read coverage under the present conditions involves a substantial number of CpG sites for an accurate determination of differential methylation between the two cell types.

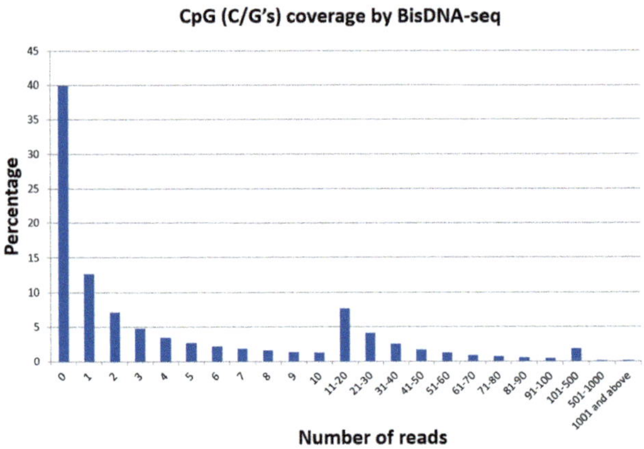

Coverage	Number of C/G
0	22,521,884
1	7,133,032
2	3,988,961
3	2,664,936
4	1,941,351
5	1,503,202
6	1,213,003
7	1,010,358
8	867,251
9	758,113
10	671,356
11-20	4,295,110
21-30	2,291,839
31-40	1,425,473
41-50	954,159
51-60	680,555
61-70	506,886
71-80	390,333
81-90	306,786
91-100	245,770
101-500	1,039,877
501-1000	10,464
1001 and above	14,197
Total	56,434,896

Figure 18.3 CpG coverage by BisDNA-seq. Reads from BisDNA-seq were organized into the amount of reads at each C/G site (predominantly CpG sites). Of the total 56,434,806 possible C/G sites in hg19 (both strands), 33,913,012 reads had at least 1 ≥ read, while 22,521,884 had zero reads (top of adjoining table). Some C/G sites had more reads than others, as shown in the C/G coverage histogram, which suggests enrichment of C/G sites by MBD2 capture.

Read counts containing individual CpG sites on hg19 from both cell types were used to call 1,094,643 bases as significant ($p < 0.01$) and differentially methylated (DMBs) using the Rao-Scott likelihood ratio test. There was a slightly higher number of DMBs, which amounted to 54.3% (594,451 DMBs), by comparison with 45.7% (500,192 DMBs) of hypomethylated DMBs in both cell types. Although methylation at individual bases is important, we also considered differential methylation in terms of the proximity of DMBs to one another. In other words, several DMBs in proximity, or in a cluster, could have a greater impact on transcription than a single DMB. We refer to ≥ 5 significant DMBs within a 300 bp region as a differentially methylated cluster (DMC). By this criterion for grouping DMBs into discrete genomic regions with substantial methylation changes, we were able to designate 6,915 DMCs from both cell types. We then separated the DMCs into hypermethylated clusters and hypomethylated clusters, according to the two cell types, and counted them. As shown in Table 18.1, there were four times more hypermethylated DMCs than hypomethylated DMCs in CAsE-PE cells. As for downstream analysis, we focused our attention only on those DMCs with a ≥ 50% average methylation difference when CAsE-PE were compared to RWPE-1 cells. Here the roughly 4 : 1 ratio of hypermethylated (5,589 DMCs in CAsE-PE) versus hypomethylated (1,226 DMCs in CAsE-PE) DMCs was observed for CAsE-PE in comparison with RWPE-1 cells.

Proximity of Hg19 DMCs to the Gene Structure

We found 6,915 differentially methylated clusters (≥ 50% methylation) on hg19 that excluded repeat and sequence-redundant regions. We were interested in the proximity of DMC changes around and within the gene structure (Figure 18.4), since these epigenetic events would be important for associating potential effects upon transcription. Notable gene features are clusters of DNA methylation that occur in exons, introns, 1 kB distance from the transcriptional start site (TSS) or transcriptional end site (TES), intergenic regions, and CpG islands (CGIs) and shores.

Table 18.1 Summary of differentially methylated clusters.

Methylation Type	# DMCs	# DMBs in DMCs	# DMCs at ≥ 50% Δ in Methylation	# DMBs in DMCs at ≥ 50% Δ methylation
Hypermethylation	11,061 (83.7%)	86,203 (86.1%)	5,689 (82.3%)	44,817 (84.8%)
Hypomethylation	2,147 (16.3%)	13,867 (13.9%)	1,226 (17.7%)	8,018 (15.2%)
Total	13,208 (100%)	100,070 (100%)	6,915 (100%)	52,835 (100%)

The numbers of differentially methylated clusters (DMCs) on the hg19 genome after BisDNA-seq are reported for transformed CAsE-PE cells versus parent RWPE-1 prostate cell line. Differentially methylated bases (DMBs) at $p < 0.01$ in DMCs (total) and the DMCs with ≥ 50% change (Δ) methylation were counted as either hypermethylated or hypomethylated in CAsE-PE cells vs. RWPE-1 cells.

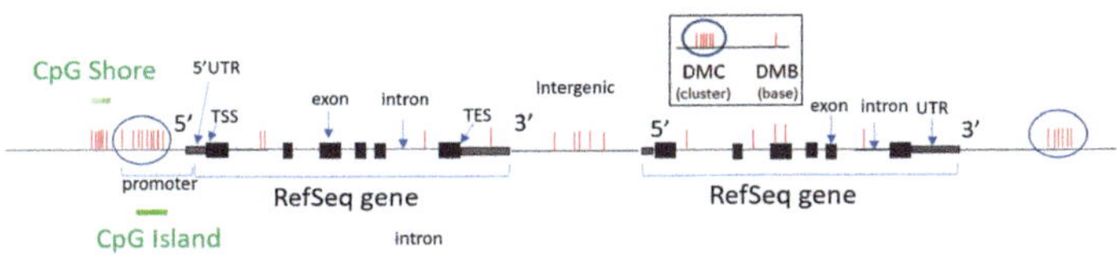

Figure 18.4 Epigenomic features of DNA methylation in model genes. Model RefSeq genes on a linear strand of DNA sequence are shown with methylation marks (red vertical bars). UTRs (untranslated regions) as smaller boxes are at the 5′ and 3′-ends of each gene. Larger boxes are exons. Introns and are the intervening sequences (line). A differentially methylated base (DMB) is a single red vertical bar, and a set of 5 DMBs within 300 nucleotides (nt) is a differentially methylated cluster (DMC). Promoter regions are sequences usually 1 kB upstream of the transcriptional start site (TSS). The transcription end site (TES) is usually followed by a regulatory 3′-UTR as part of the gene. Intergenic regions are large stretches (10–100s kB) of DNA away from the 5′- or 3′-ends of the gene. CpG islands (green bar) are segments of DNA from 500 to 1,500 nt in length, often in promoter regions, with a CpG ratio of more than 50% or 60%, and are frequent sites of regulatory methylation. CpG shores (green bar) also have concentrations of CpGs and are 2 kB upstream of islands.

Figure 18.5 shows the distribution of DMCs in and around features of gene structure such as the promoter region at 1 kB upstream of the 5′-TSS (transcriptional start site); and it also includes 1 kB of the downstream regulatory region of the 3′-TES (transcriptional end site); the region within exons and introns; and intergenic sites. There were ten times more hypermethylated clusters than hypomethylated clusters in CAsE-PE promoter regions than in RWPE-1. There was, similarly, a ten times greater number of exons in hypermethylated clusters than hypomethylated clusters, and the same was also observed for hypermethylation of intron and intergenic regions by comparison with hypomethylation. Out of the 6,915 DMCs, there were 5,689 hypermethylated clusters and 1,226 hypomethylated clusters in CAsE-PE cells. These data suggest that genomic hypermethylation occurred in CAsE-PE cells by comparison with RWPE-1 cells.

CGIs contain 50–60% GC content within a region of at least ≥ 200 bases, and often occur at promoter regions (Deaton and Bird 2011). CpG shores are of lower CpG density and occupy flanking regions outside CGIs, at a distance of up to 2kB from them (Portela and Esteller 2010). Figure 18.6 shows CAsE-PE cells that contain hypermethylated clusters in

Description	Count
Promoter(1kb upstream of TSS) hyper	636
Exon hyper	1,997
Intron hyper	1,133
Downstream (1kb downstream of TES) hyper	66
Intergenic hyper	1,857
Promoter(1kb upstream of TSS) hypo	55
Exon hypo	261
Intron hypo	513
Downstream (1kb downstream of TES) hypo	21
Intergenic hypo	376
Total	6,915

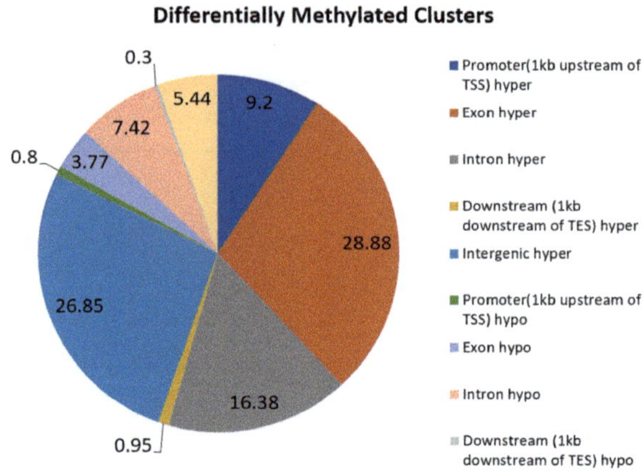

Figure 18.5 Increased gene methylation in arsenic-transformed cells (CAsE-PE) vs. parent (RWPE-1) cells. Methylated reads were organized into differentially methylated clusters (DMCs), in which we considered only the most significantly differentially methylated CpG bases (DMBs), with an average methylation difference of ≥ 50% change of 5 DMBs within 300 nt. DMCs were grouped into CAsE-PE cells according to hypermethylation (≥50% methylation) or hypomethylation (<50% methylation), by comparison with RWPE-1. More hypermethylated DMCs were found in CAsE-PE cells vs RWPE-1 in all regions of RefSeq gene structure.

Description	Count
CpG_island hyper	4,472
CpG_shore hyper	405
Not island or shore hyper	812
CpG_island hypo	334
CpG_shore hypo	216
Not island or shore hypo	676
Total	6,915

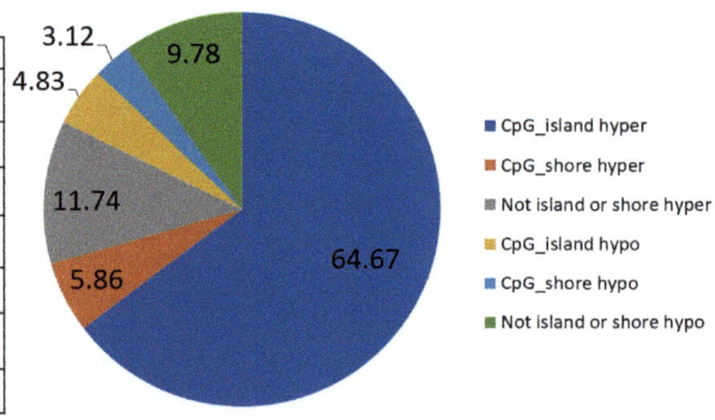

Figure 18.6 Increased methylation of CpG islands and CpG shores in CAsE-PE cells. Methylated reads were organized into differentially methylated clusters (DMCs) and mapped to known CpG islands and CpG shores. DMCs in CAsE-PE vs. RWPE-1 were grouped as hypermethylation (>50%) or hypomethylation (<50%) into CpG islands and shores. Some DMCs were in genomic regions that were considered neither islands nor shores.

CGIs and in shore regions coming in numbers almost ten times greater than those of hypomethylated islands and shores. The proportion of non-CGI/shore regions was about 22% of all DMCs (812 + 676 / 6915 total DMCs). Again, these data suggest hypermethylation in CAsE-PE cells by comparison with parent cells.

Our approach to DNA capture and the measurement of methylated cytosines by BisDNA-seq was evaluated for its accuracy via an orthogonal method, using Roche454 DNA sequencing. CpG sites near 5′-end and promoter regions were selected for validation at forty hypermethylated sites and nine hypomethylated sites. Genomic locations and gene IDs are shown in Figure 18.7. After bisulfite treatment and PCR amplification, DNA samples were sequenced on a Roche454 instrument, and results for percentage methylation by Roche454 were compared to genomic MBD2 capture and genomic BisDNA-seq estimates

Gene ID hg19	hg19 5mC Location	Site No.
CES4A:NM_001190201	chr16.67035229	1
CES4A:NM_001190201	chr16.67035231	2
FOXE1:NM_004473	chr9.100617284	3
CES4A:NM_001190201	chr16.67035225	4
FOXE1:NM_004473	chr9.100617199	5
KLF13:NM_015995	chr15.31775661	6
FOXE1:NM_004473	chr9.100617275	7
ARSI:NM_001012301	chr5.149681848	8
ARSI:NM_001012301	chr5.149681839	9
ARSI:NM_001012301	chr5.149681845	10
NES:NM_006617	chr1.156646311	11
ARSI:NM_001012301	chr5.149681854	12
SEPT9:NM_001113493	chr17.75369700	13
RUNX2:NM_004348	chr6.45631300	14
MAFB:NM_005461	chr20.39319334	15
FOXE1:NM_004473	chr9.100617297	16
RBM38:NM_017495	chr20.55964956	17
MAFB:NM_005461	chr20.39319337	18
SEPT9:NM_001113493	chr17.75369599	19
NES:NM_006617	chr1.156646468	20
NES:NM_006617	chr1.156646470	21
RBM38:NM_017495	chr20.55964963	22
SEPT9:NM_001113493	chr17.75369660	23
PRIC285:NM_033405	chr20.62197218	24
MAFB:NM_005461	chr20.39319325	25

Gene ID hg19	hg19 5mC Location	Site No.
RBM38:NM_017495	chr20.55965803	26
RBM38:NM_017495	chr20.55965865	27
RBM38:NM_017495	chr20.55964941	28
RBM38:NM_017495	chr20.55964939	29
RUNX2:NM_004348	chr6.45631486	30
RBM38:NM_017495	chr20.55965866	31
RBM38:NM_017495	chr20.55965804	32
SEPT9:NM_00111349	chr17.75369228	33
KLF13:NM_015995	chr15.31775662	34
MAFB:NM_005461	chr20.39319335	35
SEPT9:NM_001113493	chr17.75369600	36
MAFB:NM_005461	chr20.39319326	37
PMF1:NM_001199653	chr1.156186539	38
PRIC285:NM_033405	chr20.62197219	39
RUNX2:NM_004348	chr6.45631208	40
PRIC285:NM_033405	chr20.62197320	1
PRIC285:NM_033405	chr20.62197298	2
SEPT9:NM_001113493	chr17.75369251	3
SEPT9:NM_001113493	chr17.75369227	4
PRIC285:NM_0334050	chr20.62197321	5
PRIC285:NM_033405	chr20.62197299	6
SEPT9:NM_001113493	chr17.75369591	7
PMF1:NM_001199653	chr1.156186492	8
RUNX2:NM_004348	chr6.45631299	9

Figure 18.7 CpG sites for the validation of methylation. Genes from the Hg19 genome build were selected for validation of methylation at specific CpG sites. The chromosomal and nucleotide location of the 5mC (methylated cytosine) site is listed for 49 sites across 11 RefSeq genes. The site number (no.) corresponds to Figure 18.8 for sites 1–40, which had increased methylation in CAsE-PE cells vs. RWPE-1, and then the shaded sites 1–9 were found as low-methylation or hypomethylated sites.

of methylation at the same genomic 5mC site. Figure 18.8 shows excellent comparison in the proportion and direction of CAsE-PE hypermethylated sites #1 to #32 for Roche454 sequencing. However, once hypermethylation levels reached ~ 10% (#33 to #40), the proportion of methylation by BisDNA-seq became disparate and, in the case of #33, the methylation was opposite for Roche454 sequencing and for BisDNA-seq. Similarly, there was concurrence of direction and magnitude of hypomethylation at sites #1 to #4; but, when hypomethylation in CAsE-PE cells approached 10% or went further down (sites #5 to #9), BisDNA-seq became less accurate according to Roche454 sequencing.

Correlation of DNA Methylation and Gene Expression

In this study the predicted inverse relationship between DNA methylation and gene expression was inconsistent and not uniformly met. Methylation was generally observed throughout any given gene structure. Methylation differences at the promoter and 5′UTR regions were gene regions at which we expected to find the best opportunities for accompanying changes in expression. In the almost fifty CpG sites at which we performed DNA methylation validation studies for eleven genes in CAsE-PE and RWPE-1 cells, ten genes had substantial methylation changes in promoter and 5′-UTR regions. Increased methylation resulted in minor fold changes in the expression of ARSI, FOXE1, HELZ2, RBM38. CES4A

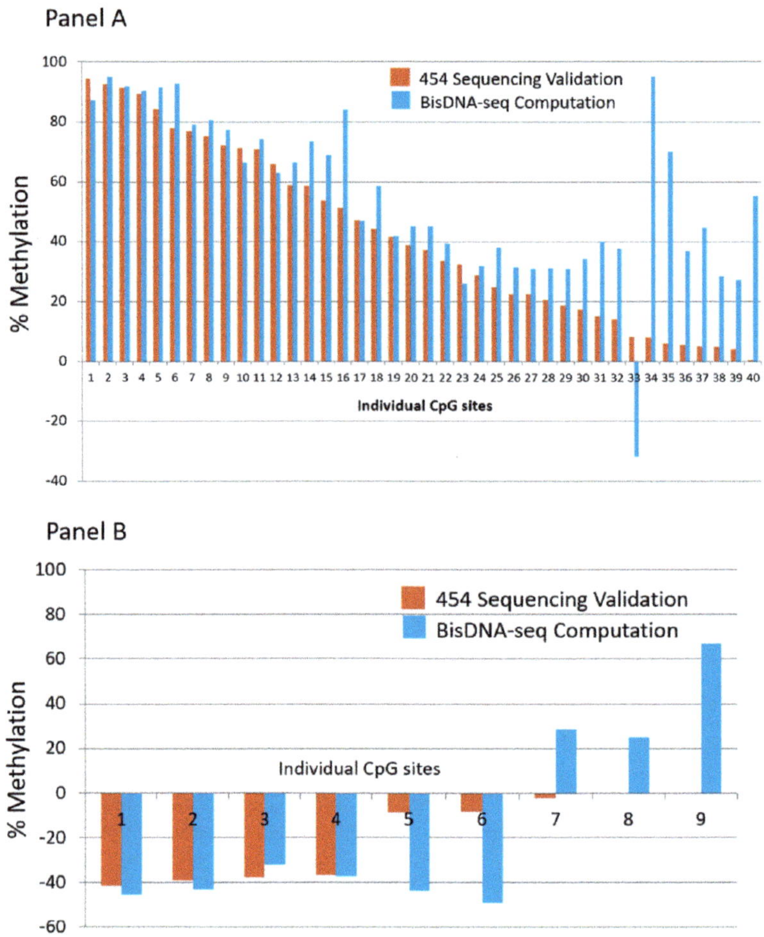

Figure 18.8 Validation of BisDNA-seq methylation CpG sites by Roche454 sequencing. Genes were selected for validation on the basis of their varying proportion of methylation (hypermethylated in Panel A; hypomethylated in Panel B). Most genes were selected on the basis of DMBs, within 5 kB upstream and 1 kB downstream of the transcriptional start site (except for RUNX2 with intergenic sites). DNA was bisulfite-treated, amplified, and Roche454-sequenced (orange bars) to determine the level of methylation for comparison with BisDNA-seq (blue bars). BisDNA-seq and Roche454 sequencing were in agreement on the amount of methylation (percentage methylation at a specific base from 1 to 100%) until a 10% methylation level occurred, where data from the two platforms began to diverge.

showed the largest expression increase—namely a 98-fold increase in CAsE-PE cells—but little methylation change at the promoter or 5′UTR. RUNX2 was included for validation because of its varying intergenic CpG methylation levels at > 100kB downstream of the transcriptional end site; even without methylation changes in promoter or 5′UTR regions, its expression was decreased 5.1-fold.

A more systematic approach was taken to evaluate the relationship between differential gene methylation and transcript expression. We considered a differentially methylated gene (DMG) as containing at least one differentially methylated cluster (5 methylated CpGs within 300nt), with an average methylation of ≥ 50% located within 1kb of its transcriptional start site (TSS). Using this criterion, the Venn diagram in Figure 18.9 shows that there were 1,627 DMGs, of which 1,476 were hypermethylated and 151 were hypomethylated in

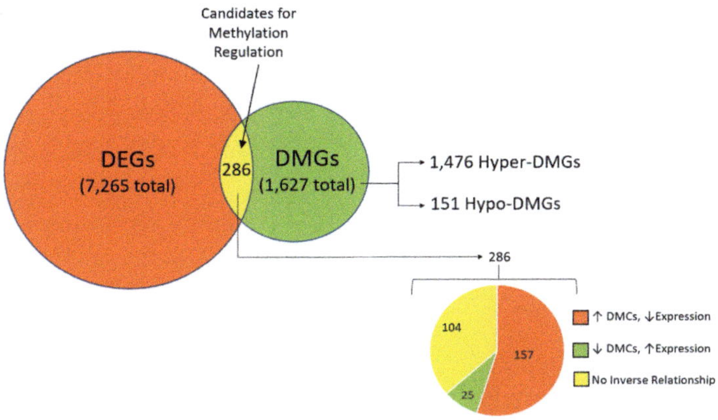

Figure 18.9 Association of gene methylation and expression. Differentially expressed genes (DEGs) were determined by RNA-seq in prior work (Merrick et al. 2019). Differentially methylated genes (DMGs) were designated as having at least ≥ 1 differentially methylated cluster (DMC) with an average methylation of $\geq 50\%$ and located within 1 kB of the transcriptional start site (TSS). By this criterion, there were 1,627 DMGs and, when these were compared with the 7,265 DEGs, only 286 were candidates for methylation regulation. The expected inverse relationship of methylation and expression was evaluated, with the result that 157 genes were found that showed increased methylation and decreased expression and 25 genes were found that had decreased methylation and increased expression.

CAsE-PE vs RWPE-1 cells. Of the 7,265 differentially expressed genes (DEGs), only 286 DMGs and DMCs overlapped as candidates for regulation. In other words, only 3.9% of DEGs (286) contained sufficient changes in methylation to be considered as potentially regulated by methylation, and just that limited number of differentially methylated genes, or 17% of total DMGs, was observed to be differentially expressed. Next, we examined each candidate gene for an inverse relationship of methylation and expression. As shown in the lower pie chart in Figure 18.9, we found 157 hypermethylated genes (↑DMCs) with decreased transcript expression and 25 hypomethylated genes (↓DMCs) with increased transcript expression. Thus, just over a half of the genes with DMCs in the promoter/5′UTR/first intron regions or 63% (157 + 25/282) conformed to an inverse relation of methylation associated with transcript expression.

Figure 18.10 shows only those methylated genes inversely associated with expression changes that had substantial read counts (≥ 0.5 FPKM) by RNA-seq. DEG fold changes were converted to the log2 scale for a more evenly scaled comparison between genes. There were 22 hypermethylated transcripts with decreased expression from 10-fold to 391-fold, and 21 hypermethylated transcripts with reduced expression from 5-fold to 10-fold. Only 7 genes (2 transcript variants of PTK6) showed increased expression, from 5-fold to 188-fold. A brief table of molecular functions shows a diversity of potential methylation effects on expression. For example, STRA6 is a binding protein for retinol transport and cellular uptake, increased 188-fold in CAsE-PE vs. RWPE-1 cells. PTK6 (with a 27-fold or 42-fold increase in each of two isoforms) is a tyrosine kinase regulating cell signaling for the differentiation, maintenance, and growth of epithelia. ZNF667-AS1, at 391-fold decrease in CAsE-PE cells, is a lncRNA operating as an antisense transcript for regulating transcription. NPTX2 plays a role in cellular plasticity and development and was decreased 290-fold in CAsE-PE cells.

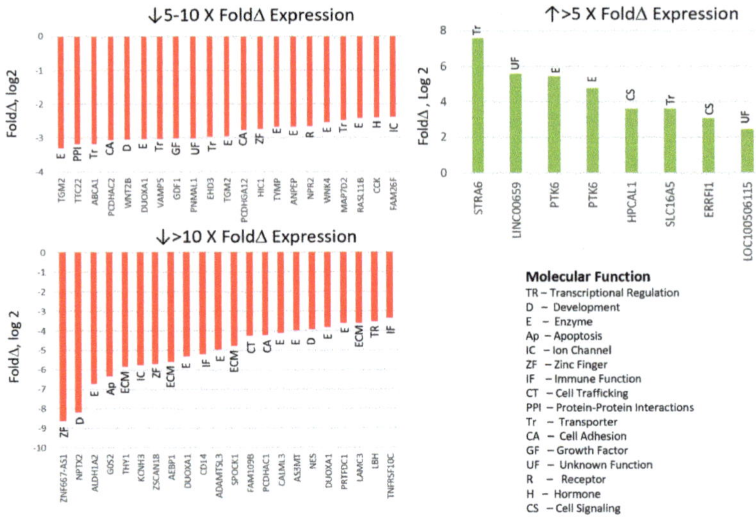

Figure 18.10 Specific genes with an inverse association between methylation and expression. Select hypermethylated and hypomethylated genes from Figure 18.9 are shown here with their relative transcript (gene) expression. RefSeq gene identities and their relative decrease or increase in expression (log2 scale) are shown for hypomethylated genes (red bars) and hypermethylated genes (green bars). There were 22 hypermethylated genes with a ≥ 10X-fold decrease expression and 21 hypermethylated genes with a 5X-fold to 10X-fold decrease in expression. By comparison, there were only 8 hypomethylated genes that had a > 5X-fold increase in expression. A legend of molecular function (from an abbreviation table) is annotated for each gene.

Examples of DNA methylation changes that were correlated to increased or decreased expression are shown in Figure 18.11. In Panel A, the gene AE Binding Protein-1 (*AEBP1*), which encodes a carboxypeptidase A transcript, showed increased methylation at the promoter, 5′UTR, and first intron in CAsE-PE cells that was associated with almost complete transcript repression (a 47-fold decreased expression in CAsE-PE cells). Alternatively, PTK6 encodes a cytoplasmic signaling kinase (protein kinase 6) where BisDNA-seq measured decreased methylation in the promoter, 5′UTR, and first intron, as shown in Panel B of Figure 18.11. Thus, decreased methylation for this gene variant of PTK6 in CAsE-PE cells was associated with a 27-fold increase in expression.

Discussion

Study of epigenomics in cancer often involves the relationship of DNA methylation to gene expression (Sharma et al. 2020). The role of DNA methylation in gene silencing has been well studied in many types of cancer (Baylin 2005). A short list of selected genes that undergo hypermethylation in sporadic cancers includes RB in retinoblastoma, APC, CDKN2A ($p14^{ARF}$) and MLH1 in colorectal tumors, CDKN2B ($p15^{INK4b}$) in leukemia, CDKN2A ($p16^{INK4a}$) in various cancers, BRCA1 in breast and ovarian cancer, VHL in renal cell cancers, and NR3A1 (ERα) in breast and other cancers (Baylin 2005). Changes in DNA methylation can affect chromosome structure, transcription, and the inactivation of tumor suppressor genes. Alternatively, a role for DNA hypomethylation has been forwarded as a means of oncogene activation in some cancer germline genes in their capacity as contributors to tumor development (Van Tongelen et al. 2017). Hypomethylated genes include SSX,

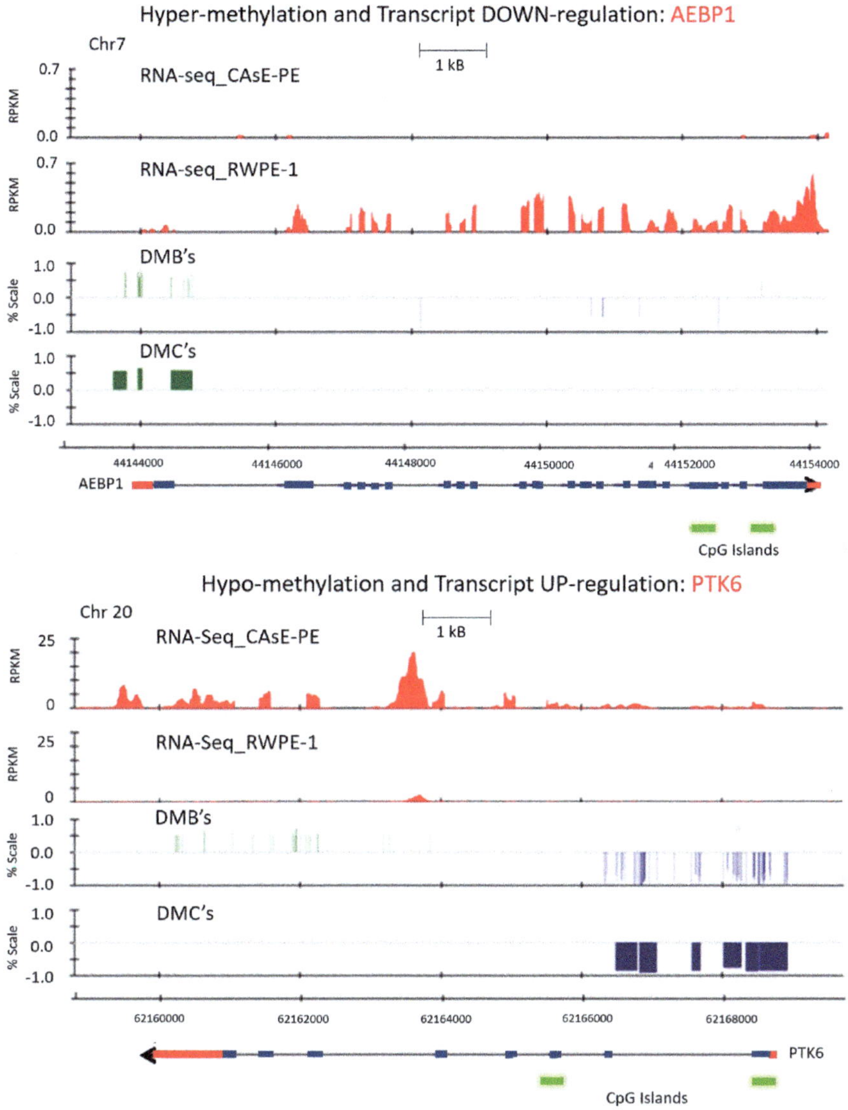

Figure 18.11 Genes showing an inverse association between methylation and expression. Comparison between the expression patterns of a hypermethylated and a hypomethylated gene are shown in UCSC genome browser format; and they conform to the expected inverse relationship between gene methylation and expression. In the top Panel A, the carboxypeptidase transcript AEBP1 (adipocyte enhancer binding protein-1) is highly expressed in normal RWPE-1 cells shown by reads in red but greatly repressed in transformed CAsE-PE cells (almost no reads). Reads on the y-axis are on equivalent scales for each cell type, and the x-axis is the genomic sequence with coordinates shown at the bottom of the panel. An increased number of DMBs (differentially methylated bases) align with DMCs (differentially methylated clusters) shown on the lower two tracks. Increased methylation (green bars) indicated by vertical lines (DMBs) and bars (DMCs) show enhanced methylation at the 5′-end of the AEBP1 gene in the promoter, 5-UTR and first exon/intron regions as shown at the bottom of the panel on the x-axis. In the bottom Panel B, PTK6 (protein tyrosine kinase 6) expression is greatly increased (red) in CAsE-PE cells in the topmost track by comparison with the low expression in RWPE-1 cells in the second RNA-seq track below. Note the decreased methylation (blue lines, DMBs; blue bars, DMCs) at the 5′-end of the gene (promoter, 5′-UTR, first exon/intron regions). A few increased methylated bases are noted toward the 3′-end of the gene, but the overall expression is increased for PTK6.

DDX43, MAEL, and PIWIL1 in cell proliferation and DDX53 in angiogenesis (Van Tongelen et al. 2017). Global hypomethylation has been observed in some human tumor types, for example lung cancer (Hong and Kim 2021) and esophageal squamous cell carcinoma (Lin et al. 2020), and some cases of prostate cancer (Zelic et al. 2015).

The effects of arsenic on DNA methylation have been approached at a global genomic level as well as through measurements at a site-specific gene locus. Findings of global DNA hypomethylation of cells or tissues with arsenic exposure in various experimental or epidemiological settings align with a hypothesis of cellular depletion of methyl donors like SAM (S-adenosyl methionine) during arsenic metabolism—depletion that may be assisted by folate deficiency (Chung et al. 2019; Mauro et al. 2016; Pilsner et al. 2009). Recent reviews have proposed that arsenic places a cellular stress on methyl donor pools, which leads to global DNA hypomethylation and hypermethylation of DNA at specific gene promoters. This in turn causes the overall dysregulation of oncogenic and tumor suppressor gene expression, which opens the way to tumor formation (Sage et al. 2017). Alternatively, others have reported a global DNA hypermethylation of peripheral blood mononuclear cells in arsenic-exposed populations (Majumdar et al. 2010; Niedzwiecki et al. 2013). Part of the challenge of studying the epigenomic effects of arsenic lies in the diversity of platforms and outputs for the measurement of DNA methylation, either at a locus-specific level or on a whole-genome basis. DNA methylation analysis reviews have covered some of the challenges in the field, and recent ones assess microarray (e.g. MethylationEPIC bead chip), affinity capture (e.g. MeDIP, methylated DNA immunoprecipitation) and sequencing (e.g. whole genome bisulfite sequencing) platforms for their robustness, throughput, and cost (Khodadadi et al. 2021).

In our study we used an affinity capture, a global bisulfite DNA sequencing method designed to enrich methylated genomic CpG sites and then to measure the proportion of methylation at each site through deep sequencing. This method has been recently reviewed and compared favorably to other methylation analysis platforms (Hsu et al. 2014; Jeltsch et al. 2019). We focused upon methylation at hg19 annotated genes and surrounding intergenic regions in a normal and arsenic-transformed cell line. We were able to account for 60% of all possible 56 million CpG sites as either methylated or non-methylated. We found over 1 million CpG bases that showed significant ($p < 0.01$) differential methylation (DMBs), and then we grouped them into 6,915 clusters that each contained more than 50% overall differential methylation of at least 5 DMBs. Our reasoning was that groups of DMBs would have a greater effect on gene transcription than differential methylation at just one individual CpG site. Further, we validated this affinity capture BisDNA-seq method by Roche454 sequencing, finding a high degree of accuracy within a 10% differential methylation level at individual CpG differential sites.

Using this analytical approach for CpG enrichment, bisulfite conversion, and measurement by BisDNA-seq, we found that the arsenic-transformed CAsE-PE cell line was globally hypermethylated in its genome by a ratio of 4 : 1 when we compared it to the RWPE-1 parent line. The distribution of DMCs demonstrated that methylation differences were a frequent occurrence in the gene structure within introns and exons, even more so than at promoter regions. This observation attests to the dynamic nature of the phenomenon of differential methylation observed in malignancy (Baylin 2005; Nishiyama and Nakanishi 2021). The organization of CpG sites into islands and shores has been helpful to epigenetic researchers by assessing the effects of differential DNA methylation at genomic loci in relation to gene structure (Nishiyama and Nakanishi 2021). The role of DNA methylation and other epigenomic modifications has been reviewed for prostate cancer (Sugiura et al. 2021). Other researchers have reported hypermethylated genes in prostate cancer that are involved in DNA repair (e.g. MGMT), cell cycle (e.g. CDKN2), cell adhesion (e.g. CDH1, CD44) or tumor suppression (e.g. APC); but the expression of these genes is not necessarily specific to the prostate gland. The hypomethylation for gene activation detected in prostate cancers so far involves a more limited set of genes, such as PLAU, HPSE, and CYP1B1 (Sugiura et al. 2021). In our study, hypermethylation in CGIs and CpG shores occurred in a proportion of almost 9 : 1 vis-à-vis hypomethylation. Thus hypermethylation in regulatory promoter regions within genes and surrounding the genomic sequence was a dominant epigenomic feature of arsenic-mediated transformation in CAsE-PE cells.

A distinguishing aspect of the current study is the availability of previously published RNA-seq data (Merrick et al. 2019) in RWPE-1 and CAsE-PE cells for determining the relationship between expression and DNA methylation changes in these two cell types. The hypothesis of the present study was that an inverse methylation–expression relationship would be manifested by the fact that gene hypermethylation leads to transcript suppression and gene hypomethylation facilitates increased gene expression critical for the malignant transformation in CAsE-PE cells vis-à-vis RWPE-1 cells. When we compared the 7,265 differentially expressed transcripts (Figure 18.9) against genes with prominent methylation changes (DMCs with ≥ 50% altered methylation), we found only 286 transcripts that contained DMCs. This amounts to only 3.9% of the changes in transcript expression of CAsE-PE cells that were accompanied by substantial methylation changes. We considered these 286 transcripts as candidates for methylation regulation, further examining whether there was an inverse relationship between methylation and expression. Our examination showed that 157 gene transcripts were hypermethylated at ± 1kB near the transcriptional start site (e.g. promoter, 5′UTR) and only 25 gene transcripts were hypomethylated. The remaining 102 transcripts did not follow an inverse relationship between expression and methylation. Examination of the most significantly expressed genes affected by methylation revealed a diversity of molecular functions of the affected genes. However, no transcriptional changes pointed consistently to a singular transcript as the one responsible for malignant transformation. It could be argued that methylated genes with altered expression (25 + 157 = 182 total) in aggregate are responsible for the CAsE-PE transformation. On the other hand, a more likely explanation seems to come from our prior work (Merrick et al. 2019); and this is the observation that KRAS was highly overexpressed in CAsE-PE cells at 400-fold increase, contained oncogenic activating mutations at codons 12 and 59, and showed downstream signaling of KRAS-dependent transcripts consistent with malignant transformation caused by mtKRAS as the driver oncogene. KRAS expression is not known to be regulated by methylation, and scrutiny of our data showed little methylation activity at the KRAS gene locus relevant to its increased expression. Therefore the current DNA methylation data do not indicate either the activation of other driver genes or the repression of tumor suppressor genes in CAsE-PE cells that might be responsible for arsenic-induced transformation.

That epigenomic alternations are a primary driver of malignant transformation remains a controversial hypothesis for explaining arsenic carcinogenesis. Epigenetic dysregulation induced by arsenic has been considered a mechanism in cancer formation, but the link between epigenetic changes and gene expression that would clearly support cell proliferation and a transformed phenotype has been difficult to establish either experimentally or epidemiologically. Several reviews have assessed the arsenic literature for specific inverse relationships between methylation and expression to the effect that hypomethylation would activate oncogene expression and hypermethylation would silence tumor suppressor genes. However, reviews of the arsenic literature have observed great ambiguity in the published data that convincingly support an arsenic epigenetic hypothesis backed up by hypo- and hypermethylation at various genomic loci (Reichard and Puga 2010). In vitro transformation models offer controlled arsenic exposures for defined periods and under defined experimental conditions. One study showed arsenic-induced transformation in human HaCaT keratinocytes after twenty-five weeks: methylation changes using a 27 K methylation array were noted in gene promoter regions in transformed cells (Barajas-Olmos et al. 2019). However, expression changes by a 47 K bead array did not produce an enrichment of specific genes identified as drivers of transformation.

In our current work and prior study, the data strongly suggest that mutated KRAS is the driver oncogene and a primary cause of transformation in CAsE-PE cells. Methylated DNA changes were documented in CAsE-PE cells versus the parent RWPE-1 cell line. Such differential methylation in CAsE-PE cells, either at the base-pair level or in clusters, may support a transformed phenotype. However, we interpret methylation-sponsored transcriptional changes in arsenic-transformed CAsE-PE cells as a consequence of altered cell signaling from the overexpression and activation of KRAS in CAsE-PE cells. It is possible that the generation of reactive oxygen species or the inhibition of DNA repair enzymes may lead to the KRAS mutations we reported and in that way make epigenetic changes be indirectly responsible for the transformation—but not by direct means, with sustained gene expression changes in our in vitro model of transformation. Before epigenetic events can be causally ascribed to arsenic-derived tumors, we suggest that a thorough genomic analysis assisted by transcriptomics should be performed in order to check cells and tissues for the kinds of genomic damage such as mutations, copy number variation, large indels, or chromosomal abnormalities that cause oncogene activation and inactivation of tumor suppressor genes. The epigenetic or genetic basis of arsenic-mediated transformation should leverage genomic, expression, and methylation analyses at a global level, with the help of the powerful sequencing technologies now available.

Acknowledgment

Funding was provided through DNTP, NIEHS intramural funds (BAM, EJT) and NIEHS contracts HHSN273201100016C (coauthors DPP, RRS, DM) and HHSN273201700001C.

Disclaimer

This article is the work product of employees of the National Institute of Environmental Health Sciences (NIEHS) and National Institutes of Health (NIH); however, the statements, opinions or conclusions contained therein do not necessarily represent the statements, opinions or conclusions of NIEHS, NIH or the United States government.

References

Achanzar, W.E., Brambila, E.M., Diwan, B.A., Webber, M.M., and Waalkes, M.P. (2002). Inorganic arsenite-induced malignant transformation of human prostate epithelial cells. *J. Natl. Cancer Inst.* 94: 1888–1891.

Arslan, B., Djamgoz, M.B.A., and Akun, E. (2017). ARSENIC: A review on exposure pathways, accumulation, mobility and transmission into the human food chain. *Rev. Environ. Contam. Toxicol.* 243: 27–51.

Barajas-Olmos, F.M., Ortiz-Sanchez, E., Imaz-Rosshandler, I., Cordova-Alarcon, E.J., Martinez-Tovar, A., Villanueva-Toledo, J., Morales-Marin, M.E., Cruz-Colin, J.L., Rangel, C., Orozco, L., and Centeno, F. (2019). Analysis of the dynamic aberrant landscape of DNA methylation and gene expression during arsenic-induced cell transformation. *Gene* 711: 143941.

Baylin, S.B. (2005). DNA methylation and gene silencing in cancer. *Nat. Clin. Pract. Oncol.* 2 (Suppl 1): S4–11.

Benbrahim-Tallaa, L., Waterland, R.A., Styblo, M., Achanzar, W.E., Webber, M.M., and Waalkes, M.P. (2005). Molecular events associated with arsenic-induced malignant transformation of human prostatic epithelial cells: Aberrant genomic DNA methylation and K-ras oncogene activation. *Toxicol. Appl. Pharmacol.* 206: 288–298.

Benbrahim-Tallaa, L., Webber, M.M., and Waalkes, M.P. (2007). Mechanisms of acquired androgen independence during arsenic-induced malignant transformation of human prostate epithelial cells. *Environ. Health Perspect.* 115: 243–247.

Bozack, A.K., Domingo-Relloso, A., Haack, K., Gamble, M.V., Tellez-Plaza, M., Umans, J.G., Best, L.G., Yracheta, J., Gribble, M.O., Cardenas, A., Francesconi, K.A., Goessler, W., Tang, W.Y., Fallin, M.D., Cole, S.A., and Navas-Acien, A. (2020). Locus-specific differential DNA methylation and urinary arsenic: An epigenome-wide association study in blood among adults with low-to-moderate arsenic exposure. *Environ. Health Perspect.* 128: 67015.

Brambila, E.M., Achanzar, W.E., Qu, W., Webber, M.M., and Waalkes, M.P. (2002). Chronic arsenic-exposed human prostate epithelial cells exhibit stable arsenic tolerance: Mechanistic implications of altered cellular glutathione and glutathione S-transferase. *Toxicol. Appl. Pharmacol.* 183: 99–107.

Bustaffa, E., Stoccoro, A., Bianchi, F., and Migliore, L. (2014). Genotoxic and epigenetic mechanisms in arsenic carcinogenicity. *Arch. Toxicol.* 88: 1043–1067.

Chen, Q.Y. and Costa, M. (2017). A comprehensive review of metal-induced cellular transformation studies. *Toxicol. Appl. Pharmacol.* 331: 33–40.

Chung, C.J., Lee, H.L., Chang, C.H., Chang, H., Liu, C.S., Jung, W.T., Liu, H.J., Liou, S.H., Chung, M.C., and Hsueh, Y.M. (2019). Measurement of urinary arsenic profiles and DNA hypomethylation in a case-control study of urothelial carcinoma. *Arch. Toxicol.* 93: 2155–2164.

Deaton, A.M. and Bird, A. (2011). CpG islands and the regulation of transcription. *Genes Dev.* 25: 1010–1022.

Ferragut Cardoso, A.P., Udoh, K.T., and States, J.C. (2020). Arsenic-induced changes in miRNA expression in cancer and other diseases. *Toxicol. Appl. Pharmacol.* 409: 115306.

Hasanvand, M., Mohammadi, R., Khoshnamvand, N., Jafari, A., Palangi, H.S., and Mokhayeri, Y. (2020). Dose-response meta-analysis of arsenic exposure in drinking water and intelligence quotient. *J. Environ. Health Sci. Eng.* 18: 1691–1697.

Hong, Y. and Kim, W.J. (2021). DNA methylation markers in lung cancer. *Curr. Genomics* 22: 79–87.

Howe, C.G. and Gamble, M.V. (2016). Influence of arsenic on global levels of histone posttranslational modifications: A review of the literature and challenges in the field. *Curr. Environ. Health Rep.* 3: 225–237.

Hsu, H.K., Weng, Y.I., Hsu, P.Y., Huang, T.H., and Huang, Y.W. (2014). Detection of DNA methylation by MeDIP and MBDCap assays: An overview of techniques. *Methods Mol. Biol.* 1105: 61–70.

Hu, Y., Li, J., Lou, B., Wu, R., Wang, G., Lu, C., Wang, H., Pi, J., and Xu, Y. (2020). The role of reactive oxygen species in arsenic toxicity. *Biomolecules* 10.

Jeltsch, A., Broche, J., Lungu, C., and Bashtrykov, P. (2019). Biotechnological applications of MBD domain proteins for DNA methylation analysis. *J. Mol. Biol.* 432 (6). doi: 10.1016/j.jmb.2019.08.020

Kapaj, S., Peterson, H., Liber, K., and Bhattacharya, P. (2006). Human health effects from chronic arsenic poisoning: A review. *J. Environ. Sci. Health A Tox Hazard Subst. Environ. Eng.* 41: 2399–2428.

Khodadadi, E., Fahmideh, L., Khodadadi, E., Dao, S., Yousefi, M., Taghizadeh, S., Asgharzadeh, M., Yousefi, B., and Kafil, H.S. (2021). Current advances in DNA methylation analysis methods. *Biomed. Res. Int.* 2021: 8827516.

Kocadal, K., Alkas, F.B., Battal, D., and Saygi, S. (2020). Cellular pathologies and genotoxic effects arising secondary to heavy metal exposure: A review. *Hum. Exp. Toxicol.* 39: 3–13.

Krueger, F. and Andrews, S.R. (2011). Bismark: A flexible aligner and methylation caller for bisulfite-seq applications. *Bioinformatics* 27: 1571–1572.

Lin, L., Cheng, X., and Yin, D. (2020). Aberrant DNA methylation in esophageal squamous cell carcinoma: Biological and clinical implications. *Front. Oncol.* 10: 549850.

Majumdar, S., Chanda, S., Ganguli, B., Mazumder, D.N., Lahiri, S., and Dasgupta, U.B. (2010). Arsenic exposure induces genomic hypermethylation. *Environ. Toxicol.* 25: 315–318.

Mandal, P. (2017). Molecular insight of arsenic-induced carcinogenesis and its prevention. *Naunyn Schmiedebergs Arch. Pharmacol.* 390: 443–455.

Markowski, V.P., Currie, D., Reeve, E.A., Thompson, D., and Wise, J.P. (2011). Tissue-specific and dose-related accumulation of arsenic in mouse offspring following maternal consumption of arsenic-contaminated water. *Basic Clin. Pharmacol. Toxicol.* 108: 326–332.

Mauro, M., Caradonna, F., and Klein, C.B. (2016). Dysregulation of DNA methylation induced by past arsenic treatment causes persistent genomic instability in mammalian cells. *Environ. Mol. Mutagen.* 57: 137–150.

Merrick, B.A., Phadke, D.P., Bostrom, M.A., Shah, R.R., Wright, G.M., Wang, X., Gordon, O., Pelch, K.E., Auerbach, S.S., Paules, R.S., Devito, M.J., Waalkes, M.P., and Tokar, E.J. (2019). Arsenite malignantly transforms human prostate epithelial cells in vitro by gene amplification of mutated KRAS. *PLoS One* 14: e0215504.

Niedzwiecki, M.M., Hall, M.N., Liu, X., Oka, J., Harper, K.N., Slavkovich, V., Ilievski, V., Levy, D., Van Geen, A., Mey, J.L., Alam, S., Siddique, A.B., Parvez, F., Graziano, J.H., and Gamble, M.V. (2013). A dose-response study of arsenic exposure and global methylation of peripheral blood mononuclear cell DNA in Bangladeshi adults. *Environ. Health Perspect.* 121: 1306–1312.

Nishiyama, A. and Nakanishi, M. (2021). Navigating the DNA methylation landscape of cancer. *Trends Genet. 37 (11): 1012–1027.*

Nordstrom, D.K. (2002). Public health: Worldwide occurrences of arsenic in ground water. *Science* 296: 2143–2145.

Pelch, K.E., Tokar, E.J., Merrick, B.A., and Waalkes, M.P. (2015). Differential DNA methylation profile of key genes in malignant prostate epithelial cells transformed by inorganic arsenic or cadmium. *Toxicol. Appl. Pharmacol.* 286: 159–167.

Pilsner, J.R., Liu, X., Ahsan, H., Ilievski, V., Slavkovich, V., Levy, D., Factor-Litvak, P., Graziano, J.H., and Gamble, M.V. (2009). Folate deficiency, hyperhomocysteinemia, low urinary creatinine, and hypomethylation of leukocyte DNA are risk factors for arsenic-induced skin lesions. *Environ. Health Perspect.* 117: 254–260.

Pogribny, I.P. and Rusyn, I. (2013). Environmental toxicants, epigenetics, and cancer. *Adv. Exp. Med. Biol.* 754: 215–232.

Portela, A. and Esteller, M. (2010). Epigenetic modifications and human disease. *Nat. Biotechnol.* 28: 1057–1068.

Rao, J.N.K. and Scott, A.J. (1987). On simple adjustments to chi-squared tests with survey data. *Ann. Stat.* 15: 385–397.

Reichard, J.F. and Puga, A. (2010). Effects of arsenic exposure on DNA methylation and epigenetic gene regulation. *Epigenomics* 2: 87–104.

Rossman, T.G., Stone, D., Molina, M., and Troll, W. (1980). Absence of arsenite mutagenicity in E coli and Chinese hamster cells. *Environ. Mutagen.* 2: 371–379.

Roy, J.S., Chatterjee, D., Das, N., and Giri, A.K. (2018). Substantial evidences indicate that inorganic arsenic is a genotoxic carcinogen: A review. *Toxicol. Res.* 34: 311–324.

Roy, N.K., Murphy, A., and Costa, M. (2020). Arsenic methyltransferase and methylation of inorganic arsenic. *Biomolecules* 10.

Sage, A.P., Minatel, B.C., Ng, K.W., Stewart, G.L., Dummer, T.J.B., Lam, W.L., and Martinez, V.D. (2017). Oncogenomic disruptions in arsenic-induced carcinogenesis. *Oncotarget* 8: 25736–25755.

Sattar, A., Xie, S., Hafeez, M.A., Wang, X., Hussain, H.I., Iqbal, Z., Pan, Y., Iqbal, M., Shabbir, M.A., and Yuan, Z. (2016). Metabolism and toxicity of arsenicals in mammals. *Environ. Toxicol. Pharmacol.* 48: 214–224.

Shao, I.H., Chang, Y.H., and Pang, S.T. (2019). Recent advances in upper tract urothelial carcinomas: From bench to clinics. *Int. J. Urol.* 26: 148–159.

Sharma, A., Sharma, K.L., Bansal, C., and Kumar, A. (2020). Updates on "Cancer genomics and epigenomics." *World J. Clin. Oncol.* 11: 890–897.

Soza-Ried, C., Bustamante, E., Caglevic, C., Rolfo, C., Sirera, R., and Marsiglia, H. (2019). Oncogenic role of arsenic exposure in lung cancer: A forgotten risk factor. *Crit. Rev. Oncol. Hematol.* 139: 128–133.

Sugiura, M., Sato, H., Kanesaka, M., Imamura, Y., Sakamoto, S., Ichikawa, T., and Kaneda, A. (2021). Epigenetic modifications in prostate cancer. *Int. J. Urol.* 28: 140–149.

Vahter, M. (2009). Effects of arsenic on maternal and fetal health. *Annu. Rev. Nutr.* 29: 381–399.

Van Tongelen, A., Loriot, A., and De Smet, C. (2017). Oncogenic roles of DNA hypomethylation through the activation of cancer-germline genes. *Cancer Lett.* 396: 130–137.

Webber, M.M., Bello, D., Kleinman, H.K., and Hoffman, M.P. (1997). Acinar differentiation by non-malignant immortalized human prostatic epithelial cells and its loss by malignant cells. *Carcinogenesis* 18: 1225–1231.

Yu, R.C., Hsu, K.H., Chen, C.J., and Froines, J.R. (2000). Arsenic methylation capacity and skin cancer. *Cancer Epidemiol. Biomarkers Prev.* 9: 1259–1262.

Zelic, R., Fiano, V., Grasso, C., Zugna, D., Pettersson, A., Gillio-Tos, A., Merletti, F., and Richiardi, L. (2015). Global DNA hypomethylation in prostate cancer development and progression: A systematic review. *Prostate Cancer Prostatic Dis.* 18: 1–12.

Zhang, Y., Baheti, S., and Sun, Z. (2018). Statistical method evaluation for differentially methylated CpGs in base resolution next-generation DNA sequencing data. *Brief. Bioinformatics* 19: 374–386.

Zhou, X., Speer, R.M., Volk, L., Hudson, L.G., and Liu, K.J. (2021). Arsenic co-carcinogenesis: Inhibition of DNA repair and interaction with zinc finger proteins. *Semin. Cancer Biol. 76:* 86–98.

Zimta, A.A., Schitcu, V., Gurzau, E., Stavaru, C., Manda, G., Szedlacsek, S., and Berindan-Neagoe, I. (2019). Biological and molecular modifications induced by cadmium and arsenic during breast and prostate cancer development. *Environ. Res.* 178: 108700.

19

Epigenetic Regulation of Sex Determination and Toxicity in Non-mammalian Vertebrates

Genki Yamagishi[1], Taisen Iguchi[2], and Shinichi Miyagawa[1]

[1] *Faculty of Advanced Engineering, Tokyo University of Science, Tokyo, Japan.*
[2] *Graduate School of Nanobioscience, Yokohama City University, Yokohama, Kanagawa, Japan.*

Introduction

Epigenetics is defined as the set of mechanisms that regulate gene expression without changing DNA sequences. The advantages of epigenetic regulation lie in its rapid responses to environmental stimuli within a single generation. Another advantage of epigenetic change is reversibility. An epigenetic modification introduced by one stimulus can be overturned by another, enabling organisms to adapt to frequently changing environmental conditions. This reversibility of epigenetic change is behind the plasticity of sex determination in non-mammalian vertebrates.

But epigenetic changes can be heritable, because some modifications escape epigenetic reprogramming in the zygotes or during primordial germ cell development (for a review, see Marczylo et al. 2016). This means that environmentally induced adaptive characteristics can be inherited from parents and passed on to offspring. However, the heritable nature of epigenetic change is double-edged. This is because not only favorable, advantageous changes are inherited, but also unfavorable and undesired ones. Considering the epigenetic toxicity of certain chemicals and drugs, the heritable nature of epigenetic change is problematic.

There are three major mechanisms of epigenetic regulation: DNA methylation, histone modifications, and regulation by non-coding RNAs. The mechanisms underlying these regulations are briefly summarized next (see also Figure 19.1).

DNA Methylation

DNA methylation takes place at cytosines within cytosine-phosphate-guanine dinucleotides (CpGs). Methylated cytosine has an added methyl group at its C5 position and is called 5-methylcytosine (5mC). 5mC is frequently deaminated to yield thymine. Because of such instability of 5mC, CpGs are scarce in vertebrate genomes (Bird 1980; Bird and Taggart 1980). However, genomic regions rich in CpGs do exist. Such regions are called CpG islands (CGIs) and are found mostly within gene regulatory elements such as promoters.

DNA is methylated by a family of DNA methyltrasnferases (Dnmts). Among members of the Dnmt family, DNMT1 maintains the methylation status of DNA by methylating the newly synthesized strand after DNA replication. In contrast, DNMT3A and DNMT3B establish *de*

Genomic and Epigenomic Biomarkers of Toxicology and Disease: Clinical and Therapeutic Actions,
First Edition. Edited by Saura C. Sahu.

novo methylation patterns (Gujar et al. 2019). Another Dnmt family member, DNMT3L, does not possess enzymatic activity. Instead, cooperative roles of DNMT3L with DNMT3A and DNMT3B to establish methylation patterns have been suggested (Hata et al. 2002).

DNA can be demethylated by loss of maintenance by DNMT1 or by deamination of 5mC. In addition, enzymes of the Ten-eleven translocation (TET) family can demethylate 5mC through multiple reactions, without altering the DNA sequence (Messerschmidt et al. 2014).

DNA methylation is generally associated with gene silencing. As we shall see in what follows, the methylation status of CpGs within gene regulatory regions has a prominent impact on gene expression in non-mammalian vertebrates. Recent research has revealed that DNA methylation also regulates alternative splicing of target genes (Maor et al. 2015) to yield mRNA variants. In some cases, alternatively spliced mRNA variants code for non-functional proteins.

Histone Modifications

In the cellular nucleus, DNA is wrapped around histones. Histones are a group of positively charged basic proteins. Their positive charge enables them to interact with negatively charged DNA. Histones are also targeted by multiple modifications such as methylation, acetylation, phosphorylation, deamination, ADP ribosylation, ubiquitination, and sumoylation (Marczylo et al. 2016). These modifications modulate the interactions between histones and DNA, and thereby regulate the accessibility of transcription factors to gene regulatory regions within DNA. Among histone modifications, histone methylation and histone acetylation have been implicated in sex determination in non-mammalian vertebrates.

Histone is methylated by histone methyltransferases (HMTs) and demethylated by histone demethylases (HDMs) (Jambhekar et al. 2019). Polycomb repressive complex 2 (PRC2) is also involved in the establishment of histone methylation, especially at Lys 27 of histone 3 (H3K27) (Margueron and Reinberg 2011). The effect of histone methylation, unlike that of DNA methylation, is bipotential, depending on the methylation sites. For example, methylation at H3K4 (histone H3 lysine K4), H3K36, and H3K79 activates gene transcription, whereas methylation at H3K9 and H3K27 represses transcription (Jambhekar et al. 2019).

In contrast, histone acetylation is generally associated with transcription activation. Histone acetyltransferases (HATs) add acetyl groups to histones, whereas histone deacetylases (HDACs) remove this modification (Eberharter and Becker 2002).

Non-coding RNAs

Non-coding RNAs (ncRNAs) do not code for functional proteins. Instead they regulate gene expression through multiple mechanisms. ncRNAs are categorized into two types, depending on their length: those shorter than 200 nucleotides (nt) are called short non-coding RNAs (sncRNAs), whereas those longer than 200 nt are called long non-coding RNAs (lncRNAs) (Marczylo et al. 2016). sncRNAs, such as miRNAs and endo-siRNAs are involved in post-transcriptional regulation through degradation of mRNA (Hale et al. 2014). In contrast, lncRNAs are involved in both transcriptional and post-transcriptional regulations by interacting with both DNA and proteins (Marchese et al. 2017). Recent investigation in teleost fish have revealed the possible role of ncRNAs in sex determination.

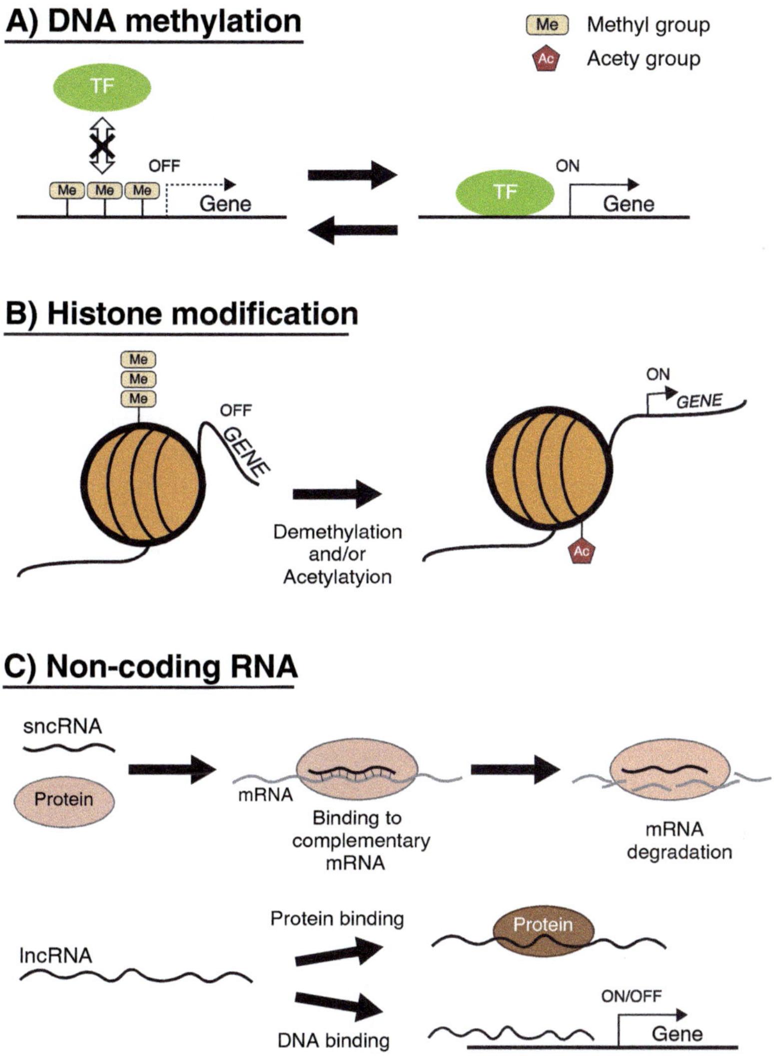

Figure 19.1 Major epigenetic machineries. (A) DNA methylation. Methylation of DNA is associated with gene silencing, because this modification represses the binding of transcription factors (TFs) to regulatory regions. (B) Histone modification. Modification of histone proteins alters DNA-histone interaction to regulate the accessibility of transcription factors to DNA. Histone methylation can either suppress or activate gene transcription depending on the site of modification. For example, H3K27me3 represses gene expression, whereas H3K4me3 activates transcription. In contrast, histone acetylation is generally associated with gene activation. (C) Non-coding RNAs. Non-coding RNAs do not code for proteins. Instead, they regulate expression of protein-coding genes through interaction with DNA and proteins. Short non-coding RNAs (sncRNAs) mainly regulate gene expression post-transcriptionally by targeting and degrading mRNAs. In contrast, long non-coding RNAs (lncRNAs) interact with DNA and proteins to regulate target genes at both transcriptional and post-transcriptional levels. Created by Genki Yamagishi.

Sex Determinig Cascades in Mammals and Non-mammals

Sex in eutherian mammals is primarily determined by the *Sry* (sex-determining region Y) gene (Koopman et al. 1991). This gene is present on the Y chromosome and activates *Sox9* (SRY-box transcription factor 9) transcription. SOX9 in turn activates downstream testicular differentiation genes such as fibroblast growth factor 9 (*Fgf9*), doublesex and mab-3-related transcription factor 1 (*Dmrt1*), and anti-Müllerian hormone (*Amh*). Among these, FGF9 maintains *Sox9* expression through a positive feedback loop between these two genes. At the same time, FGF9 represses the expression of genes involved in female pathways, such as *Wnt4* (wingless-type MMTV integration site family, member 4) and *Rspo1* (R-spondin 1). On the other hand, in female gonads, forkhead box L2 (*Foxl2*), together with *Rspo1/Wnt4* signals are activated. FOXL2 activates the cytochrome P450 family 19 subfamily A member 1 (*Cyp19a1*) gene, which encodes for aromatase; the result is estrogen synthesis, which drives ovarian differentiation (Capel 2017).

But such sex-determining (SD) cascades in mammals are not common among non-mammalian vertebrates. First of all, the *Sry* gene is absent in non-mammalian vertebrates. Instead, among teleost species, various genes are proposed as master SD genes. In some species whose sex is determined environmentally, there may not be true "master SD genes," because their role could be taken over by environmental factors (such as temperature).

Second, some components of SD cascades behave differently among vertebrate lineages. In mammals, *Dmrt1* expression is activated by the *Sry-Sox9* signaling cascade. Knockout of *Dmrt1* impairs testicular development but does not induce sex reversal in mice (Raymond et al. 2000). Thus *Dmrt1* is considered to work for the development and maintenance of the testes rather than for sex determination. However, this gene seems to play a more prominent role in sex determination in non-mammals. For example, Sun et al. (2017) investigated the function of *dmrt1* in a turtle species. According to the authors, *dmrt1* is "necessary and sufficient" to induce testicular development, which suggests a prominent role of this gene in sex determination. In birds, gene dosage (number of copies in an individual) of *dmrt1* is critical for gonadal sex differentiation (Ioannidis et al. 2021). Another example is *amh*. This gene is involved in testicular differentiation by promoting regression of the Müllerian duct (precursory structure of oviduct). Although this role is conserved among amniotes (mammals, birds, and reptiles), the regulation of this gene is somewhat different. In mammals, *Sox9* activates *Amh*. In birds, however, gonadal expression of *amh* precedes that of *sox9*, and this suggests that another factor is involved in *amh* activation. A possible role of *sf-1* (steroidogenic factor 1) has been reported instead (Takada et al. 2006). In addition, *amh* homolog is proposed as a master SD gene in a teleost fish (Hattori et al. 2012). Generalized SD pathways are presented in Figure 19.2. Although it is somewhat simplified (genes that do not appear in this chapter are not explained) and is not applicable to all vertebrate species, the authors believe that it does offer some reference for readers.

Epigenetic regulation affects gene expression at both transcriptional (through DNA methylation and histone modifications) and post-transcriptional levels (through non-coding RNAs), producing phenotypic variations. In theory, every single gene present in an individual genome can be regulated epigenetically. To date, a variety of genes in the sex determination and differentiation pathways have been identified or proposed as targets of epigenetic regulation. Among such candidate genes, two are prominent and prevailing among non-mammalian vertebrates: *cyp19a1* and *dmrt1*, whose role in non-mammalian vertebrates are explained here next.

Cyp19a1

Cyp19a1 encodes aromatase, an enzyme that irreversibly converts androgen into estrogen. Synthesis of estrogen in the gonad triggers multiple feminizing pathways, which promotes ovarian development. The expression of *cyp19a1* is activated by transcriptional factors such as *sf-1* and *foxl2*. *Cyp19a1* expression is also positively regulated by estrogen, forming

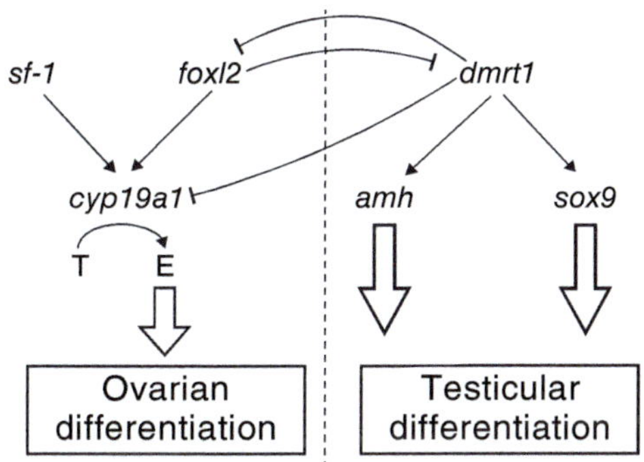

Figure 19.2 Sex determining cascades in non-mammalian vertebrates. In females, *sf-1* and *foxl2* activate the transcription of *cyp19a1*. *Cyp19a1* codes an enzyme aromatase, which converts testosterone (T) to estrogen (E). Production of estrogen is critically important for subsequent ovarian differentiation. In males, *dmrt1* expression is central to testicular differentiation, because this gene activates multiple testicular differentiation factors, such as *amh* and *sox9*. Ovarian and testicular differentiation cascades repress each other, as *foxl2* represses expression of *dmrt1*, whereas *dmrt1* represses *cyp19a1* expression directly or indirectly (through repression of *foxl2* expression). Note that the scheme represented in this figure is not applicable to all non-mammalian vertebrate species.
Created by Genki Yamagishi.

a positive feedback loop between gene expression and hormone production. This positive feedback is essential for ovarian differentiation and maintenance (Guiguen et al. 2010). Teleost fish have two isoforms of *cyp19a1*, namely *cyp19a1a* and *cyp19a1b*. The former is expressed in the gonad, whereas the latter is expressed in the brain (Ortega-Recalde et al. 2020). This chapter focuses on epigenetic regulation of *cyp19a1a*, because this isoform plays a major role in sex determination and reversal in teleost fish. Regulation of *cyp19a1b* is not discussed, because its role in sex determination and differentiation is limited.

Dmrt1

Dmrt1 drives testicular differentiation. This gene encodes a transcription factor that binds to target sequences through its zinc-finger DNA binding motif. *Dmrt1* activates and maintains testicular differentiation pathways by directly or indirectly upregulating other masculinization factors, for instance *amh*, *sox8*, and *sox9*. *Dmrt1* antagonizes feminizing pathways by downregulating *cyp19a1* expression directly (Wang et al. 2010) and indirectly, through repression of *foxl2* expression (Herpin and Schartl 2011).

As will be discussed later, gonadal expression of *cyp19a1* and *dmrt1* is up- or downregulated around the onset of gonadal differentiation. These genes also show dynamic expression changes concomitant with sex reversal. Research conducted in the past two decades has revealed that epigenetic regulation underlies the dynamism of *cyp19a1* and *dmrt1* expression during sex determination, differentiation, and reversal among non-mammals.

Sex Determination and Reversal in Fish

Sex Determination in Chondricthyan Fish

To our knowledge, the sex of chondricthyans is determined genetically, male heterogamy (XX-XY) being reported from several species. Uno et al. (2020) reported morphologically differentiated X and Y chromosomes from two shark species. However, they could not detect sex-specific chromosomal regions, such as repetitive sequences, in the Y chromosome. This suggests a transitional status for sex chromosome in the shark species. Further investigation is required for the molecular mechanisms that underlie chondrichthyan sex determination.

Diverse Sex Determination Mechanisms in Teleost Fish

Both genetic sex differentiation (GSD) and environmental sex determination (ESD) are reported from teleost fish, in contrast with chondricthyan fish. These systems are not mutually exclusive. Coexistence of both systems in single species is common, and environmental factors can override a genetic effect to produce a sex-reversed individual. In the following part, current knowledges about sex determination and sex reversal of teleosts is briefly summarized.

Modes of Sex Determination and Reversal among Teleost Species

Mammals generally reach sexual maturation as either male or female, according to chromosomal constitutions. Sex reversal does not take place throughout their lives under normal conditions. In contrast, among teleost fish sex reversal is prevalent. There are two modes of sex reversal. One occurs during developmental stages and results in a mismatch between genetic and phenotypic sex. For example, the exposure of larvae to higher temperatures induces masculinization in species such as the medaka (*Oryzias latipes*) and the Nile tilapia (*Oreochromis niloticus*). The other mode of sex reversal, called sequential hermaphroditism, takes place after sexual maturation. In this mode, individuals reach sexual maturation as either of the sexes, but reverse to the opposite sex at some point in life, and a transition of the gonad between ovary and testis takes place. The term "sequential" points to this transition. Species such as the barramundi (*Lates calcarifer*) and the ricefield eel (*Monopterus albus*) show hermaphroditism. The former reaches maturation as a male, then reverses to become a female. This direction of reversal is called protandrous. The ricefield eel shows the opposite direction of sex reversal (female-to-male), which is called protogynous. Most sequential hermaphrodites can reverse their sex only once; as a result, sex reversal is unidirectional. But in some species, for example the Okinawa rubble goby (*Trimma okinawae*), sex reversal is bidirectional (Sunobe and Nakazono 1993). This enables individuals to change sex multiple times during their lifetime (Figure 19.3).

Sex reversal in sequential hermaphrodites is induced by cues such as age and body size. In addition, population composition induces sex reversal in some coral reef fish. For example, the removal of the dominant female results in sex reversal of the largest male within the population in clown fish species (genus *Amphiprion*) (Fricke 1983; Moyer and Nakazono 1978). The opposite pattern, in which sex reversal of the largest female occurs through the removal of the dominant male, is reported in species such as the bluehead wrasse (*Thalassoma bifasciatum*) (Warner and Swearer 1991).

It should be noted that sequential hermaphrodites are distinguished from "simultaneous hermaphrodites" by the fact that, in the former, mature ovary and testis do not coexist within single individual. In contrast, simultaneous hermaphrodites such as the mangrove killifish (*Rivulus marmoratus*) possess both mature ovary and testis within a single individual and can self-fertilize to produce offspring (Harrington 1961).

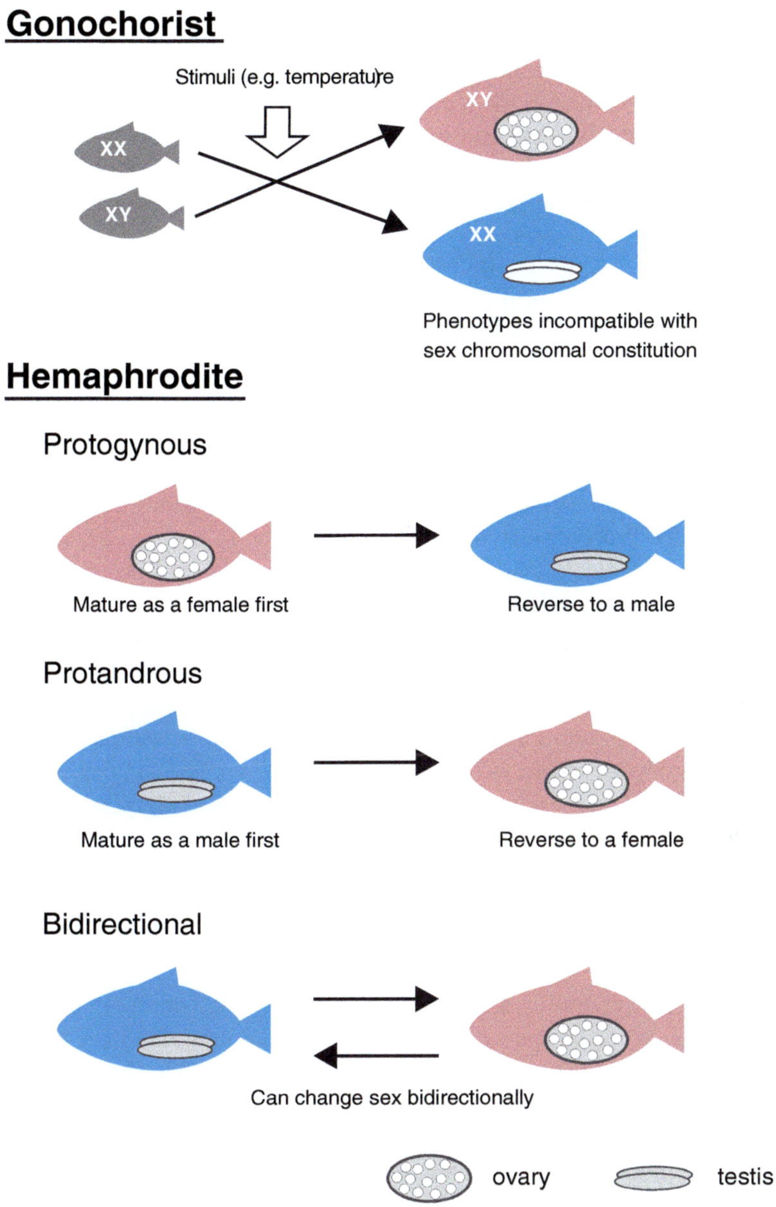

Figure 19.3 Modes of sex reversal among teleost species. Sex reversal of gonochorists takes place during early life, and thus before sexual maturation. Although the sex of gonochorists is primarily determined genetically, exposure to stimuli such as temperature during a sensitive period can induce sex reversal, resulting in discordance between genetic and phenotypic sex. Sex reversal in hermaphrodites takes place after sexual maturation. Protogynous species first mature as females, but subsequently reverse to being males. In contrast, protandrous species first mature as males, then reverse to being females. While these two modes can change sex only unidirectionally, some species, such as the Okinawa rubble goby, can change their sex bidirectionally.
Created by Genki Yamagishi.

Genetic Sex Determination

In teleost fish, GSD is dependent on sex chromosomal constitutions, just as in mammals and birds. But sex chromosomal constitutions are not limited to the classical systems of male heterogamy (XX-XY) and female heterogamy (ZZ-ZW). Systems such as XX-XO and ZZ-ZO, in which no sex chromosome is inherited from either parent, are also reported. In addition, in some species more than two sex chromosomes are involved (reviewed in Navara 2018).

Turnover from one system to another is also reported from multiple lineages; in the genus *Oryzias*, for instance, turnover from the XX-XY system to the ZZ-ZW system took place (Kikuchi and Hamaguchi 2013). This kind of turnover is made possible by processes such as translocation of SD genes, fusion of autosome and sex chromosome, and acquisition of novel SD genes (Sandra and Norma 2010).

In contrast to the morphologically distinct sex chromosomes of mammals, teleost sex chromosomes are not well differentiated. In fact only 6–11% of the species investigated possess differentiated sex chromosomes (Pandian 2012). In addition, various SD genes have been identified from teleost species. Probably the most famous teleost SD gene is *dmy* (the DM-domain gene on the Y chromosome), which was identified in medaka by Matsuda et al. (2002)—famous because it was the second SD gene to be found in vertebrates (*Sry* in mammals was the first). However, *dmy* was not the common SD gene among teleost species. In the subsequent two decades, several other SD genes have been reported in multiple teleost species, among them *amhy* (the Y chromosome-linked anti-Müllerian hormone) in Patagonian pejerrey (*Odontesthes hatcheri*) (Hattori et al. 2012), *gsdf* (gonadal soma-derived factor) in Luzon ricefish (*O. luzonensis*) (Myosho et al. 2012), *amhr2* (anti-Müllerian hormone receptor type II) in tiger puffer (*Takifugu rubripes*) (Kamiya et al. 2012), *sdy* (sexually dimorphic on the Y chromosome) in rainbow trout (*Oncorhynchus mykiss*) (Yano et al. 2012), and *sox3* (SRY-box transcription factor 3) in the Indian ricefish (*O. dancena*) (Takehana et al. 2014).

Environmental Sex Determination

Environmental factors such as temperature and pH are also involved in teleost sex determination (Sandra and Norma 2010). Temperature effects on sex determination have been well studied in a variety of fish species, including experimental models such as the medaka (Sato et al. 2005) and the zebrafish (*Danio rerio*) (Sfakianakis et al. 2012), and commercially important species like the tiger puffer (Zhou et al. 2019a).

The effect of temperature on sex determination has been reported in multiple vertebrate lineages, including teleosts, amphibians, and reptiles. In some reptile species such as the American alligator, sex is determined solely by temperature. Such a mode of sex determination is called temperature-dependent sex determination (TSD). In contrast, temperature seems not to be the sole effect that determines sex in most teleost species. Ospina-Álvarez and Piferrer (2008) revealed that "true" TSD fish, whose sex is determined solely by temperature, are limited to only a few lineages, for example *Apistogramma*. In other lineages sex is primarily determined by genetic background, but is overridden by temperature. Such species are designated as GSD + TE (temperature effect) species, as distinct from true TSD species. Most cases of sex reversal in gonochoristic teleosts fall under this designation.

To our knowledge, little is understood about molecular mechanisms underlying sex determination in true TSD teleost fish. In contrast, regulatory mechanisms underlying sex determination—or, more precisely, sex reversal—in GSD + TE species have been well studied, and prominent functions of epigenetic regulation have been reported.

Epigenetic Regulation of Sex Determination and Reversal in Gonochorists

Temperature can overwrite genotypic sex during early life in many teleost species. Epigenetic regulation plays a prominent role in sex reversal in GSD + TE species by transducing temperature effects on the expression of sex determination and differentiation genes.

Genotypic females of European seabass (*Dicentrarchus labrax*) can reverse sex when exposed to temperatures higher than 17 °C during thermally sensitive periods, which corresponds to the first sixty days after fertilization (Navarro-Martín et al. 2009, 2011). Exposure to high temperature has an effect on the genes involved in sex determination and differentiation, for instance *cyp19a1a*, which is downregulated. High temperature also promotes DNA methylation in the promoter region of *cyp19a1a* in the gonads, but not in the brain. In vitro assay using HEK293T cells revealed that the methylation of *cyp19a1a* promoter represses transcriptional activation of this gene by *foxl2* and *sf-1*, transcription factors that belong respectively to forkhead box (fox) and SRY-box (sox) families. Interestingly, two CpG sites (−431 and −13 bp upstream of the transcription start site) that showed distinct methylation patterns between genotypic females and sex-reversed females were located adjacently to putative *fox* family and *sox* family binding sites (Navarro-Martín et al. 2011). These observations suggest that high temperatures induce sex reversal by downregulating the expression of a critical feminizing gene, *cyp19a1a*, through DNA methylation. Interestingly, high temperatures can upregulate the expression of histone modification enzymes such as euchromatic *ehmt2* (histone-lysine N-methyltransferase 2), *hdac11* (histone deacetylase 11), *jarid2a* (jumonji, AT rich interactive domain 2a) and *pcgf2* (polycomb group ring finger 2) (Díaz and Piferrer 2015). Concomitant up- or downregulation of testicular (e.g. *dmrt1*) and ovarian differentiation genes (e.g. *cyp19a1a*) suggests that epigenetic changes are induced in SD and sex-differentiating genes by exposure to high temperatures (Díaz and Piferrer 2015).

Temperature-induced DNA methylation and its association with the expression of SD and sex-differentiating genes is also reported in other teleost species, such as the half-smooth tongue sole (*Cynoglossus semilaevis*), the Nile tilapia, and the tiger puffer (Shao et al. 2014; Wang et al. 2017; Zhou et al. 2019b). Among these species, the half-smooth tongue sole is especially interesting because its methylation patterns can be inherited.

Sex in the half-smooth tongue sole is primarily determined by constitutions of ZZ-ZW sex chromosomes. However, about 14% of genotypic females (ZW females) differentiate into phenotypic males, called pseudomales, even under normal conditions. Exposure of larvae to high temperatures (> 28 °C) increases this reversal rate, bringing it to around 70%.

Comparison of whole-genome methylation status between ovaries and testes from genotypic females and males revealed that regions corresponding to 4% of the whole genome were differentially methylated. In contrast, only 0.056% of the whole genome was differentially methylated between testes of males (ZZ) and pseudomales (ZW). This observation suggests the establishment of a male-like DNA methylation pattern during sex reversal in the gonads of ZW pseudomales. Among the genes associated with the differentially methylated regions between ovaries and testes (both from males and pseudomales) are sex determination and sex differentiation genes such as *amh*, *amhr2*, *dmrt1*, *gsdf* (testicular differentiation), and *cyp19a1a* (ovarian differentiation). Methylation occurs in promoter regions of these genes; methylation levels of the first four genes are lower in males and pseudomales, whereas those of the last one are lower in females. These observations suggest that the gonadal expression of SD and sex-differentiating genes is regulated through DNA methylation and that DNA methylation patterns change from female-type to male-type during sex reversal.

In addition to directly regulating gene expression, DNA methylation also plays a role in dosage compensation in pseudomales. Genotypic males possess two Z chromosomes, whereas pseudomales have only one. Such difference in sex chromosomal constitution can result in lower expression levels of Z-linked genes in pseudomales. This is problematic for species whose sex is determined by gene dosage. In the half-smooth tongue sole, genes located on a certain part of the Z chromosome are upregulated in pseudomales, and this region is differentially methylated between males and pseudomales, which suggests dosage compensation by hypomethylation in pseudomales.

It is probable that the sex-biased DNA methylation pattern just described is critical for phenotypic male differentiation. Interestingly, the evidence suggests that the DNA methylation pattern is inherited by offspring. Pseudomales are fertile and can be crossed with ZW females to produce an F1 generation. Of note, the sex reversal rate of such an F1 generation is far higher than that of the parental generation. Even under normal conditions, 94% of ZW individuals differentiate into pseudomales. Comparison of genomic methylation between testis from pseudomales and their F1 offspring revealed that 0.040% of the genome was differentially methylated. This value is close to the difference between ZZ males and ZW pseudomales (0.056% of the genome), but far smaller than the difference between ovaries and testes (4%). Such a difference suggests that the methylation pattern is either transmitted from parents to offspring or erased at some point but re-established during sex reversal. A high reversal rate among offspring of pseudomales, even without a temperature effect, indicates the former to be the case. Possibly the methylation pattern of the Z chromosome is prominent in sex reversal. Since sperms with the W chromosome are not viable, pseudomales produce sperms only with Z chromosomes. Hence their offspring with a ZW sex chromosomal constitution inherit their Z chromosome from their fathers (ZW pseudomales), but not from mothers (ZW females) (Chen et al. 2014). The result is a situation where all ZW offspring possess Z chromosomes with a pseudomale-type methylation pattern, and this could be behind the high sex reversal rate (Figure 19.4).

Epigenetic Regulation of Sex Reversal in Sequential Hermaphrodites

DNA methylation also plays a role during sex reversal in sequential hermaphrodites. This is reported in relation to the barramundi, ricefield eel, and bluehead wrasse.

The barramundi, a protandrous species, reaches sexual maturity as a male at the age of three or four, but changes its sex to female around the age of four to eight. Comparison of DNA methylation status between ovary and testis revealed significant differences in the promoter regions and first exons of sex-differentiating factors such as *cyp19a1* and *dmrt1*. However, the expression of *cyp19a1a* did not correlate with methylation levels, whereas that of *dmrt1* showed negative correlation. Interestingly, only a truncated, non-functional form of *cyp19a1a* was expressed in the testis. Similarly, *dmrt1* mRNAs present in the ovary lacked a DM-domain, which is important for DNA binding and transcriptional regulation. These observations suggest that DNA methylation plays a role in sex reversal through the regulation of alternative splicing rather than through a direct inhibition of transcription in SD genes (Domingos et al. 2018).

In contrast to the barramundi, the ricefield eel reverses from female to male. *Cyp19a1a* transcription is activated by gonadotropin through the cAMP signaling pathway in females. In summary, gonadotropin upregulates the production of cyclic AMP through the activation of adenylyl cyclase. Increased cAMP production results in turn in the phosphorylation of CRE-binding protein (CREB), which subsequently binds to the promoter region of

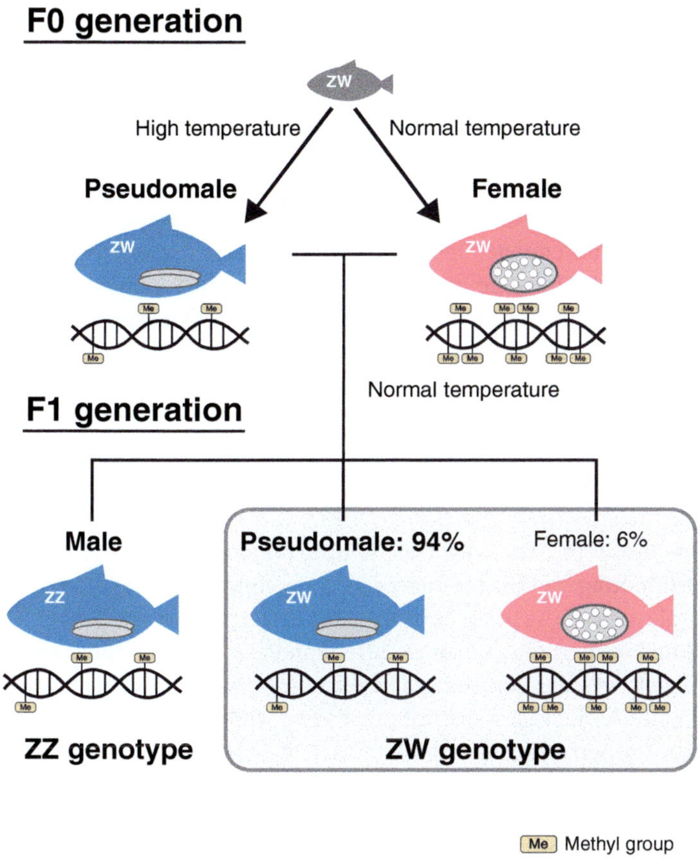

Figure 19.4 Sex reversal and inheritance of DNA methylation in the half-smooth tongue sole. In the half-smooth tongue sole, sex is determined by sex chromosomal constitutions (ZZ-ZW system). However, rearing juveniles at higher temperatures (around 28 °C) induces masculinization of ZW individuals and yields phenotypic males, called pseudomales. During this sex reversal, a male-like DNA methylation pattern is established in the pseudomales. Pseudomales are fertile and can be crossed to ZW females to produce an F1 generation. Since sperms with the W chromosome are not viable, only ZZ and ZW genotypes are present among the F1 generation. Even under normal temperatures, most of the F1 offspring with ZW genotypes gets differentiated into pseudomales. Such a high rate of sex reversal is explained through the inheritance of a male-like DNA methylation pattern, which passes from F0 pseudomales to F1 offspring. Created by Genki Yamagishi.

cyp19a1a and upregulates transcription of this gene. Two DNA regions enriched with CpG sites, called region I and II, are located upstream of ricefield eel *cyp19a1a* and are hypermethylated in testes and transitional ovotestes by comparison with ovaries. Regions I and II contain SF-1 binding site and cAMP response element (CRE) respectively. *In vitro* experiments revealed that the methylation of these regions diminishes cAMP-mediated *cyp19a1a* promoter activity. Females administered the DNA methylation inhibitor failed to develop into males, which points to a prominent role for DNA methylation in sex reversal (Zhang et al. 2013).

In addition, histone modifications also play a part in the ricefield eel. Histone methylation at the *cyp19a1* promoter region shows sex-biased patterns. In the testis, H3K9 in regions I and II are methylated, whereas in the ovary the same sites are acetylated. These modifications are hypothesized to regulate the recruitment of RNA polymerase II, thus contributing to sex-specific transcriptional regulation, namely repression in the testis and activation in the ovary (Zhang et al. 2013).

There is evidence that socially controlled sex reversal is also regulated epigenetically in the bluehead wrasse. When the dominant male is removed, the largest female in the population transitions into a male. During sex reversal, a genome-wide reprograming of methylation takes place in the gonads, producing methylation patterns that are distinct from those of mature females and males. As is the case with female to male sex reversal in other teleost species, DNA methylation is upregulated in the *cyp19a1a* promoter, whereas in the *dmrt1* promoter it is downregulated. The gene expression of DNA methylation enzymes such as *dnmt1*, *dnmt3*, and the *tet* family of genes was also investigated. The expression of *dnmt1* was downregulated during sex reversal, whereas that of *dnmt3* paralogs was upregulated. The *tet* family of enzymes showed highest expression levels during later phases of reversal. Genes coding for histone modification enzymes, such as HATs and HDACs, also showed expression change during sex reversal. These expression changes suggest dynamic reprogramming of DNA methylation and histone modification patterns during sex reversal (Todd et al. 2019).

Among sequential hermaphrodites, sex reversal in the orange-spotted grouper (*Epinephelus coioides*) is worth noting, in that the role of a non-coding RNA (ncRNA) has been elucidated. A possible role of ncRNAs in sex reversal has been reported for GSD + TE species, the Nile tilapia, and tiger puffer, too (Tao et al. 2016; Yan et al. 2021). In these species, sex-biased expression of ncRNAs is observed in the developing gonads. Possible target genes for these sex-biased ncRNAs included critical sex-differentiation factors such as *cyp19a1a* and *dmrt1*. Thus the authors suggested the importance of ncRNAs in sex determination (or reversal). However, their role is yet to be confirmed experimentally.

In contrast, the role of a miRNA, called miR-26a-5p, in sex reversal has been confirmed in the grouper. In female groupers, the expression of miR-26a-5p is repressed by estrogen. However, its expression is upregulated during sex reversal from female to male. The transcribed miRNA product induces degradation of *cyp19a1a* mRNA by interacting with its 3′-UTR sequence, which in turn results in lower estrogen levels. Thus a positive feedback loop is formed between the expression of miR-26a-5p and *cyp19a1a* to reduce estrogen production, and therefore to drive sex reversal (Yu et al. 2020).

Primordial Germ Cell Maintenance and Its Effect on Sex Determination

Besides the mechanisms just described, which are mostly dependent on the transcriptional regulation of SD genes, epigenetic regulation may also affect sex determination through the maintenance of primordial germ cells (PGCs).

The effect of PGCs on sex determination has been investigated in the medaka. In brief, the maintenance of PGCs within individual larvae drives ovarian development, whereas a reduced number of PGCs results in testicular development (Kurokawa et al. 2007; Morinaga et al. 2007).

Recent investigation has uncovered the involvement of DNA methylation in PGC maintenance. Wang et al. (2021) examined the reprogramming of DNA methylation patterns during sex determination of larvae and during sex reversal in adult zebrafish. In the larvae, the levels of DNA methylation in PGCs decreased from thirty-six hours post-fertilization to nine days post-fertilization. During this period, expression of a DNA demethylase gene, *tet3*, was upregulated. Both knockdown and knockout of *tet3* resulted in a decreased number of PGCs, which suggests that DNA demethylation plays an important role in PGC maintenance.

In the same study, the authors revealed that larvae treated with DNMT inhibitor (5-Aza-2′-deoxycytidine) preferentially developed into females. It is possible that female differentiation was driven by the hypomethylation of genomic regions important for PGC maintenance. Further investigation may reveal the relationship between DNA methylation, PGC maintenance, and phenotypic outcome.

Epigenetic Regulation of Sex Determination in Amphibians and Reptiles

Sex Determination and Reversal in Amphibians

Amphibian sex is generally determined by sex chromosomal constitutions. But sex chromosomal constitutions can be more complicated in amphibians than in other vertebrates. For example, both male heterogamy and female heterogamy are present in one species, *Glandirana rugosa* (Nishioka et al. 1993). In addition, three types of sex chromosomes (X, Y, and Z) are present in the western clawed frog (*Silurana tropicalis*), YZ, YW, and ZZ chromosomal constitutions yielding males, whereas ZW and WW yield females (Schartl 2015).

Determination of sex by chromosomal constitutions suggests the existence of SD genes in amphibian species. However, only one master SD gene, the doublesex and mab-3-related gene W (*dmw*), has been identified among amphibians. *dmw*, which was first identified in the African clawed frog (*Xenopus laevis*), is present on the W chromosome and antagonizes *dmrt1* paralogs to suppress male differentiation (Yoshimoto et al. 2010; Yoshimoto and Ito 2011). According to Mawaribuchi et al. (2017), *dmw* arose only after the divergence of the *Xenopus* lineage from its ancestral lineage, which suggests that no *dmw* is present among other amphibians.

As in the case of teleost species, here too temperature effects can override genetic sex to produce sex-reversed individuals. According to Ruiz-García et al. (2021), a higher temperature induces masculinization in most species, but it also induces feminization in a few. The developmental stages during which sex reversal can be induced are limited, and this window is known as a temperature-sensitive (or thermosensitive) period (TSP).

Currently, little information is available regarding the epigenetic regulation of amphibian sex (Ruiz-García et al. 2021). But temperature-dependent sex reversal suggests that epigenetic mechanisms similar to those observed in reptiles are present in amphibians, too. Considering the widespread application of amphibians as a biomarker for environmental pollution, together with their importance in food chains, the gap in our knowledge of the epigenetic regulation of amphibian sex is one that needs to be fulfilled.

Sex Determination in Reptiles

Two modes of sex determination exist among reptiles: GSD and temperature-dependent sex determination (TSD). In species with GSD, sex is determined by sex chromosomal constitution, while in species with TSD it is determined by the temperature at which eggs are incubated. Among reptiles, all snakes have GSD. In contrast, all extant crocodilian species and tuatara have TSD. Species with both GSD and TSD are present among lizards and turtles (Figure 19.5). As in teleost fish, in some lizard species, too, the effect of sex chromosomes is overridden by the incubation temperature. Overall, sex determination in reptiles is similar to sex determination in teleost fish.

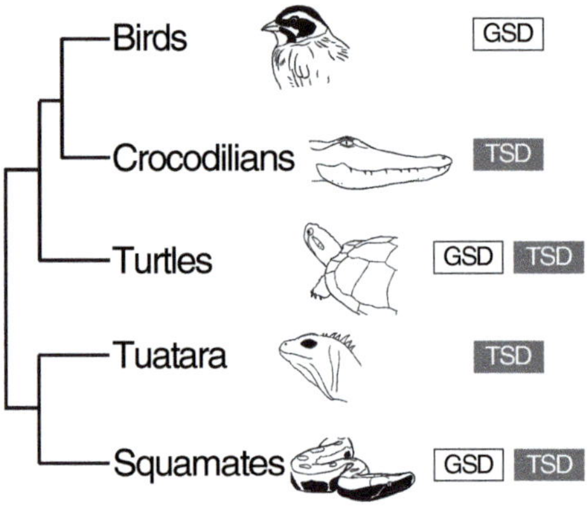

Figure 19.5 Phylogenetic relationship between reptiles and sex-determining mechanisms. Among the reptilian taxon, all crocodilians are TSD species, whereas their sister group, birds, is comprised exclusively of GSD species. In contrast, both GSD and TSD species are reported in turtles. Tuatara, the only extant species of Rhynchocephalia, is a TSD species. In contrast, both GSD and TSD species are reported in its sister taxon, squamates (lizards and snakes). Phylogenetic relationships of reptilian groups and distribution of sex-determining mechanisms suggest that shifts from TSD to GSD and vice-versa have taken place multiple times. Created by Genki Yamagishi.

Both male heterogamy (XX-XY) and female heterogamy (ZZ-ZW) are present among turtles and lizards with GSD. The phylogenetic relationship between XX-XY and ZZ-ZW species suggests that, in both turtles and lizards, turnover between the two systems took place several times (Bista and Valenzuela 2020; Gamble 2010). Snakes were long considered to be ZZ-ZW species. However, Gamble et al. (2017) has demonstrated that the XX-XY system is present in the boa (*Boa imperator*) and in the python (*Python bivittatus*).

Reptiles with TSD are classified into three categories: type Ia, type Ib, and type II. Type Ia is characterized by a higher rate of feminization at higher temperatures. Type Ib shows the opposite pattern—in other words, lower temperatures produce more females. In contrast to these patterns, in type II both higher and lower temperatures produce females, while intermediate temperatures produce males (Valenzuela and Lance 2004). Type Ia is is found in turtles, whereas type Ib is reported in tuatara. Type II is represented in crocodilian species and in a species of lizard, leopard gecko (*Eublepharis macularius*). The phylogenetic relationship between species with GSD and species with TSD suggests that the latter have evolved many times from the former, and vice-versa (Bista and Valenzuela 2020; Gamble 2010). Turnover between TSD and GSD is observed even within a genus. For example, the Japanese gecko (*Gekko japonicus*) is a species with TSD (Tokunaga 1985), whereas its close relative, the hokou gecko (*G. hokouensis*), is a ZZ-ZW species with GSD (Kawai et al. 2009).

In species with TSD, sex is determined by the temperatures to which eggs (or, more accurately, embryos) are exposed during the critical TSP. The TSP corresponds to developmental stages 15–24 in the American alligator, stages 15–20 in the red-eared slider turtle (Shoemaker and Crews 2009), and stages 32–37 in the leopard gecko (Bull 1987). In alligators and turtles, TSPs overlap with the onset of morphological differentiation of gonads. In addition, leopard gecko embryos at stage 32 are morphologically equivalent to alligator embryos at stages 13–14, as well as to embryos of the snapping turtle at stage 15, which corresponds to the onset of TSP in this species (Yntema 1968). Such overlap of TSPs among species with TSD raises the possibility that molecular mechanisms that underlie thermosensitive sex determination are conserved among reptiles.

In some lizard species, genotypic sex can be overwritten by incubation temperature. The three-lined skink (*Bassiana duperreyi*) is a GSD species with male-heterogamy (XX-XY system). However, the incubation of XX embryos at lower temperatures produced phenotypic males (Shine et al. 2002).

The Australian central bearded dragon (*Pogona vitticeps*) is another species with temperature-induced sex reversal. Sex is primarily determined by the ZZ-ZW system in this species. However, sex-reversed ZZ females were obtained experimentally by incubating eggs at temperatures higher than 32 °C (Quinn et al. 2007). Investigations into field populations revealed that sex reversal does indeed occur under natural conditions (Holleley et al. 2015). Sex-reversed ZZ female dragons are fertile, and no difference in hatching success rate was observed between ZW and ZZ females. Interestingly, ZZ females produced more eggs than ZW females. The background to this phenomenon remains to be investigated. The offspring of ZZ females were all males when the eggs were incubated at lower (28°C) temperatures, whereas higher incubation temperatures (> 32°C) produced phenotypic females (i.e. ZZ females). Overall, the relationship between incubation temperature and the proportion of female offspring followed the type Ia pattern. Such observations suggest that the bearded dragon has a transitional status between GSD and TSD.

Epigenetic Regulation of Sex Determination in Reptiles

Molecular mechanisms underlying sex determination in reptiles with TSD have long remained a mystery. However, recent advances in this field are uncovering the role of epigenetic regulation. In the following part of this section, currently available knowledges about TSD systems in reptiles are summarized.

Differential Methylation of Genomes by Incubation Temperatures

As already explained, the methylation status of DNA is critically important for sex determination and reversal in teleost fish. Among reptiles, sex-specific (or, more precisely, temperature-specific) methylation patterns have been reported in turtle and crocodilian species.

When the embryo is incubated at female-producing temperatures (FPTs), gonadal expression of *cyp19a1* in the red-eared slider turtle (*Trachemys scripta*) is rapidly upregulated after stage 16, which corresponds to the onset of TSP (Ramsey et al. 2007). This observation suggests that *cyp19a1* is involved in the feminization of FPT-incubated gonads. Matsumoto et al. (2013) reported that the promoter region of *cyp19a1* is differentially methylated between gonads incubated at FPT and gonads incubated at male-producing temperatures (MPT) in the slider turtle. Differentially methylated CpG sites were located adjacently to TATA box, which is required for RNA polymerase II recruitment, and to binding sites for FOX and SF-1. These sites were more hypomethylated in FPT-incubated gonads than in MPT-incubated ones, which suggests that female-specific upregulation is controlled epigenetically.

A significant difference in methylation status between FPT-incubated and MPT-incubated gonads was detected in stage 19 and stage 21 (each corresponding to a middle and a late TSP), but not at stage16 (early TSP). This observation suggests that a temperature-specific methylation pattern is formed after the onset of the TSP. Interestingly, epigenetic regulators differentially expressed in the gonad between FPTs and MPTs are more abundant at stage 15 than at stage 12 in the western painted turtle (*Chrysemys picta*) (Radhakrishnan et al. 2018), suggesting that a temperature-specific epigenome is actively formed around the onset of the TSP.

These sex-biased methylation patterns in the *cyp19a1* promoter can be reversed by shifting incubation temperatures between the female-producing and the male-producing ones. A shift from a MPT to a FPT during the TSP resulted in the hypomethylation of CpG sites within the *cyp19a1* promoter, whereas a shift from a FPT to a MPT resulted in hypermethylation (Matsumoto et al. 2016).

In the American alligator (*Alligator mississippiensis*), sex-specific methylation was present in the promoter regions of *cyp19a1* and *sox9* (Parrott et al. 2014), when embryos incubated at FPT (30°C) and at MPT (33.5°C) during the TSP were compared. The same study also reported an inverse relationship between methylation and gene expression. Although this study was limited in scope, in that the epigenomic status was investigated only after the TSP (stage 27), it did show the presence of temperature-dependent methylation patterns and their relationship with gene expression in the alligator, too. In addition, three *dnmt* family DNA methyltransferase genes, *dnmt1*, *dnmt3a* and *dnmt3b*, were expressed in the developing gonads, although no temperature-biased expression pattern was formed. It is possible that these genes are regulated post-transcriptionally to form sex-biased DNA methylation patterns.

Comparison between the alligator and the slider turtle suggests that methylation patterns correlate not with the absolute values of incubation temperatures, but with the sexual fate that embryos follow. This suggests plasticity of epigenetic regulation among reptiles with TSD, which allows turnover between type Ia and type II TSD patterns. Although it is unclear whether the type Ia pattern in turtles reflects the ancestral status of reptiles, such plasticity in theory allows alligators to acquire type II TSD after divergence from turtles.

It should be noted that sex-biased methylation takes place not only in DNA but also in histone proteins. In the slider turtle, trimethylation of histone 3 at Lys 4 (H3K4me3) was enriched in the *cyp19a1* promoter region when embryos were incubated at FPT but not at MPT. A shift from MPT to FPT enriched H3K4me3, whereas a shift from FPT to MPT diminished this modification. Since H3K4me3 is associated with transcriptional activation, this histone modification is probably involved in the temperature-specific regulation of *cyp19a1* transcription (Matsumoto et al. 2016).

Although the reports presented here focus on the regulation of *cyp19a1* and *dmrt1*, other sex determination and sex differentiation regulators also show sex-biased methylation status in the slider turtle (Radhakrishnan et al. 2017). This suggests that epigenetic regulation is involved in TSD at multiple points along the sex determination and differentiation pathways.

Epigenetic Regulation of Sex Determination in Alligators and Turtles

Recent research in the American alligator and red-eared slider turtle, both species with TSD, has revealed the involvement of epigenetic machinery in TSD. In the last decade, two important discoveries were made in these species: (1) identification of thermosensory molecules; and (2) identification of thermosensitive regulators of SD genes.

Considering the nature of TSD, the existence of a molecular sensor that transduces a temperature stimulus into SD pathways is probable. Yatsu et al. (2015) identified transient receptor potential vanilloid 4 (TRPV4) as just such a sensory molecule in the American alligator. TRPV4 is a member of the transient receptor potential (TRP) cation channel superfamily. The *trpv4* gene expression in the developing alligator gonads revealed clear sexual dimorphism during TSPs. Inhibition of TRPV4 activation through antagonist treatment resulted in an impaired expression of male-differentiation genes such as *amh* and *sox9* whereas the expression of *cyp19a1*, a prominent feminizing gene, was upregulated at MPTs. On the contrary, treatment with TRPV4 agonist failed to induce an aberrant expression of those genes in embryos incubated at FPT. These results suggest that TRPV4 is a sensor that primarily drives masculinization. Other than *trpv4*, four *trp* family genes are expressed in the alligator gonad at stage 21 (during the TSP). It is possible that these *trp* channels also play a thermosensory role.

Research in alligators also sowed the seed for a second breakthrough. In the American alligator, TSP corresponds to developmental stages 15–24. Among these stages, sex-specific gonadal development and gene expression occur during stages 21–24. Yatsu et al. (2016) investigated gonadal gene expression profiles during this TSP. Comparison between gonads incubated at MPT (33.5°C) and gonads incubated at FPT (30°C) revealed forty-one differentially expressed genes. Among these genes was a histone demethylase gene called *kdm6b* (lysine demethylase 6B). Ge et al. (2018) revealed the involvement of *kdm6b* in sex determination in the slider turtle. As in the alligator, gonadal expression of *kdm6b* in the turtle was dependent on incubation temperatures, showing upregulation at MPT (26°C). Knockdown of *kdm6b* through RNA interference induced female-like gonadal morphology in embryos incubated at MPT. The expression of *dmrt1*, a prominent testis-differentiating factor, was also downregulated by *kdm6b* knockdown. Interestingly, an aberrant phenotype induced by *kdm6b* knockdown was reversed through overexpression of *dmrt1*, which suggests that *kdm6b* is involved in the activation of *dmrt1* in MPT-incubated embryos. Further investigations involving immunofluorescence, chromatin immunoprecipitation, and quantitative PCR revealed that *kdm6b* removes a repressive histone modification, H3K27 trimethylation (H3K27me3), from the promoter of *dmrt1*. The investigations above revealed that testis differentiation at MPT is induced by the expression of *kdm6b*, which subsequently demethylates histone proteins to activate transcription of *dmrt1*. *Kdm6b* is also involved in ovarian differentiation at FPT. Weber et al. (2020) revealed that the signal transducer and activator of transcription 3 (STAT3) regulates gonadal *kdm6b* expression. Gene expression and protein levels of STAT3 are similar at MPT and FPT. However, STAT3 protein is phosphorylated at FPT. This phosphorylated protein binds to *kdm6b* locus and represses transcription of this gene. This transcription regulation of *kdm6b* results in silencing of *dmrt1*, because the repressive H3K27me3 modification at the *dmrt1* locus cannot be removed. As a result, testis differentiation pathways are not activated at FPT (Figure 19.6).

The series of reports presented here have uncovered molecular mechanisms that underlie TSD in reptiles. However, *kdm6b* seems not to be involved in the regulation of sex-differentiating factors other than *dmrt1*—factors such as *amh* and *cyp19a1* (Ge et al. 2018). Since *cyp19a1* was one of the differentially methylated genes between FPT and MPT in the American alligator and slider turtle (Matsumoto et al. 2013; Parrott et al. 2014), this gene is possibly regulated by another epigenetic regulator. In addition, the "link" that mediates between thermosensors (i.e. TRP channels) and epigenetic machinery (i.e. *kdm6b*) is yet to be discovered. Since TRPV4 activation increases the intracellular concentration of calcium ion, Ca^{2+}-mediated intracellular signaling is a possible mediator. It is also possible that other epigenetic regulators, such as histone modification enzymes and non-coding RNAs, are involved through the regulation of sex-determination and sex-differentiation genes such as *cyp19a1*.

Comparison between GSD and TSD Species

It is probable that gene expression of TSD regulators is up- or downregulated by incubation temperatures. Therefore temperature-responsive SD genes are likely to be present among differentially expressed genes (DEGs) between FPT-incubated and MPT-incubated gonads of species with TSD. The problem is that regulators purely related to temperature change and not involved in sex determination also respond to incubation temperatures. This can be the case with stress responders such as heat shock chaperones. Such problems can be overcome by using species with GSD as a reference. Because sex determination regulators are unlikely to respond to incubation temperatures in such species, temperature-responsive sex determination factors can be identified by "subtracting" the DEGs of these species from the DEGs of species with TSD.

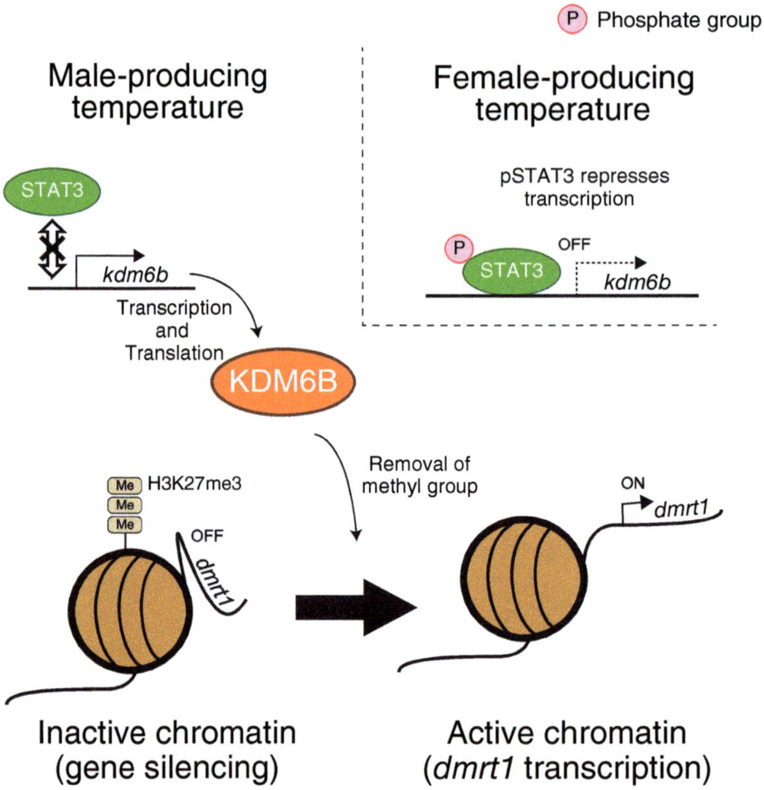

Figure 19.6 Temperature-dependent regulation of *dmrt1* in the slider turtle. Under male-producing temperatures (MPT), STAT3 is not phosphorylated and thus cannot bind to the *kdm6b* regulatory element. Under such conditions, *kdm6b* transcription is activated and the protein product removes the methyl group (H3K27me3) from histone proteins. Removal of H3K27me3 results in conformational change from inactive chromatin to active chromatin, producing the activation of *dmrt1* transcription. In contrast, STAT3 protein is phosphorylated under the female-producing temperature (FPT) conditions and binds to the regulatory element to suppress *kdm6b* transcription. Created by Genki Yamagishi.

Such an approach is reported by Radhakrishnan et al. (2018). The authors investigated gonadal expression of epigenetic regulators in a species with TSD, the western painted turtle, and in a species with GSD, the spiny softshell turtle (*Apalone spinifera*), during TSD and corresponding stages. The eggs of these turtles were incubated at MPT (26°C) and FPT (31°C), and transcripts were characterized, quantified, and compared between these two incubation schemes. Although the authors did not go so far as to reach a clear-cut conclusion about which gene is responsible for sex determination and which gene is not, they did offer some interesting insights into TSD and GSD systems.

In the western painted turtle, differential expression of multiple genes, including 3 methyltransferases (*nsd1*, *dnmt3b*, *setd1a*), 2 acetyltransferases (*ncoa6*, *ep300*), and 1 deacetylase (*hdac2*), was observed at stage 9, before the onset of the TSP. In contrast, none of these genes showed a temperature-biased pattern in the softshell turtle at the corresponding stage. Such a difference suggests that the TSD-specific epigenome is already formed before the onset of the TSP. Although this hypothesis contradicts the reports discussed earlier, which suggest that the formation of sex-biased methylation patterns occurs after the onset of TSP, it is nevertheless still possible that a basal epigenomic status unique to species with

TSD is present before the onset of the TSP—namely a status from which a male- or female-specific epigenome is formed.

GSD-specific regulation may also exist. *ncoa1* and *ncoa6* showed a temperature-biased expression in the softshell turtle, but not in the western painted turtle. Since these genes regulate or interact with nuclear receptors such as estrogen receptor α (Oñate et al. 1995; Tong et al. 2019), GSD-specific regulation is possibly present in hormonal pathways.

Epigenetic Machinery and Sex Reversal in an Australian Lizard

The Australian central bearded dragon is unique among reptiles. Although sex in this lizard is primarily determined by the constitution of sex chromosomes (ZZ yields males, whereas ZW females), egg incubation at temperatures higher than 32°C yields ZZ females. Since sex is determined by type Ia TSD in the offspring of ZZ females, this species may represent a transitional status between GSD and TSD. Deveson et al. (2017) revealed that two epigenetic regulators, *jmjd3* (alias of *kdm6b*) and *jarid2*, underwent alternative splicing when embryos were incubated at high temperatures.

As already explained, *kdm6b* is a histone demethylase. *Jarid* 2 is a member of the *jumonji* family, members of which are characterized by possession of the jumonji C domain. The *Jumonji* family is involved in histone demethylation. However, *jarid2* is thought to have lost histone demethylase activity because of mutations in its catalytic site (Klose et al. 2006); *in vitro* studies have revealed its involvement in histone methylation (Shirato et al. 2009). *Jarid2* also is a component of the polycomb repressive complex 2 (PRC2), which is involved in the establishment of H3K27me3 modification (Sanulli et al. 2015)

In the gonads of ZZ males and ZW females, both *kdm6b* and *jarid2* are transcribed normally. However, the retention of an intron (intron 19 of *kdm6b* and intron 11of *jarid2*) occurs in sex-reversed ZZ females, producing abnormal mRNA (Deveson et al. 2017). According to Deveson and colleagues, such abnormal intron retention leads either to the prevention of translation or to the production of non-functional proteins. Under such conditions, epigenetic regulation, normally under *kdm6b* and *jarid2*, is likely to be impaired, and the effect would be a blockade of default sex differentiation pathways and sex reversal (Figure 19.7).

Unfortunately the detailed mechanisms by which sex is reversed in the bearded dragon are yet to come out, although it is probable that such mechanisms are present downstream of *kdm6b* and *jarid2*. Interestingly, abnormal intron retention of *kdm6b* and *jarid2* was also observed in the gonads of TSD reptiles, the American alligator (FPT), and slider turtle (mostly MPT). Since the bearded dragon possibly represents a transitional status between GSD and TSD, abnormal intron retention of *kdm6b* and *jarid2* might be a prerequisite for turnover from GSD to TSD.

Epigenetic Regulation of Sex Determination in Birds

Sex Determination and Gonadal Differentiation in Birds

In birds sex is determined genetically, as all species reported possess the ZZ-ZW system. In most species, Z and W chromosomes are morphologically distinguishable. However, sex chromosomes in flightless ratites (such as ostriches) are not morphologically distinct (Takagi and Sasaki 1974).

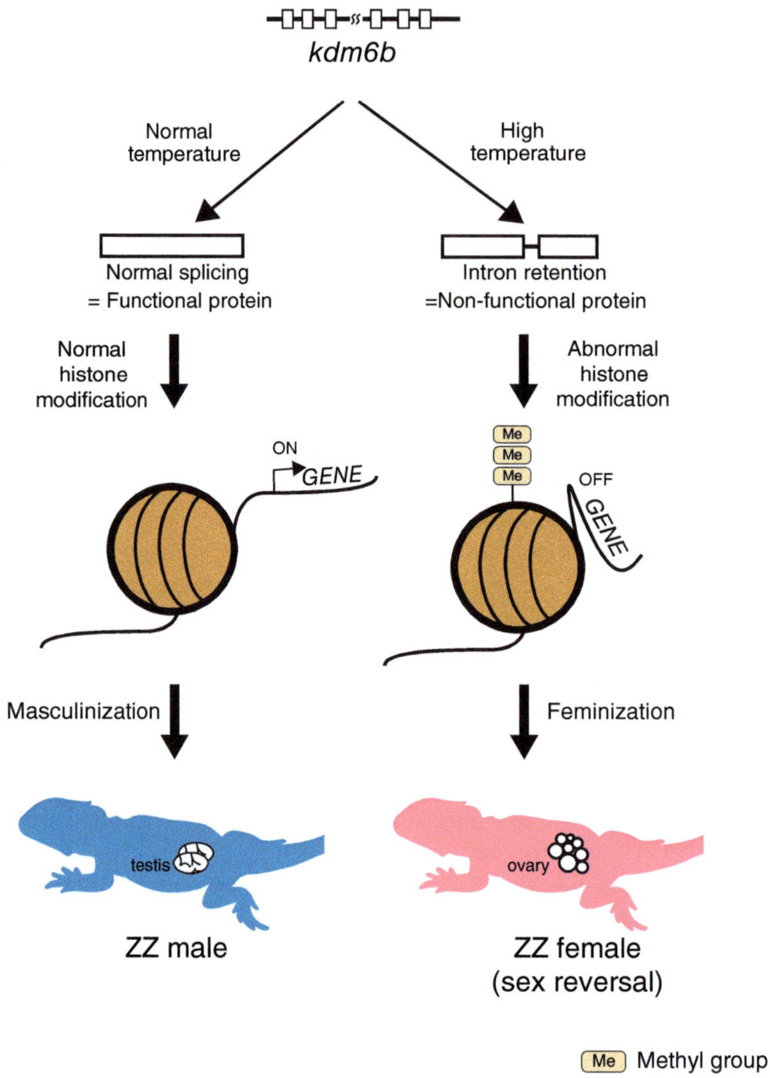

Figure 19.7 Temperature-induced sex reversal in the bearded dragon. Sex in the bearded dragon is primarily determined genetically (ZZ-ZW system). However, sex reversal takes place if ZZ embryos are incubated at temperatures higher than 32°C. Under normal temperatures (22°C to 32°C), epigenetic machineries such as *kdm6b* and *jarid2* are transcribed normally and functional enzymes are produced. Histone modification by the enzymes drives masculinization. However, incubation at higher temperatures induces alternative splicing, by which an intron is retained in the mRNA. As a result, protein products are non-functional and abnormal histone modification takes place. Under such conditions, embryos differentiate into females. Note that the *kdm6b* removes methyl groups from histone proteins, as depicted in the figure, whereas *jarid2* plays a part in the establishment of H3K27me3. Created by Genki Yamagishi.

While chicken gonads develop into bilateral testes in males, only the left gonad develops in females. In the developing embryo, urogenital tissues arise at embryonic day 3.5 (E3.5). Differentiation of gonadal morphology is first observed at E6.5. Around this time, development of the medulla and coalescence of seminiferous tubules take place in male gonads. In female gonads, the outer cortical layer starts to proliferate at this time point. On the left side of the body, the gonad continues to develop toward becoming a functional ovary.

However, proliferation stops in the right gonad and only a flattened epithelial monolayer can be observed by E8.5 (this is reviewed in Kuroiwa 2018; Smith et al. 2007).

Dimorphic expression of SD and sex-differentiating genes precedes morphological differentiation. The expression of the masculinizing gene, *dmrt1*, is already detected in male gonads at E4.5. Z-linked female marker genes such as *hintw* (W-linked histidine triad nucleotide binding protein) are also detected in female gonads at the same time point. Other sex regulators, such as *sox9* (in males) and *cyp19a1* (in females), are detected by E6.0–6.5, concomitantly with morphological differentiation (Smith et al. 2007). The expression of *amh*, a testicular differentiation factor, is also detected at E4.5, but in both males and females. The expression of *amh* in females seems to be related to regression of the right gonad (Hutson et al. 1981). The expression of estrogen receptor α (*esr1*) is also important for the development of female gonads. In a female chicken (*Gallus gallus*) embryo, *esr1* is expressed in the cortical layer of the left gonad, but not in the right one (Nakabayashi et al. 1998). This asymmetric expression of *esr1* suggests that estrogen signaling protects against *amh*-mediated regression in the left gonad.

Hypotheses Regarding Sex Determination in Birds

There are two hypotheses regarding sex determination in birds. One is the Z dosage model. According to this hypothesis, individual sex is determined by the dosage of a certain Z-linked gene. In other words, by comparison with female development, male development requires a double dose, or expression level, of a certain gene. This hypothesis is supported by the lack of dosage compensation in birds (Ellegren et al. 2007; Graves 2016; Itoh et al. 2007).

The other hypothesis is the W dominant model. In this model, avian sex is determined by a W-linked SD gene, which drives female development. The observation that only female-to-male sex reversal occurs in birds supports this hypothesis (Kuroiwa 2018), because males with ZZ chromosomal constitution can never acquire W-linked genes. Three candidates have been proposed as the W-linked sex determining gene: *hintw*, *faf* (female-associated factor), and *fet1* (female expressed transcript 1). However, overexpression of *hintw* failed to induce feminization in ZZ chicken embryos (Smith et al. 2009b). In addition, the expression of none of these three was altered during fadrozole (aromatase inhibitor)-induced sex reversal in the chicken. Furthermore, HINTW protein and mRNAs coding for *faf* or *fet1* could not be detected in embryonic gonads of the zebra finch (*Taeniopygia guttata*) (Hirst et al. 2017). These observations speak clearly against the W-linked hypothesis.

On the contrary, accumulating reports support the Z-dosage hypothesis. Knockdown of the Z-linked *dmrt1* gene in ZZ chicken embryos feminized their gonads (Smith et al. 2009a), whereas overexpression of *dmrt1* in ZW embryos induced masculinization (Lambeth et al. 2014). More recently, knockdown of one *dmrt1* copy in ZZ males ($Z^{D+}Z^{D-}$) was shown to induce feminization, causing the development of functional ovaries with follicles (Ioannidis et al. 2021). Thus gene dosage of *dmrt1* seems to be a critical factor of sex determination in birds.

Sex Steroids and Avian Sex Differentiation

The role of sex steroids in avian sex determination is somewhat complicated. Generally, vertebrate sex differentiation requires gonadal production of sex steroids such as estrogen and testosterone. However, the existence of gynandromorphs in birds speaks against such a generalization. A gynandromorph is an individual with both male and female characteristics.

Among vertebrates, birds are the sole group in which gynandromorphism occurs under natural conditions and either the left or right side of the body is feminized, while the other side is masculinized. This asymmetric sexual development cannot be explained by hormonal regulation, because hormones are equally distributed to the whole body through the bloodstream (Kuroiwa 2018).

Experimental manipulation of avian embryos also undermines the importance of hormonal regulation in avian sex differentiation. Zhao et al. (2010) demonstrated through cell transplantation that both host and transplanted cells differentiate according to their genotypic sex and that this results in the development of both ovary and testis.

Still, the various roles that sex hormones play for birds should not be disregarded. According to Ioannidis et al. (2021), *dmrt1* knockdown embryos ($Z^{D+}Z^{D-}$) treated with aromatase inhibitor develop testes, which suggests that gene dosage alone is insufficient for a complete sex differentiation and that gonadal estrogen synthesis is a prerequisite for ovarian development. As will be discussed in what follows, epigenetic regulation could play a role in avian sex determination and sex differentiation through hormone production.

The Epigenetic Regulation of Avian Sex Determination

Among avian species, the epigenetic regulation of sex determination has been studied almost exclusively in the chicken. As in teleost fish and reptiles, here too DNA methylation and histone modifications directly regulate gene expression of *cyp19a1*. Interestingly, the expression of *dmrt1* seems to be regulated by non-coding RNAs, whose own expression is in turn regulated by DNA methylation.

DNA Methylation and Histone Modifications

DNA methylation seems to silence transcription of avian genes; we have seen equivalent processes in other vertebrates. Comparison between whole-genome methylation status and transcriptome data from the chicken brain revealed that methylation levels of CpG sites within the gene or promoter regions were negatively correlated with gene expression levels (Sun et al. 2019). This observation is in agreement with a general role of DNA methylation in gene silencing. The same team also investigated the methylation status of the Z chromosomes in the chicken and white-throated sparrow (*Zonotrichia albicollis*) and found that the methylation pattern of the Z chromosome is generally the same between females and males. This observation is in agreement with the lack of gene dosage compensation in birds.

Sex-biased DNA methylation has been reported from the upstream region of the *cyp19a1* gene. According to Ellis et al. (2012), three CpG sites, located −955, −869, and −789 bp upstream of the transcription start site (TSS), were hypermethylated in males as compared to females in embryonic gonads. Sex-specific histone modifications were also observed. In male gonads, anti-transcriptional H3K27me3 was detected in the region adjacent to the three CpG sites, whereas pro-transcriptional H3K4me3 was detected in female gonads. In line with the role of DNA methylation and each histone modification, RNA polymerase II recruitment to this region was detected in female, but not in male gonads.

Interestingly, this epigenetic status around *cyp19a1* gene is reversed only partially in sex-reversed gonads. Ethynylestradiol, a synthetic estrogen, can induce sex reversal in the chicken, and the sex-reversed ZZ gonads develop into ovaries indistinguishable from ZW ovaries. However, methylation levels of the three CpG sites do not reach the value in females, although they were lowered to the intermediate level between males and females. Both H3K4me3 and H3K27me3 were detected around the CpG sites in sex-reversed gonads.

In contrast, the *cyp19a1* expression level was close to that of ZW ovaries. Recruitment of RNA polymerase II was also detected, which suggests transcriptional activation of *cyp19a1* in sex-reversed gonads. It could be that subtle changes in epigenetic status result in morphological and functional sex reversal in birds.

Regulation of Dmrt1 Expression by Non-coding RNAs

Although DNA methylation patterns are similar between Z chromosomes in female and male gonads, male-specific hypermethylation occurs in a region that spans 460 kilobase pairs (kps) on the chicken Z chromosome (Teranishi et al. 2001). This region is called the male hypermethylated (MHM) region. CpG sites within the MHM region are hypermethylated in males by comparison with females. This sex-specific methylation status is established within one day after fertilization. The MHM region codes for a non-coding RNA and is transcribed only in females. This female-specific transcription is due not only to hypomethylation, but also to the acetylation of H4K16 in females (Bisoni et al. 2005).

According to Teranishi et al. (2001), the ncRNAs transcribed from the MHM region accumulates in the nucleus. A closer look at the nucleus revealed that the ncRNAs are localized adjacently to the locus that harbors *dmrt1*. Overexpression of MHM ncRNA in male embryos results in a lower expression of *dmrt1* (Roeszeler et al. 2012; Yang et al. 2010). These observations suggest that the expression of *dmrt1* is downregulated by ncRNAs transcribed from the MHM region. Considering the importance of *dmrt1* gene dosage, the dual epigenetic control of this gene—first through DNA methylation of the MHM region, then through ncRNAs transcribed from this region—seems to play a central role in avian sex determination (Figure 19.8). Since *esr1* expression was also upregulated through MHM ncRNA overexpression (Roeszler et al. 2012), this dual epigenetic regulation could also be involved in sex determination, albeit through currently unidentified mechanisms.

However, it should also be noted that overexpression of the ncRNAs did not induce *cyp19a1* expression or ovarian development (Roeszler et al. 2012). This observation is in agreement with the importance of estrogen synthesis for feminization. Therefore the MHM region alone seems to be insufficient for sex determination. In addition, the MHM region is not common among avian lineages. Although this region is present in Galloanserae species (such as chicken), it is absent from its sister group, Neoaves, and from the more basal group, the Palaeognathae (Sun et al. 2019). In these circumstances, the dual epigenetic regulation of sex determination seems to be a derived character among birds.

Recently another region with male-specific hypermethylation was reported in the chicken (Sun et al. 2019). This region, designated MHM2, does not show homology with the classic MHM (re-designated MHM1), and its role in sex determination is yet to be investigated. However, MHM2 also codes for non-coding RNAs. In addition, genes located around either MHM1 or MHM2 showed lower expression in males than in females. These observations suggest a similar role for MHM1 and MHM2 in gene expression regulation.

Possible Effects of Epigenetic Toxicants on Sex Determination of Non-mammalian Vertebrates

Epigenetic Toxicants and Their Implications for Non-mammalian Vertebrates

As described here, epigenetic regulation allows non-mammalian vertebrates to determine or reverse their sex according to the surrounding environment. Such flexibility enables

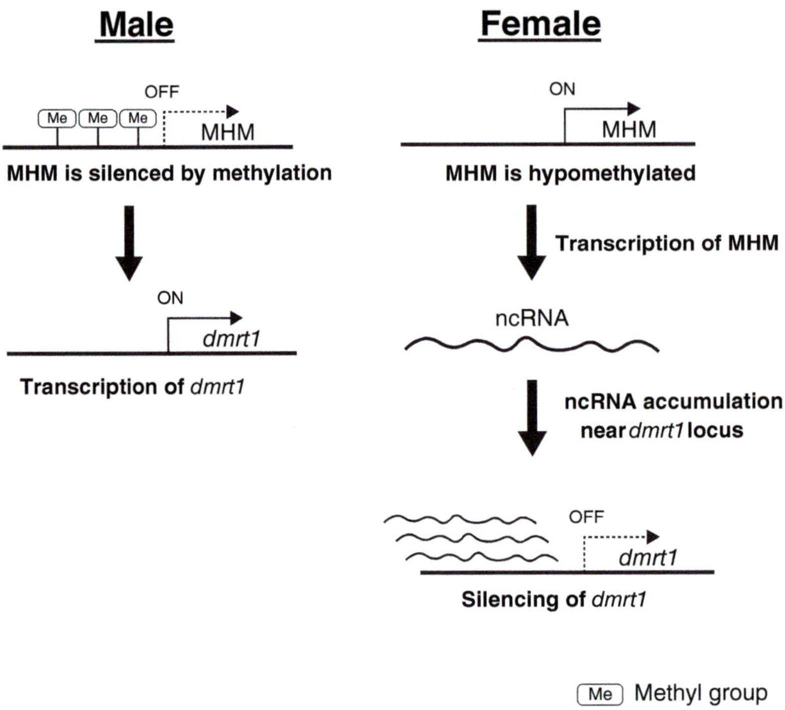

Figure 19.8 Epigenetic regulation of *dmrt1* expression in the chicken. In the chicken, *dmrt1* expression is critical for male differentiation. The expression of *dmrt1* is under dual control of epigenetic regulation, first, by the methylation status of the MHM region and, second, by transcriptional regulation of the *dmrt1* gene by ncRNAs encoded by the MHM region. In male chickens, the MHM region is hypermethylated and thus its transcription is silenced. In contrast, the MHM region is hypomethylated in females and thus transcription proceeds. Non-coding RNAs transcribed from the MHM region accumulate near the *dmrt1* locus to suppress expression of this gene. Created by Genki Yamagishi.

organisms to adapt to environmental fluctuations. At the same time, however, the same feature is liable to disruption by exogenous toxicants. Recently various chemical substances, metals, and pharmaceuticals have been found to affect the epigenetic programming of organisms (Marczylo et al. 2016). Aberrant epigenetic programs induced by such substances are said to possess epigenetic toxicity.

The effect of epigenetic toxicants on non-mammalian vertebrates is best investigated in the zebrafish, and is reviewed in Cavalieri and Spinelli (2017) and Kamstra et al. (2015). According to these review articles, all three major epigenetic mechanisms, namely DNA methylation, histone modification, and non-coding RNAs, are potentially affected by pollutants. In addition to the effect of industrial and agricultural pollutants, that of pharmaceuticals is also concerning. Epigenetic regulation is an important factor for tumorigenesis, and thus increasing attention has been paid to it as a potential therapeutic target (the topic is reviewed in Cheng et al. 2019; Patnaik and Anupriya School 2019). It is possible that such pharmaceuticals affect sex in wildlife by altering epigenetic programming.

Case Studies

The effects of substances on DNA methylation and on the expression of epigenetic machineries have been reported in teleost species (Lin et al. 2020; Pierron et al. 2014). Possible effects of pollutants in the field are also reported. For example, differential DNA methylation has been reported about alligators that inhabit environmentally contaminated and non-contaminated lakes (Guillette et al. 2016). Such reports suggest that certain contaminants do affect epigenetic programming in wildlife.

However, little has been investigated about effects on sex determination. But direct relationships between sex determination and epigenetic changes induced by pollutants have been studied in two non-mammalian vertebrates: the zebrafish and red-eared slider turtle.

Pierron et al. (2021) investigated the effect of cadmium (Cd) on sex determination in zebrafish. Although no change in the sex ratio was observed in the exposed generation (F0), the proportion of females in the population progressively increased from generation F1 to generation F3, none of which were exposed to Cd. Cd exposure in the F0 generation also resulted in the hypomethylation of the *foxl2* promoter and the upregulation of *cyp19a1a* expression in the F0, F1, and F2 generations. In addition, methylation levels of the *foxl2* promoter in the parental generation were positively correlated with *cyp19a1a* methylation levels in the next generation. This observation suggests that the hypomethylated status of *foxl2* was inherited and that it altered the sex ratio of the next generation, possibly by affecting the expression of the downstream SD gene *cyp19a1a*.

In the red-eared slider turtle, the effect of the endocrine disruptors, PCBs (polychlorinated biphenyls), has been investigated (Matsumoto et al. 2014). PCBs are reported to affect epigenetic programming, for example DNA and histone methylation, possibly by altering gene expression of DNA methyltransferases and chromatin-modifying enzymes (Casati et al. 2013; Desaulniers et al. 2009). The sex of turtle embryos incubated at MPT was successfully reversed to female through treatment with PCBs; such treatment also upregulated ovarian markers (*cyp19a1*, *foxl2* and *rspo1*) and downregulated testicular markers (*dmrt1* and *sox9*). However, DNA methylation patterns at the *cyp19a1* promoter were similar to those of embryos incubated at MPT without treatment. This observation suggests that PCBs affect the expression of sex determination genes downstream of epigenetic regulation. Whether this upstream regulation involves epigenetic changes is unconfirmed but is a possibility, because the methylation status of the *dmrt1* promoter is a key to sex determination in the slider turtle (Ge et al. 2018; Weber et al. 2020).

Possible Application of Non-mammalian Vertebrates as Biomarkers for Epigenetic Toxicants

Unlike classic endocrine disrupting chemicals (EDCs), which agonize or antagonize hormone–receptor interactions, epigenetic toxicants affect epigenetic programs, making them induce an aberrant gene expression. Serious concern arises from the nature of such epigenetic toxicity. Maintenance of PGCs is important for sex determination in teleost fish (Kurokawa et al. 2007; Morinaga et al. 2007). As already described, DNA demethylation during embryonic development is important for the maintenance of PGCs in the zebrafish, and the inhibition of demethylation results in sex reversal (Wang et al. 2021). In other words, DNA demethylation inhibitors can reverse sex completely independently of hormone–receptor interactions. Because of the existence of such substance and action mechanisms, the effectiveness of detection systems centered on hormone–receptor interaction could be impaired.

In this regard, non-mammalian vertebrates such as teleost fish and reptiles could be useful as biomarkers for epigenetic toxicity. Their sex is labile to exogenous stimulus during embryonic development. In some hermaphrodite species, sexually mature individuals could be affected, too. Thus, the effect of a pollutant could be reflected in the sex ratio of the exposed population. In addition, epigenetic changes in a small number of genes would be sufficient to evaluate the effect of pollutants. As explained earlier, sex in non-mammalian vertebrates is commonly determined by *cyp19a1* and *dmrt1*. Thus epigenetic toxicity can be evaluated by investigating epigenetic programming related to these genes.

The need for non-mammalian models also arises from epigenetic inheritance. In mammals, genome-wide epigenetic reprogramming, or “reset,” takes place in PGCs and zygotes. Hence epigenetic programs are not inherited by offspring, except in specific regions of the genome. In contrast, inheritance of a paternal epigenetic program has been reported in teleost fish (Jiang et al. 2013; Potok et al. 2013). Epigenetic inheritance has also been implicated in sex reversal of the zebrafish and half-smooth tongue sole (Pierron et al. 2021; Shao et al. 2014). Consequently, epigenetic inheritance is likely to have a more pronounced effect on non-mammalian vertebrates than on mammals. Thus the use of non-mammalian models is vital for the evaluation of effects on wildlife.

Among non-mammalian vertebrates, the zebrafish seems to be best suited as an epigenetic biomarker. This species is well established as an experimental model. Besides, epigenetic regulation of sex has been extensively investigated in this species.

The red-eared slider turtle may also offer a good experimental model. Although adult turtles are too bulky for an experimental model, their eggs are easy to handle. Turtle eggs can be purchased from farms, and an egg incubator (or two, one for FPT incubation and the other for MPT incubation) is sufficient to conduct experiments. In addition, the slider turtle is distributed worldwide (and listed as one of the world’s worst 100 invasive species), and epigenetic regulation of sex determination in this species has been well studied. Because of these features, this turtle species could be used as a standard experimental model around the world.

The weakness of non-mammalian vertebrates is that the identification of epigenetic changes induced by pollutants requires biochemical experiments dedicated to the detection of certain epigenetic modifications, for example bisulfite sequencing for DNA methylation and chromatin immunoprecipitation for histone modifications. Since such experiments involve time-consuming steps such as DNA purification, processing, and sequencing, a simple and rapid detection system is urgently needed. Recently, a mouse strain that expresses reporter gene in response to certain epigenetic changes has been developed (Ideta-Otsuka et al. 2017). In the long term, establishment of strains equivalent to such a “reporter mouse” among non-mammalian vertebrates might help with the understanding of the action mechanisms of epigenetic toxicants.

Acknowledgment

The authors would like to thank Dr. Mike Roberts, Independent Consultant, UK, for his critical readings of this manuscript. This work was supported by Grants-in-Aid for Scientific Research from the Ministry of Education, Culture, Sports, Science and Technology of Japan [17H06432, 18H02474 (SM), 20K15835 (GY)] and a grant from the Ministry of the Environment, Japan (SM, TI).

References

Bird, A.P. (1980). DNA methylation and the frequency of CpG in animal DNA. *Nucleic Acids Res.* 8 (7): 1499–1504. doi: 10.1093/nar/8.7.1499.

Bird, A.P. and Taggart, M.H. (1980). Variable patterns of total DNA and rDNA methylation in animals. *Nucleic Acids Res.* 8 (7): 1485–1497. doi: 10.1093/nar/8.7.1485.

Bisoni, L., Batlle-Morera, L., Bird, A.P., Suzuki, M., and McQueen, H.A. (2005). Female-specific hyperacetylation of histone H4 in the chicken Z chromosome. *Chromosome Res.* 13 (2): 205–214. doi: 10.1007/s10577-005-1505-4.

Bista, B. and Valenzuela, N. (2020). Turtle insights into the evolution of the reptilian karyotype and the genomic architecture of sex determination. *Genes* 11 (4): 416. doi: 10.3390/genes11040416.

Bull, J.J. (1987). Temperature-sensitive periods of sex determination in a lizard: Similarities with turtles and crocodilians. *J. Exp. Zool.* 241 (1): 143–148. doi: 10.1002/jez.1402410118.

Capel, B. (2017). Vertebrate sex determination: Evolutionary plasticity of a fundamental switch. *Nat. Rev. Genet.* 18 (11): 675–689. doi: 10.1038/nrg.2017.60.

Casati, L., Sendra, R., Poletti, A., Negri-Cesi, P., and Celotti, F. (2013). Androgen receptor activation by polychlorinated biphenyls: Epigenetic effects mediated by the histone demethylase Jarid1b. *Epigenetics* 8 (10): 1061–1068. doi: 10.4161/epi.25811.

Cavalieri, V. and Spinelli, G. (2017). Environmental epigenetics in zebrafish. *Epigenetics Chromatin* 10 (1): 1–11. doi: 10.1186/s13072-017-0154-0.

Chen, S., Zhang, G., Shao, C., Huang, Q., Liu, G., Zhang, P., and Wang, J. (2014). Whole-genome sequence of a flatfish provides insights into ZW sex chromosome evolution and adaptation to a benthic lifestyle. *Nat. Genet.* 46 (3): 253–260. doi: 10.1038/ng.2890.

Cheng, Y., He, C., Wang, M., Ma, X., Mo, F., Yang, S., and Wei, X. (2019). Targeting epigenetic regulators for cancer therapy: Mechanisms and advances in clinical trials. *Signal Transduct. Target Ther.* 4 (1): 1–39. doi: 10.1038/s41392-019-0095-0.

Desaulniers, D., Xiao, G.H., Lian, H., Feng, Y.L., Zhu, J., Nakai, J., and Bowers, W.J. (2009). Effects of mixtures of polychlorinated biphenyls, methylmercury, and organochlorine pesticides on hepatic DNA methylation in prepubertal female Sprague-Dawley rats. *Int. J. Toxicol.* 28 (4): 294–307. doi: 10.1177/1091581809337918.

Deveson, I.W., Holleley, C.E., Blackburn, J., Graves, J.A.M., Mattick, J.S., Waters, P.D., and Georges, A. (2017). Differential intron retention in *Jumonji* chromatin modifier genes is implicated in reptile temperature-dependent sex determination. *Sci. Adv.* 3 (6): e1700731. doi: 10.1126/sciadv.1700731.

Díaz, N. and Piferrer, F. (2015). Lasting effects of early exposure to temperature on the gonadal transcriptome at the time of sex differentiation in the European sea bass, a fish with mixed genetic and environmental sex determination. *BMC Genomics* 16 (1): 1–16. doi: 10.1186/s12864-015-1862-0.

Domingos, J.A., Budd, A.M., Banh, Q.Q., Goldsbury, J.A., Zenger, K.R., and Jerry, D.R. (2018). Sex-specific *dmrt1* and *cyp19a1* methylation and alternative splicing in gonads of the protandrous hermaphrodite barramundi. *PLoS One* 13 (9): e0204182. doi: 10.1371/journal.pone.0204182.

Eberharter, A. and Becker, P.B. (2002). Histone acetylation: A switch between repressive and permissive chromatin. *EMBO Rep.* 3 (3): 224–229. doi: 10.1093/embo-reports/kvf053.

Ellegren, H., Hultin-Rosenberg, L., Brunström, B., Dencker, L., Kultima, K., and Scholz, B. (2007). Faced with inequality: Chicken do not have a general dosage compensation of sex-linked genes. *BMC Biol.* 5 (1): 1–12. doi: 10.1186/1741-7007-5-40.

Ellis, H.L., Shioda, K., Rosenthal, N.F., Coser, K.R., and Shioda, T. (2012). Masculine epigenetic sex marks of the *CYP19A1/aromatase* promoter in genetically male chicken embryonic gonads are resistant to estrogen-induced phenotypic sex conversion. *Biol. Reprod.* 87 (1): 23. doi: 10.1095/biolreprod.112.099747.

Fricke, H.W. (1983). Social control of sex: Field experiments with the anemonefish *Amphiprion bicinctus*. *Zeitschrift für Tierpsychologie* 61 (1): 71–77. doi: 10.1111/j.1439-0310.1983.tb01327.x.

Gamble, T. (2010). A review of sex determining mechanisms in geckos (Gekkota: Squamata). *Sex. Dev.* 4 (1–2): 88–103. doi: 10.1159/000289578.

Gamble, T., Castoe, T.A., Nielsen, S.V., Banks, J.L., Card, D.C., Schield, D.R., and Booth, W. (2017). The discovery of XY sex chromosomes in a boa and python. *Curr. Biol.* 27 (14): 2148–2153. doi: 10.1016/j.cub.2017.06.010.

Ge, C., Ye, J., Weber, C., Sun, W., Zhang, H., Zhou, Y., and Capel, B. (2018). The histone demethylase KDM6B regulates temperature-dependent sex determination in a turtle species. *Science* 360 (6389): 645–648. doi: 10.1126/science.aap8328.

Graves, J.A.M. (2016). Evolution of vertebrate sex chromosomes and dosage compensation. *Nat. Rev. Genet.* 17 (1): 33–46. doi: 10.1038/nrg.2015.2.

Guiguen, Y., Fostier, A., Piferrer, F., and Chang, C.F. (2010). Ovarian aromatase and estrogens: A pivotal role for gonadal sex differentiation and sex change in fish. *Gen. Comp. Endocrinol.* 165 (3): 352–366. doi: 10.1016/j.ygcen.2009.03.002.

Guillette, L.J., Jr, Parrott, B.B., Nilsson, E., Haque, M.M., and Skinner, M.K. (2016). Epigenetic programming alterations in alligators from environmentally contaminated lakes. *Gen. Comp. Endocrinol.* 238: 4–12. doi: 10.1016/j.ygcen.2016.04.012.

Gujar, H., Weisenberger, D.J., and Liang, G. (2019). The roles of human DNA methyltransferases and their isoforms in shaping the epigenome. *Genes* 10 (2): 172. doi: 10.3390/genes10020172.

Hale, B.J., Yang, C.X., and Ross, J.W. (2014). Small RNA regulation of reproductive function. *Mol. Reprod. Dev.* 81 (2): 148–159. doi: 10.1002/mrd.22272.

Harrington, R.W. (1961). Oviparous hermaphroditic fish with internal self-fertilization. *Science* 134 (3492): 1749–1750. doi: 10.1126/science.134.3492.1749.

Hata, K., Okano, M., Lei, H., and Li, E. (2002). Dnmt3L cooperates with the Dnmt3 family of de novo DNA methyltransferases to establish maternal imprints in mice. *Devel.* 129 (8): 1983–1993. doi: 10.1242/dev.129.8.1983.

Hattori, R.S., Murai, Y., Oura, M., Masuda, S., Majhi, S.K., Sakamoto, T., and Strüssmann, C.A. (2012). A Y-linked anti-Müllerian hormone duplication takes over a critical role in sex determination. *Proceedings of the National Academy of Sciences, USA* 109 (8), 2955–2959. doi:10.1073/pnas.1018392109x

Herpin, A. and Schartl, M. (2011). Sex determination: Switch and suppress. *Curr. Biol.* 21 (17): R656–R659. doi: 10.1016/j.cub.2011.07.026.

Hirst, C.E., Major, A.T., Ayers, K.L., Brown, R.J., Mariette, M., Sackton, T.B., and Smith, C.A. (2017). Sex reversal and comparative data undermine the W chromosome and support Z-linked *DMRT1* as the regulator of gonadal sex differentiation in birds. *Endocrinology* 158 (9): 2970–2987. doi:10.1210/en.2017-00316.

Holleley, C.E., O'Meally, D., Sarre, S.D., Graves, J.A.M., Ezaz, T., Matsubara, K., and Georges, A. (2015). Sex reversal triggers the rapid transition from genetic to temperature-dependent sex. *Nature* 523 (7558): 79–82. doi: 10.1038/nature14574.

Hutson, J., Ikawa, H., and Donahoe, P.K. (1981). The ontogeny of Mullerian inhibiting substance in the gonads of the chicken. *J. Pediatr. Surg.* 16 (6): 822–827. doi: 10.1016/s0022-3468(81)80827-5.

Ideta-Otsuka, M., Igarashi, K., Narita, M., and Hirabayashi, Y. (2017). Epigenetic toxicity of environmental chemicals upon exposure during development-Bisphenol A and valproic acid may have epigenetic effects. *Food Chem. Toxicol.* 109: 812–816. doi: 10.1016/j.fct.2017.09.014.

Ioannidis, J., Taylor, G., Zhao, D., Liu, L., Idoko-Akoh, A., Gong, D., and Clinton, M. (2021). Primary sex determination in birds depends on DMRT1 dosage, but gonadal sex does not determine adult secondary sex characteristics. *Proceedings of the National Academy of Sciences, USA* 118 (10). doi:10.1073/pnas.2020909118

Itoh, Y., Melamed, E., Yang, X., Kampf, K., Wang, S., Yehya, N., and Arnold, A.P. (2007). Dosage compensation is less effective in birds than in mammals. *J. Biol.* 6 (1): 1–15. doi: 10.1186/jbiol53.

Jambhekar, A., Dhall, A., and Shi, Y. (2019). Roles and regulation of histone methylation in animal development. *Nat. Rev. Mol. Cell Biol.* 20 (10): 625–641. doi: 10.1038/s41580-019-0151-1.

Jiang, L., Zhang, J., Wang, J.J., Wang, L., Zhang, L., Li, G., and Liu, J. (2013). Sperm, but not oocyte, DNA methylome is inherited by zebrafish early embryos. *Cell* 153 (4): 773–784. doi: 10.1016/j.cell.2013.04.041.

Kamiya, T., Kai, W., Tasumi, S., Oka, A., Matsunaga, T., Mizuno, N., and Kikuchi, K. (2012). A trans-species missense SNP in *Amhr2* is associated with sex determination in the tiger pufferfish, *Takifugu rubripes* (fugu). *PLoS Genet.* 8 (7): e1002798. doi: 10.1371/journal.pgen.1002798.

Kamstra, J.H., Aleström, P., Kooter, J.M., and Legler, J. (2015). Zebrafish as a model to study the role of DNA methylation in environmental toxicology. *Environ. Sci. Pollut. Res.* 22 (21): 16262–16276. doi: 10.1007/s11356-014-3466-7.

Kawai, A., Ishijima, J., Nishida, C., Kosaka, A., Ota, H., Kohno, S.I., and Matsuda, Y. (2009). The ZW sex chromosomes of *Gekko hokouensis* (Gekkonidae, Squamata) represent highly conserved homology with those of avian species. *Chromosoma* 118 (1): 43–51. doi: 10.1007/s00412-008-0176-2.

Kikuchi, K. and Hamaguchi, S. (2013). Novel sex-determining genes in fish and sex chromosome evolution. *Dev. Dyn.* 242 (4): 339–353. doi: 10.1002/dvdy.23927.

Klose, R.J., Kallin, E.M., and Zhang, Y. (2006). JmjC-domain-containing proteins and histone demethylation. *Nat. Rev. Genet.* 7 (9): 715–727. doi: 10.1038/nrg1945.

Koopman, P., Gubbay, J., Vivian, N., Goodfellow, P., and Lovell-Badge, R. (1991). Male development of chromosomally female mice transgenic for *Sry*. *Nature* 351 (6322): 117–121. doi: 10.1038/351117a0.

Kuroiwa, A. (2018). Sex determination and differentiation in birds. In *Reproductive and Developmental Strategies*, ed. K. Kobayashi, T. Kitano, Y. Iwao, and M. Kondo, 391–405. Tokyo: Springer.

Kurokawa, H., Saito, D., Nakamura, S., Katoh-Fukui, Y., Ohta, K., Baba, T., and Tanaka, M. (2007). Germ cells are essential for sexual dimorphism in the medaka gonad. *Proceedings of the National Academy of Sciences, USA* 104 (43), 16958–16963. doi: 10.1073/pnas.0609932104.

Lambeth, L.S., Raymond, C.S., Roeszler, K.N., Kuroiwa, A., Nakata, T., Zarkower, D., and Smith, C.A. (2014). Over-expression of DMRT1 induces the male pathway in embryonic chicken gonads. *Dev. Biol.* 389 (2): 160–172. doi: 10.1016/j.ydbio.2014.02.012.

Lin, W., Yan, Y., Ping, S., Li, P., Li, D., Hu, J., and Ren, Y. (2020). Metformin-induced epigenetic toxicity in zebrafish: Experimental and molecular dynamics simulation studies. *Environ. Sci. Technol.* 55 (3): 1672–1681. doi: 10.1021/acs.est.0c06052.

Maor, G.L., Yearim, A., and Ast, G. (2015). The alternative role of DNA methylation in splicing regulation. *Trends Genet.* 31 (5): 274–280. doi: 10.1016/j.tig.2015.03.002.

Marchese, F.P., Raimondi, I., and Huarte, M. (2017). The multidimensional mechanisms of long noncoding RNA function. *Genome Biol.* 18 (1): 1–13. doi: 10.1186/s13059-017-1348-2.

Marczylo, E.L., Jacobs, M.N., and Gant, T.W. (2016). Environmentally induced epigenetic toxicity: Potential public health concerns. *Crit. Rev. Toxicol.* 46 (8): 676–700. doi: 10.1080/10408444.2016.1175417.

Margueron, R. and Reinberg, D. (2011). The polycomb complex PRC2 and its mark in life. *Nature* 469 (7330): 343–349. doi: 10.1038/nature09784.

Matsuda, M., Nagahama, Y., Shinomiya, A., Sato, T., Matsuda, C., Kobayashi, T., and Sakaizumi, M. (2002). *DMY* is a Y-specific DM-domain gene required for male development in the medaka fish. *Nature* 417 (6888): 559–563. doi: 10.1038/nature751.

Matsumoto, Y., Buemio, A., Chu, R., Vafaee, M., and Crews, D. (2013). Epigenetic control of gonadal aromatase (*cyp19a1*) in temperature-dependent sex determination of red-eared slider turtles. *PLoS One* 8 (6): e63599. doi: 10.1371/journal.pone.0063599.

Matsumoto, Y., Hannigan, B., and Crews, D. (2014). Embryonic PCB exposure alters phenotypic, genetic, and epigenetic profiles in turtle sex determination, a biomarker of environmental contamination. *Endocrinology* 155 (11): 4168–4177. doi: 10.1210/en.2014-1404.

Matsumoto, Y., Hannigan, B., and Crews, D. (2016). Temperature shift alters DNA methylation and histone modification patterns in gonadal aromatase (*cyp19a1*) gene in species with temperature-dependent sex determination. *PLoS One* 11 (11): e0167362. doi: 10.1371/journal.pone.0167362.

Mawaribuchi, S., Takahashi, S., Wada, M., Uno, Y., Matsuda, Y., Kondo, M., and Ito, M. (2017). Sex chromosome differentiation and the W-and Z-specific loci in *Xenopus laevis*. *Dev. Biol.* 426 (2): 393–400. doi: 10.1016/j.ydbio.2016.06.015.

Messerschmidt, D.M., Knowles, B.B., and Solter, D. (2014). DNA methylation dynamics during epigenetic reprogramming in the germline and preimplantation embryos. *Genes Dev.* 28 (8): 812–828. doi: 10.1101/gad.234294.113.

Morinaga, C., Saito, D., Nakamura, S., Sasaki, T., Asakawa, S., Shimizu, N., and Kondoh, H. (2007). The *hotei* mutation of medaka in the anti-Müllerian hormone receptor causes the dysregulation of germ cell and sexual development. *Proceedings of the National Academy of Sciences, USA* 104 (23): 9691–9696. doi: 10.1073/pnas.0611379104.

Moyer, J.T. and Nakazono, A. (1978). Protandrous hermaphroditism in six species of the anemonefish genus *Amphiprion* in Japan. *Japanese J. Ichthyol.* 25 (2): 101–106. doi: 10.11369/jji1950.25.101.

Myosho, T., Otake, H., Masuyama, H., Matsuda, M., Kuroki, Y., Fujiyama, A., and Sakaizumi, M. (2012). Tracing the emergence of a novel sex-determining gene in medaka. *Oryzias Luzonensis. Genetics* 191 (1): 163–170. doi: 10.1534/genetics.111.137497.

Nakabayashi, O., Kikuchi, H., Kikuchi, T., and Mizuno, S. (1998). Differential expression of genes for aromatase and estrogen receptor during the gonadal development in chicken embryos. *J. Mol. Endocrinol.* 20 (2): 193–202. doi: 10.1677/jme.0.0200193.

Navara, K.J. (2018). The truth about Nemo's dad: Sex-changing behaviors in fishes. In K.J. Navara, *Choosing Sexes*, 183–212. Cham: Springer.

Navarro-Martín, L., Blázquez, M., Viñas, J., Joly, S., and Piferrer, F. (2009). Balancing the effects of rearing at low temperature during early development on sex ratios, growth and maturation in the European sea bass (*Dicentrarchus labrax*): Limitations and opportunities for the production of highly female-biased stocks. *Aquaculture* 296 (3-4): 347–358. doi: 10.1016/j.aquaculture.2009.07.022.

Navarro-Martín, L., Viñas, J., Ribas, L., Díaz, N., Gutiérrez, A., Di Croce, L., and Piferrer, F. (2011). DNA methylation of the gonadal aromatase (*cyp19a*) promoter is involved in temperature-dependent sex ratio shifts in the European sea bass. *PLoS Genet.* 7 (12): e1002447. doi: 10.1371/journal.pgen.1002447.

Nishioka, M., Kodama, Y., Sumida, M., and Ryuzaki, M. (1993). Systematic evolution of 40 populations of *Rana rugosa* distributed in Japan elucidated by electrophoresis. *Sci. Rep. Lab. Amphib. Biol.* 12: 83–131. doi: 10.15027/14532.

Oñate, S.A., Tsai, S.Y., Tsai, M.J., and O'Malley, B.W. (1995). Sequence and characterization of a coactivator for the steroid hormone receptor superfamily. *Science* 270 (5240): 1354–1357. doi: 10.1126/science.270.5240.1354.

Ortega-Recalde, O., Goikoetxea, A., Hore, T.A., Todd, E.V., and Gemmell, N.J. (2020). The genetics and epigenetics of sex change in fish. *Annu. Rev. Anim. Biosci.* 8: 47–69. doi: 10.1146/annurev-animal-021419-083634.

Ospina-Álvarez, N. and Piferrer, F. (2008). Temperature-dependent sex determination in fish revisited: Prevalence, a single sex ratio response pattern, and possible effects of climate change. *PLoS One* 3 (7): e2837. doi: 10.1371/journal.pone.0002837.

Pandian, T. (2012). Differentiation in gonochores. In T. Pandian, *Genetic Sex Differentiation in Fish*, Vol. 1: 17–102. Boca Raton: CRC Press.

Parrott, B.B., Kohno, S., Cloy-McCoy, J.A., and Guillette, L.J., Jr. (2014). Differential incubation temperatures result in dimorphic DNA methylation patterning of the *SOX9* and aromatase promoters in gonads of alligator (*Alligator mississippiensis*) embryos. *Biol. Reprod.* 90 (1): 2–1. doi: 10.1095/biolreprod.113.111468.

Patnaik, S. and Anupriya School. (2019). Drugs targeting epigenetic modifications and plausible therapeutic strategies against colorectal cancer. *Front Pharmacol* 10: 588. doi: 10.3389/fphar.2019.00588.

Pierron, F., Bureau du Colombier, S., Moffett, A., Caron, A., Peluhet, L., Daffe, G., and Baudrimont, M. (2014). Abnormal ovarian DNA methylation programming during gonad maturation in wild contaminated fish. *Environ. Sci. Technol.* 48 (19): 11688–11695. doi: 10.1021/es503712c.

Pierron, F., Lorioux, S., Héroin, D., Daffe, G., Etcheverria, B., Cachot, J., and Gonzalez, P. (2021). Transgenerational epigenetic sex determination: Environment experienced by female fish affects offspring sex ratio. *Environ. Pollut.* 277: 116864. doi: 10.1016/j.envpol.2021.116864.

Potok, M.E., Nix, D.A., Parnell, T.J., and Cairns, B.R. (2013). Reprogramming the maternal zebrafish genome after fertilization to match the paternal methylation pattern. *Cell* 153 (4): 759–772. doi: 10.1016/j.cell.2013.04.030.

Quinn, A.E., Georges, A., Sarre, S.D., Guarino, F., Ezaz, T., and Graves, J.A.M. (2007). Temperature sex reversal implies sex gene dosage in a reptile. *Science* 316 (5823): 411. doi: 10.1126/science.1135925.

Radhakrishnan, S., Literman, R., Neuwald, J., Severin, A., and Valenzuela, N. (2017). Transcriptomic responses to environmental temperature by turtles with temperature-dependent and genotypic sex determination assessed by RNAseq inform the genetic architecture of embryonic gonadal development. *PLoS One* 12 (3): e0172044. doi: 10.1371/journal.pone.0172044.

Radhakrishnan, S., Literman, R., Neuwald, J.L., and Valenzuela, N. (2018). Thermal response of epigenetic genes informs turtle sex determination with and without sex chromosomes. *Sex. Dev.* 12 (6): 308–319. doi: 10.1159/000492188.

Ramsey, M., Shoemaker, C., and Crews, D. (2007). Gonadal expression of *Sf1* and aromatase during sex determination in the red-eared slider turtle (*Trachemys scripta*), a reptile with temperature-dependent sex determination. *Differentiation* 75 (10): 978–991. doi: 10.1111/j.1432-0436.2007.00182.x.

Raymond, C.S., Murphy, M.W., O'Sullivan, M.G., Bardwell, V.J., and Zarkower, D. (2000). *Dmrt1*, a gene related to worm and fly sexual regulators, is required for mammalian testis differentiation. *Genes Dev.* 14 (20): 2587–2595. doi: 10.1101/gad.834100.

Roeszler, K.N., Itman, C., Sinclair, A.H., and Smith, C.A. (2012). The long non-coding RNA, *MHM*, plays a role in chicken embryonic development, including gonadogenesis. *Dev. Biol.* 366 (2): 317–326. doi: 10.1016/j.ydbio.2012.03.025.

Ruiz-García, A., Roco, Á.S., and Bullejos, M. (2021). Sex differentiation in amphibians: Effect of temperature and its influence on sex reversal. *Sex. Dev.* 1–11. doi: 10.1159/000515220.

Sandra, G.E. and Norma, M.M. (2010). Sexual determination and differentiation in teleost fish. *Rev. Fish Biol. Fish.* 20 (1): 101–121. doi: 10.1007/s11160-009-9123-4.

Sanulli, S., Justin, N., Teissandier, A., Ancelin, K., Portoso, M., Caron, M. et al. (2015). Jarid2 methylation via the PRC2 complex regulates H3K27me3 deposition during cell differentiation. *Mol. Cell* 57 (5): 769–783. doi: 10.1016/j.molcel.2014.12.020.

Sato, T., Endo, T., Yamahira, K., Hamaguchi, S., and Sakaizumi, M. (2005). Induction of female-to-male sex reversal by high temperature treatment in medaka. *Oryzias Latipes. Zool. Sci.* 22 (9): 985–988. doi: 10.2108/zsj.22.985.

Schartl, M. (2015). Sex determination by multiple sex chromosomes in *Xenopus tropicalis. Proceedings of the National Academy of Sciences, USA* 112 (34): 10575–10576. doi: 10.1073/pnas.1513518112.

Sfakianakis, D.G., Leris, I., Mylonas, C.C., and Kentouri, M. (2012). Temperature during early life determines sex in zebrafish, *Danio rerio* (Hamilton, 1822). *J. Biol. Res.* 17: 68.

Shao, C., Li, Q., Chen, S., Zhang, P., Lian, J., Hu, Q., and Zhang, G. (2014). Epigenetic modification and inheritance in sexual reversal of fish. *Genome Res.* 24 (4): 604–615. doi: 10.1101/gr.162172.113.

Shine, R., Elphick, M.J., and Donnellan, S. (2002). Co-occurrence of multiple, supposedly incompatible modes of sex determination in a lizard population. *Ecol. Lett.* 5 (4): 486–489. doi: 10.1046/j.1461-0248.2002.00351.x.

Shirato, H., Ogawa, S., Nakajima, K., Inagawa, M., Kojima, M., Tachibana, M., and Takeuchi, T. (2009). A jumonji (Jarid2) protein complex represses *cyclin D1* expression by methylation of histone H3-K9. *J. Biol. Chem.* 284 (2): 733–739. doi: 10.1074/jbc.M804994200.

Shoemaker, C.M. and Crews, D. (2009). Analyzing the coordinated gene network underlying temperature-dependent sex determination in reptiles. *Semin. Cell Dev. Biol.* 20 (3): 293–303. doi: 10.1016/j.semcdb.2008.10.010.

Smith, C.A., Roeszler, K.N., Hudson, Q.J., and Sinclair, A.H. (2007). Avian sex determination: What, when and where? *Cytogenet. Genome Res.* 117 (1–4): 165–173. doi: 10.1159/000103177.

Smith, C.A., Roeszler, K.N., Ohnesorg, T., Cummins, D.M., Farlie, P.G., Doran, T.J., and Sinclair, A.H. (2009a). The avian Z-linked gene *DMRT1* is required for male sex determination in the chicken. *Nature* 461 (7261): 267–271. doi: 10.1038/nature08298.

Smith, C.A., Roeszler, K.N., and Sinclair, A.H. (2009b). Genetic evidence against a role for *W-linked histidine triad nucleotide binding protein (HINTW)* in avian sex determination. *Int. J. Dev. Biol.* 53 (1): 59–67. doi: 10.1387/ijdb.082742cs.

Sun, D., Maney, D.L., Layman, T.S., Chatterjee, P., and Yi, S.V. (2019). Regional epigenetic differentiation of the Z chromosome between sexes in a female heterogametic system. *Genome Res.* 29 (10): 1673–1684. doi: 10.1101/gr.248641.119.

Sun, W., Cai, H., Zhang, G., Zhang, H., Bao, H., Wang, L., Ye, J., Qian, G., and Ge, C. (2017). *Dmrt1* is required for primary male sexual differentiation in Chinese soft-shelled turtle *Pelodiscus sinensis. Sci. Rep.* 7 (1): 4433. doi: 10.1038/s41598-017-04938-5.

Sunobe, T. and Nakazono, A. (1993). Sex change in both directions by alteration of social dominance in *Trimma okinawae* (Pisces: Gobiidae). *Ethology* 94 (4): 339–345. doi: 10.1111/j.1439-0310.1993.tb00450.x.

Takada, S., Wada, T., Kaneda, R., Choi, Y.L., Yamashita, Y., and Mano, H. (2006). Evidence for activation of *Amh* gene expression by steroidogenic factor 1. *Mech. Dev.* 123 (6): 472–480. doi: 10.1016/j.mod.2006.04.004.

Takagi, N. and Sasaki, M. (1974). A phylogenetic study of bird karyotypes. *Chromosoma* 46 (1): 91–120. doi: 10.1007/bf00332341.

Takehana, Y., Matsuda, M., Myosho, T., Suster, M.L., Kawakami, K., Shin, T., and Naruse, K. (2014). Co-option of *Sox3* as the male-determining factor on the Y chromosome in the fish *Oryzias dancena*. *Nat. Commun.* 5 (1): 1–10. doi: 10.1038/ncomms5157.

Tao, W., Sun, L., Shi, H., Cheng, Y., Jiang, D., Fu, B., and Wang, D. (2016). Integrated analysis of miRNA and mRNA expression profiles in tilapia gonads at an early stage of sex differentiation. *BMC Genomics* 17 (1): 1–13. doi: 10.1186/s12864-016-2636-z.

Teranishi, M., Shimada, Y., Hori, T., Nakabayashi, O., Kikuchi, T., Macleod, T., and Mizuno, S. (2001). Transcripts of the MHM region on the chicken Z chromosome accumulate as non-coding RNA in the nucleus of female cells adjacent to the *DMRT1* locus. *Chromosome Res.* 9 (2): 147–165. doi: 10.1023/a:1009235120741.

Todd, E.V., Ortega-Recalde, O., Liu, H., Lamm, M.S., Rutherford, K.M., Cross, H., and Gemmell, N.J. (2019). Stress, novel sex genes, and epigenetic reprogramming orchestrate socially controlled sex change. *Sci. Adv.* 5 (7): eaaw7006. doi: 10.1126/sciadv.aaw7006.

Tokunaga, S. (1985). Temperature-dependent sex determination in *Gekko japonicus* (Gekkonidae, Reptilia). *Dev. Growth Differ.* 27 (2): 117–120. doi: 10.1111/j.1440-169X.1985.00117.x.

Tong, Z., Liu, Y., Yu, X., Martinez, J.D., and Xu, J. (2019). The transcriptional co-activator NCOA6 promotes estrogen-induced GREB1 transcription by recruiting ERα and enhancing enhancer–promoter interactions. *J. Biol. Chem.* 294 (51): 19667–19682. doi: 10.1074/jbc.RA119.010704.

Uno, Y., Nozu, R., Kiyatake, I., Higashiguchi, N., Sodeyama, S., Murakumo, K., and Kuraku, S. (2020). Cell culture-based karyotyping of orectolobiform sharks for chromosome-scale genome analysis. *Commun. Biol.* 3 (1): 1–10. doi: 10.1038/s42003-020-01373-7.

Valenzuela, N. and Lance, V. (2004). Temperature-dependent sex determination. In: *Reptilian Incubation: Environment, Evolution and Behaviour*, (ed. D.C. Deeming), 211–227. Nottingham: Nottingham University Press.

Wang, D.S., Zhou, L.Y., Kobayashi, T., Matsuda, M., Shibata, Y., Sakai, F., and Nagahama, Y. (2010). *Doublesex*- and *Mab-3*-related transcription factor-1 repression of aromatase transcription, a possible mechanism favoring the male pathway in tilapia. *Endocrinology* 151 (3): 1331–1340. doi: 10.1210/en.2009-0999.

Wang, X., Ma, X., Wei, G., Ma, W., Zhang, Z., Chen, X., and Liu, J. (2021). The role of DNA methylation reprogramming during sex determination and transition in zebrafish. *Genomics Proteomics Bioinformatics* S1672-0229 (21): 00023–1. Advance online publication. doi: 10.1016/j.gpb.2020.10.004.

Wang, Y.Y., Sun, L.X., Zhu, J.J., Zhao, Y., Wang, H., Liu, H.J., and Ji, X.S. (2017). Epigenetic control of *cyp19a1a* expression is critical for high temperature induced Nile tilapia masculinization. *J. Therm. Biol.* 69: 76–84. doi: 10.1016/j.jtherbio.2017.06.006.

Warner, R.R. and Swearer, S.E. (1991). Social control of sex change in the bluehead wrasse, *Thalassoma bifasciatum* (Pisces: Labridae). *Biol. Bull.* 181 (2): 199–204. doi: 10.2307/1542090.

Weber, C., Zhou, Y., Lee, J.G., Looger, L.L., Qian, G., Ge, C., and Capel, B. (2020). Temperature-dependent sex determination is mediated by pSTAT3 repression of *Kdm6b*. *Science* 368 (6488): 303–306. doi: 10.1126/science.aaz4165.

Yan, H., Liu, Q., Jiang, J., Shen, X., Zhang, L., Yuan, Z., and Liu, Y. (2021). Identification of sex differentiation-related microRNA and long non-coding RNA in *Takifugu rubripes* gonads. *Sci. Rep.* 11 (1): 1–13. doi: 10.1038/s41598-021-83891-w.

Yang, X., Zheng, J., Xu, G., Qu, L., Chen, S., Li, J., and Yang, N. (2010). Exogenous cMHM regulates the expression of *DMRT1* and *ERα* in avian testes. *Mol. Biol. Rep.* 37 (4): 1841–1847. doi: 10.1007/s11033-009-9619-y.

Yano, A., Guyomard, R., Nicol, B., Jouanno, E., Quillet, E., Klopp, C., and Guiguen, Y. (2012). An immune-related gene evolved into the master sex-determining gene in rainbow trout. *Oncorhynchus Mykiss. Curr. Biol.* 22 (15): 1423–1428. doi: 10.1016/j.cub.2012.05.045.

Yatsu, R., Miyagawa, S., Kohno, S., Parrott, B.B., Yamaguchi, K., Ogino, Y., and Iguchi, T. (2016). RNA-seq analysis of the gonadal transcriptome during *Alligator mississippiensis* temperature-dependent sex determination and differentiation. *BMC Genomics* 17 (1): 1–13. doi: 10.1186/s12864-016-2396-9.

Yatsu, R., Miyagawa, S., Kohno, S., Saito, S., Lowers, R.H., Ogino, Y., and Iguchi, T. (2015). TRPV4 associates environmental temperature and sex determination in the American alligator. *Sci. Rep.* 5 (1): 1–10. doi: 10.1038/srep18581.

Yntema, C. (1968). A series of stages in the embryonic development of *Chelydra serpentina*. *J. Morphol.* 125 (2): 219–251. doi: 10.1002/jmor.1051250207.

Yoshimoto, S., Ikeda, N., Izutsu, Y., Shiba, T., Takamatsu, N., and Ito, M. (2010). Opposite roles of *DMRT1* and its W-linked paralogue, *DM-W*, in sexual dimorphism of *Xenopus laevis*: Implications of a ZZ/ZW-type sex-determining system. *Development* 137 (15): 2519–2526. doi: 10.1242/dev.048751.

Yoshimoto, S. and Ito, M. (2011). A ZZ/ZW-type sex determination in *Xenopus laevis*. *FEBS J.* 278 (7): 1020–1026. doi: 10.1111/j.1742-4658.2011.08031.x.

Yu, Q., Peng, C., Ye, Z., Tang, Z., Li, S., Xiao, L., and Lin, H. (2020). An estradiol-17β/miRNA-26a/*cyp19a1a* regulatory feedback loop in the protogynous hermaphroditic fish. *Epinephelus Coioides. Mol Cell Endocrinol* 504: 110689. doi: 10.1016/j.mce.2019.110689.

Zhang, Y., Zhang, S., Liu, Z., Zhang, L., and Zhang, W. (2013). Epigenetic modifications during sex change repress gonadotropin stimulation of *cyp19a1a* in a teleost ricefield eel (*Monopterus albus*). *Endocrinology* 154 (8): 2881–2890. doi: 10.1210/en.2012-2220.

Zhao, D., McBride, D., Nandi, S., McQueen, H.A., McGrew, M.J., Hocking, P.M., and Clinton, M. (2010). Somatic sex identity is cell autonomous in the chicken. *Nature* 464 (7286): 237–242. doi: 10.1038/nature08852.

Zhou, H., Zhuang, Z., Zhang, R., Xu, Q., Liang, Y., Jiang, Z., and Li, Y. (2019a). Temperature-control-induced masculinization in tiger puffer *Takifugu rubripes*. *J. Oceanol. Limnol.* 37 (3): 1125–1135. doi: 10.1007/s00343-019-7382-z.

Zhou, H., Zhuang, Z.X., Sun, Y.Q., Chen, Q., Zheng, X.Y., Liang, Y.T., and Li, Y.J. (2019b). Changes in DNA methylation during epigenetic-associated sex reversal under low temperature in *Takifugu rubripes*. *PLos One* 14 (8): e0221641. doi: 10.1371/journal.pone.0221641.

20

Characterization of Genomic and Epigenomic Biomarkers of Nanoparticle Toxicity Using the Zebrafish Model System

Athira Sairanthry Suku[1,2], *Parayanthala Valappil Mohanan*[1], *and Jennifer L. Freeman*[2]

[1] *Toxicology Division, Biomedical Technology Wing, Sree Chitra Tirunal Institute for Medical Sciences and Technology, Poojapura, Kerala, India*
[2] *School of Health Sciences, Purdue University, West Lafayette, IN*

The term "nanotoxicology" was coined in 2004 and designates the toxic impacts of nanoparticles on living systems. The levels of human exposure to nanoparticles have seen a hike in the past sixteen years, the primary route of ingress being medical and commercial sources. This increase in nanoparticle exposure arose mainly due to the increasing presence of nanomaterials and their derivatives in the various dimensions of our daily life. Here are just a few examples: silver (Ag) nanoparticles in textile industries, titanium dioxide (TiO_2)-based nanoparticles in cosmetics and surface coatings, zinc oxide (ZnO) nanoparticles in opto-electronics, and carbon nanoparticles from automobiles [1]. The advancements in nanotechnology-related innovations led to an increase in toxicity research, in an attempt to identify safe levels of nanoparticles on the basis of their predetermined physical–chemical properties and the systemic response to toxicity in different model systems [2]. Newly developed nanomaterials must be subjected to safety assessments in order to certify whether they are able to cause any significant health problems. A set of study rules have been proposed for nanotoxicity assessments, with numerous prerequisites that depend on whether the study follows in vitro or in vivo routes. Marquis et al. (2009) reported in comprehensive detail on various analytical methods for the evaluation of nanotoxicity in in vitro systems including methods that involve transmission electron microscopic (TEM) observation of cellular uptake of nanoparticles, elemental analysis for nanoparticle quantification, fluorescence microscopic observation of nanoparticle exposed and tagged tissues and cells, cellular proliferation studies, live–dead pathway studies, subcellular toxicity investigation, DNA damage analysis, and other investigations [3]. But researchers suggest that in vivo models provide far more reliable results in terms of resemblance to the human system [4]. Mainstream in vivo models for nanotoxicity studies extend to rats [5–7] and mice [8–10]. Despite the ethical considerations related to the selection of models, in vivo systems basically allow chronic exposure studies, ADMR/E (absorption, distribution, metabolism, and retention/excretion) evaluations, and more significant observations, which are not normally possible via in vitro studies.

Nevertheless, conventional toxicity studies using animal or human cell lines and in vivo models such as rats, mice, and rabbits present considerable drawbacks, for instance maintenance difficulty, higher economic costs, and wider data variation. In search of an effective substitute, George Streisinger from the University of Oregon identified the freshwater zebrafish (*Danio rerio*) in the mid-1970s as a potential complementary whole-animal model

Genomic and Epigenomic Biomarkers of Toxicology and Disease: Clinical and Therapeutic Actions,
First Edition. Edited by Saura C. Sahu.

for mammalian species. George Streisinger's research group was much in awe of the transparent nature of the fish embryo, which could serve as a superlative option for neurological studies and other early developmental outcomes [11]. The impulse for a wider application of the zebrafish came into its own during the 1990s, after the identification of the prospect of induced genetic mutation and after the zebrafish was chosen as one of the organisms that were to have their genome sequenced. A major attractive feature of the zebrafish model is the degree of similarity of this fish with the human genetic constitution and physiological characteristics. More than 80% of the disease genes from humans remain in close proximity with that of zebrafish. In addition to this genetic profile, the molecular mechanisms and the patterns of cell development are also highly comparable with those of humans [12]. This vertebrate fish is 3–5 cm in length, reproduces very rapidly, is diploid in nature, and is easy to handle and propagate. Lower demand of culturing space and relatively infrequent contamination probabilities make zebrafish an ideal model for toxicity evaluations.

The use of zebrafish as a model for studies of the toxicity of nanomaterials has attracted significant attention among researchers. There has been a huge spike in zebrafish-related nanomaterial research in which studies have focused on materials of varying composition and origin. Metals and their derivatives, carbon and polymer-based nanocomposites, and also nanomaterials of different dimensions are some of the examples [13–15]. According to study reports, most of the engineered nanomaterials share common toxicity pathways with the zebrafish. For instance, Kteeba et al. demonstrated that administration of 50 parts per million (ppm, mg/L) ZnO nanoparticles to zebrafish embryos resulted in delayed hatching rate, mortality, and malformations at 72 hours post-fertilization (hpf) [16]. Similar developmental abnormalities were observed by Wang et al. while exposing zebrafish embryos to 0.1 ppm of nanoTiO_2 in which a significant reduction in cumulative number of viable embryos was noted [17]. The specific cellular mechanism of cytotoxicity also remained in harmony among certain nanomaterials during zebrafish exposure. Intracellular oxidative stress originating via reactive oxygen species (ROS) formation is one of the common toxicity mechanisms induced by nanomaterials [18,19].

Regardless of the model system, genotoxicity encompasses the foremost parameter to be analyzed in toxicological research, as it is the fundamental entity that defines every organism. Concerns have been mounting about the genotoxic response of nanomaterials, as they turned into an indispensable factor in day-to-day life. Owing to ease of research and reliability of results, the zebrafish has largely been accepted as a standard test system for nanomaterial-driven genotoxic response in different lineages of embryonic development [20]. The first article on zebrafish as a model system for genotoxicity was published in 2001 and investigated DNA damage that occurs in hepatocytes and gill cells upon exposure, with water samples from different sites in Germany [21]. From this point on, genotoxic research with zebrafish has developed at a rapid pace: about 71% of the articles were published during the period 2014–2020. Interestingly, comprehensive reviews of the genetic markers of nanoparticle toxicity in the zebrafish system appear to be scarce. This is a matter of concern and should be discussed in detail, since what is at stake here is the route source of toxicity, regardless of the tissue of origin. Bridging over this fact, the primary goal of this chapter is to address the genetic and epigenetic biomarkers and the underlying mechanisms of nanoparticle toxicity as identified via the zebrafish model system. This chapter offers a brief look at the zebrafish genome and at the possibilities of incorporating multi-omics and general analysis endpoints in the zebrafish for nanotoxicological research.

Zebrafish: A Popular Experimental Animal Model

Studies have demonstrated that the zebrafish can be considered one of the in vivo animal models for the toxicity assessment of potential agents such as environmental pollutants, drugs, chemicals, food adulterants, pesticides, and nanomaterials and their derivatives. The zebrafish embryo presents an excellent alternative for examining innumerable physiological responses of test agents. Thus the zebrafish represents an applicable, economic, and convenient model of a testing system designed to effectively reduce the number of procedures for mammalian species [22]. For instance, the developmental toxicity and altered genomic expression pattern of a potent environmental organic pollutant called perflurooctanesulfonate (PFOS) was analyzed using the zebrafish embryo [23]. Similarly, Lieschke et al. demonstrated that the early embryonic stage of the zebrafish exhibits the generation of circulating immature heterophils. They were also able to isolate a homologue of myeloperoxidase gene from embryonic zebrafish. Both of these features suggest that early embryonic granulopoiesis in zebrafish has numerous functional similarities with human granulocytes and macrophagic functionalities. Thus the embryonic stage in zebrafish can be employed as a genetically tractable system for toxicity studies [24].

The applicability of zebrafish in the context of nanomaterial toxicity studies has emerged as a latent field in toxicology research. Since most of the synthesis protocols for nanomaterials involve more than one hazardous chemical, these protocols have become inevitable in toxicity assessments required for the validation of the applicability of nanomaterials in different fields. The combination of the aforementioned advantages of the zebrafish system and the toxicity profiling of nanomaterials represents a meaningful attempt. Conventional routes for nanomaterial toxicity studies involve in vitro, in vivo, and *in silico* systems. In vitro systems use various cell lines for toxicity studies, whereas in vivo systems use animal models, for example models of zebrafish, medaka, rats, mice, rabbits, pigs, and non-human primates. *In silico* toxicity studies involve computational modelling and simulations. Overall, whole-animal model studies are still preferred, as they represent full physiological systems for toxicity assessment. While historically animal studies have primarily used mammals, the zebrafish proves to yield a reliable complementary vertebrate animal model [25]. Strengths include a high level of evolutionary conservation of multiple organ systems analogous with those of humans (e.g. heart, liver, kidney, and brain). In fact research supports the zebrafish model as a relevant animal model for toxicity research that should be used in the assessment of various categories of nanomaterials, spanning metal and metal oxide nanoparticles, carbon-based nanoparticles, nanocomposites, polymer-based nanoparticles, quantum dots, and functionalized nanomaterials (Table 20.1).

The Zebrafish Genome

Important Features of the Zebrafish Genome

The zebrafish genome is about half the size of the human genome (~1.7×109 base pairs (bp) of DNA). The zebrafish genome is also twice and respectively four times the size that of the medaka (the Japanese rice fish) genome and that of the Putterfish genome. Most of the teleosts display submetacentric chromosomes. Similarly, zebrafish have twenty-five such chromosomes, in which the centromere is placed nearly at the center, with differently sized chromosomal arms [49]. The zebrafish genome is haploid in nature, and stark conservation

Table 20.1 A summary of selected nanomaterial toxicity studies using the zebrafish.

	Type of nanomaterial	Zebrafish life stage & exposure	Study highlights	Reference
Metal and metal oxide nanoparticles	Cu^{2+}, Ag^{3+} and TiO_2	96 hpf & 4 h	Rheotaxis as an endpoint in NP toxicity & importance of behaviour study in metal NP toxicity	[26]
	Aluminium Oxide (Al_2O_3), Aluminium titanate ($Al_2O_3.TiO_2$), Chromium Oxide (Cr_2O_3), CopperOxide (CuO), Gold (Au), Iron Nickel Oxide ($NiFe_2O_4$)	2.5–3 hpf, 48 & 96 h	96 h exposure as a choice of teratogenic toxicity response by NPs & hatching time as one of the reliable indicators of toxicity in zebrafish	[27]
	Silver (Ag) NPs	120 d & 24 h	Oxidative stress, DNA damage & apoptosis are associated with induced hepatotoxicity	[28]
	Nano and bulk Zinc Oxide (ZnO) and TiO_2 NPs	120 d & 96 h	Metal ions released by NPs contributed partly to toxicity and NPs were shown to be more toxic than bulk	[29]
	Zinc Oxide (ZnO) NPs and PFOS	2 hpf & 120 d	Chronic co-exposure adversely affected growth, reproduction & transgenerational toxicity	[30]
	Gold (Au) NPs	6 to 120 hpf & 120d	NP exposure during development cause lasting impacts on CNS	[31]
	Copper NPs	48 hpf & 48 h	Morphological changes and global gene expression patterns in gills noted	[32]
	Nickel (Ni) NPs	24 hpf & 5d	NP configuration changes affected fish morphology	[33]
	Fullerenes [C60, C70 and $C60(OH)_{24}$]	24 hpf & 96 h	Changes in toxicity mechanism based on functional groups	[34]

(Continued)

Table 20.1 (Continued)

	Type of nanomaterial	Zebrafish life stage & exposure	Study highlights	Reference
Carbon based nanoparticles	Caron, Carbon nanotube (CNT) and Fullerene (C60)	2.5–3 hpf, 48 & 96 h	Cumulative mortality and teratogenicity induced by NPs	[27]
	Single walled carbon nanotubes (SWCNT), Double walled carbon nanotubes (DWCNT) and Carbon black	52 to 72 hpf & 4 h	Metal ions release caused significant alterations with larval hatching	[35]
	Graphene oxide (GO), Reduced GO and Multiwalled	2 hpf & 96 h	Induced toxicity affecting embryonic development and larval growth without any morphological changes	[36]
	Nano GO	24 hpf & 96 h	Induced morphology changes, oxidative stress and DNA methylation damage	[37]
	Buckminster fullerene (nC60) aggregates and Fullerol	1.5 hpf & 96 h	Larval development, survival rate, hatching rate and cellular functionality are affected by NP chemical composition	[38]
Polymer based nanoparticles	Chitosan NPs	120 hpf & 72 h	Dose dependent toxicity observed in terms of malformation, mortality and larval hatching rate	[39]
	Chitosan NPs	96 hpf & 4 h	Potential toxicity of NPs generated from biodegradable materials was demonstrated	[40]
	Polymeric nanocapsules	4–5 hpf & 72 h	Applicability zebrafish model for screening of nanocarriers	[41]
	Dendrimers (Polyamidoamine and Polypropylenimine)	120 d & 96 hpf	Charge and size dependent toxicity	[42]

(*Continued*)

Table 20.1 (Continued)

	Type of nanomaterial	Zebrafish life stage & exposure	Study highlights	Reference
	Up conversion nanoparticles (UCNPs)	7–8 months, 120 hpf	Malformations, delayed hatching rate, embryonic and larval development at higher levels of NPs	[43]
Luminescent nanoparticles	Nitrogen doped Carbon dots (N@CDs)	120 d & 0.5 h	Zebrafish as a potential model for analysing N@CDs as bioagents for biomedical	[44]
	Core-shell Silica nanoparticles (FSNP)	<6 hpf & 96 h	Chorion accumulation without significant toxic response	[45]
	Carbon Quantum dots (CQDs)	24 hpf & 4 h	QDs exhibited tissue dependent affinity for zebrafish imaging	[46]
	$CuInS_2$-ZnS nanocrystals	6 hpf & 2 h	Zebrafish based validation of bright luminescent nanocrystals for medical applications	[47]
	Silicon Quantum dots nanoparticles (SiQDNPs)	26 hpf & 4 h	Zebrafish as a model to evaluate photosensitizing property of SiQDNPs in photodynamic therapy	[48]

is observed between zebrafish and humans. This proximity to humans is far higher than the similarity of sequencing between zebrafish and mice or rats [50], and it signifies evolutionary lineage, indicating that the zebrafish genome emerged from a rodent rather than from primates. Higher karyotypic resemblance between the zebrafish and human genomes allows us to determine the approximate position of genes in humans. The zebrafish genome is sequenced and has proven to be highly relevant to researchers who investigate genetic disorders and other diseases. Zebrafish also have gene sequences homologous with those found in humans (there is ~75% similarity between the two sets). This similarity further increases to above 80% when we specifically focus on disease-associated genes. Many zebrafish genes are paralogs (i.e. genes or chromosomes that originated from a similar ancestor and have identical functionality). For example, humans have two HOX retinoic acid receptor genes: *RARA* and *RARG*. (HOX genes are the ones responsible for the axial positioning of an organism.) These two genes correspond to the syntenic genes *HOXB* and *HOXC*. Zebrafish possess *hoxbb*, which is syntenic with *rara2b* (instead of *RARA*). These genes have similar functionalities, sequence distributions, and positions in the chromosomes of the two species [51].

Zebrafish often exhibit two orthologs of gene sequences, which exist as a single gene in mammalian genomes. For instance, *shh* (sonic-class genes) and *twhh* (tiggywinkle) in zebrafish have proven similarity with the *SHH* gene in humans [52]. *SHH* is the gene responsible for Sonic hedgehog protein synthesis, which in turn is involved in embryonic development. This structure permits the identification of the precise role of gene duplication and of its relevance in terms of an evolutionary origin of phenotypes. In addition, gene duplication often allows for non-lethal genetic zebrafish knockout models for targets that, when knocked out in rodents, are lethal. A variety of advanced genome editing techniques have been applied in the zebrafish, including gene knockout experiments. For example, Ablain et al. developed a high-throughput vector system for in vivo screening strategies and gene editing procedures. This group developed a simple gene-targeting vector that was able to accomplish knockout zebrafish models by injecting guide RNA (gRNA) and Cas9 mRNA at the single-cell stage. In this study, the vector specifically targeted the silencing of the *urod* gene, which is involved in heme biosynthesis. The study aimed to expand the scope of using gene silencing research possibilities in zebrafish [53]. Similarly, a RNA-guided endonuclease (RGEN)-based study for gene knockout in zebrafish has also been reported [54]. Using the zebrafish embryo at the single-cell stage, this work highlighted the importance of RGEN-mediated in vivo mutagenesis and its relevance to the development of genetically engineered organisms. Also, the authors of the same study could identify the germline transmission of the RGEN-induced mutation through F1 progeny. Thus knowledge of the zebrafish genome helps enlarge the scope of genetic engineering and of other branches of molecular medicine in order to establish human disease models [54].

Zebrafish and "-omics" Approaches

The quantitative measurement, by various -omics strategies, of the changes that occur in a living system during a toxic encounter constitutes a useful tool of analysis in many fields of research. The term "toxicomics" encompasses a combination of toxicology with several -omics procedures. Toxicomic approaches are applied in order to assess the mechanism of action of toxicants at systems level. These approaches extend to genomics, proteomics, metabolomics, transcriptomics, and other fields (Figure 20.1) [55]. The evolutionary conservation of molecular structures translates into an immense potential for application in toxicology studies that use the zebrafish.

Proteins constitute a group of functional molecules that demonstrate the precise consequences of chemical and nanoparticle exposure for a biological system. Proteomics approaches are employed in toxicity studies because they detect alterations in the levels of protein profiles in

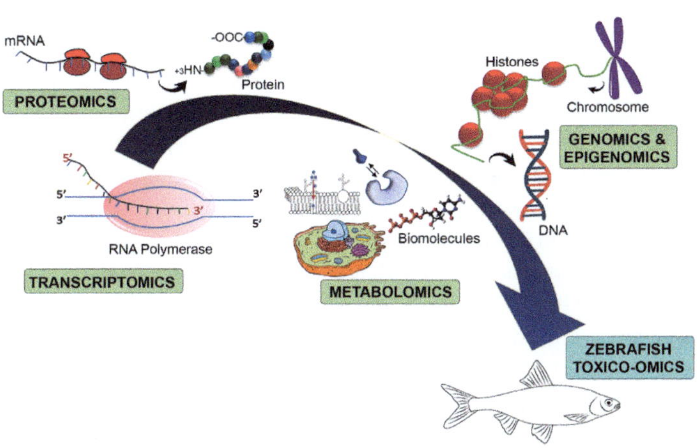

Figure 20.1 Different -omics approaches applied in zebrafish toxicology studies (zebrafish toxico-omics).

response to these exposure events. Toxicological studies in the area of proteomics using zebrafish are currently limited, and the limitation is mainly attributed to the abundance of protein (e.g. vitellogenin) accumulated in the yolk sac during the early developmental stages of the zebrafish embryo. This abundance diminishes the capacity to detect proteins at subsequent stages in early life, when they are in lower supply. Some researchers put forward a concept of manually removing the yolk sac during early embryonic stages in order to overcome this limitation; but such a process can cause stress-induced responses, which in turn compete with the proteome changes directly related to chemical and nanoparticle exposure [56]. Other researchers have waited until later in development (e.g. at least five or seven days after fertilization) to alleviate the competition of yolk sac proteins. Once the yok sac is absorbed, this concern is no longer present and proteome analysis, including that of specific organs, is much easier at later stages. A set of globally accepted zebrafish embryonic proteome identification protocols and a detailed protein expression profile for zebrafish are published [56,57]. Proteomic toxicity studies in the zebrafish continue to expand and there are recent evaluations of nanomaterials (Table 20.2).

Just like proteome identification, zebrafish metabolomics created huge possibilities by allowing the detection of changes in levels of metabolites during xenobiotic exposure or

Table 20.2 Hallmark features of nanoparticle toxicity identified via various -omics approaches in zebrafish.

Class of -omics technique	Identified toxicity features following nanoparticle exposure	Reference
Proteomics	– Dysregulated actin cytoskeleton development • Pericardial edema • Spinal curvature • Yolk sac edema • Dysregulated cell signaling pathways • Intracellular stress events	[64–67]
Transcriptomics	– Dysregulated gene clusters involved in • DNA and RNA processing • Hydrolase enzyme activity • Methyl transferase activity • Heat shock protein binding • Developmental and morphogenesis processes	[60,64,68,69]
Metabolomics	• Disturbed fatty acid and amino acid metabolism • Downregulation of TCA cycle • Perturbed energy regulation • Inhibition of ATP production	[65,70–72]
Genomics	• Altered locomotor behavior • Impact on vision and sensory perception • Perturbed nervous system development • DNA modification due to oxidative stress • Dysregulated genes associated with mitochondrial respiration, metal detoxification, and DNA repair	[31,73,74]
Epigenomics	• DNA hypermethylation • Histone (H3 & H4) acetylation • Condensed chromatin status • Transcription repression • mRNA instability • Posttranslational modifications of histone	[75–77]

during encounter with a xenobiotic. Several different techniques are useful in this regard: nuclear magnetic resonance (NMR), liquid chromatography coupled with mass spectroscopy (LC-MS), and high-resolution mass spectroscopy (HRMS) are just three examples. Studies have employed an open-field test paradigm for chemometric endpoints [58] and liquid–liquid extraction or solid-phase extraction methods [59].

Transcriptomics allows the identification, as transcripts, of gene expression patterns of mRNAs that correspond to a specific toxicity outcome of xenobiotic exposure. Transcriptomics permits the assessment of all genes in the genome in order to determine expression perturbations of the chemical and of the nanoparticle using array-based or sequencing approaches. This method ensures precise and reliable knowledge of toxicity mechanisms via databases such as the Kyoto Encyclopedia of Genes and Genome Pathways and Gene Ontology (the KEGG Pathway database) or Ingenuity Pathway Analysis, which integrate the altered genes so as to produce an understanding of functions, pathways, and disease outcomes [60]. Out of -omic approaches, transcriptomics is currently the one most applied in toxicity studies; and there are several reports on exposures to nanomaterials (Table 20.2).

Zebrafish are considered to offer a profuse model for genomic research, as it presents a classical system, which offers several significant advantages not usually available in traditional vertebrate models. Regarding genomics, studies at the level of DNA mutagenesis in zebrafish contribute profusely to the development of models for human diseases such as Parkinson's, Alzheimer's, cancers, and many others. A recent and commendable contribution to zebrafish genome research is antisense sequence-based target gene inhibition or "knockdown screening." This approach subsidizes higher levels for the functional identification of unknown vertebrate genome subsets in vivo. With the invention of chemical genetics in zebrafish, it became possible to locate drug targets and their biology, and hence to develop innovative pharmaceuticals in the future [61]. In the field of epigenetics, zebrafish are considered to yield an excellent research model, as it presents a system that is closely related to mammalian development. Mammalian embryogenesis exhibits close proximity in various respects with zebrafish epigenetic mechanisms; for instance DNA methylation, demethylation, histone modification, and non-coding RNAs. However, in contrast with what happens in mammals, in zebrafish DNA methylation in sperm is inherited during embryogenesis. Also, most of the histone methylations are not involved in zygotic genome maturation. The remaining histone methylations or other modified loci are observed to be retained in the zebrafish genome, which interacts with the promoter regions of the developmental genes and further determines the transcriptional activation of the respective genes [62]. The global histone methylation pattern and the sperm-inherited DNA methylation in the zebrafish genome give us an understanding of the mechanism through which inherited DNA methylations prime zygotic transcription events. More than 75% of the epigenetic genomic incidents during germ cell programming in zebrafish are comparable with those of mammals; and this includes certain key epigenetic mediator molecules. During zygote genome activation in zebrafish, a set of epigenetics events such as DNA methylation, histone modifications, and non-coding RNAs integrate for the fine-tuning of gene expression [63].

General Molecular Determinants and Genomic and Epigenomic Biomarkers of Nanoparticle Toxicity in Zebrafish

Owing to the rapid and passive diffusion of nanoparticles into the yolk sac and into the vascular compartment, zebrafish are considered to yield one of the foremost vertebrate models in point of toxicity assessment. In terms of toxicity mechanisms in zebrafish, nanoparticles differ according to chemical composition, which includes the type of functional group(s) and inherent physicochemical characteristics, as well as the route of synthesis (Figure 20.2). Moreover, the severity of toxicity and its mode vary according to the site of accumulation and the average size or the hydrodynamic size of the particles (or both) [33,45,78,79]. Nanoparticles elicit toxicity in early embryonic stages of larval development, primarily through the expulsion of free ions either into the microenvironment or into the embryonic chorion. Since these ions have a charge, they are prone to disrupting the electron transport chain (ETC) in cells; hence they produce oxidative stress. Alterations in the expression of genes associated with ETC during events of exposure to nanoparticles have been reported earlier in zebrafish. For example, Van Aerle et al. demonstrated via sequencing-based transcription profiling that silver nanoparticles (AgNPs) are capable of causing perturbations in oxidative phosphorylation in zebrafish embryos owing to the presence of $Ag+$ ions. Embryonic exposure at 24 hpf was followed by a selective hike in intracellular ROS levels. Among the thirteen genes the research group identified, cryptochrome 1a (*cry1a*) was found to be downregulated at 24 and 48 hpf. In addition, elevated protein turnover was observed through the upregulation of proteasome subunit beta type-1 (*psmb1*) and ribosomal protein S6 modification-like protein B (*zgc:92164*), which led to the compensation of the protein damage associated with ROS generation.[79]

In another study, Kim et al. explored the mechanism of toxicity induced by gold nanoparticles (AuNPs) in zebrafish larvae [80]. They found that AuNPs induced cellular apoptosis and DNA damage in the eye, which was apparent from the reduction in eye pigmentation levels (Figure 20.2). Both the TUNEL assay and flow cytometry revealed that exposure to AuNP resulted in a concentration-dependent increase in apoptotic cells ranging from 0.9% to 5.0% for 0.5 mg/L to 30 mg/L concentrations. Further study revealed that genes regulating apoptosis (*p53* and *bax*) were expressed in greater amounts, while significant downregulation of transcription factors for eye development (*pax6a, pax6b, otx2*, and *rx1*) was observed. Transcription factors for lens pigmentation, *sox10*, was also diminished. The study linked these abnormalities with motor neuron axonal deficits and subsequent behavioral transformations [85].

In an attempt to evaluate the influence of the surface functionality of nanoparticles on fish larval viability, Paatero et al. found that nanoparticles follow individual toxicity

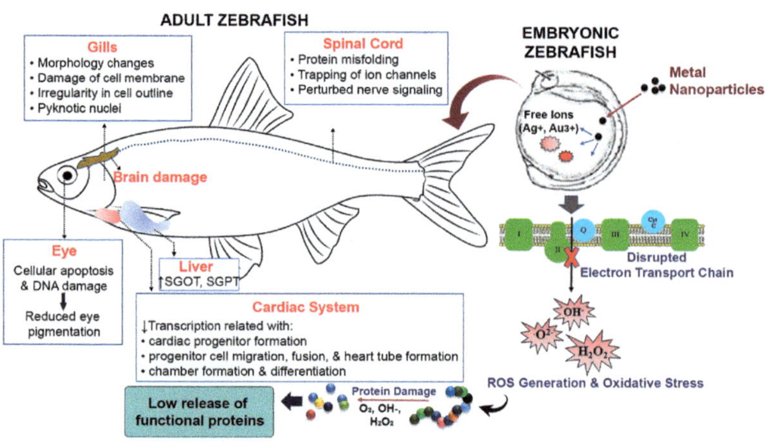

Figure 20.2 General outcomes of nanoparticle toxicity in multiple organ systems and at different life stages in zebrafish.

pathways, depending on the site of administration, the dosage used, and the stage of embryonic or larval development [81]. The researchers analyzed the toxicity of biomedically relevant mesoporous silica nanoparticles with different surface functional groups. Apart from the expected morphological and behavioral changes, the nanoparticles induced significant alterations in the blood circulation patterns under stereomicroscopic observation, thus confirming cardiovascular toxicity [81]. Reports suggest that the cardiac toxicity of nanoparticles can be due to the dysregulation of genes and transcription factors that control different stages of heart development (Figure 20.2). For example, genes involved in cardiac progenitor formation (*gata5, hand2, tbx20, smarcd3, rtf1*), genes involved in progenitor cell migration, fusion, and heart tube formation (*gata5, hand2*), and genes involved in the formation and differentiation of the heart chamber (*gata5, hand2, nkx2.5, ctr9, cdc73*) [82].

Because of the homologous nature of the zebrafish's immune system and associated parameters with that of humans, zebrafish are extensively used to test immunological anomalies caused by deleterious agents, including nanoparticles. Studies provide adequate evidence to attest to the presence of cells homologous to human monocytes, tissue macrophages, granulocytes, and lymphocytes [83]. Significant variations for inflammatory genes are reported in studies of nanoparticle toxicity in zebrafish. For instance, Truong et al. investigated gene expression profiles of zebrafish embryos after administration of AuNPs with different functional groups [84]. A large fraction of the dysregulated genes—*ELF4, RUNX3, c-FOS, STAT4, PAK2, PP2A cat-α*, and *ubiquitin*—were associated with the inflammatory pathway. All of these genes were observed to be nanoparticle functional group-specific. Other altered genes corresponded to protein folding, proliferation, and G-protein coupled receptor (GCR) protein signalling. Also, the study put forward the notion that AuNPs are able to disrupt the cell surface transport pathways selectively, by trapping the ion channels on the cell surface. This was possible at 24 to 48 hours post-fertilization, when embryos were routed through altered GCR protein expression, and hence it imbalanced the ionic potential across the cell membrane. This perturbation leads to interrupted physiological response and in turn altered metabolic rate in fish [84].

Nanoparticles generated via green synthesis protocols (e.g. those from plant extract) can have deleterious consequences too, as evidenced by some in vivo studies that use zebrafish. For example, Krishnaraj et al. showed that AgNPs synthesized from plant extract selectively accumulated in zebrafish gills and liver tissue [85]. In addition, serious cellular anomalies were observed, for example intrahepatic oxidative stress, immunotoxicity, and genotoxic responses. In the genotoxic perspective, AgNPs caused abnormalities in the nuclear compartment, including the presence of micronuclei in peripheral monocytic cells. The loss of cell contact caused by the occurrence of an irregular cell shape was also noted. Upon accumulation in gills, AgNPs produced a number of morphological changes such as damage to the cell membrane, irregularity in cell outline, pyknotic nuclei, and broader damage to gill cells (Figure 20.2). Krishnaraj et al.'s study claims that such morphological changes are possible during oxidative stress in cells via the generation of ROS, including hydroxyl and peroxyl radicals. Oxidative damage in the liver was confirmed through the identification of the elevated threshold for SGOT and SGPT enzymes (Figure 20.2). mRNA profiling of oxidative stress and inflammatory pathways also revealed that AgNPs caused serious incongruity in intracellular enzyme cascades [85]. This effect can be exemplified by the altered regulation of metal response element-binding Transcription Factor-1 (MTF-1) in response to cellular stress events. These events are further regulated by the selective expression of allothionein genes (MT genes). The expression of these MT genes was noticeably downregulated as a result of AgNP exposure, which underlies induced oxidative damage and subsequent hepatic toxicity. Also, these alterations led to dysregulated transcription of genes that encode for apoptosis, storing, and detoxification of metallic residues (e.g., Ag^+ ions). Most of the nanoparticle exposure event resulted in the expression of heat shock protein genes as an immediate cell response to the foreign agent [86].

Magnetic nanoparticles are indicated to have toxic impacts on zebrafish, without significant induction in the oxidative stress pathway; instead, magnetic nanoparticles have lethal consequences. They harm cell growth patterns and synthesis and the release of new functional proteins. Villacis et al. analyzed toxic insults caused by iron oxide nanoparticles (IONPs) in zebrafish; and they did so through a classic combination of genotoxicity, oxidative stress, and transcriptomic methodologies [87]. These researchers found that IONPs accumulate in significant amounts in gut and liver tissue, which suggests a preference for the digestive tract as the major route of exposure and toxicity. Mutagenic effects of IONPs were evident from selective DNA fragmentation and chromosomal damage induced at higher concentrations. In contrast to earlier findings, Villacis et al.'s study confirmed that higher levels of lipid peroxidation underlie cell growth obstruction instead of ROS release. Genomic analysis implied that metallic iron released from IONPs led to the elevation of metal binding genes, and that this ionic imbalance in cells resulted in a minimal release of newly translated proteins and abolished associated cellular functions. The presence of micronuclei also signified genotoxicity associated with IONPs in aquatic fauna [87]. Apart from all these toxic consequences, certain metal nanoparticles are observed to induce oocyte maturation in zebrafish by increasing the apoptotic death of ovarian follicle cells. In particular, AgNPs are reported to be capable of inducing germinal vesicle breakdown, which is a key step on the pathway to oocyte maturation. Genomic analysis indicated that AgNPs secured this effect through the selective inhibition of cyclic AMP (cAMP) levels in treated oocytes and the ensuing resumption of meiosis [88].

Epigenomic alterations are also being identified after nanoparticle exposure using the zebrafish (Figure 20.3). An example of a nanoparticle-induced alteration in an epigenetic event in zebrafish is the elevation in gene expression for the nucleotide excision repair (NER) mechanism upon the onset of an oxidative stress damage in DNA. The main genes involved in the NER pathway are *ercc5* and *baxa* [89]. Indications of homologous end joining (HEJ) and homologous recombination (HR) were also evident in zebrafish DNA samples during post-nanoparticle exposure events. This suggests possible double-stranded DNA breaks caused by nanoparticles in the genetic material of fish—a possibility that is particularly

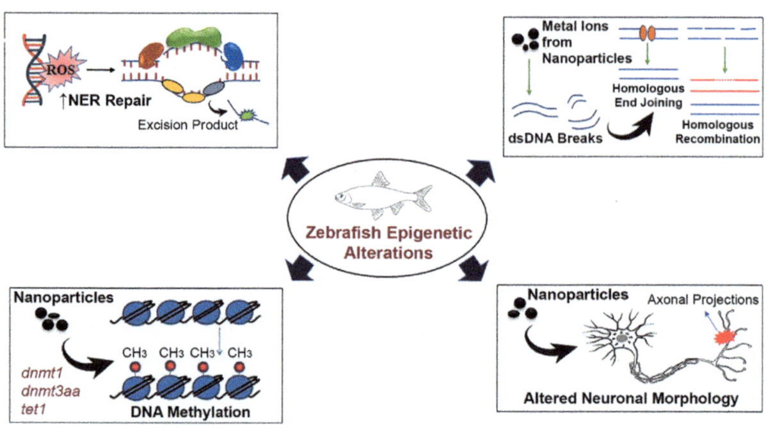

Figure 20.3 Major epigenetic alterations identified in the zebrafish induced by nanoparticle exposure and subsequent outcomes.

strong during interaction with dissolved metal ions released from nanoparticles. The applicability of microRNAs (miRNAs) to the prediction of nanoparticle-induced toxicity in zebrafish has been studied [90]. *In silico* analysis was performed using mRNA-miRNA regulatory networks as potential markers of nanoparticle toxicity. The networks developed were *dre-miR-124b* (a gene involved in cell cycle and apoptosis), *dre-miR-144* (a regulator of immune response), *dre-miR-155* (a regulator of inflammatory response), *dre-miR-19a* (a regulator of cancer development and inflammation), *dre-miR-148* (a gene related to the activation of innate immunity), and *dre-miR-223* (a gene for immune cell differentiation and tumor suppression). Each of the constructs triggered a gene expression pattern that indicated nanoparticle-induced toxic outcomes in the corresponding signalling pathways of zebrafish [90].

Post-exposure elevation in the cardiac DNA methylation rate was observed in the zebrafish after sixty days of exposure to carbon nanoparticles (CNPs). A significant remodelling of the zebrafish methylome was evident from the elevation in responsive genes in cardiac tissue (e.g., *dnmt1*, *dnmt3aa*, and *tet1*) (Figure 20.3). Also, CNPs executed the methylation modification in a concentration-dependent manner. On the contrary, CNPs induced a significant reduction in gene expression that regulates the demethylation of pro-inflammatory genes and of genes related to tissue homeostasis. This contradictory toxic outcome of CNPs was also visually evident from the targeted accumulation in the cardiac endothelium, not in macrophages or mononuclear phagocytes. The study provides useful information on ROS-induced DNA methylation and subsequent chronic toxic effects on the cardiac tissue of zebrafish [91]. The activation of metal-responsive DNA damage repair genes is also identified during long-term exposure of aquatic life to AgNPs. Silver ions (Ag^+) interact with the gill epithelium and cause a blockage of Na^+-K^+ ATPase pump, thereby inhibiting ion transportation across the plasma membrane. Also, these ions inhibit the carbonic anhydrase enzyme involved in the respiratory pathway [92]. ZnO NP-driven abnormal phenotypes were studied by Kteeba et al. in the zebrafish [93]. ZnO NPs at 0.5 ppm altered motor neuron axonal morphology, showing dorsal axonal projections (Figure 20.3). The exposure also resulted in disruption of vasculature development. All the phenotypic changes identified in the study were attributed to epigenetic changes associated with the tissue's DNA. Moreover, anomalies noted in generation F0 were transferred to progeny F1 [93].

Genome and Genetic Biomarkers of Nanoparticle Toxicity Associated with Different Systems in Zebrafish

The past decade has seen a sharp increase in translational research on sex dimorphism among vertebrate species. Generally vertebrates exhibit a wide difference between sexes in genotypic and phenotypic behavior, and this is an important endpoint to consider and scrutinize in study protocols [94]. Just like other vertebrates, zebrafish exhibit two sexes and present functional and anatomical dimorphism. Moreover, the organization of the genome of zebrafish displays in various ways close similarities with the organization of the mammalian genome. Beginning with the gonads, zebrafish possess Sertoli and Leydig cells in testes, whereas the ovaries contain thecal and granulosa cells. Steroid synthesis is regulated by hypothalamus–pituitary–interrenal (HPI) and hypothalamus–pituitary–gonadal axes in both sexes. As in mammals, the process of reproduction in zebrafish is regulated by the release of estradiol hormone into the ovary, which takes place at the onset of ovulation. This hormone in turn regulates vitellogenesis (yolk deposition), primarily by acting on the corresponding brain loci [95]. Furthermore, genetic sequencing and mutational studies of the zebrafish genome provide insight into a large number of homologous genes, which are conserved among other vertebrates, including humans, and allow us to identify them [96]. Since numerous mutational phenotypes of zebrafish resemble characteristics of human diseases, zebrafish are successfully used in research as in vivo disease models. Alongside the advancement in sophisticated analysis tools, it became easier to recognize and characterize mutant phenotypic models of zebrafish, including the use of transplantation, cell labelling, and microinjection [97]. Overall, zebrafish experiments are being completed for most human genetic diseases and provide a map of the activity of individual genes with higher accuracy than other in vivo models.

The zebrafish genome-sequencing project was introduced in 2001 at the Wellcome Trust Sanger Institute, Hinxton, United Kingdom. A rapid growth in zebrafish genetic research occurred, since the zebrafish was one of the vertebrates chosen for genome sequencing. Further, the development and application of several genomic mutation technologies propelled the zebrafish as one of the genetic animal models for human development and human disease. During the early 1990s, high efficiency chemical mutagenesis protocols were developed for the zebrafish genome. These techniques all comprised N-ethyl-N-Nitrosourea (ENU), an active mutant that could unleash thousands of mutated genes that coded for embryonic development [96]. This tool was then followed by the emergence of transgenics, which used retroviruses via pseudotyping [98]. The retroviral strategy opened wider possibilities for understanding developmental defects with the help of the zebrafish. Since the 1990s, several resources have been developed or refined for the zebrafish research community; one of them is the Zebrafish Information Network (ZFIN), which constitutes the basic zebrafish genetic information repository. Information housed by ZFIN includes gene sequences and their coded molecules, possible numbers of alleles, genome targeting agents such as CRISPR guide RNA sequences, transgenics, and other numerous tools. This major database gives zebrafish researchers access to a seamless source of genetic information on models of human diseases. Other foremost resources for the study of the zebrafish genome are the Genome Wiki, Genome Browsers, stock centers, the Zebrafish International Resource Center (ZIRC), the European Zebrafish Resource Center (EZRC), and the China Zebrafish Resource Center (CZRC) [99].

Overall, the zebrafish is considered to supply a complementary vertebrate animal model to mammals for environmental and human safety. In this regard, the zebrafish provides an excellent platform for nanomaterial toxicity studies. As mentioned earlier, nanoparticles can cause tissue system-based or organ system-based toxic insults, whose toxic mechanism can be seen at work in zebrafish in genetic deviations associated with nanoparticle invasion. Since most of the cellular metabolic reactions are interrelated, alterations induced by nanoparticles in the zebrafish genome can in fact pose a cascade of toxic events in various physiological systems [100]. Examples of neuronal and immune lineage markers and biomarkers of reproductive health and development follow in the next section.

Neuronal Lineage Genetic Biomarkers of Nanoparticle Toxicity

The brain is one of the pivotal organs as it drives bodily movements, complex behaviors, and regulation of endocrine functions. Studies suggest that nanoparticles can have selective toxic impacts on neuronal tissue by interfering with the genes associated with it (Figure 20.4).

According to their size, nanoparticles can enter the interstitial space of neuronal tissue and tend to reside in various loci. Their presence in the brain does not always imply that they are able to cross the blood–brain barrier unambiguously. Instead, it is possible for nanoparticles to distribute themselves throughout the body during early embryonic stages within 3–6 hpf. Nevertheless, a neuronal deposition of nanoparticles can initiate a torrent of toxic reactions mediated through oxidative stress, DNA damage, behavioral changes, and developmental defects [101]. The most significant impacts of nanoparticles on the brain are alterations in neurotransmission and metabolic and physiological homeostasis. The invasion of nanoparticles into the brain's tissue can be discernibly detected through the upregulation and the downregulation of certain key genes. For example, Dedeh et al. demonstrated the impact of contaminated sediment containing 14 nm-Gold NPs (AuNPs) in zebrafish after twenty days of exposure [102]. AuNPs triggered a modification in gene expression related to acetylcholine esterase activity (AchE), oxidative stress, mitochondrial metabolism, DNA repair, and detoxification. Relative gene expression variation was analyzed through random amplified polymorphic DNA (RAPD)-based methodology. There occurred a marked deviation of expression titres for genes such as *sod1*, *sod2*, *cox1*, *mt2*, *hsp70*, *gad*, *rad51*, and *ache*, corresponding to oxidative stress, mitochondrial metabolism, detoxification, DNA repair, and neurotransmission (Figure 20.4). All these genes showed remarkable upregulation in response to nanoparticle invasion. Quantitative RAPD-PCR measurements recorded AuNP-induced genotoxicity, as observed from the number of OPB7 hybridization sites in the genomic DNA of the fish [102]. In fact functionalized AuNPs were previously reported to cause conformational changes to DNA, which is attributable to the conformational homogeneity of nanoparticles with histone octamer in DNA motifs [103].

A contradictory report suggests an inhibitory effect of other metal particles—such as copper, iron, lead, and cadmium—against *AchE* activity [104]. According to this report, the inhibitory effect was contributed by a specific binding of these particles to a thiol group near the enzyme active site and may in turn have contributed to the termination of the hydrolysis of acetylcholine into choline and acetic acid in post-synaptic neuromuscular junctions, a process that indicates blocked neurotransmission and altered locomotor, behavioral, or other responses [104]. Other metal and metal oxide nanoparticles are also found to induce a series of genetic damages that affect neuronal functions. For example, the subchronic exposure of zebrafish to titanium dioxide (TiO_2) nanoparticles induced a

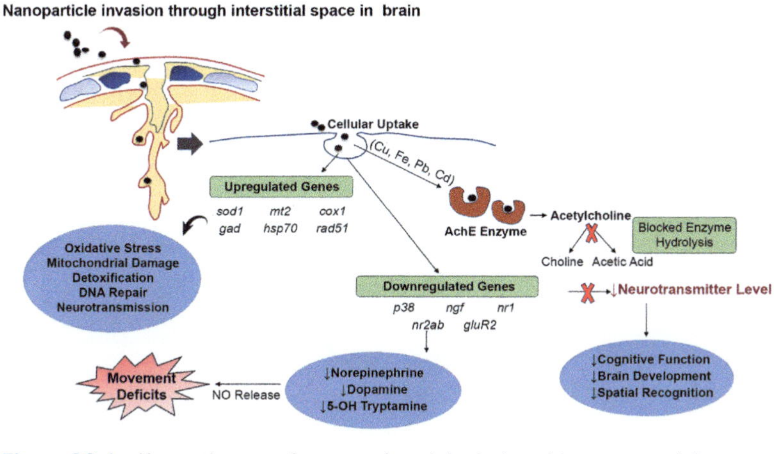

Figure 20.4 Key pathways of neuronal toxicity induced by nanoparticle exposure in studies using the zebrafish.

significant reduction in transcriptional activation of spatial recognition memory and in the levels of neurotransmitters like norepinephrine, dopamine, and 5-hydroxy tryptamine, along with a simultaneous increase in nitric oxide (NO) levels. Also, TiO_2 NPs at higher concentrations have induced expression of learning and memory behavior genes such as *c-fos*, *c-jun*, and *bdnf* and have downregulated *p38*, *ngf*, *creb*, *nr1*, *nr2ab*, and *gluR2* (Figure 20.4). Cognitive functions and brain development were adversely affected by the selective inhibition of neurotransmitter production and release [105]. Nanoparticles often diminish retinal photoreceptor activity, and this is followed by impaired color preference in zebrafish. Conceivably this is the result of targeted damage in highly organized neurons and primary sensory cells in the retina of zebrafish. Also, the reduced locomotor activity under the influence of nanoparticles is possible to explain as a result of targeted loss of dopaminergic neurons in the substantia nigra of amniotes, which may lead to difficulties in the execution of movement that are similar to what we witness in Parkinsonism [79].

Immune Lineage Genetic Biomarkers Associated with Nanoparticle Toxicity

The immunotoxicity of nanoparticles, as manifested in zebrafish, has been a topic of interest to researchers for many years. Several studies suggest that nanoparticles are capable of inducing cytokine production and release by promoting the transcriptional activation of oxidative stress-related genes (Figure 20.5). Certain nanoparticles accumulate selectively in fish lymph nodes, where they are processed by dendritic cells and modify their own antigenicity, triggering an immune response. Depending on their physico-chemical properties, nanoparticles vary widely in terms of an immunotoxicity pathway. For example, nanoparticles such as nickel, cobalt, and palladium stimulate Th1 cytokine release, whereas others, for instance titanium, single-wall carbon nanotubes (SWCNTs), and multi-wall carbon nanotubes (MWCNTs), induce Th2 cytokines. Some nanoparticles can also induce an allergic sensitization in zebrafish indirectly, namely a sensitization mediated through acting as an adjuvant. This response directs a specific pattern of cytokines and immune cell production for an allergic response. Primary pro-inflammatory mediators activated by nanoparticles are Interleukin-1 Beta (*IL-1β*), MIP-1 Alpha (*MIP-1α*), *MCP-1*, and *MIP-2*, along with stress-activated MAPKs *p38* and *JNKs* (Figure 20.5) [106,107]. Brun et al. reported that zinc oxide

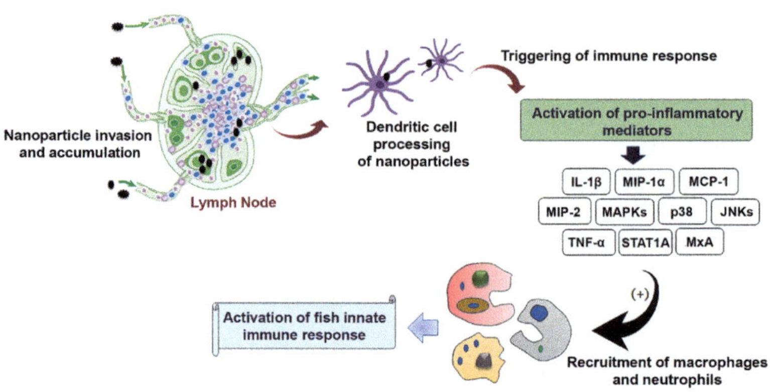

Figure 20.5 General routes of immunotoxicity following nanoparticle exposure as identified in zebrafish studies.

(ZnO) nanoparticles caused transcriptional alterations for pro-inflammatory cytokines, *IL-1β* and tumor necrosis factor alpha or *TNF-α* [108]. Additionally, genes such as *c-jun* proto-oncogene, which is a member of the c-jun N-terminal protein kinase (JNK) signalling pathway, *Stat1a*, and *MxA* were also significantly upregulated (Figure 20.5). These findings suggest a significant immune modulation of ZnO nanoparticles in zebrafish [108].

In another study, the immunomodulatory effects of graphene oxide (GO) nanoparticles in zebrafish were evaluated after a fourteen-day exposure. Evidence of immunotoxicity was noted; as part of it, a significant elevation in the transcriptional activation of pro-inflammatory genes was observed for *TNF-α*, *IL-1β*, and *IL-6* (Figure 20.5). The expression titre was seen to be markedly deviating from the control spleen sample [109]. Furthermore, in order to understand the genetic mechanism of immunotoxicity induced by copper (Cu) nanoparticles, Aksakal et al. conducted a toxicity analysis in zebrafish [110]. Gene expression evaluation indicated that there was a momentous increase in the expression of *IL-1β*, which is a key mediator in the innate immune system as well as a regulator in the aggregation of phagocytes at the site of infection. Also, *IL-1β* helps recruit activated macrophages and neutrophils to the site. Instantaneously, Cu nanoparticles downregulated the expression of some other immune-related genes such as *TLR-4*, *TLR-22*, and *CEBP*, which can be attributed to the nanoparticle-induced stress and subsequent suppression of the innate immune system in zebrafish [110]. Variation of mRNA levels for the immune-related genes was also reported to be induced by GO nanoparticles in zebrafish embryos, where the upregulation of *IL-6*, *IL-8*, *TNF-α*, and *Interferon-gamma (IFN-γ)* was observed [111]. Although nanoparticles are seen to be a causative agent in immune system stress and oxidative stress in zebrafish, the wide-ranging expression pattern and underlying signalling pathways are largely unknown. Generally speaking, most of the identified genes associated with nanoparticle-induced immunotoxicity belong to pattern recognition receptors (PRRs), Toll-like receptor (TLR), adapter molecules and the TLR signalling pathway, cytokines, the MAPK-signalling pathway, and the mTOR signalling pathway.

Genetic Biomarkers of Nanoparticle Toxicity Associated with Reproductive Health

Partial or whole life cycle tests can be carried out to analyze the reproductive toxicity of nanoparticles in the zebrafish. Researchers commonly rely on toxicity indicators such as vitellogenin (the precursor of egg yolk), the histological examination of gonads, the gonadal index, sperm plasma membrane integrity, embryo survival, and ovary histology (Figure 20.6). Wang et al. conducted a chronic toxicity study of TiO_2 nanoparticles in zebrafish through ninety-one days of exposure. Global gene expression in ovarian tissue by microarray analysis confirmed that the maturation and functionality of the zebrafish ovary were significantly affected. qPCR analysis confirmed that key genes regulating oocyte maturation (e.g. *cyp11a*, *tgfb1*, and *egf*) were remarkably reduced (Figure 20.6). These findings further indicated a drop in the maturation of stage 1 follicles. Most of the upregulated genes were responsible for the degradation of proteins, which all play crucial roles in ovarian growth and functionality [17]. The loss of fecundity in adult zebrafish as a result of AgNPs is reported by Ma et al. [18]. These researchers claimed that AgNPs induced apoptotic loss of ovarian and testicular cells via a mitochondrial-dependent pathway. A transcriptional deviation of genes (e.g. *bax*, *bcl-2*, *caspase-3*, and *caspase-9*) occurred, and it caused oxidative stress and apoptotic loss of germ cells (Figure 20.6). The abnormal expression patterns of *caspase-9* and *caspase-3* were said to be responsible for

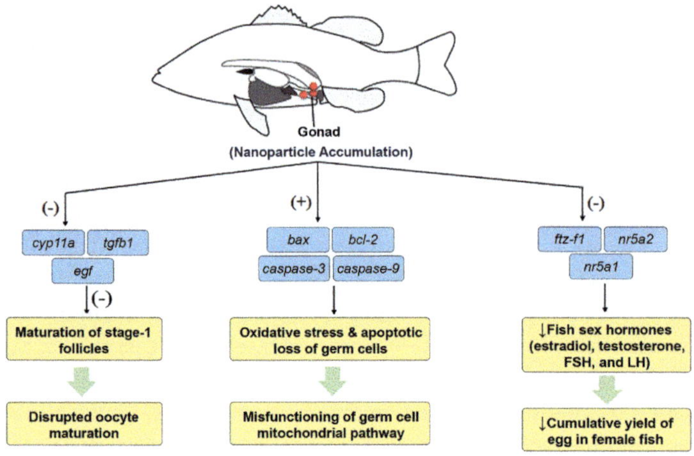

Figure 20.6 Reproductive toxicity outcomes identified in studies using the zebrafish following.

the malfunctioning of the mitochondrial pathway. The adverse outcome pathway identified in this study supports the idea of reproductive system consequences of AgNPs [18].

Apart from causing transcriptional alterations in the genes involved in germ cell lineage development, certain nanoparticles can act as mediators of genetic abnormalities related to reproductive enzymes. In particular, TiO_2 NPs, together with a potent environmental contaminant, bisphenol A (BPA), caused a noteworthy level of reproductive toxicity in zebrafish. The co-exposure showed selective accumulation of nanoparticles in the brain, liver, and gonad. The plasma concentration of steroidogenic fish sex hormones (including estradiol, testosterone, follicle-stimulating hormone, and luteinizing hormone) was significantly reduced through the transcriptional inactivation of the responsive genes (Figure 20.6). Although the co-exposure did not elicit any morphological alterations in gonad tissue, the cumulative yield of egg production in female zebrafish was reduced. The same study put forward the concept that TiO_2 NPs can be a carrier of BPA in aquatic bodies, heightening its bioconcentration in these bodies and subsequently causing an impairment of reproductive health via transcriptional alteration of endocrine functioning [112].

Genetic Biomarkers of Nanoparticle Toxicity Associated with Embryonic Development

The central markers of developmental toxicity in zebrafish are hatching rate, mortality, and teratogenicity. The in vitro fertilization and transparency of embryos make it easier to detect teratogenicity hallmarks such as cell movement, heartbeat, blood circulation, and neurovascular development. Wang et al. studied the developmental and behavioral toxicity of oxine copper (OxCu) nanoparticles in embryonic–larval stages of zebrafish for eighteen to twenty-four hours of exposure [113]. Gene expression studies indicated that the exposure caused a significant reduction in gene expression levels for neonatal neurons in embryos. Notably *hoxb1a*, *krox-neurolin*, *pax2*, and *pax6* were upregulated, indicating impaired embryonic brain development. Upregulation in the astrocyte marker gene *gfap* was also observed, implying damage in proliferating neuronal progenitor cells. Increased apoptotic cell death in the embryo leading to vascular development obstruction was also noticed, along with larval behavior [113].

Metal nanoparticles are marked by an ability to induce embryonic developmental toxicity in zebrafish. Most of the available literature appends the information that metal nanoparticles cause pericardial edema, spinal cord curvature, yolk sac edema, twisted tail endings, irregular eye shape with reduced pigmentation, notochord deformity, abnormal blood circulation, and retarded embryonic growth and development. For example, Cu(OH)2 nanoparticles caused altered transcriptional regulation of the genes that code for yolk sac development and hatching at 96 hpf [114]. Another case study was carried out by Xu et al. using CuO nanoparticles in zebrafish embryos. After thirteen hours of exposure, these nanoparticles caused a reduced expression of genes related to the development of the anteroposterior axis [115]. The genes that caused these insults included *ntl* (no tail), *dlx3* (distal-less homeobox 3), and *hgg* (hedgehog), which are responsible for normalizing cell movement during gastrulation. CuO nanoparticles reduced the expression of *ntl* genes and led to a poor development of posterior structures and tailbud formation in the embryo. Perturbed expression of *hgg* caused an underdeveloped prechordal mesoderm, which is the most likely origin of the rostral cranial mesoderm. CuO nanoparticle exposure in the zebrafish embryo also resulted in the interruption of neural plate enlargement and convergence [115]. Additionally, certain luminescent nanoparticles are reported to result in embryonic toxicity. For instance, different surface-modified CdSe/ZnS quantum dots caused embryonic pericardial edema through the selective disruption of gene expression patterns related to cardiac tissue (*tbx5*, *gata4*, *myh6*, *tnnt2*) development. An intracellular accumulation of Cd^{2+} ions was attributed to these genetic changes and was partially removed by the addition of the Cd^{2+} chelator, diethylene triamine pentaacetic acid [116].

Future Directions of Zebrafish Research for Informing Genomic and Epigenomic Biomarkers of Nanoparticle Toxicity through Novel Genetic Models and Integrated -omic Approaches to Clinical and Therapeutic Action

The development of groundbreaking genome-editing techniques, such as clustered regularly interspaced short palindromic repeats (CRISPR) and CRISPR-associated (Cas) systems, permits advancements in the identification of toxicity mechanisms and therapeutic molecular targets. These systems use guide RNA (gRNA) to target specific short DNA sequences of interest and to make them functionally inactive via the selective activation of the enzyme Cas9. This technique proves to be an exceptional tool for zebrafish research that targets DNA sequences associated with transcription, translation, and epigenomic regulation. Overall, the generation of knockout and knock-in zebrafish models through precise point mutations using short oligonucleotide sequences represents a set of promising techniques for the development of zebrafish models for the understanding of nanoparticle toxicity mechanisms and therapeutic actions [117].

Another important technique for disease model screening using the zebrafish is the xenotransplantation method. Here cancer cells derived from patient samples (e.g. from bone marrow biopsy samples) can be xenografted to transgenic zebrafish models (e.g. to those that have fluorescent immune cells or vasculature). The mechanism of action of anti-cancer drugs can thus be detected easily and in a more convenient way through these zebrafish models than through regular in vitro ones [118]. Such a system provides an excellent platform for cancer biologists to apply longitudinal analysis to tumor growth, metastasis, and angiogenesis using a transparent model like zebrafish in real time. Since adult zebrafish serve as a unique system for the evaluation of stem cell interactions with the in vivo microenvironment, analysis procedures that combine genetic tools can generate greater outputs in the context of human biology. Studies have successfully demonstrated, via genetic approaches, the usefulness of zebrafish as a potent hematopoietic model. Many of the phenotypic zebrafish mutants resemble single or multiple stages of hematopoiesis such as iron metabolism, heme synthesis, and globin protein expression [119]. Similarly, these zebrafish models can be applied in nanoparticle toxicity research for mechanistic identification and therapeutics.

In addition to these novel genetic models, advances in gene sequencing, transcriptomics, bioinformatics, and similar fields have contributed greatly to the flourishing of genetic and epigenetic toxicity research. Recent trends in biomedical and analytical research create far-reaching possibilities for regulatory toxicology and for the development of clinical and therapeutic actions. The incorporation of multiomics or integrative -omics can provide a comprehensive assessment of the molecular determinants of induced toxicity in experimental models [120]. Although genomic and epigenomic tools have found wide applicability in research on various stages of cancer, diabetes, obesity, neuronal degeneration, and other human diseases, their role in environmental safety as well as in toxicology is quite recent. These tools promise an accurate targeting of disease-causing gene sequences and in this way nurture the development of specific therapeutic agents related to nanoparticle toxicity.

Conclusions

Traditional toxicity assessment requires intensive testing resources, which include large numbers of mammalian models; and these increase the economic burden of research, along with other associated concerns. The establishment of the zebrafish as a complementary whole-animal model for toxicity testing and for the study of human disease has lessened this burden somewhat and provides alternative approaches and disease models that further our knowledge of health risks and therapeutics. While all in vitro and in vivo models of human disease have limitations, recognizing the strengths of each one and expanding our toolbox gives us greater confidence in the study of human health. The zebrafish is a new item in this toolbox: an additional whole-animal model on which we can study the toxicity mechanisms of nanoparticles. In fact, as already discussed, zebrafish are already being widely used in studies that attempt to find genetic and epigenetic biomarkers of nanoparticle toxicity. Future research, taking advantage of the most recent advances in zebrafish genetic models and integrated -omics approaches, will refine our knowledge of nanoparticle toxicity mechanisms, to the benefit of regulatory toxicology and therapeutics.

Funding

Funding support was provided by the Science and Engineering Research Board—Purdue University Overseas Visiting Doctoral Fellowship (SERB-OVDF).

References

1 Love, S.A., Maurer-Jones, M.A., Thompson, J.W., Lin, Y.S., and Haynes, C.L. (2012). Assessing nanoparticle toxicity. *Annu. Rev. Anal. Chem.* 5: 181–205.

2 Grassian, V.H., Haes, A.J., Mudunkotuwa, I.A., Demokritou, P., Kane, A.B., Murphy, C.J., Hutchison, J.E., Isaacs, J.A., Jun, Y.S., Karn, B. and Khondaker, S.I. (2016). NanoEHS–defining fundamental science needs: No easy feat when the simple itself is complex. *Environmental Science: Nano* 3(1): 15–27.

3 Marquis, B.J., Love, S.A., Braun, K.L., and Haynes, C.L. (2009). Analytical methods to assess nanoparticle toxicity. *Analyst* 134 (3): 425–439.

4 Yang, Y., Qin, Z., Zeng, W., Yang, T., Cao, Y., Mei, C., and Kuang, Y. (2017). Toxicity assessment of nanoparticles in various systems and organs. *Nanotechnol. Rev.* 6 (3): 279–289.

5 Kim, W.Y., Kim, J., Park, J.D., Ryu, H.Y., and Yu, I.J. (2009). Histological study of gender differences in accumulation of silver nanoparticles in kidneys of Fischer 344 rats. *J. Toxicol. Environ. Health Part A* 72 (21–22): 1279–1284.

6 Lee, J.H., Kim, Y.S., Song, K.S., Ryu, H.R., Sung, J.H., Park, J.D., Park, H.M., Song, N.W., Shin, B.S., Marshak, D., and Ahn, K. (2013). Bio persistence of silver nanoparticles in tissues from Sprague–Dawley rats. *Part. Fibre Toxicol.* 10 (1): 1–14.

7 Hamidi, M., Azadi, A., Rafiei, P., and Ashrafi, H. (2013). A pharmacokinetic overview of nanotechnology-based drug delivery systems: An ADME-oriented approach. *Cri. Rev.™ Ther. Drug Carrier Sys.* 30 (5).

8 Kim, J.S., Yoon, T.J., Yu, K.N., Kim, B.G., Park, S.J., Kim, H.W., Lee, K.H., Park, S.B., Lee, J.K., and Cho, M.H. (2006). Toxicity and tissue distribution of magnetic nanoparticles in mice. *Toxicol. Sci.* 89 (1): 338–347.

9 Chen, J., Dong, X., Zhao, J., and Tang, G. (2009). In vivo acute toxicity of titanium dioxide nanoparticles to mice after intraperitoneal injection. *J Appl Toxicol* 29 (4): 330–337.

10 Cho, Y.M., Mizuta, Y., Akagi, J.I., Toyoda, T., Sone, M., and Ogawa, K. (2018). Size-dependent acute toxicity of silver nanoparticles in mice. *J. Toxicol. Pathol.* 31 (1): 73–80.

11 Khan, F.R. and Alhewairini, S. (2019). Zebrafish (*Danio rerio*) as a model organism. In *Current Trends in Cancer Management*, ed. L. Streba, D. Gheonea, and M. Schenker, 3–18. London: IntechOpen. https://www.intechopen.com/books/7054.

12 Santoriello, C. and Zon, L.I. (2012). Hooked! Modeling human disease in zebrafish. *J. Clin. Invest.* 122 (7): 2337–2343.

13 Haque, E. and Ward, A.C. (2018). Zebrafish as a model to evaluate nanoparticle toxicity. *Nanomaterials* 8 (7): 561.

14 Wang, R.,.N., Meredith, A., Lee, M., Jr, Deutsch, D., Miadzvedskaya, L., Braun, E., Pantano, P., Harper, S., and Draper, R. (2016). Toxicity assessment and bioaccumulation in zebrafish embryos exposed to carbon nanotubes suspended in Pluronic® F-108. *Nanotoxicol.* 10 (6): 689–698.

15 Ramachandran, R., Krishnaraj, C., Sivakumar, A.S., Prasannakumar, P., Kumar, V.A., Shim, K.S., Song, C.G., and Yun, S.I. (2017). Anticancer activity of biologically synthesized silver and gold nanoparticles on mouse myoblast cancer cells and their toxicity against embryonic zebrafish. *Mat. Sci. Eng.: C* 73: 674–683.

16 Kteeba, S.M., El-Adawi, H.I., El-Rayis, O.A., El-Ghobashy, A.E., Schuld, J.L., Svoboda, K.R., and Guo, L. (2017). Zinc oxide nanoparticle toxicity in embryonic zebrafish: Mitigation with different natural organic matter. *Environ. Pollut.* 230: 1125–1140.

17 Wang, J., Zhu, X., Zhang, X., Zhao, Z., Liu, H., George, R., Wilson-Rawls, J., Chang, Y., and Chen, Y. (2011). Disruption of zebrafish (Danio rerio) reproduction upon chronic exposure to TiO_2 nanoparticles. *Chemosphere* 83 (4): 461–467.

18 Ma, Y.B., Lu, C.J., Junaid, M., Jia, P.P., Yang, L., Zhang, J.H., and Pei, D.S. (2018). Potential adverse outcome pathway (AOP) of silver nanoparticles mediated reproductive toxicity in zebrafish. *Chemosphere* 207: 320–328.

19 Palaniappan, P.R. and Pramod, K.S. (2011). The effect of titanium dioxide on the biochemical constituents of the brain of zebrafish (Danio rerio): An FT-IR study. *Spectrochim. Acta A Mol. Biomol. Spectrosc.* 79 (1): 206–212.

20 Canedo, A. and Rocha, T.L. (2020). Zebrafish (Danio rerio) using as model for genotoxicity and DNA repair assessments: Historical review, current status and trends. *Science of the Total Environment* 762: 144084.

21 Schnurstein, A. and Braunbeck, T. (2001). Tail moment versus tail length—application of an in vitro version of the comet assay in biomonitoring for genotoxicity in native surface waters using primary hepatocytes and gill cells from zebrafish (Danio rerio). *Ecotoxicol. Environ. Saf.* 49 (2): 187–196.

22 Jeevanandam, J., San Chan, Y., and Danquah, M.K. (2019). Zebrafish as a model organism to study nanomaterial toxicity. *Emerg. Sci. J.* 3 (3): 195–208.

23 Shi, X., Du, Y., Lam, P.K., Wu, R.S., and Zhou, B. (2008). Developmental toxicity and alteration of gene expression in zebrafish embryos exposed to PFOS. *Toxicol. Appl. Pharmacol.* 230 (1): 23–32.

24 Lieschke, G.J., Oates, A.C., Crowhurst, M.O., Ward, A.C., and Layton, J.E. (2001). Morphologic and functional characterization of granulocytes and macrophages in embryonic and adult zebrafish. *Blood* 98 (10): 3087–3096.

25 Cheng, D., Shami, G.J., Morsch, M., Chung, R.S., and Braet, F. (2016). Ultrastructural mapping of the zebrafish gastrointestinal system as a basis for experimental drug studies. *Biomed. Res. Int.* 2016: https://doi.org/10.1155/2016/8758460.

26 McNeil, P.L., Boyle, D., Henry, T.B., Handy, R.D., and Sloman, K.A. (2014). Effects of metal nanoparticles on the lateral line system and behaviour in early life stages of zebrafish (Danio rerio). *Aquat. Toxicol.* 152: 318–323.

27 Kovrižnych, J.A., Sotníková, R., Zeljenková, D., Rollerová, E., Szabova, E., and Wimmerová, S. (2013). Acute toxicity of 31 different nanoparticles to zebrafish (Danio rerio) tested in adulthood and in early life stages–comparative study. *Interdiscip Toxicol* 6 (2): 67.

28 Choi, J.E., Kim, S., Ahn, J.H., Youn, P., Kang, J.S., Park, K., Yi, J., and Ryu, D.Y. (2010). Induction of oxidative stress and apoptosis by silver nanoparticles in the liver of adult zebrafish. *Aquat. Toxicol.* 100 (2): 151–159.

29 Xiong, D., Fang, T., Yu, L., Sima, X., and Zhu, W. (2011). Effects of nano-scale TiO_2, ZnO and their bulk counterparts on zebrafish: Acute toxicity, oxidative stress and oxidative damage. *Sci. Total Environ.* 409 (8): 1444–1452.

30 Du, J., Tang, J., Xu, S., Ge, J., Dong, Y., Li, H., and Jin, M. (2018). Parental transfer of perfluorooctane sulfonate and ZnO nanoparticles chronic co-exposure and inhibition of growth in F1 offspring. *Regul. Toxicol. Pharmacol.* 98: 41–49.

31 Truong, L., Saili, K.S., Miller, J.M., Hutchison, J.E., and Tanguay, R.L. (2012). Persistent adult zebrafish behavioral deficits results from acute embryonic exposure to gold nanoparticles. *Comp. Biochem. Physiol. C Toxicol. Pharmacol.* 155 (2): 269–274.

32 Griffitt, R.J., Weil, R., Hyndman, K.A., Denslow, N.D., Powers, K., Taylor, D., and Barber, D.S. (2007). Exposure to copper nanoparticles causes gill injury and acute lethality in zebrafish (Danio rerio). *Environ. Sci. Technol.* 41 (23): 8178–8186.

33 Ispas, C., Andreescu, D., Patel, A., Goia, D.V., Andreescu, S., and Wallace, K.N. (2009). Toxicity and developmental defects of different sizes and shape nickel nanoparticles in zebrafish. *Environ. Sci. Technol.* 43 (16): 6349–6356.

34 Usenko, C.Y., Harper, S.L., and Tanguay, R.L. (2007). In vivo evaluation of carbon fullerene toxicity using embryonic zebrafish. *Carbon* 45 (9): 1891–1898.

35 Cheng, J., Flahaut, E., and Cheng, S.H. (2007). Effect of carbon nanotubes on developing zebrafish (Danio rerio) embryos. *Environ. Toxicol. Chem. Int. J.* 26 (4): 708–716.

36 Liu, X. T., Mu, X. Y., Wu, X. L., Meng, L. X., Guan, W. B., Ma, Y. Q., Sun, H., Wang, C. J., and Li, X. F. (2014). Toxicity of multi-walled carbon nanotubes, graphene oxide, and reduced graphene oxide to zebrafish embryos. *Biomed. Environ. Sci.* 27 (9): 676–683.

37 Chen, Y., Hu, X., Sun, J., and Zhou, Q. (2016). Specific nanotoxicity of graphene oxide during zebrafish embryogenesis. *Nanotoxicol.* 10 (1): 42–52.

38 Zhu, X., Zhu, L., Li, Y., Duan, Z., Chen, W., and Alvarez, P.J. (2007). Developmental toxicity in zebrafish (Danio rerio) embryos after exposure to manufactured nanomaterials: Buckminsterfullerene aggregates (nC60) and fullerol. *Environ. Toxicol. Chem. Int. J.* 26 (5): 976–979.

39 Wang, Y., Zhou, J., Liu, L., Huang, C., Zhou, D., and Fu, L. (2016). Characterization and toxicology evaluation of chitosan nanoparticles on the embryonic development of zebrafish, Danio rerio. *Carbohydr. Polym.* 141: 204–210.

40 Hu, Y.L., Qi, W., Han, F., Shao, J.Z., and Gao, J.Q. (2011). Toxicity evaluation of biodegradable chitosan nanoparticles using a zebrafish embryo model. *Int. J. Nanomedicine* 6: 3351.

41 Teijeiro-Valiño, C., Yebra-Pimentel, E., Guerra-Varela, J., Csaba, N., Alonso, M.J., and Sánchez, L. (2017). Assessment of the permeability and toxicity of polymeric nanocapsules using the zebrafish model. *Nanomedicine* 12 (17): 2069–2082.

42 Bodewein, L., Schmelter, F., Di Fiore, S., Hollert, H., Fischer, R., and Fenske, M. (2016). Differences in toxicity of anionic and cationic PAMAM and PPI dendrimers in zebrafish embryos and cancer cell lines. *Toxicol. Appl. Pharmacol.* 305: 83–92.

43 Wang, K., Ma, J., He, M., Gao, G., Xu, H., Sang, J., Wang, Y., Zhao, B., and Cui, D. (2013). Toxicity assessments of near-infrared upconversion luminescent LaF3: Yb, Er in early development of zebrafish embryos. *Theranostics* 3 (4): 258.

44 Khajuria, D.K., Kumar, V.B., Karasik, D., and Gedanken, A. (2017). Fluorescent nanoparticles with tissue-dependent affinity for live zebrafish imaging. *ACS Appl. Mater. Interfaces* 9 (22): 18557–18565.

45 Fent, K., Weisbrod, C.J., Wirth-Heller, A., and Pieles, U. (2010). Assessment of uptake and toxicity of fluorescent silica nanoparticles in zebrafish (Danio rerio) early life stages. *Aquat. Toxicol.* 100 (2): 218–228.

46 Kang, Y.F., Li, Y.H., Fang, Y.W., Xu, Y., Wei, X.M., and Yin, X.B. (2015). Carbon quantum dots for zebrafish fluorescence imaging. *Sci Rep* 5 (1): 1–12.

47 Chetty, S.S., Praneetha, S., Basu, S., Sachidanandan, C., and Murugan, A.V. (2016). Sustainable, rapid synthesis of bright-luminescent CuInS 2-ZnS alloyed nanocrystals: Multistage nano-xenotoxicity assessment and intravital fluorescence bioimaging in zebrafish-embryos. *Sci Rep* 6 (1): 1–15.

48 Srivastava, P.K., Han, S., Tu, C.C., and Jing, L. (2019). Phototoxicity generated by silicon quantum dot nanoparticles on zebrafish embryos. *ACS Applied Bio Mat.* 2 (7): 2872–2878.

49 Amores, A. and Postlethwait, J.H. (1998). Banded chromosomes and the zebrafish karyotype. *Methods Cell Biol.* 60: 323–338.

50 Morizot, D.C. (1990). Use of fish gene maps to predict ancestral vertebrate genome organization. *Prog. Clin. Biol. Res.* 344: 207–234.

51 Risinger, C., Deitcher, D.L., Lundell, I., Schwarz, T.L., and Larhammar, D. (1997). Complex gene organization of synaptic protein SNAP-25 in Drosophila melanogaster. *Gene* 194 (2): 169–177.

52 Postlethwait, J., Amores, A., Force, A., and Yan, Y.L. (1998). The zebrafish genome. *Methods Cell Biol.* 60: 149–163.

53 Ablain, J., Durand, E.M., Yang, S., Zhou, Y., and Zon, L.I. (2015). A CRISPR/Cas9 vector system for tissue-specific gene disruption in zebrafish. *Dev. Cell* 32 (6): 756–764.

54 Sung, Y.H., Kim, J.M., Kim, H.T., Lee, J., Jeon, J., Jin, Y., Choi, J.H., Ban, Y.H., Ha, S.J., Kim, C.H., and Lee, H.W. (2014). Highly efficient gene knockout in mice and zebrafish with RNA-guided endonucleases. *Genome Res.* 24 (1): 125–131.

55 Merrick, B.A. and Bruno, M.E. (2004). Genomic and proteomic profiling for biomarkers and signature profiles of toxicity. *Curr. Opin. Mol. Ther.* 6 (6): 600–607.

56 Link, V., Shevchenko, A., and Heisenberg, C.P. (2006). Proteomics of early zebrafish embryos. *BMC Dev. Biol.* 6 (1): 1–9.

57 Lucitt, M.B., Price, T.S., Pizarro, A., Wu, W., Yocum, A.K., Seiler, C., Pack, M.A., Blair, I.A., FitzGerald, G.A., and Grosser, T. (2008). Analysis of the zebrafish proteome during embryonic development. *Mol. Cell. Proteom.* 7 (5): 981–994.

58 Gómez-Canela, C., Prats, E., Tauler, R., and Raldúa, D. (2017). Analysis of neurobehavioural data by chemometric methods in ecotoxicological studies. *Ecotoxicol. Environ. Saf.* 145: 583–590.

59 Zhou, B., Xiao, J.F., Tuli, L., and Ressom, H.W. (2012). LC-MS-based metabolomics. *Mol. Biosyst.* 8 (2): 470–481.

60 Veneman, W.J., Spaink, H.P., Brun, N.R., Bosker, T., and Vijver, M.G. (2017). Pathway analysis of systemic transcriptome responses to injected polystyrene particles in zebrafish larvae. *Aquat. Toxicol.* 190: 112–120.

61 Chen, E. and Ekker, S.C. (2004). Zebrafish as a genomics research model. *Curr. Pharm. Biotechnol.* 5 (5): 409–413.

62 Hammoud, S.S., Nix, D.A., Zhang, H., Purwar, J., Carrell, D.T., and Cairns, B.R. (2009). Distinctive chromatin in human sperm packages genes for embryo development. *Nature* 460 (7254): 473–478.

63 Kimmel, C.B., Ballard, W.W., Kimmel, S.R., Ullmann, B., and Schilling, T.F. (1995). Stages of embryonic development of the zebrafish. *Dev. Dyn.* 203 (3): 253–310.

64 Sun, J., Zhou, Q., and Hu, X. (2019). Integrating multi-omics and regular analyses identifies the molecular responses of zebrafish brains to graphene oxide: Perspectives in environmental criteria. *Ecotoxicol. Environ. Saf.* 180: 269–279.

65 Pomés, A. and Arruda, L.K. (2014). Investigating cockroach allergens: Aiming to improve diagnosis and treatment of cockroach allergic patients. *Methods* 66 (1): 75–85.

66 Dong, X., Qiu, X., Meng, S., Xu, H., Wu, X., and Yang, M. (2018). Proteomic profile and toxicity pathway analysis in zebrafish embryos exposed to bisphenol A and di-n-butyl phthalate at environmentally relevant levels. *Chemosphere* 193: 313–320.

67 Du, B., Liu, G., Ke, M., Zhang, Z., Zheng, M., Lu, T., Sun, L., and Qian, H. (2019). Proteomic analysis of the hepatotoxicity of Microcystis aeruginosa in adult zebrafish (Danio rerio) and its potential mechanisms. *Environ. Pollut.* 254: 113019.

68 Griffitt, R.J., Lavelle, C.M., Kane, A.S., Denslow, N.D., and Barber, D.S. (2013). Chronic nanoparticulate silver exposure results in tissue accumulation and transcriptomic changes in zebrafish. *Aquat. Toxicol.* 130: 192–200.

69 Zheng, M., Lu, J., and Zhao, D. (2018). Toxicity and transcriptome sequencing (RNA-seq) analyses of adult zebrafish in response to exposure carboxymethyl cellulose stabilized iron sulfide nanoparticles. *Sci Rep* 8 (1): 1–11.

70 Lindeque, J.Z., Matthyser, A., Mason, S., Louw, R., and Taute, C.J.F. (2018). Metabolomics reveals the depletion of intracellular metabolites in HepG2 cells after treatment with gold nanoparticles. *Nanotoxicol.* 12 (3): 251–262.

71 Lee, S.H., Wang, T.Y., Hong, J.H., Cheng, T.J., and Lin, C.Y. (2016). NMR-based metabolomics to determine acute inhalation effects of nano-and fine-sized ZnO particles in the rat lung. *Nanotoxicol.* 10 (7): 924–934.

72 Chen, L., Hu, Y., He, J., Chen, J., Giesy, J.P., and Xie, P. (2017). Responses of the proteome and metabolome in livers of zebrafish exposed chronically to environmentally relevant concentrations of microcystin-LR. *Environ. Sci. Technol.* 51 (1): 596–607.

73 De Souza Filho, J., Matsubara, E.Y., Franchi, L.P., Martins, I.P., Rivera, L.M.R., Rosolen, J.M., and Grisolia, C.K. (2014). Evaluation of carbon nanotubes network toxicity in zebrafish (Danio rerio) model. *Environ. Res.* 134: 9–16.

74 Dedeh, A., Ciutat, A., Treguer-Delapierre, M., and Bourdineaud, J.P. (2015). Impact of gold nanoparticles on zebrafish exposed to a spiked sediment. *Nanotoxicol.* 9 (1): 71–80.

75 Bonadio, R.S., Cunha, M.C.P.C.D., Longo, J.P.F., Azevedo, R.B., and Poças-Fonseca, M.J. (2020). Exposure to maghemite nanoparticles induces epigenetic alterations in human submandibular gland cells. *J Nanosci Nanotechnol* 20 (3): 1454–1462.

76 Cao, D., Bromberg, P.A., and Samet, J.M. (2007). COX-2 expression induced by diesel particles involves chromatin modification and degradation of HDAC1. *Am. J. Respir. Cell Mol. Biol.* 37 (2): 232–239.

77 Lawrence, M., Daujat, S., and Schneider, R. (2016). Lateral thinking: How histone modifications regulate gene expression. *Trends in Genetics* 32 (1): 42–56.

78 Nelson, S.M., Mahmoud, T., Beaux, M., II, Shapiro, P., McIlroy, D.N., and Stenkamp, D.L. (2010). Toxic and teratogenic silica nanowires in developing vertebrate embryos. *Nanomedicine: Nanotechnol. Biol. Med.* 6 (1): 93–102.

79 Li, X., Liu, B., Li, X.L., Li, Y.X., Sun, M.Z., Chen, D.Y., Zhao, X., and Feng, X.Z. (2014). SiO 2 nanoparticles change colour preference and cause Parkinson's-like behaviour in zebrafish. *Sci Rep* 4 (1): 1–9.

80 Kim, K.T., Zaikova, T., Hutchison, J.E., and Tanguay, R.L. (2013). Gold nanoparticles disrupt zebrafish eye development and pigmentation. *Toxicol. Sci.* 133 (2): 275–288.

81 Paatero, I., Casals, E., Niemi, R., Özliseli, E., Rosenholm, J.M., and Sahlgren, C. (2017). Analyses in zebrafish embryos reveal that nanotoxicity profiles are dependent on surface-functionalization controlled penetrance of biological membranes. *Sci Rep* 7 (1): 1–13.

82 Zakaria, Z.Z., Benslimane, F.M., Nasrallah, G.K., Shurbaji, S., Younes, N.N., Mraiche, F., Da'as, S.I., and Yalcin, H.C. (2018). Using zebrafish for investigating the molecular mechanisms of drug-induced cardiotoxicity. *Biomed. Res. Int.* 2018: 1642684.

83 Traver, D., Herbomel, P., Patton, E.E., Murphey, R.D., Yoder, J.A., Litman, G.W., Catic, A., Amemiya, C.T., Zon, L.I., and Trede, N.S. (2003). The zebrafish as a model organism to study development of the immune system. *Adv. Immunol.* 81: 253–330.

84 Truong, L., Tilton, S.C., Zaikova, T., Richman, E., Waters, K.M., Hutchison, J.E., and Tanguay, R.L. (2013). Surface functionalities of gold nanoparticles impact embryonic gene expression responses. *Nanotoxicol.* 7 (2): 192–201.

85 Krishnaraj, C., Harper, S.L., and Yun, S.I. (2016). In vivo toxicological assessment of biologically synthesized silver nanoparticles in adult zebrafish (Danio rerio). *J. Hazard. Mater.* 301: 480–491.

86 Van Aerle, R., Lange, A., Moorhouse, A., Paszkiewicz, K., Ball, K., Johnston, B.D., de Bastos, E., Booth, T., Tyler, C.R., and Santos, E.M. (2013). Molecular mechanisms of toxicity of silver nanoparticles in zebrafish embryos. *Environ. Sci. Technol.* 47 (14): 8005–8014.

87 Villacis, R.A., José Filho, S., Piña, B., Azevedo, R.B., Pic-Taylor, A., Mazzeu, J.F., and Grisolia, C.K. (2017). Integrated assessment of toxic effects of maghemite (γ-Fe2O3) nanoparticles in zebrafish. *Aquat. Toxicol.* 191: 219–225.

88 Chen, S.X., Yang, X.Z., Deng, Y., Huang, J., Li, Y., Sun, Q., Yu, C.P., Zhu, Y., and Hong, W.S. (2017). Silver nanoparticles induce oocyte maturation in zebrafish (Danio rerio). *Chemosphere* 170: 51–60.

89 Hanawalt, P.C. and Spivak, G. (2008). Transcription-coupled DNA repair: Two decades of progress and surprises. *Nat. Rev. Mol. Cell Biol.* 9 (12): 958–970.

90 Hu, M., Jovanovic, B., and Palic, D. (2019). In silico prediction of microRNA role in regulation of zebrafish (Danio rerio) responses to nanoparticle exposure. *Toxicol. Vitro* 60: 187–202.

91 Zhou, W., Tian, D., He, J., Yan, X., Zhao, J., Yuan, X., and Peng, S. (2019). Prolonged exposure to carbon nanoparticles induced methylome remodeling and gene expression in zebrafish heart. *J. Appl. Toxicol.* 39 (2): 322–332.

92 Poynton, H.C., Lazorchak, J.M., Impellitteri, C.A., Blalock, B.J., Rogers, K., Allen, H.J., Loguinov, A., Heckman, J.L., and Govindasmawy, S. (2012). Toxicogenomic responses of nanotoxicity in Daphnia magna exposed to silver nitrate and coated silver nanoparticles. *Environ. Sci. Technol.* 46 (11): 6288–6296.

93 Kteeba, S.M., El-Ghobashy, A.E., El-Adawi, H.I., El-Rayis, O.A., Sreevidya, V.S., Guo, L., and Svoboda, K.R. (2018). Exposure to ZnO nanoparticles alters neuronal and vascular development in zebrafish: Acute and transgenerational effects mitigated with dissolved organic matter. *Environ. Pollut.* 242: 433–448.

94 Ngun, T.C., Ghahramani, N., Sánchez, F.J., Bocklandt, S., and Vilain, E. (2011). The genetics of sex differences in brain and behavior. *Front Neuroendocrinol* 32 (2): 227–246.

95 Busby, E.R., Roch, G.J., and Sherwood, N.M. (2010). Endocrinology of zebrafish: A small fish with a large gene pool. *Fish Physiology* 29: 173–247.

96 Driever, W., Solnica-Krezel, L., Schier, A.F., Neuhauss, S.C., Malicki, J., Stemple, D.L., Stainier, D.Y., Zwartkruis, F., Abdelilah, S., Rangini, Z., and Belak, J. (1996). A genetic screen for mutations affecting embryogenesis in zebrafish. *Development* 123 (1): 37–46.

97 Melby, A.E., Warga, R.M., and Kimmel, C.B. (1996). Specification of cell fates at the dorsal margin of the zebrafish gastrula. *Development* 122 (7): 2225–2237.

98 Amsterdam, A., Burgess, S., Golling, G., Chen, W., Sun, Z., Townsend, K., Farrington, S., Haldi, M., and Hopkins, N. (1999). A large-scale insertional mutagenesis screen in zebrafish. *Genes Dev.* 13 (20): 2713–2724.

99 Varshney, G.K., Sood, R., and Burgess, S.M. (2015). Understanding and editing the zebrafish genome. *Adv. Genet.* 92: 1–52.

100 Jia, H.R., Zhu, Y.X., Duan, Q.Y., Chen, Z., and Wu, F.G. (2019). Nanomaterials meet zebrafish: Toxicity evaluation and drug delivery applications. *J. Control. Release* 311: 301–318.

101 Xia, T., Kovochich, M., Brant, J., Hotze, M., Sempf, J., Oberley, T., Sioutas, C., Yeh, J.I., Wiesner, M.R., and Nel, A.E. (2006). Comparison of the abilities of ambient and manufactured nanoparticles to induce cellular toxicity according to an oxidative stress paradigm. *Nano Lett.* 6 (8): 1794–1807.

102 Dedeh, A., Ciutat, A., Treguer-Delapierre, M., and Bourdineaud, J.P. (2015). Impact of gold nanoparticles on zebrafish exposed to a spiked sediment. *Nanotoxicol.* 9 (1): 71–80.

103 Goodman, C.M., Chari, N.S., Han, G., Hong, R., Ghosh, P., and Rotello, V.M. (2006). DNA-binding by functionalized gold nanoparticles: Mechanism and structural requirements. *Chem. Biol. Drug Des.* 67 (4): 297–304.

104 De Lima, D., Roque, G.M., and De Almeida, E.A. (2013). In vitro and in vivo inhibition of acetylcholinesterase and carboxylesterase by metals in zebrafish (Danio rerio). *Mar. Environ. Res.* 91: 45–51.

105 Sheng, L., Wang, L., Su, M., Zhao, X., Hu, R., Yu, X., Hong, J., Liu, D., Xu, B., Zhu, Y., and Wang, H. (2016). Mechanism of TiO_2 nanoparticle-induced neurotoxicity in zebrafish (D anio rerio). *Environ. Toxicol.* 31 (2): 163–175.

106 Di Gioacchino, M., Petrarca, C., Lazzarin, F., Di Giampaolo, L., Sabbioni, E., Boscolo, P., Mariani-Costantini, R., and Bernardini, G. (2011). Immunotoxicity of nanoparticles. *Int. J. Immunopathol. Pharmacol.* 24 (1 Suppl): 65S–71S.

107 Lankveld, D.P.K., Van Loveren, H., Baken, K.A., and Vandebriel, R.J. (2010). In vitro testing for direct immunotoxicity: State of the art. *Methods Mol Biol.* 598: 401–423.

108 Brun, N.R., Lenz, M., Wehrli, B., and Fent, K. (2014). Comparative effects of zinc oxide nanoparticles and dissolved zinc on zebrafish embryos and eleuthero-embryos: Importance of zinc ions. *Sci. Total Environ.* 476: 657–666.

109 Chen, M., Yin, J., Liang, Y., Yuan, S., Wang, F., Song, M., and Wang, H. (2016). Oxidative stress and immunotoxicity induced by graphene oxide in zebrafish. *Aquat. Toxicol.* 174: 54–60.

110 Aksakal, F.I. and Ciltas, A. (2019). Impact of copper oxide nanoparticles (CuO NPs) exposure on embryo development and expression of genes related to the innate immune system of zebrafish (Danio rerio). *Comp. Biochem. Physiol. C Toxicol. Pharmacol.* 223: 78–87.

111 Yang, X., Yang, Q., Zheng, G., Han, S., Zhao, F., Hu, Q., and Fu, Z. (2019). Developmental neurotoxicity and immunotoxicity induced by graphene oxide in zebrafish embryos. *Environ. Toxicol.* 34 (4): 415–423.

112 Fang, Q., Shi, Q., Guo, Y., Hua, J., Wang, X., and Zhou, B. (2016). Enhanced bioconcentration of bisphenol A in the presence of nano-TiO2 can lead to adverse reproductive outcomes in zebrafish. *Environ. Sci. Technol.* 50 (2): 1005–1013.

113 Wang, H., Zhou, L., Liao, X., Meng, Z., Xiao, J., Li, F., Zhang, S., Cao, Z., and Lu, H. (2019). Toxic effects of oxine-copper on development and behavior in the embryo-larval stages of zebrafish. *Aquat. Toxicol.* 210: 242–250.

114 Aksakal, F.I. and Sisman, T. (2020). Developmental toxicity induced by Cu (OH) 2 nanopesticide in zebrafish embryos. *Environ. Toxicol.* 35 (12): 1289–1298.

115 Xu, J., Zhang, Q., Li, X., Zhan, S., Wang, L., and Chen, D. (2017). The effects of copper oxide nanoparticles on dorsoventral patterning, convergent extension, and neural and cardiac development of zebrafish. *Aquat. Toxicol.* 188: 130–137.

116 Zheng, N., Yan, J., Qian, W., Song, C., Zuo, Z., and He, C. (2021). Comparison of developmental toxicity of different surface modified CdSe/ZnS QDs in zebrafish embryos. *J. Environ. Sci.* 100: 240–249.

117 Prykhozhij, S.V., Fuller, C., Steele, S.L., Veinotte, C.J., Razaghi, B., Robitaille, J.M., McMaster, C.R., Shlien, A., Malkin, D., and Berman, J.N. (2018). Optimized knock-in of point mutations in zebrafish using CRISPR/Cas9. *Nucleic Acids Res.* 46 (17): 102–e102.

118 White, R.M., Sessa, A., Burke, C., Bowman, T., LeBlanc, J., Ceol, C., Bourque, C., Dovey, M., Goessling, W., Burns, C.E., and Zon, L.I. (2008). Transparent adult zebrafish as a tool for in vivo transplantation analysis. *Cell Stem Cell* 2 (2): 183–189.

119 Amatruda, J.F. and Zon, L.I. (1999). Dissecting hematopoiesis and disease using the zebrafish. *Dev. Biol.* 216 (1): 1–15.

120 Van Aerle, R., Lange, A., Moorhouse, A., Paszkiewicz, K., Ball, K., Johnston, B.D., de Bastos, E., Booth, T., Tyler, C.R., and Santos, E.M. (2013). Molecular mechanisms of toxicity of silver nanoparticles in zebrafish embryos. *Environ. Sci. Technol.* 47 (14): 8005–8014.

Index

Genomic and Epigenomic Biomarkers of Toxicology and Disease: Clinical and Therapeutic Actions,
First Edition. Edited by Saura C. Sahu.